AF510811

Current Topics in Behavioral Neurosciences

Volume 73

Series Editors

Bart A. Ellenbroek, School of Psychology, Victoria University of Wellington, Wellington, New Zealand

Thomas R. E. Barnes, Division of Psychiatry, Imperial College London, London, UK

Martin P. Paulus, Laureate Institute for Brain Research, Tulsa, USA

Jocelien Olivier, GELIFES, University of Groningen, Groningen, The Netherlands

Current Topics in Behavioral Neurosciences provides critical and comprehensive discussions of the most significant areas of behavioral neuroscience research, written by leading international authorities. Each volume in the series represents the most informative and contemporary account of its subject available, making it an unrivalled reference source. Each volume will be made available in both print and electronic form.

With the development of new methodologies for brain imaging, genetic and genomic analyses, molecular engineering of mutant animals, novel routes for drug delivery, and sophisticated cross-species behavioral assessments, it is now possible to study behavior relevant to psychiatric and neurological diseases and disorders on the physiological level. The Behavioral Neurosciences series focuses on translational medicine and cutting-edge technologies. Preclinical and clinical trials for the development of new diagnostics and therapeutics as well as prevention efforts are covered whenever possible. Special attention is also drawn on epigenetical aspects, especially in psychiatric disorders.

CTBN series is indexed in PubMed and Scopus.

Founding Editors:
Emeritus Professor Mark A. Geyer
Department of Psychiatry, University of California San Diego, La Jolla, USA
Emeritus Professor Charles A. Marsden
Institute of Neuroscience, School of Biomedical Sciences, University of Nottingham Medical School Queen's Medical Centre, Nottingham, UK
Professor Bart A. Ellenbroek
School of Psychology, Victoria University of Wellington, Wellington, New Zealand

Jennifer Urbano Blackford • Mohammed R. Milad
Editors

New Discoveries in the Brain Sciences of Fear and Anxiety - From Basic to Clinical Neuroscience

Springer

Editors
Jennifer Urbano Blackford
Munroe-Meyer Institute
University of Nebraska Medical Center
Omaha, NE, USA

Mohammed R. Milad
Psychiatry and Behavioral Sciences
UTHealth Houston
Houston, TX, USA

ISSN 1866-3370 ISSN 1866-3389 (electronic)
Current Topics in Behavioral Neurosciences
ISBN 978-3-031-98401-3 ISBN 978-3-031-98402-0 (eBook)
https://doi.org/10.1007/978-3-031-98402-0

© The Editor(s) (if applicable) and The Author(s), under exclusive license to Springer Nature Switzerland AG 2025

This work is subject to copyright. All rights are solely and exclusively licensed by the Publisher, whether the whole or part of the material is concerned, specifically the rights of translation, reprinting, reuse of illustrations, recitation, broadcasting, reproduction on microfilms or in any other physical way, and transmission or information storage and retrieval, electronic adaptation, computer software, or by similar or dissimilar methodology now known or hereafter developed.
The use of general descriptive names, registered names, trademarks, service marks, etc. in this publication does not imply, even in the absence of a specific statement, that such names are exempt from the relevant protective laws and regulations and therefore free for general use.
The publisher, the authors and the editors are safe to assume that the advice and information in this book are believed to be true and accurate at the date of publication. Neither the publisher nor the authors or the editors give a warranty, expressed or implied, with respect to the material contained herein or for any errors or omissions that may have been made. The publisher remains neutral with regard to jurisdictional claims in published maps and institutional affiliations.

This Springer imprint is published by the registered company Springer Nature Switzerland AG
The registered company address is: Gewerbestrasse 11, 6330 Cham, Switzerland

If disposing of this product, please recycle the paper.

Introduction

This volume is part of a series on Current Topics in Behavioral Neurosciences and focuses on the latest neuroscience findings about the neurobiology of fear and anxiety. The last Current Topics in Behavioral Neurosciences volume on anxiety was in 2010, led by Murray Stein and Thomas Steckler. Over the past 15 years, significant progress has been achieved toward our understanding of the neurobiology of anxiety and its relevance to psychopathology. This volume includes a collection of reviews written by an international group of scientific leaders in their respective domains that highlight some of the most recent advances in behavioral and psychiatric neuroscience pertaining to fear and anxiety. These reviews are highly relevant for both scientists and clinicians interested in the brain mechanisms and associated psychopathologies related to fear and anxiety.

Several key points highlight the significance of the contributions in this volume. First, fear is a highly conserved and evolutionarily adaptive response to threats in the environment. Second, everyone experiences fear at some point during our lives and must learn to regulate fear responses. Third, if left unchecked, these responses can lead to the development of fear- and anxiety-based disorders that negatively impact our daily lives. Fear- and anxiety-based disorders often begin in childhood, highlighting the importance of genetics, early environment, and risk factors. Importantly, anxiety disorders are often co-occurring with other psychiatric disorders including alcohol and substance use disorders, depressive disorders, autism spectrum disorders, and schizophrenia.

Decades of research have identified the basic underlying neurocircuitry of fear and cutting-edge methods used in both animal models and humans continue to refine and shape our understanding of fear and anxiety. The contributions included in this volume will highlight and summarize some of the recent advances made in some of the domains noted above. We selected topics for this book based on the following core principles:

1. Translational research is critical for advancing cutting-edge discoveries and improving human health. To achieve this goal, the reviews in this book span

the continuum from basic to clinical neuroscience with representation across species. We hope that these reviews will provide an accessible way to bridge the gap between basic and clinical neuroscience.

2. Bringing together findings across different disciplines is critical for a full understanding of fear and anxiety. To achieve this goal the contributors to this book are international experts in a wide range of fields, including behavioral neuroscience, cognitive neuroscience, clinical neuroscience, psychiatry, and psychology. Collectively, they provide a range of perspectives on the neurobiology of fear and anxiety.

In summary, the sections of this volume: (1) review the basic mechanisms underlying fear and anxiety, (2) review the factors that contribute to anxiety, including developmental stage, risk factors, sex differences, and genetics, (3) elucidate how current treatments work and describe novel treatments, and (4) demonstrate the role of anxiety in other commonly co-occurring disorders, including autism, substance use disorders, and schizophrenia.

We end with an expression of gratitude and appreciation to all contributing authors to this volume for their generosity with their time and effort. We also acknowledge the outstanding research and publications of many other outstanding scientists and clinicians in the fields of behavioral and psychiatric neuroscience that were not included in this volume.

Contents

Part I
Basic Mechanisms

Animal Models

Disentangling the Neural Circuits of Arousal and Anxiety-Like Behavior

Antonio Florido, Vincent R. Curtis, Nicolas C. Pégard, and Jose Rodriguez-Romaguera

Contents

A. Florido
Department of Psychiatry, University of North Carolina, Chapel Hill, NC, USA

Department of Applied Physical Sciences, University of North Carolina, Chapel Hill, NC, USA

V. R. Curtis
Department of Applied Physical Sciences, University of North Carolina, Chapel Hill, NC, USA

N. C. Pégard (✉)
Department of Applied Physical Sciences, University of North Carolina, Chapel Hill, NC, USA

Department of Biomedical Engineering, University of North Carolina, Chapel Hill, NC, USA

Neuroscience Center, University of North Carolina, Chapel Hill, NC, USA

Carolina Stress Initiative, University of North Carolina, Chapel Hill, NC, USA
e-mail: pegard@unc.edu

J. Rodriguez-Romaguera (✉)
Department of Psychiatry, University of North Carolina, Chapel Hill, NC, USA

Department of Applied Physical Sciences, University of North Carolina, Chapel Hill, NC, USA

Neuroscience Center, University of North Carolina, Chapel Hill, NC, USA

Carolina Stress Initiative, University of North Carolina, Chapel Hill, NC, USA

Carolina Institute for Developmental Disorders, University of North Carolina, Chapel Hill, NC, USA

Department of Cell Biology and Physiology, University of North Carolina, Chapel Hill, NC, USA
e-mail: jose_rodriguezromaguera@med.unc.edu

© The Author(s), under exclusive license to Springer Nature Switzerland AG 2024

Curr Topics Behav Neurosci (2025) 73: 3–32
https://doi.org/10.1007/7854_2024_539
Published Online: 24 November 2024

Abstract Anxiety disorders are prevalent and debilitating conditions characterized by excessive concern and fear, affecting thoughts, behaviors, and sensations. A critical component of anxiety is arousal, a complex process involving alertness regulation and stimulus salience modulation. While arousal is adaptive in normal circumstances, dysregulation can lead to hypoarousal or hyperarousal, affecting response selection and threat perception. This chapter reviews challenges in studying arousal in preclinical anxiety models, emphasizing the need for multicomponent measurement and analysis. Novel methodologies integrating physiological measurement with activity tracking of neurons with single-cell resolution in awake animals are discussed, with emphasis in current challenges. Understanding these mechanisms is crucial for developing effective treatments for anxiety disorders.

Keywords Anxiety · Arousal · Behavior · Internal physiology · Stress

1 Neural Circuits of Arousal and Anxiety

Anxiety is a state characterized by excessive concern over a potential threat that may or may not arise in the near future (Association AP 2015; Penninx et al. 2021; Szuhany and Simon 2022). It is marked by feelings of unease or fear that significantly affect one's thoughts, behavior, and sensations, eliciting avoidance or escape responses to safeguard the individual's physical and mental integrity (Schoenfeld 1950; Sege et al. 2018). Currently, the Diagnostic and Statistical Manual of Mental Disorders (DSM-5) distinguishes 11 categories of anxiety disorders, including generalized anxiety disorder (GAD) and panic disorder (PD) (American Psychiatric Association 2013). These conditions are highly prevalent worldwide, with an estimated lifetime prevalence ranging from 4.8% to 10.9% (Kessler et al. 2012; Bandelow and Michaelis 2015), with a higher prevalence in women compared to men (Bekker and van Mens-Verhulst 2007; McLean et al. 2011; Maeng and Milad 2015). Among the various components of the anxious response, arousal has emerged as a critical construct for understanding physiological dysregulations, such as hypervigilance, that persist in anxiety disorders (Kimble et al. 2014; Wessing et al. 2017; Wermes et al. 2018; Hur et al. 2019). Arousal encompasses a multifaceted array of executive processes aimed at maintaining an organism's state of readiness to respond to stimuli crucial for its integrity (Mayes 2000; VaezMousavi et al. 2007; Arrieta-Cruz and Pfaff 2009). Existing literature indicates that arousal is a composite process, involving the regulation of the organism's overall alertness state, intricately linked with the cyclical nature of sleep and wakefulness, as well as the rapid modulation of the biological salience of prevailing stimuli to facilitate swift and efficient response execution (Tyree et al. 2018; Jones 2020; Sotelo et al. 2020). This modulation of stimulus salience is particularly significant for survival, as the need to swiftly identify potential threats emphasizes the necessity for rapid response initiation to preserve life.

Under physiological conditions, arousal serves as an adaptive mechanism, enabling the selection of the most effective response within a short time frame. However, under pathological circumstances, the identification of dangerous stimuli may be compromised, leading to hypoarousal characterized by reduced responsiveness to salient cues (Gostisha et al. 2014; Hegerl and Ulke 2016), or hyperarousal, triggering avoidance or flight responses to stimuli predictive of imminent threat through direct or associative pathways (Bonnet and Arand 1997; McCurry et al. 2020; Morabito et al. 2020). This chapter provides a review of current challenges in studying arousal in preclinical models of anxiety disorders. It highlights available methods for measuring arousal as a construct and discusses the limitations of current arousal research across species to guide future research toward multicomponent measurement and analysis of arousal-related processes.

The anxious phenotype is characterized by two core traits: (1) fear responses upon imminent threats to safeguard one's integrity and (2) excessive and uncontrolled worry that is inflexible to safety cues (American Psychiatric Association 2013). Multiple brain regions have been described to be involved in these processes. Here, we will describe the amygdala as one of the principal hubs in the brain for the acquisition, maintenance, and recall of fear memories. In addition, we will discuss how other critical hubs across the cortex, striatum, and thalamus interact with the salience network to generate excessive worry.

1.1 Neurobiology of Fear and Stress in Anxiety Disorders

The amygdalar complex is a key region for fear processing in the brain located in the deep temporal lobe (Davis 1992). It is conformed by multiple structures, with the most common subdivisions being the lateral (LA), basal (BA), medial (MeA), and central amygdala (CeA) (Sah et al. 2003; Dalla Corte et al. 2024) (Fig. 1a). It is supported by an additional network usually referred to as the extended amygdala, which includes projections from the bed nucleus of the stria terminalis (BNST) and the nucleus accumbens (NAc). Functionally and anatomically, these structures are highly preserved across mammals, permitting a high degree of translationability of results between species (Roxo et al. 2011; Barger et al. 2014). In humans, lesions of the amygdala have been linked to an inability to recognize fearful face expressions through impaired sight fixation on the eyes area of another person's face (Adolphs et al. 2005). However, it has also been reported that patients with amygdala lesions

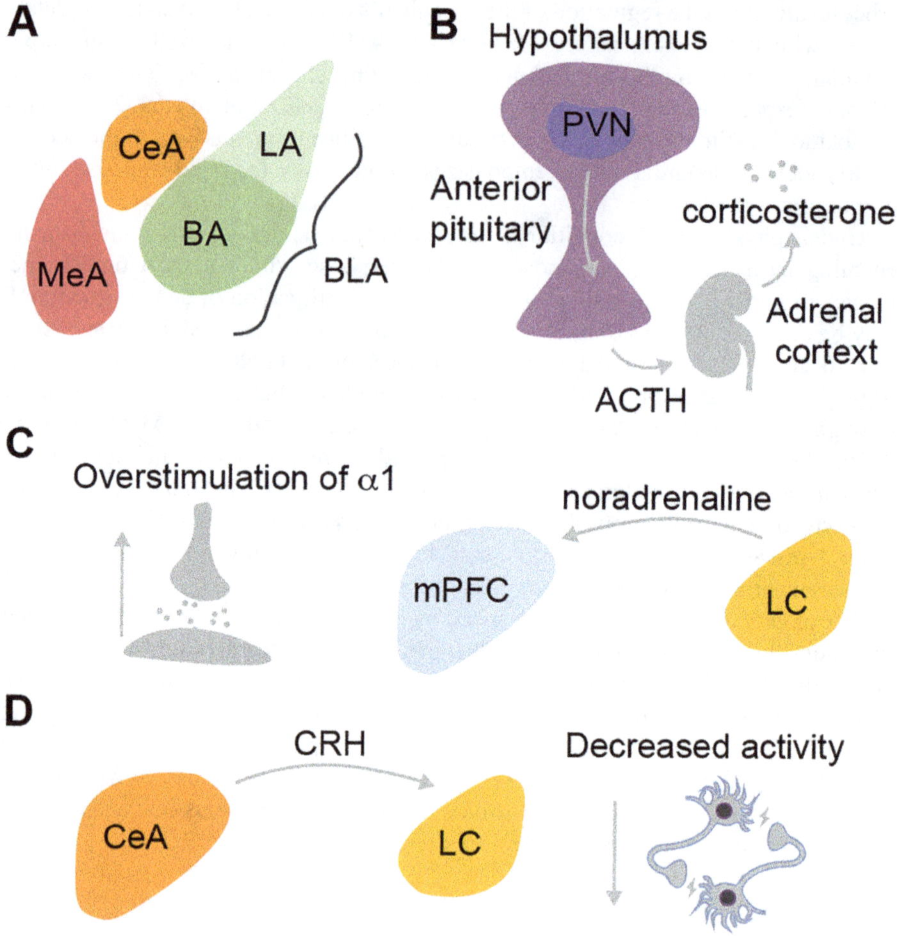

Fig. 1 A comprehensive overview of neural circuitry associated with arousal and anxiety. (**a**) The amygdalar complex, located in the deep temporal lobe, plays a pivotal role in fear processing. (**b**) The corticotropin-releasing factor neurons in the paraventricular nucleus (PVN) stimulate the anterior pituitary to release adrenocorticotropic hormone (ACTH), initiating the stress response cascade. (**c**) The noradrenergic system, characterized by increased firing in the locus coeruleus (LC) and norepinephrine (NE) release, influences cognitive performance via overstimulation of α-1 receptors in the medial prefrontal cortex (mPFC). (**d**) During stressful conditions, corticotropin-releasing hormone (CRH) release from the central amygdala (CeA) can dampen neural firing in the LC. CeA = central amygdala, MeA = medial amygdala, BA = basal amygdala, LA = lateral amygdala, BLA = basolateral amygdala, mPFC = medial prefrontal cortex

are able to preserve aspects of the subjective experience of fear (Feinstein et al. 2011; Fullana et al. 2016).

When we detect the presence of a threat, information travels from the sensory cortices and the sensory thalamus to the basolateral amygdala (BLA) (Hintiryan et al. 2021). The convergence of sensory inputs onto these neurons is subject to synaptic plasticity, as neutral stimuli that initially do not elicit a fear response can have their synaptic connections with the BLA strengthened through an experience-dependent process known as long-term potentiation (Maren 1999; Dityatev and Bolshakov 2005). This process entails a series of molecular and structural changes in cortico-amygdalar and thalamo-amygdalar fibers to generate sufficient synaptic strength to depolarize the BLA without the need for a stimulus that, without prior conditioning, would achieve this depolarization. Through various processes, the central nucleus of the amygdala (CeA) is the ultimate effector of the fear response via direct inputs to autonomic, endocrine, cognitive, or behavioral processing regions (Ciocchi et al. 2010; Duvarci et al. 2011; Keifer et al. 2015). Among these inputs, projections to the paraventricular nucleus of the hypothalamus (PVN) are particularly relevant for understanding the etiology of anxiety disorders, as these projections regulate the hypothalamic-pituitary–adrenal axis response to perceived threats (whether external through our senses or internal through thoughts or proprioception) (Herman et al. 2008; Jiang et al. 2019) (Fig. 1b). Then, corticotropin-releasing factor neurons in the PVN stimulate the anterior pituitary, a region interconnected with the general blood circulation through a *porta* vessel system, to release adrenocorticotropic hormone (ACTH) into the bloodstream. ACTH stimulates adrenal glands to produce glucocorticoids, whose actions through their receptors spread all through the body to ready the organism to fight or flight (Stephens and Wand 2012; Herman et al. 2016).

The noradrenergic system from the locus coeruleus (LC) has also been shown to play a pivotal role in the regulation of the stress response to perceived threats through direct modulation of arousal (Winklewski et al. 2017). Acute stress has been shown to increase LC firing and norepinephrine (NE) release (Gubits et al. 1989; Bing et al. 1991, 1992). Stressful events have shown to correlate positively with high arousal and affect cognitive performance through overstimulation of α-1 receptors in the medial prefrontal cortex (mPFC) (Fig. 1c) (Berridge and Waterhouse 2003). During stressful conditions, corticotropin-releasing hormone (CRH) afferents into the LC have shown to decrease neuronal firing of the LC induced by weakening of synaptic connections coming from sensory inputs to increase arousal and therefore perceptual efficiency (Fig. 1d) (Devilbiss et al. 2012). Interestingly, prevailing theories suggest an effect of chronic stress in accumulation of Aβ42 peptides in the LC NE system, causing brain damage and correlating with aged-related cognitive decline and dementia (Berridge and Waterhouse 2003).

1.2 *The Bed Nucleus of the Stria Terminalis (BNST)*

For years, research on anxiety disorders has been focused on amygdalar responses upon perceived threats. Notwithstanding, the functional interconnections between the amygdalar complex and the BNST have been demonstrated to play a key role in the development of anxiety disorders (Lebow and Chen 2016). The BNST is a region

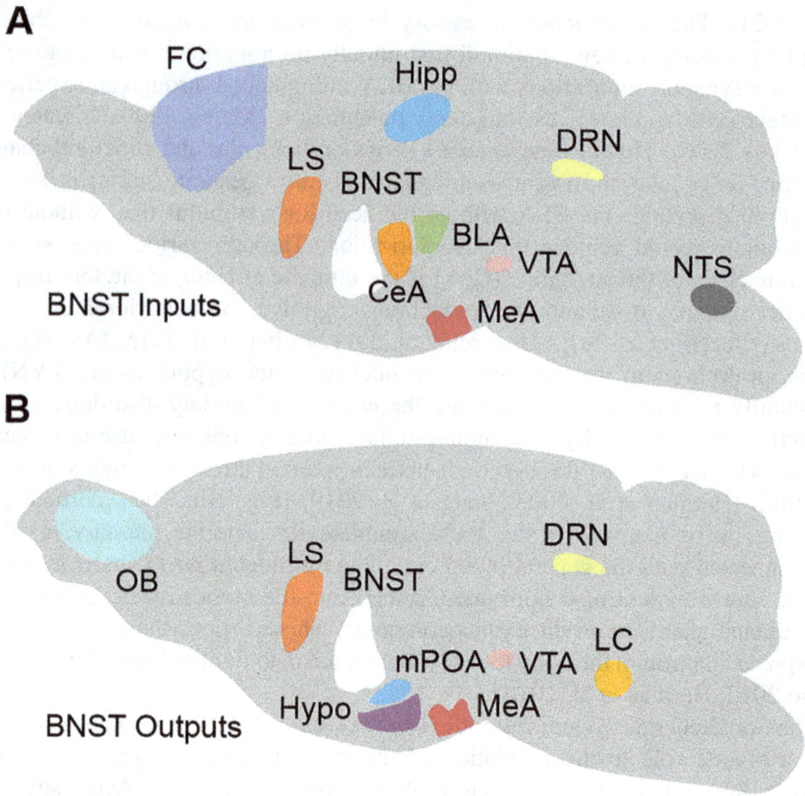

Fig. 2 The BNST is highly connected with regions associated with arousal and anxiety across the brain. (**a**) It serves as an integration hub for limbic inputs and (**b**) a regulator region for other processes through projections to hypothalamic and brainstem regions. FC = frontal cortex, LS = lateral septum, BNST = bed nucleus of the stria terminalis, CeA = central amygdala, BLA = basolateral amygdala, MeA = medial amygdala, Hipp = hippocampus, VTA = ventral tegmental area, DRN = dorsal raphe nuclei, OB = olfactory bulb, mPOA = medial preoptic, Hypo = hypothalamic area, LC = locus coeruleus

of the basal forebrain, commonly referred to as the extended amygdala given its anatomical proximity and relationship, that serves as an integration hub for limbic inputs (Fig. 2a) and a regulator region for other processes through projections to the hypothalamus, the hippocampus, or the midbrain (Fig. 2b) (Hammack et al. 2021). It has been vastly associated with the detection of threats, anxiety, and its characteristic hypervigilance. While the amygdala has been related to defensive responses upon imminent threats, the BNST has been mostly related to the fear of encountering that threat again (Knight and Depue 2019). The BNST also contains an important population of CRF neurons that project to the amygdala to stimulate GABAergic transmission and CRF release at the PVN and induce a stress response (Partridge et al. 2016). Although the BNST has been involved in fear conditioning, it is particularly sensitive to contextual information and it has a crucial role in the discrimination of safety cues from threatening ones (Radke 2009; Foilb et al. 2021). Therefore, it has been hypothesized that the BNST exerts a modulatory role over fear responses related to the induction of a threat-responding state based on environmental cues. In recent years, the BNST has emerged as a critical translational hub for anxiety research (Lebow and Chen 2016). Although researchers have put effort into designing complex behavioral paradigms to differentiate the role of the amygdala and the BNST in anxiety, it is hard to imagine a situation in real life that isolates one of the two components.

1.3 The Human Salience Network and Anxiety Disorders

The salience network or midcingulo-insular network is an intricate cortical system principally conformed by the anterior insula (AI) and the dorsal anterior cingulate cortex (dACC), with influence of temporal, parietal, and prefrontal cortices, as well as subthalamic regions like the amygdala (Chen et al. 2016; Seeley 2019; Schimmelpfennig et al. 2023). It is known for regulating the relevance of incoming stimuli and it also plays the role of an integration hub for different cognitive and emotional processes such as communication and social interactions (Uddin 2017). It is especially relevant in anxiety disorders since it has been involved in impaired detection of threats and overgeneralization of fear memories (Berg et al. 2021). Functional connectivity of the salience network has shown to be stronger under hypervigilance states, as suggested by a strong correlation with anxiety states, and its activity has been considered aberrant in anxiety disorders (Chaposhloo et al. 2023) and depression (Menon 2011). It has been studied in response to emotionally salient stimuli as well as in resting state. When exposed to negative stimuli, anxiety and depressive patients present abnormally enhanced activity of the AI, dACC, and amygdala (Fonzo and Etkin 2017; Wang et al. 2018b; He et al. 2019; Bertocci et al. 2023). In resting state, this connectivity is stronger in anxiogenic situations as demonstrated by a positive correlation with reported pretest anxiety (Seeley et al. 2017). Altered connectivity of the dACC and frontal, parietal, and occipital areas has been reported in panic disorder (Pannekoek et al. 2013). Additionally, functional imaging studies have reported an activation of the dACC, AI, and BNST in phobic patients upon presentation of images related to the threat (Straube et al. 2007) with no involvement of the amygdala.

Several factors have been suggested to contribute to the hyperactivity of the salience network in anxiety disorders. Among these, genetic polymorphisms in the catechol-*O*-methyltransferase (COMT) gene have been strongly associated with anxious personality traits and anxiety disorders (Stein et al. 2005), as they increase dopamine availability in the striatum, leading to difficulty in inhibiting prepotent responses (Simpson et al. 2014). Specifically, the rs737865 and rs4680 polymorphisms of the COMT gene have been associated with increased anxiety, neuroticism, and variations in extraversion (Stein et al. 2005; Montag et al. 2012; Klein et al.

2016). Other genetic polymorphisms have been linked to increased levels of anxiety, specifically those affecting the brain-derived neurotrophic factor (BDNF) (Frustaci et al. 2009), the serotonin transporter 5-HTTLPR (Schinka et al. 2004; Kenna 2012), GABA receptors' subunits (Pham et al. 2009; Olivier and Olivier 2020), and the Hsp90 complex co-chaperone Fkbp5 (Kang et al. 2012; Scheuer et al. 2016; Bryant et al. 2016). Taken together, multiple genetic biomarkers have been identified as predictors of abnormal processing of salient cues, which often lead to heightened anxious states in patients suffering for anxiety-related disorders. Since many of these processes are conserved across species, rodent model can shed light into the intricate mechanisms involved.

2 Rodent Models for Anxiety-Like Behavior

Rodent models for anxiety are prevalent in the scientific literature, and multiple methodological variations exist. These models are characterized by inducing conflict in mice, either through natural circumstances or via training protocols (Rodgers 1997). Unconditioned conflict paradigms, which do not necessitate prior training of the animal to elicit an anxious response, are often referred to as ethological models due to their naturalistic relevance (Carobrez and Bertoglio 2005). Within this category, anxiety measurement tests are typically classified into exploratory behaviors (such as the open field (Walsh and Cummins 1976) and elevated plus maze (Walf and Frye 2007) tests), dark/light transitions (dark/light box test (Bourin and Hascoët 2003)), social-induced conflicts (maternal separation (Vetulani 2013)), defensive behaviors (predator odor avoidance (Staples 2010; Albrechet-Souza and Gilpin 2019)), and physiological responses (stress-induced hyperthermia (Adriaan Bouwknecht et al. 2007) and freezing (Hagenaars et al. 2014)). In contrast, learned conflict tests require prior training of the animals. In these paradigms, animals are faced with a conflict where they must choose between executing a trained response (e.g., drinking sucrose water from a bottle in the Vogel test (Millan and Brocco 2003)) or avoiding an aversive event associated with that response (e.g., foot shock in the active avoidance models (Spiegler et al. 2020)).

Typically, these tests enable the assessment of behavioral and physiological responses in animals, allowing inference about anxiety states (such as avoidance time, freezing behavior, blood corticosterone levels, etc.). However, the manifestation of certain components of the anxious state requires the tracking of preparatory behavioral and physiological responses, which determine the animal's survival within fractions of a second. Currently, such measurements involve tracking heart rate, respiratory rate, pupil size, and/or skin electrodermal responses. Nevertheless, the current state-of-the-art technology significantly limits the combination of these data collection techniques with rodent anxiety models. Primarily, this limitation arises from the requirement for the animal to be fixed or motion-restricted to utilize these devices, greatly restricting the applicability of these measures. Next, we will describe the main methodologies for measuring arousal in rodent anxiety models, considering two modalities: ocular and visceral dynamics.

3 Strategies for Monitoring Arousal in Mice

The initial strategies that have allowed us to obtain measurements of these rapid and dynamic arousal processes in anxiogenic settings are based on electrical or mechanical recordings of signals emitted by the organism, obtained through electrodes or piezoelectric sensors placed at strategic locations on the animal's body. However, these techniques often do not support long-term monitoring, require animal habituation to mitigate stress effects, may necessitate invasive surgeries to implant electrodes, and are prone to artifacts from the animal's unpredictable movements (Ho et al. 2011; Calvet and Seebeck 2023). The diversity of methodologies and devices complicates data comparison across different systems, necessitating cautious interpretation (Calvet and Seebeck 2023). Among these measurements, pupil size, heart rate, breathing rate, skin temperature, or blood pressure are the most abundant in the literature, as they have shown to be reliably affected by manipulations on arousal hubs.

3.1 Ocular Dynamics

Changes in pupil size and facial movements around the eye such as blinks can be associated with both rapid arousal responses as well as long term state changes (Wang et al. 2018a; Kim et al. 2023). Pupil size and blinking frequency have been found to increase with increased mental effort or arousal (Gorin et al. 2024). Other more complex pupillary outputs such as tracing the behavior of saccades or counting pupillary spikes can reveal additional information related to environmental uncertainty and fatigue (Mahanama et al. 2022; Chen et al. 2022; Fink et al. 2023). To capture all relevant information from the pupil, camera-based pupillometry is commonly used (Fig. 3a). This technique involves placing a camera in front of the eye and imaging the surrounding area (Manno 2009; Privitera et al. 2020; Lara-Doña et al. 2021). Beyond ensuring sufficient spatial resolution to capture complex pupillary dynamics and temporal resolution to capture rapid events like blinks (which occur within 100 ms), cameras used in these systems have minimal hardware requirements. Head-fixation, while enabling the collection of stable data in a controlled environment, produces data that does not often generalize well to naturalistic behavior. Elevated corticosterone and decreased body weight, two indicators of anxiogenesis, have been observed in animals undergoing head fixation even after 25 days of habituation (Juczewski et al. 2020). Freely moving experiments offer animals the ability to perform their full repertoire of natural behavior; however, the cameras used in traditional pupillometry have historically restricted the technique to

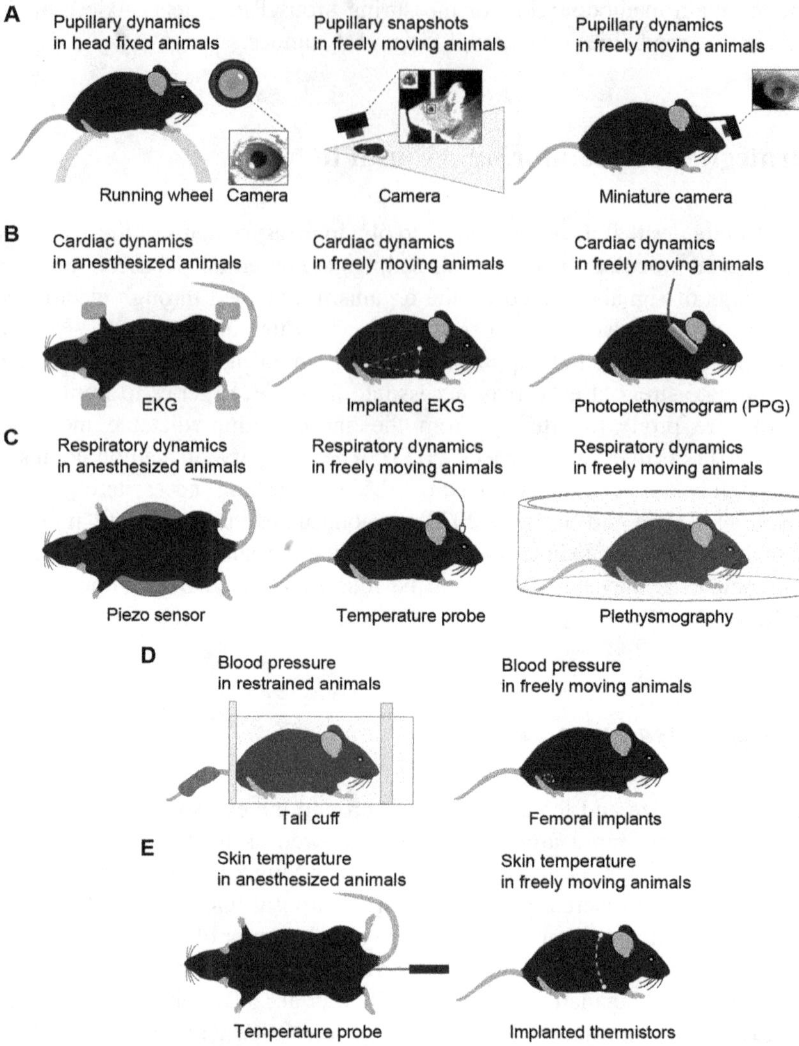

Fig. 3 Techniques for the measurement of pupillary and visceral arousal dynamics in mice. (**a**) Examples of experimental approaches to measure pupillary responses in head-fixed and freely moving setups. (**b**) Examples of experimental approaches to measure heartbeat dynamics in anesthetized and freely moving mice using electrocardiography (EKG) or photoplethysmography (PPG). Traditional electrocardiogram (EKG) setup involves electrodes placed on the skin surface of the mouse to record electrical signals of the heart, providing valid measurements of HR and cardiac physiology. Telemetry-based EKG allows for long-term HR monitoring in freely moving mice, requiring surgical implantation of telemetry systems under the skin. Photoplethysmography (PPG) collars provide a noninvasive method for monitoring HR, with sensors placed around the neck to detect blood volume changes in superficial vessels. Both EKG and PPG methods enable the extraction of two key metrics: HR, the frequency of heartbeats; and heart rate variability (HRV), reflecting fluctuations in time between consecutive heartbeats. (**c**) Setups for the measurement of breath rate dynamics in anesthetized and freely moving mice using piezoelectric sensors, temperature sensors, or PPG. Methods for measuring blood pressure (**d**) and skin temperature (**e**) in mice. Tail cuff method measures blood pressure by inflating a cuff around the tail and recording arterial

head-fixed experimentation. One way around the hardware constraints is to position the camera in an arena and record the mouse's pupil as it passes into frame. Stimulus can be introduced near the camera, and the arena can be designed to encourage the mouse to remain in focus for as long as possible to capture the pupil at a critical experimental moment. Even with these enhancements, the pupil information gathered is still often limited to a few in-focus frames losing important pupillary dynamics. Alternatively, the camera can be placed on the mouse to record the pupil continuously as it navigates its environment. Cameras have been successfully mounted directly in front of the eye as well as on the head, with a mirror used to direct the image of the pupil to the camera without obscuring the mouse's field of view (Meyer et al. 2018).

3.2 *Visceral Dynamics*

Heart rate is perhaps one of the most classical methodologies to assess rapid arousal responses to anxiogenic stimuli. Stressful stimuli induce an increase in heart rate through a sympathetic activation, parasympathetic withdrawal, and stimulation of glucocorticoid receptors in the heart muscle (Nicolaides et al. 2014; Lowrance et al. 2016). The first methodology for an electrocardiogram (EKG) in mice was described by Goldbarg et al. in 1968 in C57BL/10 and SEC/1 mice (Goldbarg et al. 1968) (Fig. 3b). Briefly, using electrodes strategically placed on the skin surface of the animal, electrical signals of the heart are continuously recorded (Ho et al. 2011). It provides valid measurements of heart rate and the physiology of heart contraction and dilation, with the main caveat that requires a steady placement of the electrodes to avoid interference in the signal. Novel methodologies based on telemetry have been developed to permit the applicability of the EKG to freely moving settings. Beyond electrical methods such as EKG, there are also optical methods like photoplethysmography (PPG). PPG works by emitting light into the skin, and measuring the amount of light absorbed or reflected by the blood vessels. Whether electrical or optical, the readouts most relevant for arousal are heart rate (HR) and heart rate variability (HRV). HR is a critical arousal read-out with both tachycardia and bradycardia being associated with high arousal behavioral states. While high HR is classically associated with anxiety, decreases in HR has also been found to be associated with high fear states. Heart rate variability (HRV), which gives even more detailed information about the variation in time intervals between heartbeats, offers deeper insights into physiological responses and overall health.

Fig. 3 (continued) pressure changes. Femoral implantation involves surgically implanting a sensor near the femoral artery to directly measure blood pressure. Rectal temperature probe assesses core temperature by inserting a probe into the rectum, while implanted temperature probe monitors core temperature continuously

HRV, understood as the fluctuations in time between consecutive heartbeats, is reflective of the relationship between sympathetic activity (SA) and parasympathetic activity (PSA), offering deeper insights into physiological arousal than HR alone (Shaffer and Ginsberg 2017). While sympathetic autonomous activity is mostly associated with the fight or flight response (Jansen et al. 1995), the parasympathetic autonomic nervous system has been related to the rest and digest response (McCorry 2007; Tindle and Tadi 2024), thus considered to be antagonistic to keep a balanced state of awareness. PSA indices in HRV measurements have been shown to be decreased in chronic anxiety and have been proposed as an endophenotype for chronic anxiety disorders. During resting state, PSA is dominant over SA on the control of heartbeat, and the vagus nerve is the anatomical connection between PSA-generating activity centers in the brain and the heart, therefore making HRV an ideal measure of cardiac vagal tone (Shaffer et al. 2014). Low HRV has been related to impoverished cardiac health, cardiac death, and chronic anxiety. Anxiety disorders are considered one of the main psychological risk factors for cardiac disease, and it is associated with death in those with a preexisting cardiac condition (Emdin et al. 2016). Since HRV has proven to be affected not just in anxiety disorders but also in a greater number of psychopathologies across the different spectra (Beauchaine and Thayer 2015; Heiss et al. 2021), it has been proposed as a biomarker of emotional self-regulation under stressful conditions (Beauchaine and Thayer 2015), a hallmark that is disrupted in anxiety disorders as well as in other kind of mental conditions such as major depressive disorder and schizophrenia (Montaquila et al. 2015; Hartmann et al. 2019; Dell'Acqua et al. 2020; Liu et al. 2021).

Respiratory dynamics, including respiratory rate (RR), are also highly linked with both neural activity and behavioral state. Patterns of respiratory oscillation were recently identified that match activity pulses in the prelimbic prefrontal cortex and occur only during conditioned fear-induced freezing behavior (Moberly et al. 2018; Bagur et al. 2021). In restrained animals, techniques such as using piezoelectric sensors, which rely on the contraction movement that occurs with every breath, or computer vision methods to track chest wall contraction and expansion are commonly employed (Fig. 3c). Due to respiration cycles occupying a similar frequency band to motion artifacts, few techniques exist for measuring respiration in freely moving animals (Grimaud and Murthy 2018; Pais-Roldán et al. 2018). Plethysmography chambers are an alternative option for measuring respiration that requires no invasive surgeries (Glaab et al. 2007; Prada-Dacasa et al. 2020). These chambers monitor respiratory parameters by measuring pressure changes inside an airtight chamber as the subject breathes. While accurate, these chambers often must be extremely small to derive respiratory information, restricting the repertoire of behavioral experimentation that can be performed.

Recently, another method was introduced for simultaneously monitoring both respiration and cardiac dynamics in freely moving mice using a mechanico-acoustic approach (Lee et al. 2020), offering a more comprehensive understanding of physiological responses. This involved implanting an inertial measurement unit into the chest wall to collect the small movements and sounds associated with viscera. The rationale behind tracking multiple biometrics, such as HR and RR, simultaneously is rooted in the understanding that each individual metric can be influenced by processes parallel to arousal but not directly related. For instance, respiration may increase due to sniffing a novel odor, while HR may increase solely due to movement. By collecting data on both respiratory and cardiac dynamics simultaneously, we gain a more comprehensive overview of the arousal state of the animal, allowing us to discern between various physiological responses and external stimuli. Apart from benefiting from a rapid identification of anxiety states that could drive to prompt intervention, cardiac and respiratory monitoring techniques form the basis of biofeedback that has also been proven to be a valuable resource in the self-management of anxiety or panic attacks (Banerjee and Argáez 2017; Alneyadi et al. 2021).

Similar to HRV, blood pressure variability (BPV) has shown to be associated with cardiovascular disease and anxiety states (de la Sierra 2023). Using pulse intervals and systolic arterial pressure, acute stress has been characterized by an increase in BPV and the power of low frequency oscillations, while chronic stress is characterized by a marked decrease in BPV (Farah et al. 2004). While there is not a clear association between hypertension and mood disorders, the long-term exposure to autonomic alterations typical of chronic anxiety may cause a long-term impact on blood pressure physiology, leading to hypertensive syndromes in a considerable proportion of patients (Wu et al. 2014). Measuring blood pressure in restrained animals normally involves a cuff placed on the tail or limbs (Fig. 3d). The cuff inflates and deflates, intermittently occluding blood flow in the tail artery. Sensors detect the pulsatile blood flow, allowing for determination of systolic and diastolic pressures. While this method is noninvasive and suitable for longitudinal studies, it may cause stress to the animal and yield less accurate results compared to invasive methods. Femoral artery implantation involves surgically exposing the femoral artery in the mouse and inserting a catheter connected to pressure transducers. This method provides continuous and direct measurements of arterial blood pressure, offering high accuracy and reliability. However, it requires surgical expertise and may introduce confounding factors due to the invasiveness of the procedure.

One of the components of the stress response is a peripheral vasoconstriction that drives peripheral blood into the core (Kim et al. 2019). Vasoconstriction, in conjunction with stress-induced thermogenesis, have a temperature rising effect in the organism in the presence of a threat that is observable seconds after the threat appearance (Meyer et al. 2010; Bastías-Pérez et al. 2020). The stress-induced hyperthermia (SIH) is a paradigm with multiple variations, in which the temperature change from a basal condition is assessed in response to an acute stressor, in a degree that is proportional to the stressor intensity (Adriaan Bouwknecht et al. 2007). During acute stress, sympathetic activation leads to the release of epinephrine and norepinephrine that stimulate thermogenesis through activation beta-3 adrenergic receptor in adipocytes in the brown adipose tissue to facilitate the catabolism of fats and the release of free fatty acids (FFA) into the bloodstream (Guilherme et al. 2020; Auger and Kajimura 2021; Cero et al. n.d.). Once these FFA reach the skeletal muscle, stimulation of beta-3 adrenergic receptors in the myocytes increases mitochondrial activity through cAMP- and PKA-mediated mechanisms. Additionally, the increased expression of uncoupling protein 1 (UCP1) in the brown adipose tissue uncouples oxidative phosphorylation from ATP formation, leading to thermogenesis instead of ATP synthesis (Fedorenko et al. 2012).

Different methods have been employed to measure this temperature rise (Barrows and Dodds 1942; Kawakami et al. 2018; Takahashi et al. 2023) (Fig. 3e). The use of probes to assess core temperature is probably the most widespread methodology for its low cost and reproducibility, but it requires a high degree of manipulation of the animal since it needs to be restrained to prevent the probe from moving. Besides, the mere insertion of the probe into the animal's rectum supposes an invasive form of stress that generates itself an increase in temperature and sympathetic tone. In contrast, measurements of skin temperature using infrared thermography (IRT) are noninvasive and continuous, allowing a real-time monitoring of the animal temperature during baseline periods and during the stress response (Gjendal et al. 2018; Blenkuš et al. 2022). The use of IRT has been demonstrated to accurately track patterns of thermal changes in a within-subject manner, with reduced reliability when assessing absolute temperature values (van der Vinne et al. 2020). Besides, novel methodologies are capturing a differential thermal regulation in the head and back compared to the paws and tail upon stress, concordant with previous knowledge describing peripheral vasoconstriction in the presence of threats (Usamentiaga et al. 2014).

4 Single-Cell Strategies to Monitor and Manipulate the Neural Circuits of Arousal and Anxiety

To ascertain the role of circuitry in driving changes in arousal, researchers have both neuroimaging and stimulation techniques at their disposal. Two-photon (2P) imaging offers enhanced resolution, particularly at specific *z* planes (Fig. 4a). Utilizing a 2P excitation process, photons are absorbed simultaneously by a fluorophore, enabling imaging at greater depths with reduced scattering. Traditionally confined to head-fixed setups, recent advancements include the development of miniaturized head-mounted 2P systems, with minimal behavioral interference. One-photon miniscopes are more prevalent for monitoring neural activity in freely moving mice due to their simpler light path and ease of miniaturization. This setup involves a single-photon excitation process, allowing for compact designs (Fig. 4b). However, because light from the visible spectrum is used, scattering is more prevalent reducing *z* specificity. Additionally, more light is required to reach areas of interest, leading to an increased risk of photobleaching cells. Despite this, miniscopes have been pivotal in uncovering the circuitry of arousal and motivation (Rodriguez-Romaguera et al. 2022).

While imaging can unveil correlations between neural activity and behaviors, it falls short in establishing direct causal relationships (Stamatakis et al. 2021). One

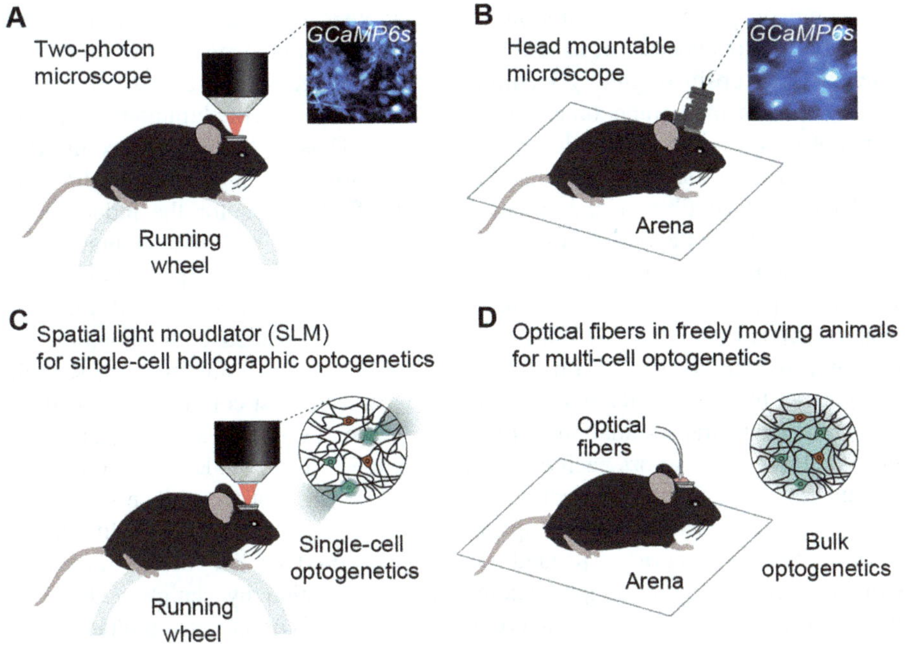

Fig. 4 Optical techniques for stimulating and imaging neurons in mice. (**a**) Two-photon imaging of a head-fixed mouse allows high-resolution imaging of neural activity in specific brain regions. (**b**) An unrestrained mouse with a head-mounted miniscope enables imaging of neuronal activity during natural behavior. (**c**) Holographic stimulation targets single neurons within the same genetic subgroup, enabling investigation of neural function. (**d**) Bulk optogenetic stimulation involves stimulating a large area of the brain, activating specific genetic subgroups of neurons

notable approach of achieving this goal is via the manipulation of genetically labeled individual neurons. This approach involves selecting a population of neurons relevant to behavior, such as arousal and anxiety, based on the expression of a specific peptide (Madisen et al. 2010; Taniguchi et al. 2011; Gerfen et al. 2013). To achieve this, the selected neurons are targeted by a viral vector containing the mechanism to induce the expression of a protein of interest in the target cells. These proteins of interest can have various functions, but in systems neuroscience, two main approaches are employed: chemogenetics and optogenetics. In chemogenetics, the protein expressed in the target cells is a modified version of a receptor, typically a muscarinic receptor coupled to a G protein (Campbell and Marchant 2018; Kang et al. 2023). It is hypothesized that this receptor lacks an endogenous ligand potent enough to activate it naturally but can be activated by exogenous ligands such as clozapine-*N*-oxide (CNO). Notwithstanding CNO is reverse-metabolized into clozapine producing some discreet effects on the rodent central nervous system (Manvich et al. 2018). To overcome this difficulty, other ligands with lower side effects have been developed for the same purpose (Bonaventura et al. 2019). However, the activation of this receptor depends on the pharmacokinetics of the exogenous ligand, limiting the temporal resolution of the manipulation and making it dependent on plasma concentration and metabolism of the compound (Jendryka et al. 2019; Bonaventura et al. 2019). In contrast, in optogenetics, the protein of interest is an opsin originally identified in algae (Fig. 4c, d) (Deisseroth 2015). Opsins are transmembrane ion channels that respond to specific luminous stimuli in terms of power and wavelength (Guru et al. 2015; Deisseroth 2015). Once inserted into the target cells, these proteins can respond to a luminous stimulus from an implanted optical fiber in the infected region. This allows for the induction of responses such as depolarization or hyperpolarization of neurons, facilitating the study of the necessity and sufficiency of a neuron population in specific behaviors with high spatial and temporal resolution (Guru et al. 2015; Rodriguez-Romaguera et al. 2022). However, relying solely on genetic subtypes to dictate functionality may oversimplify neural circuits' complexity. Even neurons expressing the same specific peptide can have dramatically different functions. The most common approach for more targeted neural stimulation employs two-photon excitation, scanning a focused infrared laser beam across each cell body of interest (Tong et al. 2023). While the nonlinearity of technique restricts the laser to a small spot to achieve single-cell resolution, many photosensitive opsins deactivate rapidly, making it difficult to achieve the photocurrent needed to reliably produce an action potential. Computer-generated holography addresses this limitation by sculpting the laser beam into a custom 3D pattern that is delivered to all cells of interest simultaneously, in contrast to bulk optogenetic approaches that modulate all neurons that genetically express a particular opsin (Szabo et al. 2014; Pégard et al. 2017; Eybposh et al. 2022). This technique offers unprecedented control, allowing researchers to dissect neural circuits with precision and uncover their contributions to arousal regulation.

Optogenetic and chemogenetic manipulations in mouse models of anxiety have identified key circuits involved in increased arousal and anxiety. The fine dissection of neuronal circuits allowed by optogenetics has fueled the identification of functional units within the brain, referring to specific neuronal populations with an identitary neuropeptidergic, anatomical and functional correlation that support a specific function (Nieh et al. 2013). Using channelrhodopsin 2 (ChR2)-induced optogenetic stimulation and halorhodopsin (NpHR)-induced inhibition of BLA pyramidal neurons projections to the centrolateral amygdala (CeL), Tye et al. (2011) uncovered the anxiolytics role of these neurons during the elevated plus maze and open field tests. The bed nucleus of the stria terminalis (BNST) has shown to be involved in contextual fear memory and anxiety states, through inputs from the BLA and the CeA, as well as regulating fear states by its projections to the CeA (Dong et al. 2001; Alheid 2003; Dong and Swanson 2004). ChR2 stimulation of CeA GABAergic projections into the BNST increases the inhibitory postsynaptic currents in the BNST as measured by whole-cell patch-clamp, an effect that is reduced by kappa opioid receptor agonist (KOR) norbinaltorphimine (Li et al. 2012). Using chemogenetics, Milosavljevic et al. (2016) found that, in dark-housed mice, stimulation of melanopsin retinal ganglion cells induces an arousal and anxiety that correlate with neural signatures of these states like hypothalamic or limbic cFos expression (Milosavljevic et al. 2016). Chemogenetic stimulation of BLA excitatory β-adrenergic signaling induces acute and social anxiety (Siuda et al. 2016), similar to the chemogenetic stimulation of CaMKIIα forebrain excitatory cells (Salvi et al. 2019). These approaches have uncovered the specific role of functional units in the rodent brain involved in the development and maintenance of anxiety disorders.

More recently, Hsueh et al. (2023) developed an optical strategy to manipulate the heart cells in order to test a possible cardiogenic control of anxiety-like behaviors in mice. In this study, the researchers investigate the relationship between emotional states and physiological processes, particularly focusing on the influence of heart rate on anxiety and fear responses. While it is commonly understood that anxiety can cause the heart to beat faster, it is less clear whether an increased heart rate could itself induce anxiety or fear. The researchers developed a noninvasive optogenetic pacemaker that allows precise control of cardiac rhythms in mice. Optically induced tachycardia significantly increased anxiety-like behavior in mice, but only in risky contexts, suggesting that central (brain) and peripheral (body) are integrated to create the subjective experience of emotion.

5 The Bed Nucleus of the Stria Terminalis (BNST) as a Hub to Integrate Arousal and Anxiety-Like Behavior

The bed nucleus of the stria terminalis (BNST) is a deep brain region widely implicated in the regulation of motivation and reward. Its complex network of neuropeptidergic populations reflects the various processes integrated within it to regulate motivated responses (Ortiz-Juza et al. 2021). Recent research has underscored the fundamental role of the BNST in dynamically regulating arousal processes in response to emotionally salient stimuli (Rodriguez-Romaguera et al. 2020). Among the different neuronal populations in the BNST, neurons expressing the *prepronociceptin* (*Pnoc*) gene have shown to play a significant role in arousal responses to motivational stimuli. $BNST^{Pnoc}$ neurons are a subset of GABAergic neurons primarily located in the anterodorsal region of the BNST, with dense projections to the MeA and mPOA. We recently studied the role of $BNST^{Pnoc}$ neurons in regulating rapid arousal responses using trimethylthiazoline (TMT), a compound present in fox urine that elicits fear and avoidance responses in rodents. Employing two-photon calcium imaging approaches, Rodriguez-Romaguera et al. (2020) found robust correlations between the activity of $BNST^{Pnoc}$ neurons and pupil size dynamics (Fig. 5a). Interestingly, this correlation was observed only in the presence of aversive salient odor TMT (or rewarding odor peanut oil), suggesting a possible role of these neurons in regulating rapid arousal responses to motivationally salient cues. Furthermore, optogenetic activation of $BNST^{Pnoc}$ neurons led to an increase in physiological arousal measurements (pupillary response and heart rate) and modulation of anxiety-like behavior without inducing approach, avoidance, or locomotion (Fig. 5b). These findings suggest a crucial role for $BNST^{\mathrm{Pnoc}}$ neurons in coordinating arousal-related responses associated with motivationally salient stimuli (Rodriguez-Romaguera et al. 2020).

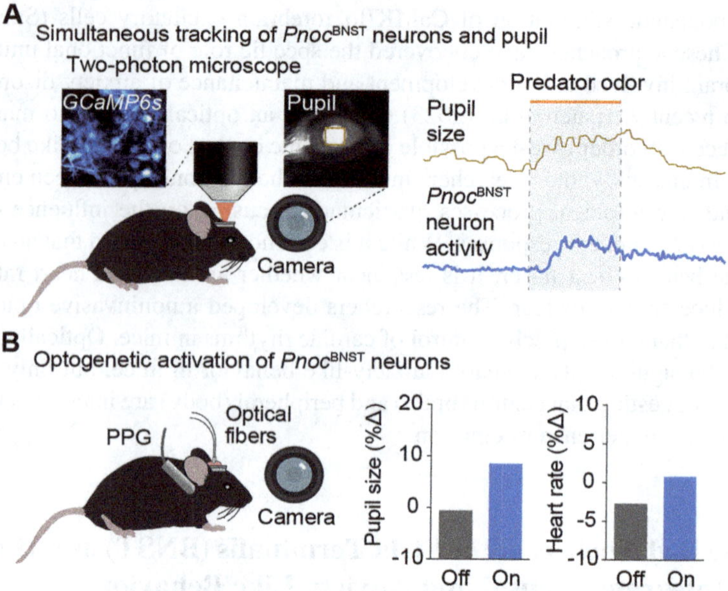

Fig. 5 Tracking and manipulating *prepronociceptin* (*Pnoc*) neurons in the BNST. (**a**) Two-photon imaging of a head-fixed mouse allows the analysis of BNST^{Pnoc} neuron activity with pupillary dynamics in head-fixed setups. (**b**) The use of optical fibers to optogenetically activate or silence specific functional units in the brain is compatible with technology for arousal tracking as PPG or camera-based pupillometry. Data depicting published findings from Rodriguez-Romaguera et al. (2020)

Additionally, using neuronal labeling studies, researchers have isolated the role of certain projections in states of hypervigilance and anxiety. For example, projections from the BNST to the lateral hypothalamic area have been shown to be necessary for the anxiety-like response in the elevated plus maze (EPM), whereas projections to the parabrachial nucleus are necessary for the regulation of autonomic arousal (Kim et al. 2013; Kodani et al. 2017). Importantly, BNST interactions with brainstem nuclei such as the locus coeruleus (LC) or the nucleus tractus solitarius (NTS) have emerged as promising candidates to understand the complex interaction between motivational states and rapid changes in arousal (Scheuer et al. 2017; Kawai 2018), although these interactions remain elusive.

6 Concluding Remarks: Translating Studies Across Species

Technological advancements in animal research have paralleled the formulation of increasingly precise questions, allowing us to detect processes that, within fractions of a second, determine an organism's readiness to respond to relevant stimuli (Rodriguez-Romaguera et al. 2022). This process of arousal or preparatory response to potent stimuli involves not a singular mechanism, but rather a multitude of distinct processes. These processes are primarily orchestrated by norepinephrinergic nuclei in the brainstem, with the additional involvement of multiple catecholaminergic and neuropeptidergic systems. These neural circuits interact with deep brain nuclei, including the amygdala complex, the BNST, and the hypothalamic–pituitary–adrenal axis. Additionally, cortical regions regulate subsequent fear responses and anxiety-like behaviors.

The methodologies available to study arousal processes in humans are extensive, owing to the versatility of techniques that can be applied not only in research but also for clinical and diagnostic purposes (Coull 1998; Zsidó 2024; Kiss and Linnell 2024). The fundamental principles characterizing these techniques are their low invasiveness and external approach, which minimize harm and make them highly ecological (Tinga et al. 2020). Among these methods, electrocardiographic, thermographic, and blood pressure measurements are relatively established techniques that have been adapted for various applications, ranging from clinical diagnostics to wearable devices (Guk et al. 2019). These devices, such as smartwatches, can automatically record large amounts of data without requiring active participation from the user.

Pupillometric measurements, on the other hand, have benefited greatly from advancements in eye-tracking technology. Eye trackers have revolutionized research in neuroscience, psychology, and even other disciplines like marketing by providing insights into both pupil size and gaze patterns (Holmqvist et al. 2011; Hutton 2019). The evolution of eye trackers can be divided into three stages. Initially, stationary eye trackers that require the use of chin rests to minimize movement. The next generation that allows for head movement while still requiring the subject to remain relatively still. Lately, the most recent advancements have focused on developing wearable and mobile eye trackers, such as those integrated into glasses, capable of measuring both pupil size and gaze in real-world settings (Valliappan et al. 2020; Ivanchenko et al. n.d.).

While wearable eye-tracking devices still face challenges such as high costs and limited accessibility to the general population, hemodynamic measurement devices (e.g., blood pressure monitors, electrocardiograms, and thermometers) have been successfully integrated into more accessible and affordable wearables like smartwatches. These devices produce reliable and efficient measurements, making them widely available to the public (Konstantinidis et al. 2022; Pan et al. 2024). The use of such devices in clinical settings, particularly for anxiety disorders, is invaluable. They provide real-time information about autonomic arousal, a variable that can be directly linked to the onset of anxiety or panic attacks. The convergence of similar techniques in animal models of anxiety and the increasingly wearable, noninvasive, and cost-effective devices for measuring these variables in humans enhances the apparent validity of translational research (Marková 2018; Pratt et al. 2022). This alignment between animal and human studies is pivotal for the advancement of translational neuroscience.

In the study of physiological responses as indicators of arousal, a multitude of methodologies are employed. While these techniques are not typically associated with arousal research, they provide relevant information. Notably, measuring heart rate and blood pressure has shed light on the relationship between their variability and the presence of anxiety disorders. Although these methodologies are highly informative, their application to animal models for anxiety remains challenging due to the technical need for electrode implantation or restriction of animal movements. More adapted techniques include pupillometry in head-fixed mice. Specifically, the combination of optogenetic activation with pupillometry has revealed that driving the activity of neurons in the BNST is sufficient to drive arousal in the absence of motivated behaviors (Rodriguez-Romaguera et al. 2020), suggesting that these two processes are distinct.

This chapter underscores the necessity of technological development focused on integrating manipulation and neural recording techniques with specific measurements of rapid physiological processes. Efforts to combine these techniques in anxiety disorder research represent a crucial step toward identifying new functional units that disentangle the neural circuits that regulate arousal from those that regulate anxiety-like behaviors.

References

Adolphs R, Gosselin F, Buchanan TW et al (2005) A mechanism for impaired fear recognition after amygdala damage. Nature 433:68–72. https://doi.org/10.1038/nature03086

Adriaan Bouwknecht J, Olivier B, Paylor RE (2007) The stress-induced hyperthermia paradigm as a physiological animal model for anxiety: a review of pharmacological and genetic studies in the mouse. Neurosci Biobehav Rev 31:41–59. https://doi.org/10.1016/j.neubiorev.2006.02.002

Albrechet-Souza L, Gilpin NW (2019) The predator odor avoidance model of post-traumatic stress disorder in rats. Behav Pharmacol 30:105–114. https://doi.org/10.1097/FBP.0000000000000460

Alheid GF (2003) Extended amygdala and basal forebrain. Ann N Y Acad Sci 985:185–205. https://doi.org/10.1111/j.1749-6632.2003.tb07082.x

Alneyadi M, Drissi N, Almeqbaali M, Ouhbi S (2021) Biofeedback-based connected mental health interventions for anxiety: systematic literature review. JMIR Mhealth Uhealth 9:e26038. https://doi.org/10.2196/26038

American Psychiatric Association (2013) Diagnostic and statistical manual of mental disorders. In: DSM Libr. https://dsm.psychiatryonline.org/doi/book/10.1176/appi.books.9780890425596. Accessed 31 Mar 2024

Arrieta-Cruz I, Pfaff DW (2009) Definition of arousal and mechanistic studies in intact and brain-damaged mice. Ann N Y Acad Sci 1157:24–31. https://doi.org/10.1111/j.1749-6632.2008.04118.x

Association AP (2015) Anxiety disorders: DSM-5® selections. American Psychiatric Publishing

Auger C, Kajimura S (2021) Detouring adrenergic stimulation to induce adipose thermogenesis. Nat Rev Endocrinol 17:579–580. https://doi.org/10.1038/s41574-021-00546-6

Bagur S, Lefort JM, Lacroix MM et al (2021) Breathing-driven prefrontal oscillations regulate maintenance of conditioned-fear evoked freezing independently of initiation. Nat Commun 12: 2605. https://doi.org/10.1038/s41467-021-22798-6

Bandelow B, Michaelis S (2015) Epidemiology of anxiety disorders in the 21st century. Dialogues Clin Neurosci 17:327–335

Banerjee S, Argáez C (2017) Neurofeedback and biofeedback for mood and anxiety disorders: a review of clinical effectiveness and guidelines. Canadian Agency for Drugs and Technologies in Health, Ottawa

Barger N, Hanson KL, Teffer K et al (2014) Evidence for evolutionary specialization in human limbic structures. Front Hum Neurosci 8:277. https://doi.org/10.3389/fnhum.2014.00277

Barrows EF, Dodds H (1942) Body temperature of mice during anesthesia. Am J Physiol-Leg Content 137:259–262. https://doi.org/10.1152/ajplegacy.1942.137.2.259

Bastías-Pérez M, Zagmutt S, Soler-Vázquez MC et al (2020) Impact of adaptive thermogenesis in mice on the treatment of obesity. Cells 9:316. https://doi.org/10.3390/cells9020316

Beauchaine TP, Thayer JF (2015) Heart rate variability as a transdiagnostic biomarker of psychopathology. Int J Psychophysiol Off J Int Organ Psychophysiol 98:338–350. https://doi.org/10.1016/j.ijpsycho.2015.08.004

Bekker MHJ, van Mens-Verhulst J (2007) Anxiety disorders: sex differences in prevalence, degree, and background, but gender-neutral treatment. Gend Med 4:S178–S193. https://doi.org/10.1016/S1550-8579(07)80057-X

Berg H, Ma Y, Reuter A et al (2021) Salience and central executive networks track overgeneralization of conditioned-fear in post-traumatic stress disorder. Psychol Med 51: 2610–2619. https://doi.org/10.1017/S0033291720001166

Berridge CW, Waterhouse BD (2003) The locus coeruleus-noradrenergic system: modulation of behavioral state and state-dependent cognitive processes. Brain Res Brain Res Rev 42:33–84. https://doi.org/10.1016/s0165-0173(03)00143-7

Bertocci MA, Afriyie-Agyemang Y, Rozovsky R et al (2023) Altered patterns of central executive, default mode and salience network activity and connectivity are associated with current and future depression risk in two independent young adult samples. Mol Psychiatry 28:1046–1056. https://doi.org/10.1038/s41380-022-01899-8

Bing G, Stone EA, Zhang Y, Filer D (1992) Immunohistochemical studies of noradrenergic-induced expression of c-fos in the rat CNS. Brain Res 592:57–62. https://doi.org/10.1016/0006-8993(92)91658-2

Bing GY, Filer D, Miller JC, Stone EA (1991) Noradrenergic activation of immediate early genes in rat cerebral cortex. Brain Res Mol Brain Res 11:43–46. https://doi.org/10.1016/0169-328x(91)90019-t

Blenkuš U, Gerós AF, Carpinteiro C et al (2022) Non-invasive assessment of mild stress-induced hyperthermia by infrared thermography in laboratory mice. Anim Open Access J MDPI 12:177. https://doi.org/10.3390/ani12020177

Bonaventura J, Eldridge MAG, Hu F et al (2019) High-potency ligands for DREADD imaging and activation in rodents and monkeys. Nat Commun 10:4627. https://doi.org/10.1038/s41467-019-12236-z

Bonnet MH, Arand DL (1997) Hyperarousal and insomnia. Sleep Med Rev 1:97–108. https://doi.org/10.1016/s1087-0792(97)90012-5

Bourin M, Hascoët M (2003) The mouse light/dark box test. Eur J Pharmacol 463:55–65. https://doi.org/10.1016/s0014-2999(03)01274-3

Bryant RA, Felmingham KL, Liddell B et al (2016) Association of FKBP5 polymorphisms and resting-state activity in a frontotemporal–parietal network. Transl Psychiatry 6:e925–e925. https://doi.org/10.1038/tp.2016.149

Calvet C, Seebeck P (2023) What to consider for ECG in mice – with special emphasis on telemetry. Mamm Genome 34:166–179. https://doi.org/10.1007/s00335-023-09977-0

Campbell EJ, Marchant NJ (2018) The use of chemogenetics in behavioural neuroscience: receptor variants, targeting approaches and caveats. Br J Pharmacol 175:994–1003. https://doi.org/10.1111/bph.14146

Carobrez AP, Bertoglio LJ (2005) Ethological and temporal analyses of anxiety-like behavior: the elevated plus-maze model 20 years on. Neurosci Biobehav Rev 29:1193–1205. https://doi.org/10.1016/j.neubiorev.2005.04.017

Cero C, Lea HJ, Zhu KY et al (n.d.) β3-Adrenergic receptors regulate human brown/beige adipocyte lipolysis and thermogenesis. JCI Insight 6:e139160. https://doi.org/10.1172/jci.insight.139160

Chaposhloo M, Nicholson AA, Becker S et al (2023) Altered resting-state functional connectivity in the anterior and posterior hippocampus in post-traumatic stress disorder: the central role of the anterior hippocampus. NeuroImage Clin 38:103417. https://doi.org/10.1016/j.nicl.2023.103417

Chen J-T, Kuo Y-C, Hsu T-Y, Wang C-A (2022) Fatigue and arousal modulations revealed by saccade and pupil dynamics. Int J Environ Res Public Health 19:9234. https://doi.org/10.3390/ijerph19159234

Chen T, Cai W, Ryali S et al (2016) Distinct global brain dynamics and spatiotemporal organization of the salience network. PLoS Biol 14:e1002469. https://doi.org/10.1371/journal.pbio.1002469

Ciocchi S, Herry C, Grenier F et al (2010) Encoding of conditioned fear in central amygdala inhibitory circuits. Nature 468:277–282. https://doi.org/10.1038/nature09559

Coull JT (1998) Neural correlates of attention and arousal: insights from electrophysiology, functional neuroimaging and psychopharmacology. Prog Neurobiol 55:343–361. https://doi.org/10.1016/s0301-0082(98)00011-2

Dalla Corte A, Pinzetta G, Ruwel AG et al (2024) Anatomical organization of the amygdala: a brief visual review. Cogn Behav Neurol Off J Soc Behav Cogn Neurol 37:13–22. https://doi.org/10.1097/WNN.0000000000000364

de la Sierra A (2023) Blood pressure variability as a risk factor for cardiovascular disease: which antihypertensive agents are more effective? J Clin Med 12:6167. https://doi.org/10.3390/jcm12196167

Deisseroth K (2015) Optogenetics: 10 years of microbial opsins in neuroscience. Nat Neurosci 18:1213–1225. https://doi.org/10.1038/nn.4091

Dell'Acqua C, Dal Bò E, Messerotti Benvenuti S, Palomba D (2020) Reduced heart rate variability is associated with vulnerability to depression. J Affect Disord Rep 1:100006. https://doi.org/10.1016/j.jadr.2020.100006

Devilbiss DM, Waterhouse BD, Berridge CW, Valentino R (2012) Corticotropin-releasing factor acting at the locus coeruleus disrupts thalamic and cortical sensory-evoked responses. Neuropsychopharmacology 37:2020–2030. https://doi.org/10.1038/npp.2012.50

Dityatev AE, Bolshakov VY (2005) Amygdala, long-term potentiation, and fear conditioning. Neurosci Rev J Bringing Neurobiol Neurol Psychiatry 11:75–88. https://doi.org/10.1177/1073858404270857

Dong H-W, Petrovich GD, Swanson LW (2001) Topography of projections from amygdala to bed nuclei of the stria terminalis. Brain Res Rev 38:192–246. https://doi.org/10.1016/S0165-0173(01)00079-0

Dong H-W, Swanson LW (2004) Projections from bed nuclei of the stria terminalis, posterior division: implications for cerebral hemisphere regulation of defensive and reproductive behaviors. J Comp Neurol 471:396–433. https://doi.org/10.1002/cne.20002

Duvarci S, Popa D, Paré D (2011) Central Amygdala activity during fear conditioning. J Neurosci 31:289–294. https://doi.org/10.1523/JNEUROSCI.4985-10.2011

Emdin CA, Odutayo A, Wong CX et al (2016) Meta-analysis of anxiety as a risk factor for cardiovascular disease. Am J Cardiol 118:511–519. https://doi.org/10.1016/j.amjcard.2016.05.041

Eybposh MH, Curtis VR, Rodríguez-Romaguera J, Pégard NC (2022) Advances in computer-generated holography for targeted neuronal modulation. Neurophotonics 9:041409. https://doi.org/10.1117/1.NPh.9.4.041409

Farah VMA, Joaquim LF, Bernatova I, Morris M (2004) Acute and chronic stress influence blood pressure variability in mice. Physiol Behav 83:135–142. https://doi.org/10.1016/j.physbeh.2004.08.004

Fedorenko A, Lishko PV, Kirichok Y (2012) Mechanism of fatty-acid-dependent UCP1 uncoupling in Brown fat mitochondria. Cell 151:400–413. https://doi.org/10.1016/j.cell.2012.09.010

Feinstein JS, Adolphs R, Damasio AR, Tranel D (2011) The human amygdala and the induction and experience of fear. Curr Biol 21:34–38. https://doi.org/10.1016/j.cub.2010.11.042

Fink L, Simola J, Tavano A et al (2023) From pre-processing to advanced dynamic modeling of pupil data. Behav Res Methods. https://doi.org/10.3758/s13428-023-02098-1

Foilb AR, Sansaricq GN, Zona EE et al (2021) Neural correlates of safety learning. Behav Brain Res 396:112884. https://doi.org/10.1016/j.bbr.2020.112884

Fonzo GA, Etkin A (2017) Affective neuroimaging in generalized anxiety disorder: an integrated review. Dialogues Clin Neurosci 19:169–179. https://doi.org/10.31887/DCNS.2017.19.2/gfonzo

Frustaci A, Pozzi G, Gianfagna F et al (2009) Meta-analysis of the brain-derived neurotrophic factor gene (BDNF) Val66Met polymorphism in anxiety disorders and anxiety-related personality traits. Neuropsychobiology 58:163–170. https://doi.org/10.1159/000182892

Fullana MA, Harrison BJ, Soriano-Mas C et al (2016) Neural signatures of human fear conditioning: an updated and extended meta-analysis of fMRI studies. Mol Psychiatry 21:500–508. https://doi.org/10.1038/mp.2015.88

Gerfen CR, Paletzki R, Heintz N (2013) GENSAT BAC cre-recombinase driver lines to study the functional organization of cerebral cortical and Basal Ganglia circuits. Neuron 80:1368–1383. https://doi.org/10.1016/j.neuron.2013.10.016

Gjendal K, Franco NH, Ottesen JL et al (2018) Eye, body or tail? Thermography as a measure of stress in mice. Physiol Behav 196:135–143. https://doi.org/10.1016/j.physbeh.2018.08.022

Glaab T, Taube C, Braun A, Mitzner W (2007) Invasive and noninvasive methods for studying pulmonary function in mice. Respir Res 8:63. https://doi.org/10.1186/1465-9921-8-63

Goldbarg AN, Hellerstein HK, Bruell JH, Daroczy AF (1968) Electrocardiogram of the normal mouse, Mus Musculus*: general considerations and genetic aspects. Oxford Academic

Gorin H, Patel J, Qiu Q et al (2024) A review of the use of gaze and pupil metrics to assess mental workload in gamified and simulated sensorimotor tasks. Sensors 24:1759. https://doi.org/10.3390/s24061759

Gostisha AJ, Vitacco MJ, Dismukes AR et al (2014) Beyond physiological hypoarousal: the role of life stress and callous-unemotional traits in incarcerated adolescent males. Horm Behav 65:469–479. https://doi.org/10.1016/j.yhbeh.2014.03.016

Grimaud J, Murthy VN (2018) How to monitor breathing in laboratory rodents: a review of the current methods. J Neurophysiol 120:624–632. https://doi.org/10.1152/jn.00708.2017

Gubits RM, Smith TM, Fairhurst JL, Yu H (1989) Adrenergic receptors mediate changes in c-fos mRNA levels in brain. Brain Res Mol Brain Res 6:39–45. https://doi.org/10.1016/0169-328x(89)90026-0

Guilherme A, Yenilmez B, Bedard AH et al (2020) Control of adipocyte thermogenesis and lipogenesis through β3-adrenergic and thyroid hormone signal integration. Cell Rep 31:107598. https://doi.org/10.1016/j.celrep.2020.107598

Guk K, Han G, Lim J et al (2019) Evolution of wearable devices with real-time disease monitoring for personalized healthcare. Nanomaterials 9:813. https://doi.org/10.3390/nano9060813

Guru A, Post RJ, Ho Y-Y, Warden MR (2015) Making sense of Optogenetics. Int J Neuropsychopharmacol 18:pyv079. https://doi.org/10.1093/ijnp/pyv079

Hagenaars MA, Oitzl M, Roelofs K (2014) Updating freeze: aligning animal and human research. Neurosci Biobehav Rev 47:165–176. https://doi.org/10.1016/j.neubiorev.2014.07.021

Hammack SE, Braas KM, May V (2021) Chemoarchitecture of the BNST: neurophenotypic diversity and function. Handb Clin Neurol 179:385–402. https://doi.org/10.1016/B978-0-12-819975-6.00025-X

Hartmann R, Schmidt FM, Sander C, Hegerl U (2019) Heart rate variability as indicator of clinical state in depression. Front Psych 9:735. https://doi.org/10.3389/fpsyt.2018.00735

He C, Gong L, Yin Y et al (2019) Amygdala connectivity mediates the association between anxiety and depression in patients with major depressive disorder. Brain Imaging Behav 13:1146–1159. https://doi.org/10.1007/s11682-018-9923-z

Hegerl U, Ulke C (2016) Fatigue with up- vs downregulated brain arousal should not be confused. Prog Brain Res 229:239–254. https://doi.org/10.1016/bs.pbr.2016.06.001

Heiss S, Vaschillo B, Vaschillo EG et al (2021) Heart rate variability as a biobehavioral marker of diverse psychopathologies: a review and argument for an "ideal range". Neurosci Biobehav Rev 121:144–155. https://doi.org/10.1016/j.neubiorev.2020.12.004

Herman JP, Flak J, Jankord R (2008) Chronic stress plasticity in the hypothalamic paraventricular nucleus. Prog Brain Res 170:353–364. https://doi.org/10.1016/S0079-6123(08)00429-9

Herman JP, McKlveen JM, Ghosal S et al (2016) Regulation of the hypothalamic-pituitary-adrenocortical stress response. Compr Physiol 6:603–621. https://doi.org/10.1002/cphy.c150015

Hintiryan H, Bowman I, Johnson DL et al (2021) Connectivity characterization of the mouse basolateral amygdalar complex. Nat Commun 12:2859. https://doi.org/10.1038/s41467-021-22915-5

Ho D, Zhao X, Gao S et al (2011) Heart rate and electrocardiography monitoring in mice. Curr Protoc Mouse Biol 1:123–139. https://doi.org/10.1002/9780470942390.mo100159

Holmqvist K, Nystrom M, Andersson R et al (2011) Eye tracking: a comprehensive guide to methods and measures. Oxford University Press, Oxford

Hsueh B, Chen R, Jo Y et al (2023) Cardiogenic control of affective behavioural state. Nature 615: 292–299. https://doi.org/10.1038/s41586-023-05748-8

Hur J, Stockbridge MD, Fox AS, Shackman AJ (2019) Chapter 16 – dispositional negativity, cognition, and anxiety disorders: an integrative translational neuroscience framework. In: Srinivasan N (ed) Progress in brain research. Elsevier, pp 375–436

Hutton SB (2019) Eye tracking methodology. In: Klein C, Ettinger U (eds) Eye movement research: an introduction to its scientific foundations and applications. Springer, Cham, pp 277–308

Ivanchenko D, Rifai K, Hafed ZM, Schaeffel FA (n.d.) Low-cost, high-performance video-based binocular eye tracker for psychophysical research. J Eye Mov Res 14. https://doi.org/10.16910/jemr.14.3.3

Jansen AS, Nguyen XV, Karpitskiy V et al (1995) Central command neurons of the sympathetic nervous system: basis of the fight-or-flight response. Science 270:644–646. https://doi.org/10.1126/science.270.5236.644

Jendryka M, Palchaudhuri M, Ursu D et al (2019) Pharmacokinetic and pharmacodynamic actions of clozapine-N-oxide, clozapine, and compound 21 in DREADD-based chemogenetics in mice. Sci Rep 9:4522. https://doi.org/10.1038/s41598-019-41088-2

Jiang Z, Rajamanickam S, Justice NJ (2019) CRF signaling between neurons in the paraventricular nucleus of the hypothalamus (PVN) coordinates stress responses. Neurobiol Stress 11:100192. https://doi.org/10.1016/j.ynstr.2019.100192

Jones BE (2020) Arousal and sleep circuits. Neuropsychopharmacology 45:6–20. https://doi.org/10.1038/s41386-019-0444-2

Juczewski K, Koussa JA, Kesner AJ et al (2020) Stress and behavioral correlates in the head-fixed method: stress measurements, habituation dynamics, locomotion, and motor-skill learning in mice. Sci Rep 10:12245. https://doi.org/10.1038/s41598-020-69132-6

Kang HJ, Minamimoto T, Wess J, Roth BL (2023) Chemogenetics for cell-type-specific modulation of signalling and neuronal activity. Nat Rev Methods Primer 3:1–17. https://doi.org/10.1038/s43586-023-00276-1

Kang JI, Chung HC, Jeung H-C et al (2012) FKBP5 polymorphisms as vulnerability to anxiety and depression in patients with advanced gastric cancer: a controlled and prospective study. Psychoneuroendocrinology 37:1569–1576. https://doi.org/10.1016/j.psyneuen.2012.02.017

Kawakami Y, Sielski R, Kawakami T (2018) Mouse body temperature measurement using infrared thermometer during passive systemic anaphylaxis and food allergy evaluation. J Vis Exp:58391. https://doi.org/10.3791/58391

Keifer OP, Hurt RC, Ressler KJ, Marvar PJ (2015) The physiology of fear: reconceptualizing the role of the central amygdala in fear learning. Physiology 30:389–401. https://doi.org/10.1152/physiol.00058.2014

Kenna G (2012) Association of the 5-HTT gene-linked promoter region (5-HTTLPR) polymorphism with psychiatric disorders: review of psychopathology and pharmacotherapy. Pharmacogenomics Pers Med 19. https://doi.org/10.2147/PGPM.S23462

Kessler RC, Petukhova M, Sampson NA et al (2012) Twelve-month and lifetime prevalence and lifetime morbid risk of anxiety and mood disorders in the United States. Int J Methods Psychiatr Res 21:169–184. https://doi.org/10.1002/mpr.1359

Kim JH, Almuwaqqat Z, Hammadah M et al (2019) Peripheral vasoconstriction during mental stress and adverse cardiovascular outcomes in patients with coronary artery disease. Circ Res 125:874–883. https://doi.org/10.1161/CIRCRESAHA.119.315005

Kim JH, Yin C, Merriam EP, Roth ZN (2023) Pupil size is sensitive to low-level stimulus features, independent of arousal-related modulation. eNeuro 10. https://doi.org/10.1523/ENEURO.0005-23.2023

Kim S-Y, Adhikari A, Lee SY et al (2013) Diverging neural pathways assemble a behavioural state from separable features in anxiety. Nature 496:219–223. https://doi.org/10.1038/nature12018

Kimble M, Boxwala M, Bean W et al (2014) The impact of hypervigilance: evidence for a forward feedback loop. J Anxiety Disord 28:241–245. https://doi.org/10.1016/j.janxdis.2013.12.006

Kiss L, Linnell KJ (2024) The role of mood and arousal in the effect of background music on attentional state and performance during a sustained attention task. Sci Rep 14:9485. https://doi.org/10.1038/s41598-024-60218-z

Klein M, Schmoeger M, Kasper S, Schosser A (2016) Meta-analysis of the COMT Val158Met polymorphism in major depressive disorder: the role of gender. World J Biol Psychiatry 17:147–158. https://doi.org/10.3109/15622975.2015.1083615

Knight LK, Depue BE (2019) New Frontiers in anxiety research: the translational potential of the bed nucleus of the stria terminalis. Front Psych 10:510. https://doi.org/10.3389/fpsyt.2019.00510

Kodani S, Soya S, Sakurai T (2017) Excitation of GABAergic neurons in the bed nucleus of the stria terminalis triggers immediate transition from non-rapid eye movement sleep to wakefulness in mice. J Neurosci 37:7164–7176. https://doi.org/10.1523/JNEUROSCI.0245-17.2017

Konstantinidis D, Iliakis P, Tatakis F et al (2022) Wearable blood pressure measurement devices and new approaches in hypertension management: the digital era. J Hum Hypertens 36:945–951. https://doi.org/10.1038/s41371-022-00675-z

Lara-Doña A, Torres-Sanchez S, Priego-Torres B et al (2021) Automated mouse pupil size measurement system to assess Locus Coeruleus activity with a deep learning-based approach. Sensors 21:7106. https://doi.org/10.3390/s21217106

Lebow MA, Chen A (2016) Overshadowed by the amygdala: the bed nucleus of the stria terminalis emerges as key to psychiatric disorders. Mol Psychiatry 21:450–463. https://doi.org/10.1038/mp.2016.1

Lee K, Ni X, Lee JY et al (2020) Mechanoacoustic sensing of physiological processes and body motions via a soft wireless device placed at the suprasternal notch. Nat Biomed Eng 4:148–158. https://doi.org/10.1038/s41551-019-0480-6

Li C, Pleil KE, Stamatakis AM et al (2012) Presynaptic inhibition of GABA release in the BNST by kappa opioid receptor signaling. Biol Psychiatry 71:725–732. https://doi.org/10.1016/j.biopsych.2011.11.015

Liu Y, Huang Y, Zhou J et al (2021) Altered heart rate variability in patients with schizophrenia during an autonomic nervous test. Front Psych 12. https://doi.org/10.3389/fpsyt.2021.626991

Lowrance SA, Ionadi A, McKay E et al (2016) Sympathetic nervous system contributes to enhanced corticosterone levels following chronic stress. Psychoneuroendocrinology 68:163–170. https://doi.org/10.1016/j.psyneuen.2016.02.027

Madisen L, Zwingman TA, Sunkin SM et al (2010) A robust and high-throughput Cre reporting and characterization system for the whole mouse brain. Nat Neurosci 13:133–140. https://doi.org/10.1038/nn.2467

Maeng LY, Milad MR (2015) Sex differences in anxiety disorders: interactions between fear, stress, and gonadal hormones. Horm Behav 76:106–117. https://doi.org/10.1016/j.yhbeh.2015.04.002

Mahanama B, Jayawardana Y, Rengarajan S et al (2022) Eye movement and pupil measures: a review. Front Comput Sci 3. https://doi.org/10.3389/fcomp.2021.733531

Manno F (2009) Pupillometry in mice: sex and strain-dependent phenotypes of pupillary functioning. Optom Vis Sci Off Publ Am Acad Optom 86:895–899. https://doi.org/10.1097/OPX.0b013e3181adfde9

Manvich DF, Webster KA, Foster SL et al (2018) The DREADD agonist clozapine N-oxide (CNO) is reverse-metabolized to clozapine and produces clozapine-like interoceptive stimulus effects in rats and mice. Sci Rep 8:3840. https://doi.org/10.1038/s41598-018-22116-z

Maren S (1999) Long-term potentiation in the amygdala: a mechanism for emotional learning and memory. Trends Neurosci 22:561–567. https://doi.org/10.1016/s0166-2236(99)01465-4

Marková IS (2018) Translational neuroscience and psychiatry: a conceptual analysis. J Eval Clin Pract 24:791–796. https://doi.org/10.1111/jep.12914

Mayes LC (2000) A developmental perspective on the regulation of arousal states. Semin Perinatol 24:267–279. https://doi.org/10.1053/sper.2000.9121

McCorry LK (2007) Physiology of the autonomic nervous system. Am J Pharm Educ 71:78

McCurry KL, Frueh BC, Chiu PH, King-Casas B (2020) Opponent effects of hyperarousal and re-experiencing on affective habituation in posttraumatic stress disorder. Biol Psychiatry Cogn Neurosci Neuroimaging 5:203–212. https://doi.org/10.1016/j.bpsc.2019.09.006

McLean CP, Asnaani A, Litz BT, Hofmann SG (2011) Gender differences in anxiety disorders: prevalence, course of illness, comorbidity and burden of illness. J Psychiatr Res 45:1027–1035. https://doi.org/10.1016/j.jpsychires.2011.03.006

Menon V (2011) Large-scale brain networks and psychopathology: a unifying triple network model. Trends Cogn Sci 15:483–506. https://doi.org/10.1016/j.tics.2011.08.003

Meyer AF, Poort J, O'Keefe J et al (2018) A head-mounted camera system integrates detailed behavioral monitoring with multichannel electrophysiology in freely moving mice. Neuron 100: 46–60.e7. https://doi.org/10.1016/j.neuron.2018.09.020

Meyer CW, Willershäuser M, Jastroch M et al (2010) Adaptive thermogenesis and thermal conductance in wild-type and UCP1-KO mice. Am J Physiol-Regul Integr Comp Physiol 299:R1396–R1406. https://doi.org/10.1152/ajpregu.00021.2009

Millan MJ, Brocco M (2003) The Vogel conflict test: procedural aspects, gamma-aminobutyric acid, glutamate and monoamines. Eur J Pharmacol 463:67–96. https://doi.org/10.1016/s0014-2999(03)01275-5

Milosavljevic N, Cehajic-Kapetanovic J, Procyk CA, Lucas RJ (2016) Chemogenetic activation of Melanopsin retinal ganglion cells induces signatures of arousal and/or anxiety in mice. Curr Biol CB 26:2358–2363. https://doi.org/10.1016/j.cub.2016.06.057

Moberly AH, Schreck M, Bhattarai JP et al (2018) Olfactory inputs modulate respiration-related rhythmic activity in the prefrontal cortex and freezing behavior. Nat Commun 9:1528. https://doi.org/10.1038/s41467-018-03988-1

Montag C, Jurkiewicz M, Reuter M (2012) The role of the catechol-O-methyltransferase (COMT) gene in personality and related psychopathological disorders. CNS Neurol Disord Drug Targets 11:236–250

Montaquila JM, Trachik BJ, Bedwell JS (2015) Heart rate variability and vagal tone in schizophrenia: a review. J Psychiatr Res 69:57–66. https://doi.org/10.1016/j.jpsychires.2015.07.025

Morabito DM, Boffa JW, Bedford CE et al (2020) Hyperarousal symptoms and perceived burdensomeness interact to predict suicidal ideation among trauma-exposed individuals. J Psychiatr Res 130:218–223. https://doi.org/10.1016/j.jpsychires.2020.07.029

Nicolaides N, Kyratzi E, Lamprokostopoulou A et al (2014) Stress, the stress system and the role of glucocorticoids. Neuroimmunomodulation 22:6–19. https://doi.org/10.1159/000362736

Nieh EH, Kim S-Y, Namburi P, Tye KM (2013) Optogenetic dissection of neural circuits underlying emotional valence and motivated behaviors. Brain Res 1511:73–92. https://doi.org/10.1016/j.brainres.2012.11.001

Olivier JDA, Olivier B (2020) Translational studies in the complex role of neurotransmitter systems in anxiety and anxiety disorders. In: Kim Y-K (ed) Anxiety disorders: rethinking and understanding recent discoveries. Springer, Singapore, pp 121–140

Ortiz-Juza MM, Alghorazi RA, Rodriguez-Romaguera J (2021) Cell-type diversity in the bed nucleus of the stria terminalis to regulate motivated behaviors. Behav Brain Res 411:113401. https://doi.org/10.1016/j.bbr.2021.113401

Pais-Roldán P, Biswal B, Scheffler K, Yu X (2018) Identifying respiration-related aliasing artifacts in the rodent resting-state fMRI. Front Neurosci 12:788. https://doi.org/10.3389/fnins.2018.00788

Pan H-Y, Lee C-K, Liu T-Y et al (2024) The role of wearable home blood pressure monitoring in detecting out-of-office control status. Hypertens Res 47:1033–1041. https://doi.org/10.1038/s41440-023-01539-w

Pannekoek JN, Van Der Werff SJA, Stein DJ, Van Der Wee NJA (2013) Advances in the neuroimaging of panic disorder. Hum Psychopharmacol Clin Exp 28:608–611. https://doi.org/10.1002/hup.2349

Partridge JG, Forcelli PA, Luo R et al (2016) Stress increases GABAergic neurotransmission in CRF neurons of the central amygdala and bed nucleus stria terminalis. Neuropharmacology 107:239–250. https://doi.org/10.1016/j.neuropharm.2016.03.029

Pégard NC, Mardinly AR, Oldenburg IA et al (2017) Three-dimensional scanless holographic optogenetics with temporal focusing (3D-SHOT). Nat Commun 8:1228. https://doi.org/10.1038/s41467-017-01031-3

Penninx BW, Pine DS, Holmes EA, Reif A (2021) Anxiety disorders. Lancet Lond Engl 397:914–927. https://doi.org/10.1016/S0140-6736(21)00359-7

Pham X, Sun C, Chen X et al (2009) Association study between GABA receptor genes and anxiety spectrum disorders. Depress Anxiety 26:998–1003. https://doi.org/10.1002/da.20628

Prada-Dacasa P, Urpi A, Sánchez-Benito L et al (2020) Measuring breathing patterns in mice using whole-body plethysmography. Bio-Protoc 10:e3741

Pratt JA, Robinson ESJ, Fernandes C et al (2022) BAP editorial: improving the validity and translation of preclinical research. J Psychopharmacol Oxf Engl 36:779–780. https://doi.org/10.1177/02698811221104064

Privitera M, Ferrari KD, Von Ziegler LM et al (2020) A complete pupillometry toolbox for real-time monitoring of locus coeruleus activity in rodents. Nat Protoc 15:2301–2320. https://doi.org/10.1038/s41596-020-0324-6

Radke AK (2009) The role of the bed nucleus of the stria terminalis in learning to fear. J Neurosci 29:15351–15352. https://doi.org/10.1523/JNEUROSCI.5194-09.2009

Rodgers RJ (1997) Animal models of "anxiety": where next? Behav Pharmacol 8:477

Rodriguez-Romaguera J, Namboodiri VMK, Basiri ML et al (2022) Developments from bulk optogenetics to single-cell strategies to dissect the neural circuits that underlie aberrant motivational states. Cold Spring Harb Perspect Med 12:a039792. https://doi.org/10.1101/cshperspect.a039792

Rodriguez-Romaguera J, Ung RL, Nomura H et al (2020) Prepronociceptin-expressing neurons in the extended amygdala encode and promote rapid arousal responses to motivationally salient stimuli. Cell Rep 33:108362. https://doi.org/10.1016/j.celrep.2020.108362

Roxo MR, Franceschini PR, Zubaran C et al (2011) The limbic system conception and its historical evolution. Sci World J 11:2428–2441. https://doi.org/10.1100/2011/157150

Sah P, Faber ESL, Lopez De Armentia M, Power J (2003) The amygdaloid complex: anatomy and physiology. Physiol Rev 83:803–834. https://doi.org/10.1152/physrev.00002.2003

Salvi SS, Pati S, Chaudhari PR et al (2019) Acute chemogenetic activation of CamKIIα-positive forebrain excitatory neurons regulates anxiety-like behaviour in mice. Front Behav Neurosci 13:249. https://doi.org/10.3389/fnbeh.2019.00249

Scheuer S, Ising M, Uhr M et al (2016) FKBP5 polymorphisms moderate the influence of adverse life events on the risk of anxiety and depressive disorders in preschool children. J Psychiatr Res 72:30–36. https://doi.org/10.1016/j.jpsychires.2015.10.009

Schimmelpfennig J, Topczewski J, Zajkowski W, Jankowiak-Siuda K (2023) The role of the salience network in cognitive and affective deficits. Front Hum Neurosci 17:1133367. https://doi.org/10.3389/fnhum.2023.1133367

Schinka JA, Busch RM, Robichaux-Keene N (2004) A meta-analysis of the association between the serotonin transporter gene polymorphism (5-HTTLPR) and trait anxiety. Mol Psychiatry 9:197–202. https://doi.org/10.1038/sj.mp.4001405

Schoenfeld WN (1950) An experimental approach to anxiety, escape and avoidance behavior. In: Anxiety. Grune & Stratton, New York City, pp 70–101

Seeley JR, Manitsas T, Gau JM (2017) Feasibility study of a peer-facilitated low intensity cognitive-behavioral intervention for mild to moderate depression and anxiety in older adults. Aging Ment Health 21:968–974. https://doi.org/10.1080/13607863.2016.1186152

Seeley WW (2019) The salience network: a neural system for perceiving and responding to homeostatic demands. J Neurosci 39:9878–9882. https://doi.org/10.1523/JNEUROSCI.1138-17.2019

Sege CT, Bradley MM, Lang PJ (2018) Avoidance and escape: defensive reactivity and trait anxiety. Behav Res Ther 104:62–68. https://doi.org/10.1016/j.brat.2018.03.002

Shaffer F, Ginsberg JP (2017) An overview of heart rate variability metrics and norms. Front Public Health 5:258. https://doi.org/10.3389/fpubh.2017.00258

Shaffer F, McCraty R, Zerr CL (2014) A healthy heart is not a metronome: an integrative review of the heart's anatomy and heart rate variability. Front Psychol 5. https://doi.org/10.3389/fpsyg.2014.01040

Simpson EH, Morud J, Winiger V et al (2014) Genetic variation in COMT activity impacts learning and dopamine release capacity in the striatum. Learn Mem 21:205–214. https://doi.org/10.1101/lm.032094.113

Siuda ER, Al-Hasani R, McCall JG et al (2016) Chemogenetic and optogenetic activation of Gαs signaling in the basolateral Amygdala induces acute and social anxiety-like states. Neuropsychopharmacol Off Publ Am Coll Neuropsychopharmacol 41:2011–2023. https://doi.org/10.1038/npp.2015.371

Sotelo MI, Tyan J, Dzera J, Eban-Rothschild A (2020) Sleep and motivated behaviors, from physiology to pathology. Curr Opin Physio 15:159–166. https://doi.org/10.1016/j.cophys.2020.01.006

Spiegler KM, Palmieri J, Pang KCH, Myers CE (2020) A reinforcement-learning model of active avoidance behavior: differences between Sprague Dawley and Wistar-Kyoto rats. Behav Brain Res 393:112784. https://doi.org/10.1016/j.bbr.2020.112784

Stamatakis AM, Resendez SL, Chen K-S et al (2021) Miniature microscopes for manipulating and recording in vivo brain activity. Microscopy 70:399–414. https://doi.org/10.1093/jmicro/dfab028

Staples LG (2010) Predator odor avoidance as a rodent model of anxiety: learning-mediated consequences beyond the initial exposure. Neurobiol Learn Mem 94:435–445. https://doi.org/10.1016/j.nlm.2010.09.009

Stein MB, Fallin MD, Schork NJ, Gelernter J (2005) COMT polymorphisms and anxiety-related personality traits. Neuropsychopharmacol Off Publ Am Coll Neuropsychopharmacol 30:2092–2102. https://doi.org/10.1038/sj.npp.1300787

Stephens MAC, Wand G (2012) Stress and the HPA axis. Alcohol Res Curr Rev 34:468–483

Straube T, Mentzel H-J, Miltner WHR (2007) Waiting for spiders: brain activation during anticipatory anxiety in spider phobics. NeuroImage 37:1427–1436. https://doi.org/10.1016/j.neuroimage.2007.06.023

Szabo V, Ventalon C, De Sars V et al (2014) Spatially selective holographic photoactivation and functional fluorescence imaging in freely behaving mice with a fiberscope. Neuron 84:1157–1169. https://doi.org/10.1016/j.neuron.2014.11.005

Szuhany KL, Simon NM (2022) Anxiety disorders: a review. JAMA 328:2431–2445. https://doi.org/10.1001/jama.2022.22744

Takahashi TM, Sakurai T, Hirano A (2023) Measuring body temperature of freely moving mice under an optogenetics-induced long-term hypothermic state. STAR Protoc 4:102321. https://doi.org/10.1016/j.xpro.2023.102321

Taniguchi H, He M, Wu P et al (2011) A resource of Cre driver lines for genetic targeting of GABAergic neurons in cerebral cortex. Neuron 71:995–1013. https://doi.org/10.1016/j.neuron.2011.07.026

Tindle J, Tadi P (2024) Neuroanatomy, parasympathetic nervous system. In: StatPearls. StatPearls Publishing, Treasure Island

Tinga AM, de Back TT, Louwerse MM (2020) Non-invasive neurophysiology in learning and training: mechanisms and a SWOT analysis. Front Neurosci 14:589. https://doi.org/10.3389/fnins.2020.00589

Tong L, Han S, Xue Y et al (2023) Single cell in vivo optogenetic stimulation by two-photon excitation fluorescence transfer. iScience 26:107857. https://doi.org/10.1016/j.isci.2023.107857

Tye KM, Prakash R, Kim S-Y et al (2011) Amygdala circuitry mediating reversible and bidirectional control of anxiety. Nature 471:358–362. https://doi.org/10.1038/nature09820

Tyree SM, Borniger JC, de Lecea L (2018) Hypocretin as a hub for arousal and motivation. Front Neurol 9:413. https://doi.org/10.3389/fneur.2018.00413

Uddin LQ (2017) Chapter 3 – functions of the salience network. In: Uddin LQ (ed) Salience network of the human brain. Academic Press, San Diego, pp 11–16

Usamentiaga R, Venegas P, Guerediaga J et al (2014) Infrared thermography for temperature measurement and non-destructive testing. Sensors 14:12305–12348. https://doi.org/10.3390/s140712305

VaezMousavi SM, Barry RJ, Rushby JA, Clarke AR (2007) Arousal and activation effects on physiological and behavioral responding during a continuous performance task. Acta Neurobiol Exp (Wars) 67:461–470

Valliappan N, Dai N, Steinberg E et al (2020) Accelerating eye movement research via accurate and affordable smartphone eye tracking. Nat Commun 11:4553. https://doi.org/10.1038/s41467-020-18360-5

van der Vinne V, Pothecary CA, Wilcox SL et al (2020) Continuous and non-invasive thermography of mouse skin accurately describes core body temperature patterns, but not absolute core temperature. Sci Rep 10:20680. https://doi.org/10.1038/s41598-020-77786-5

Vetulani J (2013) Early maternal separation: a rodent model of depression and a prevailing human condition. Pharmacol Rep 65:1451–1461. https://doi.org/10.1016/s1734-1140(13)71505-6

Walf AA, Frye CA (2007) The use of the elevated plus maze as an assay of anxiety-related behavior in rodents. Nat Protoc 2:322–328. https://doi.org/10.1038/nprot.2007.44

Walsh RN, Cummins RA (1976) The open-field test: a critical review. Psychol Bull 83:482–504

Wang C-A, Baird T, Huang J et al (2018a) Arousal effects on pupil size, heart rate, and skin conductance in an emotional face task. Front Neurol 9. https://doi.org/10.3389/fneur.2018.01029

Wang H-Y, Zhang X-X, Si C-P et al (2018b) Prefrontoparietal dysfunction during emotion regulation in anxiety disorder: a meta-analysis of functional magnetic resonance imaging studies. Neuropsychiatr Dis Treat 14:1183–1198. https://doi.org/10.2147/NDT.S165677

Wermes R, Lincoln TM, Helbig-Lang S (2018) Anxious and alert? Hypervigilance in social anxiety disorder. Psychiatry Res 269:740–745. https://doi.org/10.1016/j.psychres.2018.08.086

Wessing I, Romer G, Junghöfer M (2017) Hypervigilance-avoidance in children with anxiety disorders: magnetoencephalographic evidence. J Child Psychol Psychiatry 58:103–112. https://doi.org/10.1111/jcpp.12617

Winklewski PJ, Radkowski M, Wszedybyl-Winklewska M, Demkow U (2017) Stress response, brain noradrenergic system and cognition. Adv Exp Med Biol 980:67–74. https://doi.org/10.1007/5584_2016_204

Wu E-L, Chien I-C, Lin C-H (2014) Increased risk of hypertension in patients with anxiety disorders: a population-based study. J Psychosom Res 77:522–527. https://doi.org/10.1016/j.jpsychores.2014.10.006

Zsidó AN (2024) The effect of emotional arousal on visual attentional performance: a systematic review. Psychol Res 88:1–24. https://doi.org/10.1007/s00426-023-01852-6

The Emerging Role of Brain Mitochondria in Fear and Anxiety

Alessandro Chioino and Carmen Sandi

Contents

Abstract The functional complexity of brain circuits underlies the broad spectrum of behaviors, cognitive functions, and their associated disorders. Mitochondria, traditionally known for their role in cellular energy metabolism, are increasingly recognized as central to brain function and behavior. This review examines how mitochondria are pivotal in linking cellular energy processes with the functioning of neural circuits that govern fear and anxiety. Following an introductory section in which we summarize current knowledge about fear and anxiety neural circuits, we

A. Chioino and C. Sandi (✉)
Laboratory of Behavioral Genetics, Brain Mind Institute, School of Life Sciences, École Polytechnique Fédérale de Lausanne (EPFL), Lausanne, Switzerland

Synapsy Center for Neuroscience and Mental Health Research, School of Life Sciences, École Polytechnique Fédérale de Lausanne (EPFL), Lausanne, Switzerland
e-mail: carmen.sandi@epfl.ch

© The Author(s), under exclusive license to Springer Nature Switzerland AG 2024
Curr Topics Behav Neurosci (2025) 73: 33–54
https://doi.org/10.1007/7854_2024_537
Published Online: 07 November 2024

provide a brief summary of mitochondria fundamental roles (e.g., from energy production and calcium buffering to their involvement in reactive oxygen species (ROS) generation, mitochondrial dynamics, and signaling), particularly emphasizing their contribution to synaptic plasticity, neurodevelopment, and stress response mechanisms. The review's core focuses on the current state of knowledge regarding how mitochondrial function and dysfunction impact the neural substrates of fear and anxiety. Furthermore, we explore the implications of mitochondrial alterations in the context of posttraumatic stress disorder (PTSD) and anxiety disorders, underscoring the potential of mitochondrial pathways as new therapeutic targets. Integrating insights from genetic, biochemical, neurobiological, behavioral, and clinical studies, we propose a model in which mitochondrial function is critical for regulating the neural circuits that underpin fear and anxiety behaviors, highlighting how mitochondrial dysfunction can lead to their pathological manifestations. This integration emphasizes the potential for developing novel treatments targeting the biological roots of fear, anxiety, and related disorders. By merging mitochondrial biology with behavioral and circuit neuroscience, we enrich our neurobiological understanding of fear and anxiety, uncovering promising avenues for therapeutic intervention.

Keywords Anxiety · Fear · Mitochondria · Neural circuits · Nutritional interventions

1 Introduction

The interplay between brain mitochondria and the regulation of fear and anxiety unveils a fascinating frontier in cognitive, behavioral, and systems neuroscience, challenging traditional perspectives on the etiology and management of these complex emotional states. Fear and anxiety, essential for survival, can escalate into debilitating disorders, highlighting a critical gap in our understanding of their neurobiological underpinnings. This review focuses on the emerging role of mitochondria – beyond their well-recognized role in energy production – as pivotal regulators within the neural circuits governing fear and anxiety behaviors. By examining mitochondrial dysfunction's contribution to variations in these behaviors, we aim to illuminate novel pathways implicated in the pathophysiology of anxiety disorders, offering novel therapeutic insights. The economic and human burden of fear and anxiety disorders underscores the urgent need for this exploration, promising to expand our understanding of their biology and foster the development of targeted interventions.

2 Fear, Anxiety, and Their Brain Circuits

Fear and anxiety represent fundamental emotional states characterized by a complex set of physiological, behavioral, and cognitive responses. Fear is a response to immediate threats, facilitating survival by orchestrating a range of behaviors, including defense or avoidance. It encompasses both innate reactions to danger and learned responses, with its biological underpinnings highly conserved across species. Anxiety, while closely related to fear, is an evolutionary conserved, more pervasive emotional state of heightened vigilance and apprehension, often occurring in the absence of a direct threat. When excessive or prolonged, it can become maladaptive, leading to anxiety disorders (Price 2003). Similarly, the beneficial and transient emotional and motivational aspects of "state" anxiety are clear when facing uncertainty. However, the excessive and prolonged activation of anxiety response, the so-called "trait" anxiety, can lead to significant impairments in life quality.

2.1 *The Neural Circuitry of Fear*

Fear, a fundamental emotional response to perceived threats, is mediated by a conserved set of neural circuits across species. It encompasses both innate and learned aspects, facilitating survival through immediate behavioral adaptations. Central to our understanding of fear mechanisms is fear conditioning research, which has elucidated the pivotal roles of specific brain regions in fear acquisition, expression, and extinction (Herry and Johansen 2014).

The amygdala, particularly the central (CE) and lateral (LA) nuclei, plays a central role in fear processing. The LA serves as the initial processing hub for fear stimuli, integrating conditioned and unconditioned stimuli to facilitate the learning of fear responses through synaptic plasticity. Information from the LA is then relayed to the basal amygdala (BA) and subsequently to the CE, which coordinates the execution of fear responses by projecting to brainstem areas like the periaqueductal gray (PAG) and the hypothalamus. This pathway initiates autonomic and behavioral fear responses while also engaging cortical areas to modulate the cognitive and emotional aspects of fear (Ciocchi et al. 2010; Herry and Johansen 2014; Duvarci and Pare 2014). Furthermore, the HPC encodes contextual information vital for triggering conditioned fear responses (Namkung et al. 2022).

The medial prefrontal cortex (mPFC) plays a critical role in modulating fear acquisition (particularly its prelimbic part; PL) and extinction (particularly its infralimbic part; IL), as well as the retrieval of fear memories, through its interactions with the basolateral amygdala (BLA) and hippocampus (HPC). The PL regulates fear expression as fear conditioning increases the strength of the PL–BLA pathway (Giustino and Maren 2015). The mechanism through which the IL promotes fear suppression is still under debate. Some studies have reported the influence of the BA and HPC inputs on the IL, while others have highlighted the indirect action of IL on the intercalated cells (ITC) via BA (Herry and Johansen 2014; Bloodgood et al. 2018; Namkung et al. 2022).

Recent research has highlighted the critical role of thalamic nuclei, particularly the mediodorsal thalamus (MD) and the paraventricular nucleus of the thalamus (PVT), and nucleus reuniens (NR) in fear processing through their interactions with the mPFC and HPC. The MD thalamus regulates fear retrieval and extinction through its connections with the mPFC (Lee et al. 2012; Do-Monte et al. 2015). Similarly, the PVT modulates fear and stress responses through its projections to both the mPFC and HPC (Do-Monte et al. 2015; Ratigan et al. 2023). In addition, emerging research highlights a multifaceted role for the NR in contextual fear memory encoding (Ramanathan et al. 2018, Ramanathan and Maren 2019) as well as in the extinction of contextual fear through its interactions with the hippocampal CA1 region (Ratigan et al. 2023). Specifically, the NR coordinates prefrontal–hippocampal synchrony to suppress extinguished fear memories (Totty et al. 2023). Furthermore, the NR plays a critical role in the extinction of remote fear through its interactions with the basolateral amygdala (BLA) (Silva et al. 2021).

Additionally, the nucleus accumbens (NAc) has been identified as a key structure in fear generalization and stress susceptibility. The NAc integrates signals from the amygdala, hippocampus, and mPFC, modulating the emotional and motivational aspects of fear responses. Studies have shown that the NAc is involved in the regulation of fear and stress responses through its dopaminergic and glutamatergic inputs, which affect learning and expression of fear responses, and its dysfunctional activity has been associated with heightened fear generalization and increased vulnerability to stress-related disorders, such as post-traumatic stress disorder (PTSD) (Fan et al. 2022). Moreover, the NAc has been implicated in fear extinction through the modulation of activity in the mPFC and amygdala (Rodriguez-Romaguera et al. 2012; Correia et al. 2016). This highlights the importance of the NAc in both the generalization of fear responses and the development of stress-related psychopathologies, underscoring its potential as a therapeutic target for fear-related disorders.

2.2 *Anxiety and its Neural Substrates*

Anxiety, characterized by a pervasive sense of apprehension and heightened vigilance, engages a widespread neural network encompassing – among other brain regions – the amygdala, bed nucleus of the stria terminalis (BNST), lateral septum, prefrontal cortex (PFC), and ventral tegmental area (VTA). This network integrates internal and external cues to modulate defensive behaviors and states of alertness (Lüthi and Lüscher 2014; Calhoon and Tye 2015).

Recent findings emphasize the role of corticolimbic and corticostriatal circuits in mediating anxiety, where disruptions within these circuits are linked to the emergence of anxiety disorders. Specifically, the NAc, receiving inputs from the mPFC, ventral hippocampus (vHPC), and BLA, is emerging as a significant player in adapting behavior and stress responses, interacting with the putative anxiety circuitries at the intersection of the interplay between anxiety and motivation (Britt et al. 2012; Lüthi and Lüscher 2014; Calhoon and Tye 2015; Burkhouse et al. 2020; Pignatelli et al. 2021).

Despite the booming in understanding the neural circuits encoding for fear and anxiety, little was known on their metabolic substrates. As our understanding for molecular and mechanistic insights into psychiatric disorders progresses, attention has shifted toward mitochondria function (Andreazza and Nierenberg 2018). The upcoming sections will delve into the specific role of mitochondria in neural circuits regulating behaviors with crucial implications for mental health, with a particular focus on fear and anxiety.

3 Mitochondrial Biology and Their Central Role in Brain Function

Mitochondria, the powerhouses of the cell, are dynamic organelles essential for eukaryotic cell survival and functionality, particularly in the brain, which has high energy demands. Despite comprising only 2% of the body's mass, the brain consumes about 20% of the oxygen and 25% of the glucose used by the body. Beyond energy production, mitochondria play crucial roles in bioenergetics, metabolism, cellular signaling, and gene regulation (Picard et al. 2015; Latorre-Pellicer et al. 2016; Palikaras et al. 2018; Monzel et al. 2023; Chen et al. 2023). This section summarizes some of the key roles of mitochondria in cellular functions and their contributions to brain function (Palikaras et al. 2018; Monzel et al. 2023; Chen et al. 2023), laying the foundations for subsequent inquiry into their involvement in fear and anxiety.

3.1 Energy Metabolism and ROS Production

Mitochondria produce ATP through the citric acid cycle (TCA) and oxidative phosphorylation (OXPHOS), which is critical for brain functions like synaptic transmission and plasticity (Picard et al. 2015) (Fig. 1A). At activated synapses, presynaptic ATP is necessary for neurotransmitter transport, release, and recycling, while postsynaptic ATP and calcium buffering are key for synaptic plasticity and memory formation (Rangaraju et al. 2014; Pathak et al. 2015; Devine and Kittler 2018; Rossi and Pekkurnaz 2019; Datta and Jaiswal 2021).

During ATP generation, mitochondria also produce reactive oxygen species (ROS) (Fig. 1A). While ROS are essential for cell signaling, excess ROS can cause oxidative stress, damaging cells and contributing to diseases. The brain is particularly vulnerable to oxidative damage, exacerbated by stress (Schiavone et al. 2012;

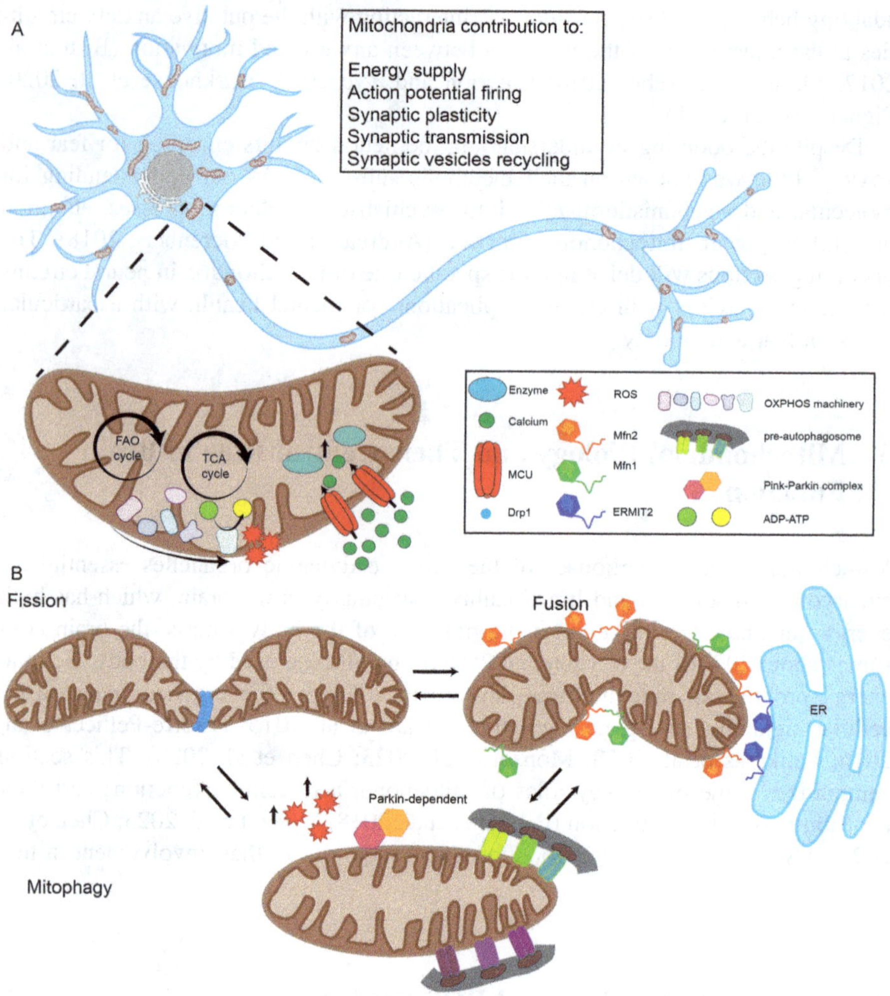

Fig. 1 (**a**) Representative figure of a neuron and mitochondria. Mitochondria are localized in dendrites, soma, and axons, although they are present in different shapes and distributions, with more mitochondria present at the somatic and dendritic levels. A zoom-in on a single mitochondrion shows that beta-oxidation of fatty acids (FAO) and TCA cycles take place inside the mitochondrion. Mitochondria play a role in buffering calcium into the matrix from the cytoplasm through the mitochondrial calcium uniporter (MCU). Increases in calcium concentration activate several enzymes that can support the TCA cycle. The OXPHOS chain is localized on the inner mitochondrial membrane (IMM) and produces energy in the form of ATP. OXPHOS leads to ROS production as a byproduct of the energy reactions. (**b**) Mitochondria are dynamic organelles that can be found in different biological states. Fission is a multistep event that requires the oligomerization of Drp1 on the outer mitochondrial membrane (OMM). Fusion initiates with the fusion of the OMM of two adjacent mitochondria through the homotypic and/or heterotypic interaction of Mfn2 and Mfn1. Fission allows the recycling of old, suboptimal mitochondria, while fusion allows the genesis of new, functional mitochondria. Mitochondria tether with the endoplasmic reticulum (ER) through Mfn2. Mitochondrial damage, such as a pathological increase in ROS, results in the removal of suboptimal mitochondria via mitophagy in a Parkin-dependent or Parkin-independent

Filiou and Sandi 2019; Zalachoras et al. 2020). Mitochondrial quality control mechanisms, including mitophagy, maintain the balance between energy production and ROS generation, preventing neuronal damage (Twig and Shirihai 2011; Mergenthaler et al. 2013). Dysregulation in mitochondrial energy metabolism can impair cognitive functions and is implicated in various neuropsychiatric conditions.

3.2 *Mitochondrial Signaling and Calcium Buffering*

Mitochondria regulate calcium, which is critical for synaptic activity, neurotransmitter release, and plasticity. The mitochondrial calcium uniporter (MCU) mediates calcium uptake, activating enzymes involved in ATP production and maintaining energy balance necessary for cognitive processes (Duchen 2000; Brand et al. 2004) (Fig. 1A). Precise execution of these mitochondrial functions is essential for synaptic health and cognitive function, with dysregulation leading to neuronal damage (Woods and Wilson 2020; Ryan et al. 2020).

Mitochondria also influence gene expression and mediate intercellular communication, including the transfer of mitochondrial DNA (mtDNA) between cells, which impacts development and stress responses (Kraus et al. 2021; Liu et al. 2021). Furthermore, the presence of mtDNA in the bloodstream underscores a systemic role in organismal health and disease.

3.3 *Mitochondrial Dynamics*

Mitochondria undergo fusion and fission, maintaining cellular health and optimizing energy distribution. Fusion, mediated by Mitofusin 1 and 2 (Mfn1, Mfn2) and Optic atrophy 1 (OPA1), supports an interconnected network and cellular energy needs, while fission, driven by dynamin-related protein 1 (Drp1), is essential for mitochondrial replication and removal of damaged mitochondria through mitophagy (Schrepfer and Scorrano 2016; Kraus et al. 2021). Interactions of mitochondria with the endoplasmic reticulum (ER) support calcium homeostasis and oxidative stress regulation (Patergnani et al. 2011; Krols et al. 2016). These interactions support synaptic function and plasticity, essential for neuronal function and cognitive processes (Fig. 1B).

Fig. 1 (continued) cascade. Note that the three states are tightly linked to each other as suboptimal mitochondria can clear out their damaged parts via fission, generating two new organelles. The part of the mitochondrion that expresses the proper markers undergoes mitophagy, while the functional part can undergo fusion

3.4 Mitochondrial Quality Control: Mitophagy

Mitophagy is crucial for maintaining mitochondrial integrity by removing damaged mitochondria, preventing ROS accumulation and protecting against oxidative stress, essential for neuronal health (Evans and Holzbaur 2020; Han et al. 2021). Disrupting the dynamic nature of mitochondrial quality control can have profound implications for brain function and the onset of neurodegenerative and psychiatric conditions (Fig. 1B).

3.5 Mitochondrial Contributions to Cellular Biosynthesis

Mitochondria are involved in synthesizing vital molecules such as fatty acids, steroids, and heme, which are crucial for cellular structure and signaling. This biosynthetic capacity supports brain development, synaptic plasticity, and resilience to stress (Tyurina et al. 2014; Spinelli and Haigis 2018). Disrupting the balance of brain bioenergetics, ROS production, mitochondrial dynamics, and quality control can lead to metabolic imbalances, affecting brain function and increasing vulnerability to disorders. The upcoming sections will delve into the specific role of mitochondria in neural circuits that regulate behaviors with crucial implications for mental health, with a particular focus on fear and anxiety.

4 Mitochondrial Involvement in Mental Health

As our understanding of mitochondrial function extends beyond their role in energy production, so do their implications in pathologies. Mitochondrial dysfunction is increasingly recognized as a key factor in the etiology and exacerbation of neuropsychiatric disorders, highlighting a complex interplay between metabolic processes and mental health (Pei and Wallace 2018; Picard and Sandi 2021). The brain's disproportionate reliance on mitochondrial energy – consuming about 20% of the body's oxygen and 25% of its glucose, despite constituting only around 2% of body mass – underscores mitochondria's critical role in maintaining the integrity and functionality of brain cells and their interactions. This high demand for energy renders the brain particularly susceptible to disruptions in mitochondrial function, leading to increased vulnerability to psychiatric conditions (Morava and Kozicz 2013).

Evidence associates mitochondrial dysfunction with a range of neuropsychiatric disorders, including depression, bipolar disorder, schizophrenia, and, more recently, anxiety disorders (Daniels et al. 2020; van de Burgt et al. 2023). Genetic, biochemical, molecular, structural, and functional studies collectively indicate that aberrations in mitochondrial function lie at the core of these conditions. For example, initial research showed that individuals with primary mitochondrial disorders exhibit higher incidences of depressive behaviors (Morava et al. 2010), and rodent studies have highlighted mitochondrial sensitivity to stress (Andreazza et al. 2010). These findings align with a growing body of evidence connecting primary mitochondrial diseases and neuropsychiatric symptoms.

Importantly, stress plays a central role in the association between mitochondrial dysfunction and psychiatric disorders. Stress can induce and exacerbate mitochondrial dysfunction, potentially leading to mtDNA mutations, oxidative stress, and, ultimately, brain dysfunction and psychiatric symptoms (Picard and McEwen 2018; Daniels et al. 2020; Weger et al. 2020).

In the following section, we will explore the connection between mitochondrial function and dysfunction and behaviors related to fear and anxiety.

5 Mitochondria and Fear

Emerging research highlights the crucial role of mitochondrial function and dysfunction in fear behaviors, particularly in learned fear. Studies increasingly link alterations in mitochondrial functions within neurons and astrocytes to fear conditioning and extinction, highlighting the interaction between mitochondrial functions and synaptic plasticity in shaping and modulating fear learning over time. This body of evidence illuminates the multifaceted contributions of mitochondria to fear memory, ranging from cellular and molecular mechanisms to potential therapeutic targets for fear-related disorders.

Early studies showed that fear conditioning induces changes in mitochondrial gene expression, identifying mitochondria-related genes that are differentially expressed in the mid-temporal regions of the brain in trained rats (Federighi et al. 2013). Further analyses of microarray data from both rodent and human studies linked mitochondrial function genes in the PFC to fear-related behavior, including key genes like monoamine oxidase B (MAOB), which is essential for the catabolism of neurotransmitters such as dopamine; NADH dehydrogenase flavoprotein 1 (NDUFV1), which is crucial for ATP synthesis; and mitochondrial uncoupling protein 5 (SLC25A14), which modulates neuronal energy metabolism and survival (Choi et al. 2012). These early findings underscored mitochondrial adaptation as vital for the regulation of fear memory, suggesting the potential for therapeutic interventions targeting mitochondrial function. For instance, methylene blue, a mitochondrial enhancer, was shown effective in improving memory retention and facilitating fear extinction post-training, highlighting the therapeutic potential of modulating mitochondrial function in treating fear-related disorders (Telch et al. 2014). More recent research, including studies on mutations in genes critical for mitochondrial protein synthesis, has emphasized the importance of mitochondrial integrity during neurodevelopment for the proper development of neural circuits critically involved in fear conditioning (Kawarada et al. 2021).

Mitochondrial dynamics, particularly the regulation of energy production within the hippocampus, have also been shown to play integral roles in fear learning and memory formation. Contextual fear conditioning has been shown to increase mitochondrial respiration, especially within the presynaptic terminals of the CA1 region of the hippocampus, likely due to the energy-intensive nature of processes such as protein synthesis, neurotransmitter release, and synapse formation essential for long-term memory consolidation (Underwood et al. 2023). This increase in energy demand has been linked to mitochondrial fission, mediated by dynamin-related protein 1 (Drp1), with inhibitors of Drp1 diminishing the enhancement in mitochondrial respiration induced by fear conditioning and impairing the consolidation of contextual fear memory (Underwood et al. 2023). Further investigations into mitochondrial dynamics have revealed their impact on fear memory processes beyond mere energy provision. Specifically, alterations in mitochondrial fission and fusion, observed in serotonin transporter knockout rats, were associated with marked deficiencies in fear extinction recall, along with changes in mitochondrial oxidative phosphorylation, biogenesis, and the production of antioxidant enzymes (Brivio et al. 2022). This deficiency suggests a mechanism whereby mitochondrial dynamics directly influence the neural circuitry involved in fear conditioning, alongside broader implications for mitochondrial functions. Moreover, balanced mitochondrial dynamics have been highlighted in studies of FOXP1 haploinsufficiency mice, where several proteins regulating mitochondrial biogenesis (e.g., Foxo1, Pgc-1α, Tfam) and dynamics (e.g., Mfn1, Opa1, Drp1, and Fis1) were significantly dysregulated, leading to alterations in mitochondrial networks and fear memory deficits (Wang et al. 2022). Additionally, the role of astrocytes and their mitochondrial function in fear conditioning has been emphasized (Shevelkin et al. 2020; Murlanova et al. 2022), with reports that impaired mitochondrial respiration in astrocytes, due to the deletion of Cox10, leads to deficiencies in fear conditioning (Murlanova et al. 2022). These studies underscore the fundamental role of mitochondrial dynamics in supporting the bioenergetic and synaptic requirements of fear learning.

Oxidative processes, closely linked to mitochondrial respiration and bioenergetics, have been identified as playing critical roles in fear learning, with antioxidant defense mechanisms crucial for maintaining cognitive functions. NRF2, a key antioxidant regulatory transcription factor involved in modulating mitochondrial biogenesis, antioxidant defenses, and cellular metabolism, has been shown to contribute to fear learning in aged mice (Zweig et al. 2020), suggesting NRF2 as a valuable target for cognitive enhancement strategies. Furthermore, oxidative stress has been directly linked to synaptic dysfunction and memory deficits in the context of neurodevelopmental disorders. For example, increased levels of superoxide were found in the hippocampus of a mouse model of Angelman syndrome, and targeting mitochondrial reactive oxygen species (ROS) with antioxidants like MitoQ ameliorated impairments in synaptic plasticity and contextual fear memory deficits (Santini et al. 2015).

Mitochondria also play key roles in the synthesis and catabolism of molecules involved in both general metabolism and neurotransmission, thereby affecting neurotransmitter balance, which is fundamental for proper brain function and behavior, including mechanisms underlying fear and anxiety. For instance, glutamate dehydrogenase 1 (GDH), a pivotal enzyme encoded by the Glud1 gene, plays a significant role in maintaining glutamate homeostasis. GDH deficiency has been shown to lead to elevated levels of glutamate and glutamine in the hippocampus and medial prefrontal cortex, altering neurotransmitter balance, affecting NMDA receptor function, and being associated with deficits in fear memory retrieval and changes in neuronal activation patterns, highlighting the vulnerability of the hippocampus to GDH deficiency (Lander et al. 2020). Moreover, monoamine oxidase A (MAOA), an enzyme involved in the metabolism of neurotransmitters such as serotonin and located on the outer mitochondrial membrane, has been found to play a role in modulating fear responses. High MAOA activity in the prelimbic cortex of rats displaying persistent fear responses suggests a potential mitochondrial link to individual differences in the processing of fearful stimuli and, by extension, to trauma-related vulnerabilities (Poirier et al. 2016). This perspective aligns with the emerging involvement of mitochondria in dysfunctions related to the development of post-traumatic stress disorder (PTSD), further underscoring the importance of mitochondrial functions in the neurobiology of fear and anxiety.

6 Mitochondria and Anxiety

The link between mitochondrial function, anxiety, and the neural circuits regulating anxiety-related behaviors has emerged as a critical area of investigation. A growing body of research underscores the pivotal role of mitochondrial bioenergetics, structural integrity, and signaling pathways in shaping these behaviors, highlighting how mitochondrial dysfunctions contribute to anxiety across various states and disorders. Advances in this field are illuminating the underlying mechanisms of vulnerability to stress-related disorders, with trait anxiety identified as a significant risk factor (Sandi and Richter-Levin 2009; Weger and Sandi 2018).

Genetic studies, both in humans and through genetic manipulation in animal models, have robustly evidenced a causal link between mitochondrial alterations and the emergence of anxiety behaviors. These investigations reveal a significant interaction between mitochondrial respiratory processes and inflammatory responses in the development of anxiety (Fan et al. 2019; Liu et al. 2023). Research comparing different mouse strains has elucidated mitochondrial metabolism's role in anxiety-related phenotypes, highlighting strain-specific differences in anxiety behaviors and their correlation with mitochondrial function (Filiou et al. 2021). Moreover, studies on mitochondrial DNA variations and the effects of mitochondrial uncoupling protein 2 gene knockout have reinforced the connection between mitochondrial function and anxiety, emphasizing how mitochondrial variations can affect stress reactivity and neurotransmitter systems (Gimsa et al. 2009, 2011). Interestingly, observations in humans, particularly in matrilineal relatives of children with presumed maternally inherited mitochondrial disorders, have suggested that mitochondrial DNA sequences may predispose individuals to mental health conditions (Boles et al. 2005). These different lines of genetic evidence offer compelling support to the view that mitochondrial alterations play an etiological role in the development and manifestation of anxiety behaviors.

Stressful life events, significant risk factors for anxiety disorders, has been linked to alterations in mitochondrial function in key brain regions associated with increased anxiety-like behaviors (Misiewicz et al. 2019; Weger et al. 2020; Zhvania et al. 2022; Morella et al. 2022; Rosenberg et al. 2023). The bed nucleus of the stria terminalis (BNST), a central node in the neural circuits regulating anxiety, has been a focus of these studies, with research revealing stress-induced alterations in mitochondrial gene expression and the identification of distinct mitochondrial phenotypes associated with anxiety-like behavior (Misiewicz et al. 2019; Rosenberg et al. 2023). A multi-omics analysis in the bed nucleus of the stria terminalis (BNST), a key brain region regulating anxiety, investigated how environmental stress affects anxiety using the chronic social defeat stress (CSDS) paradigm in mice (Misiewicz et al. 2019). Stress induced differential expression of mRNA, miRNA, and proteins, with enriched mitochondrial-related genes in stressed mice, reflecting stress-specific alterations in mitochondrial gene expression between stress-susceptible and stress-resilient mice. Rosenberg et al. (2023) further explored brain mitochondrial diversity and network organization after chronic stress and glucocorticoid treatment, identifying mitochondria-behavior associations across brain regions. The corticostriatal network, particularly, showed strong correlations with behavior, indicating the significant influence of mitochondrial capacity in these regions on anxiety-like behavior.

Oxidative stress, a frequent outcome of mitochondrial dysfunction, has been closely linked to anxiety and its development (Filiou and Sandi 2019). Increased oxidative stress has been consistently found in highly anxious individuals, particularly when compounded by inflammation and the activation of stress response (Hovatta et al. 2010; Olsen et al. 2013; Krolow et al. 2014). These humans and rodent studies showed correlations between anxiety-related behaviors and oxidative damage markers in both blood, neurons, and glia across various brain regions. Pharmacological interventions targeting oxidative stress, such as MitoQ, have demonstrated anxiolytic effects, highlighting the potential for mitochondrial modulation in anxiety treatment (Nussbaumer et al. 2016).

Mitophagy, another mitochondrial function-related mechanism, has been also implicated in anxiety regulation, with studies showing that chronic social defeat stress induces mitochondrial impairment in the amygdala that was related to the activation of the PINK1–Parkin mitophagy pathway and mitophagy activation in the amygdala, leading to increased anxiety (Duan et al. 2021). Such mitochondrial deficiency was related to impaired synaptic transmission in the basolateral amygdala-bed nucleus of the stria terminalis (BLA-BNST) pathway, which is crucial for anxiety regulation. This suggests mitophagy's importance in the regulation of neural circuits related to anxiety.

A significant line of research has focused on the NAc and its role in anxiety, particularly showing how mitochondrial functions within this neural hub relate to anxiety predisposition (i.e., trait anxiety) and the regulation of anxiety-related behaviors. This is important because anxiety trait is a known risk factor for the development of stress-related alterations, notably anxiety disorders and depression (Sandi and Richter-Levin 2009; Weger and Sandi 2018). Variations in mitochondrial function within the NAc medium spiny neurons (MSNs) have been linked to individual differences in anxiety behaviors and their interaction with motivated behaviors (Hollis et al. 2015; Van Der Kooij et al. 2018; Calarco and Lobo 2021; Gebara et al. 2021; Zalachoras et al. 2022; Ghosal et al. 2023). Mechanistic insights have revealed that decreased expression of Mfn2, a protein crucial for mitochondrial dynamics and mitochondria–endoplasmic reticulum interactions in MSNs causally links heightened anxiety levels to alterations in mitochondrial and neuronal structure and function, predisposing individuals to depression-like behaviors (Gebara et al. 2021; Ghosal et al. 2023).

Furthermore, in vivo ^{1}H-magnetic resonance spectroscopy (^{1}H-MRS) studies have provided insights into the relationship between NAc metabolites and anxiety, revealing how metabolite levels can reflect individual differences in anxiety and social rank, and how they respond to chronic stress and nutritional interventions. This research underscores the metabolic underpinnings of anxiety and the interplay between anxiety and motivation (Larrieu et al. 2017; Strasser et al. 2019, 2020; Cherix et al. 2020). Together, these findings highlight the NAc's critical role in anxiety, with mitochondrial function and metabolic state acting as key mediators.

This body of evidence not only advances our understanding of the neural circuitry underlying anxiety but also opens avenues for targeted interventions aimed at ameliorating anxiety and depression through mitochondrial modulation.

7 Mitochondria in PTSD

Post-traumatic stress disorder (PTSD) represents a complex interplay between traumatic experiences and biological predispositions, where mitochondrial dysfunction increasingly recognized as a fundamental biological underpinning. Across diverse research domains – from animal models to human studies – mitochondria are being pinpointed as key players in physiological processes integral to PTSD, including abnormal fear learning, brain network activation, synaptic plasticity, steroidogenesis, and inflammation (Preston et al. 2018). This section synthesizes findings that associate mitochondrial dysfunction with PTSD, both in humans and animal models.

Evidence suggests mitochondrial dysfunction might contributes to PTSD's etiology. Human genetic studies have revealed altered mitochondrial gene expression in the prefrontal cortex, decreased mitochondrial DNA copy numbers in granulocytes, and a link between PTSD severity and mitochondrial DNA (mtDNA) single nucleotide polymorphism heteroplasmy levels (reviewed in Preston et al., 2018). These findings point at a connection between mitochondrial dysfunction and PTSD development, implicating alterations in oxidative phosphorylation, the tricarboxylic acid

(TCA) cycle, and pyruvate metabolism. Beyond genetics, a recent study highlighted the exacerbation of PTSD symptoms linked to adolescent sexual trauma in women, evidenced by increased fear-potentiated startle responses and skin conductance, alongside significant rises in circulating cell-free mitochondrial DNA (ccf-mtDNA) levels in adulthood (Morrison et al. 2022). This timing-specific association underscores the enduring impact of trauma on mitochondrial function.

Animal studies further bolster the link between mitochondrial dysfunction and PTSD-like behaviors in animal models for PTSD, showing that even mild acute stress can inhibit mitochondrial complex I activity (Batandier et al. 2014). Environmental (Li et al. 2014) and genetically based (Ma et al. 2020) rodent models have detailed how PTSD-like conditions precipitate mitochondrial dysregulation across key neural regions, including disruptions in oxidative phosphorylation, β-oxidation of fatty acids, mitochondrial dynamics, and the TCA cycle, leading to neural oxidative stress and subsequent cellular damage (Kaplan et al. 2023). Such trauma-induced mitochondrial alterations extend to structural changes, replication issues, neuroinflammatory responses, signal transduction abnormalities, and apoptosis, underlining mitochondrial dysfunction's profound impact on PTSD pathology and its potential as a therapeutic target.

The potential therapeutic benefits of targeting mitochondrial dysfunction in PTSD are highlighted by the efficacy of antidepressants in reversing stress-induced behavioral and mitochondrial changes (Preston et al. 2018). For example, a study utilizing a foot shock-induced PTSD mouse model showed that fluoxetine, a common antidepressant, rectifies alterations in energy metabolism pathways within the neural circuitry underlying PTSD-like behaviors (Kao et al. 2016). Specifically, it modulates the TCA cycle in different brain regions and ameliorates conditioned fear responses. This emerging evidence advocates for further research focusing on mitochondria in both the understanding and treatment of PTSD.

8 Toward Novel Therapeutic Approaches

The recent elucidation of neurometabolic mechanisms underlying anxiety opens promising avenues for innovative treatments. This research underscores the critical role of mitochondrial function and metabolism in modulating anxiety-related behaviors and points toward pharmacological and metabolic interventions as potential therapeutic strategies.

Pharmacological approaches targeting mitochondrial dysfunction, such as the use of MitoQ and other antioxidants (Zalachoras et al. 2020), have shown promise in alleviating anxiety-like behaviors in animal models. MitoQ, a mitochondria-targeted antioxidant, has been shown to decrease anxiety-related behavior in high anxiety-related behavior (HAB) mice by altering the brain metabolome and inducing distinct molecular signatures (Nussbaumer et al. 2016). The role of glutathione (GSH), the brain's principal endogenous antioxidant, in anxiety-related deficits in motivation further illustrates the importance of mitochondrial and antioxidant function in the

NAc's regulation of anxiety and motivation. Studies have established a causal role between accumbal GSH levels and steady performance in effort-related tasks, and showed that improving accumbal GSH levels and antioxidant function through nutritional supplementation with *N*-acetyl-cysteine could ameliorate anxiety-related deficits in motivated performance (Zalachoras et al. 2022). eNAMPT, the extracellular form of the rate-limiting enzyme NAMPT in the nicotinamide adenine dinucleotide (NAD+) synthesis, is emerging as a potential candidate in modulating brain behavior and body-to-brain signaling, through its action on NAD+/SIRT+ pathway, specifically operating in the nucleus accumbens (Morató et al. 2022).

Additionally, interventions aimed at enhancing mitochondrial function through the modulation of metabolic pathways have emerged as viable strategies. Dietary supplementation with medium-chain triglycerides (MCT) has been found to reduce anxiety-like behaviors and enhance social competitiveness in rats, potentially through the enhancement of mitochondrial energy metabolism (Hollis et al. 2018). Similarly, acetyl-L-carnitine (LAC), a mitochondria-boosting supplement, has exhibited antidepressant-like effects in stressed mice by restoring neuroenergetic alterations in the NAc (Cherix et al. 2020). Urolithin A, a compound that induces mitophagy, has also been spotlighted for its lifespan-prolonging and muscle function-enhancing effects in animal models (Ryu et al. 2016), hinting at its potential for addressing anxiety-related disorders by improving mitochondrial quality control.

These insights into the neurometabolic underpinnings of anxiety, underlining the therapeutic potential of enhancing mitochondrial function and antioxidant defenses, offer a promising avenue for developing effective treatments. By targeting mitochondrial dysfunction and metabolic pathways, these novel therapeutic approaches could provide a much-needed alternative to current treatments, potentially offering relief to those with anxiety disorders and related psychopathologies.

9 Outstanding Questions

While the existing body of research has begun to unveil the connection between mitochondria and fear and anxiety-related behaviors, several critical questions remain unanswered. Foremost, the precise neurobiological mechanisms through which mitochondrial dysfunction precipitates the onset of anxiety disorders are yet to be fully elucidated. How do alterations in mitochondrial bioenergetics and ROS production influence neural circuits specifically associated with fear and anxiety? What role do mitochondrial dynamics, including fission, fusion, and mitophagy, play in the regulation of these behaviors? Another key and rather unexplored question is to what extent potential sex-related differences in mitochondrial function and dysfunction contribute to differential susceptibility to develop PTSD and its neurobiological underpinnings. In this connection, it is also important to explore whether changes in mitochondrial function during the reproductive lifespan in women contribute to differential susceptibility to develop PTSD. Moreover, at the moment, it is unclear whether brain regions underlying fear and anxiety are particularly susceptible to mitochondrial changes under exposure to traumatic experiences or whether changes are distributed across the brain, and if not, what could be the reasons underlying those differences. Furthermore, the potential of mitochondrial targeting as a therapeutic strategy raises important questions: Can enhancing mitochondrial function ameliorate anxiety disorders and PTSD? Additionally, the regional specificity of mitochondrial dysfunction across key neural areas involved in anxiety, as well as the interaction between genetic and environmental factors in mitochondrial perturbations, warrants further investigation. Addressing these questions demands integrated, multidisciplinary approaches, promising to unravel the intricate mitochondrial underpinnings of anxiety and fear.

10 Conclusions

In this review, we have detailed the significant role of mitochondrial function in the neural mechanisms underlying fear and anxiety, highlighting its importance across a spectrum of psychiatric conditions. Mitochondria are central to both the regulation of normal behavioral responses and the pathophysiology of anxiety disorders. By focusing on how these organelles influence brain circuits involved in fear and anxiety, we offer new insights into potential therapeutic targets for anxiety disorders. This growing body of evidence not only enriches our understanding of the neurobiology of anxiety but also suggests that targeting mitochondrial dysfunction could provide effective strategies for treating these complex conditions. Future research should continue to explore the intricate relationship between mitochondrial function, brain circuitry, and behavior, with the aim of developing novel interventions that can improve treatment outcomes for individuals with anxiety disorders. This approach holds the promise of advancing our management of anxiety and related disorders, moving toward interventions that address the underlying biological mechanisms.

References

Andreazza AC, Andersen ML, Alvarenga TA et al (2010) Impairment of the mitochondrial electron transport chain due to sleep deprivation in mice. J Psychiatr Res 44:775–780. https://doi.org/10.1016/j.jpsychires.2010.01.015

Andreazza AC, Nierenberg AA (2018) Mitochondrial dysfunction: at the Core of psychiatric disorders? Biol Psychiatry 83:718–719

Batandier C, Poulet L, Hininger I et al (2014) Acute stress delays brain mitochondrial permeability transition pore opening. J Neurochem 131:314–322. https://doi.org/10.1111/jnc.12811

Bloodgood DW, Sugam JA, Holmes A, Kash TL (2018) Fear extinction requires infralimbic cortex projections to the basolateral amygdala. Transl Psychiatry 8:60. https://doi.org/10.1038/s41398-018-0106-x

Boles RG, Burnett BB, Gleditsch K et al (2005) A high predisposition to depression and anxiety in mothers and other matrilineal relatives of children with presumed maternally inherited mitochondrial disorders. Am J Med Genet B Neuropsychiatr Genet 137B:20–24. https://doi.org/10.1002/ajmg.b.30199

Brand MD, Affourtit C, Esteves TC et al (2004) Mitochondrial superoxide: production, biological effects, and activation of uncoupling proteins. Free Radic Biol Med 37:755–767

Britt JP, Benaliouad F, McDevitt RA et al (2012) Synaptic and behavioral profile of multiple glutamatergic inputs to the nucleus Accumbens. Neuron 76:790–803. https://doi.org/10.1016/j.neuron.2012.09.040

Brivio P, Gallo MT, Karel P et al (2022) Alterations of mitochondrial dynamics in serotonin transporter knockout rats: a possible role in the fear extinction recall mechanisms. Front Behav Neurosci 16. https://doi.org/10.3389/fnbeh.2022.957702

Burkhouse KL, Jimmy J, Defelice N et al (2020) Nucleus accumbens volume as a predictor of anxiety symptom improvement following CBT and SSRI treatment in two independent samples. Neuropsychopharmacology 45:561–569. https://doi.org/10.1038/s41386-019-0575-5

Calarco CA, Lobo MK (2021) The individualized powerhouse: Mitofusin-2 regulates nucleus Accumbens mitochondrial influence on individual differences in trait anxiety. Biol Psychiatry 89:1024–1026. https://doi.org/10.1016/j.biopsych.2021.03.031

Calhoon GG, Tye KM (2015) Resolving the neural circuits of anxiety. Nat Neurosci 18:1394–1404

Chen W, Zhao H, Li Y (2023) Mitochondrial dynamics in health and disease: mechanisms and potential targets. Signal Transduct Target Ther 8:333. https://doi.org/10.1038/s41392-023-01547-9

Cherix A, Larrieu T, Grosse J et al (2020) Metabolic signature in nucleus accumbens for anti-depressant-like effects of acetyl-l-carnitine. Elife 9:1–19. https://doi.org/10.7554/eLife.50631

Choi KH, Le T, McGuire J et al (2012) Expression profiles of mitochondrial genes in the frontal cortex and the caudate nucleus of developing humans and mice selectively bred for high and low fear. PLoS One 7:e49183. https://doi.org/10.1371/journal.pone.0049183

Ciocchi S, Herry C, Grenier F et al (2010) Encoding of conditioned fear in central amygdala inhibitory circuits. Nature 468:277–282. https://doi.org/10.1038/nature09559

Correia SS, McGrath AG, Lee A et al (2016) Amygdala-ventral striatum circuit activation decreases long-term fear. Elife 5. https://doi.org/10.7554/eLife.12669

Daniels TE, Olsen EM, Tyrka AR (2020) Stress and psychiatric disorders: the role of mitochondria. Annu Rev Clin Psychol 16:165–186. https://doi.org/10.1146/annurev-clinpsy-082719-104030

Datta S, Jaiswal M (2021) Mitochondrial calcium at the synapse. Mitochondrion 59135-153. https://doi.org/10.1016/j.mito.2021.04.006

Devine MJ, Kittler JT (2018) Mitochondria at the neuronal presynapse in health and disease. Nat Rev Neurosci 19(2):63–80. https://doi.org/10.1038/nrn.2017.170

Do-Monte FH, Quiñones-Laracuente K, Quirk GJ (2015) A temporal shift in the circuits mediating retrieval of fear memory. Nature 519:460–463. https://doi.org/10.1038/nature14030

Duan K, Gu Q, Petralia RS et al (2021) Mitophagy in the basolateral amygdala mediates increased anxiety induced by aversive social experience. Neuron 109:3793–3809.e8. https://doi.org/10.1016/j.neuron.2021.09.008

Duchen MR (2000) Mitochondria and calcium: from cell signalling to cell death. J Physiol 529:57–68. https://doi.org/10.1111/j.1469-7793.2000.00057.x

Duvarci S, Pare D (2014) Amygdala microcircuits controlling learned fear. Neuron 82:966–980

Evans CS, Holzbaur ELF (2020) Quality control in neurons: mitophagy and other selective autophagy mechanisms. J Mol Biol 432:240–260. https://doi.org/10.1016/j.jmb.2019.06.031

Fan B, Hao B, Dai Y et al (2022) Deficiency of Tet3 in nucleus accumbens enhances fear generalization and anxiety-like behaviors in mice. Brain Pathol 32. https://doi.org/10.1111/bpa.13080

Fan K, Li Y, Wang H et al (2019) Stress-induced metabolic disorder in peripheral CD4+ T cells leads to anxiety-like behavior. Cell 179:864–879.e19. https://doi.org/10.1016/j.cell.2019.10.001

Federighi G, Traina G, Macchi M et al (2013) Modulation of gene expression in contextual fear conditioning in the rat. PLoS One 8:e80037. https://doi.org/10.1371/journal.pone.0080037

Filiou MD, Nussbaumer M, Teplytska L, Turck CW (2021) Behavioral and metabolome differences between C57BL/6 and DBA/2 mouse strains: implications for their use as models for depression- and anxiety-like phenotypes. Metabolites 11:128. https://doi.org/10.3390/metabo11020128

Filiou MD, Sandi C (2019) Anxiety and brain mitochondria: a bidirectional crosstalk. Trends Neurosci 42:573–588

Gebara E, Zanoletti O, Ghosal S et al (2021) Mitofusin-2 in the nucleus Accumbens regulates anxiety and depression-like behaviors through mitochondrial and neuronal actions. Biol Psychiatry 89:1033–1044. https://doi.org/10.1016/j.biopsych.2020.12.003

Ghosal S, Gebara E, Ramos-Fernández E et al (2023) Mitofusin-2 in nucleus accumbens D2-MSNs regulates social dominance and neuronal function. Cell Rep 42. https://doi.org/10.1016/j.celrep.2023.112776

Gimsa U, Kanitz E, Otten W et al (2011) Alterations in anxiety-like behavior following knockout of the uncoupling protein 2 (ucp2) gene in mice. Life Sci 89:677–684. https://doi.org/10.1016/j.lfs.2011.08.009

Gimsa U, Kanitz E, Otten W, Ibrahim SM (2009) Behavior and stress reactivity in mouse strains with mitochondrial DNA variations. Ann N Y Acad Sci 1153:131–138. https://doi.org/10.1111/j.1749-6632.2008.03960.x

Giustino TF, Maren S (2015) The role of the medial prefrontal cortex in the conditioning and extinction of fear. Front Behav Neurosci 9. https://doi.org/10.3389/fnbeh.2015.00298

Han S, Zhang M, Jeong YY et al (2021) The role of mitophagy in the regulation of mitochondrial energetic status in neurons. Autophagy 17:4182–4201. https://doi.org/10.1080/15548627.2021.1907167

Herry C, Johansen JP (2014) Encoding of fear learning and memory in distributed neuronal circuits. Nat Neurosci 17:1644–1654

Hollis F, Mitchell ES, Canto C et al (2018) Medium chain triglyceride diet reduces anxiety-like behaviors and enhances social competitiveness in rats. Neuropharmacology 138:245–256. https://doi.org/10.1016/j.neuropharm.2018.06.017

Hollis F, Van Der Kooij MA, Zanoletti O et al (2015) Mitochondrial function in the brain links anxiety with social subordination. Proc Natl Acad Sci U S A 112:15486–15491. https://doi.org/10.1073/pnas.1512653112

Hovatta I, Juhila J, Donner J (2010) Oxidative stress in anxiety and comorbid disorders. Neurosci Res 68:261–275. https://doi.org/10.1016/j.neures.2010.08.007

Kao C-Y, He Z, Henes K et al (2016) Fluoxetine treatment rescues energy metabolism pathway alterations in a posttraumatic stress disorder mouse model. Complex Psychiatry 2:46–59. https://doi.org/10.1159/000445377

Kaplan GB, Dadhi NA, Whitaker CS (2023) Mitochondrial dysfunction in animal models of PTSD: relationships between behavioral models, neural regions, and cellular maladaptation. Front Physiol 14. https://doi.org/10.3389/fphys.2023.1105839

Kawarada L, Fukaya M, Saito R et al (2021) Telencephalon-specific *Alkbh1* conditional knockout mice display hippocampal atrophy and impaired learning. FEBS Lett 595:1671–1680. https://doi.org/10.1002/1873-3468.14098

Kraus F, Roy K, Pucadyil TJ, Ryan MT (2021) Function and regulation of the divisome for mitochondrial fission. Nature 590:57–66

Krolow R, Arcego DM, Noschang C et al (2014) Oxidative imbalance and anxiety disorders. Curr Neuropharmacol 12:193–204. https://doi.org/10.2174/1570159X11666131120223530

Krols M, van Isterdael G, Asselbergh B et al (2016) Mitochondria-associated membranes as hubs for neurodegeneration. Acta Neuropathol 131:505–523. https://doi.org/10.1007/s00401-015-1528-7

Lander SS, Chornyy S, Safory H et al (2020) Glutamate dehydrogenase deficiency disrupts glutamate homeostasis in hippocampus and prefrontal cortex and impairs recognition memory. Genes Brain Behav 19. https://doi.org/10.1111/gbb.12636

Larrieu T, Cherix A, Duque A et al (2017) Hierarchical status predicts behavioral vulnerability and nucleus Accumbens metabolic profile following chronic social defeat stress. Curr Biol 27:2202–2210.e4. https://doi.org/10.1016/j.cub.2017.06.027

Latorre-Pellicer A, Moreno-Loshuertos R, Lechuga-Vieco A et al (2016) Mitochondrial and nuclear DNA matching shapes metabolism and healthy ageing. Nature 535(7613):561–565. https://doi.org/10.1038/nature18618

Lee S, Ahmed T, Lee S et al (2012) Bidirectional modulation of fear extinction by mediodorsal thalamic firing in mice. Nat Neurosci 15:308–314. https://doi.org/10.1038/nn.2999

Li H, Li X, Smerin SE et al (2014) Mitochondrial gene expression profiles and metabolic pathways in the Amygdala associated with exaggerated fear in an animal model of PTSD. Front Neurol 5. https://doi.org/10.3389/fneur.2014.00164

Liu D, Gao Y, Liu J et al (2021) Intercellular mitochondrial transfer as a means of tissue revitalization. Signal Transduct Target Ther 6:65. https://doi.org/10.1038/s41392-020-00440-z

Liu L, Cheng S, Qi X et al (2023) Mitochondria-wide association study observed significant interactions of mitochondrial respiratory and the inflammatory in the development of anxiety and depression. Transl Psychiatry 13:216. https://doi.org/10.1038/s41398-023-02518-y

Lüthi A, Lüscher C (2014) Pathological circuit function underlying addiction and anxiety disorders. Nat Neurosci 17:1635–1643

Ma L, Chen X, Zhao B et al (2020) Enhanced apoptosis and decreased AMPA receptors are involved in deficit in fear memory in rin1 knockout rats. J Affect Disord 268:173–182. https://doi.org/10.1016/j.jad.2020.02.040

Mergenthaler P, Lindauer U, Dienel GA, Meisel A (2013) Sugar for the brain: the role of glucose in physiological and pathological brain function. Trends Neurosci 36:587–597. https://doi.org/10.1016/j.tins.2013.07.001

Misiewicz Z, Iurato S, Kulesskaya N et al (2019) Multi-omics analysis identifies mitochondrial pathways associated with anxiety-related behavior. PLoS Genet 15:e1008358. https://doi.org/10.1371/journal.pgen.1008358

Monzel AS, Enríquez JA, Picard M (2023) Multifaceted mitochondria: moving mitochondrial science beyond function and dysfunction. Nat Metab 5:546–562. https://doi.org/10.1038/s42255-023-00783-1

Morató L, Astori S, Zalachoras I et al (2022) eNAMPT actions through nucleus accumbens NAD $^{+}$ / SIRT1 link increased adiposity with sociability deficits programmed by peripuberty stress. Sci Adv 8. https://doi.org/10.1126/sciadv.abj9109

Morava E, Gardeitchik T, Kozicz T et al (2010) Depressive behaviour in children diagnosed with a mitochondrial disorder. Mitochondrion 10:528–533. https://doi.org/10.1016/j.mito.2010.05.011

Morava É, Kozicz T (2013) Mitochondria and the economy of stress (mal)adaptation. Neurosci Biobehav Rev 37:668–680

Morella IM, Brambilla R, Morè L (2022) Emerging roles of brain metabolism in cognitive impairment and neuropsychiatric disorders. Neurosci Biobehav Rev 142:104892. https://doi.org/10.1016/j.neubiorev.2022.104892

Morrison KE, Stenson AF, Marx-Rattner R et al (2022) Developmental timing of Trauma in women predicts unique extracellular vesicle proteome signatures. Biol Psychiatry 91:273–282. https://doi.org/10.1016/j.biopsych.2021.08.003

Murlanova K, Jouroukhin Y, Huseynov S et al (2022) Deficient mitochondrial respiration in astrocytes impairs trace fear conditioning and increases naloxone-precipitated aversion in morphine-dependent mice. Glia 70:1289–1300. https://doi.org/10.1002/glia.24169

Namkung H, Thomas KL, Hall J, Sawa A (2022) Parsing neural circuits of fear learning and extinction across basic and clinical neuroscience: towards better translation. Neurosci Biobehav Rev 134

Nussbaumer M, Asara JM, Teplytska L et al (2016) Selective mitochondrial targeting exerts anxiolytic effects in vivo. Neuropsychopharmacology 41:1751–1758. https://doi.org/10.1038/npp.2015.341

Olsen RKJ, Cornelius N, Gregersen N (2013) Genetic and cellular modifiers of oxidative stress: what can we learn from fatty acid oxidation defects? Mol Genet Metab 110:S31–S39. https://doi.org/10.1016/j.ymgme.2013.10.007

Palikaras K, Lionaki E, Tavernarakis N (2018) Mechanisms of mitophagy in cellular homeostasis, physiology and pathology. Nat Cell Biol 20:1013–1022. https://doi.org/10.1038/s41556-018-0176-2

Patergnani S, Suski JM, Agnoletto C et al (2011) Calcium signaling around mitochondria associated membranes (MAMs). Cell Commun Signal 9:19. https://doi.org/10.1186/1478-811X-9-19

Pathak RK, Kolishetti N, Dhar S (2015) Targeted nanoparticles in mitochondrial medicine. WIREs Nanomed Nanobiotechnol 7(3):315–329. https://doi.org/10.1002/wnan.2015.7.issue-3 10.1002/wnan.1305

Pei L, Wallace DC (2018) Mitochondrial etiology of neuropsychiatric disorders. Biol Psychiatry 83: 722–730

Picard M, McEwen BS (2018) Psychological stress and mitochondria: a systematic review. Psychosom Med 80:141–153

Picard M, Meagan J, McManus Jason D et al (2015) Mitochondrial functions modulate neuroendocrine metabolic inflammatory and transcriptional responses to acute psychological stress. Significance Proceedings Nat Academy Sci 112(48). https://doi.org/10.1073/pnas.1515733112

Picard M, Sandi C (2021) The social nature of mitochondria: implications for human health. Neurosci Biobehav Rev 120:595–610. https://doi.org/10.1016/j.neubiorev.2020.04.017

Pignatelli M, Tejeda HA, Barker DJ et al (2021) Cooperative synaptic and intrinsic plasticity in a disynaptic limbic circuit drive stress-induced anhedonia and passive coping in mice. Mol Psychiatry 26:1860–1879. https://doi.org/10.1038/s41380-020-0686-8

Poirier GL, Hitora-Imamura N, Sandi C (2016) Emergence in extinction of enhanced and persistent responding to ambiguous aversive cues is associated with high MAOA activity in the prelimbic cortex. Neurobiol Stress 5:1–7. https://doi.org/10.1016/j.ynstr.2016.08.005

Preston G, Kirdar F, Kozicz T (2018) The role of suboptimal mitochondrial function in vulnerability to post-traumatic stress disorder. J Inherit Metab Dis 41:585–596. https://doi.org/10.1007/s10545-018-0168-1

Price JS (2003) Evolutionary aspects of anxiety disorders. Dialogues Clin Neurosci 5:223–236. https://doi.org/10.31887/DCNS.2003.5.3/jprice

Ramanathan KR, Maren S (2019) Nucleus reuniens mediates the extinction of contextual fear conditioning. Behav Brain Res 374:112114. https://doi.org/10.1016/j.bbr.2019.112114

Ramanathan KR, Ressler RL, Jin J, Maren S (2018) Nucleus Reuniens is required for encoding and retrieving precise, hippocampal-dependent contextual fear memories in rats. J Neurosci 38: 9925–9933. https://doi.org/10.1523/JNEUROSCI.1429-18.2018

Rangaraju V, Calloway N, Ryan TA (2014) Activity-Driven Local ATP Synthesis Is Required for Synaptic Function. Cell 156(4):825–835. https://doi.org/10.1016/j.cell.2013.12.042

Ratigan HC, Krishnan S, Smith S, Sheffield MEJ (2023) A thalamic-hippocampal CA1 signal for contextual fear memory suppression, extinction, and discrimination. Nat Commun 14:6758. https://doi.org/10.1038/s41467-023-42429-6

Rodriguez-Romaguera J, Do Monte FHM, Quirk GJ (2012) Deep brain stimulation of the ventral striatum enhances extinction of conditioned fear. Proc Natl Acad Sci 109:8764–8769. https://doi.org/10.1073/pnas.1200782109

Rosenberg AM, Saggar M, Monzel AS et al (2023) Brain mitochondrial diversity and network organization predict anxiety-like behavior in male mice. Nat Commun 14. https://doi.org/10.1038/s41467-023-39941-0

Rossi MJ, Pekkurnaz G (2019) Powerhouse of the mind: mitochondrial plasticity at the synapse. Curr Opin Neurobiol 57149-155. https://doi.org/10.1016/j.conb.2019.02.001

Ryan KC, Ashkavand Z, Norman KR (2020) The role of mitochondrial calcium homeostasis in Alzheimer's and related diseases. Int J Mol Sci 21:9153. https://doi.org/10.3390/ijms21239153

Ryu D, Mouchiroud L, Andreux PA et al (2016) Urolithin A induces mitophagy and prolongs lifespan in C. elegans and increases muscle function in rodents. Nat Med 22:879–888. https://doi.org/10.1038/nm.4132

Sandi C, Richter-Levin G (2009) From high anxiety trait to depression: a neurocognitive hypothesis. Trends Neurosci 32:312–320. https://doi.org/10.1016/j.tins.2009.02.004

Santini E, Turner KL, Ramaraj AB et al (2015) Mitochondrial superoxide contributes to hippocampal synaptic dysfunction and memory deficits in Angelman syndrome model mice. J Neurosci 35:16213–16220. https://doi.org/10.1523/JNEUROSCI.2246-15.2015

Schiavone S, Jaquet V, Trabace L, Krause KH (2012) Severe life stress and oxidative stress in the brain: from animal models to human pathology. Antioxid Redox Signal 12:1475–1490. https://doi.org/10.1089/ars.2012.4720

Schrepfer E, Scorrano L (2016) Mitofusins, from mitochondria to metabolism. Mol Cell 61:683–694

Shevelkin AV, Terrillion CE, Hasegawa Y et al (2020) Astrocyte DISC1 contributes to cognitive function in a brain region-dependent manner. Hum Mol Genet 29:2936–2950. https://doi.org/10.1093/hmg/ddaa180

Silva BA, Astori S, Burns AM et al (2021) A thalamo-amygdalar circuit underlying the extinction of remote fear memories. Nat Neurosci 24:964–974. https://doi.org/10.1038/s41593-021-00856-y

Spinelli JB, Haigis MC (2018) The multifaceted contributions of mitochondria to cellular metabolism. Nat Cell Biol 20:745–754. https://doi.org/10.1038/s41556-018-0124-1

Strasser A, Luksys G, Xin L et al (2020) Glutamine-to-glutamate ratio in the nucleus accumbens predicts effort-based motivated performance in humans. Neuropsychopharmacology 45:2048–2057. https://doi.org/10.1038/s41386-020-0760-6

Strasser A, Xin L, Gruetter R, Sandi C (2019) Nucleus accumbens neurochemistry in human anxiety: a 7 T 1 H-MRS study. Eur Neuropsychopharmacol 29:365–375. https://doi.org/10.1016/j.euroneuro.2018.12.015

Telch MJ, Bruchey AK, Rosenfield D et al (2014) Effects of post-session administration of methylene blue on fear extinction and contextual memory in adults with claustrophobia. Am J Psychiatry 171:1091–1098. https://doi.org/10.1176/appi.ajp.2014.13101407

Totty MS, Tuna T, Ramanathan KR et al (2023) Thalamic nucleus reuniens coordinates prefrontal-hippocampal synchrony to suppress extinguished fear. Nat Commun 14:6565. https://doi.org/10.1038/s41467-023-42315-1

Twig G, Shirihai OS (2011) The interplay between mitochondrial dynamics and mitophagy. Antioxid Redox Signal 14:1939–1951. https://doi.org/10.1089/ars.2010.3779

Tyurina YY, Poloyac SM, Tyurin VA et al (2014) A mitochondrial pathway for biosynthesis of lipid mediators. Nat Chem 6:542–552. https://doi.org/10.1038/nchem.1924

Underwood EL, Redell JB, Hood KN et al (2023) Enhanced presynaptic mitochondrial energy production is required for memory formation. Sci Rep 13. https://doi.org/10.1038/s41598-023-40877-0

van de Burgt N, van Doesum W, Grevink M et al (2023) Psychiatric manifestations of inborn errors of metabolism: a systematic review. Neurosci Biobehav Rev 144:104970. https://doi.org/10.1016/j.neubiorev.2022.104970

Van Der Kooij MA, Hollis F, Lozano L et al (2018) Diazepam actions in the VTA enhance social dominance and mitochondrial function in the nucleus accumbens by activation of dopamine D1 receptors. Mol Psychiatry 23:569–578. https://doi.org/10.1038/mp.2017.135

Wang J, Fröhlich H, Torres FB et al (2022) Mitochondrial dysfunction and oxidative stress contribute to cognitive and motor impairment in FOXP1 syndrome. Proc Natl Acad Sci 119. https://doi.org/10.1073/pnas.2112852119

Weger M, Alpern D, Cherix A et al (2020) Mitochondrial gene signature in the prefrontal cortex for differential susceptibility to chronic stress. Sci Rep 10:18308. https://doi.org/10.1038/s41598-020-75326-9

Weger M, Sandi C (2018) High anxiety trait: a vulnerable phenotype for stress-induced depression. Neurosci Biobehav Rev 87:27–37. https://doi.org/10.1016/j.neubiorev.2018.01.012

Woods JJ, Wilson JJ (2020) Inhibitors of the mitochondrial calcium uniporter for the treatment of disease. Curr Opin Chem Biol 55:9–18. https://doi.org/10.1016/j.cbpa.2019.11.006

Zalachoras I, Astori S, Meijer M et al (2022) Opposite effects of stress on effortful motivation in high and low anxiety are mediated by CRHR1 in the VTA

Zalachoras I, Hollis F, Ramos-Fernández E et al (2020) Therapeutic potential of glutathione-enhancers in stress-related psychopathologies. Neurosci Biobehav Rev 114:134–155. https://doi.org/10.1016/j.neubiorev.2020.03.015

Zhvania M, Japaridze N, Tizabi Y et al (2022) Anxiety and ultrastructural consequences of chronic mild stress in rats. Neurosci Lett 771:136390. https://doi.org/10.1016/j.neulet.2021.136390

Zweig JA, Caruso M, Brandes MS, Gray NE (2020) Loss of NRF2 leads to impaired mitochondrial function, decreased synaptic density and exacerbated age-related cognitive deficits. Exp Gerontol 131:110767. https://doi.org/10.1016/j.exger.2019.110767

Humans

Behavioral and Brain Mechanisms of Active Avoidance and Their Relevance to Anxiety Disorders

Hannah McManus and Mohammed R. Milad

Contents

Abstract Avoidance behaviors are paramount for the survival of all species. While adaptive in response to genuine danger, avoidance can become maladaptive when generalized or persistent in the absence of threat. In anxiety disorders, maladaptive avoidance behaviors frequently arise and impede everyday life for individuals with these mental health conditions. Given the central role of avoidance in anxiety disorders, understanding the neural circuits that underly these behaviors is essential for developing both targeted and effective treatments. This review explores avoidance through both a behavioral and neural mechanistic lens, while also examining the current behavioral and pharmacological treatments aimed at addressing avoidance. Understanding of the neural underpinnings of avoidance, psychotherapy, and pharmacotherapy may improve clinical care and outcomes for those suffering with anxiety disorders.

Keywords Active avoidance · Anxiety · Behavioral mechanisms · Neural correlates

H. McManus
New York State Psychiatric Institute, New York, NY, USA

M. R. Milad (✉)
The University of Texas Health Science Center at Houston, Houston, TX, USA
e-mail: Mohammed.R.Milad@uth.tmc.edu

© The Author(s), under exclusive license to Springer Nature Switzerland AG 2025

Curr Topics Behav Neurosci (2025) 73: 57–72
https://doi.org/10.1007/7854_2025_593
Published Online: 11 June 2025

1 Background

Avoidance is critical to the survival of all species. From an evolutionary perspective, avoidance has an adaptive purpose, and individuals learned to avoid true threats in their environment to keep themselves and their families safe. However, avoidance can be maladaptive and interfere with everyday functioning and engagement in value-driven behaviors. For example, when an individual avoids a hot stove, avoidance serves as an adaptive behavior, protecting the person from burning themselves. Though often within mental health conditions, avoidance is experienced as maladaptive and unhelpful, limiting individuals' behaviors and interfering with their everyday lives, for example, a student avoiding taking a college course because of fear of public speaking or avoiding public spaces in fear of germs.

Avoidance is highly prevalent within anxiety disorders, affecting approximately 30% of adults over the course of their lifetime (Michael et al. 2007), and prevalence rates continue to rise (Xiong et al. 2022). There are many conditions that fall under the umbrella of "anxiety disorders," defined as mental health conditions characterized by unrestrained fear or avoidance of actual or perceived threats that lead to a modification of behavior and likely physical symptoms (Craske et al. 2011). Of the eleven anxiety disorders classified by the DSM-5, engagement in avoidance behaviors is a component of all conditions.

Avoidance behaviors refer to actions taken to avoid or anticipate troublesome stimuli, events, or situations (Elliot 1999). Within the context of anxiety disorders, avoidance of anxiety provoking stimuli continues to be negatively reinforced, which leads to a significant impact on their social, interpersonal, and occupational functioning (American Psychiatric Association 2013). For example, someone with social anxiety disorder (SAD) may avoid interviewing for jobs or attending medical appointments in fear of socializing with others. The avoidance of these anxiety provoking events negatively reinforces the avoidance cycle, thus making one less and less likely to deviate from their comfort zone and may result in the further restriction of their outings, work, and social life.

To develop effective treatments for individuals with anxiety disorders, a mechanistic understanding of avoidance behaviors and their underlying neural circuits is essential. These behaviors manifest in various forms – some are passive and some are active avoidance responses, and it is crucial for researchers, clinicians, and patients alike to distinguish between the adaptive and maladaptive patterns. For example, an adaptive avoidance pattern for social anxiety could be thoroughly preparing for a public speaking engagement. While the individual is not avoiding public speaking entirely, they are reducing their distress by engaging in safety behaviors and continuing to undertake the challenge. A maladaptive avoidance response to this scenario would be turning down the speaking engagement, thus avoiding it and their distress entirely. Identifying and understanding maladaptive avoidance behaviors is a critical step in optimizing therapeutic interventions and improving treatment outcomes. By refining this understanding, clinical strategies can be effectively tailored to target the specific avoidance mechanisms and behaviors that perpetuate anxiety symptoms.

Current research increasingly focuses on identifying the neural circuits involved in fear and avoidance, which are key drivers in the pathophysiology of anxiety disorders. By precisely targeting and understanding the brain regions implicated in these processes, there is potential to develop medications and inform other therapeutics that can interrupt maladaptive neural activity. In parallel, advancing therapeutic interventions is essential to disrupt maladaptive cognitive patterns stemming from these neural structures. This review will examine avoidance behaviors through a behavioral and neural mechanistic lens, including a review of current treatments aimed at addressing avoidance and how treatments can restructure the brain to extinguish these maladaptive behaviors.

2 Passive Versus Active Avoidance

There are two different types of avoidance behaviors: passive and active. Passive avoidance is defined as the absence of a motor response when presented with a fearful stimulus (i.e., freezing) (Diehl et al. 2019). Conversely, active avoidance is defined as the occurrence of a motor response when presented with a fearful stimulus (i.e., running away) (Diehl et al. 2019). Humans tend to have a more variable response to aversive stimuli that is situationally dependent. For example, active avoidance might present itself as fleeing from a loud noise to prevent potential harm or altering one's route to avoid a street associated with bad memories. Conversely, passive avoidance may involve remaining indoors to avoid exposure to germs outside or not crossing the road in fear of an accident (Fig. 1).

Passive Avoidance Behaviors		Active Avoidance Behaviors	
Rodents	**Humans**	**Rodents**	**Humans**
Freezing behavior	Staying in their home to avoid anxiety-inducing situations	Pressing a lever to avoid an aversive stimulus	Crossing the street to avoid a certain place or person
Reducing their exploration	Not speaking up in classes or meetings	Running away from a predator or aversive stimulus	Calling in sick to avoid a stressful presentation
Avoiding a dark chamber associated with an aversive stimulus	Not opening emails or messages in fear of bad news	Learning to avoid a specific route in which an aversive stimulus is present	Leaving a social event early to avoid a stressful interaction

Fig. 1 Examples of passive and active avoidance behaviors in rodents and humans

Both types of avoidance behaviors have distinct implications for treatment, and the treatment approach may depend on the type of response the individual exhibits. While a passive avoider with SAD may remain in a conversation with new people, they may not speak at all. Although they are not facing their fear of socializing, these passive avoiders may have the opportunity to habituate to the situation slowly and eventually tolerate it since they did not escape the situation. While on the other hand, an active avoider with SAD is more likely to remove themselves from the situation entirely, thus not getting that same opportunity to habituate. Treatments for withholding behaviors may lean toward slow integration into the feared situation or activity, while treatments for withdrawing behaviors may start with remaining in the situation and building tolerance to engage in more social behaviors in those situations.

To further improve individualized treatment for passive versus active avoidance, we must understand if there are certain situations or environments that promote them. Mkrtchian et al. (2017) presented 101 healthy adults with a go/no-go task, where they could "go" to win, "go" to avoid losing, "no-go" to win, and "no-go" to avoid losing. When faced with these options, the participants were less accurate when told to "go" to avoid losing versus when they were told to "no-go" to avoid losing (Mkrtchian et al. 2017), thus displaying participants propensity to passively avoid harmful stimuli. Binti Affandi et al. (2022) sought to uncover if the threat of a shock (which is often used to induce stress or anxiety) promotes active or passive avoidance in 59 healthy participants. Through their avoidance task they found that threat of shock promotes passive avoidance but not active (Binti Affandi et al. 2022), suggesting that anxiety in general promotes passive avoidance strategies, possibly to conserve energy while avoiding harm (Binti Affandi et al. 2022).

3 Costs and Benefits of Avoidance

Within the context of anxiety disorders, the costs of avoidance often outweigh the benefits. For example, a patient with aerophobia (fear of flying) is very likely to avoid flying, which may result in many costs/disruptions to their life. They may not be able to see friends and family, feel trapped in their location, unable to travel for work, or have increased anxiety about the possibility of flying. By avoiding flying, the person with aerophobia does not experience the anxiety associated with it. However, they may be limiting other meaningful and important aspects of their life, thus missing the opportunity to engage in value driven behaviors associated with social and occupational obligations. This experience is common across many anxiety disorders, highlighting avoidance as a critical component and the need to further understand and examine its function and treatment implications.

Positive and negative reinforcement contribute substantially to the maladaptive cycle of avoidance. Positive reinforcement involves the introduction of a rewarding stimulus to strengthen the likelihood of a particular response, whereas negative reinforcement involves the removal of an aversive stimulus to achieve that same effect. Negative reinforcement (e.g., engaging in avoidance behaviors), resulting in the sense of relief a person feels when successfully avoiding anxiety inducing stimuli, encourages future avoidance behaviors (Leng 2024; Papalini et al. 2021a, b; San Martín et al. 2020; Vervliet et al. 2017). A study done by Leng et al. (2024) found that higher magnitudes of relief from avoiding an unpleasant stimulus and greater monetary rewards both elicit higher pleasantness ratings (Vervliet et al. 2017) in a sample of 50 healthy adults after engaging in both reward and relief tasks. These findings highlight relief as a positive reinforcer for active avoidance behaviors. Additionally, compared to healthy controls, people with anxiety disorders often feel a stronger sense of relief after participating in active avoidance behaviors (De Kleine et al. 2023; San Martín et al. 2020; Vervliet et al. 2017). These findings suggest that it is much easier for people with anxiety disorders to fall into the maladaptive cycle of active avoidance thus effecting their lives in a much greater capacity.

However, the severity of avoidance behaviors widely varies, and not all avoidance behaviors that induce relief are maladaptive. It is important to distinguish the point at which avoidance behaviors become maladaptive, and research has found that this distinction is dependent on the feeling of relief rather than the behavior itself (Vervliet et al. 2017). In an avoidance fMRI task, Vervliet and colleagues (2017) found that healthy participants with low distress tolerance scores displayed sustained levels of relief throughout the task, while participants with higher distress tolerance scores' level of relief gradually decreased with each omission of the aversive stimulus. Distress tolerance is defined as an individual's capacity to experience and persist through adverse emotional states (Simons and Gaher 2005). These findings suggest that sustained relief may over-reinforce active avoidance. A person that experiences some benefit of relief from an active avoidance behavior, but not a great deal, may be less vulnerable to continue the maladaptive cycle.

To further expand upon this finding, San Martin and colleagues (2020) examined how anxiety traits affect the dynamics of experiencing relief during avoidance learning. In a sample of 101 college students who underwent a fear conditioning and avoidance paradigm, the researchers found that anxiety traits (measured by distress tolerance using the Distress Tolerance Scale and intolerance of uncertainty using the Intolerance of Uncertainty Scale (de Bruin et al. 2006; Simons and Gaher 2005)) were linked to increased levels of avoidance behavior and perceived relief (San Martín et al. 2020). They also found that participants experienced and generalized relief and avoidance behaviors based on whether they perceived the situation as controllable (avoidable) or safe. Taken together, these findings indicate that heightened experiences of relief may contribute to the onset and generalization of avoidance behaviors.

4 Neural Correlates of Avoidance

Understanding the neural correlates that drive avoidance behaviors provides critical insight into how dysregulated neural activity can contribute to the development and maintenance of maladaptive patterns, particularly displayed in anxiety disorders. Before discussing the neurobiological underpinnings of avoidance behaviors, we first outline the neurobiological mechanisms of threat conditioning to provide essential context and foundational understanding.

4.1 Threat Conditioning and Avoidance Learning

The amygdala, particularly its lateral and central nuclei, plays a critical role in threat conditioning, the process through which neutral cues become associated with aversive stimuli (Davis 1992; LeDoux 2000), which guides avoidance behaviors. During fear conditioning, the lateral amygdala encodes the initial association between a neutral stimulus (e.g., a sound or image) and an aversive outcome (e.g., a shock or loud noise). Once this association is established, the neutral stimulus alone can activate the lateral amygdala and trigger a fear response (Davis 1992; LeDoux 2000). The central amygdala then orchestrates these fear responses by signaling downstream structures, such as the lateral hypothalamus and periaqueductal gray, to produce behavioral and physiological reactions to threat (Davis 1992; Kim and Jung 2006; LeDoux 2000). These processes collectively ensure that learned threats are effectively recognized and avoided.

While the amygdala is primarily responsible for forming the association between the stimulus and the aversive outcome, the hippocampus works with the amygdala to process the contextual information surrounding threat conditioning (Maren 2001; Phillips and LeDoux 1992). During both the learning and retrieval phases of fear conditioning, the hippocampus helps encode the environment in which the aversive stimulus occurs and tags it as threatening if the stimulus is consistently present (Maren 2001; Phillips and LeDoux 1992). This contextual encoding allows the hippocampus to activate the amygdala when a previously threatening environment is re-encountered, prompting an appropriate fear response/avoidance behavior (Maren 2001). Importantly, the hippocampus also generalizes fear to similar contexts while maintaining specificity, preventing unwarranted fear responses in safe environments (Phillips and LeDoux 1992). Dysfunctional hippocampal processing, as seen in anxiety disorders and post-traumatic stress disorder (PTSD), can lead to overgeneralization of fear and avoidance of benign contexts or cues, exacerbating maladaptive behaviors (Phillips and LeDoux 1992; Shin et al. 2006). Together, the amygdala and hippocampus work in tandem to regulate the balance between adaptive fear responses and maladaptive avoidance behaviors.

In coordination with the amygdala and hippocampus, the dorsal anterior cingulate cortex (dACC) plays a key role in regulating and expressing fear responses and resisting fear extinction (Dunsmoor et al. 2008; Linnman et al. 2011; Milad et al. 2007). Responsive to both conditioned and unconditioned stimuli, the dACC remains active even when an expected aversive stimulus is omitted, contributing to the persistence of fear and hindering extinction (Dunsmoor et al. 2008; Linnman et al. 2011; Milad et al. 2007). This function underscores the dACC's role in sustaining maladaptive avoidance behaviors and its potential contribution to the presentation of anxiety disorders.

4.2 *Active Avoidance*

Once fear has been conditioned, various regions of the brain will activate to avoid experiencing the aversive stimuli. Specifically, regions in the fronto-limbic-striatal network (i.e., insula, amygdala, striatum, and the ventromedial prefrontal cortex (vmPFC)) work in tandem to both produce avoidance behaviors and make them hard to deviate from. It has been shown that the insular cortex plays a key role in emotional awareness and the integration of internal body states which helps individuals evaluate risks and threats (Barrett et al. 2004; Barrett and Bar 2009; Xue et al. 2010). This has many implications for research findings regarding the insula's activation or deactivation during avoidance learning and active avoidance. A study done in children found that avoidance response rates were negatively correlated with insular cortex activation, while avoidance cues elicited insula, amygdala, striatum, and thalamus activation (Schlund et al. 2010). The observed activation in the amygdala and the insula indicates that threat-related actions can persist even when negative experiences are continuously avoided. This may explain why avoidance-based coping strategies remain prevalent in children with anxiety disorders, as the brain maintains the fear response despite the absence of direct threat (Schlund et al. 2010). Reduced insula activation during active avoidance suggests that higher rates of avoidance responses may also be associated with less emotional awareness, a less accurate perception of the body's internal signals, and a decreased ability to evaluate risks and threats.

In addition to the insula, the ventral striatum has been found to be related to the acquisition of an avoidance response and active coping (Boeke et al. 2017; Delgado et al. 2009). Experiencing control over a threat activates the striatum and helps transition from automatic, instinctual defensive reactions to more flexible and learned responses tailored to the specific threat (Boeke et al. 2017). This shift enables individuals to engage in behaviors that are better suited to managing or avoiding danger (Boeke et al. 2017). Delgado et al. (2009) found that the activation patterns in the amygdala during a classical conditioning paradigm with the option to avoid delivery of a shock were closely aligned with the activity in the striatum. These findings suggest that these two brain regions may interact during the process of avoidance learning (Delgado et al. 2009).

Rodent studies have also found that the striatum is engaged in anticipating an aversive stimulus, regardless of whether there is a chance to avoid the negative outcome (Delgado et al. 2008; Jensen et al. 2003). This suggests that the striatum processes the expectation of a threat both in situations where avoidance is possible and where it is not. Therefore, the striatum is activated by the amygdala and works to invoke control over the feared stimulus through avoidance and over time learns when and how to avoid the stimulus in future contexts.

The ventromedial prefrontal cortex (vmPFC) plays a unique role in fear responding, as it has been shown in rodent models to toggle between different output paths of the amygdala, specifically the central amygdala pathway and basolateral amygdala pathway (Moscarello and LeDoux 2013; Quirk et al. 2003; Ramirez et al. 2015). An individual's ability to shift from reactive to proactive responses implies a mechanism that alternates between different defensive behaviors. The vmPFC (specifically the infralimbic PFC in rodents) can inhibit the central amygdala through a feedforward mechanism, anticipating the aversive stimulus and thus taking action to prevent it, therefore reducing the expression of automatic defensive behaviors like freezing (Moscarello and LeDoux 2013; Sotres-Bayon and Quirk 2010). This regulatory function of the vmPFC allows for a more flexible behavioral response to threats. Building off of rodent models, Boeke et al. (2017) later confirmed that vmPFC in humans plays a key role in avoidance learning. This study compared active avoidance, where the participant could prevent receiving a shock through action, with yoked extinction where the shock presentation mirrored that of the active condition but without the active avoidance option (Boeke et al. 2017). The researchers found between-group differences in vmPFC BOLD signal during cue presentation in the later stages of avoidance/extinction and retrieval in a sample of 56 adult healthy controls. These findings underscore the vmPFC's pivotal role in modulating defensive behaviors, facilitating a shift from automatic reactions to more adaptive or avoidant responses to threats by regulating amygdala output and supporting flexible behavior.

Working in tandem with the vmPFC to shift from passive to active avoidance behaviors is the nucleus accumbens (NAcc) (Levita et al. 2012), located in the ventral striatum. Many studies have supported the NAcc's key role in active avoidance (Delgado et al. 2009; Jensen et al. 2003; Niznikiewicz and Delgado 2011). In a pilot study, Sturm et al. (2003) found evidence through deep-brain stimulation that the NAcc plays a large role in information relay to the amygdaloid complex. The amygdaloid complex is a group of over ten nuclei (Sah et al. 2003) widely recognized for its role in anxiety and fear responses (LeDoux 2000; Shumyatsky et al. 2002). This deep brain stimulation resulted in significant symptom reduction in participants with treatment resistant obsessive-compulsive disorder (OCD) (Sturm et al. 2003), a condition in which avoidance behaviors are extensive and significant. Thus, as deep brain stimulation to the NAcc had a direct effect on OCD symptom reduction, it likely had a significant influence on the expression of active avoidance. Levita and colleagues (2012) furthered this work through conducting an avoidance task in a sample of 20 adult healthy controls and found that during active avoidance, the NAcc displayed greater activation, whereas passive avoidance actions triggered greater deactivation in the NAcc (Levita et al. 2012). This finding further suggests that the NAcc's role to one that can both emit or withhold a response to avoid harmful stimuli. Additionally, the researchers found that participants with higher anxiety levels showed greater activation of the NAcc while performing avoidance behaviors, thus the level of activation or deactivation of the NAcc during avoidance behaviors correlated with individual anxiety levels, suggesting that the NAcc is crucial in the development and persistence of maladaptive avoidance behaviors in anxiety disorders (Levita et al. 2012). This association indicates that variations in NAcc activity may underlie the severity and maintenance of anxiety-related avoidance, pointing to its potential role in anxiety disorders' neural mechanisms (Levita et al. 2012).

5 Dysfunction of Neural Circuitry and Anxiety Disorders

The study of avoidance responses within patients diagnosed with anxiety disorders is an understudied area of research. What is currently known, however, is that as the brain regions just reviewed above and noted to play a critical role in facilitating the avoidance behaviors are also implicated in the psychopathology of several anxiety disorders. For example, given the insula's role in the perception of internal body states as well as the evaluation of risks and threats (Barrett et al. 2004; Barrett and Bar 2009; Xue et al. 2010), it may contribute to the development of panic disorder. You et al. (2024) conducted an fMRI study on participants with panic disorder and healthy controls to investigate possible brain network abnormalities in those with panic disorder. This study found that patients with panic disorder have decreased functional connectivity and nodal efficiency within the right insula, suggesting that impaired information integration plays a significant role in the disorder (You et al. 2024). Furthermore, Simmons et al. (2006) conducted an fMRI study in anxiety-prone (defined as participants who scored in the top 15 percentile for trait anxiety) and anxiety normative participants (participants who scored from the 40th to 60th percentile for trait anxiety) and found increased insula activation during visual anticipation of an aversive stimulus in anxiety-prone individuals, indicating that insula hyperactivity may be a common feature in people with anxiety disorders (Simmons et al. 2006).

Additionally, findings across anxiety and related disorders and fear conditioning literature indicate widespread insula (Chavanne and Robinson 2021) and amygdala hyperactivity, reinforcing the notion that these regions are consistently implicated in anxiety disorders (Etkin and Wager 2007). Chavanne and Robinson (2021), performed a meta-analysis of whole-brain fMRI studies contrasting anxious patients (with PTSD, SAD, specific phobias, generalized anxiety disorder, and panic disorder) with healthy controls or healthy participants that underwent an unpredictable threat condition versus a safe condition. Analyses found that patients with induced anxiety and pathological anxiety had increased activation in the basolateral insula and medial PFC (Chavanne and Robinson 2021). A meta-analysis of fMRI and PET studies on PTSD, SAD, specific phobias, and fear conditioning found heightened activation in the insula and amygdala across these anxiety-related disorders (Etkin and Wager 2007). This elevated activity likely reflects the amygdala's role in associating a stimulus with a fear response. However, unique to PTSD, reduced activation was observed in the dACC and vmPFC (Etkin and Wager 2007). This hypoactivity suggests disrupted regulatory function within these regions. Given the dACC's role in modulating and expressing fear responses, diminished activation may contribute to exaggerated and contextually inappropriate fear seen in PTSD, leading to exaggerated and contextually inappropriate avoidance responses as well. Similarly, the vmPFC, involved in anticipatory processing of aversive stimuli, likely underperforms, leading to the maladaptive avoidance patterns common in PTSD patients. These findings support a differential neural profile for PTSD in which the amygdala remains hyperactive (Etkin and Wager 2007), while dACC and vmPFC activity is notably reduced, aligning with the persistent, generalized fear responses characteristic of the disorder.

As for the striatum, dysfunction in the fronto-striatal circuitry has been found to be less active in patients with OCD, specifically when planning future actions (van den Heuvel et al. 2005). These findings align with what has been shown in rodent studies, specifically that the striatum plays a key role in the expectation of threat (Delgado et al. 2008; Jensen et al. 2003) and, therefore, participants with OCD may have difficulty planning because they are caught in an obsessive thought cycle. Additionally, in a sample of 22 OCD patients and 22 healthy controls who underwent a Tower of London task (a task commonly used to examine planning processes) in an MR machine, van den Heuvel et al. (2005) found that regions associated with performance monitoring and short-term memory processing (the ACC, ventrolateral prefrontal, and parahippocampal cortices) showed increased activity. These findings suggest that executive function impairment may be a central feature of OCD. Taken together, differing rates of over-activation or deactivation in the amygdala, hippocampus, dACC, IC, striatum, vmPFC, and NAcc may play a role in the prominent avoidance behaviors that are prevalent in anxiety disorders.

6 Psychotherapy's Effects on Avoidance

Therapeutic interventions have been designed to target fears and modify maladaptive avoidance behaviors, aiming to enhance the quality of life for individuals affected by anxiety disorders. Exposure therapy, working with individuals to face fearful stimuli, aims to extinguish the fear response associated with the feared stimulus (Abramowitz 2013) and is one of the most effective treatments for targeting maladaptive avoidance behaviors (Hofmann and Hay 2018). Exposure therapy is also a core component of cognitive behavioral therapy (CBT), which is commonly used for the treatment of anxiety disorders, OCD, and PTSD (Abramowitz 2013) and has shown to be an effective treatment for the reduction of avoidance behaviors (Hofmann and Hay 2018). Examining the neurobiological changes that occur post successful psychotherapy treatment is critical for understanding the mechanisms at play in maladaptive anxiety symptoms such as avoidance. Although research focused on avoidance specifically is limited, there is a growing body of literature on brain changes subsequent to psychotherapy treatment (Barsaglini et al. 2014; Brooks and Stein 2015) providing implications on how these treatments may impact neural mechanisms of avoidance behaviors.

Previous literature suggests that psychotherapy (i.e., cognitive behavioral interventions) leads to either normalization of previously dysregulated brain activity, activation of previously inactive regions, or a combination of both (Barsaglini et al. 2014). A meta-analysis done by Schrammen et al. (2022) examined neural changes post psychotherapy in participants with anxiety disorders. Analyses of 18 studies revealed that activation in the right IC, ACC, and dorsolateral PFC decreased after psychotherapy intervention (Schrammen et al. 2022). A subgroup of studies that utilized CBT or stand-alone exposure found decreased activity in the supplementary motor area as well (Schrammen et al. 2022). These results display that cognitive and behavioral interventions impact brain areas associated with the creation of avoidance behaviors in the limbic system. Through participation in CBT, individuals are encouraged to engage in new learning experiences that challenge and modify their previously held beliefs and fears (Kaczkurkin and Foa 2022). These analyses indicate that engaging in these new learning experiences may be a successful path to normalize the previously overactive regions that control the perception of internal and external threats, the allocation of cognitive resources, and changes in cognitive control that are responsible for maladaptive avoidance behaviors.

7 Pharmacotherapy's Effects on Avoidance

While exposure therapy and CBT are highly effective in the treatment of anxiety disorders, not all individuals achieve a significant reduction in their symptoms after a course of treatment (Norton and Price 2007). For these individuals, psychotropic medication may be considered to help reduce their symptoms and achieve remission. Research on the effects of pharmacotherapy on the neural mechanisms underlying avoidance remains limited; however, an expanding body of literature is investigating how medication trials influence the brain's connectivity and activation patterns and can further inform what we know about the mechanisms of avoidance behaviors.

Pharmacological treatments have demonstrated measurable effects on brain structure and function in anxiety and related disorders. Bernstein et al. (2019) found increased striatal connectivity after a 12-week course of sertraline in a pediatric sample with OCD. Resting state connectivity increased specifically between the right putamen and the left frontal cortex and insula, as well as the left putamen and regions including the frontal cortex and pre- and post-central cortices (Bernstein et al. 2019). Increased striatal activation has been associated with enhanced planning abilities and improved threat expectation in individuals with OCD (Delgado et al. 2008; Jensen et al. 2003; van den Heuvel et al. 2005). Based on these findings, increased connectivity in these areas may assist in diminishing the reliance on avoidance behaviors if the participant feels safer encountering these experiences.

Similarly, Shin et al. (2014) found a correlation between resting state connectivity in the right vmPFC and improved symptoms after a 16-week Selective Serotonin Reuptake Inhibitor (SSRI) trial in adults with OCD (Shin et al. 2014). Since the vmPFC has been found to play a role in the modulation of defensive behaviors like avoidance (Boeke et al. 2017), increased connectivity within this region may enhance behavioral flexibility, thereby facilitating a shift away from maladaptive avoidance behaviors.

Beyond connectivity, structural MRI studies reveal that antidepressant treatment is also associated with gray matter volume changes in multiple regions, including the thalamus and the amygdala (Seyedmirzaei et al. 2024). In a meta-analysis of 13 studies, Seyedmirzaei et al. (2024) found that antidepressant use in participants with OCD was associated with a decrease in thalamic gray matter volume, and one particular study found reductions in left amygdala volume in children with OCD after a trial of paroxetine (Szeszko et al. 2004). Previous findings have shown that participants with OCD have an increased thalamic volume as compared to healthy controls (Atmaca et al. 2007), suggesting this reduction could reflect how antidepressant use reverses this formerly dysfunctional sensory-motor gateway contributing to symptom reduction (Seyedmirzaei et al. 2024) and in turn a reduction in avoidance behaviors. This change in thalamus and amygdala volumes may be the result of increased serotonin available in these regions (Chugani et al. 1998; Oke et al. 1997), thereby reducing the excessive excitatory signaling associated with OCD. This inhibition of the amygdala may directly impact patients' fear responses to aversive stimuli, reducing their perceived threat and, consequently, thereby diminishing the likelihood of engaging in avoidance behaviors.

8 Conclusion

Taken together, maladaptive avoidance behaviors play a critical role in the development and maintenance of anxiety disorders. Examining the nuances of avoidance behaviors is essential for understanding how quickly it can transition from adaptive to maladaptive and interfere with one's daily life. While avoidance may offer immediate relief from distress, it often reinforces maladaptive patterns that can exacerbate anxiety symptoms over time. The neural correlates of avoidance are intricately tied to key brain regions such as the amygdala, hippocampus, striatum, and ACC, which are involved in threat assessment and avoidance learning. Dysfunction within these circuits can lead to heightened anxiety responses and ultimately manifest as anxiety disorders.

Therapeutic approaches such as cognitive behavioral therapy, and specifically exposure therapy, offer promising avenues for addressing avoidance behaviors. The underlying mechanisms of these interventions offer insight into brain regions that are most impacted by cognitive and behavioral therapies, which can inform clinical care. Lastly, pharmacological interventions play an important role in managing avoidance behaviors, particularly when patients are unresponsive to therapeutic interventions alone. Understanding medications' ability to modulate neural circuits and impact clinical symptoms is of critical importance to obtain a holistic view of the mechanisms underlying avoidance behaviors. By merging all these insights, we can better comprehend the mechanisms of avoidance and develop more effective treatment strategies for those affected by anxiety disorders. Overall, a multifaceted approach that combines an understanding of the neural underpinnings of avoidance, psychotherapy, and pharmacotherapy may improve clinical care and outcomes for those suffering with anxiety disorders.

References

Abramowitz JS (2013) The practice of exposure therapy: relevance of cognitive-behavioral theory and extinction theory. Behav Ther 44(4):548–558. https://doi.org/10.1016/j.beth.2013.03.003

American Psychiatric Association (2013) Diagnostic and statistical manual of mental disorders, vol 5. American Psychiatric Association. https://doi.org/10.1176/appi.books.9780890425596

Atmaca M, Yildirim H, Ozdemir H, Tezcan E, Kursad Poyraz A (2007) Volumetric MRI study of key brain regions implicated in obsessive–compulsive disorder. Prog Neuro-Psychopharmacol Biol Psychiatry 31(1):46–52. https://doi.org/10.1016/j.pnpbp.2006.06.008

Barrett LF, Bar M (2009) See it with feeling: affective predictions during object perception. Philos Trans R Soc B Biol Sci 364(1521):1325–1334. https://doi.org/10.1098/rstb.2008.0312

Barrett LF, Quigley KS, Bliss-Moreau E, Aronson KR (2004) Interoceptive sensitivity and self-reports of emotional experience. J Pers Soc Psychol 87(5):684–697. https://doi.org/10.1037/0022-3514.87.5.684

Barsaglini A, Sartori G, Benetti S, Pettersson-Yeo W, Mechelli A (2014) The effects of psychotherapy on brain function: a systematic and critical review. Prog Neurobiol 114:1–14. https://doi.org/10.1016/j.pneurobio.2013.10.006

Bernstein GA, Cullen KR, Harris EC, Conelea CA, Zagoloff AD, Carstedt PA, Lee SS, Mueller BA (2019) Sertraline effects on striatal resting-state functional connectivity in youth with obsessive-compulsive disorder: a pilot study. J Am Acad Child Adolesc Psychiatry 58(5):486–495. https://doi.org/10.1016/j.jaac.2018.07.897

Binti Affandi AH, Pike AC, Robinson OJ (2022) Threat of shock promotes passive avoidance, but not active avoidance. Eur J Neurosci 55(9–10):2571–2580. https://doi.org/10.1111/ejn.15184

Boeke EA, Moscarello JM, LeDoux JE, Phelps EA, Hartley CA (2017) Active avoidance: neural mechanisms and attenuation of Pavlovian conditioned responding. J Neurosci 37(18): 4808–4818. https://doi.org/10.1523/JNEUROSCI.3261-16.2017

Brooks SJ, Stein DJ (2015) A systematic review of the neural bases of psychotherapy for anxiety and related disorders. Dialogues Clin Neurosci 17(3):261–279. https://doi.org/10.31887/DCNS.2015.17.3/sbrooks

Chavanne AV, Robinson OJ (2021) The overlapping neurobiology of induced and pathological anxiety: a meta-analysis of functional neural activation. Am J Psychiatry 178(2):156–164. https://doi.org/10.1176/appi.ajp.2020.19111153

Chugani DC, Muzik O, Chakraborty P, Mangner T, Chugani HT (1998) Human brain serotonin synthesis capacity measured in vivo with α-[C-11]methyl-L-tryptophan. Synapse 28(1):33–43. https://doi.org/10.1002/(SICI)1098-2396(199801)28:1<33::AID-SYN5>3.0.CO;2-D

Craske MG, Rauch SL, Ursano R, Prenoveau J, Pine DS, Zinbarg RE (2011) What is an anxiety disorder? Focus 9(3):369–388. https://doi.org/10.1176/foc.9.3.foc369

Davis M (1992) The role of the amygdala in fear and anxiety. Ann Rev Neurosci 15(Volume 15, 1992):353–375. https://doi.org/10.1146/annurev.ne.15.030192.002033

de Bruin GO, Rassin E, van der Heiden C, Muris P (2006) Psychometric properties of a Dutch version of the intolerance of uncertainty scale. Neth J Psychol 62(2):87–92. https://doi.org/10.1007/BF03061055

De Kleine RA, Hutschemaekers MHM, Hendriks GJ, Kampman M, Papalini S, Van Minnen A, Vervliet B (2023) Impaired action-safety learning and excessive relief during avoidance in patients with anxiety disorders. J Anxiety Disord 96:102698. https://doi.org/10.1016/j.janxdis.2023.102698

Delgado MR, Li J, Schiller D, Phelps EA (2008) The role of the striatum in aversive learning and aversive prediction errors. Philos Trans R Soc B Biol Sci 363(1511):3787–3800. https://doi.org/10.1098/rstb.2008.0161

Delgado MR, Jou RL, LeDoux JE, Phelps EA (2009) Avoiding negative outcomes: tracking the mechanisms of avoidance learning in humans during fear conditioning. Front Behav Neurosci 3:33. https://doi.org/10.3389/neuro.08.033.2009

Diehl MM, Bravo-Rivera C, Quirk GJ (2019) The study of active avoidance: a platform for discussion. Neurosci Biobehav Rev 107:229–237. https://doi.org/10.1016/j.neubiorev.2019.09.010

Dunsmoor JE, Bandettini PA, Knight DC (2008) Neural correlates of unconditioned response diminution during Pavlovian conditioning. NeuroImage 40(2):811–817. https://doi.org/10.1016/j.neuroimage.2007.11.042

Elliot AJ (1999) Approach and avoidance motivation and achievement goals. Educ Psychol 34(3):169–189. https://doi.org/10.1207/s15326985ep3403_3

Etkin A, Wager TD (2007) Functional neuroimaging of anxiety: a meta-analysis of emotional processing in PTSD, social anxiety disorder, and specific phobia. Am J Psychiatry 164(10):1476–1488. https://doi.org/10.1176/appi.ajp.2007.07030504

Hofmann SG, Hay AC (2018) Rethinking avoidance: toward a balanced approach to avoidance in treating anxiety disorders. J Anxiety Disord 55:14–21. https://doi.org/10.1016/j.janxdis.2018.03.004

Jensen J, McIntosh AR, Crawley AP, Mikulis DJ, Remington G, Kapur S (2003) Direct activation of the ventral striatum in anticipation of aversive stimuli. Neuron 40(6):1251–1257. https://doi.org/10.1016/S0896-6273(03)00724-4

Kaczkurkin AN, Foa EB (2022) Cognitive-behavioral therapy for anxiety disorders: an update on the empirical evidence. J Consult Clin Psychol 17:337. https://www.tandfonline.com/doi/full/10.31887/DCNS.2015.17.3/akaczkurkin

Kim JJ, Jung MW (2006) Neural circuits and mechanisms involved in Pavlovian fear conditioning: a critical review. Neurosci Biobehav Rev 30(2):188–202. https://doi.org/10.1016/j.neubiorev.2005.06.005

LeDoux JE (2000) Emotion circuits in the brain. Annu Rev Neurosci 23(1):155–184. https://doi.org/10.1146/annurev.neuro.23.1.155

Leng L, Beckers T, Vervliet B (2024) What a relief! The pleasure of threat avoidance. Emotion 24(3):539–550. https://doi.org/10.1037/emo0001312

Levita L, Hoskin R, Champi S (2012) Avoidance of harm and anxiety: a role for the nucleus accumbens. NeuroImage 62(1):189–198. https://doi.org/10.1016/j.neuroimage.2012.04.059

Linnman C, Rougemont-Bücking A, Beucke JC, Zeffiro TA, Milad MR (2011) Unconditioned responses and functional fear networks in human classical conditioning. Behav Brain Res 221(1):237–245. https://doi.org/10.1016/j.bbr.2011.02.045

Maren S (2001) Neurobiology of Pavlovian fear conditioning. Annu Rev Neurosci 24(1):897–931. https://doi.org/10.1146/annurev.neuro.24.1.897

Michael T, Zetsche U, Margraf J (2007) Epidemiology of anxiety disorders. Psychiatry 6(4):136–142. https://doi.org/10.1016/j.mppsy.2007.01.007

Milad MR, Wright CI, Orr SP, Pitman RK, Quirk GJ, Rauch SL (2007) Recall of fear extinction in humans activates the ventromedial prefrontal cortex and hippocampus in concert. Biol Psychiatry 62(5):446–454. https://doi.org/10.1016/j.biopsych.2006.10.011

Mkrtchian A, Aylward J, Dayan P, Roiser JP, Robinson OJ (2017) Modeling avoidance in mood and anxiety disorders using reinforcement learning. Biol Psychiatry 82(7):532–539. https://doi.org/10.1016/j.biopsych.2017.01.017

Moscarello JM, LeDoux JE (2013) Active avoidance learning requires prefrontal suppression of amygdala-mediated defensive reactions. J Neurosci 33(9):3815–3823. https://doi.org/10.1523/JNEUROSCI.2596-12.2013

Niznikiewicz MA, Delgado MR (2011) Two sides of the same coin: learning via positive and negative reinforcers in the human striatum. Dev Cognit Neurosci 1(4):494–505. Scopus. https://doi.org/10.1016/j.dcn.2011.07.006

Norton PJ, Price EC (2007) A meta-analytic review of adult cognitive-behavioral treatment outcome across the anxiety disorders. J Nerv Ment Dis 195(6):521–531. https://doi.org/10.1097/01.nmd.0000253843.70149.9a

Oke AF, Carver LA, Gouvion CM, Adams RN (1997) Three-dimensional mapping of norepinephrine and serotonin in human thalamus. Brain Res 763(1):69–78. https://doi.org/10.1016/S0006-8993(97)00404-6

Papalini S, Ashoori M, Zaman J, Beckers T, Vervliet B (2021a) The role of context in persistent avoidance and the predictive value of relief. Behav Res Ther 138:103816. https://doi.org/10.1016/j.brat.2021.103816

Papalini S, Beckers T, Claes L, Vervliet B (2021b) The drive for thinness: towards a mechanistic understanding of avoidance behaviors in a non-clinical population. Behav Res Ther 142:103868. https://doi.org/10.1016/j.brat.2021.103868

Phillips RG, LeDoux JE (1992) Differential contribution of amygdala and hippocampus to cued and contextual fear conditioning. Behav Neurosci 106(2):274–285. https://doi.org/10.1037/0735-7044.106.2.274

Quirk GJ, Likhtik E, Pelletier JG, Paré D (2003) Stimulation of medial prefrontal cortex decreases the responsiveness of central amygdala output neurons. J Neurosci 23(25):8800–8807. https://doi.org/10.1523/JNEUROSCI.23-25-08800.2003

Ramirez F, Moscarello JM, LeDoux JE, Sears RM (2015) Active avoidance requires a serial basal amygdala to nucleus Accumbens Shell circuit. J Neurosci 35(8):3470–3477. https://doi.org/10.1523/JNEUROSCI.1331-14.2015

Sah P, Faber ESL, Lopez De Armentia M, Power J (2003) The Amygdaloid complex: anatomy and physiology. Physiol Rev 83(3):803–834. https://doi.org/10.1152/physrev.00002.2003

San Martín C, Jacobs B, Vervliet B (2020) Further characterization of relief dynamics in the conditioning and generalization of avoidance: effects of distress tolerance and intolerance of uncertainty. Behav Res Ther 124:103526. https://doi.org/10.1016/j.brat.2019.103526

Schlund MW, Siegle GJ, Ladouceur CD, Silk JS, Cataldo MF, Forbes EE, Dahl RE, Ryan ND (2010) Nothing to fear? Neural systems supporting avoidance behavior in healthy youths. NeuroImage 52(2):710–719. https://doi.org/10.1016/j.neuroimage.2010.04.244

Schrammen E, Roesmann K, Rosenbaum D, Redlich R, Harenbrock J, Dannlowski U, Leehr EJ (2022) Functional neural changes associated with psychotherapy in anxiety disorders—a meta-analysis of longitudinal fMRI studies. Neurosci Biobehav Rev 142:104895. https://doi.org/10.1016/j.neubiorev.2022.104895

Seyedmirzaei H, Bayan N, Ohadi MAD, Cattarinussi G, Sambataro F (2024) Effects of antidepressants on brain structure and function in patients with obsessive-compulsive disorder: a review of neuroimaging studies. Psychiatry Res Neuroimaging 342:111842. https://doi.org/10.1016/j.pscychresns.2024.111842

Shin LM, Rauch SL, Pitman RK (2006) Amygdala, medial prefrontal cortex, and hippocampal function in PTSD. Ann N Y Acad Sci 1071:67–79. https://doi.org/10.1196/annals.1364.007

Shin D-J, Jung WH, He Y, Wang J, Shim G, Byun MS, Jang JH, Kim SN, Lee TY, Park HY, Kwon JS (2014) The effects of pharmacological treatment on functional brain connectome in obsessive-compulsive disorder. Biol Psychiatry 75(8):606–614. https://doi.org/10.1016/j.biopsych.2013.09.002

Shumyatsky GP, Tsvetkov E, Malleret G, Vronskaya S, Hatton M, Hampton L, Battey JF, Dulac C, Kandel ER, Bolshakov VY (2002) Identification of a signaling network in lateral nucleus of amygdala important for inhibiting memory specifically related to learned fear. Cell 111(6): 905–918. https://doi.org/10.1016/S0092-8674(02)01116-9

Simmons A, Strigo I, Matthews SC, Paulus MP, Stein MB (2006) Anticipation of aversive visual stimuli is associated with increased insula activation in anxiety-prone subjects. Biol Psychiatry 60(4):402–409. https://doi.org/10.1016/j.biopsych.2006.04.038

Simons JS, Gaher RM (2005) The distress tolerance scale: development and validation of a self-report measure. Motiv Emot 29(2):83–102. https://doi.org/10.1007/s11031-005-7955-3

Sotres-Bayon F, Quirk GJ (2010) Prefrontal control of fear: more than just extinction. Curr Opin Neurobiol 20(2):231–235. https://doi.org/10.1016/j.conb.2010.02.005

Sturm V, Lenartz D, Koulousakis A, Treuer H, Herholz K, Klein JC, Klosterkötter J (2003) The nucleus accumbens: a target for deep brain stimulation in obsessive-compulsive- and anxiety-disorders. J Chem Neuroanatomy 26(4):293–299. Scopus. https://doi.org/10.1016/j.jchemneu.2003.09.003

Szeszko PR, MacMillan S, McMeniman M, Lorch E, Madden R, Ivey J, Banerjee SP, Moore GJ, Rosenberg DR (2004) Amygdala volume reductions in pediatric patients with obsessive–compulsive disorder treated with paroxetine: preliminary findings. Neuropsychopharmacology 29(4):826–832. https://doi.org/10.1038/sj.npp.1300399

van den Heuvel OA, Veltman DJ, Groenewegen HJ, Cath DC, van Balkom AJLM, van Hartskamp J, Barkhof F, van Dyck R (2005) Frontal-striatal dysfunction during planning in obsessive-compulsive disorder. Arch Gen Psychiatry 62(3):301–309. https://doi.org/10.1001/archpsyc.62.3.301

Vervliet B, Lange I, Milad MR (2017) Temporal dynamics of relief in avoidance conditioning and fear extinction: experimental validation and clinical relevance. Behav Res Ther 96:66–78. https://doi.org/10.1016/j.brat.2017.04.011

Xiong P, Liu M, Liu B, Hall BJ (2022) Trends in the incidence and DALYs of anxiety disorders at the global, regional, and national levels: estimates from the global burden of disease study 2019. J Affect Disord 297:83–93. https://doi.org/10.1016/j.jad.2021.10.022

Xue G, Lu Z, Levin IP, Bechara A (2010) The impact of prior risk experiences on subsequent risky decision-making: the role of the insula. NeuroImage 50(2):709–716. https://doi.org/10.1016/j.neuroimage.2009.12.097

You L, Jiang W, Zhang X, Li Y, Wei J, Zhou Y, Chen S, Lu N, Yue Y, Kong Y, Yuan Y (2024) Decreased temporal variabilities of functional connectivities in insula and lingual gyrus are associated with better early treatment response in patients with panic disorder. J Psychiatry Neurosci 49(6):E377–E387. https://doi.org/10.1503/jpn.240016

Current State of the Neuroscience of Fear Extinction and Its Relevance to Anxiety Disorders

Elena Andres, Benjamin Meyer, Kenneth S. L. Yuen, and Raffael Kalisch

Contents

Abstract The elucidation of the functional neuroanatomy of human fear, or threat, extinction has started in the 2000s by a series of enthusiastically greeted functional magnetic resonance imaging (fMRI) studies that were able to translate findings from rodent research about an involvement of the ventromedial prefrontal cortex (vmPFC)

E. Andres
Leibniz Institute for Resilience Research (LIR), Mainz, Germany

Donders Centre for Cognitive Neuroimaging, Radboud University, Nijmegen, The Netherlands

B. Meyer, K. S. L. Yuen, and R. Kalisch (✉)
Leibniz Institute for Resilience Research (LIR), Mainz, Germany

Neuroimaging Center (NIC), Focus Program Translational Neuroscience (FTN), Johannes Gutenberg University Medical Center, Mainz, Germany
e-mail: rkalisch@uni-mainz.de

© The Author(s), under exclusive license to Springer Nature Switzerland AG 2024

Curr Topics Behav Neurosci (2025) 73: 73–92
https://doi.org/10.1007/7854_2024_555
Published Online: 03 January 2025

and the hippocampus in fear extinction into human models. Enthusiasm has been painfully dampened by a meta-analysis of human fMRI studies by Fullana and colleagues in 2018 who showed that activation in these areas is inconsistent, sending shock waves through the extinction research community. The present review guides readers from the field (as well as non-specialist readers desiring safe knowledge about human extinction mechanisms) during a series of exposures with corrective information. New information about extinction-related brain activation not considered by Fullana et al. will also be presented. After completion of this exposure-based fear reduction program, readers will trust that the reward learning system, the cerebellum, the vmPFC, the hippocampus, and a wider brain network are involved in human fear extinction, along with the neurotransmitters dopamine and noradrenaline. Specific elements of our exposure program include exploitation of the temporal dynamics of extinction, of the spatial heterogeneity of extinction-related brain activation, of functional connectivity methods, and of large sample sizes. Implications of insights from studies in healthy humans for the understanding and treatment of anxiety-related disorders are discussed.

Keywords Dopamine · fMRI · Hippocampus · Noradrenaline · Reward system · Ventromedial prefrontal cortex

1 Introduction

In fear, or threat, conditioning, an initially neutral stimulus (conditioned stimulus, CS) is paired with an aversive stimulus (unconditioned stimulus, US) and, after one or several pairings, comes to elicit a conditioned fear reaction (CR) (Pavlov 1927). CRs in human fear conditioning have been observed in electrodermal activity (skin conductance responses, SCR), startle reactions to other sudden and intense stimuli, including sounds or airblasts (fear-potentiated startle, FPS), heart rate (HR), other physiological indices, and also self-report measures of US expectancy, fear, arousal, or CS valence (Beckers et al. 2023). In fear extinction, the CS is repeatedly presented in the absence of the US, and CRs decline (Pavlov 1927). Retention of the extinction learning experience can be tested by presenting the extinguished CS again at some later time point. In these situations, it is sometimes observed that CRs come back. This "return of fear" at a post-extinction test indicates that extinction does not erase the CS-US association (the fear memory) that has been formed during conditioning but rather complements it with a fear-inhibitory CS-noUS association (the extinction memory), which then competes with the CS-US association for behavioral expression at testing (Bouton 2004). The likelihood of fear returning is the higher the less the test context resembles the context of the extinction learning. This in turn indicates that context importantly biases the fear vs. extinction memory competition and that the extinction memory is relatively strongly bound to its learning context, in contrast to the fear memory, which tends to prevail outside extinction contexts (Bouton 2004).

While early research has considered conditioning as a suitable laboratory model for fear-related disorders, in particular post-traumatic stress disorder (PTSD), an emerging consensus is that fear disorders are less related to abnormal responses to threat than to abnormal responses to safety. Next to the US-paired CS ("CS+"), human conditioning studies typically employ a non-paired comparison stimulus ("CS−"), which comes to signal absence of threat, or safety, during the procedure. Meta-analysis has shown that patients with fear- and anxiety-related disorders do not exhibit increased CS+ responses relative to healthy controls, but increased CS− responses (Duits et al. 2015). That is, they fail to recognize the CS− as a source of safety. This deficit in safety learning extends to extinction, where patients show relatively heightened CRs to the now-safe CS+ (Duits et al. 2015). Deficient safety learning may not just be a corollary of fear disorders, but a predisposing risk factor (Beckers et al. 2023; Kalisch et al. 2024). These observations indicate that safety learning is a naturally occurring adaptive behavior that provides the benefit of saving the individual from producing costly stress reactions when they are not necessary and of instead providing space for restoration and resource-building (Kalisch et al. 2015). Extinction in particular is also of interest because it can be considered a laboratory analog of exposure-based therapies, where patients are repeatedly confronted with the object of their fear, to make the experience that the consequences of the confrontation are not as dramatic as they expected (Craske et al. 2022).

The present review will focus on fear extinction as one key facet of safety learning and specifically on the macroanatomical neural correlates of this behavior in healthy adult humans. Defining the neurocircuitry that mediates human extinction is not only of academic interest but may provide anchor points for neuroscience-based methods to boost extinction in a clinical context, for instance with non-invasive transcranial neurostimulation methods (Marković et al. 2021; Becker and Milad 2023; Webler et al. 2023), with biofeedback methods targeting neural signatures (Chiba et al. 2019), or also with transcutaneous vagal nerve stimulation (Bierwirth and Stockhorst 2022). Further, knowing how extinction works in normal brains is essential to identify and interpret deviations from normality in at-risk individuals and patients.

2 The Fullana Shock

The history of human neuroscientific extinction research can be read as a collective fear conditioning experience, where findings of non-replicability figure as the US. Early fMRI studies had celebrated observations of CS + -induced activations of the vmPFC during extinction learning and/or testing (Phelps et al. 2004; Kalisch et al. 2006; Milad et al. 2007) as a translation of corresponding results in rodent models of extinction, where a homologous PFC area, the infralimbic cortex (IL), had been found to be necessary for the formation and consolidation of the extinction memory and for its later expression in test situations (Quirk et al. 2000; Milad and Quirk 2002; Santini et al. 2004; Sierra-Mercado et al. 2006; reviewed in Milad and Quirk

2012). Also, findings that vmPFC activity during learning or testing in these early studies correlated with SCR-based metrics of extinction success (reduction of CRs) (Phelps et al. 2004; Milad et al. 2007) were taken to reflect the observation that the rodent IL can suppress CRs via inhibition of CS-induced amygdala activity (Royer and Paré 2002; Quirk et al. 2003; Milad et al. 2004; Vidal-Gonzalez et al. 2006).

A meta-analysis by Fullana et al. (2018) could not confirm these initial results. There was no consistent vmPFC activation when comparing CS+ and CS− fMRI responses during extinction learning (in 31 studies with 1,074 participants). Null results were also reported for separate analyses of late extinction learning trials (where the CS+ might already be recognized as safe and the formation of the extinction memory might have progressed to some extent), of conditioning vs. extinction learning phases (which might be a more sensitive contrast to detect threat vs. safety processing), and of the test phase (where rodent studies suggested IL/vmPFC activation is necessary to suppress CRs). Only in a post-hoc analysis of nine studies in altogether 161 participants that had used a paradigm developed by Milad and colleagues (2007) could an effect be detected. In this paradigm, conditioning occurs with two CS + s, of which one is then extinguished (CS + E), while the other (unextinguished CS+, CS + U) is only presented again later during the test phase. This permits a contrast CS + E > CS + U at test, where extinction memory expression, CR suppression, and vmPFC activation should be associated specifically with the CS + E. However, other studies using this paradigm had subsequently not been able to reproduce this finding (Kampa et al. 2018) in $N = 50$; Kampa et al. 2020 in $N = 103$; Wen et al. 2021 in $N = 137$; Young et al. 2021 in $N = 272$). Generally, between-sample replicability of fMRI spatial activation patterns from an identical extinction task was found to be poor (Kampa et al. 2020).

Similarly disappointing results (Fullana et al. 2018) were obtained for two other regions strongly implicated in extinction by animal research, the hippocampus and the amygdala (Bouton et al. 2021). The only well-replicable activations during extinction learning and testing are in brain regions that also activate during fear conditioning and therefore presumably reflect the retrieval and expression of the fear memory formed during conditioning, most prominently the dorsal anterior cingulate and adjacent dorsal medial prefrontal cortices (dACC, dmPFC) and the anterior insula (Etkin et al. 2011; Fullana et al. 2018).

To complement the Fullana shock, animal studies, too, raised doubts about the role of the IL/vmPFC in extinction memory expression. It is unambiguously established that a functioning IL is not needed during extinction learning for CRs to decrease, but that the extinction memory will not be retrieved and/or the amygdala will not be suppressed at testing without intact IL function during the learning and subsequent post-extinction consolidation phases (Bouton et al. 2021; Plas et al. 2024). This extinction memory retrieval/amygdala suppression (that is, in short: extinction memory expression) was supposed to necessitate CS-related IL activity at test. Introducing the new optogenetics methods to the field, Do-Monte et al. (2015) could not find an effect of IL silencing at CS presentations during testing in three experiments; these null findings were corroborated with two pharmacological inhibition experiments. In the same year, Bukalo et al. (2015) reported null effects of optogenetic silencing. Hence, the very rationale for expecting CS+ > CS− or CS + E > CS-U activation differences in the vmPFC in human extinction expression experiments appeared to crumble.

3 Problem Analysis and Development of Problem-Solving Strategies

Conditioning is unpleasant but adaptive. An intense debate following the Fullana publication worked out several important problem diagnostics and potential solutions. One, it is challenging to find a neural trace of safety associated with an extinguished CS+ by comparing it to a CS−, which itself has acquired safety properties during prior conditioning (Fullana et al. 2018). We here extend this argument to CS + E > CS + U comparisons, based on the observation that CS + U-induced CRs typically disappear quickly during test (e.g., Milad et al. 2007; Kampa et al. 2018), that is, the CS + U, too, becomes quickly recognized as safe. Hence, even with this potentially more sensitive paradigm, there are only very few CS trials during which differential expression-related activity might be observed. This also shows that the comparison problem becomes amplified by the dynamic nature of extinction learning (Morriss et al. 2018; LaBar 2023), which typically occurs over only a couple of trials and is not restricted to the learning phase. Relevant learning-related activity may thus occur only during the first few trials of the learning phase (Raczka et al. 2011; Gerlicher et al. 2018; Andres et al. 2024), making it hard to capture; it is hard to determine when during learning a representation of the CS+ as safe may arise; and testing is confounded by renewed extinction. The *comparison problem* is even further amplified by the ethical necessity to restrict the aversiveness of conditioning in human experiments, such that the threat value of the CS+ early in extinction learning (against which the CS+ during late extinction is sometimes compared in order to identify safety-related processing), or also the threat value of the CS + U during testing (against which the CS + E can be compared) are also limited (e.g., Fullana et al. 2018; Hennings et al. 2022a).

One solution to this agglomeration of problems may be to abandon standard CS + vs. CS− comparisons across arbitrarily fixed long temporal intervals (many numbers of trials) and instead embrace the temporal dynamic of extinction, by either allowing exploratory analysis of fMRI signal dynamics in a temporally highly resolved fashion or by recurring to computational models of associative learning that explicitly try to explain temporal changes in responding, e.g., in a trial-by-trial fashion. Temporal dynamics can also be incorporated into fMRI analyses by differentiating activity during a CS from activity at the offset of a CS, which is the time point in most experimental paradigms where the unexpected omission of the US that drives extinction learning occurs. We will come back to exemplary solutions below.

Another solution that can be applied when a test has low sensitivity is raising the numbers. This approach, too, has born first fruits.

The comparison problem is not the only one. In the context of a similar debate in the field of conditioning, where amygdala activation has been inconsistently observed (Mechias et al. 2010; Fullana et al. 2016; Visser et al. 2021), it has been pointed out that the amygdala is a heterogeneous structure consisting of several subnuclei, each of which subserves different functions and itself exhibits considerable microanatomical and functional heterogeneity (Wen et al. 2022a). So, animal studies have shown that the amygdala is necessary for both conditioning and extinction and that a sub-nucleus like the basolateral amygdala possesses both cell populations responding to threat and cell populations responding to safety (Tovote et al. 2015). This constellation will make it challenging (but not impossible; Wen et al. 2022a) to find conditioning-related activity with a comparatively gross, spatially poorly resolved technique like fMRI. By extension, fMRI will be insensitive to find extinction-related activity in the amygdala.

We think that the *heterogeneity problem* also applies, though maybe to a lesser extent, to the vmPFC. This area has become so much tagged as mediating extinction, positive affect, and regulation of negative emotion in general that observations of functional heterogeneity in this area, too, have simply been forgotten. For example, an early study in the rat did find negative effects of vmPFC inactivation on CR expression (Sierra-Mercado et al. 2006). In one of the early reviews of the human literature (Etkin et al. 2011), the authors found a preponderance of fear-, conditioning, and arousal-related activations in dorsal parts of the ACC and the mPFC as well as a preponderance of extinction- and anti-arousal-related activations in ventral parts. Nevertheless, this pattern was not "clean", and both classes of experimental conditions also were linked with activations in the "wrong" parts of the ACC/mPFC. This extended to studies employing measures of functional connectivity (FC). Recently, Battaglia and colleagues observed impairment of fear conditioning in vmPFC-lesioned patients (Battaglia et al. 2020) and summarized evidence for processing of both positively and negatively valenced information in this area (Battaglia et al. 2022). Hence, similar to the amygdala, fMRI studies of vmPFC activation during extinction may suffer from extinction-related activity being canceled out by fear- or threat-related activity simultaneously going on in the same area, in particular in situations where fear and extinction memories are both cued by the CS+ and compete for expression.

One solution to the heterogeneity problem may be to complement analyses searching for average activation in a large amygdala, vmPFC, or also hippocampus region of interest (ROI) or for large homogeneous activation clusters (the standard mass-univariate approach to fMRI analysis) with analyses that exploit spatial signal heterogeneity, by determining patterns of activation across multiple voxels in a brain region, set of brain regions, or in the whole brain. Such multi-voxel pattern analysis (MVPA) approaches (Haxby et al. 2001; LaBar 2023) have already been proven to be more sensitive in detecting brain activity related to fear learning and memory than univariate approaches (e.g., Bach et al. 2011; Visser et al. 2013, 2015). We will also come back to this issue.

Finally, standard fMRI activation studies assume that the key neural processes that drive a behavior of interest consist in the massive activity of neurons in glutamatergic projections or circuits, associated with the generation of action potentials and excitatory post-synaptic potentials, processes for which the blood oxygenation-level dependent (BOLD) signal exploited in fMRI is primarily sensitive (Howarth et al. 2021). This could be called the *BOLD problem*. Not all extinction-related neural processes may be of this nature. A particular type of neuron with special relevance for extinction (see below) are dopaminergic neurons originating in the ventral tegmental area (VTA) of the midbrain and projecting to the ventral striatum in the basal ganglia, where they synapse onto GABAergic medium spiny neurons (de Jong et al. 2022). It is not clear whether dopamine release from these neurons causes the same kind of neurovascular changes that make glutamatergic synaptic transmission visible, indirectly, to fMRI (Brocka et al. 2018; Thiele et al. 2021). Another type of neural process are the cascades of molecular changes occurring inside cells and in synapses after a learning event and which are crucial for the consolidation of newly learned material (Izquierdo et al. 2006). These plasticity processes are unlikely to be directly associated with glutamatergic transmission, but they may well change the representation of information in a given area or network of areas, and they may change the functional coupling of cell ensembles and brain areas that participate in a common behavior. Here, MVPA and also FC analyses, which trace subtle co-fluctuations of the brain signal between voxels or areas (Friston et al. 1993), may be a preferential tool.

In the following, we will highlight the more refined approaches to the macroanatomical study of human extinction that were spurred by the Fullana shock, with a focus on those findings whose replicability has either already been demonstrated or is likely, due to large sample sizes.

4 Step-Wise Exposure-Based Therapy

4.1 Step 1: Exploiting the Temporal Dynamics of Extinction (Prediction Error-Based Extinction Learning in the Reward System and the Cerebellum)

Fear extinction, as fear conditioning, is driven by unexpected outcomes. In fear conditioning, the first presentation of the CS does not signal threat, and the concurrent or subsequent presentation of the US is therefore unexpected and violates the initial "no-US" prediction. This prediction error (PE) is a teaching signal that leads to an update of the CS− associated US prediction toward higher values. Error-based updating continues until expectation and outcome frequency match, such that the CS carries an accurate US prediction, and US presentation no longer induces a PE (Rescorla et al. 1972). In extinction, the PE is elicited by the unexpected omission of the US and leads to an update of the US prediction toward lower values.

Three observations suggest that the extinction PE (EPE) may not be instantiated in the same neural substrates as the conditioning PE. First, as has already been discussed, the fear and extinction memories that result from PE-based learning are behaviorally separable. Second, the neural correlates of fear and extinction learning known from animal studies only partially overlap. And third, surprisingly not receiving an aversive stimulus is a very different experience from surprisingly receiving it, the former being pleasant, the latter unpleasant. In a rare case of human extinction research going ahead of animal extinction research, these considerations have motivated Raczka et al. (2011) to posit that fear extinction is not mediated by the aversive system but rather constitutes an appetitive learning process mediated by the reward system. In particular, they suggested that the EPE is encoded by the same phasic release of dopamine from terminals of dopaminergic VTA neurons in the ventral striatum that also encodes reward PEs (RPE) (Schultz 2006; de Jong et al. 2022).

US omission-induced EPEs are high in the beginning of extinction and then quickly decline with every unpaired CS trial. The resulting update of the US prediction becomes apparent at every subsequent CS trial as decreased CR. CR reduction over trials is also often fast. Raczka et al. used a simple computational model of associative learning (Rescorla et al. 1972) to fit the decrease in subjective fear of the CS they observed in their $N = 69$ participants and derive a trial-by-trial dynamic EPE estimate. Because during conditioning the US was presented at CS offsets, they expected a neural EPE correlate at CS offsets, precisely when the expectation of the US was violated. By contrast, they expected a neural correlate of the emotional value of the CS, determining the magnitude of fear, at CS onsets. They observed an EPE-related signal in the ventral striatum. There was no signal conforming to a decline of an aversive value of the CS or the fear reaction. By contrast, at the same rate that fear declined, there was an increase in CS onset-related signal in orbitofrontal areas, also consistent with an appetitive account of extinction.

Finally, the authors separated their participants into those carrying one or two 9R variants of the *dopamine transporter 1 (DAT1)* gene and those without any 9R variant. The dopamine transporter (DAT) is nearly exclusively expressed in the striatum and there clears released dopamine from the extracellular space. At the time of the study, it was considered that carriers of the 9R variant have less DAT expression and therefore higher extracellular DA levels in the striatum when there is phasic dopamine release. Indeed, 9R carriers learned extinction faster than non9R carriers and also showed a higher ventral striatal EPE signal, yielding a first indication that the EPE might be mediated by the mesolimbic dopamine system (Raczka et al. 2011).

While in hindsight, analyses betting on effects of genetic variants in small samples may not be considered reliable anymore, the authors took care to replicate their main finding of an EPE-related signal in the ventral striatum (Thiele et al. 2021). Importantly, this signal was observed in an area previously linked with RPE signaling in human reward studies (O'Doherty et al. 2002; Garrison et al. 2013). CS + offset-related activity during extinction in the ventral striatum has since also been reported in a non-computational analysis (that is, assuming temporally invariant activation) in $N = 103$ participants by Kampa et al. (2023).

Since Raczka et al., backtranslational animal studies have demonstrated activity increases in dopaminergic VTA neurons projecting to the nucleus accumbens and phasic release of dopamine in the nucleus accumbens precisely at CS offsets, both of which decline over the course of extinction (Luo et al. 2018; Salinas-Hernández et al. 2018, 2023). The same dopamine neurons take part in a reward learning task, but are not sensitive to aversive stimulation (Salinas-Hernández et al. 2023). Their activity is both necessary and sufficient to induce extinction learning (Luo et al. 2018; Salinas-Hernández et al. 2018, 2023). Also, systemic administration of the dopamine precursor L-DOPA before extinction in mice with an extinction learning deficit has been reproducibly shown to rescue extinction (Whittle et al. 2016; Sartori et al. 2024), further corroborating a role of dopamine in fear extinction.

Evidence for a comparable effect of L-DOPA in humans has not been obtained so far (Esser et al. 2021) (own publication in preparation), presumably because participants' normal extinction abilities did not leave room for improvement by the drug. However, anhedonia, characterized by a reduced motivation to obtain reward and reduced pleasure in anticipating and consuming rewards, has been associated with higher activity of areas of the fear circuitry to CS + E vs. CS− stimuli during extinction (Young et al. 2021), and in the same sample extinction fMRI data have been used to predict anhedonia from a whole-brain MVPA-based decoder (Rosenberg et al. 2023). These results further suggest a link between extinction and the reward system.

Another brain area that has been consistently shown to activate to CS+ offsets in early extinction in humans is the cerebellum (Ernst et al. 2019; Batsikadze et al. 2022). Although there is currently no corresponding work in animal models that has been able to elucidate a neurophysiological basis for these observations, a general role for the cerebellum in computing predictions and prediction errors about sensory states (Popa and Ebner 2018) and its direct excitatory projections to the VTA (Pisano et al. 2021) make it conceivable that the cerebellum interacts with the reward circuitry in generating EPEs.

4.2 Step 2: Exploiting Spatial Heterogeneity in Extinction-Related Brain Activation (Extinction Memory Consolidation through Spontaneous vmPFC Pattern Reactivations)

Other human fMRI studies also turned the temporal dynamics of extinction into an advantage and combined this with a solution to address the heterogeneity problem. Based on the idea that the vmPFC is important for extinction memory formation already during extinction learning and also during the subsequent memory consolidation phase (see above), Gerlicher et al. (2018) asked whether the state of the vmPFC at the time of the EPE would be relevant for memory expression at a test 1 day later. Specifically, they indexed the EPE-related vmPFC state by the average spatial pattern of activation across vmPFC voxels at CS+ offset during the first five (of fifteen) CS+ trials and then tested, one, whether this multi-voxel pattern (MVP) would be spontaneously reactivated during consolidation and, two, whether the number of reactivations would in turn predict extinction memory expression at the test. Spontaneous MVP reactivations during consolidation were captured by resting-state fMRI scans conducted 10, 45, and 90 min after extinction. Predictions were confirmed. Results were recently reproduced in a preregistered direct replication study by Andres et al. (2024). vmPFC MVPs obtained from CS+ offsets during later CS+ trials, from CS+ onsets, or from CS− onsets or offsets did not show the effect in both studies. Also, the effect was only observed in the vmPFC. These findings clearly implicate the human vmPFC in the plasticity processes occurring during extinction learning and consolidation. Although it remains unclear what exactly the vmPFC MVPs express (perhaps a subjective state of pleasant surprise or relief?), the studies by Gerlicher et al. and Andres et al. demonstrate that combining a dynamic perspective with advanced analysis approaches, moving away from averaging fMRI signals across large regions, can reliably establish a macroanatomical correlate of human fear extinction.

4.3 Step 3: Exploiting Functional Connectivity and Large Sample Sizes (Extinction-Induced Plasticity in a Wider Network)

Another therapeutic result was reported by Wen et al. (2021), who approached the question of dynamic plasticity- and memory-related brain changes during extinction from a different angle. Specifically, the authors calculated average FC between 432 brain regions in response to CS+ and CS− trials during two trial blocks of conditioning, four trial blocks of extinction, and two trial blocks of an extinction memory test, conducted 1 day later. Mean FC to the CS+ increased relative to FC to the CS− during the last two blocks of extinction and was still higher during the first block of the test. The FC increase on day 1 predicted extinction expression on day 2. Interestingly, CS+ > CS− activation differences observed in the same areas that contributed to the FC change were temporally dissociated from the FC changes, being observed only during the first block of extinction, and they did not predict extinction expression. This suggests that connectivity measures may be more suitable to capture plasticity occurring during extinction, and rough activation measures may indeed not capture essential extinction processes. Due to the large sample size ($N = 137$) and the extremely robust nature of the analysis, the finding can be considered reliable.

The areas that contributed to the FC-related memory effect were widely distributed across the brain and included the dACC/dmPFC, the precuneus, the insula, the thalamus, and parts of the vmPFC and the orbitofrontal, lateral frontal, temporal, parietal, and occipital cortices. This might indicate that extinction memory formation is not an isolated task of the vmPFC (Wen et al. 2021), in line with animal studies (e.g., Senn et al. 2014), although post-extinction processes in this area may eventually be key for longer-term consolidation (Gerlicher et al. 2018; Andres et al. 2024).

4.4 Step 4: Exploiting Spatial Heterogeneity in Extinction-Related Brain Activation II (Extinction Memory Expression Carried by Pattern Reactivations in vmPFC and Hippocampus)

Our therapy program so far consisted in exposing the reader to corrective information indicating that neural activity related to extinction learning and extinction memory consolidation can be reliably detected with fMRI and cannot simply be explained by the expression of the fear memory that is induced by CS+ presentations during extinction. But can we also see activity related to extinction memory expression at post-extinction tests?

To address this issue, Hennings and colleagues developed a variant of conditioning where pictures from one category of objects (e.g., tools) are paired with a US while pictures from another category (e.g., animals) are not, such that participants learn to see the paired category as CS+ and the unpaired category as CS−. Other exemplars from these categories are then presented unpaired during extinction and yet others during test a day later (Hennings et al. 2020). In two studies, the authors observed that an MVP associated with the CS+ during extinction in the vmPFC was reactivated in this region at CS+ presentation during testing (Hennings et al. 2020, 2022b). Interestingly, while a dACC ROI showed selective test reactivation of an MVP associated with the CS+ during conditioning, suggesting a role in fear memory expression, the vmPFC showed not only extinction-related, but also conditioning-related MVP reactivation, though to a smaller extent (Hennings et al. 2022b). The latter result is in line with functional heterogeneity within the vmPFC, as discussed above (Sierra-Mercado et al. 2006; Etkin et al. 2011; Battaglia et al. 2022), and further underlines the limitations of mass-univariate analysis.

The Hennings et al. studies also demonstrated extinction-related reactivation in the hippocampus, predominantly in its anterior portion (while the posterior hippocampus may be more specialized on reactivating conditioning-related MVPs) (Hennings et al. 2022b). Also, hippocampus reactivations do not appear to differentiate between CS+ and CS− (Hennings et al. 2022b), indicating the hippocampus may be more sensitive to context than cue information. The amygdala did not show comparable effects.

The Hennings et al. studies are the last step of our exposure-based therapy, and the authors hope that the reader will at this stage share their relieving conclusion that we can safely assume that the reward learning system, the vmPFC, and the hippocampus are involved in human extinction, along with a broader network of brain regions. The reader is also pointed to an exciting development in the field of non-invasive brain stimulation, which suggests that methods such as transcranial magnetic stimulation (TMS) can be used to probe the causal role of areas such as the vmPFC (Raij et al. 2018).

5 Neurotransmitter Systems: Focus on Dopamine and Noradrenaline

In the last part of this review, we will try to convince the reader that we can also make relatively safe assumptions about the contribution of two neurotransmitter systems to human fear extinction. We have already cited preliminary evidence that dopaminergic projections from the VTA to the ventral striatum carry error learning signals during extinction (Raczka et al. 2011; Thiele et al. 2021). A different series of studies implicate dopamine also in human extinction memory consolidation. Notably, systemic administration of L-DOPA immediately after extinction learning has been shown to improve extinction expression and vmPFC activity a day later, though this was only observed when using a simple cue conditioning paradigm (Haaker et al. 2013; Gerlicher et al. 2018). In a more complex combined cue and context conditioning paradigm, only enhanced vmPFC activity was observed (Haaker et al. 2015). Since L-DOPA has a short half-life, it can be excluded that these effects are caused by acute L-DOPA actions at test. Because one behavioral study outside the scanner (Gerlicher et al. 2022) failed to reproduce the L-DOPA effect and two other behavioral studies saw the effect only in participants with successful extinction learning (Gerlicher et al. 2019), Andres et al. (2024) wondered whether the high arousal states that scanning was reported to cause in participants make observation of L-DOPA-induced improvements of extinction consolidation more likely. Although they did not observe a moderating effect of acute state arousal, they found that average baseline levels of salivary alpha-amylase (sAA), a stable and heritable individual-differences marker (Nater et al. 2007; Out et al. 2011, 2013), interacted with L-DOPA in determining extinction expression. Specifically, L-DOPA was more effective in individuals with higher trait sAA. High trait sAA in itself was associated with worse extinction expression.

Trait sAA can be used to index basal activity of the sympathetic nervous system (SNS), associated with the neurotransmitter noradrenaline (norepinephrine) (Andres et al. 2024). The finding that high trait sAA was associated with a weaker extinction memory in the study by Andres et al. resonates with the proposal, based on pharmacological and optogenetic studies in rodents, that high levels of arousal and/or noradrenergic activity impair extinction, presumably via direct release of noradrenaline from locus coeruleus projections to the IL and indirectly via amygdala-mediated IL suppression resulting from enhanced amygdala activity (Giustino and Maren 2018; Bierwirth and Stockhorst 2022; Maren 2022). In the rodent, noradrenergic arousal and extinction appear to exhibit an inverse U-shaped relationship: arousal is detrimental for extinction at high and low levels, but good at intermediate levels (Bierwirth and Stockhorst 2022). As a consequence, effects of noradrenergic pharmacological manipulations (or also other methods, such as transcutaneous vagal nerve stimulation) are likely to depend on an individual's baseline arousal levels. This complicates human investigations and may have contributed to the inconclusive picture in the human literature on noradrenaline and extinction (Bierwirth and Stockhorst 2022).

Nevertheless, the rodent data offer a possible interpretation for the observed trait sAA x L-DOPA interaction. Extinction in rodents, like other stressful situations (Abercrombie et al. 1989; Pascucci et al. 2007), leads to dopamine efflux in the frontal cortex that outlasts the extinction session (Feenstra et al. 2001; Hugues et al. 2007; Sartori et al. 2024). This effect is amplified by systemic L-DOPA (Sartori et al. 2024). There is also a clear link between post-extinction dopaminergic transmission in the IL and extinction consolidation in rodents (Sartori et al. 2024). Hence, L-DOPA may boost an endogenous, dopaminergic pro-consolidation process in the IL/vmPFC, counteracting the dampening influence of noradrenaline. In the absence of high prefrontal noradrenaline levels, when extinction works well, the L-DOPA boost may not be needed. L-DOPA may also decrease extracellular prefrontal noradrenaline by increasing its turnover or reuptake or by activating inhibitory $\alpha 2$-adrenoceptors (Dayan and Finberg 2003), and it may directly reduce amygdala-mediated inhibition of IL/vmPFC neurons (Floresco and Tse 2007).

Together, these combined observations from human and animal studies on extinction indicate that human extinction is under opposing influences from two major neurotransmitter systems, the noradrenergic system, which promotes extinction at intermediate levels and impairs it at high levels, and the dopaminergic system, which promotes extinction and counteracts detrimental influences of high noradrenaline. Major crossroads, where these two systems meet, appear to be the amygdala and the vmPFC.

Next to fast increases in central noradrenaline levels and locus coeruleus activity, stress reactions are characterized by delayed increases in central cortisol levels, derived from the adrenal glands via the bloodstream (de Kloet and Joëls 2023). Cortisol, too, affects human extinction. Cortisol administered before extinction reduces fear memory expression during extinction and facilitates later extinction expression, to the extent that return of fear becomes weak even when the CS+ is presented in the conditioning context (that is, the extinction memory becomes relatively context-independent). Cortisol administered after extinction also facilitates extinction expression, but not to the extent that it makes the memory context independent. Finally, cortisol administered before the test impairs extinction expression. For a comprehensive review, see Merz and Wolf (2022). No replicated or large-sized studies of the neuroanatomical correlates of these effects are yet available.

6 Summary of Findings in Healthy Adult Humans

Figure 1 summarizes the results of this review and can serve as an extinction reminder.

7 Implications for the Understanding and Treatment of Anxiety-Related Disorders

Extinction learning and probably also extinction memory retrieval is impaired in patients suffering in disorders characterized by pathological fear and/or anxiety; this includes patients with post-traumatic stress disorder (PTSD), where the implication of conditioning processes in the etiology of the disease is most obvious (Duits et al. 2015; Zuj et al. 2016; Beckers et al. 2023). There is also some evidence that impaired extinction may enhance individual vulnerability to developing an anxiety disorder following exposure to adverse experiences (Beckers et al. 2023; Kalisch et al. 2024).

A methodological challenge the study of individual and patient-to-control differences in fear extinction is confronted with is the low stability (that is, test-retest reliability) of conditioning measures (Cooper et al. 2022), meaning that showing even small effects requires many hundreds of participants (Kalisch et al. 2024). Task-based fMRI metrics exhibit similarly poor stability (Elliott et al. 2020). Coupled with the high cost of MRI, this has made the identification of abnormalities in the fear extinction circuitry in vulnerable individuals and patients difficult (for a recent overview of task-based fMRI studies, see Webler et al. 2023).

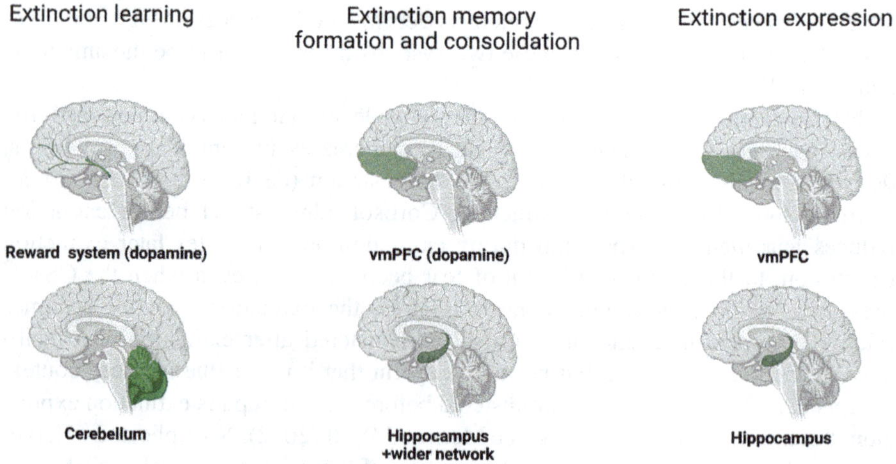

Fig. 1 Overview of current knowledge about the functional neuroanatomy of human fear extinction

FC metrics are generally superior to task-based metrics in terms of reliability (Elliott et al. 2020). This was exploited by Wen et al. (2022b) who applied their described robust mean FC analysis (see Sect. 4.3) to compare 91 anxiety patients and 81 PTSD patients with 77 healthy controls. Both patient groups failed to show the CS + -associated FC increase over the course of the extinction characteristic for the healthy controls (Wen et al. 2021); instead, FC decreased. The observation of this pattern in two independent patient groups indicates that mean FC decreases in extinction are a general feature of anxiety disorders. The strongest overlap in this pattern between the two patients groups was observed in the dorsolateral PFC (DLPFC), which was in turn less connected than in the controls with many other brain areas, including the insula, the thalamus, and cingulate and medial PFC areas, known to be involved in emotional processing. This might point to a failure in the patients to link the extinction learning experience and the resulting safety memory with a higher-order processing and regulation circuitry that may support memory recall and fear inhibition at later CS confrontations.

While the DLPFC has not been in the focus of the fMRI studies summarized in this review, a role for the DLPFC in modulating extinction is supported by the observation that transcranial magnetic stimulation (TMS) of the DLPFC during extinction improved extinction memory retrieval (Raij et al. 2018).

Like the pharmacological findings presented in Sect. 5, this latter result opens up avenues toward improved treatment of anxiety disorders. If the assumption that fear extinction is a major working mechanism of exposure-based therapy (Craske et al. 2022) is correct, therapy augmentation with a drug such as L-DOPA, other memory-enhancing drugs (Singewald et al. 2023), or also targeted TMS might enhance success rates or reduce relapse. Imaging-based biofeedback (Chiba et al. 2019) might be another approach making use of our increasing knowledge about the neural mechanisms of fear extinction. Further work at the intersection of neuroimaging and neuromodulation will be needed to fully exploit these possibilities.

Acknowledgments This work was funded by Deutsche Forschungsgemeinschaft (DFG grant KA 1623/11-1).

References

Abercrombie ED, Keefe KA, DiFrischia DS, Zigmond MJ (1989) Differential effect of stress on in vivo dopamine release in striatum, nucleus accumbens, and medial frontal cortex. J Neurochem 52:1655–1658

Andres E, Chuan-Peng H, Gerlicher AMV, Meyer B, Tüscher O, Kalisch R (2024) Replication study on the role of dopamine-dependent prefrontal reactivations in human extinction memory retrieval. Nat Commun 15:2699

Bach DR, Weiskopf N, Dolan RJ (2011) A stable sparse fear memory trace in human amygdala. J Neurosci 31:9383–9389

Batsikadze G, Diekmann N, Ernst TM, Klein M, Maderwald S, Deuschl C, Merz CJ, Cheng S, Quick HH, Timmann D (2022) The cerebellum contributes to context-effects during fear extinction learning: a 7T fMRI study. Neuroimage 253:119080

Battaglia S, Garofalo S, di Pellegrino G, Starita F (2020) Revaluing the role of vmPFC in the acquisition of Pavlovian threat conditioning in humans. J Neurosci 40:8491–8500

Battaglia S, Harrison BJ, Fullana MA (2022) Does the human ventromedial prefrontal cortex support fear learning, fear extinction or both? A commentary on subregional contributions. Mol Psychiatry 27:784–786

Becker CR, Milad MR (2023) Contemporary approaches toward neuromodulation of fear extinction and its underlying neural circuits. Curr Top Behav Neurosci 64:353–387

Beckers T, Hermans D, Lange I, Luyten L, Scheveneels S, Vervliet B (2023) Understanding clinical fear and anxiety through the lens of human fear conditioning. Nat Rev Psychol 2:233–245

Bierwirth P, Stockhorst U (2022) Role of noradrenergic arousal for fear extinction processes in rodents and humans. Neurobiol Learn Mem 194:107660

Bouton ME (2004) Context and behavioral processes in extinction. LearnMem 11:485–494

Bouton ME, Maren S, McNally GP (2021) Behavioral and neurobiological mechanisms of PAVLOVIAN and instrumental extinction learning. Physiol Rev 101:611–681

Brocka M, Helbing C, Vincenz D, Scherf T, Montag D, Goldschmidt J, Angenstein F, Lippert M (2018) Contributions of dopaminergic and non-dopaminergic neurons to VTA-stimulation induced neurovascular responses in brain reward circuits. Neuroimage 177:88–97

Bukalo O, Pinard CR, Silverstein S, Brehm C, Hartley ND, Whittle N, Colacicco G, Busch E, Patel S, Singewald N, Holmes A (2015) Prefrontal inputs to the amygdala instruct fear extinction memory formation. Sci Adv 1:e1500251

Chiba T, Kanazawa T, Koizumi A, Ide K, Taschereau-Dumouchel V, Boku S, Hishimoto A, Shirakawa M, Sora I, Lau H, Yoneda H, Kawato M (2019) Current status of neurofeedback for post-traumatic stress disorder: a systematic review and the possibility of decoded neurofeedback. Front Hum Neurosci 13:233

Cooper SE, Dunsmoor JE, Koval KA, Pino ER, Steinman SA (2022) Test-retest reliability of human threat conditioning and generalization across a 1-to-2-week interval. Psychophysiology: e14242

Craske MG, Sandman CF, Stein MB (2022) How can neurobiology of fear extinction inform treatment? Neurosci Biobehav Rev 143:104923

Dayan L, Finberg JPM (2003) L-DOPA increases noradrenaline turnover in central and peripheral nervous systems. Neuropharmacology 45:524–533

de Jong JW, Fraser KM, Lammel S (2022) Mesoaccumbal dopamine heterogeneity: what do dopamine firing and release have to do with it? Annu Rev Neurosci 45:109–129

de Kloet ER, Joëls M (2023) The cortisol switch between vulnerability and resilience. Mol Psychiatry

Do-Monte FH, Manzano-Nieves G, Quiñones-Laracuente K, Ramos-Medina L, Quirk GJ (2015) Revisiting the role of infralimbic cortex in fear extinction with optogenetics. J Neurosci 35: 3607–3615

Duits P, Cath DC, Lissek S, Hox JJ, Hamm AO, Engelhard IM, van den Hout MA, Baas JMP (2015) Updated meta-analysis of classical fear conditioning in the anxiety disorders. Depress Anxiety 32:239–253

Elliott ML, Knodt AR, Ireland D, Morris ML, Poulton R, Ramrakha S, Sison ML, Moffitt TE, Caspi A, Hariri AR (2020) What is the test-retest reliability of common task-functional MRI measures? New empirical evidence and a meta-analysis. Psychol Sci 31:792–806

Ernst TM, Brol AE, Gratz M, Ritter C, Bingel U, Schlamann M, Maderwald S, Quick HH, Merz CJ, Timmann D (2019) The cerebellum is involved in processing of predictions and prediction errors in a fear conditioning paradigm. eLife 8:e46831

Esser R, Korn CW, Ganzer F, Haaker J (2021) L-DOPA modulates activity in the vmPFC, nucleus accumbens, and VTA during threat extinction learning in humans. eLife 10:e65280

Etkin A, Egner T, Kalisch R (2011) Emotional processing in anterior cingulate and medial prefrontal cortex. Trends Cogn Sci 15:85–93

Feenstra MGP, Vogel M, Botterblom MHA, Joosten RNJMA, De Bruin JPC (2001) Dopamine and noradrenaline efflux in the rat prefrontal cortex after classical aversive conditioning to an auditory cue. Eur J Neurosci 13:1051–1054

Floresco SB, Tse MT (2007) Dopaminergic regulation of inhibitory and excitatory transmission in the basolateral amygdala-prefrontal cortical pathway. J Neurosci 27:2045–2057

Friston KJ, Frith CD, Liddle PF, Frackowiak RS (1993) Functional connectivity: the principal-component analysis of large (PET) data sets. J Cereb Blood Flow Metab 13:5–14

Fullana MA, Harrison BJ, Soriano-Mas C, Vervliet B, Cardoner N, Àvila-Parcet A, Radua J (2016) Neural signatures of human fear conditioning: an updated and extended meta-analysis of fMRI studies. Mol Psychiatry 21:500–508

Fullana MA, Albajes-Eizagirre A, Soriano-Mas C, Vervliet B, Cardoner N, Benet O, Radua J, Harrison BJ (2018) Fear extinction in the human brain: a meta-analysis of fMRI studies in healthy participants. Neurosci Biobehav Rev 88:16–25

Garrison J, Erdeniz B, Done J (2013) Prediction error in reinforcement learning: a meta-analysis of neuroimaging studies. Neurosci Biobehav Rev 37:1297–1310

Gerlicher AMV, Tüscher O, Kalisch R (2018) Dopamine-dependent prefrontal reactivations explain long-term benefit of fear extinction. Nat Commun 9:4294

Gerlicher AMV, Tüscher O, Kalisch R (2019) L-DOPA improves extinction memory retrieval after successful fear extinction. Psychopharmacology (Berl) 236:3401–3412

Gerlicher A, Ilhan-Bayrakcı M, Tüscher O, Kalisch R (2022) Post-extinction L-DOPA administration to reduce the return of fear in humans. Available at: https://osf.io/rhnsa. Accessed 29 Apr 2024

Giustino TF, Maren S (2018) Noradrenergic modulation of fear conditioning and extinction. Front Behav Neurosci 12:43

Haaker J, Gaburro S, Sah A, Gartmann N, Lonsdorf TB, Meier K, Singewald N, Pape H-C, Morellini F, Kalisch R (2013) Single dose of l-dopa makes extinction memories context-independent and prevents the return of fear. Proc Natl Acad Sci U S A 110:E2428–E2436

Haaker J, Lonsdorf TB, Kalisch R (2015) Effects of post-extinction l-DOPA administration on the spontaneous recovery and reinstatement of fear in a human fMRI study. Eur Neuropsychopharmacol

Haxby JV, Gobbini MI, Furey ML, Ishai A, Schouten JL, Pietrini P (2001) Distributed and overlapping representations of faces and objects in ventral temporal cortex. Science 293: 2425–2430

Hennings AC, McClay M, Lewis-Peacock JA, Dunsmoor JE (2020) Contextual reinstatement promotes extinction generalization in healthy adults but not PTSD. Neuropsychologia 147: 107573

Hennings AC, Cooper SE, Lewis-Peacock JA, Dunsmoor JE (2022a) Pattern analysis of neuroimaging data reveals novel insights on threat learning and extinction in humans. Neurosci Biobehav Rev 142:104918

Hennings AC, McClay M, Drew MR, Lewis-Peacock JA, Dunsmoor JE (2022b) Neural reinstatement reveals divided organization of fear and extinction memories in the human brain. Curr Biol 32:304–314.e5

Howarth C, Mishra A, Hall CN (2021) More than just summed neuronal activity: how multiple cell types shape the BOLD response. Philos Trans R Soc Lond B Biol Sci 376:20190630

Hugues S, Garcia R, Léna I (2007) Time course of extracellular catecholamine and glutamate levels in the rat medial prefrontal cortex during and after extinction of conditioned fear. Synapse 61: 933–937

Izquierdo I, Bevilaqua LR, Rossato JI, Bonini JS, Medina JH, Cammarota M (2006) Different molecular cascades in different sites of the brain control memory consolidation. Trends Neurosci 29:496–505

Kalisch R, Korenfeld E, Stephan KE, Weiskopf N, Seymour B, Dolan RJ (2006) Context-dependent human extinction memory is mediated by a ventromedial prefrontal and hippocampal network. J Neurosci 26:9503–9511

Kalisch R, Müller MB, Tüscher O (2015) A conceptual framework for the neurobiological study of resilience. Behav Brain Sci:e92

Kalisch R, Russo SJ, Müller MB (2024) Neurobiology and systems biology of stress resilience. Physiol Rev. Available at: https://journals.physiology.org. Accessed 19 Apr 2024. https://doi.org/10.1152/physrev.00042.2023

Kampa M, Schick A, Yuen K, Sebastian A, Chmitorz A, Saase V, Wessa M, Tüscher O, Kalisch R (2018) A combined behavioral and neuroimaging battery to test positive appraisal style theory of resilience in longitudinal studies. bioRxiv

Kampa M, Schick A, Sebastian A, Wessa M, Tüscher O, Kalisch R, Yuen K (2020) Replication of fMRI group activations in the neuroimaging battery for the Mainz resilience project (MARP). Neuroimage 204:116223

Kampa M, Hermann A, Stark R, Klucken T (2023) Neural correlates of immediate versus delayed extinction when simultaneously varying the time of the test in humans. Cereb Cortex 33:9325–9338

LaBar KS (2023) Neuroimaging of fear extinction. Curr Top Behav Neurosci 64:79–101

Luo R, Uematsu A, Weitemier A, Aquili L, Koivumaa J, McHugh TJ, Johansen JP (2018) A dopaminergic switch for fear to safety transitions. Nat Commun 9:2483

Maren S (2022) Unrelenting fear under stress: neural circuits and mechanisms for the immediate extinction deficit. Front Syst Neurosci 16:888461

Marković V, Vicario CM, Yavari F, Salehinejad MA, Nitsche MA (2021) A systematic review on the effect of transcranial direct current and magnetic stimulation on fear memory and extinction. Front Hum Neurosci 15:655947

Mechias ML, Etkin A, Kalisch R (2010) A meta-analysis of instructed fear studies: implications for conscious appraisal of threat. Neuroimage 49:1760–1768

Merz CJ, Wolf OT (2022) How stress hormones shape memories of fear and anxiety in humans. Neurosci Biobehav Rev 142:104901

Milad MR, Quirk GJ (2002) Neurons in medial prefrontal cortex signal memory for fear extinction. Nature 420:70–74

Milad MR, Quirk GJ (2012) Fear extinction as a model for translational neuroscience: ten years of progress. Annu Rev Psychol 63:129–151

Milad MR, Vidal-Gonzalez I, Quirk GJ (2004) Electrical stimulation of medial prefrontal cortex reduces conditioned fear in a temporally specific manner. Behav Neurosci 118:389–394

Milad MR, Wright CI, Orr SP, Pitman RK, Quirk GJ, Rauch SL (2007) Recall of fear extinction in humans activates the ventromedial prefrontal cortex and hippocampus in concert. Biol Psychiatry 62:446–454

Morriss J, Hoare S, van Reekum CM (2018) It's time: a commentary on fear extinction in the human brain using fMRI. Neurosci Biobehav Rev 94:321–322

Nater UM, Rohleder N, Schlotz W, Ehlert U, Kirschbaum C (2007) Determinants of the diurnal course of salivary alpha-amylase. Psychoneuroendocrinology 32:392–401

O'Doherty JP, Deichmann R, Critchley HD, Dolan RJ (2002) Neural responses during anticipation of a primary taste reward. Neuron 33:815–826

Out D, Bakermans-Kranenburg MJ, Granger DA, Cobbaert CM, van Ijzendoorn MH (2011) State and trait variance in salivary α-amylase: a behavioral genetic study. Biol Psychol 88:147–154

Out D, Granger DA, Sephton SE, Segerstrom SC (2013) Disentangling sources of individual differences in diurnal salivary α-amylase: reliability, stability and sensitivity to context. Psychoneuroendocrinology 38:367–375

Pascucci T, Ventura R, Latagliata EC, Cabib S, Puglisi-Allegra S (2007) The medial prefrontal cortex determines the accumbens dopamine response to stress through the opposing influences of norepinephrine and dopamine. Cereb Cortex 17:2796–2804

Pavlov IP (1927) Conditioned reflexes: an investigaton of the physiological activity of the cerebral cortex. Oxford University Press, Oxford

Phelps EA, Delgado MR, Nearing KI, LeDoux JE (2004) Extinction learning in humans. Neuron 43:897–905

Pisano TJ, Dhanerawala ZM, Kislin M, Bakshinskaya D, Engel EA, Hansen EJ, Hoag AT, Lee J, de Oude NL, Venkataraju KU, Verpeut JL, Hoebeek FE, Richardson BD, Boele H-J, Wang SS-H (2021) Homologous organization of cerebellar pathways to sensory, motor, and associative forebrain. Cell Rep 36:109721

Plas SL, Tuna T, Bayer H, Juliano VAL, Sweck SO, Arellano Perez AD, Hassell JE, Maren S (2024) Neural circuits for the adaptive regulation of fear and extinction memory. Front Behav Neurosci 18:1352797

Popa LS, Ebner TJ (2018) Cerebellum, predictions and errors. Front Cell Neurosci 12:524

Quirk GJ, Russo GK, Barron JL, Lebron K (2000) The role of ventromedial prefrontal cortex in the recovery of extinguished fear. J Neurosci 20:6225–6231

Quirk GJ, Likhtik E, Pelletier JG, Pare D (2003) Stimulation of medial prefrontal cortex decreases the responsiveness of central amygdala output neurons. J Neurosci 23:8800–8807

Raczka KA, Mechias ML, Gartmann N, Reif A, Deckert J, Pessiglione M, Kalisch R (2011) Empirical support for an involvement of the mesostriatal dopamine system in human fear extinction. Transl Psychiatry 1:e12. https://doi.org/10.1038/tp.2011.10

Raij T, Nummenmaa A, Marin M-F, Porter D, Furtak S, Setsompop K, Milad MR (2018) Prefrontal cortex stimulation enhances fear extinction memory in humans. Biol Psychiatry 84:129–137

Rescorla RA, Wagner AR, Black AH, Prokasy WF (1972) A theory of Pavlovian conditioning: variations in the effectiveness of reinforcement and nonreinforcement. In: Classical conditioning II: current research and theory. Appleton-Century-Crofts, New York, pp 64–99

Rosenberg BM, Taschereau-Dumouchel V, Lau H, Young KS, Nusslock R, Zinbarg RE, Craske MG (2023) A multivoxel pattern analysis of anhedonia during fear extinction: implications for safety learning. Biol Psychiatry Cogn Neurosci Neuroimaging 8:417–425

Royer S, Paré D (2002) Bidirectional synaptic plasticity in intercalated amygdala neurons and the extinction of conditioned fear responses. Neuroscience 115:455–462

Salinas-Hernández XI, Vogel P, Betz S, Kalisch R, Sigurdsson T, Duvarci S (2018) Dopamine neurons drive fear extinction learning by signaling the omission of expected aversive outcomes. eLife 7:e38818

Salinas-Hernández XI, Zafiri D, Sigurdsson T, Duvarci S (2023) Functional architecture of dopamine neurons driving fear extinction learning. Neuron. S0896-6273(23)00636-0

Santini E, Ge H, Ren K, Peña de Ortiz S, Quirk GJ (2004) Consolidation of fear extinction requires protein synthesis in the medial prefrontal cortex. J Neurosci 24:5704–5710

Sartori SB, Keil TMV, Kummer KK, Murphy CP, Gunduz-Cinar O, Kress M, Ebner K, Holmes A, Singewald N (2024) Fear extinction rescuing effects of dopamine and L-DOPA in the ventromedial prefrontal cortex. Transl Psychiatry 14:11

Schultz W (2006) Behavioral theories and the neurophysiology of reward. Annu Rev Psychol 57: 87–115

Senn V, Wolff SBE, Herry C, Grenier F, Ehrlich I, Gründemann J, Fadok JP, Müller C, Letzkus JJ, Lüthi A (2014) Long-range connectivity defines behavioral specificity of Amygdala neurons. Neuron 81:428–437

Sierra-Mercado D, Corcoran KA, Lebron-Milad K, Quirk GJ (2006) Inactivation of the ventromedial prefrontal cortex reduces expression of conditioned fear and impairs subsequent recall of extinction. Eur J Neurosci 24:1751–1758

Singewald N, Sartori SB, Reif A, Holmes A (2023) Alleviating anxiety and taming trauma: novel pharmacotherapeutics for anxiety disorders and posttraumatic stress disorder. Neuropharmacology 226:109418

Thiele M, Yuen KSL, Gerlicher AVM, Kalisch R (2021) A ventral striatal prediction error signal in human fear extinction learning. Neuroimage 229:117709

Tovote P, Fadok JP, Lüthi A (2015) Neuronal circuits for fear and anxiety. Nat Rev Neurosci 16: 317–331

Vidal-Gonzalez I, Vidal-Gonzalez B, Rauch SL, Quirk GJ (2006) Microstimulation reveals opposing influences of prelimbic and infralimbic cortex on the expression of conditioned fear. Learn Mem 13:728–733

Visser RM, Scholte HS, Beemsterboer T, Kindt M (2013) Neural pattern similarity predicts long-term fear memory. Nat Neurosci 16:388–390

Visser RM, Kunze AE, Westhoff B, Scholte HS, Kindt M (2015) Representational similarity analysis offers a preview of the noradrenergic modulation of long-term fear memory at the time of encoding. Psychoneuroendocrinology 55:8–20

Visser RM, Bathelt J, Scholte HS, Kindt M (2021) Robust BOLD responses to faces but not to conditioned threat: challenging the Amygdala's reputation in human fear and extinction learning. J Neurosci 41:10278–10292

Webler RD, Oathes DJ, van Rooij SJH, Gewirtz JC, Nahas Z, Lissek SM, Widge AS (2023) Causally mapping human threat extinction relevant circuits with depolarizing brain stimulation methods. Neurosci Biobehav Rev 144:105005

Wen Z, Chen ZS, Milad MR (2021) Fear extinction learning modulates large-scale brain connectivity. Neuroimage 238:118261

Wen Z, Raio CM, Pace-Schott EF, Lazar SW, LeDoux JE, Phelps EA, Milad MR (2022a) Temporally and anatomically specific contributions of the human amygdala to threat and safety learning. Proc Natl Acad Sci U S A 119:e2204066119

Wen Z, Seo J, Pace-Schott EF, Milad MR (2022b) Abnormal dynamic functional connectivity during fear extinction learning in PTSD and anxiety disorders. Mol Psychiatry 27:2216–2224

Whittle N, Maurer V, Murphy C, Rainer J, Bindreither D, Hauschild M, Scharinger A, Oberhauser M, Keil T, Brehm C, Valovka T, Striessnig J, Singewald N (2016) Enhancing dopaminergic signaling and histone acetylation promotes long-term rescue of deficient fear extinction. Transl Psychiatry 6:e974

Young KS, Bookheimer SY, Nusslock R, Zinbarg RE, Damme KSF, Chat IK-Y, Kelley NJ, Vinograd M, Perez M, Chen K, Cohen AE, Craske MG (2021) Dysregulation of threat neurocircuitry during fear extinction: the role of anhedonia. Neuropsychopharmacology 46:1650–1657

Zuj DV, Palmer MA, Lommen MJJ, Felmingham KL (2016) The centrality of fear extinction in linking risk factors to PTSD: a narrative review. Neurosci Biobehav Rev 69:15–35

The Organization of Anxiety Symptoms Along the Threat Imminence Continuum

Ya'ira Somerville and Rany Abend

Contents

Abstract Pathological anxiety is highly prevalent, impairing, and often chronic. Yet, despite considerable research, mechanistic understanding of anxiety and its translation to clinical practice remain limited. Here, we first highlight two foundational complications that contribute to this gap: a reliance on a phenomenology-driven definition of pathological anxiety in neurobiological mechanistic research, and a limited understanding of the chronicity of anxiety symptom expression. We then posit that anxiety symptoms may reflect aberrant expression of otherwise normative defensive responses. Accordingly, we propose that threat imminence, an organizing dimension for normative defensive responses observed across species, may be applied to organize and understand anxiety symptoms along a temporal dimension of expression. Empirical evidence linking distinct anxiety symptoms and the aberrant expression of imminence-dependent defensive responses is reviewed, alongside the neural

Y. Somerville and R. Abend (✉)
Baruch Ivcher School of Psychology, Reichman University, Herzliya, Israel
e-mail: rany.abend@runi.ac.il

© The Author(s), under exclusive license to Springer Nature Switzerland AG 2024
Curr Topics Behav Neurosci (2025) 73: 93–121
https://doi.org/10.1007/7854_2024_548
Published Online: 24 November 2024

mechanisms which may underpin these cognitive, physiological, and behavioral responses. Drawing from extensive translational and clinical research, we suggest that understanding anxiety symptoms through this neurobiologically-informed framework may begin to overcome the conceptual complications hindering advancement in mechanistic research and clinical interventions for pathological anxiety.

Keywords Anxiety · Anxiety disorders · Anxiety symptoms · Avoidance · Defensive response · Fear · Neural circuitry · Neurobiological mechanisms · Physiological response · Threat imminence · Worry

1 Introduction

Anxiety disorders are among the most prevalent psychiatric disorders (Bandelow et al. 2017), and are often chronic, debilitating, and associated with a multitude of adverse outcomes (Kessler and Wang 2008; Shackman and Fox 2021). Several distinct diagnoses comprise the wide umbrella of anxiety disorders in psychiatric nosology (APA 2013; World Health Organization 2019). These diagnoses share symptoms that reflect excessive, distressing, and impairing responses to potential threats, but are distinguished primarily by the types of objects or settings that are perceived as threatening and elicit these responses (World Health Organization 2019). Despite considerable research, mechanistic understanding of anxiety disorders and its translation to clinical practice remain limited. In this article we highlight two foundational complications that contribute to these shortcomings: a reliance on a phenomenology-driven definition of pathological anxiety in neurobiological mechanistic research, and a limited understanding of the chronicity of anxiety symptom expression. We then propose that *threat imminence*, an organizing dimension for defensive responses observed across species (Fanselow and Lester 1988), may be applied to likewise organize anxiety symptoms along a temporal dimension of expression. Through this neurobiologically informed framework, we may understand anxiety symptoms as aberrant expressions of otherwise normative, threat imminence-dependent defensive responses, and begin to overcome conceptual complications impeding progress in mechanistic research and clinical interventions for pathological anxiety.

2 Limitations in Mechanistic Understanding of Anxiety and Translation from Research to Clinic

According to psychiatric nosology, anxiety disorders share features of excessive, persistent, and impairing physiological (e.g., autonomic arousal), cognitive (e.g., vigilance and worrying), and behavioral (e.g., avoidance) responses to real or perceived potential threats (APA 2013; World Health Organization 2019). Current first-line guidelines recommend treatment through psychotherapy (e.g., cognitive behavioral therapy), pharmacotherapy (e.g., selective serotonin reuptake inhibitors), or a combination of the two (Bandelow et al. 2017; Wehry et al. 2015). However, given the considerable distress experienced, treatment-seeking persons frequently seek immediate relief, often leading to the prescription of fast-acting anxiolytic agents such as benzodiazepines (Roy-Byrne 2015), which can result in significant difficulties associated with long-term use (Soyka 2017). Furthermore, despite available intervention options backed by clinical guidelines (Bandelow et al. 2023), incidence cases have increased dramatically while existing treatments remain inconsistently effective (Shackman and Fox 2021) and a significant proportion of individuals with anxiety disorders display treatment resistance (Domschke et al. 2024), indicating the need for improved intervention options.

Despite considerable research investment and ever-improving scientific methodologies, our mechanistic understanding of pathological anxiety and its translation from the lab to the clinic remains limited (Scull 2021; Cuthbert 2022). Next, we briefly highlight two core obstacles contributing to this challenge.

Phenomenological Descriptions in Neurobiological Research

For decades, clinical research on psychopathology and its treatment has depended primarily on phenomenology-driven categorical diagnoses outlined in psychiatric nosology such as the Diagnostic and Statistical Manual of Mental Disorders (DSM) and the International Classification of Diseases (ICD) (Conway et al. 2019). The clinical diagnosis of anxiety disorders is defined in terms of observable symptoms and signs, with different disorders distinguished primarily by the types of stimuli or situations that appear to elicit these overt reactions (APA 2013; World Health Organization 2019).

However, while classification of pathological anxiety into distinct diagnoses may hold clinical value in some settings, the diagnostic boundaries that are drawn are blurry. Thus, considerable data from clinical observations and empirical research consistently indicate limited disorder specificity in anxiety symptom presentation, pathophysiology, familial aggregation and genetic loci, pathophysiology, diagnostic reliability and validity, and treatment targets or response (Cuthbert and Insel 2013; Kotov et al. 2022). In other words, diagnostic boundaries do not clearly align with clinical presentation and biological boundaries. Yet, much of clinical practice and research continue to depend on categorical, phenomenological classification. As a result, despite substantial research efforts, only limited gains have been made in improving the well-being of affected individuals through the development of targeted treatments, drawing growing criticism and increased calls for development of alternative conceptualizations of anxiety and psychopathology as a whole (Cuthbert 2022).

Chronicity of Anxiety Symptom Expression

A second obstacle to effective translation is insufficient understanding of temporal variation in anxiety symptom presentation. While different anxiety symptoms, such as excessive worry cognitions, physiological arousal, and behavioral avoidance, are fairly distinct and discernable, a clear framework describing precisely which and when different symptoms are expressed is lacking. For example, diagnostic criteria for social anxiety disorder (SAD) indicate marked fear or anxiety about social situations (APA 2013), but provide limited guidance as to which specific symptoms are expected at different time points relative to the anxiogenic situation. Indeed, empirical data indicate that anxiety symptom presentation, much like responses to aversive events, shows substantial temporal variability (Schoevers et al. 2021), in line with influential theories on the dynamic nature of emotion and the emergence of denser clinical assessment strategies in which symptoms are repeatedly probed throughout the day (Trull et al. 2015). However, within current clinical conceptualizations of anxiety, different anxiety symptoms appear to emerge and dissolve without an organizing, explanatory framework that can contextualize their expression over time (Insel et al. 2010; Kotov et al. 2017). Consequently, treatment efforts to target and alleviate specific anxiety symptoms are hindered by limited ability to anticipate when specific symptoms are expected to present. Similarly, without a clear understanding of symptom chronicity, attempts to systematically evoke specific symptoms in experimental settings may conflate distinct evoked responses. For example, responses to emotional faces are often used to experimentally assess hypervigilance in anxiety; however, the common practice of indexing responses as an aggregate across a time frame of several seconds makes it difficult to differentiate initial attentional capture from other symptom-relevant processes such as preparation for avoidance behavior (Abend et al. 2021a). This may complicate efforts to capture fundamental underlying mechanisms of dysfunction and potentially contribute to inconsistent findings observed across studies (Abend 2023).

These forms of imprecise description of pathological anxiety may arise in part from terminology. The DSM highlights reported symptoms pertaining to "excessive fear and anxiety" as defining features and diagnostic criteria shared by anxiety disorders (APA 2013). The terms "fear" and "anxiety" are used to broadly distinguish emotional, cognitive, physiological, and behavioral responses to near-future- vs. far-future threats, respectively (Barlow 2002; Perusini and Fanselow 2015). However, given that the DSM does not clearly define these terms, an ongoing debate remains regarding the precise nature of this distinction in terms of theoretical considerations, underlying neural circuitry, and clinical expression (Perusini and Fanselow 2015; Tovote et al. 2015). This elusive distinction must be noted, as it hinders the utility of these terms as a foundation for describing anxiety symptomatology, especially with regard to clinical care and linking quantitative, psychobiological research in lab models to clinical classification (Perusini and Fanselow 2015). Here, rather than attempting to delineate these terms or formally use them in definitions, "fear" and "anxiety" symptoms may instead be understood as commonly reflecting maladaptive defensive responses to potential threat observed across anxiety disorders.

3 Addressing Limitations in Nosology Through a Neurobiologically Informed Framework of Anxiety Symptoms

Given the substantial role that accurate and comprehensive description of psychopathology plays in guiding both clinical research and practice (Conway et al. 2019), a refined conceptualization of anxiety may be necessary in order to bridge existing gaps. The converging lines of evidence noted above suggest that the different anxiety disorders may be characterized more by similar, rather than distinct, etiology, mechanistic perturbations, and patterns of expression, potentially arising from a shared biological or psychosocial diathesis (Barlow 2002; Chavanne and Robinson 2021). Further, the conserved nature of responses to threat, the central feature of anxiety symptoms, suggests that key aspects of pathological anxiety may be understood in terms of the function of adaptive neurobiological circuitry (Fanselow and Lester 1988; LeDoux 2012; Cisek 2019). Pathological anxiety therefore lends itself to an alternative conceptualization that draws from conserved neurobiological processes that promote survival – to explain shared "core" symptomatology. Importantly, as will be described, the proposed framework serves to specify which anxiety symptoms are expected to present at which times, in order to promote an enhanced understanding of pathological anxiety and its treatment (Xia and Kheirbek 2020).

Substantial translational research indicates that objects or settings which signal potential threat, such as danger to one's life and health, or loss of social standing, caregivers, or resources (Marks and Nesse 1994; Beauchaine and Zisner 2017), elicit a cascade of defensive responses in anticipation of the threat. This cascade can be observed on molecular, synaptic, physiological, and cognitive levels, subserved by partially conserved neural circuitry, and operating to promote survival by driving defensive behaviors (Cisek 2019; Rosen and Schulkin 1998). This wealth of responses appears to reflect an expanding evolutionary process whereby dedicated systems that initially emerged to support simple, defensive reflexes to immediate physical threats evolved to become increasingly complex, promoting survival when faced with a seemingly infinite variety of threats via nuanced detection, assessment, and behavior selection and mobilization processes (LeDoux 2012; Cisek 2019; Fox and Shackman 2019; Dranias et al. 2008; Adolphs 2009; Blanchard et al. 2011).

Arising from these evolutionary observations is the notion that, while humans appear to cope with different types of threats that call for diverging behavioral outputs, the neurobiological processes subserving the anticipation of threat (i.e., preparation, detection, evaluation, and subsequent behavior selection) may be shared. Emerging evidence begins to support this notion, indicating the centrality of excessive cognitive (e.g., vigilance, worry cognitions) and physiological (autonomic arousal) defensive responses in anxiety symptom presentation, regardless of specific source of potential threat (Insel et al. 2010; Abend 2023; Craske et al. 2009). The excessive expression of these threat-anticipatory responses may subsequently promote exaggerated execution of defensive behaviors (e.g., avoidance) frequently observed across anxiety disorders (Craske et al. 2009; Bijsterbosch et al. 2014;

Abend et al. 2021b). Thus, while the specific objects or situations perceived as threatening may differ, the exaggerated expression of otherwise normative cognitive, physiological, and behavioral defensive responses may be shared across disorders (APA 2013; Marks and Nesse 1994; Rosen and Schulkin 1998; Blanchard et al. 2011; Gray and McNaughton 2000; Etkin and Wager 2007). Taken together, it may be posited that pathological anxiety reflects a tendency for enhanced defensive responsivity, prominently manifesting in physiological, cognitive, and behavioral domains (Craske et al. 2009). If these responses are expressed pervasively, or with such a magnitude that they become maladaptive, distressing, and impairing, they constitute the symptoms described in clinical anxiety (Abend 2023; Rosen and Schulkin 1998; Blanchard 2017).

We may therefore begin to link specific clinical symptoms of anxiety to specific patterns of exaggerated and pervasive expression of otherwise normative defensive responses to anticipated potential threat (Shackman and Fox 2021; Rosen and Schulkin 1998; Craske et al. 2009; Mobbs et al. 2020; Kenwood et al. 2022; Mineka and Oehlberg 2008; Grupe and Nitschke 2013). While evolution has favored a low threshold for the initiation of threat-anticipatory defensive responses (Nesse 2005), a high rate of false alarms confers the risk for excessive and persistent execution of responses that can lead to pathological anxiety (Kenwood et al. 2022; Grupe and Nitschke 2013; Nesse 2005). Anxiety symptoms may therefore reflect aberrant expression of otherwise normative defensive responses. This notion is further supported by recent work showing an association between pathological anxiety and aberrant anticipation, rather than ultimate experience, of aversive events (Abend et al. 2020, 2021b; Grupe and Nitschke 2013), underscoring the importance of focusing on responses to potential threats.

4 Organization of Defensive Responses Along a Temporal Continuum

Substantial cross-species research reveals that defensive responses to threat are organized along a threat imminence continuum, typically comprising pre-encounter, post-encounter, and circa-strike phases (Fanselow and Lester 1988; Perusini and Fanselow 2015; Mobbs et al. 2020; Blanchard and Blanchard 1989; McNaughton and Corr 2004). This organization reflects the formation of discrete responses, observed across multiple, coordinated functional levels (e.g., autonomic reactions supporting defensive behaviors) that have evolved to anticipate threat and minimize harm and which are differentially and dynamically expressed as a function of physical or temporal threat proximity, and mobilized by partially conserved neural circuitry (LeDoux 2012; Fox and Shackman 2019; Mobbs et al. 2020; McNaughton and Corr 2018). Recent work has begun to extend these insights to humans, providing a strong translational, empirical and conceptual framework for describing human defensive responding (Blanchard 2017; Mobbs 2018; Hamm 2020; Mobbs and Kim 2015).

If, as proposed prior, pathological anxiety reflects a propensity for exaggerated expression of otherwise normative responses to potential threat (Rosen and Schulkin 1998; Blanchard 2017; Kenwood et al. 2022), and these responses can be organized by threat imminence (Fanselow and Lester 1988; Blanchard 2017; Mobbs 2018; Adolphs 2013), we may begin to systematically link distinct anxiety symptoms to aberrant expression of specific, imminence-dependent defensive responses, providing a theory-driven model for dynamically contextualizing symptoms within a wider, translational framework of response to threat (Abend 2023; Craske et al. 2009; McNaughton and Corr 2004; Hamm 2020; Krueger and DeYoung 2020). In other words, this conceptualization of pathological anxiety posits that as a certain threat becomes increasingly proximal, a relatively fixed set of distinct, conserved defensive responses are differentially expressed, regardless of the specific nature of the threat. A tendency for facilitated execution of defensive responding then manifests as excessive expression of each of these responses, which we clinically refer to as anxiety symptoms (Abend 2023).

By applying this organizing framework, we may begin to address some of the challenges posed by phenomenology-based classification of anxiety disorders, and limited understanding of symptom chronicity (Insel et al. 2010; Kotov et al. 2017). First, this framework grounds symptoms in conserved psychobiological mechanisms that support a common pattern of imminence-dependent defensive responses shared by different species in response to various threats (Fanselow and Lester 1988; Mobbs 2018; Adolphs 2013). As such, it lends itself to, and may propel, cross-species, translational mechanistic research on neurobiological determinants of pathological anxiety. Importantly, by focusing on common patterns of expression of conserved defensive responses, as opposed to weaker diagnostic distinctions (Cuthbert 2022), we may be able to identify more robust biomarkers necessary for improving translation of neuroscientific insight on pathophysiological mechanisms to clinical practice (Xia and Kheirbek 2020).

Second, in anchoring anxiety symptom expression to defensive responses evoked by the imminence of specific perceived threats, we may resolve the challenge posed by temporally dynamic symptom expression. Acknowledging a threat imminence-based temporal organization of anxiety symptoms, such as excessive vigilance, worry cognitions, physiological arousal, or avoidance behavior, could help clinicians inform patients on when different symptoms may be expected to present, and facilitate more targeted, efficacious treatment practices. This may also serve to ultimately attenuate seemingly high levels of comorbidity, as changes in threat imminence, rather than multiple anxiety disorders, could explain the expression of different "disorder-specific" symptoms within individuals. Mechanistic research will likewise benefit from such organization as experimental paradigms may be designed to elicit symptom-relevant responses in a more precise manner (Abend 2023).

5 The Phases of the Threat Imminence Continuum: Characteristic Defensive Responses, Anxiety Symptoms, and Neural Circuitry

Extensive translational research delineates several phases of encounter with potential threat (Fanselow and Lester 1988; Mobbs et al. 2020; Blanchard and Blanchard 1989; Adolphs 2013). The following framework extends prior work by identifying four such phases; see Fig. 1. The first phase identified is *Pre-Encounter*, characterized by presence in a context assessed to entail an increased likelihood of encounter with threat. Next, the *Encounter* phase may be initiated when an object conferring potential threat is spotted, acting to interrupt ongoing activities and inhibit behavior in favor of threat assessment. If the encountered object is assessed to confer sufficient threat, state transition to the *Post-Encounter* phase takes place, entailing preparation and execution of appropriate defensive behaviors according to threat proximity, current goals, and available options. Finally, the *Contact* phase is initiated if preventative defensive behaviors fail, and a threat is in direct contact, leading to

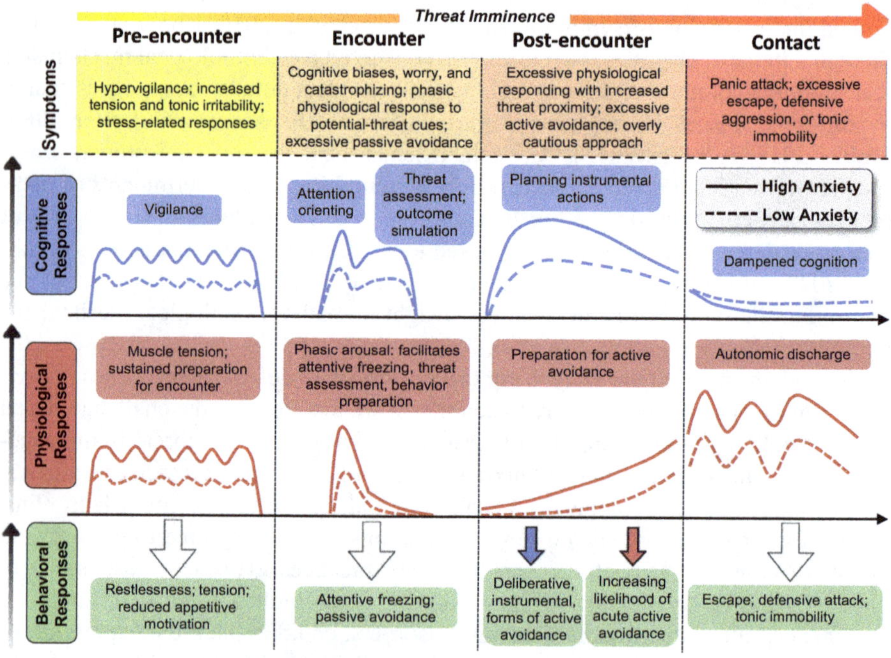

Fig. 1 Each of the four phases of threat imminence (left to right) is associated with cognitive (blue) and physiological (red) defensive responses, and the behaviors they drive (green); response magnitude is expressed from low to high along the vertical axis. Anxiety symptoms correspond to exaggerated expression of these responses. Solid and dashed lines describe expected response intensity for individuals with higher or lower anxious tendencies, respectively (but note that anxiety severity, and magnitude of defensive responses, likely reflect a continuum)

immediate execution of acute, stereotypic, and coarse behavioral responses. In this section, we delineate our framework grounding anxiety symptoms in imminence-dependent defensive responding – by reviewing extant literature concerned with three aspects of each phase of the continuum. First, we describe phase-specific characteristic patterns of normative cognitive and physiological responses, and the defensive behaviors which they drive. Next, we link specific anxiety symptoms to the aberrant expression of these phase-specific responses and review corroborating evidence. Finally, given that this framework is rooted in neurobiological perspectives of conserved defensive responding, we describe key elements in the hypothesized neural circuitry underlying anxiety symptoms.

5.1 Pre-Encounter

Pre-encounter reflects a context-activated state, characterized by increased possibility of encountering a threat, although one has yet to be spotted (Fiddick 2011; Hamm et al. 2016). It is initiated in species-specific settings, such as exposure to predator odor in rodents, or contexts conferring perceived threat to humans, such as presence in a war zone in soldiers, walking down a dark alley at night, or arriving at social functions (Grillon 2008). Vigilance is the primary cognitive response during *pre-encounter* (Hamm 2020), mediated by increased sensory gain to facilitate detection of the potential threat (Cornwell et al. 2017). Simultaneously, sustained physiological responses are expressed, such as elevated skin conductance levels, muscle tone, and startle reflexes, to facilitate rapid responding when needed (*encounter* phase, see below), given the increased likelihood of threat presence (Grillon 2008). As such, *pre-encounter* defensive responding serves to promote threat detection and rapid execution of *encounter* responses (Cornwell et al. 2017; Grillon 2002; Mkrtchian et al. 2017). This prepotent effect is inherently adaptive as the energetic cost of false-positive, downstream defensive-response cascade initiation is low relative to that of threat contact (e.g., attack) (Marks and Nesse 1994; Nesse 2005).

Anxiety Symptoms

In human studies, the *pre-encounter* state is typically modeled with context-conditioning and threat-of-shock paradigms, whereby participants are experientially or verbally informed about the potential occurrence of context-dependent, temporally unpredictable aversive stimuli such as electric shocks (Chavanne and Robinson 2021; Grillon et al. 2019). Such research robustly links pathological anxiety to enhanced and persistent expression of *pre-encounter* defensive responses, with findings consistently demonstrating exaggerated cognitive (e.g., hypervigilance) and physiological (elevated autonomic levels) responding in such contexts, likely reflecting excessive efforts to predict, detect, and prepare for aversive events (Grupe and Nitschke 2013; Grillon 2008; Armstrong and Olatunji 2012; Abend et al. 2022). These responses correspond to several key clinical features of pathological anxiety, including excessive vigilance and tension "in preparation for future danger" (APA 2013).

Behaviorally, persistent and pervasive preparation for potential threat occurrence may manifest in restlessness and tension, as well as irritability and sleep disturbances, all commonly observed symptoms of pathological anxiety (APA 2013; Silver et al. 2021). Moreover, excessive behavioral avoidance of contexts perceived as conferring increased likelihood of threat (context or situational avoidance) is prominent in pathological anxiety, such as avoidance of social settings, separation from caregivers, and agoraphobia (APA 2013; Barlow 2002; Grillon et al. 2006). When prolonged, *pre-encounter* defensive responses may be accompanied by "stress"-induced reduction in motivation to seek out appetitive and hedonic behaviors such as food intake or sexual activities as these may be deprioritized in threat contexts (Mobbs et al. 2020; Hamm 2020; Grillon et al. 2006).

Neural Circuitry

Cross-species research links presence in *pre-encounter* contexts to activation in the hippocampus, amygdala, and contiguous cortices, comprising a conserved network strongly implicated in associations between environmental, interoceptive, and temporal contexts, and aversive outcomes (Chavanne and Robinson 2021; Blanchard et al. 2011; Maren et al. 2013). Considerable research in humans and animal models additionally implicates bed nucleus of stria terminalis (BNST) responding during states of anticipation of potential threat, particularly for sustained periods (Fox and Shackman 2019; Torrisi et al. 2015). Tracing studies have demonstrated projections between sensory and context-encoding regions and amygdala and BNST, which then trigger somatomotor and neuroendocrine signals via additional projections to subcortical effector regions (Fox and Shackman 2019; Fox et al. 2015). These pathways may therefore play a key role in mediating the elevated central and peripheral physiological tone which characterizes *pre-encounter* states (Alvarez et al. 2015). BNST function during states relevant to *pre-encounter* has been further linked to nodes of the dorsal attention network (DAN), such as dorsolateral PFC and parietal cortex, and of the salience network (SN), such as anterior insula and dorsal anterior cingulate cortex (dACC) (Clauss et al. 2019; Feola et al. 2021). These networks are strongly involved in spatial attention processes (Dixon et al. 2017) and detection of salient stimuli (Menon and Uddin 2010), and as such may mediate heightened *pre-encounter* vigilance. Additional *pre-encounter* findings in the anterior and mid-cingulate cortices and inferior frontal gyrus (Chavanne and Robinson 2021) may correspond to aspects of selection, preparation, and control of behavior (Cavanagh and Shackman 2015). The amygdala and BNST are therefore well-positioned to modulate distributed neural circuits subserving physiological and cognitive responding in ways that promote threat detection and rapid behavior execution (Fox and Shackman 2019; Hamm 2020; Grillon 2008; Torrisi et al. 2015; Fox et al. 2015; Shackman et al. 2016; Robinson et al. 2019).

Pathological anxiety has been robustly linked to aberrant function within and between these circuits during experimentally induced *pre-encounter* states (Chavanne and Robinson 2021; Robinson et al. 2019; Lago et al. 2017; Torrisi et al. 2019). For example, during such states, individuals with anxiety demonstrate aberrant amygdala and BNST activation and connectivity with dACC (Avery et al. 2016), as well as aberrant anterior insula activity (Etkin and Wager 2007; Robinson et al. 2019) and connectivity with dACC and DAN nodes, which has been further linked to modulation of physiological responding (Robinson et al. 2019; Balderston et al. 2020). Together, cross-species research identifies *pre-encounter* recruitment of regions promoting and maintaining vigilance and preparation for action, while their aberrant function is observed in pathological anxiety and may underlie symptoms characterizing this phase of threat encounter (Chavanne and Robinson 2021; Kenwood et al. 2022; Davis et al. 2010; Robinson et al. 2013).

5.2 Encounter

The *encounter* phase is rapidly initiated when an object conferring potential threat is spotted during the *pre-encounter* phase. Encounter responses serve to acutely interrupt and "reset" ongoing behavior (Dayan and Yu 2006; Roelofs and Dayan 2022), and facilitate computation of potential threat value to determine whether subsequent defensive responses are required (Blanchard et al. 2011; Lang et al. 2000; Bolles and Fanselow 1980; Gross and Canteras 2012; Corr 2013; Livermore et al. 2021). As such, *encounter* determines potential state transition from *pre-* to *post-encounter* with a specific threat. *Encounter* responses are prominently characterized by phasic physiological arousal (e.g., skin conductance response, tachycardia) to potential-threat cues, promoting attentive freezing (behavior inhibition) and evasion of detection, and facilitating attention allocation and threat assessment (appraisal) processes (Blanchard et al. 2011; Gray and McNaughton 2000; Hamm 2020; Abend et al. 2022; Roelofs and Dayan 2022; LeDoux 2000; Wendt et al. 2017). For example, a person walking down a dark alley at night notices a sudden movement, and must immediately gauge whether it signifies threat. Threat assessment, and the subsequent selection of *post-encounter* defensive behavior, is determined through computation of potential threat magnitude, accounting for estimated factors such as strike probability (given type of detected object, proximity, speed, and intent) and cost, as well as context and available behavioral options. An object determined to confer high threat value will lead to initiation of the *post-encounter* phase (see below) while low threat value will result in a return to *pre-encounter*.

Humans' evolved ability to generate and maintain powerful, nuanced, and realistic internal representations of future situations holds substantial adaptive value as threats can be "encountered" via mental simulation (LeDoux and Daw 2018; Riskind and Calvete 2020). Invoked imagery of possible negative outcomes and action outcomes may serve to increase survival by generating considerable time and space for carrying out a wide variety of proactive defensive actions to mitigate the threat. "Mentally encountered" threat still necessitates assessment of strike probability and cost to guide subsequent defensive responding, if needed, and therefore may engage similar threat assessment mechanisms (Lang et al. 2000; Moulton and

Kosslyn 2009; Ottaviani et al. 2016). Experimentally, paradigms often examine threat imagery by instructing participants to imagine perceiving an event or object conferring potential danger (e.g., witnessing bodily harm or receiving embarrassing social feedback), while measuring physiological responding (Ji et al. 2016). In such tasks, individuals with anxiety may display significant neural and physiological activity reflecting heightened defensive responding (Ji et al. 2016), while in treatment settings, patients are encouraged to use imaginal exposure as a robust way to evoke powerful trauma-related and anxiogenic states in order to modify pathological defensive responsivity (Minnen and Foa 2006).

More broadly, the *encounter* phase encompasses the resolution of approach-avoid conflicts, in which potential threats and defensive behaviors are considered alongside contrasting approach behaviors invoked by rewarding objects (Cisek 2019; Dranias et al. 2008). Such conflicts are ubiquitous in life, and their resolution entails the selection and pursuit of an adaptive motivated behavior determined by comparing estimated potential threat and reward outcomes (Gray and McNaughton 2000; McNaughton and Corr 2018; Livermore et al. 2021; Averbeck and Murray 2020). A situation characterized by greater estimated threat value may result in passive avoidance, a defensive behavioral response which prioritizes inaction at the cost of forgone rewards; greater reward value will elicit *post-encounter* cautious approach behavior (see below). As noted above, simulation of situations and action outcomes enables the resolution of approach-avoid conflicts far before they physically materialize, allowing individuals to passively avoid rather than engage in situations which confer greater negative relative to positive outcomes (Gray and McNaughton 2000; McNaughton and Corr 2018).

Anxiety Symptoms

Observations of threat-related biases in cognitive processes subserving threat detection and assessment have inspired influential cognitive models of pathological anxiety (Grupe and Nitschke 2013; Eysenck 1992). Empirical findings suggest that anxiety severity may relate to prioritized allocation of attention to, and visual processing of, encountered threat cues (Armstrong and Olatunji 2012; Bar-Haim et al. 2007; Clauss et al. 2022). Such biases toward threat in early attention processes have been framed as hypervigilance (Armstrong and Olatunji 2012; Bar-Haim et al. 2007), although a clearer operational distinction is needed to distinguish *encounter*-related attentional biases from *pre-encounter* vigilance and *post-encounter* sustained attention to threat (Mogg et al. 2004).

Pathological anxiety is also characterized by biases in threat valuation processes, such as inflated estimates of the cost and likelihood of negative outcome occurrence, and the interpretation of ambiguous cues as conferring greater threat value (Grupe and Nitschke 2013; Hazlett-Stevens and Borkovec 2004; Lissek et al. 2014). Such biases may clinically present as pessimistic expectations or catastrophizing cognitions prominently associated with excessive worry and observed across anxiety disorders (Barlow 2002; Grupe and Nitschke 2013; Moulton and Kosslyn 2009; Wells 1997), as well as trigger unnecessary *post-encounter* responding (Marks and Nesse 1994; Nesse 2005) (see below).

Such cognitive biases may extend to mentally simulated, distal threats (Moulton and Kosslyn 2009; Ottaviani et al. 2016) and exacerbated by difficulty in accurate estimation of their potential magnitude and probability (especially in situations of significant uncertainty (Grupe and Nitschke 2013)), relative to situations in which the threat is physically perceived. Extended temporal distance and associated uncertainty may also delay achieving sufficient confidence in threat valuation and selection of effective actions, perpetuating repeated initiation of these processes and contributing to the persistent and intrusive nature of worry in pathological anxiety (Barlow 2002; Wells 1997).

Major learning theories posit that aberrant fear conditioning and extinction of defensive responses to threat cues contribute to pathological anxiety (Mineka and Oehlberg 2008; LeDoux 2000; Milad and Quirk 2012). Such theories are typically examined using fear learning paradigms, which reveal enhanced phasic physiological response and attentive freezing in response to cues signaling potential threat in individuals with elevated anxiety (Roelofs and Dayan 2022; Duits et al. 2015; Ojala and Bach 2020). It should be noted, however, that these models do not typically enable the analytical distinction between *encounter* and *post-encounter* responses; recent studies providing clearer phase separation have provided preliminary evidence for stronger encounter-specific physiological responding in pathological anxiety, occurring prior to threat assessment (Abend et al. 2022).

Finally, excessive passive avoidance is a prominent and impairing feature of pathological anxiety, whereby individuals prefer to relinquish potential rewards (e.g., positive social interactions) given the risk of aversive outcomes (e.g., social rejection) (APA 2013; Livermore et al. 2021; Aupperle and Paulus 2010; McNaughton et al. 2016). These maladaptive behavioral tendencies (especially toward novel objects or situations) are frequently observed as *behavioral inhibition* temperament in children (Kalin 2017), and confer a noted increased risk for pathological anxiety later in life (Clauss and Blackford 2012).

Neural Circuitry

Cross-species research suggests that initial detection of potential threats elicits rapid activation of the sympathetic branch of the autonomic nervous system (Roelofs and Dayan 2022; Ojala and Bach 2020; Bradley 2009), primarily via widespread ascending and descending noradrenergic projections of the locus coeruleus (LC) (Dayan and Yu 2006; Roelofs and Dayan 2022). This phasic physiological responding manifests as increased perspiration, heart rate, breathing, and muscle tone (Mineka and Oehlberg 2008; LeDoux 2000; LeDoux and Daw 2018; Maren 2001), facilitating preparation for immediate mobilization of acute defensive behaviors through LC and amygdala interactions with the hypothalamus, dorsal periaqueductal gray (PAG), and other downstream structures (Dayan and Yu 2006; Roelofs and Dayan 2022; Silva et al. 2016; Fullana et al. 2016). LC and amygdala nuclei further modulate cognitive functions by promoting general cortical excitability, including in visual cortex regions, facilitating perception and attention orientation toward the encountered object (Blanchard et al. 2011; Dayan and Yu 2006; Roelofs and Dayan 2022; Pessoa 2010; Mogg and Bradley 2018).

These activating responses are concomitantly counteracted by the parasympathetic branch, involving basal forebrain, hippocampus, and ventral PAG pathways which facilitate rapid response inhibition and attentive freezing behavior (Roelofs and Dayan 2022; Tovote et al. 2016). This offers the opportunity for more elaborate feature processing and nuanced computation of threat value toward selection of appropriate action, especially when the perceived threat is ambiguous or during approach-avoid conflicts (Blanchard et al. 2011; Blanchard 2017; McNaughton and Corr 2018; Hamm 2020; Roelofs and Dayan 2022; Drabant et al. 2011).

Cross-species research suggests a key role for basolateral and central amygdala, particularly in conjunction with ventromedial prefrontal cortex (vmPFC), in processing of threat-relevant (e.g., conditioned) cues (Fox et al. 2015; Shackman et al. 2016; LeDoux and Daw 2018; Milad and Quirk 2012; Fullana et al. 2016). This circuit is part of a broader network mediating threat valuation and selection of motivated behaviors given internal and external contexts, with key subcortical nodes including the ventral striatum, hypothalamus, and ventral hippocampus (LeDoux 2000; Averbeck and Murray 2020; Pessoa 2010). Extensive research in humans and animal models has also provided evidence for the integration of subcortical input in vmPFC/orbitofrontal cortex and septum, cingulate cortex, and anterior insula during valuation-driven cognitive functions (e.g., encoding of future aversive and appetitive outcomes, conflict monitoring, mediation of approach-avoid conflict resolution, and broader decision making) (Grupe and Nitschke 2013; McNaughton and Corr 2018; Fiddick 2011; Corr 2013; Drabant et al. 2011; Keefer et al. 2021).

Aberrant function in this circuitry has been linked to *encounter*-phase threat-related biases in cognitive processes and pathological anxiety (Grupe and Nitschke 2013; Aupperle and Paulus 2010; McNaughton et al. 2016; Pelletier and Fellows 2021). Most notably, abnormal amygdala connectivity with vmPFC and dorsal PFC has been suggested to underlie perturbed threat assessment (Kenwood et al. 2022; Aupperle and Paulus 2010; Kalin 2017; Etkin et al. 2011; Myers-Schulz and Koenigs 2012; Gold et al. 2016). Additionally, atypical reinforcement-based updating of threat value, in part facilitated by amygdala-vmPFC function, has been linked to anxiety, hypothesized to play a key role in its etiology as well as its treatment (Kenwood et al. 2022; Milad and Quirk 2012).

Mental simulation and imagery of threats have been shown to activate sensory-motor brain regions, impact neural plasticity, and elicit powerful cognitive and emotional responses in a way that at least partially overlaps with the effects of perceived threats (Holmes and Mathews 2010; Reddan et al. 2018; Greening et al. 2022). Extant research on negative imagery and worry cognitions has additionally suggested the involvement of the "default mode network" (DMN), associated with self-referential emotional processes, and its interaction with cortical regions associated with emotion regulation such as the lateral PFC (Weber-Goericke and Muehlhan 2019; Makovac et al. 2020). Additional work implicates amygdala and PAG activation and functional connectivity with DMN hubs and lateral PFC during induced threat imagery and worry (Meeten et al. 2016; Shuhama et al. 2016). Thus, internally generated imagery of negative outcomes appears to involve some of the brain circuitry contributing to processing of perceived threat during *encounter*, but additional research on threat imagery is needed to more strongly link responses to simulated and perceived threats.

5.3 Post-Encounter

Following *encounter*-phase determination of sufficient threat value conferred by an identified threat object, execution of appropriate defensive behaviors to mitigate the danger, given available behavioral options and current goals, takes place during the *post-encounter* phase (Fanselow and Lester 1988; Blanchard et al. 2011; Mobbs et al. 2020). This phase comprises preparation for and eventual execution of the chosen behavior, with cross-species research indicating that the proximity of the threat object plays a key role in determining when behavior is ultimately initiated (Mobbs et al. 2020; Mobbs 2018; Hamm 2020; Wendt et al. 2017; Choi and Kim 2010; Hashemi et al. 2019). Accordingly, as a threat becomes increasingly proximal, changes in physiological responding (e.g., increased skin conductance, heart rate) are observed as the likelihood of behavior initiation rises (Hamm 2020; Abend et al. 2022; Wendt et al. 2017).

A concrete threat of high proximity or magnitude will likely engage acute active avoidance, a prototypical, highly-conserved defensive behavior leading to reduced likelihood of harm through active removal of oneself from the source of imminent danger (Cisek 2019; Gray and McNaughton 2000; Choi and Kim 2010; LeDoux et al. 2017). For example, humans will often create distance from a potential attacker; given adequate space and escape routes, animals will flee an approaching predator (Choi and Kim 2010). During motivational conflict, reward-driven approach behaviors may be intertwined with avoidance, generating cautious approach or risky foraging behavior (Gray and McNaughton 2000; Mobbs and Kim 2015; Choi and Kim 2010; Chou and Smith n.d.).

As noted, humans' evolved ability to mentally simulate distal threats provides opportunities for the execution of more instrumental and deliberative actions to avoid danger (Mobbs et al. 2020; Roelofs and Dayan 2022; LeDoux and Daw 2018; Moulton and Kosslyn 2009; Shuhama et al. 2019; Schubert et al. 2020; Pittig et al. 2020). These behaviors may be slower, more elaborate and nuanced, requiring action simulation and planning to determine an expected outcome (LeDoux and Daw 2018; Pittig et al. 2020). For example, repeated rehearsal of an upcoming public presentation is an active behavior intended to avoid mistakes perceived to result in social embarrassment or rejection. Other complex forms of instrumental avoidance, such as defensive aggression or submissive behaviors, likewise share a goal of diminishing harm by a specific threat. Precursors for such behaviors can be seen in animals, as well as defensive burrowing and other ways to deter threat (Berridge 2019). Extended time and space further allow for iterative initiation of *encounter* and *post-encounter* responses as the threat gradually looms and different mitigating actions are taken; eventually, a shift to acute active avoidance is executed if danger reduction was not achieved and the threat has become highly proximal (Roelofs and Dayan 2022; Livermore et al. 2021).

Anxiety Symptoms
Excessive expression of *post-encounter* defensive responses and their behavioral outcomes is a central feature across anxiety disorders (APA 2013; Barlow 2002; Craske et al. 2009; Davey and Wells 2006), potentially resulting from *encounter*-phase biases, such as inflated threat cost and probability (Grupe and Nitschke 2013), which lead to stronger downstream mobilization of defensive behaviors (Nesse 2005; Arnaudova et al. 2017). Excessive physiological responding to proximal threats is a defining feature of pathological anxiety (APA 2013), and may correspond to instances of preparation for acute active avoidance. Experimental evidence from manipulations of *post-encounter* threat proximity indicates that approaching interoceptive threats (e.g., dyspnea) and exteroceptive threats (e.g., noxious thermal stimulation) induce exaggerated increases in physiological responses and execution of active avoidance in individuals with high relative to low anxiety (Hamm 2020; Abend et al. 2022; Benke et al. 2017).

Pathological anxiety is likewise associated with excessive expression of deliberative, planning-driven forms of active avoidance. For example, excessive preparation for a public speaking event (if one could not be passively avoided) characterizes social anxiety symptoms (APA 2013; Vassilopoulos 2008). Similarly, execution of other forms of avoidance such as defensive aggression or submissive behaviors is observed in excess in pathological anxiety, sometimes to an extent that it disrupts or restricts other normative behaviors (APA 2013; Davis et al. 2010; Ottaviani et al. 2016; Schubert et al. 2020; Russell et al. 2011). Alternatively, rather than actively utilizing instrumental actions to mitigate future threat, individuals may instead engage strategies for effortful emotion regulation (Gross 1998). This may reflect "internal" forms of *post-encounter* active avoidance, to attenuate distress associated with defensive responsivity to simulated, imagined negative outcomes, rather than actions to cope with the threat itself. The increased use of maladaptive regulation strategies (e.g., distraction, emotional avoidance, and suppression) is a characteristic feature of anxiety (Kelly and Forsyth 2009; Schäfer et al. 2017).

Finally, learning theories of anxiety have proposed that excessive and pervasive use of avoidance behavior (both in passive and active forms) prevents determination of (likely) eventual safety, and the integration of this information into future threat assessments (Mowrer and Lamoreaux 1946; Krypotos et al. 2015; Craske et al. 2018). This impedes normative extinction of learned threat contingencies, and the correction of inaccurate beliefs, contributing to the persistence and maintenance of heightened threat responding in pathological anxiety (Blanchard et al. 2011; Pittig et al. 2020; Bouton et al. 2001).

Neural Circuitry
The likelihood of active avoidance execution, scaling with proximity of the threat, is accompanied by changes in autonomic responding, reflecting sympathetic activation (e.g., increasing skin conductance, and muscle tone) alongside inhibitory parasympathetic activation (e.g., hypoventilation, and motor inhibition) (Hamm 2020;

Abend et al. 2022; Wendt et al. 2017; Benke et al. 2017; Klumpers et al. 2017). Parasympathetic dominance acts as a brake to facilitate increased action preparation, while a threshold-based shift to sympathetic dominance (and parasympathetic withdrawal) occurs when behavior execution is deemed necessary, enabling its initiation (Roelofs and Dayan 2022; Drabant et al. 2011; Roelofs 2017; Evans et al. 2018). The maintenance of parasympathetic dominance has been associated with ACC and adjacent mPFC regions (Roelofs and Dayan 2022; Hashemi et al. 2019; Bagur et al. 2021). Cross-species research suggests that the shift from parasympathetic to sympathetic dominance to execute acute active avoidance is mediated by the amygdala and perigenual cingulate (Roelofs and Dayan 2022; Livermore et al. 2021; Hashemi et al. 2019).

Further research in humans and non-human primates suggests a causal link between vmPFC function and the expression of physiological responses to a specific, proximal threat (Kenwood et al. 2022; Kalin 2017; Myers-Schulz and Koenigs 2012; Alexander et al. 2020) and the gating of acute active avoidance, via projections to the ventral striatum, a key structure implicated in executing motivated behaviors, as well as amygdala nuclei, anterior insula, multiple cingulate sectors, and premotor cortices (Tovote et al. 2015; Mobbs et al. 2020; Shackman et al. 2016; Livermore et al. 2021; Diehl et al. 2019). The function of various nodes in this network has been shown to be dynamic and sensitive to threat proximity (Mobbs et al. 2020; Hamm 2020; Adolphs 2013; Meyer et al. 2019), with amygdala, BNST, anterior/mid-cingulate and motor cortices showing increased engagement with perceived increases in threat proximity (Hamm 2020; Wendt et al. 2017; Meyer et al. 2019; Hur et al. 2020; Mobbs et al. 2007; Mobbs et al. 2009; Qi et al. 2018). Furthermore, anxiety severity has been found to moderate amygdala, BNST, and vmPFC connectivity during cued anticipation of outcomes (Clauss et al. 2019; Williams et al. 2015), as well as the association between amygdala-vmPFC intrinsic functional connectivity and the magnitude of proximity-dependent increases in physiological response (Abend et al. 2022).

5.4 Contact

It has been postulated that a key impetus for the substantial cortical expansion across evolution is to increase survival by enabling a greater capacity to anticipate and address primary motivational needs, including coping with threats (Murray et al. 2017). However, when all preventative defensive behaviors failed, and a threat is already in direct contact, the immediate execution of a restricted set of highly conserved, acute, and stereotypic defensive behaviors is invoked, such as disorganized evasive responses, tonic immobility, and defensive attack (Gray and McNaughton 2000; Mobbs et al. 2020; Blanchard and Blanchard 1989; McNaughton and Corr 2004; Hamm 2020; Kalin 2017). While *post-encounter* responses appear to involve proximity-dependent, increasing preparation for active avoidance execution, the acute activation of *contact*-driven defensive behaviors is distinctly all-or-none, implying a threshold that is crossed following only swift and coarse computation (Mobbs et al. 2020; Evans et al. 2018). *Contact* behaviors are accompanied by phasic and systemic sympathetic discharge of physiological responses (e.g., perspiration, hyperventilation, and tachycardia) to support the immediate execution of behavior, supported by only limited cognition (Marks and Nesse 1994; Craske et al. 2009; Hamm et al. 2016).

Anxiety Symptoms
Acute and sudden physiological discharge (e.g., palpitations, accelerated heart rate, dyspnea) and stereotypic defensive behavior (e.g., escape) associated with situations in which danger is perceived as immediate and inevitable are often defined in clinical terms as a panic attack (APA 2013; Barlow 2002; Blanchard 2017; Hamm 2020; White et al. 2006). Panic attacks are a common feature of anxiety disorders (APA 2013), and given the conditions that evoke them and their features, they may reflect the exaggerated expression of *contact* responses (Nesse 2005).

Given that physiological responding is the most prominent clinical feature of panic attacks (APA 2013), it may be conjectured that behavioral aspects are often inhibited. In situations where *Contact* defensive behaviors are carried out, they are differentially expressed as a function of available behavioral options. Studies in individuals with anxiety have demonstrated excessive rates of escape behavior in the face of immediate but escapable threat (Hamm 2020; Hamm et al. 2016), whereas tonic immobility (or defensive immobilization) has been seen in anxiety in situations where threat contact is perceived as imminent and inescapable (Moskowitz 2004). Defensive attack may correspond to triggered reactive aggression and rage episodes observed in individuals with anxiety and related conditions such as post-traumatic stress disorder (APA 2013), and may reflect instances where attack is perceived to be potentially effective in the immediate term.

Neural Circuitry
Cross-species expression of *contact* phase behaviors suggests the involvement of highly conserved circuitry. Research primarily implicates specific PAG columns (via amygdala and hypothalamus paths) in directly generating escape, tonic immobility, and aggression behaviors across species (Mobbs et al. 2020; McNaughton and Corr 2004; Evans et al. 2018; Silva and McNaughton 2019). Electrical stimulation of the PAG in humans evokes acute physiological responding, as well as symptoms which closely resemble those of panic attacks such as feelings of fear of death and desire to flee (Schenberg et al. 2014). Further, both structural and neurochemical PAG abnormalities are associated with a propensity for panic attacks in patients (Del-Ben and Graeff 2009). The PAG also receives projections from the ACC and insula, regions implicated in threat assessment (Faull and Pattinson 2017), potentially enabling cortical modulation of *contact* behavioral responding.

6 Summary

Considerable cross-species research provides evidence for a dynamic organization of defensive responses by threat imminence, subserved by partially conserved neural circuitry. According to the proposed framework, pathological anxiety reflects a tendency for exaggerated expression of what would otherwise be normative defensive responses (Fanselow and Lester 1988; Rosen and Schulkin 1998). As such, the expression of specific anxiety symptoms is not necessarily linked to *what* the threat is (which guides current nosology) but rather to *where* the threat is relative to the individual, in terms of physical or temporal proximity. This framework therefore provides a plausible organization of anxiety symptom expression that is grounded in extensively studied, conserved neurobiological mechanisms (Fig. 1).

7 Implications

The limitations of phenomenology-driven, category-based anxiety diagnosis and insufficient understanding of symptom chronicity pose a significant challenge for clinical application and its improvement through systematic mechanistic research. Grounding anxiety symptoms in conserved, threat imminence-dependent defensive responding can begin to address some of these concerns.

Clinically, a temporal organization of anxiety symptoms that derives from a conserved pattern of imminence-dependent defensive responding may attenuate comorbidity and other issues related to diagnostic boundaries. If pathological anxiety reflects a general tendency for enhanced defensive responsivity (Rosen and Schulkin 1998), then different symptoms are expected to be temporarily expressed at different imminence phases, while the temporal organization of symptoms is expected to be relatively similar for different types of threats. This reduces the need for multiple diagnoses and instead highlights "core" anxiety symptomatology, in line with recent findings (Abend et al. 2021c). Further, a clearer understanding of symptom dynamics enables both clinicians and patients to better anticipate and thus target specific cognitive, physiological, and behavioral symptoms in accordance with their expected presentation as a function of the imminence of a relevant, identified threat. This may serve to complement and enhance existing treatments, in line with models that integrate different treatment modules according to symptom presentation (Weisz et al. 2012; Evans et al. 2021).

Finally, the proposed view of anxiety is intended to complement and integrate existing models of pathological anxiety across different disciplines. As such, it complements biologically focused, evolutionary approaches that emphasize observable cross-species behavior and neurobiological measures such as Research Domain Criteria (RDoC) (Cuthbert 2022; Insel et al. 2010), reductionist classification approaches seeking to identify shared, core symptomatology such as the Hierarchical Taxonomy of Psychopathology (HiTOP) (Kotov et al. 2021), cognitive and motivational-behavioral models of anxiety (Barlow 2002; Gray and McNaughton 2000; Wells 1997; McNaughton et al. 2016), as well as classic psychodynamic theories on the role of fear and defense mechanisms in psychopathology (Freud 1935).

From a research perspective, limited mechanistic insight on pathophysiological bases for anxiety symptoms may be due, at least in part, to experimental paradigms or common analytic approaches that conflate evoked, phase-specific responses (Abend 2023; Abend et al. 2022). For example, paradigms designed to elicit "threat processing processes" often quantify responses to threat cues through aggregate indices (e.g., of physiological or neural responses) derived across a window of several seconds (Lonsdorf et al. 2017); such indices may actually encompass elements of *pre-encounter*, *encounter*, and *post-encounter*-driven responses, and obscure links to specific symptoms (Abend 2023). Experimentally accounting for distinct patterns of defensive responses along the threat imminence continuum may help bridge this gap, enabling more precise quantification of symptom-relevant responses and addressing inconsistent findings observed across studies (Abend 2023; Abend et al. 2020; Duits et al. 2015; Lissek et al. 2005). Importantly, grounding symptoms in specific, conserved defensive responses allows for more effective leveraging of translational mechanistic research, as conserved patterns of defensive responses may track more closely with function in specific neural circuits (Kenwood et al. 2022; Robinson et al. 2019). The isolation of reliable cross-species biomarkers may subsequently facilitate neuroscience-guided development of novel, targeted, therapeutic interventions (Xia and Kheirbek 2020).

References

Abend R (2023) Understanding anxiety symptoms as aberrant defensive responding along the threat imminence continuum. Neurosci Biobehav Rev 152:105305. https://doi.org/10.1016/j.neubiorev.2023.105305

Abend R, Gold AL, Britton JC et al (2020) Anticipatory threat responding: associations with anxiety, development, and brain structure. Biol Psychiatry 87(10):916–925. https://doi.org/10.1016/j.biopsych.2019.11.006

Abend R, Bajaj MA, Matsumoto C et al (2021a) Converging multi-modal evidence for implicit threat-related bias in pediatric anxiety disorders. Res Child Adolesc Psychopathol 49(2): 227–240. https://doi.org/10.1007/s10802-020-00712-w

Abend R, Bajaj MA, Harrewijn A et al (2021b) Threat-anticipatory psychophysiological response is enhanced in youth with anxiety disorders and correlates with prefrontal cortex neuroanatomy. J Psychiatry Neurosci 46(2). https://doi.org/10.1503/jpn.200110

Abend R, Bajaj MA, Coppersmith DDL et al (2021c) A computational network perspective on pediatric anxiety symptoms. Psychol Med 51(10):1752–1762. https://doi.org/10.1017/S0033291720000501

Abend R, Ruiz SG, Bajaj MA et al (2022) Threat imminence reveals links among unfolding of anticipatory physiological response, cortical-subcortical intrinsic functional connectivity, and anxiety. Neurobiol Stress 16:100428. https://doi.org/10.1016/j.ynstr.2022.100428

Adolphs R (2009) The social brain: neural basis of social knowledge. Annu Rev Psychol 60(1): 693–716. https://doi.org/10.1146/annurev.psych.60.110707.163514

Adolphs R (2013) The biology of fear. Curr Biol 23(2):R79–R93. https://doi.org/10.1016/j.cub.2012.11.055

Alexander L, Wood CM, Gaskin PLR et al (2020) Over-activation of primate subgenual cingulate cortex enhances the cardiovascular, behavioral and neural responses to threat. Nat Commun 11(1):5386. https://doi.org/10.1038/s41467-020-19167-0

Alvarez RP, Kirlic N, Misaki M et al (2015) Increased anterior insula activity in anxious individuals is linked to diminished perceived control. Transl Psychiatry 5(6):e591–e591. https://doi.org/10.1038/tp.2015.84

APA (2013) Diagnostic and statistical manual of mental disorders: DSM-5, 5th edn. American Psychiatric Association

Armstrong T, Olatunji BO (2012) Eye tracking of attention in the affective disorders: a meta-analytic review and synthesis. Clin Psychol Rev 32(8):704–723. https://doi.org/10.1016/j.cpr.2012.09.004

Arnaudova I, Kindt M, Fanselow M, Beckers T (2017) Pathways towards the proliferation of avoidance in anxiety and implications for treatment. Behav Res Ther 96:3–13. https://doi.org/10.1016/j.brat.2017.04.004

Aupperle RL, Paulus MP (2010) Neural systems underlying approach and avoidance in anxiety disorders. Dialogues Clin Neurosci 12(4):517–531. https://doi.org/10.31887/DCNS.2010.12.4/raupperle

Averbeck BB, Murray EA (2020) Hypothalamic interactions with large-scale neural circuits underlying reinforcement learning and motivated behavior. Trends Neurosci 43(9):681–694. https://doi.org/10.1016/j.tins.2020.06.006

Avery SN, Clauss JA, Blackford JU (2016) The human BNST: functional role in anxiety and addiction. Neuropsychopharmacology 41(1):126–141. https://doi.org/10.1038/npp.2015.185

Bagur S, Lefort JM, Lacroix MM et al (2021) Breathing-driven prefrontal oscillations regulate maintenance of conditioned-fear evoked freezing independently of initiation. Nat Commun 12(1):2605. https://doi.org/10.1038/s41467-021-22798-6

Balderston NL, Beydler EM, Roberts C et al (2020) Mechanistic link between right prefrontal cortical activity and anxious arousal revealed using transcranial magnetic stimulation in healthy subjects. Neuropsychopharmacology 45(4):694–702. https://doi.org/10.1038/s41386-019-0583-5

Bandelow B, Michaelis S, Wedekind D (2017) Treatment of anxiety disorders. Dialogues Clin Neurosci 19(2):93–107. https://doi.org/10.31887/DCNS.2017.19.2/bbandelow

Bandelow B, Allgulander C, Baldwin DS et al (2023) World Federation of Societies of biological psychiatry (WFSBP) guidelines for treatment of anxiety, obsessive-compulsive and posttraumatic stress disorders – version 3. Part I: anxiety disorders. World J Biol Psychiatry 24(2):79–117. https://doi.org/10.1080/15622975.2022.2086295

Bar-Haim Y, Lamy D, Pergamin L, Bakermans-Kranenburg MJ, van IJzendoorn MH. (2007) Threat-related attentional bias in anxious and nonanxious individuals: a meta-analytic study. Psychol Bull 133(1):1–24. https://doi.org/10.1037/0033-2909.133.1.1

Barlow DH (2002) Anxiety and its disorders: the nature and treatment of anxiety and panic. The Guilford Press

Beauchaine TP, Zisner A (2017) Motivation, emotion regulation, and the latent structure of psychopathology: an integrative and convergent historical perspective. Int J Psychophysiol 119:108–118. https://doi.org/10.1016/j.ijpsycho.2016.12.014

Benke C, Hamm AO, Pané-Farré CA (2017) When dyspnea gets worse: suffocation fear and the dynamics of defensive respiratory responses to increasing interoceptive threat. Psychophysiology 54(9):1266–1283. https://doi.org/10.1111/psyp.12881

Berridge KC (2019) A liking versus wanting perspective on emotion and the brain. In: Gruber J (ed) The Oxford handbook of positive emotion and psychopathology. Oxford University Press, pp 184–196

Bijsterbosch J, Smith S, Forster S, John OP, Bishop SJ (2014) Resting state correlates of subdimensions of anxious affect. J Cogn Neurosci 26(4):914–926. https://doi.org/10.1162/jocn_a_00512

Blanchard DC (2017) Translating dynamic defense patterns from rodents to people. Neurosci Biobehav Rev 76(Pt A):22–28. https://doi.org/10.1016/j.neubiorev.2016.11.001

Blanchard RJ, Blanchard DC (1989) Attack and defense in rodents as ethoexperimental models for the study of emotion. Prog Neuropsychopharmacol Biol Psychiatry 13(Suppl):S3–S14. https://doi.org/10.1016/0278-5846(89)90105-x

Blanchard DC, Griebel G, Pobbe R, Blanchard RJ (2011) Risk assessment as an evolved threat detection and analysis process. Neurosci Biobehav Rev 35(4):991–998. https://doi.org/10.1016/j.neubiorev.2010.10.016

Bolles RC, Fanselow MS (1980) A perceptual-defensive-recuperative model of fear and pain. Behav Brain Sci 3(2):291–301. https://doi.org/10.1017/S0140525X0000491X

Bouton ME, Mineka S, Barlow DH (2001) A modern learning theory perspective on the etiology of panic disorder. Psychol Rev 108(1):4–32. https://doi.org/10.1037/0033-295x.108.1.4

Bradley MM (2009) Natural selective attention: orienting and emotion. Psychophysiology 46(1):1–11. https://doi.org/10.1111/j.1469-8986.2008.00702.x

Cavanagh JF, Shackman AJ (2015) Frontal midline theta reflects anxiety and cognitive control: meta-analytic evidence. J Physiol Paris 109(1–3):3–15. https://doi.org/10.1016/j.jphysparis.2014.04.003

Chavanne AV, Robinson OJ (2021) The overlapping neurobiology of induced and pathological anxiety: a meta-analysis of functional neural activation. Am J Psychiatry 178(2):156–164. https://doi.org/10.1176/appi.ajp.2020.19111153

Choi JS, Kim JJ (2010) Amygdala regulates risk of predation in rats foraging in a dynamic fear environment. Proc Natl Acad Sci U S A 107(50):21773–21777. https://doi.org/10.1073/pnas.1010079108

Chou K, Smith R (n.d.) Computational mechanisms of information-seeking in anxiety

Cisek P (2019) Resynthesizing behavior through phylogenetic refinement. Atten Percept Psychophys 81(7):2265–2287. https://doi.org/10.3758/s13414-019-01760-1

Clauss JA, Blackford JU (2012) Behavioral inhibition and risk for developing social anxiety disorder: a meta-analytic study. J Am Acad Child Adolesc Psychiatry 51(10):1066–1075.e1. https://doi.org/10.1016/j.jaac.2012.08.002

Clauss JA, Avery SN, Benningfield MM, Blackford JU (2019) Social anxiety is associated with BNST response to unpredictability. Depress Anxiety 36(8):666–675. https://doi.org/10.1002/da.22891

Clauss K, Gorday JY, Bardeen JR (2022) Eye tracking evidence of threat-related attentional bias in anxiety- and fear-related disorders: a systematic review and meta-analysis. Clin Psychol Rev 93:102142. https://doi.org/10.1016/j.cpr.2022.102142

Conway CC, Forbes MK, Forbush KT et al (2019) A hierarchical taxonomy of psychopathology can transform mental health research. Perspect Psychol Sci J Assoc Psychol Sci 14(3):419–436. https://doi.org/10.1177/1745691618810696

Cornwell BR, Garrido MI, Overstreet C, Pine DS, Grillon C (2017) The unpredictive brain under threat: a neurocomputational account of anxious hypervigilance. Biol Psychiatry 82(6):447–454. https://doi.org/10.1016/j.biopsych.2017.06.031

Corr PJ (2013) Approach and avoidance behaviour: multiple systems and their interactions. Emot Rev 5(3):285–290. https://doi.org/10.1177/1754073913477507

Craske MG, Rauch SL, Ursano R, Prenoveau J, Pine DS, Zinbarg RE (2009) What is an anxiety disorder? Depress Anxiety 26(12):1066–1085. https://doi.org/10.1002/da.20633

Craske MG, Hermans D, Vervliet B (2018) State-of-the-art and future directions for extinction as a translational model for fear and anxiety. Philos Trans R Soc Lond B Biol Sci 373(1742):20170025. https://doi.org/10.1098/rstb.2017.0025

Cuthbert BN (2022) Research domain criteria (RDoC): Progress and potential. Curr Dir Psychol Sci 31(2):107–114. https://doi.org/10.1177/09637214211051363

Cuthbert BN, Insel TR (2013) Toward the future of psychiatric diagnosis: the seven pillars of RDoC. BMC Med 11:126. https://doi.org/10.1186/1741-7015-11-126

Davey GCL, Wells A (eds) (2006) Worry and its psychological disorders: theory, assessment and treatment, 1st edn. Wiley. https://doi.org/10.1002/9780470713143

Davis M, Walker DL, Miles L, Grillon C (2010) Phasic vs sustained fear in rats and humans: role of the extended amygdala in fear vs anxiety. Neuropsychopharmacology 35(1):105–135. https://doi.org/10.1038/npp.2009.109

Dayan P, Yu AJ (2006) Phasic norepinephrine: a neural interrupt signal for unexpected events. Netw Bristol Engl 17(4):335–350. https://doi.org/10.1080/09548980601004024

Del-Ben CM, Graeff FG (2009) Panic disorder: is the PAG involved? Neural Plast 2009:108135. https://doi.org/10.1155/2009/108135

Diehl MM, Bravo-Rivera C, Quirk GJ (2019) The study of active avoidance: a platform for discussion. Neurosci Biobehav Rev 107:229–237. https://doi.org/10.1016/j.neubiorev.2019.09.010

Dixon ML, Andrews-Hanna JR, Spreng RN et al (2017) Interactions between the default network and dorsal attention network vary across default subsystems, time, and cognitive states. Neuroimage 147:632–649. https://doi.org/10.1016/j.neuroimage.2016.12.073

Domschke K, Seuling PD, Schiele MA et al (2024) The definition of treatment resistance in anxiety disorders: a Delphi method-based consensus guideline. World Psychiatry 23(1):113–123. https://doi.org/10.1002/wps.21177

Drabant EM, Kuo JR, Ramel W et al (2011) Experiential, autonomic, and neural responses during threat anticipation vary as a function of threat intensity and neuroticism. Neuroimage 55(1):401–410. https://doi.org/10.1016/j.neuroimage.2010.11.040

Dranias MR, Grossberg S, Bullock D (2008) Dopaminergic and non-dopaminergic value systems in conditioning and outcome-specific revaluation. Brain Res 1238:239–287. https://doi.org/10.1016/j.brainres.2008.07.013

Duits P, Cath DC, Lissek S et al (2015) Updated meta-analysis of classical fear conditioning in the anxiety disorders. Depress Anxiety 32(4):239–253. https://doi.org/10.1002/da.22353

Etkin A, Wager TD (2007) Functional neuroimaging of anxiety: a meta-analysis of emotional processing in PTSD, social anxiety disorder, and specific phobia. Am J Psychiatry 164(10):1476–1488. https://doi.org/10.1176/appi.ajp.2007.07030504

Etkin A, Egner T, Kalisch R (2011) Emotional processing in anterior cingulate and medial prefrontal cortex. Trends Cogn Sci 15(2):85–93. https://doi.org/10.1016/j.tics.2010.11.004

Evans DA, Stempel AV, Vale R, Ruehle S, Lefler Y, Branco T (2018) A synaptic threshold mechanism for computing escape decisions. Nature 558(7711):590–594. https://doi.org/10.1038/s41586-018-0244-6

Evans SC, Wei MA, Harmon SL, Weisz JR (2021) Modular psychotherapy outcomes for youth with different latent profiles of irritability and emotion dysregulation. Front Psych 12:618455. https://doi.org/10.3389/fpsyt.2021.618455

Eysenck MW (1992) Anxiety: the cognitive perspective. Psychology Press

Fanselow MS, Lester LS (1988) A functional behavioristic approach to aversively motivated behavior: predatory imminence as a determinant of the topography of defensive behavior. In: Bolles RC, Beecher MD (eds) Evolution and learning. Lawrence Erlbaum Associates, Inc, pp 185–212

Faull OK, Pattinson KT (2017) The cortical connectivity of the periaqueductal gray and the conditioned response to the threat of breathlessness. Elife 6:e21749. https://doi.org/10.7554/eLife.21749

Feola B, McHugo M, Armstrong K et al (2021) BNST and amygdala connectivity are altered during threat anticipation in schizophrenia. Behav Brain Res 412:113428. https://doi.org/10.1016/j.bbr.2021.113428

Fiddick L (2011) There is more than the amygdala: potential threat assessment in the cingulate cortex. Neurosci Biobehav Rev 35(4):1007–1018. https://doi.org/10.1016/j.neubiorev.2010.09.014

Fox AS, Shackman AJ (2019) The central extended amygdala in fear and anxiety: closing the gap between mechanistic and neuroimaging research. Neurosci Lett 693:58–67. https://doi.org/10.1016/j.neulet.2017.11.056

Fox AS, Oler JA, Tromp DPM, Fudge JL, Kalin NH (2015) Extending the amygdala in theories of threat processing. Trends Neurosci 38(5):319–329. https://doi.org/10.1016/j.tins.2015.03.002

Freud S (1935) Inhibitions, symptoms and anxiety. Psychoanal Q 4(4):616–625. https://doi.org/10.1080/21674086.1935.11925261

Fullana MA, Harrison BJ, Soriano-Mas C et al (2016) Neural signatures of human fear conditioning: an updated and extended meta-analysis of fMRI studies. Mol Psychiatry 21(4):500–508. https://doi.org/10.1038/mp.2015.88

Gold AL, Shechner T, Farber MJ et al (2016) Amygdala-cortical connectivity: associations with anxiety, development, and threat. Depress Anxiety 33(10):917–926. https://doi.org/10.1002/da.22470

Gray LA, McNaughton N (2000) The neuropsychology of anxiety. Oxford University Press

Greening SG, Lee TH, Burleigh L et al (2022) Mental imagery can generate and regulate acquired differential fear conditioned reactivity. Sci Rep 12(1):997. https://doi.org/10.1038/s41598-022-05019-y

Grillon C (2002) Startle reactivity and anxiety disorders: aversive conditioning, context, and neurobiology. Biol Psychiatry 52(10):958–975. https://doi.org/10.1016/s0006-3223(02)01665-7

Grillon C (2008) Models and mechanisms of anxiety: evidence from startle studies. Psychopharmacology (Berl) 199(3):421–437. https://doi.org/10.1007/s00213-007-1019-1

Grillon C, Baas JMP, Cornwell B, Johnson L (2006) Context conditioning and behavioral avoidance in a virtual reality environment: effect of predictability. Biol Psychiatry 60(7):752–759. https://doi.org/10.1016/j.biopsych.2006.03.072

Grillon C, Robinson OJ, Cornwell B, Ernst M (2019) Modeling anxiety in healthy humans: a key intermediate bridge between basic and clinical sciences. Neuropsychopharmacology 44(12):1999–2010. https://doi.org/10.1038/s41386-019-0445-1

Gross JJ (1998) Antecedent- and response-focused emotion regulation: divergent consequences for experience, expression, and physiology. J Pers Soc Psychol 74(1):224–237. https://doi.org/10.1037/0022-3514.74.1.224

Gross CT, Canteras NS (2012) The many paths to fear. Nat Rev Neurosci 13(9):651–658. https://doi.org/10.1038/nrn3301

Grupe DW, Nitschke JB (2013) Uncertainty and anticipation in anxiety: an integrated neurobiological and psychological perspective. Nat Rev Neurosci 14(7):488–501. https://doi.org/10.1038/nrn3524

Hamm AO (2020) Fear, anxiety, and their disorders from the perspective of psychophysiology. Psychophysiology 57(2):e13474. https://doi.org/10.1111/psyp.13474

Hamm AO, Richter J, Pané-Farré C et al (2016) Panic disorder with agoraphobia from a behavioral neuroscience perspective: applying the research principles formulated by the research domain criteria (RDoC) initiative. Psychophysiology 53(3):312–322. https://doi.org/10.1111/psyp.12553

Hashemi MM, Gladwin TE, de Valk NM et al (2019) Neural dynamics of shooting decisions and the switch from freeze to fight. Sci Rep 9(1):4240. https://doi.org/10.1038/s41598-019-40917-8

Hazlett-Stevens H, Borkovec TD (2004) Interpretive cues and ambiguity in generalized anxiety disorder. Behav Res Ther 42(8):881–892. https://doi.org/10.1016/S0005-7967(03)00204-3

Holmes EA, Mathews A (2010) Mental imagery in emotion and emotional disorders. Clin Psychol Rev 30(3):349–362. https://doi.org/10.1016/j.cpr.2010.01.001

Hur J, Smith JF, DeYoung KA et al (2020) Anxiety and the neurobiology of temporally uncertain threat anticipation. J Neurosci 40(41):7949–7964. https://doi.org/10.1523/JNEUROSCI.0704-20.2020

Insel T, Cuthbert B, Garvey M et al (2010) Research domain criteria (RDoC): toward a new classification framework for research on mental disorders. Am J Psychiatry 167(7):748–751. https://doi.org/10.1176/appi.ajp.2010.09091379

Ji JL, Heyes SB, MacLeod C, Holmes EA (2016) Emotional mental imagery as simulation of reality: fear and beyond-a tribute to Peter Lang. Behav Ther 47(5):702–719. https://doi.org/10.1016/j.beth.2015.11.004

Kalin NH (2017) Mechanisms underlying the early risk to develop anxiety and depression: a translational approach. Eur Neuropsychopharmacol 27(6):543–553. https://doi.org/10.1016/j.euroneuro.2017.03.004

Keefer SE, Gyawali U, Calu DJ (2021) Choose your path: divergent basolateral amygdala efferents differentially mediate incentive motivation, flexibility and decision-making. Behav Brain Res 409:113306. https://doi.org/10.1016/j.bbr.2021.113306

Kelly MM, Forsyth JP (2009) Associations between emotional avoidance, anxiety sensitivity, and reactions to an observational fear challenge procedure. Behav Res Ther 47(4):331–338. https://doi.org/10.1016/j.brat.2009.01.008

Kenwood MM, Kalin NH, Barbas H (2022) The prefrontal cortex, pathological anxiety, and anxiety disorders. Neuropsychopharmacology 47(1):260–275. https://doi.org/10.1038/s41386-021-01109-z

Kessler RC, Wang PS (2008) The descriptive epidemiology of commonly occurring mental disorders in the United States. Annu Rev Public Health 29:115–129. https://doi.org/10.1146/annurev.publhealth.29.020907.090847

Klumpers F, Kroes MCW, Baas JMP, Fernández G (2017) How human amygdala and bed nucleus of the Stria Terminalis may drive distinct defensive responses. J Neurosci 37(40):9645–9656. https://doi.org/10.1523/JNEUROSCI.3830-16.2017

Kotov R, Krueger RF, Watson D et al (2017) The hierarchical taxonomy of psychopathology (HiTOP): a dimensional alternative to traditional nosologies. J Abnorm Psychol 126(4): 454–477. https://doi.org/10.1037/abn0000258

Kotov R, Krueger RF, Watson D et al (2021) The hierarchical taxonomy of psychopathology (HiTOP): a quantitative nosology based on consensus of evidence. Annu Rev Clin Psychol 17: 83–108. https://doi.org/10.1146/annurev-clinpsy-081219-093304

Kotov R, Cicero DC, Conway CC et al (2022) The hierarchical taxonomy of psychopathology (HiTOP) in psychiatric practice and research. Psychol Med 52(9):1666–1678. https://doi.org/10.1017/S0033291722001301

Krueger RF, DeYoung CG (2020) The indispensable value of a coherent phenotypic model of psychopathology. Biol Psychiatry 88(1):6–8. https://doi.org/10.1016/j.biopsych.2020.04.010

Krypotos AM, Effting M, Kindt M, Beckers T (2015) Avoidance learning: a review of theoretical models and recent developments. Front Behav Neurosci 9:189. https://doi.org/10.3389/fnbeh.2015.00189

Lago T, Davis A, Grillon C, Ernst M (2017) Striatum on the anxiety map: small detours into adolescence. Brain Res 1654(Pt B):177–184. https://doi.org/10.1016/j.brainres.2016.06.006

Lang PJ, Davis M, Ohman A (2000) Fear and anxiety: animal models and human cognitive psychophysiology. J Affect Disord 61(3):137–159. https://doi.org/10.1016/s0165-0327(00)00343-8

LeDoux JE (2000) Emotion circuits in the brain. Annu Rev Neurosci 23:155–184. https://doi.org/10.1146/annurev.neuro.23.1.155

LeDoux J (2012) Rethinking the emotional brain. Neuron 73(4):653–676. https://doi.org/10.1016/j.neuron.2012.02.004

LeDoux J, Daw ND (2018) Surviving threats: neural circuit and computational implications of a new taxonomy of defensive behaviour. Nat Rev Neurosci 19(5):269–282. https://doi.org/10.1038/nrn.2018.22

LeDoux JE, Moscarello J, Sears R, Campese V (2017) The birth, death and resurrection of avoidance: a reconceptualization of a troubled paradigm. Mol Psychiatry 22(1):24–36. https://doi.org/10.1038/mp.2016.166

Lissek S, Powers AS, McClure EB et al (2005) Classical fear conditioning in the anxiety disorders: a meta-analysis. Behav Res Ther 43(11):1391–1424. https://doi.org/10.1016/j.brat.2004.10.007

Lissek S, Kaczkurkin AN, Rabin S, Geraci M, Pine DS, Grillon C (2014) Generalized anxiety disorder is associated with overgeneralization of classically conditioned fear. Biol Psychiatry 75(11):909–915. https://doi.org/10.1016/j.biopsych.2013.07.025

Livermore JJA, Klaassen FH, Bramson B et al (2021) Approach-avoidance decisions under threat: the role of autonomic psychophysiological states. Front Neurosci 15:621517. https://doi.org/10.3389/fnins.2021.621517

Lonsdorf TB, Menz MM, Andreatta M et al (2017) Don't fear "fear conditioning": methodological considerations for the design and analysis of studies on human fear acquisition, extinction, and return of fear. Neurosci Biobehav Rev 77:247–285. https://doi.org/10.1016/j.neubiorev.2017.02.026

Makovac E, Fagioli S, Rae CL, Critchley HD, Ottaviani C (2020) Can't get it off my brain: meta-analysis of neuroimaging studies on perseverative cognition. Psychiatry Res Neuroimaging 295:111020. https://doi.org/10.1016/j.pscychresns.2019.111020

Maren S (2001) Neurobiology of Pavlovian fear conditioning. Annu Rev Neurosci 24:897–931. https://doi.org/10.1146/annurev.neuro.24.1.897

Maren S, Phan KL, Liberzon I (2013) The contextual brain: implications for fear conditioning, extinction and psychopathology. Nat Rev Neurosci 14(6):417–428. https://doi.org/10.1038/nrn3492

Marks IFM, Nesse RM (1994) Fear and fitness: an evolutionary analysis of anxiety disorders. Ethol Sociobiol 15(5–6):247–261. https://doi.org/10.1016/0162-3095(94)90002-7

McNaughton N, Corr PJ (2004) A two-dimensional neuropsychology of defense: fear/anxiety and defensive distance. Neurosci Biobehav Rev 28(3):285–305. https://doi.org/10.1016/j.neubiorev.2004.03.005

McNaughton N, Corr PJ (2018) Survival circuits and risk assessment. Curr Opin Behav Sci 24:14–20. https://doi.org/10.1016/j.cobeha.2018.01.018

McNaughton N, DeYoung CG, Corr PJ (2016) Approach/avoidance. In: Neuroimaging personality, social cognition, and character. Elsevier, pp 25–49. https://doi.org/10.1016/B978-0-12-800935-2.00002-6

Meeten F, Davey GCL, Makovac E et al (2016) Goal directed worry rules are associated with distinct patterns of amygdala functional connectivity and vagal modulation during perseverative cognition. Front Hum Neurosci 10:553. https://doi.org/10.3389/fnhum.2016.00553

Menon V, Uddin LQ (2010) Saliency, switching, attention and control: a network model of insula function. Brain Struct Funct 214(5–6):655–667. https://doi.org/10.1007/s00429-010-0262-0

Meyer C, Padmala S, Pessoa L (2019) Dynamic threat processing. J Cogn Neurosci 31(4):522–542. https://doi.org/10.1162/jocn_a_01363

Milad MR, Quirk GJ (2012) Fear extinction as a model for translational neuroscience: ten years of progress. Annu Rev Psychol 63:129–151. https://doi.org/10.1146/annurev.psych.121208.131631

Mineka S, Oehlberg K (2008) The relevance of recent developments in classical conditioning to understanding the etiology and maintenance of anxiety disorders. Acta Psychol (Amst) 127(3):567–580. https://doi.org/10.1016/j.actpsy.2007.11.007

Minnen AV, Foa EB (2006) The effect of imaginal exposure length on outcome of treatment for PTSD. J Trauma Stress 19(4):427–438. https://doi.org/10.1002/jts.20146

Mkrtchian A, Roiser JP, Robinson OJ (2017) Threat of shock and aversive inhibition: induced anxiety modulates Pavlovian-instrumental interactions. J Exp Psychol Gen 146(12):1694–1704. https://doi.org/10.1037/xge0000363

Mobbs D (2018) The ethological deconstruction of fear(s). Curr Opin Behav Sci 24:32–37. https://doi.org/10.1016/j.cobeha.2018.02.008

Mobbs D, Kim JJ (2015) Neuroethological studies of fear, anxiety, and risky decision-making in rodents and humans. Curr Opin Behav Sci 5:8–15. https://doi.org/10.1016/j.cobeha.2015.06.005

Mobbs D, Petrovic P, Marchant JL et al (2007) When fear is near: threat imminence elicits prefrontal-periaqueductal gray shifts in humans. Science 317(5841):1079–1083. https://doi.org/10.1126/science.1144298

Mobbs D, Marchant JL, Hassabis D et al (2009) From threat to fear: the neural organization of defensive fear systems in humans. J Neurosci 29(39):12236–12243. https://doi.org/10.1523/JNEUROSCI.2378-09.2009

Mobbs D, Headley DB, Ding W, Dayan P (2020) Space, time, and fear: survival computations along defensive circuits. Trends Cogn Sci 24(3):228–241. https://doi.org/10.1016/j.tics.2019.12.016

Mogg K, Bradley BP (2018) Anxiety and threat-related attention: cognitive-motivational framework and treatment. Trends Cogn Sci 22(3):225–240. https://doi.org/10.1016/j.tics.2018.01.001

Mogg K, Bradley B, Miles F, Dixon R (2004) Brief report time course of attentional bias for threat scenes: testing the vigilance-avoidance hypothesis. Cognit Emot 18(5):689–700. https://doi.org/10.1080/02699930341000158

Moskowitz AK (2004) "Scared stiff": catatonia as an evolutionary-based fear response. Psychol Rev 111(4):984–1002. https://doi.org/10.1037/0033-295X.111.4.984

Moulton ST, Kosslyn SM (2009) Imagining predictions: mental imagery as mental emulation. Philos Trans R Soc Lond B Biol Sci 364(1521):1273–1280. https://doi.org/10.1098/rstb.2008.0314

Mowrer OH, Lamoreaux RR (1946) Fear as an intervening variable in avoidance conditioning. J Comp Psychol 39:29–50. https://doi.org/10.1037/h0060150

Murray EA, Wise SP, Graham KS (2017) The evolution of memory systems: ancestors, anatomy, and adaptations. First published in paperback. Oxford University Press

Myers-Schulz B, Koenigs M (2012) Functional anatomy of ventromedial prefrontal cortex: implications for mood and anxiety disorders. Mol Psychiatry 17(2):132–141. https://doi.org/10.1038/mp.2011.88

Nesse RM (2005) Natural selection and the regulation of defenses. Evol Hum Behav 26(1):88–105. https://doi.org/10.1016/j.evolhumbehav.2004.08.002

Ojala KE, Bach DR (2020) Measuring learning in human classical threat conditioning: translational, cognitive and methodological considerations. Neurosci Biobehav Rev 114:96–112. https://doi.org/10.1016/j.neubiorev.2020.04.019

Ottaviani C, Thayer JF, Verkuil B et al (2016) Physiological concomitants of perseverative cognition: a systematic review and meta-analysis. Psychol Bull 142(3):231–259. https://doi.org/10.1037/bul0000036

Pelletier G, Fellows LK (2021) Viewing orbitofrontal cortex contributions to decision-making through the lens of object recognition. Behav Neurosci 135(2):182–191. https://doi.org/10.1037/bne0000447

Perusini JN, Fanselow MS (2015) Neurobehavioral perspectives on the distinction between fear and anxiety. Learn Mem 22(9):417–425. https://doi.org/10.1101/lm.039180.115

Pessoa L (2010) Emotion and cognition and the amygdala: from "what is it?" to "what's to be done?". Neuropsychologia 48(12):3416–3429. https://doi.org/10.1016/j.neuropsychologia.2010.06.038

Pittig A, Wong AHK, Glück VM, Boschet JM (2020) Avoidance and its bi-directional relationship with conditioned fear: mechanisms, moderators, and clinical implications. Behav Res Ther 126:103550. https://doi.org/10.1016/j.brat.2020.103550

Qi S, Hassabis D, Sun J, Guo F, Daw N, Mobbs D (2018) How cognitive and reactive fear circuits optimize escape decisions in humans. Proc Natl Acad Sci U S A 115(12):3186–3191. https://doi.org/10.1073/pnas.1712314115

Reddan MC, Wager TD, Schiller D (2018) Attenuating neural threat expression with imagination. Neuron 100(4):994–1005.e4. https://doi.org/10.1016/j.neuron.2018.10.047

Riskind JH, Calvete E (2020) Anxiety and the dynamic self as defined by the prospection and mental simulation of looming future threats. J Pers 88(1):31–44. https://doi.org/10.1111/jopy.12465

Robinson OJ, Krimsky M, Grillon C (2013) The impact of induced anxiety on response inhibition. Front Hum Neurosci 7:69. https://doi.org/10.3389/fnhum.2013.00069

Robinson OJ, Pike AC, Cornwell B, Grillon C (2019) The translational neural circuitry of anxiety. J Neurol Neurosurg Psychiatry 90(12):1353–1360. https://doi.org/10.1136/jnnp-2019-321400

Roelofs K (2017) Freeze for action: neurobiological mechanisms in animal and human freezing. Philos Trans R Soc Lond B Biol Sci 372(1718):20160206. https://doi.org/10.1098/rstb.2016.0206

Roelofs K, Dayan P (2022) Freezing revisited: coordinated autonomic and central optimization of threat coping. Nat Rev Neurosci 23(9):568–580. https://doi.org/10.1038/s41583-022-00608-2

Rosen JB, Schulkin J (1998) From normal fear to pathological anxiety. Psychol Rev 105(2): 325–350. https://doi.org/10.1037/0033-295x.105.2.325

Roy-Byrne P (2015) Treatment-refractory anxiety; definition, risk factors, and treatment challenges. Dialogues Clin Neurosci 17(2):191–206. https://doi.org/10.31887/DCNS.2015.17.2/proybyrne

Russell JJ, Moskowitz DS, Zuroff DC, Bleau P, Pinard G, Young SN (2011) Anxiety, emotional security and the interpersonal behavior of individuals with social anxiety disorder. Psychol Med 41(3):545–554. https://doi.org/10.1017/S0033291710000863

Schäfer JÖ, Naumann E, Holmes EA, Tuschen-Caffier B, Samson AC (2017) Emotion regulation strategies in depressive and anxiety symptoms in youth: a meta-analytic review. J Youth Adolesc 46(2):261–276. https://doi.org/10.1007/s10964-016-0585-0

Schenberg LC, Schimitel FG, Armini R d S et al (2014) Translational approach to studying panic disorder in rats: hits and misses. Neurosci Biobehav Rev 46(Pt 3):472–496. https://doi.org/10.1016/j.neubiorev.2014.10.002

Schoevers RA, van Borkulo CD, Lamers F et al (2021) Affect fluctuations examined with ecological momentary assessment in patients with current or remitted depression and anxiety disorders. Psychol Med 51(11):1906–1915. https://doi.org/10.1017/S0033291720000689

Schubert T, Eloo R, Scharfen J, Morina N (2020) How imagining personal future scenarios influences affect: systematic review and meta-analysis. Clin Psychol Rev 75:101811. https://doi.org/10.1016/j.cpr.2019.101811

Scull A (2021) American psychiatry in the new millennium: a critical appraisal. Psychol Med 51(16):2762–2770. https://doi.org/10.1017/S0033291721001975

Shackman AJ, Fox AS (2021) Two decades of anxiety neuroimaging research: new insights and a look to the future. Am J Psychiatry 178(2):106–109. https://doi.org/10.1176/appi.ajp.2020.20121733

Shackman AJ, Stockbridge MD, Tillman RM et al (2016) The neurobiology of dispositional negativity and attentional biases to threat: implications for understanding anxiety disorders in adults and youth. J Exp Psychopathol 7(3):311–342. https://doi.org/10.5127/jep.054015

Shuhama R, Rondinoni C, de Araujo DB et al (2016) Behavioral and neuroimaging responses induced by mental imagery of threatening scenarios. Behav Brain Res 313:358–369. https://doi.org/10.1016/j.bbr.2016.06.059

Shuhama R, Blanchard DC, Graeff FG, Del-Ben CM (2019) What we imagine is what we do? A critical overview about mental imagery as a strategy to study human defensive responses. Behav Brain Res 357-358:18–28. https://doi.org/10.1016/j.bbr.2017.07.012

Silva C, McNaughton N (2019) Are periaqueductal gray and dorsal raphe the foundation of appetitive and aversive control? A comprehensive review. Prog Neurobiol 177:33–72. https://doi.org/10.1016/j.pneurobio.2019.02.001

Silva BA, Gross CT, Gräff J (2016) The neural circuits of innate fear: detection, integration, action, and memorization. Learn Mem 23(10):544–555. https://doi.org/10.1101/lm.042812.116

Silver J, Carlson GA, Olino TM et al (2021) Differential outcomes of tonic and phasic irritability in adolescent girls. J Child Psychol Psychiatry 62(10):1220–1227. https://doi.org/10.1111/jcpp.13402

Soyka M (2017) Treatment of benzodiazepine dependence. N Engl J Med 376(12):1147–1157. https://doi.org/10.1056/NEJMra1611832

Torrisi S, O'Connell K, Davis A et al (2015) Resting state connectivity of the bed nucleus of the stria terminalis at ultra-high field. Hum Brain Mapp 36(10):4076–4088. https://doi.org/10.1002/hbm.22899
Torrisi S, Alvarez GM, Gorka AX et al (2019) Resting-state connectivity of the bed nucleus of the stria terminalis and the central nucleus of the amygdala in clinical anxiety. J Psychiatry Neurosci 44(5):313–323. https://doi.org/10.1503/jpn.180150
Tovote P, Fadok JP, Lüthi A (2015) Neuronal circuits for fear and anxiety. Nat Rev Neurosci 16(6): 317–331. https://doi.org/10.1038/nrn3945
Tovote P, Esposito MS, Botta P et al (2016) Midbrain circuits for defensive behaviour. Nature 534(7606):206–212. https://doi.org/10.1038/nature17996
Trull TJ, Lane SP, Koval P, Ebner-Priemer UW (2015) Affective dynamics in psychopathology. Emot Rev 7(4):355–361. https://doi.org/10.1177/1754073915590617
Vassilopoulos SP (2008) Coping strategies and anticipatory processing in high and low socially anxious individuals. J Anxiety Disord 22(1):98–107. https://doi.org/10.1016/j.janxdis.2007.01.010
Weber-Goericke F, Muehlhan M (2019) A quantitative meta-analysis of fMRI studies investigating emotional processing in excessive worriers: application of activation likelihood estimation analysis. J Affect Disord 243:348–359. https://doi.org/10.1016/j.jad.2018.09.049
Wehry AM, Beesdo-Baum K, Hennelly MM, Connolly SD, Strawn JR (2015) Assessment and treatment of anxiety disorders in children and adolescents. Curr Psychiatry Rep 17(7):52. https://doi.org/10.1007/s11920-015-0591-z
Weisz JR, Chorpita BF, Palinkas LA et al (2012) Testing standard and modular designs for psychotherapy treating depression, anxiety, and conduct problems in youth: a randomized effectiveness trial. Arch Gen Psychiatry 69(3):274–282. https://doi.org/10.1001/archgenpsychiatry.2011.147
Wells A (1997) Cognitive therapy of anxiety disorders: a practice manual and conceptual guide. Wiley
Wendt J, Löw A, Weymar M, Lotze M, Hamm AO (2017) Active avoidance and attentive freezing in the face of approaching threat. Neuroimage 158:196–204. https://doi.org/10.1016/j.neuroimage.2017.06.054
White KS, Brown TA, Somers TJ, Barlow DH (2006) Avoidance behavior in panic disorder: the moderating influence of perceived control. Behav Res Ther 44(1):147–157. https://doi.org/10.1016/j.brat.2005.07.009
Williams LE, Oler JA, Fox AS et al (2015) Fear of the unknown: uncertain anticipation reveals amygdala alterations in childhood anxiety disorders. Neuropsychopharmacology 40(6): 1428–1435. https://doi.org/10.1038/npp.2014.328
World Health Organization (2019) International classification of diseases eleventh revision (ICD-11). World Health Organization (WHO). https://icd.who.int/browse11
Xia F, Kheirbek MA (2020) Circuit-based biomarkers for mood and anxiety disorders. Trends Neurosci 43(11):902–915. https://doi.org/10.1016/j.tins.2020.08.004

Structural Brain Correlates of Anxiety During Development

E. Leighton Durham and Antonia N. Kaczkurkin

Contents

Abstract Anxiety and related disorders are prevalent across the lifespan, often have their onset during youth, and are associated with notable levels of impairment and burden across multiple domains. Elucidating the associations between differential patterns of neurodevelopment and anxiety in youth is a promising approach for developing deeper insights regarding the neurobiological etiologies and maintenance factors associated with anxiety and related disorders. A growing body of literature has yielded evidence of associations between patterns of brain structure (i.e., volume, cortical thickness, and cortical surface area) and anxiety. Here, we present a review and synthesis of the existing body of literature surrounding neurostructural correlates of anxiety in youth spanning multiple anxiety presentations and three neurostructural modalities. We reveal substantially more research focusing on brain volume than cortical thickness or surface area and a greater

---

E. L. Durham and A. N. Kaczkurkin (✉)
Department of Psychological Sciences, Vanderbilt University, Nashville, TN, USA
e-mail: antonia.kaczkurkin@Vanderbilt.Edu

© The Author(s), under exclusive license to Springer Nature Switzerland AG 2024
Curr Topics Behav Neurosci (2025) 73: 123–153
https://doi.org/10.1007/7854_2024_541
Published Online: 08 November 2024

number of studies examining anxiety broadly defined, obsessive-compulsive disorder, or posttraumatic stress disorder. There is also evidence of considerable variability in the brain regions implicated and the direction of associations across studies. Finally, we discuss the gaps and limitations in this body of work, which suggest avenues for future directions.

Keywords Anxiety · Brain structure · Cortical surface area · Cortical thickness · Obsessive-compulsive disorder · Posttraumatic stress disorder · Volume · Youth

1 Introduction

Anxiety and related disorders are some of the most prevalent and impairing forms of psychopathology worldwide (de Sousa Júnior et al. 2021; Roza et al. 2003; Kessler et al. 2005; Remes et al. 2016). Research shows that the onset of anxiety and related disorders occurs predominantly during youth (encompassing childhood, adolescence, and young adulthood) with symptoms increasing across adolescence into young adulthood (Roza et al. 2003). Further, it has been found that experiencing psychopathology during childhood and adolescence, including anxiety, serves as a risk factor for experiencing psychopathology during adulthood (Reef et al. 2010). Youth is also a time characterized by extensive brain development. The brain undergoes substantial remodeling through both progressive shifts (i.e., increased myelination) and regressive shifts (i.e., increased synaptic pruning), as well as continued maturation of neural networks (Giedd et al. 1999; Stiles and Jernigan 2010).

Differences in the development of structural aspects of the brain during youth have been found to be associated with a range of negative consequences across multiple domains including deficits in cognitive functioning (Nave et al. 2019) and higher levels of psychopathology (Kaczkurkin et al. 2019, 2020; Durham et al. 2021). Gray matter volume, cortical thickness, and cortical surface area are each distinct, but related, facets of brain structure. Gray matter volume is generally considered to be the product of cortical thickness (distance between the pial surface and the white matter surface beneath it) and surface area (exterior plane of the cortex) (Wierenga et al. 2014; Winkler et al. 2010). Despite their relation to each other, gray matter volume, cortical thickness, and surface area offer unique information and have differing normative developmental trajectories (Wierenga et al. 2014; Winkler et al. 2010).

Understanding the associations between patterns of neurodevelopment and anxiety in youth is critical to gaining further insight about the neurobiological etiologies of anxiety and related disorders and to enhancing early identification capabilities and prevention efforts at a young age. While the existing body of research investigating neural mechanisms of anxiety in youth tends to favor a functional (i.e., neural activation) focus over a structural focus (Ashworth et al. 2021; Pine 2007; Sylvester et al. 2012), a fair amount of evidence surrounding neurostructural correlates of anxiety in youth has accumulated. In studying the neurobiology of psychopathology, structural and functional approaches are complementary to each other and both offer unique insight into the neural mechanisms underlying psychological symptoms and disorders. This is, in part, due to the fact that structural and functional imaging techniques are susceptible to different types of confounds (Madan 2017). For instance, structural imaging data tends to be more robust than functional data, as it not task-dependent and is generally less sensitive to potential artifacts (Madan 2017). Therefore, the convergence of both structural and functional techniques is the most comprehensive approach to understanding the neurobiology underlying psychopathology.

There is a paucity of literature synthesizing neurostructural findings associated with youth anxiety, particularly in comparison to the functional literature. Thus, in the current review, we provide an overview of the existing literature organized by symptom classes. First, we begin with a brief description of our methods; then, we describe the neurostructural differences associated with anxiety, starting from the broadest conceptualizations (anxiety broadly defined and transdiagnostic anxiety) and transitioning to more precise conceptualizations (specific disorders). Finally, we end with a discussion of the gaps and limitations in this body of literature, which provide insights into future directions.

2 Methods

Research studies included in the current review were identified with a series of searches in both the PubMed and PsycINFO databases using combinations of search terms (*anxiety, phobia, panic, agoraphobia, selective mutism*), youth samples (*children, adolescence, youth*), and aspects of brain structure (*brain structure, brain volume, gray matter, cortical thickness, surface area*). Studies were selected for inclusion if the sample was comprised of youth (defined as age 25 or younger) and the study examined relationships between brain structure (defined as global brain volume, gray matter volume, cortical thickness, or cortical surface area) and anxiety symptoms or disorders. Articles were excluded if brain structure and internalizing problems were not measured in the same individuals (i.e., maternal anxiety symptoms being associated with offspring brain structure).

3 Neurostructural Differences Associated with Anxiety Symptoms in Youth

Increasing our understanding of the specific ways in which brain volume, cortical thickness, and cortical surface area relate to different forms of anxiety in youth will promote further insight into the etiology and maintenance of anxiety and related disorders and contribute to the advancement of more precise early identification, prevention, and intervention efforts. Here, we outline the existing literature regarding associations between brain structure and anxiety-related phenotypes in youth. The following sections will be ordered from a broader focus (e.g., continuous measures of anxiety symptoms) to a narrower focus (e.g., specific disorders) in terms of the conceptualization and measurement of symptoms. Details of the studies included in this review are summarized in Table 1.

3.1 Anxiety Symptoms Broadly Defined

Some studies investigating associations between brain structure and anxiety have measured symptoms in a transdiagnostic, continuous way rather than focusing on one particular type of anxiety or implementing a case-control design based on diagnostic criteria for an anxiety or related disorder. This section will review studies which have measured broadly defined anxiety symptoms and related them to brain structure in youth samples.

3.1.1 Differences in Brain Volume Implicated in Studies Measuring Continuous Anxiety Symptoms

While there does not seem to be much evidence of global associations between gray matter volume and anxiety symptoms broadly defined in youth, some associations with regional gray matter volumes have been identified. The amygdala is the region that has most consistently been found to have a relationship with anxiety symptoms broadly defined in youth; however, the direction of this association has not been entirely consistent. One prior study found a positive relationship between anxiety symptoms based on the anxiety subscale of the Child Behavior Checklist (CBCL) (Achenbach 2009) and amygdala volume (Qin et al. 2014). Similarly, another prior study demonstrated that the physical anxiety subscale of the Multidimensional Anxiety Scale for Children (MASC) (March et al. 1997) was associated with greater left amygdala volume, while the social anxiety subscale of the MASC was associated with greater right hippocampal volume (Castagna et al. 2018).

Alternatively, using the Screen for Child Anxiety-Related Emotional Disorders (SCARED) (Birmaher et al. 1997), a prior study found that anxiety symptoms were *inversely* associated with amygdala gray matter volume, but only for children younger than 11 years old within the sample (Warnell et al. 2018). Further, others using the Screen for Child Anxiety-Related Emotional Disorders-Revised (SCARED-R) (Muris 2002) and the CBCL anxiety subscale (Achenbach 2009) have found no associations between regional subcortical volumes and anxiety symptoms (Vargas and Mittal 2022; Merz et al. 2018). These mixed findings may be due to differences across study samples (e.g., age ranges, anxiety levels) and study designs (e.g., anxiety measures, statistical methods). As for cortical regions, there are not as many reported findings around gray matter volume and youth anxiety symptoms; however, one recent study found higher trait anxiety, measured with the State-Trait Anxiety Inventory for Children (STAI-C) (Spielberger et al. 1973), was inversely associated with middle frontal gyrus gray matter volume in a sample of adolescents (Schuttenberg et al. 2022). In sum, limited prior research has found evidence of differences in amygdala, hippocampal, and middle frontal gyrus volumes associated with anxiety symptoms broadly defined in youth. For the hippocampus and middle frontal gyrus, greater volume was associated with less anxiety. For the amygdala, the direction of the association has been inconsistent.

3.1.2 Associations Between Broad Anxiety Symptoms and Cortical Thickness Are Mixed

Studies relating anxiety symptoms with cortical thickness in youth samples have yielded evidence of both global and regional associations. In a longitudinal study, greater youth anxiety symptoms, measured with the anxiety subscale of the Trauma Symptom Checklist for Children (Briere 1996), were associated with smaller average cortical thickness at baseline, but also with slower cortical thinning over a 3-year period (Taylor et al. 2021). Likewise, another prior study found anxiety symptoms, measured with the SCARED-R (Muris 2002), were inversely associated with global cortical thickness; however, the association was stronger in the younger children within the sample and diminished with age (Newman et al. 2016).

In terms of regional associations between cortical thickness and youth anxiety symptoms, several studies have implicated a variety of regions. However, similar to the volume findings described in the prior section, the directionality of these regional associations has not always been consistent. In one longitudinal study, youth anxiety symptoms during preschool years, measured with the Preschool-Age Psychiatric Assessment (Egger et al. 1999), were found to predict thinner cortices of the insula, subgenual anterior cingulate cortex, and inferior lateral parietal cortex in later childhood; however, the anxiety symptoms of those children in middle childhood did not predict differences in cortical thickness (Luking et al. 2022). Additionally, less superior frontal thickness, but greater precuneus and isthmus cingulate thickness has been associated with greater youth anxiety symptoms, measured with the anxiety subscale of the CBCL (Achenbach 2009) cross-sectionally, while thinner cortices in the inferior parietal cortex, rostral middle frontal gyrus, and fusiform gyrus, but greater thickness in the caudal middle frontal gyrus all predicted increased anxiety symptoms over time during youth (Vargas and Mittal 2022). Of note, in another study, no associations were found between anxiety symptoms, as measured by the SCARED-R (Muris 2002), and cortical thickness across a number of a priori-defined regions (Merz et al. 2018).

Interestingly, one study examined how relationships between youth anxiety symptoms and cortical thickness differed across the physical anxiety and social anxiety subscales of the MASC (March et al. 1997) and found that both subscales were positively associated with widespread regional cortical thickness. Additionally, they found that there was a stronger relationship between physical anxiety symptoms and cortical thickness in the right anterior insula, right somatosensory area, left supramarginal gyrus, and bilateral precentral gyrus, as compared to social anxiety symptoms, and that no associations between social anxiety symptoms and cortical thickness were significantly stronger than the associations with physical anxiety symptoms (Castagna et al. 2018). This suggests that heightened physical anxiety (e.g., heart racing, shakiness) may have a stronger association with cortical thickness than social anxiety symptoms across a number of brain regions. The authors note that this may be due, in part, to the involvement of many of the identified regions in sensation and interoception processes (Castagna et al. 2018).

In sum, there is evidence of both global and regional cortical thickness associations with anxiety symptoms broadly defined in youth. Some have found anxiety to be associated with thinner global cortical thickness and slower cortical thinning over time. Prior studies have also identified a variety of regional associations between cortical thickness and youth anxiety, particularly in the frontal and parietal lobes and the insula, among other areas; though, the direction of these findings is not always consistent.

3.1.3 Minimal Research on Associations Between Broad Anxiety Symptoms and Surface Area

Studies examining the relationship between cortical surface area and anxiety symptoms broadly defined in youth are sparse. One study found an association between greater anxiety symptoms, measured with the SCARED-R (Muris 2002), and less cortical surface area of the ventromedial prefrontal cortex in younger children, which seemed to diminish in older children (Newman et al. 2016). Additionally, a longitudinal study found that greater surface area in the entorhinal cortex and posterior cingulate cortex, but less surface area in the temporal pole were associated with more youth anxiety symptoms at baseline, measured with the CBCL anxiety subscale, while greater surface area in the supramarginal gyrus predicted increased anxiety symptoms over time from childhood into adolescence (Vargas and Mittal 2022). Of note, another prior study reported that no associations were found between anxiety symptoms measured with the SCARED-R (Muris 2002) and cortical surface area across a variety of regions defined a priori (Merz et al. 2018). In sum, the limited prior research on associations between cortical surface area and anxiety symptoms broadly defined in youth indicate inverse associations with the ventromedial prefrontal cortex and temporal pole, as well as positive associations with the entorhinal cortex, posterior cingulate cortex, and supramarginal gyrus.

3.1.4 Summary of Findings Related to Youth Brain Structure and Continuous Anxiety Symptoms

Together, these findings suggest that broad anxiety symptoms in youth may be associated with differences in brain volume, cortical thickness, and cortical surface area. In terms of volume, differences in amygdala and hippocampal volumes, as well as in the middle frontal gyrus, have been implicated with a positive association for the hippocampus, an inverse association for the middle frontal gyrus, and mixed directionality for the amygdala. In terms of cortical thickness, multiple studies have suggested anxiety symptoms may be associated with globally thinner cortices. Cortical thickness of specific regions has also been implicated across studies focusing on anxiety symptoms broadly defined; however, the directionality of those associations appears to be dependent on the brain area and sample. Additionally, there is substantial variability in how continuous anxiety symptoms have been measured across the reviewed studies, which likely contributes to the inconsistency and variability of findings.

3.2 *Transdiagnostic Anxiety Disorders*

Several prior studies investigating relationships between brain structure and anxiety in youth have focused on samples which consist of several anxiety or related disorders, rather than focusing only on one specific disorder. This section will review studies which have analyzed data from samples which include multiple anxiety disorder diagnoses.

3.2.1 Mixed Evidence of Brain Volume Differences Associated with Transdiagnostic Anxiety Disorders

There is evidence of associations between brain volume and anxiety yielded by studies which include samples characterized by a mix of anxiety diagnoses. Prior studies found that youth who met the criteria for generalized anxiety disorder (GAD), separation anxiety disorder, and/or social anxiety disorder exhibited smaller right hippocampal volume (Gold et al. 2017) and smaller left amygdala volume (Milham et al. 2005), as compared to control groups. Likewise, in another study, youth with a current diagnosis of GAD, panic disorder, separation anxiety disorder, and/or specific phobia showed smaller amygdala gray matter volume compared to a healthy control group (Suffren et al. 2019). Interestingly, one longitudinal study also found that larger nucleus accumbens gray matter volume at baseline predicted greater decreases in anxiety symptoms from pre- to post-treatment with medication and therapy in youth with GAD or social anxiety disorder (Burkhouse et al. 2020). Finally, in addition to subcortical findings, one study reported that youth with GAD, separation anxiety disorder, and/or social anxiety disorder had larger gray matter volumes in the dorsal anterior cingulate, and smaller gray matter volumes in the ventrolateral prefrontal cortex, postcentral gyrus, and cuneus/precuneus areas (Strawn et al. 2015). In sum, identified studies relating brain volume to transdiagnostic anxiety in youth indicate inverse associations with hippocampus, amygdala, ventrolateral prefrontal cortex, postcentral gyrus, and cuneus/precuneus volumes, as well as a positive association with dorsal anterior cingulate volume.

3.2.2 Paucity of Studies on Relationships Between Cortical Thickness and Transdiagnostic Anxiety Disorders

A few studies also identify associations between anxiety and cortical thickness in samples characterized by transdiagnostic anxiety disorders. In one study, youth diagnosed with any anxiety disorder showed greater widespread regional cortical thickness as compared to healthy controls (Yin et al. 2022). Additionally, another study found that youth who met the criteria for GAD, separation anxiety disorder, and/or social anxiety disorder exhibited greater cortical thickness in the left ventromedial prefrontal cortex and left precentral gyrus than healthy controls (Gold et al. 2017). Finally, in a third study, youth with current GAD, panic disorder, separation anxiety disorder, and/or specific phobia showed thinner cortices in the rostral anterior cingulate cortex and insula compared to healthy controls (Suffren et al. 2019).

3.2.3 Summary of Findings Related to Brain Structure and Transdiagnostic Anxiety Disorders in Youth

In sum, in youth samples characterized by more than one anxiety disorder, a relatively small number of studies have yielded evidence of regional volume and cortical thickness associations which vary in terms of direction and are not yet widely replicated. While inverse relationships have been found between transdiagnostic anxiety disorders and volumes in the hippocampus, amygdala, ventrolateral prefrontal cortex, postcentral gyrus, and cuneus/precuneus, positive relationships have been identified with volumes of the nucleus accumbens and anterior cingulate. Very few studies have reported findings on differences in cortical thickness or surface area associated with transdiagnostic anxiety disorders in youth. Notably, the most commonly combined disorders in these transdiagnostic studies include GAD, social anxiety disorder, and separation anxiety disorder; however, there is an emerging body of evidence suggesting that social anxiety and separation anxiety cluster together into the "fear" disorders while GAD may be better described as a "distress" disorder (Conway et al. 2022), which implies potentially different underlying mechanisms. Thus, combining distress and fear disorders may obscure important structural distinctions between these two classes of symptoms. In addition, due to inconsistent neurostructural findings in transdiagnostic anxiety disorders in youth, there has been a shift in the field towards using data-driven approaches to provide greater clarity. However, a large ENIGMA study using supervised machine learning models with cortical thickness, cortical surface area, and subcortical volume concluded that only modest classification performances can be achieved with such approaches (Bruin et al. 2024).

3.3 *Generalized Anxiety Disorder*

3.3.1 Limited Evidence of Differences in Brain Volumes Associated with Generalized Anxiety Disorder

Prior literature on the relationship between brain volume and GAD in youth is limited. Youth with GAD have been found to have significantly larger gray matter volume in the right putamen (Liao et al. 2013), amygdala (De Bellis et al. 2000), superior temporal gyrus (De Bellis et al. 2002a), right precuneus, and right precentral gyrus (Strawn et al. 2013). Youth GAD is also associated with smaller gray matter volumes in the left orbital frontal gyrus, posterior cingulate cortex (Strawn et al. 2013), right inferior frontal gyrus, ventral pre- and post-central gyri, insular cortex, and superior temporal gyrus (Chaarani et al. 2020).

3.3.2 Research on Cortical Thickness and Generalized Anxiety Disorder Is Limited

Studies investigating cortical thickness and youth GAD are also limited and no studies focusing on cortical surface area and youth GAD were identified. The few studies that found associations between cortical thickness and GAD in youth were limited to adolescent samples. In one of these studies, adolescents with GAD exhibited greater cortical thickness in the right inferolateral and ventromedial prefrontal cortex, left inferior and middle temporal cortex, and right lateral occipital cortex, as compared to healthy control participants (Strawn et al. 2014). In another study, greater cortical thickness of the middle temporal gyrus predicted increased symptoms of GAD over a two-year period during adolescence (Busso et al. 2017).

3.3.3 Summary of Findings Related to Youth Brain Structure and Generalized Anxiety Disorder

In sum, the small body of literature investigating brain volume and GAD in youth suggests that youth GAD may be associated with volumetric differences, including positive associations with volume in the right putamen, amygdala, superior temporal gyrus, right precuneus, and right precentral gyrus, as well as inverse associations with the left orbitofrontal cortex, posterior cingulate cortex, right inferior frontal gyrus, ventral pre- and post-central gyrus, insula, and superior temporal gyrus. For cortical thickness, studies have identified positive associations between youth GAD and thickness of the right inferolateral and ventromedial prefrontal cortex, left inferior and middle temporal cortex, right lateral occipital cortex, and middle temporal gyrus.

3.4 Social Anxiety Disorder

3.4.1 Differences in Brain Volume Associated with Social Anxiety Disorder

There is a small amount of prior literature which identified associations between social anxiety and brain volume. Associations have been found between youth social anxiety and smaller volumes in the amygdala, hippocampus (Mueller et al. 2013), and ventrolateral prefrontal cortex (Auday and Pérez-Edgar 2019), as well as larger volumes in the right middle temporal gyrus (Wang et al. 2021), orbitofrontal cortex, and insula (Liu et al. 2021). In the latter study, it was also demonstrated that adolescents with social anxiety disorder exhibited less covariance of gray matter volume between the orbitofrontal cortex and amygdala and between the orbitofrontal cortex and insula than a healthy control group, which suggests that social anxiety disorder may be associated with altered maturational coupling between those regions (Liu et al. 2021). Lastly, another study found that youth with social anxiety disorder exhibited significantly greater right thalamic volume compared to healthy controls, which was a stronger association in adolescents than in younger youth within the sample (Zhang et al. 2020). In sum, studies examining relationships between youth social anxiety and brain volume have indicated positive associations with the right middle temporal gyrus, orbitofrontal cortex, insula, and right thalamus, as well as inverse associations with the amygdala, hippocampus, and ventrolateral prefrontal cortex.

3.4.2 Summary of Findings Related to Brain Structure and Social Anxiety Disorder in Youth

Overall, there is uncertainty regarding the relationship between social anxiety disorder and brain structure in youth. A handful of studies have yielded evidence of positive associations between volume and social anxiety disorder in some frontal, temporal, and insular areas, as well as in the thalamus, while other studies have provided preliminary evidence of negative associations between social anxiety disorder and volume in the amygdala, hippocampus, and ventromedial prefrontal cortex. However, most of these findings are cross-sectional and have not been replicated. Additionally, the samples in these studies were largely limited to adolescent age ranges. No studies specifically examining patterns of cortical thickness or surface area associated with youth social anxiety disorder were identified. There is substantial room for advancement and additional research in the area of youth brain structure and social anxiety.

3.5 *Obsessive-Compulsive Disorder*

3.5.1 Mixed Evidence of Brain Volume Differences Associated with Obsessive-Compulsive Disorder

Prior research investigating brain volume correlates of youth OCD have identified both global and regional associations. Specifically, youth OCD symptoms have been found to be associated with smaller global volume (Carmona et al. 2007; Chen et al. 2013; Weeland et al. 2021). Smaller regional volumes associated with youth OCD have also been found for the putamen (Rosenberg et al. 1997), cingulate cortex, frontal lobe (Carmona et al. 2007; Chen et al. 2013; Jayarajan et al. 2015), temporal and occipital lobes (Chen et al. 2013), and parietal lobe (Chen et al. 2013; Lázaro et al. 2009; Cheng et al. 2016). Another study found that youth OCD primarily characterized by compulsive acts and repeated thoughts was associated with smaller volume of the parietal lobe, precentral gyrus, and precuneus, while youth OCD primarily characterized by perfectionism and anxiety was not (Wu et al. 2022).

Other prior studies have found *positive* associations between youth OCD and regional volumes, including the thalamus (Weeland et al. 2021; Ivanov et al. 2022), hippocampus (Vattimo et al. 2021), putamen, and orbitofrontal cortex (Cheng et al. 2016). Youth OCD was also found to be associated with an increase in orbitofrontal cortex volume over time in one study (Huyser et al. 2014). Of note, another study found different volumetric associations for particular aspects of OCD symptomology in youth. Specifically, ordering symptoms were inversely associated with caudate volume; obsessing symptoms were inversely associated with right temporal pole volume; and hoarding symptoms were positively associated with left inferior frontal gyrus volume (Suñol et al. 2018). Finally, some investigations of brain structure in youth OCD have found no associations with regional gray matter volume (Lázaro et al. 2014; Pagliaccio et al. 2020, 2021).

In sum, studies investigating the relationship between youth OCD and brain volume have found inverse associations with global volume, as well as with regional volume of the putamen, cingulate, and frontal, temporal, parietal, and occipital lobes. At the same time, positive associations have been found for regional volumes of the thalamus, hippocampus, putamen, and orbitofrontal cortex. Additionally, there is evidence to suggest that volumetric associations with youth OCD may differ across symptom clusters within OCD.

3.5.2 Inconclusive Evidence of Associations between Cortical Thickness and Obsessive-Compulsive Disorder

Some differences in cortical thickness patterns in youth OCD have also been found, including thinner cortex in the inferior and superior parietal cortices (Boedhoe et al. 2018; Fallucca et al. 2011). However, some investigations of brain structure in youth OCD have found no associations with regional cortical thickness (Pagliaccio et al. 2020, 2021; Boedhoe et al. 2020). Of note, in one of those studies, while no neurostructural differences were associated with youth OCD at a baseline timepoint, thinner cortex in the left parietal lobe, right insular cortex, left middle and superior frontal gyri, left precentral gyrus, left superior temporal gyrus, and left and right lingual gyri predicted a greater decrease in symptoms following OCD treatment (Pagliaccio et al. 2020).

3.5.3 Summary of Findings Related to Brain Structure and Obsessive-Compulsive Disorder in Youth

In sum, while some have found associations between brain structure and youth OCD, there is also notable uncertainty regarding these relationships. Some studies have yielded evidence of negative associations between volume and OCD throughout the brain, while other studies have found positive associations between youth OCD and volume in several regions, including the thalamus, hippocampus, putamen, and some frontal lobe regions. For cortical thickness, there is preliminary evidence that youth OCD is associated with thinner cortices in the inferior and superior parietal cortex, as well as evidence of differences in regional cortical thickness predicting treatment response over time. Of note, some studies have also found no associations between brain structure and youth OCD and others found neurostructural differences associated with only particular symptom clusters within OCD. OCD is a multifaceted disorder with heterogenous obsessions, compulsions, severity profiles, and levels of insight amongst patients with this disorder (Broekhuizen et al. 2023; Yovel et al. 2012), which may impede the ability to reliably find common neurostructural correlates of OCD.

3.6 *Posttraumatic Stress Disorder*

3.6.1 Posttraumatic Stress Disorder Associated with Differences in Brain Volume

For youth posttraumatic stress disorder (PTSD), a variety of both global and regional brain structure associations have been identified. Youth PTSD has been associated with smaller overall volume across multiple studies (Carrion et al. 2001; De Bellis et al. 1999, 2002b). Further, youth PTSD has also been associated with smaller regional volumes in the hippocampus (Mutluer et al. 2018), amygdala (Mutluer et al. 2018; Ross et al. 2021), cerebellum (De Bellis et al. 2015; De Bellis and Kuchibhatla 2006), right temporal lobe (De Bellis et al. 2002b), areas within the prefrontal cortex (De Bellis et al. 2002b; Kedin and Herringa 2015; Morey et al. 2016), right precuneus (Garrett et al. 2021), and areas within the cingulate cortex (Mutluer et al. 2018; Garrett et al. 2021; Rinne-Albers et al. 2017).

Some studies have identified associations between youth PTSD and larger regional volumes as well, including in the hippocampus (Tupler and De Bellis 2006), right amygdala (Weems et al. 2015), prefrontal cortex (Carrion et al. 2009; Richert et al. 2006), superior temporal gyrus (De Bellis et al. 2002c), right medial orbitofrontal cortex, and left posterior cingulate (Garrett et al. 2021). Of note, in the latter study, a variety of neurostructural differences associated with treatment response in youth PTSD were also identified (Garrett et al. 2021). In sum, prior research on the relationship between PTSD in youth and brain volume has indicated inverse associations with global volume and regional volumes of the hippocampus, amygdala, cerebellum, precuneus, and areas within the prefrontal cortex, cingulate cortex, and temporal lobe. However, the opposite effect has also been reported in some of the same regions. Positive associations have been indicated with regional volume of the hippocampus, amygdala, prefrontal cortex, superior temporal gyrus, medial orbital cortex, and posterior cingulate cortex.

3.6.2 Paucity of Evidence of Associations Between Cortical Thickness and Posttraumatic Stress Disorder

The literature regarding associations between youth PTSD and cortical thickness is very limited. However, persistent PTSD during youth has been found to be associated with thinner cortices in the prefrontal cortex (Mutluer et al. 2018), while PTSD remission has been found to be related to increased cortical thickness in the ventromedial prefrontal cortex over time (Heyn and Herringa 2019).

3.6.3 Minimal Research on Surface Area Associations with Posttraumatic Stress Disorder

Finally, there is also minimal research surrounding associations between PTSD during youth and cortical surface area. One study found that persistent youth PTSD was associated with decreased surface area of the ventromedial prefrontal cortex over time, while PTSD remission was associated with surface area expansion in the frontal pole over time (Heyn and Herringa 2019).

3.6.4 Summary of Findings Related to Brain Structure and Posttraumatic Stress Disorder in Youth

Overall, prior literature has yielded mixed findings with regard to brain structure associations with youth PTSD. Some studies have found that smaller overall volume is associated with youth PTSD, as well as smaller regional volumes in both subcortical and cortical regions (e.g., hippocampus, amygdala, cerebellum, precuneus, prefrontal cortex, cingulate cortex, and temporal lobe). Other studies have identified associations between youth PTSD and larger regional volumes in many of the same regions (e.g., hippocampus, amygdala, prefrontal cortex, superior temporal gyrus, medial orbital cortex, and posterior cingulate cortex). In terms of other neurostructural modalities, there is some evidence that youth PTSD is associated with thinner cortices and smaller surface area in prefrontal areas. The inconsistencies across the studies relating youth PTSD to brain structure in terms of the regions implicated and direction of associations may be due to PTSD's heterogeneous nature where the type of trauma experienced may impact the physiological, cognitive, and emotional symptoms experienced, which could result in divergent neurostructural differences (Lanius et al. 2006; DiMauro et al. 2014).

4 Discussion

The present review presents the existing body of literature regarding neurostructural correlates of anxiety and related disorders in youth spanning multiple phenotypes and multiple neurostructural modalities. This review of empirical studies focusing on the neurostructural modalities of brain volume, cortical thickness, and cortical surface area revealed that the most commonly reported anxiety phenotypes studied included anxiety symptoms broadly defined, transdiagnostic anxiety disorders, generalized anxiety disorder, social anxiety disorder, obsessive-compulsive disorder, and posttraumatic stress disorder. Of note, literature searches around panic, specific phobia, agoraphobia, and selective mutism yielded a dearth of empirical studies examining associations with brain structure in youth. Separation anxiety disorder was commonly subsumed under the transdiagnostic studies.

Overall, the body of literature investigating associations between brain structure and anxiety symptoms in youth shows notable variability across anxiety phenotype and neurostructural modality (Fig. 1). Further, there is also considerable variability in the brain regions implicated and the direction of associations across studies. Additionally, the amount of research focusing on brain volume is notably more substantial than that of cortical thickness or surface area. Broad measures of anxiety symptoms as well as the specific disorders of OCD and PTSD were represented in the largest portion of studies, while other individual anxiety disorders have fewer studies by comparison. This body of literature is characterized by a number of gaps and limitations which have likely contributed to the inconsistency and variability of findings and are important considerations that may inform future research directions. We discuss each consideration, organized around the topics of sample characteristics/study design, symptom measurement, and neuroimaging approaches.

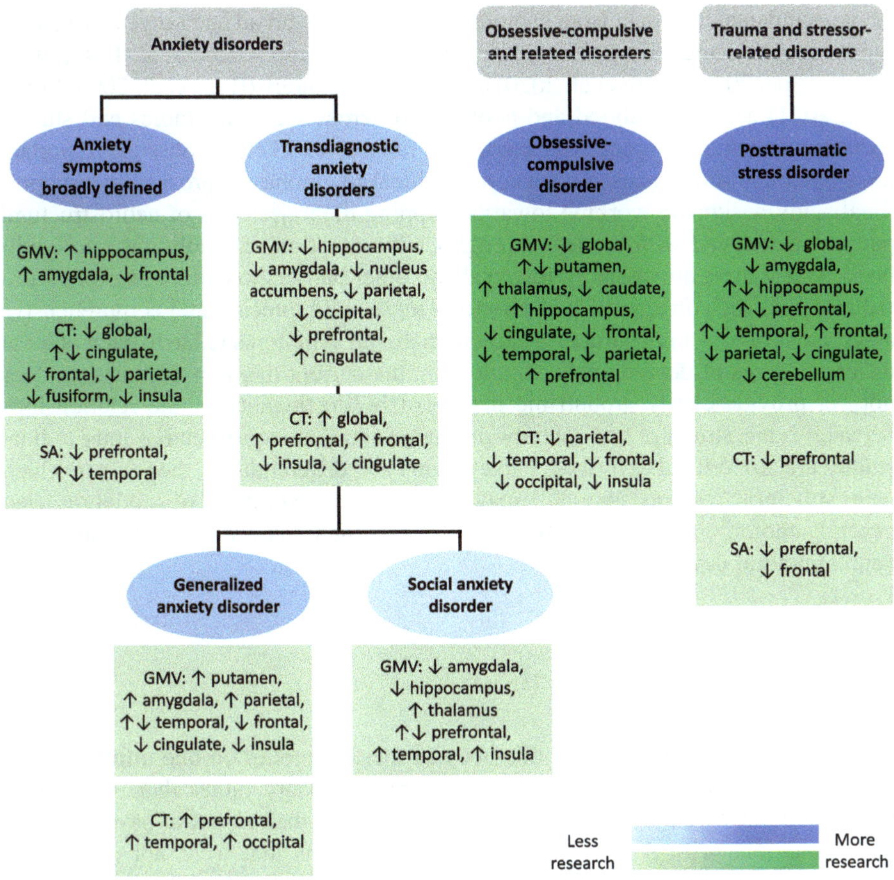

Fig. 1 *Synthesis of Findings Across Anxiety Phenotypes and Structural Modalities.* Blue circles represent the internalizing phenotype and green boxes represent the structural modality. The shade of the shapes represents the amount of published literature in that area with lighter colors corresponding to fewer studies and darker colors corresponding to more studies. Gray matter volume, cortical thickness, and surface area are referred to as GMV, CT, and SA, respectively. The direction of the association is indicated with arrows with the upward arrow representing a positive association between anxiety and structure and the downward arrow representing an inverse association. This figure reflects a synthesis of the prevailing trends in the reviewed literature. More detailed information about the reviewed findings can be found in Table 1

5 Sample Characteristics and Study Design

This review highlights considerable variation in the sample size examined across studies, with some samples being quite small. Small sample sizes may fail to detect effects or may detect spurious effects. Additionally, the majority of investigations on relationships between brain structure and anxiety in youth have used cross-sectional designs and analyzed data from samples with relatively broad age ranges. Given the rapid rates of change in the development of both the brain (Giedd et al. 1999; Stiles and Jernigan 2010; Baum et al. 2020) and psychopathology (Roza et al. 2003; Solmi et al. 2021) across childhood and adolescence, cross-sectional studies and studies focusing on youth samples with broad age ranges are likely to obscure important developmental differences. As such, it is critical to emphasize prospective longitudinal study designs in working towards mapping the trajectories of neurostructural patterns associated with anxiety throughout the duration of youth and into adulthood. While there are certainly financial and other barriers to conducting longitudinal research, such an approach allows for a more valid, precise, and comprehensive assessment of how brain structure and internalizing symptoms relate to each other at different stages of development. In addition, the current literature in this area does not yet provide clarity around the degree of bidirectionality of the relationships between brain structure and anxiety during youth. Thus, prospective longitudinal studies are also critical to further characterizing and differentiating the ways in which neurostructural patterns in youth may affect the development of anxiety-related psychopathology, and conversely, the ways in which anxiety in youth may affect neurostructural development.

6 Symptom Measurement

The vast majority of research in this body of literature relies on one informant and most studies used a caregiver-report measure of anxiety rather than self-report. Research has shown that caregivers and youth differ in their report of psychopathology symptoms, with some youth reporting symptoms that were not noticed by caregivers (Bonadio et al. 2022). Importantly, the reporting of symptoms by youth but not caregivers was specific to internalizing symptoms and not found for externalizing problems, which is consistent with the notion that internalizing symptoms are less outwardly apparent and may be harder to detect by caregivers (Bonadio et al. 2022). In addition, the variability in results across studies may be due to differences in the measurement of anxiety symptoms. The two most common approaches to defining anxiety in these studies were categorical approaches and dimensional approaches. The phenotypes of anxiety in categorical studies are primarily based on diagnostic classification systems like the *Diagnostic and Statistical Manual of Mental Disorders* (DSM), where patients who meet diagnostic criteria for an anxiety or related disorder are compared to heathy controls. While this is a commonly used framework for studying anxiety, categorical diagnoses have a number of limitations including relying on a relatively arbitrary cutoff to define clinical versus non-clinical levels of symptoms, ignoring subthreshold symptoms, and inflating effect sizes by comparing the extremes (clinically impaired vs. healthy). Alternatively, a number of studies included in this review used a dimensional approach to measure anxiety symptoms that takes into account the continuum from no symptoms to many symptoms, which is more representative of how symptoms appear in the general population. While dimensional approaches are an improvement over categorical approaches, there is a movement in the field to reconceptualize psychopathology as hierarchically organized. Using this approach, an overarching internalizing dimension is subdivided into subdimensions such as distress and fear (Kotov et al. 2017; Lahey et al. 2021). Although studies relating distress and fear dimensions to brain structure are sparse, such an approach may be useful for capturing the hierarchical and dimensional nature of anxiety symptoms and may lead to new insights when examining structural differences in youth.

7 Neuroimaging Approaches

Finally, this review revealed limitations in the neuroimaging approaches used in studies of neurostructural correlates of anxiety symptoms. The majority of studies examining associations between brain structure and youth anxiety focus on one neurostructural modality and many of these studies focus on specific regions only. The modality of focus in this body of research has most frequently been volume. Studies focusing on cortical thickness and youth anxiety are more limited and those focusing on or including cortical surface area are even more rare. This is a limitation, as it has been found that while different aspects of brain structure are interrelated, they also offer unique contributions and have varied developmental patterns (Wierenga et al. 2014; Winkler et al. 2010). Thus, examining multiple structural modalities simultaneously can offer a more thorough insight into how neurostructural development relates to youth anxiety, as well as into which particular facets of brain structure may be most relevant for anxiety outcomes. In addition, many of the studies reviewed use a region of interest (ROI) approach, in which researchers pre-select particular brain regions to investigate. In contrast, others use a whole-brain approach to examine global structural measures or all regions across the brain. While ROI approaches increase the power to detect significant associations, they may miss important relationships in regions that were not examined. Whole-brain approaches are useful for ensuring that no important associations are overlooked but come with the caveat of requiring correction for multiple comparisons, which can render small but meaningful effects nonsignificant. Thus, different decisions regarding neuroimaging approaches may also contribute to the substantial variability of findings that we see in this body of literature.

8 Conclusion

Together, the literature reviewed illustrates that researchers have made substantial progress in identifying and characterizing neurostructural patterns which are related to the development and experience of anxiety in youth over the past several decades. However, given the multifaceted nature of brain development and the complexities of the development and maintenance of psychopathology, there is still much to uncover. There is an accumulating body of evidence suggesting that anxiety in youth has associations with differences in brain volume, cortical thickness, and cortical surface area with many studies pointing to key subcortical and cortical brain regions. However, there is considerable variability across studies, particularly in terms of the implicated regions and direction of effects. In response to these mixed findings, future research in this area could prioritize employing prospective longitudinal designs to capture all stages of youth, using multi-informant and hierarchical methods for measuring anxiety symptoms, and integrating multiple neurostructural modalities in a whole-brain approach. By doing so, our field can begin to clarify the neurostructural correlates of anxiety in youth, with the hope of advancing our understanding of the pathophysiology of anxiety and ameliorating the long-term negative effects of these symptoms.

Appendix

Table 1 Investigations of neurostructural differences associated with youth anxiety and related disorders (NS = nonsignificant)

Citation	N	Age	Direction/region/ modality	Relevant findings
Anxiety symptoms broadly defined				
Castagna et al. (2018)	35	8–17 years	↑ amygdala volume; ↑ hippocampus volume; ↑ global ct	Physical anxiety symptoms positively associated with left amygdala volume; social anxiety symptoms positively associated with right hippocampal volume; both physical and social anxiety positively associated with widespread regional cortical thickness; correlations between anxiety symptoms and regional cortical thickness generally stronger for physical anxiety than social anxiety
Luking et al. (2022)	138	3–5 years	↓ insula ct; ↓ subgenual anterior cingulate cortex ct; ↓ inferior lateral parietal cortex ct	Anxiety symptoms predicted cortical thinning over time in the insula, subgenual anterior cingulate cortex, and inferior lateral parietal cortex
Merz et al. (2018)	328	7–21 years	NS volume, ct, sa results	Anxiety symptoms not associated with a prior defined regional volume, cortical thickness, or cortical surface area
Newman et al. (2016)	287	7–20 years	↓ global cortical thickness; ↓ ventromedial prefrontal cortex sa	Anxiety symptoms inversely associated with global cortical thickness; anxiety symptoms inversely associated with cortical surface area of the ventromedial prefrontal cortex
Qin et al. (2014)	76	7–9 years	↑ amygdala volume	Anxiety symptoms positively associated with amygdala volume
Schuttenberg et al. (2022)	64	13–14 years	↓ middle frontal gyrus volume	Anxiety symptoms inversely associated with middle frontal gyrus volume

(continued)

Table 1 (continued)

Citation	N	Age	Direction/region/ modality	Relevant findings
Taylor et al. (2021)	205	9–15 years	↓ global cortical thickness	Anxiety symptoms inversely associated with average cortical thickness; anxiety symptoms associated with slower cortical thinning over time
Vargas and Mittal (2022)	8,645	9–10 years	NS volume results; ↓ superior frontal cortex ct; ↑ precuneus ct; ↑ isthmus cingulate ct; ↓ inferior parietal cortex ct; ↓ rostral middle frontal gyrus ct; ↓ fusiform gyrus ct; ↑ caudal middle frontal gyrus ct; ↑ entorhinal cortex sa; ↑ posterior cingulate cortex sa; ↓ temporal pole sa; ↑ supramarginal gyrus sa	Anxiety symptoms not associated with subcortical regional volumes; anxiety symptoms inversely associated with superior frontal cortex thickness and positively associated with precuneus and isthmus cingulate cortex thickness at baseline; thinner cortices in the inferior parietal cortex, rostral middle frontal gyrus, and fusiform gyrus predicted increased anxiety symptoms over time; thicker cortices in the caudal middle frontal gyrus predicted increased anxiety symptoms over time; anxiety symptoms positively associated with cortical surface area in the entorhinal cortex and posterior cingulate cortex; anxiety symptoms inversely associated with cortical surface area in the temporal pole at baseline; greater surface area of the supramarginal gyrus predicted increased anxiety symptoms over time
Warnell et al. (2018)	72	6–13 years	↓ amygdala volume	Anxiety symptoms inversely associated with amygdala volume
Transdiagnostic anxiety				
Burkhouse et al. (2020)	55	7–19 years	↓ nucleus accumbens volume	Larger nucleus accumbens volume predicted greater decrease in anxiety symptoms in response to intervention

(continued)

Table 1 (continued)

Citation	N	Age	Direction/region/ modality	Relevant findings
Gold et al. (2017)	151	8–18 years	↓ hippocampus volume; ↑ ventromedial prefrontal cortex ct; ↑ precentral gyrus ct	Transdiagnostic anxiety disorders inversely associated with right hippocampal volume; transdiagnostic anxiety disorders positively associated with cortical thickness in the left ventromedial prefrontal cortex and left precentral gyrus
Milham et al. (2005)	51	Mean age 12 years, range not reported	↓ amygdala volume	Transdiagnostic anxiety disorders inversely associated with left amygdala volume
Strawn et al. (2015)	65	7–19 years	↑ dorsal anterior cingulate cortex volume; ↓ cuneus/precuneus volume, ↓ ventrolateral prefrontal cortex volume; ↓ postcentral gyrus volume	Transdiagnostic anxiety disorders positively associated with volume of the dorsal anterior cingulate cortex and inversely associated with volume of the cuneus/precuneus, ventrolateral prefrontal cortex, and postcentral gyrus
Suffren et al. (2019)	68	10–17 years	↓ amygdala volume; ↓ insula ct; ↓ rostral anterior cingulate cortex ct	Transdiagnostic anxiety disorders inversely associated with amygdala volume; transdiagnostic anxiety disorders inversely associated with cortical thickness in the insula and rostral anterior cingulate cortex
Yin et al. (2022)	79	Under 18 years; mean age 12 years, range not reported	↑ global ct	Transdiagnostic anxiety disorders positively associated with widespread cortical thickness
Generalized anxiety disorder				
Busso et al. (2017)	51	13–20 years	↑ middle temporal gyrus ct	Thicker cortices in the middle temporal gyrus predicted increased GAD symptoms over time
Chaarani et al. (2020)	160	14 years	↓ inferior frontal gyrus volume; ↓ ventral pre- and post-central gyri volume; ↓ insula volume; ↓ superior temporal gyrus volume	GAD inversely associated with volume in the right inferior frontal gyrus, ventral pre- and post-central gyri, insula, and superior temporal gyrus

(continued)

Table 1 (continued)

Citation	N	Age	Direction/region/ modality	Relevant findings
De Bellis et al. (2000)	36	8–16 years	↑ amygdala volume	GAD positively associated with amygdala volume
De Bellis et al. (2002a, b, c)	111	8–15 years	↑ superior temporal gyrus volume	GAD positively associated with superior temporal gyrus volume
Liao et al. (2013)	51	16–17 years	↑ putamen volume	GAD positively associated with right putamen volume
Strawn et al. (2014)	32	10–17 years	↑ inferolateral and ventromedial prefrontal cortex ct; ↑ inferior and middle temporal cortex ct; ↑ lateral occipital cortex ct	GAD positively associated with cortical thickness in the right inferolateral and ventromedial prefrontal cortex, left inferior and middle temporal cortex, and right lateral occipital cortex
Strawn et al. (2013)	43	10–18 years	↑ precuneus volume; ↑ precentral gyrus volume; ↓ orbital gyrus volume; ↓ posterior cingulate cortex volume	GAD positively associated with right precuneus and right precentral gyrus volumes and inversely associated with left orbital gyrus and posterior cingulate cortex volumes
Social anxiety disorder				
Auday and Pérez-Edgar (2019)	130	9–12 years	↓ ventrolateral prefrontal cortex volume	Social anxiety inversely associated with ventrolateral prefrontal cortex volume
Liu et al. (2021)	73	15–17 years	↑ orbitofrontal cortex volume; ↑ insula volume	Social anxiety positively associated with orbitofrontal cortex and insula volumes
Mueller et al. (2013)	102	Mean age 13 years, range not reported	↓ amygdala volume; ↓ hippocampus	Social anxiety inversely associated with amygdala and hippocampal volumes
Wang et al. (2021)	204	16–20 years	↑ middle temporal gyrus volume	Social anxiety positively associated with right middle temporal gyrus volume
Zhang et al. (2020)	106	8–17 years	↑ thalamus volume	Social anxiety positively associated with right thalamus volume
Obsessive-compulsive disorder				
Boedhoe et al. (2018)	731	Under 18 years, mean age 13 years, range not reported	↓ inferior and superior parietal cortex ct	OCD inversely associated with cortical thickness of inferior and superior parietal cortices
Boedhoe et al. (2020)	3,457	Under 18 years,	NS volume, ct, sa results	No associations found between OCD and regional

(continued)

Table 1 (continued)

Citation	N	Age	Direction/region/ modality	Relevant findings
		mean age 12 years, range not reported		subcortical volume, regional cortical thickness, or regional cortical surface area
Carmona et al. (2007)	36	10–15 years	↓ global volume; ↓ cingulate cortex volume; ↓ frontal cortex volume	OCD inversely associated with global volume and volume of areas in the cingulate cortex and frontal cortex
Chen et al. (2013)	20	8–15 years	↓ global volume; ↓ widespread regional volumes	OCD inversely associated with global volume, cingulate cortex volume, and volume of areas in the frontal, temporal, parietal, and occipital lobes
Cheng et al. (2016)	60	8–14 years	↑ putamen volume; ↑ orbitofrontal cortex volume; ↓ inferior parietal lobule volume	OCD positively associated with putamen and orbitofrontal cortex volumes, and inversely with inferior parietal lobule volume
Fallucca et al. (2011)	54	8–18 years	↓ supramarginal gyrus ct	OCD inversely associated with cortical thickness in the left supramarginal gyrus
Huyser et al. (2014)	37	8–19 years	↑ orbitofrontal cortex volume	OCD associated with increased orbitofrontal cortex volume over time
Ivanov et al. (2022)	577	6–19 years	↑ thalamus volume	OCD positively associated with right thalamus
Jayarajan et al. (2015)	30	Mean age 14 years, range not reported	↓ anterior cingulate cortex volume	OCD inversely associated with volume of the left anterior cingulate cortex
Lázaro et al. (2009)	30	9–17 years	↓ parietal lobe volume	OCD inversely associated with volume of the parietal lobe
Lázaro et al. (2014)	108	11–18 years	NS volume results	No associations found between OCD and regional gray matter volume
Pagliaccio et al. (2020)	55	7–18 years	↓ parietal lobe ct; ↓ insula ct; ↓ middle and superior frontal gyri ct; ↓ precentral gyrus ct; ↓ superior temporal gyrus ct; ↓ lingual gyrus ct	No associations found between OCD and regional subcortical volume or regional cortical thickness at baseline; thinner cortex in left parietal lobe, right insula, left middle and

(continued)

Table 1 (continued)

Citation	N	Age	Direction/region/ modality	Relevant findings
				superior frontal gyri, left precentral gyrus, left superior temporal gyrus, and lingual gyrus predicted greater symptom decrease with treatment
Pagliaccio et al. (2021)	9,475	9–10 years	NS volume, ct results	No associations found between OCD and regional subcortical volume or regional cortical thickness
Rosenberg et al. (1997)	38	7–18 years	↓ putamen volume	OCD inversely associated with putamen volume
Suñol et al. (2018)	255	8–12 years	↓ caudate volume; ↓ temporal pole volume; ↑ inferior frontal gyrus volume	OCD ordering symptoms inversely associated with caudate volume; OCD obsessing symptoms inversely associated with right temporal pole volume; OCD hoarding symptoms positively associated with volume of the left inferior frontal gyrus
Vattimo et al. (2021)	57	7–17 years	↑ hippocampus volume	OCD positively associated with hippocampus volume
Weeland et al. (2021)	2,551	9–12 years	↓ global volume; ↑ thalamus volume	OCD inversely associated with global volume and positively associated with thalamus volume
Wu et al. (2022)	5,256	9–10 years	↓ parietal lobe volume; ↓ precentral gyrus volume; ↓ precuneus volume	OCD compulsive acts and repeated thoughts symptoms inversely associated with volume of the parietal lobe, precentral gyrus, precuneus
Posttraumatic stress disorder				
Carrion et al. (2001)	24	7–14 years	↓ global volume	PTSD inversely associated with global volume
Carrion et al. (2009)	48	7–14 years	↑ prefrontal cortex volume	PTSD positively associated with prefrontal cortex volume
De Bellis et al. (2015)	132	6–16 years	↓ cerebellum volume	PTSD inversely associated with cerebellum volume
De Bellis et al. (1999)	105	6–17 years	↓ global volume	PTSD inversely associated with global volume
De Bellis et al. (2002a, b, c)	104	6–17 years	↑ superior temporal gyrus volume	PTSD positively associated with superior temporal gyrus volume

(continued)

Table 1 (continued)

Citation	N	Age	Direction/region/ modality	Relevant findings
De Bellis et al. (2002a, b, c)	94	4–17 years	↓ global volume; ↓ prefrontal cortex volume; ↓ temporal lobe volume	PTSD inversely associated with global volume and volume of the prefrontal cortex and right temporal lobe
De Bellis and Kuchibhatla (2006)	169	Mean age 12 years, range not reported	↓ cerebellum volume	PTSD inversely associated with cerebellum volume
Garrett et al. (2021)	40	10–17 years	↓ precuneus volume; ↓ midcingulate cortex volume; ↑ medial orbital cortex volume; ↑ posterior cingulate cortex volume	PTSD inversely associated with right precuneus and midcingulate cortex volumes, and positively associated with right medial orbital cortex and left posterior cingulate cortex volumes; structural differences also were associated with treatment response
Heyn and Herringa (2019)	55	8–18 years	↓ ventromedial prefrontal cortex sa; ↓ frontal pole sa	PTSD associated with decreased surface area of ventromedial prefrontal cortex over time; PTSD remission associated with increased frontal pole surface area and increased ventromedial prefrontal cortex thickness over time
Kedin and Herringa (2015)	54	8–18 years	↓ ventromedial prefrontal cortex volume	PTSD inversely associated with ventromedial prefrontal cortex volume
Morey et al. (2016)	120	6–16 years	↓ ventromedial prefrontal cortex volume	PTSD inversely associated with ventromedial prefrontal cortex volume
Mutluer et al. (2018)	44	13–18 years	↓ hippocampus volume; ↓ amygdala volume; ↓ anterior cingulate cortex volume; ↓ prefrontal cortex ct	PTSD inversely associated with hippocampus, amygdala, and anterior cingulate cortex volumes; PTSD inversely associated with cortical thickness in the prefrontal cortex
Richert et al. (2006)	47	7–14 years	↑ middle inferior and ventral prefrontal cortex volume	PTSD positively associated with middle inferior and ventral prefrontal cortex volume
Rinne-Albers et al. (2017)	46	12–21 years	↓ dorsal anterior cingulate cortex volume	PTSD inversely associated with dorsal anterior cingulate cortex volume

(continued)

Table 1 (continued)

Citation	N	Age	Direction/region/ modality	Relevant findings
Ross et al. (2021)	98	11–17 years	↓ amygdala volume	PTSD inversely associated with amygdala volume
Tupler and De Bellis (2006)	183	4–17 years	↑ hippocampus volume	PTSD positively associated with hippocampus volume
Weems et al. (2015)	54	9–17 years	↑ amygdala volume	PTSD positively associated with right amygdala volume

References

Achenbach TM (2009) The Achenbach system of empirically based assessment (ASEBA): development, findings, theory, and applications. University of Vermont Research Center for Children, Youth, and Families, Burlington

Ashworth E, Brooks SJ, Schiöth HB (2021) Neural activation of anxiety and depression in children and young people: a systematic meta-analysis of fMRI studies. Psychiatry Res Neuroimaging 311. https://doi.org/10.1016/J.PSCYCHRESNS.2021.111272

Auday ES, Pérez-Edgar KE (2019) Limbic and prefrontal neural volume modulate social anxiety in children at temperamental risk. Depress Anxiety 36:690–700. https://doi.org/10.1002/da.22941

Baum GL, Cui Z, Roalf DR et al (2020) Development of structure–function coupling in human brain networks during youth. Proc Natl Acad Sci U S A 117:771–778. https://doi.org/10.1073/PNAS.1912034117

Birmaher B, Khetarpal S, Brent D et al (1997) The screen for child anxiety related emotional disorders (SCARED): scale construction and psychometric characteristics. J Am Acad Child Adolesc Psychiatry 36:545–553. https://doi.org/10.1097/00004583-199704000-00018

Boedhoe PSW, Schmaal L, Abe Y et al (2018) Cortical abnormalities associated with pediatric and adult obsessive-compulsive disorder: findings from the ENIGMA obsessive-compulsive disorder working group. Am J Psychiatry 175:453–462. https://doi.org/10.1176/appi.ajp.2017.17050485

Boedhoe PSW, van Rooij D, Hoogman M et al (2020) Subcortical brain volume, regional cortical thickness, and cortical surface area across disorders: findings from the ENIGMA ADHD, ASD, and OCD working groups. Am J Psychiatry 177:834–843. https://doi.org/10.1176/appi.ajp.2020.19030331

Bonadio FT, Evans SC, Cho GY et al (2022) Whose outcomes come out? Patterns of caregiver- and youth-reported outcomes based on caregiver-youth baseline discrepancies. J Clin Child Adolesc Psychol 51:469–483. https://doi.org/10.1080/15374416.2021.1955367

Briere J (1996) Trauma symptom checklist for children. In: Psychological assessment resources. Odessa, pp 00253–00258

Broekhuizen A, Vriend C, Wolf N et al (2023) Poor insight in obsessive-compulsive disorder as a multifaceted phenomenon: evidence from brain activation during symptom provocation. Biol Psychiatry Cogn Neurosci Neuroimaging. https://doi.org/10.1016/J.BPSC.2023.04.006

Bruin WB, Zhutovsky P, van Wingen GA et al (2024) Brain-based classification of youth with anxiety disorders: transdiagnostic examinations within the ENIGMA-anxiety database using machine learning. Nat Mental Health 2:104–118. https://doi.org/10.1038/s44220-023-00173-2

Burkhouse KL, Jimmy J, Defelice N et al (2020) Nucleus accumbens volume as a predictor of anxiety symptom improvement following CBT and SSRI treatment in two independent samples. Neuropsychopharmacology 45:561–569. https://doi.org/10.1038/s41386-019-0575-5

Busso DS, McLaughlin KA, Brueck S et al (2017) Child abuse, neural structure, and adolescent psychopathology: a longitudinal study. J Am Acad Child Adolesc Psychiatry 56:321–328. https://doi.org/10.1016/j.jaac.2017.01.013

Carmona S, Bassas N, Rovira M et al (2007) Pediatric OCD structural brain deficits in conflict monitoring circuits: a voxel-based morphometry study. Neurosci Lett 421:218–223. https://doi.org/10.1016/j.neulet.2007.05.047

Carrion VG, Weems CF, Eliez S et al (2001) Attenuation of frontal asymmetry in pediatric posttraumatic stress disorder. Biol Psychiatry 50:943–951. https://doi.org/10.1016/S0006-3223(01)01218-5

Carrion VG, Weems CF, Watson C et al (2009) Converging evidence for abnormalities of the prefrontal cortex and evaluation of midsagittal structures in pediatric posttraumatic stress disorder: an MRI study. Psychiatry Res Neuroimaging 172:226–234. https://doi.org/10.1016/j.pscychresns.2008.07.008

Castagna PJ, Roye S, Calamia M et al (2018) Parsing the neural correlates of anxious apprehension and anxious arousal in the grey-matter of healthy youth. Brain Imaging Behav 12:1084–1098. https://doi.org/10.1007/s11682-017-9772-1

Chaarani B, Kan KJ, Mackey S et al (2020) Neural correlates of adolescent irritability and its comorbidity with psychiatric disorders. J Am Acad Child Adolesc Psychiatry 59:1371–1379. https://doi.org/10.1016/j.jaac.2019.11.028

Chen J, Silk T, Seal M et al (2013) Widespread decreased grey and white matter in paediatric obsessive-compulsive disorder (OCD): a voxel-based morphometric MRI study. Psychiatry Res Neuroimaging 213:11–17. https://doi.org/10.1016/j.pscychresns.2013.02.003

Cheng B, Cai W, Wang X et al (2016) Brain gray matter abnormalities in first-episode, treatment-naive children with obsessive-compulsive disorder. Front Behav Neurosci 10. https://doi.org/10.3389/fnbeh.2016.00141

Conway CC, Forbes MK, South SC (2022) A hierarchical taxonomy of psychopathology (HiTOP) primer for mental health researchers. Clin Psychol Sci 10:236–258. https://doi.org/10.1177/21677026211017834

De Bellis MD, Kuchibhatla M (2006) Cerebellar volumes in pediatric maltreatment-related posttraumatic stress disorder. Biol Psychiatry 60:697–703. https://doi.org/10.1016/j.biopsych.2006.04.035

De Bellis MD, Keshavan MS, Clark DB et al (1999) Developmental traumatology: II. Brain development. Biol Psychiatry 45:1271–1284. https://doi.org/10.1016/S0006-3223(99)00045-1

De Bellis MD, Casey BJ, Dahl RE et al (2000) A pilot study of amygdala volumes in pediatric generalized anxiety disorder. Biol Psychiatry 48:51–57. https://doi.org/10.1016/S0006-3223(00)00835-0

De Bellis MD, Keshavan MS, Shifflett H et al (2002a) Superior temporal gyrus volumes in pediatric generalized anxiety disorder. Biol Psychiatry 51:553–562. https://doi.org/10.1016/S0006-3223(01)01375-0

De Bellis MD, Keshavan MS, Shifflett H et al (2002b) Brain structures in pediatric maltreatment-related posttraumatic stress disorder: a sociodemographically matched study. Biol Psychiatry 52:1066–1078. https://doi.org/10.1016/S0006-3223(02)01459-2

De Bellis MD, Keshavan MS, Frustaci K et al (2002c) Superior temporal gyrus volumes in maltreated children and adolescents with PTSD. Biol Psychiatry 51:544–552. https://doi.org/10.1016/S0006-3223(01)01374-9

De Bellis MD, Hooper SR, Chen S et al (2015) Posterior structural brain volumes differ in maltreated youth with and without chronic posttraumatic stress disorder. Dev Psychopathol 27:1555–1576. https://doi.org/10.1017/S0954579415000942

de Sousa Júnior GM, Tavares VD d O, de Meiroz Grilo MLP et al (2021) Mental health in COVID-19 pandemic: a meta-review of prevalence meta-analyses. Front Psychol 12. https://doi.org/10.3389/FPSYG.2021.703838/BIBTEX

DiMauro J, Carter S, Folk JB, Kashdan TB (2014) A historical review of trauma-related diagnoses to reconsider the heterogeneity of PTSD. J Anxiety Disord 28:774–786. https://doi.org/10.1016/J.JANXDIS.2014.09.002

Durham EL, Jeong HJ, Moore TM et al (2021) Association of gray matter volumes with general and specific dimensions of psychopathology in children. Neuropsychopharmacology 46:1333–1339. https://doi.org/10.1038/s41386-020-00952-w

Egger HL, Ascher BH, Angold A (1999) The preschool age psychiatric assessment (PAPA). Duke University Medical Center, Durham

Fallucca E, MacMaster FP, Haddad J et al (2011) Distinguishing between major depressive disorder and obsessive-compulsive disorder in children by measuring regional cortical thickness. Arch Gen Psychiatry 68:527–533. https://doi.org/10.1001/archgenpsychiatry.2011.36

Garrett AS, Abazid L, Cohen JA et al (2021) Changes in brain volume associated with trauma-focused cognitive behavioral therapy among youth with posttraumatic stress disorder. J Trauma Stress. https://doi.org/10.1002/jts.22678

Giedd JN, Blumenthal J, Jeffries NO et al (1999) Brain development during childhood and adolescence: a longitudinal MRI study. Nat Neurosci 2:861–863. https://doi.org/10.1038/13158

Gold AL, Steuber ER, White LK et al (2017) Cortical thickness and subcortical gray matter volume in pediatric anxiety disorders. Neuropsychopharmacology 42:2423–2433. https://doi.org/10.1038/npp.2017.83

Heyn SA, Herringa RJ (2019) Longitudinal cortical markers of persistence and remission of pediatric PTSD. Neuroimage Clin 24:102028. https://doi.org/10.1016/j.nicl.2019.102028

Huyser C, OA van den H, Wolters L et al (2014) A longitudinal VBM study in paediatric obsessive-compulsive disorder at 2-year follow-up after cognitive behavioural therapy. World J Biol PsychiatryBiol Psychiatry 15:443–452. https://doi.org/10.3109/15622975.2013.819122

Ivanov I, Boedhoe PSW, Abe Y et al (2022) Associations of medication with subcortical morphology across the lifespan in OCD: results from the international ENIGMA consortium. J Affect Disord 318:204–216. https://doi.org/10.1016/j.jad.2022.08.084

Jayarajan RN, Agarwal SM, Viswanath B et al (2015) A voxel based morphometry study of brain gray matter volumes in juvenile obsessive compulsive disorder. J Can Acad Child Adolesc Psychiatry 24:84–91

Kaczkurkin AN, Park SS, Sotiras A et al (2019) Evidence for dissociable linkage of dimensions of psychopathology to brain structure in youths. Am J Psychiatry 17:1000–1009. https://doi.org/10.1176/appi.ajp.2019.18070835

Kaczkurkin AN, Sotiras A, Baller EB et al (2020) Neurostructural heterogeneity in youths with internalizing symptoms. Biol Psychiatry 87:473–482. https://doi.org/10.1016/j.biopsych.2019.09.005

Kedin TJ, Herringa RJ (2015) Abnormal structure of fear circuitry in pediatric post-traumatic stress disorder. Neuropsychopharmacology 40:537–545. https://doi.org/10.1038/npp.2014.239

Kessler RC, Berglund P, Demler O et al (2005) Lifetime prevalence and age-of-onset distributions of DSM-IV disorders in the national comorbidity survey replication. Arch Gen Psychiatry 62:593–602. https://doi.org/10.1001/archpsyc.62.6.593

Kotov R, Waszczuk MA, Krueger RF et al (2017) The hierarchical taxonomy of psychopathology (HiTOP): a dimensional alternative to traditional nosologies. J Abnorm Psychol 126:454–477. https://doi.org/10.1037/abn0000258

Lahey BB, Moore TM, Kaczkurkin AN, Zald DH (2021) Hierarchical models of psychopathology: empirical support, implications, and remaining issues. World Psychiatry 20:57–63. https://doi.org/10.1002/wps.20824

Lanius RA, Bluhm R, Lanius U, Pain C (2006) A review of neuroimaging studies in PTSD: heterogeneity of response to symptom provocation. J Psychiatr Res 40:709–729. https://doi.org/10.1016/J.JPSYCHIRES.2005.07.007

Lázaro L, Bargalló N, Castro-Fornieles J et al (2009) Brain changes in children and adolescents with obsessive–compulsive disorder before and after treatment: a voxel-based morphometric

MRI study. Psychiatry Res Neuroimaging 172:140–146. https://doi.org/10.1016/j.pscychresns.2008.12.007

Lázaro L, Ortiz AG, Calvo A et al (2014) White matter structural alterations in pediatric obsessive–compulsive disorder: relation to symptom dimensions. Prog Neuropsychopharmacol Biol Psychiatry 54:249–258. https://doi.org/10.1016/j.pnpbp.2014.06.009

Liao M, Yang F, Zhang Y et al (2013) Childhood maltreatment is associated with larger left thalamic gray matter volume in adolescents with generalized anxiety disorder. PLoS One 8. https://doi.org/10.1371/journal.pone.0071898

Liu Z, Hu Y, Zhang Y et al (2021) Altered gray matter volume and structural co-variance in adolescents with social anxiety disorder: evidence for a delayed and unsynchronized development of the fronto-limbic system. Psychol Med 51:1742–1751. https://doi.org/10.1017/S0033291720000495

Luking KR, Jirsaraie RJ, Tillman R et al (2022) Timing and type of early psychopathology symptoms predict longitudinal change in cortical thickness from middle childhood into early adolescence. Biol Psychiatry Cogn Neurosci Neuroimaging 7:397–405. https://doi.org/10.1016/j.bpsc.2021.06.013

Madan CR (2017) Advances in studying brain morphology: the benefits of open-access data. Front Hum Neurosci 11. https://doi.org/10.3389/FNHUM.2017.00405

March JS, Parker JDA, Sullivan K et al (1997) The multidimensional anxiety scale for children (MASC): factor structure, reliability, and validity. J Am Acad Child Adolesc Psychiatry 36:554–565. https://doi.org/10.1097/00004583-199704000-00019

Merz EC, He X, Noble K (2018) Anxiety, depression, impulsivity, and brain structure in children and adolescents. Neuroimage Clin 20:243–251. https://doi.org/10.1016/j.nicl.2018.07.020

Milham MP, Nugent AC, Drevets WC et al (2005) Selective reduction in amygdala volume in pediatric anxiety disorders: a voxel-based morphometry investigation. Biol Psychiatry 57:961–966. https://doi.org/10.1016/j.biopsych.2005.01.038

Morey RA, Haswell CC, Hooper SR, De Bellis MD (2016) Amygdala, hippocampus, and ventral medial prefrontal cortex volumes differ in maltreated youth with and without chronic posttraumatic stress disorder. Neuropsychopharmacology 41:791–801. https://doi.org/10.1038/npp.2015.205

Mueller SC, Aouidad A, Gorodetsky E et al (2013) Gray matter volume in adolescent anxiety: an impact of the brain-derived neurotrophic factor Val[66]Met polymorphism? J Am Acad Child Adolesc Psychiatry 52:184–195. https://doi.org/10.1016/j.jaac.2012.11.016

Muris P (2002) The revised version of the screen for anxiety related emotional disorders (SCARED-R): a new scale for measuring childhood anxiety. Int Perspect Child Adolescent Mental Health 2:257–274. https://doi.org/10.1016/S1874-5911(02)80011-2

Mutluer T, Şar V, Kose-Demiray Ç et al (2018) Lateralization of neurobiological response in adolescents with post-traumatic stress disorder related to severe childhood sexual abuse: the tri-modal reaction (T-MR) model of protection. J Trauma Dissociation 19:108–125. https://doi.org/10.1080/15299732.2017.1304489

Nave G, Jung WH, Karlsson Linnér R et al (2019) Are bigger brains smarter? Evidence from a large-scale preregistered study. Psychol Sci 30:43–54. https://doi.org/10.1177/0956797618808470

Newman E, Thompson WK, Bartsch H et al (2016) Anxiety is related to indices of cortical maturation in typically developing children and adolescents. Brain Struct Funct 221:3013–3025. https://doi.org/10.1007/s00429-015-1085-9

Pagliaccio D, Cha J, He X et al (2020) Structural neural markers of response to cognitive behavioral therapy in pediatric obsessive-compulsive disorder. J Child Psychol Psychiatry 61:1299–1308. https://doi.org/10.1111/jcpp.13191

Pagliaccio D, Durham K, Fitzgerald KD, Marsh R (2021) Obsessive-compulsive symptoms among children in the adolescent brain and cognitive development study: clinical, cognitive, and brain connectivity correlates. Biol Psychiatry Cogn Neurosci Neuroimaging 6:399–409. https://doi.org/10.1016/j.bpsc.2020.10.019

Pine DS (2007) Research review: a neuroscience framework for pediatric anxiety disorders. J Child Psychol Psychiatry 48:631–648. https://doi.org/10.1111/J.1469-7610.2007.01751.X

Qin S, Young CB, Duan X et al (2014) Amygdala subregional structure and intrinsic functional connectivity predicts individual differences in anxiety during early childhood. Biol Psychiatry 75:892–900. https://doi.org/10.1016/j.biopsych.2013.10.006

Reef J, Van Meurs I, Verhulst FC, Van Der Ende J (2010) Children's problems predict adults' DSM-IV disorders across 24 years. J Am Acad Child Adolesc Psychiatry 49:1117–1124. https://doi.org/10.1016/j.jaac.2010.08.002

Remes O, Brayne C, van der Linde R, Lafortune L (2016) A systematic review of reviews on the prevalence of anxiety disorders in adult populations. Brain Behav 6

Richert KA, Carrion VG, Karchemskiy A, Reiss AL (2006) Regional differences of the prefrontal cortex in pediatric PTSD: an MRI study. Depress Anxiety 23:17–25. https://doi.org/10.1002/da.20131

Rinne-Albers MA, Pannekoek JN, van Hoof M-J et al (2017) Anterior cingulate cortex grey matter volume abnormalities in adolescents with PTSD after childhood sexual abuse. Eur Neuropsychopharmacol 27:1163–1171. https://doi.org/10.1016/j.euroneuro.2017.08.432

Rosenberg DR, Keshavan M, O'Hearn KM et al (1997) Frontostriatal measurement in treatment-naive children with obsessive-compulsive disorder. Arch Gen Psychiatry 54:824–830. https://doi.org/10.1001/archpsyc.1997.01830210068007

Ross MC, Dvorak D, Sartin-Tarm A et al (2021) Gray matter volume correlates of adolescent posttraumatic stress disorder: a comparison of manual intervention and automated segmentation in Freesurfer. Psychiatry Res Neuroimaging 313:111297. https://doi.org/10.1016/j.pscychresns.2021.111297

Roza SJ, Hofstra MB, Van Der Ende J, Verhulst FC (2003) Stable prediction of mood and anxiety disorders based on behavioral and emotional problems in childhood: a 14-year follow-up during childhood, adolescence, and young adulthood. Am J Psychiatry 160:2116–2121. https://doi.org/10.1176/appi.ajp.160.12.2116

Schuttenberg E, Sneider JT, Rosmarin DH et al (2022) Forgiveness mediates the relationship between middle frontal gyrus volume and clinical symptoms in adolescents. Front Hum Neurosci 16. https://doi.org/10.3389/fnhum.2022.782893

Solmi M, Radua J, Olivola M et al (2021) Age at onset of mental disorders worldwide: large-scale meta-analysis of 192 epidemiological studies. Mol Psychiatry 27:281–295. https://doi.org/10.1038/s41380-021-01161-7

Spielberger CD, Edwards CD, Montouri J, Lushene R (1973) State-trait anxiety inventory for children. APA PsycTests. https://doi.org/10.1037/t06497-000

Stiles J, Jernigan TL (2010) The basics of brain development. Neuropsychol Rev 20:327–348. https://doi.org/10.1007/S11065-010-9148-4

Strawn JR, Wehry AM, Chu WJ et al (2013) Neuroanatomic abnormalities in adolescents with generalized anxiety disorder: a voxel-based morphometry study. Depress Anxiety 30:842–848. https://doi.org/10.1002/da.22089

Strawn JR, John Wegman C, Dominick KC et al (2014) Cortical surface anatomy in pediatric patients with generalized anxiety disorder. J Anxiety Disord 28:717–723. https://doi.org/10.1016/j.janxdis.2014.07.012

Strawn JR, Hamm L, Fitzgerald DA et al (2015) Neurostructural abnormalities in pediatric anxiety disorders. J Anxiety Disord 32:81–88. https://doi.org/10.1016/j.janxdis.2015.03.004

Suffren S, Chauret M, Nassim M et al (2019) On a continuum to anxiety disorders: adolescents at parental risk for anxiety show smaller rostral anterior cingulate cortex and insula thickness. J Affect Disord 248:34–41. https://doi.org/10.1016/j.jad.2019.01.028

Suñol M, Contreras-Rodríguez O, Macià D et al (2018) Brain structural correlates of subclinical obsessive-compulsive symptoms in healthy children. J Am Acad Child Adolesc Psychiatry 57:41–47. https://doi.org/10.1016/j.jaac.2017.10.016

Sylvester CM, Corbetta M, Raichle ME et al (2012) Functional network dysfunction in anxiety and anxiety disorders. Trends Neurosci 35:527–535. https://doi.org/10.1016/J.TINS.2012.04.012

Taylor BK, Eastman JA, Frenzel MR et al (2021) Subclinical anxiety and posttraumatic stress influence cortical thinning during adolescence. J Am Acad Child Adolesc Psychiatry 60:1288–1299. https://doi.org/10.1016/j.jaac.2020.11.020

Tupler LA, De Bellis MD (2006) Segmented hippocampal volume in children and adolescents with posttraumatic stress disorder. Biol Psychiatry 59:523–529. https://doi.org/10.1016/j.biopsych.2005.08.007

Vargas TG, Mittal VA (2022) Brain morphometry points to emerging patterns of psychosis, depression, and anxiety vulnerability over a 2-year period in childhood. Psychol Med 1–13. https://doi.org/10.1017/S0033291721005304

Vattimo EFQ, dos Santos AC, Hoexter MQ et al (2021) Higher volumes of hippocampal subfields in pediatric obsessive-compulsive disorder. Psychiatry Res Neuroimaging 307:111200. https://doi.org/10.1016/j.pscychresns.2020.111200

Wang S, Zhao Y, Wang X et al (2021) Emotional intelligence mediates the association between middle temporal gyrus gray matter volume and social anxiety in late adolescence. Eur Child Adolesc Psychiatry 30:1857–1869. https://doi.org/10.1007/s00787-020-01651-z

Warnell KR, Pecukonis M, Redcay E (2018) Developmental relations between amygdala volume and anxiety traits: effects of informant, sex, and age. Dev Psychopathol 30:1503–1515. https://doi.org/10.1017/S0954579417001626

Weeland CJ, White T, Vriend C et al (2021) Brain morphology associated with obsessive-compulsive symptoms in 2,551 children from the general population. J Am Acad Child Adolesc Psychiatry 60:470–478. https://doi.org/10.1016/j.jaac.2020.03.012

Weems CF, Klabunde M, Russell JD et al (2015) Post-traumatic stress and age variation in amygdala volumes among youth exposed to trauma. Soc Cogn Affect Neurosci 10:1661–1667. https://doi.org/10.1093/scan/nsv053

Wierenga LM, Langen M, Oranje B, Durston S (2014) Unique developmental trajectories of cortical thickness and surface area. Neuroimage 87:120–126. https://doi.org/10.1016/J.NEUROIMAGE.2013.11.010

Winkler AM, Kochunov P, Blangero J et al (2010) Cortical thickness or grey matter volume? The importance of selecting the phenotype for imaging genetics studies. Neuroimage 53:1135–1146. https://doi.org/10.1016/J.NEUROIMAGE.2009.12.028

Wu X, Yu G, Zhang K et al (2022) Symptom-based profiling and multimodal neuroimaging of a large Preteenage population identifies distinct obsessive-compulsive disorder-like subtypes with neurocognitive differences. Biol Psychiatry Cogn Neurosci Neuroimaging 7:1078–1089. https://doi.org/10.1016/j.bpsc.2021.06.011

Yin S, Hong SJ, Di Martino A et al (2022) Shared and distinct patterns of atypical cortical morphometry in children with autism and anxiety. Cereb Cortex 32:4565–4575. https://doi.org/10.1093/cercor/bhab502

Yovel I, Gershuny BS, Steketee G et al (2012) Idiosyncratic severity profiles of obsessive compulsive dysfunction: a detailed self report assessment of a multifaceted syndrome. Cognit Ther Res 36:694–713. https://doi.org/10.1007/S10608-011-9427-2/TABLES/6

Zhang Y, Liu W, Lebowitz ER et al (2020) Abnormal asymmetry of thalamic volume moderates stress from parents and anxiety symptoms in children and adolescents with social anxiety disorder. Neuropharmacology 180:8. https://doi.org/10.1016/j.neuropharm.2020.108301

Computational Mechanisms of Information-Seeking in Anxiety

Ko-Ping Chou and Ryan Smith

Contents

Abstract The drive to seek information through exploratory behavior is widespread in both humans and other animals. This can be adaptive in reducing uncertainty about the best course of action within novel or changing environments. However, exploratory behaviors can also become maladaptive if subjective uncertainty levels remain too high or too low, as may happen in states of elevated anxiety. In this article, we review recent studies investigating the influence of anxiety on information-seeking behavior. We focus primarily on studies using cognitive computational models and associated behavioral tasks designed to test specific exploratory strategies, which could each be affected by anxiety in distinct ways. Results of current studies remain mixed and highlight the importance of distinguishing potential effects of task, state vs. trait anxiety, somatic vs. cognitive anxiety, and clinical vs. sub-clinical anxiety. There are also a range of different information-seeking strategies that are necessary to consider. At present, many findings could be taken to support a picture in which cognitive anxiety, and/or trait anxiety more broadly, may increase information-seeking, while somatic and/or state anxiety could have opposing effects. However, a number of previous results also appear inconsistent or task-dependent. Future studies are needed to resolve these apparent

K.-P. Chou and R. Smith (✉)
Laureate Institute for Brain Research, Tulsa, OK, USA
e-mail: rssmith@laureateinstitute.org

© The Author(s), under exclusive license to Springer Nature Switzerland AG 2024

Curr Topics Behav Neurosci (2025) 73: 155–177
https://doi.org/10.1007/7854_2024_552
Published Online: 19 December 2024

inconsistencies and more directly disentangle effects of different dimensions of anxiety on the adaptive and maladaptive use of information-seeking.

Keywords Anxiety · Directed exploration · Foraging · Information-seeking

1 Introduction

Anxiety is closely linked to uncertainty (Grupe and Nitschke 2013). Anxiety disorders are characterized by persistent and exaggerated responses to uncertain outcomes and a lack of confidence in how to respond – resulting in a set of characteristic changes in affect, cognition, and behavior. Anxious feelings can be transient and short-lived (i.e., *state anxiety*); however, some individuals may also be more pre-disposed than others to experience such feelings – so-called *trait anxiety* (Spielberger et al. 1971). Importantly, the experience of anxiety is multidimensional, often broadly divided into cognitive and somatic domains (although narrower distinctions, such as between social and non-social domains, are also commonly discussed; Ree et al. 2008; Sandin et al. 2001). For example, cognitive anxiety is more closely linked to worry, exaggerated/persistent uncertainty, and indecision, while somatic anxiety is often associated with feelings of tension, stress, and high arousal. Different aspects of anxiety may also influence behavior in distinct ways. For example, while some anxious states may drive individuals to reduce uncertainty through excessive information-seeking, others may promote avoidance of feared situations and prevent information gain (i.e., the basis of exposure therapy, which promotes information-seeking; Barlow et al. 2016). In the latter case, uncertainty may instead be reduced by staying in familiar environments believed to be safe. Thus, anxiety is a broad and multifaceted construct, and some aspects will likely be of greater relevance than others depending on the context in question. In this chapter, we review work on one specific class of behaviors aimed at uncertainty reduction – often referred to as *exploratory* behaviors – and how they may be exaggerated or diminished by different facets of anxiety. We also focus primarily on recent studies using cognitive computational modeling, which suggest specific mechanisms by which information-seeking behaviors may become maladaptive in anxiety disorders.

In psychology research, the difficult decision to seek information or to trust current knowledge and immediately seek reward is known as the "explore-exploit dilemma" (Schulz and Gershman 2019; Wilson et al. 2021). A poor balance between exploration and exploitation of current knowledge has been proposed as a potentially important factor across multiple psychiatric disorders (Addicott et al. 2017). Here, over-exploration can sacrifice time that could be used to maximize quality of life, while under-exploration can lead to uninformed, non-reflective choices – with clear links to mental health issues associated with both exaggerated uncertainty and impulsivity, respectively (Boswell et al. 2013; Moeller et al. 2001). However, multiple types of exploratory strategies have been distinguished in past literature, and some appear more relevant to anxiety than others. Here, the most important distinction has been between directed versus random exploration (Gershman 2019; Wilson et al. 2014). *Directed exploration (DE)* arises when one is more confident about the outcomes of some actions than others (e.g., perhaps one action has been chosen more often in the past, allowing more experience with the outcomes that follow). To gain information, this exploratory strategy involves choosing the action for which one is more uncertain (e.g., the one chosen less often in the past). That is, choice is directed toward the actions for which one has the least information. This competes with reward-seeking behaviors, especially when past experience suggests the uncertain action may be less rewarding (i.e., but with the possibility that past experience was misleading). As this form of exploration depends on sensitivity to relative uncertainty between actions, and has been tied to trait differences in cognitive reflectiveness (Smith et al. 2022b), it could be viewed as goal-directed. However, it should be noted that DE is not always modeled as explicitly dependent on the expected usefulness of new information and, in such cases, could instead reflect a type of motivated curiosity. The other commonly discussed strategy, *random exploration (RE)*, does not specifically target some actions over others. Instead, it simply involves choosing actions more randomly when overall uncertainty is high (e.g., being more likely to choose actions that appear less rewarding, independent of relative uncertainty). In this chapter, we focus on DE, both for the sake of space and because previous work has mainly identified relationships between anxiety and this exploration strategy.

A related type of exploration we consider is foraging. Foraging involves repeatedly choosing to stay in one location (e.g., a tree) and harvest a steadily reducing amount of reward (e.g., a slowly depleting number of apples), or to leave the current location to explore whether other locations might be better. Optimal exploration in this context entails that one should explore other options when the remaining reward at the current location drops below the expected average reward that could be found elsewhere, where this expected average reward needs to be learned through the (risky) exploratory choice of leaving the current location despite uncertainty. While this type of exploration could be seen as directed in some ways, the task and choice structure are sufficiently different that we review this literature separately.

A final type of exploration we consider is explicitly non-instrumental. Here, individuals choose whether or not to incur a cost to gain information when they know that this information cannot be used to inform future choices (and therefore cannot help them maximize long-term rewards). Thus, it simply reflects intrinsic curiosity or the drive to reduce uncertainty for its own sake. Non-instrumental information-seeking tasks therefore do not create an explore-exploit dilemma. They instead provide a way to measure the intrinsic value an individual assigns to uncertainty reduction.

In what follows, we first offer a brief primer on computational modeling for those new to this approach. We then review recent studies linking anxiety to DE. We further review related work on foraging behavior and non-instrumental information seeking. We conclude with a discussion of important future directions that will be needed to clarify underlying mechanisms and resolve apparent inconsistencies in existing results.

2 Computational Modeling

In many behavioral tasks used in psychology research, summary statistics are sufficient to capture the aspects of behavior necessary to test a prediction of interest. However, when tasks involve learning, this becomes challenging because behavior shows a dependence on outcome order (e.g., behavior should be different if a win is followed by a loss versus if a loss is followed by a win). In addition, outcome order depends on participants' choices (e.g., better choices will lead a participant to see more wins). These factors lead to important behavioral patterns that evolve over time, which can be missed when averaging over trials to get summary statistics – creating the need for more sophisticated measures. Computational modeling is an approach designed to fill this need by mathematically encoding hypothesized mechanisms of learning and choice in models that predict behavioral dynamics (Wilson and Collins 2019). Similar needs also arise in other tasks that involve dynamics in perceptual judgment, planning, and reaction times, making this type of modeling an important tool across a wide range of experimental domains. Here we provide a brief example of a computational model, which we hope will help unfamiliar readers gain a deeper understanding of the empirical results reviewed below.

One of the simplest models of learning is called the Rescorla-Wagner (RW) model. In a decision-making task with two (or more) options to choose from on each trial that may or may not deliver a reward – often called "bandit tasks" based on the analogy to choosing between slot machines – this model assumes a participant starts out on the first trial ($t = 1$) assigning an expected reward value (R_a) to each option (a). Let us assume $R_{a,t=1} = 0$, meaning there is no expected reward (positive value) or punishment (negative value). If a participant chooses Option 1 ($a = 1$) and observes a loss ($r = -1$), the RW model assumes they learn by first computing a reward prediction error (*RPE*), which is the difference between the observed and expected reward:

$$\text{RPE} = r - R_{a,t}$$

In our example, this would be:

$$\text{RPE} = -1 - 0 = -1$$

The RPE is then used to learn a new expected value for the next trial ($R_{a,t=2}$), by adding it to the old value ($R_{a,t=1}$). More generically, this can be written as:

$$R_{a,t+1} = R_{a,t} + \text{RPE}$$

Here, this would be:

$$R_{a=1,t=2} = 0 + (-1) = -1$$

This means that, on the next trial, a negative outcome will be expected if Option 1 were again chosen. Thus, Option 2 may be favored, as its expected value is still equal to 0 and therefore higher than that of Option 1. However, in probabilistic environments, it can be suboptimal to change expectations too much or too little after a single outcome. This magnitude of change can also differ between individuals. These factors motivate the inclusion of a model parameter called the *learning rate* (often denoted by the Greek letter alpha; α). This parameter is added to the previous equation and controls how strongly an *RPE* changes expectations after each trial:

$$R_{a,t+1} = R_{a,t} + \alpha(\text{RPE})$$

In our example, assuming $\alpha = .5$, this would be:

$$R_{a=1,t=2} = 0 + .5(-1) = -.5$$

This smaller change prevents a possible "jumping to conclusions" effect after a single observation. Importantly, computational modeling allows us to measure individual differences in learning rate, which can provide useful information about individual participants beyond summary statistics.

Modeling can also measure individual differences in exploratory behavior using additional parameters. While there are many approaches, we will briefly illustrate two. Both depend on modeling how people use expected rewards to choose between options. One intuitive approach is a logistic choice model, which says that choosing Option 1 over Option 2 will become more probable as the expected reward for Option 1 becomes greater than that for Option 2. Here, one can include two further parameters: an *information bonus* (A) and a *decision noise* (σ). Higher values for A increase the probability of choosing the option that was less often chosen previously, promoting DE. Higher values for σ increase the probability of choosing the option with the lower expected reward, promoting RE. The logistic choice model provides the probability of choosing Option 1 ($P_{a=1}$) as follows:

$$P_{a=1} = \frac{1}{1 + \exp\left(\frac{(R_{a=2} - R_{a=1}) + A(I_{a=2} - I_{a=1})}{\sigma}\right)}$$

The variable $I_a = 1$ for the option chosen less often in the past and $I_a = 0$ for the option chosen more often in the past. The information bonus (A) controls how strongly that difference influences choice. While this equation might look complex, it simply assigns a higher probability to Option 1 than Option 2 if the expected value

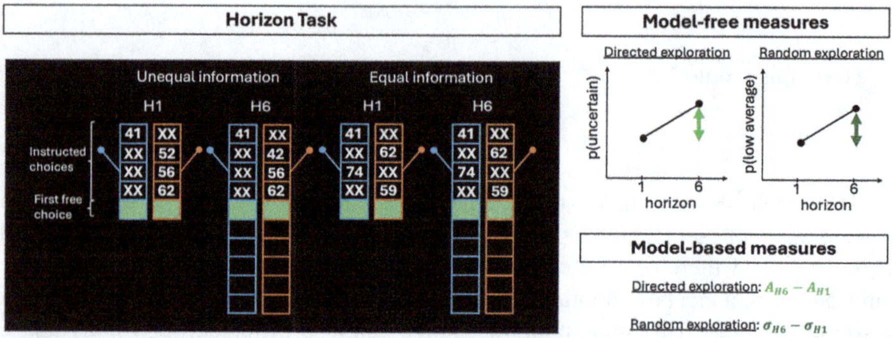

Fig. 1 The horizon task. *Left*: In this task, there are two options (bandits; orange and blue vertical stacked boxes). After choosing an option, the reward value appears in the associated box, and "XX" appears in the box for the other option. Choices go from top to bottom. Four variations of games are shown. In all cases, the outcomes of the first four (forced) choices are shown. The green filled boxes indicate the first choice participants can freely make by themselves. In some games, there is only one free choice (H1); in other games, there are six (H6). In some games, the forced choices provide two outcomes per bandit (equal information), while in other games they provide three outcomes for one bandit and only one outcome for the other (unequal information). *Right*: Directed exploration is measured in unequal information games based on how much more often the uncertain option is chosen in H6 than in H1, as can be estimated by information bonus parameter differences ($A_{H6} - A_{H1}$). Similarly, random exploration is measured based on how much more often the option with the lower observed reward average is chosen in H6 than in H1 games, as can be estimated by decision noise parameter differences ($\sigma_{H6} - \sigma_{H1}$)

of Option 1 is higher, and this effect is amplified by the information bonus if Option 1 has also been chosen less often in the past. Conversely, this probability will be reduced somewhat if the decision noise is high.

Crucially, we can use these models to estimate information bonus and decision noise values (and therefore exploration levels) for each participant based on dynamics in their choice behavior. This is done using *parameter estimation algorithms*, which are beyond the scope of this article. Briefly, these algorithms test out many different possible values for each parameter until they identify values that lead the model to best predict a participant's behavior. This approach can then be repeated for each participant in a study.

As a concrete example, we will consider the Horizon Task (Wilson et al. 2014), which is used in many of the studies we review below (Fig. 1). In each game, participants repeatedly choose between two bandits that differ in average reward values (points between 0 and 100), which need to be learned through exploration. The average reward for each bandit is stable in a game but differs between games. Critically, participants are told which option to choose in the first four choices, controlling the amount of information participants have about each bandit. In the *unequal* information condition, participants are instructed to choose one of the options three times while choosing the other option only once, resulting in less information about the latter option. There is also an *equal* information condition, where participants are instructed to choose each option twice. Orthogonally, the pragmatic benefit of information is manipulated between games by changing the number of free choices afterwards: either one free choice (H1) or six free choices (H6) per game. Here, choosing the more uncertain option (i.e., the option only chosen once during forced choices) is only useful in H6 games, because what is learned could guide the next five choices. DE levels can be estimated based on how much more often a participant chooses the uncertain option in H6 games compared to H1 games (treated as a baseline condition). More specifically, the information bonus parameter (A) described above is estimated separately in H1 and H6 games, and the difference ($A_{H6} - A_{H1}$) is taken as the level of DE. In parallel, RE can be estimated in equal information games based on how often individuals choose the option with the lower average observed reward. More specifically, the decision noise parameter (σ) described above is estimated separately in H1 and H6 games, and the difference ($\sigma_{H6} - \sigma_{H1}$) is taken as their level of RE. Learning rates can also be estimated based on how choices change over time depending on observed outcomes.

It is important to highlight that this is just one example. Many tasks and models described below are designed to measure different aspects of exploration and use different equations and parameters. For example, some tasks may have more than two options, they might have binary (win/loss) outcomes instead of continuous reward values, or they might have evolving average reward values. It is beyond the scope of this article to thoroughly review each task and model. However, this example should provide a sense of how this methodology can be used to describe mechanisms of learning and choice on a task and the unique benefits it can offer.

3 Relationships Between Information-Seeking and Anxiety

3.1 Directed Exploration

One recent study by Aberg et al. (2022) examined how *trait* anxiety (State-Trait Anxiety Inventory [STAI] Trait scale; Spielberger et al. 1971) related to behavior on a dynamic bandit task within a community sample (mean STAI score below clinically significant cutoffs; Ercan et al. 2015; see Table 1). In this task, participants repeatedly chose between three bandits with average reward values that evolved unexpectedly over time. Thus, exploratory choices were required to determine the current best choice. Here, DE was modeled as a growing drive to select an option based on how long since it was last chosen. A positive association between anxiety and DE was observed, such that more anxious individuals showed a stronger tendency to choose more uncertain options. They further observed a quadratic (inverted-U) relationship between anxiety and overall task performance, suggesting low anxiety (associated with under-exploration) and high anxiety (associated with over-exploration) both led to worse performance. This study therefore suggested that trait anxiety promoted greater DE in an environment where the best choice unpredictably changed over time, and moderate levels of anxiety appeared to optimize task performance.

Table 1 Details of reviewed studies

Reviewed studies		Task/sample information	Anxiety measures [Mean (SD)] or induction methods	
Directed exploration	Aberg et al. (2022)[a]	Dynamic bandit task • Sample 1 (behavioral): $N = 20$ • Sample 2 (fMRI): $N = 31$	• STAI-trait: – Sample 1: 38.00 (8.77) – Sample 2: 36.32 (9.98)	
	Smith et al. (2022b)	Horizon task • $N = 414$	• STAI-trait: 36.55 (10.98) • STAI-state: 30.12 (8.22)	
	Fan et al. (2023)	Two-armed bandit task • Study 1: $N = 501$ • Study 2: $N = 484$	• STAI-trait: – Study 1: 43.5 (12.9) – Study 2: 42.9 (12.7)	• STICSA-trait: – Study 1: 37.4 (15.1) – Study 2: 35.4 (13.4)
	Yan et al. (2023)[a]	Dynamic bandit task • $N = 1{,}001$	• GAD-7: 7.17 (5.54)	
	Witte et al. (2024)	Grid task • Study 1: $N = 282$ • Study 2: $N = 75$	• STICSA-trait [unavailable]	
	Li et al. (2024)	Horizon task • Anxious depression: $N = 61$ • Healthy comparisons: $N = 58$	• STAI-trait: – Clinical group: 51.87 (9.06) – Healthy group: 29.17 (7.40)	• OASIS: – Clinical group: 8.41 (3.02) – Healthy group: 1.07 (1.52)
			• Somatic anxiety induction paradigm – STAI-state	
			• Clinical group: • Pre: 42.03 (9.13) • Post: 50.06 (11.42)	• Healthy group: • Pre: 29.40 (8.00) • Post: 38.53 (10.32)
Foraging	Lenow et al. (2017)	Patch-foraging task • $N = 65$	• Acute stress induction: Cold pressor – Saliva cortisol measure • Chronic stress: PSS [unavailable]	
	Bustamante et al. (2023)	Modified patch-foraging task • $N = 678$	• GAD-7 [unavailable]	

	Lloyd et al. (2022a)	Modified patch-foraging task • $N = 181$	• GAD-7 [unavailable]
	Trier et al. (2023)	Threat-based foraging task • $N = 1,076$	• STICSA-trait somatic anxiety [unavailable]
Non-instrumental information-seeking	Charpentier et al. (2022)	Stock market task • Study 1: $N = 45$ • Study 2: $N = 48$ – Anxiety induction: $N = 24$ – Control group: $N = 24$	• Study 1: STAI-trait: 42.21(11.69) • Study 2: Modified TSST – Short State Anxiety Inventory [SSAI] (Marteau and Bekker 1992) • Anxiety induction group: • Pre: 9.79 (3.93) • Post: 16.42 (4.35) • Control group: • Pre: 10.08 (3.78) • Post: 10.46 (4.00)
	Bennett et al. (2021)	Study-specific lottery task • $N = 139$	• BFAS volatility, withdrawal subscales [unavailable] • BFI-2 anxiety, emotional volatility subscales [unavailable]
	Bode et al. (2023)	Study-specific lottery task • $N = 67$	• Socially evaluated cold pressor (stress induction) – Saliva cortisol measure
Clinical cutoff scores			
STAI: state = 40 (Emons et al. 2019); trait = 44 (Ercan et al. 2015)			OASIS: 8 (Campbell-Sills et al. 2009)
STICSA: trait = 40 (Van Dam et al. 2013); state = unavailable			BFAS: Unavailable
GAD-7: 10 (Spitzer et al. 2006)			BFI-2: Unavailable
PSS: 0–13 = low stress; 14–25 = moderate stress; 27–40 = high stress (Othman et al. 2019)			SSAI: Unavailable

[a] Explicitly excluded for previous history of psychiatric or neurological disorders

In another study, Smith et al. (2022b) used the horizon task (described above; see Fig. 1) to investigate the relationship between DE and both *state* and *trait* anxiety (STAI). Their sample included both undergraduate students and other non-diagnosed individuals from the surrounding community (mean anxiety levels below clinical cutoffs; see Table 1). They found that higher levels of state anxiety were associated with lower levels of DE (associations with trait anxiety were numerically similar but above significance thresholds). Further analyses revealed that results were not only driven by less information-seeking in H6 games, but also by *more* information-seeking in H1 games. Recall that information gain is not beneficial in H1 games, since it sacrifices reward when no further choices are allowed. Thus, high-state anxiety may also motivate uncertainty reduction when it is maladaptive. This appears consistent with the first study reviewed above (Aberg et al. 2022), and suggests a general intolerance of uncertainty (Carleton et al. 2012).

Another recent study (Fan et al. 2023) conducted two online experiments to test the relationship between DE and trait anxiety (mean anxiety scores below clinical cutoffs; see Table 1). Here, a different bandit task was used in which participants repeatedly chose between two options explicitly labeled as "Stable" (S) or "Fluctuating" (F). These options differed in the stability of their average reward values (S = constant average reward; F = evolving average reward). Participants played a series of games with four possible combinations of the two bandit types presented on the left and right sides of the screen: S/S, S/F, F/S, F/F. These combinations were designed to vary the level of total uncertainty across options and relative uncertainty between options, while also minimizing the potential effects of side bias. For example, in S/F and F/S games, there is a difference in relative uncertainty, because the fluctuating option has more uncertainty than the stable option. Thus, DE can be modeled by how often participants choose the fluctuating option. To derive finer-grained symptom measures, this study performed exploratory factor analysis across items of the State-Trait Inventory for Cognitive and Somatic Anxiety [STICSA] (Ree et al. 2008), and STAI trait scales. Trait anxiety was then divided into scores on four factors: somatic anxiety, cognitive anxiety, negative affect, and low self-esteem. Results showed that those with higher trait somatic anxiety also showed lower DE. Interestingly, follow-up analyses indicated this relationship could be explained by subjective underestimation of relative uncertainty. This was measured based on the difference in confidence ratings reported for the average reward value of each option. A final noteworthy finding was that scores for the negative affect factor instead showed positive associations with DE (i.e., more consistent with some of the results reviewed above); but this pattern was only found in one of their two datasets.

A further study applied dimensionality reduction to behavior on a dynamic bandit task (similar to Aberg et al. 2022) and derived summary measures for two unique dimensions in choice behavior (Yan et al. 2023). By mapping computational measures onto the above-mentioned behavioral dimension scores, they found that one score represented the balance between exploration and exploitation, while the other represented choice stability. Results with explore/exploit balance scores indicated positive associations between generalized anxiety symptoms (Generalized Anxiety Disorder-7 [GAD-7]; Spitzer et al. 2006) and exploration. Note, however, that

GAD-7 scores were largely within the sub-clinical range (see Table 1), and DE and RE could not be clearly distinguished. Yet, these findings appear largely consistent with those of the first two studies reviewed above (Aberg et al. 2022; Smith et al. 2022b).

Another recent study in an online sample (Witte et al. 2024) used a task called the grid task (Schulz et al. 2018; Wu et al. 2018) to test associations between exploration and anxiety/depression scores, based on the STICSA and a range of depressive symptom scales. In their task, each option corresponded to a location in an 11x11 grid (i.e., 121 options in total) with an unknown average reward value, whereas nearby options tended to have similar average reward values. In each game, full knowledge of average reward values could not be learned for all options because the number of choices was less than the number of locations to choose. The task also had safe and risky conditions. In the safe condition, options could be repeatedly chosen with relative gains and losses. In the risky condition, participants could instead lose all previously obtained rewards if choices led to reward values below a threshold value. In either condition, exploration corresponded to selecting options that had not yet been chosen. Results showed that both cognitive anxiety and depression were associated with increased exploration, particularly in the risky condition. Modeling suggested this was best explained by participants assigning greater value to unobserved options. In a second dataset, in-person participants also completed an intervention designed to reduce worry, which they found decreased exploration. These results therefore link cognitive aspects of anxiety to increased exploration, and provide initial evidence for causation.

However, when considering most studies reviewed above, state and trait anxiety could not be fully disentangled. Further, these studies have focused on non-clinical student and community samples with average anxiety scores below clinical cutoffs (Table 1), preventing a thorough evaluation of the higher anxiety levels linked to psychopathology. To address this, more recent work has combined the horizon task with a somatic (breathing-based) state anxiety induction in individuals diagnosed with anxious depression and/or methamphetamine use disorders (Goldman et al. 2024; Li et al. 2024). This anxiety induction asked individuals to breathe through an apparatus that provides moderate resistance during inspiration, creating an anxiety-promoting sensation of air hunger (Berk et al. 2015). When comparing those with anxious depression to a healthy comparison (HC) group with low trait anxiety (STAI trait scores and Overall Anxiety Severity and Impairment Scale scores [OASIS]; Norman et al. 2006), they found that those with anxious depression showed greater DE at baseline. However, after the state anxiety induction, DE was reduced in the anxious depression sample (i.e., becoming similar to HCs). No effect of anxiety induction was found in HCs or in those with methamphetamine use disorder. However, the anxiety induction only induced mild anxiety in HCs; it is possible that greater state anxiety may have been necessary to see an effect. The methamphetamine users also showed significantly lower DE in general, many of which had moderately elevated affective symptoms. Together, these results suggest that trait anxiety may promote DE, while state anxiety may instead reduce it, specifically in those with affective disorders.

From a clinical perspective, learning from unpredictable or otherwise adverse childhood experiences may also be important, given their known relationships to anxiety disorders (e.g., Chorpita and Barlow 2018), as well as to reduced cognitive performance and lower emotional awareness (e.g., Smith et al. 2022a) – each of which could influence mechanisms of exploration (Humphreys et al. 2015; Pérez-Edgar 2018). Consistent with this possibility, reduced DE on the horizon task was recently observed in children who perceived more unpredictability in their childhood environment (Xu et al. 2023), as measured by the questionnaire of unpredictability in childhood (Glynn et al. 2019). This highlights how childhood experiences may change sensitivity to uncertainty and further shape subsequent learning and exploratory behavior. However, more developmental studies are necessary to confirm these differences and better characterize underlying mechanisms.

In summary, recent studies have begun to elucidate relationships between anxiety and information-seeking. Both the severity of generalized anxiety symptoms (GAD-7) and trait cognitive anxiety (STISCA) have been positively associated with exploration. However, trait anxiety (STAI) showed positive associations with DE in one study and non-significantly negative associations in another. State anxiety (STAI) was linked to lower DE in one study, but this partially reflected greater exploration in a baseline condition where information was not useful. In contrast, another study found trait somatic anxiety (STICSA) was associated with reduced DE, with some evidence of an opposing relationship to general negative affect (STAI). Work attempting to experimentally disentangle the effects of state and trait anxiety also suggests they may have opposing effects.

Some of the apparent discrepancies above could be explained by the use of different tasks and the way they operationalize exploration. For example, not all tasks could fully disentangle DE from other strategies; nor have all studies defined DE in the same manner. The horizon task, for example, measures DE as a change in sensitivity to relative uncertainty when information gain would be beneficial. In contrast, other tasks estimate sensitivity to relative uncertainty in general. This pragmatic element of DE in the horizon task therefore sets it apart from other measures reviewed above. Additional discrepancies between results could relate to anxiety subtype or choice of measures. For example, it could be that somatic anxiety reduces DE, while more cognitive aspects of anxiety instead increase it. This would explain many results above, as both somatic anxiety induction and trait somatic anxiety appeared to reduce DE, while baseline anxiety levels and general negative affect (perhaps reflecting elevated/persistent uncertainty) were more often associated with greater exploration. Here one might speculate that high-arousal, somatic anxiety creates a greater urgency to act and reduces reflection on one's own uncertainty, therefore reducing the exploratory drive, while cognitive anxiety has greater links to elevated worry and attention to uncertainty, which could instead elevate exploratory motivation. Future work will be necessary to further test these potential distinctions.

However, the possibility that some reports reflect false positives, or false negatives should also be kept in mind, highlighting the crucial need for replication studies. This is further emphasized by recent preliminary results failing to find associations between DE and either worry or intolerance of uncertainty (Zaller et al. 2021). We return to consideration of these important future directions below. Recall also that nearly all studies reviewed here have measured anxiety levels in community samples without affective disorders. Only Li et al. (2024) made direct comparisons between non-clinical and clinical groups.

3.2 Foraging

We now move on to discuss links between anxiety and foraging behavior, which has primarily been studied using the patch-foraging task (PFT; Constantino and Daw 2015). In each round of this task, participants visit different virtual tree orchards to harvest as many apples as possible in a limited time window. On each trial, participants decide whether to continue to harvest from their current tree or to leave in hopes of finding a better tree. The number of apples on the current tree depletes further with each successive choice to harvest. If one chooses to leave the current tree, a travel time is required before reaching a new tree. As there is limited time to gather as many apples as possible, this travel time is costly. The choice to leave and explore new trees should thus depend on both the expected travel time and the expected number of apples per tree across the orchard on average. Optimal choice behavior can be judged based on the marginal value theorem (MVT), which states that participants should leave the current tree when its expected value falls below the average value of other trees in the orchard. However, this average value needs to be learned through exploration (e.g., by choosing to leave on early trials to figure out how many apples tend to be on other trees in the orchard). This task and associated models thus allow one to measure differences in exploration based on how quickly individuals choose to leave each tree after beginning to harvest (i.e., leaving rate) or the threshold of remaining apples after which an individual usually chooses to leave (i.e., leaving threshold). It should be highlighted that the PFT differs from the bandit tasks described above in that one cannot return to the current tree after choosing to leave. Exploration is therefore driven by subjective estimates of the average reward rate in the environment, as it only makes sense to risk leaving the current tree if one believes better trees are likely to be found.

Lenow et al. (2017) used this task in a community sample, finding that both acute stress (induced by a cold pressor task [CPT]; Lovallo 1975) and chronic stress (measured by the Perceived Stress Scale [PSS]; Cohen et al. 1983) biased participants' decisions towards under-exploration. Namely, participants with higher levels of stress were more likely to harvest apples from a tree longer than would be optimal. Notably this pattern of behavior was present regardless of environment quality (i.e., average reward rates), suggesting that stress downwardly modulated subjective estimates of average reward levels generally. Although not identical constructs, stress and anxiety are highly related and have overlapping neural/behavioral correlates (Daviu et al. 2019), suggesting these results are likely also relevant to anxiety.

The role of anxiety in foraging decisions was also recently investigated in an online sample using a variant of the PFT in which the cost of exploration was not passive (i.e., travel time), but instead required either cognitive or physical effort (Bustamante et al. 2023). For cognitive effort trials, participants were required to perform a difficult Multi-Source Interference Task (MSIT) (Bush and Shin 2006); whereas physical effort trials required participants to repeatedly key press at a fast rate. Results showed a lower leaving threshold when high cognitive or physical effort was required (i.e., less exploration if the cost of traveling was high). Generalized anxiety symptoms (GAD-7) were also associated with a lower error rate on the MIST, linked to a greater subjective cognitive cost in the high-effort condition. This suggested that participants with higher trait anxiety exhibited greater caution and error aversiveness, and expended greater cognitive effort to avoid mistakes. Though there was no reported relationship between anxiety and overall leaving thresholds, the greater cognitive effort cost shown by those with higher trait anxiety modulates leaving thresholds in the direction of reduced exploration.

A further question about how anxiety relates to foraging strategies involves the possible effects of environmental stability. To answer this question, a recent online study in a non-clinical sample modified the PFT such that, while travel time was fixed, the initial reward level and reward depletion rate in one condition were unstable (i.e., the volatile environment condition; Lloyd et al. 2022a). Namely, some trials unpredictably started with fewer apples on each tree and a faster depletion rate, while others started with a higher number of apples and a slower depletion rate. Results showed that participants were relatively more likely to choose options with lower observed values (greater RE) in the volatile environment, leading to better overall task performance. In other words, observing outcomes of occasional random choices helped gain information and keep track of unknown changes in depletion rate. Crucially, this study found that higher levels of generalized anxiety symptoms (GAD-7) were associated with less adjustment in RE between conditions, leading to less exploration when it would be beneficial.

Another recent study added an acute threat to a novel foraging task (i.e., a predator; Trier et al. 2023). In this task, participants could decide between (1) foraging, (2) checking for predators, and (3) hiding from predators. To avoid getting caught, one must check for the presence of predators. Thus, the time portioned between foraging, checking, and hiding is relevant to task performance. This online study found that individuals with higher trait somatic anxiety (STICSA) spent a smaller proportion of time in the task choosing to check for the predator (i.e., less information-seeking), which led to lower task performance. Self-report measures further suggested these individuals avoided checking specifically due to fear of finding the predator. These findings therefore also suggested a link between somatic anxiety and maladaptive reductions in information-seeking, due to a fear of learning undesired information. Here one might speculate that elevated physiological symptoms of somatic anxiety amplified feelings of fear and thereby increased this type of avoidance.

Pharmacological work has also manipulated anxiety in conjunction with the PFT. In particular, a recent study showed that, relative to placebo, propranolol administration (i.e., a norepinephrine antagonist that reduces heart rate and blood pressure) led to more exploratory behavior (Cremer et al. 2023). More specifically, they found that lowering norepinephrine levels with propranolol led to increased RE by selectively reducing the degree to which value information contributed to the choice to stay or leave a patch. This was consistent with another study that used a norepinephrine agonist (atomoxetine), showing selective decreases in RE on the horizon task (Warren et al. 2017). However, these results contrast with another study suggesting decreased choice randomness on the horizon task under propranolol administration (Dubois et al. 2021). They also appear at odds with a previous rodent study showing earlier patch leaving under stimulation of tonic norepinephrine activity that was best explained by an increase in choice randomness (Kane et al. 2017). It remains unclear why directional effects appear inconsistent between these studies. One suggestion is that there is a positive relationship between tonic norepinephrine levels and RE, while phasic changes instead facilitate exploitation by promoting faster commitment to the choice under consideration (Aston-Jones and Cohen 2005; Cremer et al. 2023). However, both modes of norepinephrine signaling may be influenced by pharmacological agents such as propranolol (Lawson et al. 2021), and task differences between studies are likely relevant. Thus, further work will be needed to fully disentangle these effects.

Finally, developmental studies offer some additional insights into how early life experiences could impact exploration in foraging tasks, and how relationships between anxiety and exploration could arise. For example, one study found reduced exploration during foraging in adults who previously experienced greater early adversity (Lloyd et al. 2022b). In addition, the study using the horizon task in children reviewed above (Xu et al. 2023) also had participants complete a foraging task. In this case, however, they found no relationship between early childhood unpredictability and exploration. More research is therefore required to test these relationships, as these studies differed in study population (children vs. adults), behavioral task (bandit task vs. foraging), and measures of early life experiences (adverse childhood events vs. childhood unpredictability).

In summary, when exploration is measured in terms of foraging behavior, results have been consistent in suggesting that both state and trait anxiety lead to reduced exploration. Further analyses suggested this may be explained by misrepresenting the quality of the environment, a greater subjective cost of effort, fear of detecting a threat, and maladaptive adjustment in decision strategy when the stability of the environment changes. It is currently unclear why these results differ from those of some studies above testing DE within bandit tasks. There are notable differences between bandit and foraging tasks, but it remains unclear which differences are most relevant. Gaining a better understanding of the relevant differences between the cognitive demands of these tasks will be an important direction for future work. At present, results suggest some bandit tasks may tap more into the exaggerated uncertainty associated with high cognitive anxiety and negative affect, while foraging tasks may tap into maladaptive avoidance behaviors. However, some of the bandit tasks above also suggested reductions in DE with greater somatic and state anxiety.

Although our focus is on computational modeling studies, it is worth highlighting that the foraging results also appear consistent with findings in other animal paradigms. For example, exploratory behavior has been extensively examined in rodent studies using behavioral assays such as the open-field test and the elevated plus-maze task (Prut and Belzung 2003; Walf and Frye 2007). A common feature of these assays is that animals have free access to explore open areas. However, some locations within the fields/mazes have features that influence exploratory behavior. In the open field test, a preference to stay near the borders is observed due to advantages associated with a more complete view of the surrounding environment as well as less exposure. Similarly, animals tend to avoid open areas of the elevated plus-maze since, while offering a more complete view of surrounding areas, they are less protected due to the lack of surrounding walls. The time spent in different locations within these environments is therefore an indicator of risk assessment and whether subjects are willing to fully explore the environment. Crucially, these task paradigms have consistently found a negative correlation between anxiety and exploratory behavior (Prut and Belzung 2003; Walf and Frye 2007).

Recent studies have also modified these tasks to study analogous exploratory behavior in humans. For example, in a human version of the open-field test, people with higher levels of anxiety sensitivity (measured by the Anxiety Sensitivity Index [ASI] total scores; Reiss et al. 1986) spent more time in the surrounding borders of the field, analogous to the findings in other animals (Walz et al. 2016). Similar results were found in a human version of the elevated plus-maze task, in which more anxious individuals (based on a study-specific state anxiety scale) spent less time exploring the open arm and tended to remain in the safe arms with borders (Biedermann et al. 2017). Overall, these studies therefore support a picture similar to the results within foraging tasks, suggesting that anxiety reduces exploration and promotes avoidance.

3.3 Non-instrumental Information-Seeking

Finally, we turn to studies investigating non-instrumental information-seeking. Here, information-seeking cannot inform future choices or affect observations; it is simply driven by one's desire to reduce the uncertainty of not knowing. In laboratory settings, non-instrumental information-seeking can be measured in task paradigms where participants choose whether or not to pay money to receive more detailed information about a fluctuating signal (e.g., stock prices). They have no control over the signal and simply watch it passively, where this signal can either show small or large average fluctuations over time. For example, they might simply pay to find out what their current investments are worth, despite not being able to adjust them, where the stock market may show large or small changes over time. The amount participants are willing to pay therefore serves as an indirect measure of the intrinsic value they attach to gaining information.

One recent study found no relationship between trait anxiety (STAI-trait) and non-instrumental information-seeking in this type of paradigm (Charpentier et al. 2022); nor was there a relationship found with state anxiety induced by (a modified version of) the Trier Social Stress Test [TSST] (Birkett 2011; Garrett et al. 2018; Kirschbaum et al. 1993). However, compared to those with low anxiety, participants with high anxiety (both trait and state) were more willing to pay for information when the stock market signal showed larger fluctuations. Note that mean trait anxiety scores were below clinical cutoffs (see Table 1). Using computational modeling, another study using a related paradigm also found that anxiety (using the Big Five Aspects Scale [BFAS-2] volatility and withdrawal subscales (DeYoung et al. 2007), and the Big Five Inventory-2 [BFI-2] anxiety and emotional volatility subscales (Soto and John 2017)) was associated with more willingness to pay for information (Bennett et al. 2021). However, the paradigm used here differed from the previous study in potentially meaningful ways (e.g., outcomes were categorical and fluctuated with stable, chance-level probabilities; and cost levels differed by trial type). In contrast, one further study using an anxiety induction protocol, the Socially Evaluated Cold Pressor Test [SECPT] (Schwabe et al. 2008), found no relationship between state anxiety and non-instrumental information-seeking (Bode et al. 2023). Similar to the previous study, the paradigm here also consisted of chance-level categorical outcomes. Together, these studies tend to indicate the absence of a consistent relationship between non-instrumental information-seeking and either trait or state anxiety (at least in non-clinical community samples), but also suggest behavior may be altered by anxiety when the passively observed signal is less stable over time. Here, it also appears that computational modeling may have been able to capture more subtle individual differences in non-instrumental information-seeking that were not apparent using model-free approaches.

4 Discussion

Anxiety involves maladaptive responses associated with distal threat and has been associated with intolerance/misestimation of uncertainty (Pulcu and Browning 2019). In this review, we examined altered information-seeking, as measured by exploratory behaviors that function to reduce uncertainty. We drew on a wide range of studies using different task environments to measure distinct types of exploratory strategies, including DE, foraging, and non-instrumental information-seeking. As we have seen, findings to date are somewhat mixed and consistency seems to depend on the type of exploration being measured, how it is operationalized within a given task, and how precisely different dimensions of anxiety are measured (e.g., state vs. trait, somatic vs. cognitive, etc.). Overall, most, but not all, results could jointly be taken to suggest: (1) cognitive and/or trait anxiety may increase DE, while somatic and/or state anxiety may decrease DE; (2) anxiety symptoms more consistently relate to lower foraging-related exploration, which requires taking a risk to leave a current resource in hopes of finding something better (possibly due to pessimistic expectations about the average reward available elsewhere); and (3) both state and trait anxiety are inconsistently related to non-instrumental information-seeking, and their influence may depend on the amount of environmental uncertainty that could be resolved (e.g., due to differences in the magnitude of signal fluctuations over time) or the methods used for investigation (e.g., design of paradigm or adoption of computational models).

These results paint a nuanced and incomplete picture of how anxiety influences information-seeking. More work will be necessary to disentangle the exact ways in which existing tasks differ and may produce divergent results. Greater consistency and breadth in the choice of measures would also benefit these lines of research going forward. It will also be important to better elucidate how decision-making processes in these tasks map onto the structure of real-world problems anxious individuals face in their daily lives – and how this may explain differences in the adaptiveness of each exploratory strategy. More work will also be needed to test connections between differences observed in exploration and differences in learning rates and environmental volatility estimates previously seen within anxious individuals (e.g., Browning et al. 2015).

Yet another area that may be important for further study is the neural basis of exploration and its links to anxiety. Prior work identifying brain regions linked to anxiety appears to show some overlap with brain regions previously linked to exploration, suggesting potential connections. For example, in studies focused on learning under uncertainty, a broad set of regions, including the dorsal anterior cingulate cortex (dACC), insula, ventral medial prefrontal cortex (vmPFC), and amygdala, have been identified showing patterns of activation associated with greater uncertainty and/or anxiety (e.g., Brown et al. 2018; Hein et al. 2023; Lawrance et al. 2022; Piray et al. 2019). Separately, the dACC and insula have been linked to exploration while vmPFC and amygdala have been implicated in exploitation (e.g., Aberg et al. 2022; Laureiro-Martínez et al. 2014, 2015). Evidence has also linked the right rostral prefrontal cortex (RPFC) with DE and its role in tracking relative uncertainty (e.g., Badre et al. 2012; Tomov et al. 2020), and resting state functional connectivity between RPFC and a number of other regions has been linked to exploratory behavior as well (Kayser et al. 2016; Morris et al. 2016). The causal role of RPFC in controlling exploration has further been established with noninvasive brain stimulation techniques (Beharelle et al. 2015; Zajkowski et al. 2017). Specifically, Beharelle et al. (2015) found that different patterns of transcranial direct current stimulation (TDCS) over RPFC could either increase or decrease exploratory behavior, and Zajkowski et al. (2017) found that transcranial magnetic stimulation (TMS) over RPFC could particularly modulate levels of DE. However, it remains to be tested how neural circuitry underlying different types of exploration is specifically affected by anxiety induction or how associated neural responses are linked with trait anxiety and/or dimensions of somatic and cognitive anxiety. This could shed further light on underlying mechanisms and potentially highlight novel treatment targets.

A final direction for future work of potential importance pertains to clinical intervention. At present, it is not known whether exploratory behavior can be persistently modified with training, nor is it known whether changes in exploratory behavior would lead to changes in anxiety. Multiple psychotherapies target uncertainty in various ways, but it is also not clear whether this alters information-seeking.

However, this may be of considerable clinical relevance, as common interventions, such as exposure therapy or behavioral activation therapy, could both be seen as promoting information-seeking behaviors (i.e., testing pessimistic beliefs about tolerability of feared situations or about how exercise or social interaction might improve mood). There are also some recent findings suggesting that other correlates of DE, such as cognitive reflectiveness (Smith et al. 2022b), can be improved with training (e.g., Orona and Trautwein 2023) – which suggests potential malleability and other possible intervention approaches. If information-seeking is malleable, and future work can establish causal links between this and anxiety, this knowledge could therefore offer potential clinical benefit.

In summary, there has been significant progress in recent work testing how different forms of information-seeking may be associated with different forms of anxiety. Computational modeling has also been a powerful tool in characterizing specific cognitive mechanisms. However, a wide range of tasks and anxiety measures have been used in inconsistent ways, leading to some apparent inconsistencies in results that are difficult to interpret. Given its potential clinical relevance, future progress may require a more concerted effort to increase consistency in experimental approaches and perhaps compare behavior and links to anxiety using multiple tasks and measures in the same participants. We look forward to these next steps and the additional insights they will likely offer.

References

Aberg KC, Toren I, Paz R (2022) A neural and behavioral trade-off between value and uncertainty underlies exploratory decisions in normative anxiety. Mol Psychiatry 27(3):1573–1587

Addicott MA, Pearson JM, Sweitzer MM, Barack DL, Platt ML (2017) A primer on foraging and the explore/exploit trade-off for psychiatry research. Neuropsychopharmacology 42(10): 1931–1939

Aston-Jones G, Cohen JD (2005) An integrative theory of locus coeruleus-norepinephrine function: adaptive gain and optimal performance. Annu Rev Neurosci 28:403–450

Badre D, Doll BB, Long NM, Frank MJ (2012) Rostrolateral prefrontal cortex and individual differences in uncertainty-driven exploration. Neuron 73(3):595–607

Barlow DH, Allen LB, Choate ML (2016) Toward a unified treatment for emotional disorders – republished article. Behav Ther 47(6):838–853. https://doi.org/10.1016/j.beth.2016.11.005

Beharelle AR, Polanía R, Hare TA, Ruff CC (2015) Transcranial stimulation over frontopolar cortex elucidates the choice attributes and neural mechanisms used to resolve exploration–exploitation trade-offs. J Neurosci 35(43):14544–14556

Bennett D, Sutcliffe K, Tan NP-J, Smillie LD, Bode S (2021) Anxious and obsessive-compulsive traits are independently associated with valuation of noninstrumental information. J Exp Psychol Gen 150(4):739

Berk L, Stewart JL, May AC, Wiers RW, Davenport PW, Paulus MP, Tapert SF (2015) Under pressure: adolescent substance users show exaggerated neural processing of aversive interoceptive stimuli. Addiction 110(12):2025–2036

Biedermann SV, Biedermann DG, Wenzlaff F, Kurjak T, Nouri S, Auer MK et al (2017) An elevated plus-maze in mixed reality for studying human anxiety-related behavior. BMC Biol 15: 1–13

Birkett MA (2011) The trier social stress test protocol for inducing psychological stress. J Vis Exp 56:e3238

Bode S, Jiwa M, Chum C, Frost L, Heekeren HR, Wingenfeld K, Deuter CE (2023) Non-instrumental information seeking is resistant to acute stress. Sci Rep 13(1):19505

Boswell JF, Thompson-Hollands J, Farchione TJ, Barlow DH (2013) Intolerance of uncertainty: a common factor in the treatment of emotional disorders. J Clin Psychol 69(6):630–645

Brown VM, Zhu L, Wang JM, Frueh BC, King-Casas B, Chiu PH (2018) Associability-modulated loss learning is increased in posttraumatic stress disorder. eLife 7:e30150

Browning M, Behrens TE, Jocham G, O'reilly JX, Bishop SJ (2015) Anxious individuals have difficulty learning the causal statistics of aversive environments. Nat Neurosci 18(4):590–596

Bush G, Shin LM (2006) The multi-source interference task: an fMRI task that reliably activates the cingulo-frontal-parietal cognitive/attention network. Nat Protoc 1(1):308–313

Bustamante LA, Oshinowo T, Lee JR, Tong E, Burton AR, Shenhav A et al (2023) Effort foraging task reveals positive correlation between individual differences in the cost of cognitive and physical effort in humans. Proc Natl Acad Sci 120(50):e2221510120

Campbell-Sills L, Norman SB, Craske MG, Sullivan G, Lang AJ, Chavira DA et al (2009) Validation of a brief measure of anxiety-related severity and impairment: the overall anxiety severity and impairment scale (OASIS). J Affect Disord 112(1–3):92–101

Carleton RN, Mulvogue MK, Thibodeau MA, McCabe RE, Antony MM, Asmundson GJ (2012) Increasingly certain about uncertainty: intolerance of uncertainty across anxiety and depression. J Anxiety Disord 26(3):468–479

Charpentier CJ, Cogliati Dezza I, Vellani V, Globig LK, Gädeke M, Sharot T (2022) Anxiety increases information-seeking in response to large changes. Sci Rep 12(1):7385

Chorpita BF, Barlow DH (2018) The development of anxiety: the role of control in the early environment. Neurotic Paradox 2:227–264

Cohen S, Kamarck T, Mermelstein R (1983) A global measure of perceived stress. J Health Soc Behav:385–396

Constantino SM, Daw ND (2015) Learning the opportunity cost of time in a patch-foraging task. Cogn Affect Behav Neurosci 15:837–853

Cremer A, Kalbe F, Müller JC, Wiedemann K, Schwabe L (2023) Disentangling the roles of dopamine and noradrenaline in the exploration-exploitation tradeoff during human decision-making. Neuropsychopharmacology 48(7):1078–1086

Daviu N, Bruchas MR, Moghaddam B, Sandi C, Beyeler A (2019) Neurobiological links between stress and anxiety. Neurobiol Stress 11:100191

DeYoung CG, Quilty LC, Peterson JB (2007) Between facets and domains: 10 aspects of the big five. J Pers Soc Psychol 93(5):880

Dubois M, Habicht J, Michely J, Moran R, Dolan RJ, Hauser TU (2021) Human complex exploration strategies are enriched by noradrenaline-modulated heuristics. eLife 10:e59907

Emons W, Habibović M, Pedersen S (2019) Prevalence of anxiety in patients with an implantable cardioverter defibrillator: measurement equivalence of the HADS-A and the STAI-S. Qual Life Res 28:3107–3116

Ercan I, Hafizoglu S, Ozkaya G, Kirli S, Yalcintas E, Akaya C (2015) Examining cut-off values for the state-trait anxiety inventory/Examinando los puntajes de corte para el inventario de ansiedad estado-rasgo. Revista Argentina de Clinica Psicológica 24(II):143

Fan H, Gershman SJ, Phelps EA (2023) Trait somatic anxiety is associated with reduced directed exploration and underestimation of uncertainty. Nat Hum Behav 7(1):102–113

Garrett N, González-Garzón AM, Foulkes L, Levita L, Sharot T (2018) Updating beliefs under perceived threat. J Neurosci 38(36):7901–7911

Gershman SJ (2019) Uncertainty and exploration. Decision 6(3):277

Glynn LM, Stern HS, Howland MA, Risbrough VB, Baker DG, Nievergelt CM et al (2019) Measuring novel antecedents of mental illness: the questionnaire of unpredictability in childhood. Neuropsychopharmacology 44(5):876–882

Goldman CM, Takahashi T, Lavalley CA, Li N, Taylor S, Chuning AE et al (2024) Individuals with methamphetamine use disorder show reduced directed exploration and learning rates independent of an aversive interoceptive state induction. medRxiv. 2024.2005. 2017.24307491

Grupe DW, Nitschke JB (2013) Uncertainty and anticipation in anxiety: an integrated neurobiological and psychological perspective. Nat Rev Neurosci 14(7):488–501

Hein TP, Gong Z, Ivanova M, Fedele T, Nikulin V, Herrojo Ruiz M (2023) Anterior cingulate and medial prefrontal cortex oscillations underlie learning alterations in trait anxiety in humans. Commun Biol 6(1):271

Humphreys KL, Lee SS, Telzer EH, Gabard-Durnam LJ, Goff B, Flannery J, Tottenham N (2015) Exploration – exploitation strategy is dependent on early experience. Dev Psychobiol 57(3): 313–321

Kane GA, Vazey EM, Wilson RC, Shenhav A, Daw ND, Aston-Jones G, Cohen JD (2017) Increased locus coeruleus tonic activity causes disengagement from a patch-foraging task. Cogn Affect Behav Neurosci 17:1073–1083

Kayser AS, Op de Macks Z, Dahl RE, Frank MJ (2016) A neural correlate of strategic exploration at the onset of adolescence. J Cogn Neurosci 28(2):199–209

Kirschbaum C, Pirke K-M, Hellhammer DH (1993) The 'Trier Social Stress Test' – a tool for investigating psychobiological stress responses in a laboratory setting. Neuropsychobiology 28(1–2):76–81

Laureiro-Martínez D, Canessa N, Brusoni S, Zollo M, Hare T, Alemanno F, Cappa SF (2014) Frontopolar cortex and decision-making efficiency: comparing brain activity of experts with different professional background during an exploration-exploitation task. Front Hum Neurosci 7:927

Laureiro-Martínez D, Brusoni S, Canessa N, Zollo M (2015) Understanding the exploration–exploitation dilemma: an fMRI study of attention control and decision-making performance. Strateg Manag J 36(3):319–338

Lawrance EL, Gagne CR, O'reilly JX, Bjisterbosch J, Bishop SJ (2022) The computational and neural substrates of ambiguity avoidance in anxiety. Comput Psychiatry 6(1):8

Lawson RP, Bisby J, Nord CL, Burgess N, Rees G (2021) The computational, pharmacological, and physiological determinants of sensory learning under uncertainty. Curr Biol 31(1):163–172. e164

Lenow JK, Constantino SM, Daw ND, Phelps EA (2017) Chronic and acute stress promote overexploitation in serial decision making. J Neurosci 37(23):5681–5689

Li N, Lavalley CA, Chou K-P, Chuning AE, Taylor S, Goldman CM et al (2024) Directed exploration is elevated in affective disorders but reduced by an aversive interoceptive state induction. medRxiv:2024.2006.2019.24309110. https://doi.org/10.1101/2024.06.19.24309110

Lloyd A, McKay R, Furl N (2022a) Stochastic decisions support optimal foraging of volatile environments, and are disrupted by anxiety

Lloyd A, McKay RT, Furl N (2022b) Individuals with adverse childhood experiences explore less and underweight reward feedback. Proc Natl Acad Sci 119(4):e2109373119

Lovallo W (1975) The cold pressor test and autonomic function: a review and integration. Psychophysiology 12(3):268–282

Marteau TM, Bekker H (1992) The development of a six-item short-form of the state scale of the Spielberger state – trait anxiety inventory (STAI). Br J Clin Psychol 31(3):301–306

Moeller FG, Barratt ES, Dougherty DM, Schmitz JM, Swann AC (2001) Psychiatric aspects of impulsivity. Am J Psychiatry 158(11):1783–1793

Morris LS, Baek K, Kundu P, Harrison NA, Frank MJ, Voon V (2016) Biases in the explore–exploit tradeoff in addictions: the role of avoidance of uncertainty. Neuropsychopharmacology 41(4):940–948

Norman SB, Hami Cissell S, Means-Christensen AJ, Stein MB (2006) Development and validation of an overall anxiety severity and impairment scale (OASIS). Depress Anxiety 23(4):245–249

Orona GA, Trautwein U (2023) Thinking disposition education improves cognitive reflection: experimental results from two intervention studies

Othman N, Ahmad F, El Morr C, Ritvo P (2019) Perceived impact of contextual determinants on depression, anxiety and stress: a survey with university students. Int J Ment Heal Syst 13:1–9

Pérez-Edgar K (2018) Attention mechanisms in behavioral inhibition: exploring and exploiting the environment. Behavioral inhibition: Integrating theory, research, and clinical perspectives, pp 237–261

Piray P, Ly V, Roelofs K, Cools R, Toni I (2019) Emotionally aversive cues suppress neural systems underlying optimal learning in socially anxious individuals. J Neurosci 39(8): 1445–1456

Prut L, Belzung C (2003) The open field as a paradigm to measure the effects of drugs on anxiety-like behaviors: a review. Eur J Pharmacol 463(1–3):3–33

Pulcu E, Browning M (2019) The misestimation of uncertainty in affective disorders. Trends Cogn Sci 23(10):865–875

Ree MJ, French D, MacLeod C, Locke V (2008) Distinguishing cognitive and somatic dimensions of state and trait anxiety: development and validation of the state-trait inventory for cognitive and somatic anxiety (STICSA). Behav Cogn Psychother 36(3):313–332

Reiss S, Peterson RA, Gursky DM, McNally RJ (1986) Anxiety sensitivity, anxiety frequency and the prediction of fearfulness. Behav Res Ther 24(1):1–8

Sandin B, Chorot P, McNally RJ (2001) Anxiety sensitivity index: normative data and its differentiation from trait anxiety. Behav Res Ther 39(2):213–219. https://doi.org/10.1016/s0005-7967(00)00009-7

Schulz E, Gershman SJ (2019) The algorithmic architecture of exploration in the human brain. Curr Opin Neurobiol 55:7–14

Schulz E, Wu CM, Huys QJ, Krause A, Speekenbrink M (2018) Generalization and search in risky environments. Cognit Sci 42(8):2592–2620

Schwabe L, Haddad L, Schachinger H (2008) HPA axis activation by a socially evaluated cold-pressor test. Psychoneuroendocrinology 33(6):890–895

Smith R, Chuning AE, Tidwell CA, Allen JJ, Lane RD (2022a) Psychopathic tendencies are selectively associated with reduced emotional awareness in the context of early adversity. PLoS One 17(12):e0277475

Smith R, Taylor S, Wilson RC, Chuning AE, Persich MR, Wang S, Killgore WD (2022b) Lower levels of directed exploration and reflective thinking are associated with greater anxiety and depression. Front Psych 12:782136

Soto CJ, John OP (2017) The next big five inventory (BFI-2): developing and assessing a hierarchical model with 15 facets to enhance bandwidth, fidelity, and predictive power. J Pers Soc Psychol 113(1):117

Spielberger CD, Gonzalez-Reigosa F, Martinez-Urrutia A, Natalicio LF, Natalicio DS (1971) The state-trait anxiety inventory. Int Am J Psychol 5(3 & 4)

Spitzer RL, Kroenke K, Williams JB, Löwe B (2006) A brief measure for assessing generalized anxiety disorder: the GAD-7. Arch Intern Med 166(10):1092–1097

Tomov MS, Truong VQ, Hundia RA, Gershman SJ (2020) Dissociable neural correlates of uncertainty underlie different exploration strategies. Nat Commun 11(1):2371

Trier H, O'Reilly JX, Spiering L, Ma S, Kolling N, Rushworth M, Scholl J (2023) Emotions and individual differences shape foraging under threat

Van Dam NT, Gros DF, Earleywine M, Antony MM (2013) Establishing a trait anxiety threshold that signals likelihood of anxiety disorders. Anxiety Stress Coping 26(1):70–86

Walf AA, Frye CA (2007) The use of the elevated plus maze as an assay of anxiety-related behavior in rodents. Nat Protoc 2(2):322–328

Walz N, Mühlberger A, Pauli P (2016) A human open field test reveals thigmotaxis related to agoraphobic fear. Biol Psychiatry 80(5):390–397

Warren CM, Wilson RC, van der Wee NJ, Giltay EJ, van Noorden MS, Cohen JD, Nieuwenhuis S (2017) The effect of atomoxetine on random and directed exploration in humans. PLoS One 12(4):e0176034. https://doi.org/10.1371/journal.pone.0176034

Wilson RC, Collins AG (2019) Ten simple rules for the computational modeling of behavioral data. eLife 8:e49547
Wilson RC, Geana A, White JM, Ludvig EA, Cohen JD (2014) Humans use directed and random exploration to solve the explore–exploit dilemma. J Exp Psychol Gen 143(6):2074
Wilson RC, Bonawitz E, Costa VD, Ebitz RB (2021) Balancing exploration and exploitation with information and randomization. Curr Opin Behav Sci 38:49–56
Witte K, Wise T, Huys QJ, Schulz E (2024) Exploring the unexplored: worry as a catalyst for exploratory behavior in anxiety and depression
Wu CM, Schulz E, Speekenbrink M, Nelson JD, Meder B (2018) Generalization guides human exploration in vast decision spaces. Nat Hum Behav 2(12):915–924
Xu Y, Harms MB, Green CS, Wilson RC, Pollak SD (2023) Childhood unpredictability and the development of exploration. Proc Natl Acad Sci 120(49):e2303869120
Yan X, Ebitz RB, Grissom N, Darrow DP, Herman AB (2023) A low dimensional manifold of human exploratory behavior reveals opposing roles for apathy and anxiety. bioRxiv. 2023.2006. 2019.545645
Zajkowski WK, Kossut M, Wilson RC (2017) A causal role for right frontopolar cortex in directed, but not random, exploration. eLife 6:e27430
Zaller I, Zorowitz S, Niv Y (2021) Information seeking on the horizons task does not predict anxious symptomatology. Biol Psychiatry 89(9):S217–S218

Emotion Regulation Under Stress: A Social Processing and Memory Perspective

Nicholas J. Dennis, Tasha Bulgin, Casey M. Nicastri, Cassandra Bell, and Mauricio R. Delgado

Contents

Abstract Research on emotion regulation often focuses on cognitively effortful self-regulation strategies, but exposure to stress has been shown to interfere with the underlying mechanisms supporting such processes. Understanding alternative strategies that potentially bolster emotion regulation under stress is an important topic of investigation. Two potential alternatives involve everyday occurrences of social processing and memory recall. Social support and past emotional experiences may help in guiding us toward appropriate neurophysiological responses through overlapping circuitry with stress and reward systems, while also buttressing cognitive regulation strategies by expanding one's perspective and allowing multiple opportunities to regulate retrospectively. In recognition that ongoing social and emotional events are often at the beginning of a cascade of both emotion regulation and memory processes, this chapter focuses on the emerging role of social

---

N. J. Dennis, T. Bulgin, C. M. Nicastri, C. Bell, and M. R. Delgado (✉)
Department of Psychology, Rutgers University-Newark, Newark, NJ, USA
e-mail: nicholas.dennis@rutgers.edu; delgado@psychology.rutgers.edu

© The Author(s), under exclusive license to Springer Nature Switzerland AG 2024
Curr Topics Behav Neurosci (2025) 73: 179–204
https://doi.org/10.1007/7854_2024_560
Published Online: 31 December 2024

relationships and autobiographical memory recall in regulating emotions under stress, highlighting opportunities and challenges associated with this process.

Keywords Affect · Emotion regulation · Memory · Social · Stress

1 Introduction

Stress is an evolutionarily adaptive response when expressed in appropriate contexts and magnitudes, often facilitating motivation to guide behavior in service of our goals (for review see Lang and Bradley 2010; Koolhaas et al. 2011; McEwen and Akil 2020). However, stress can also elicit emotions that, if not appropriately regulated, can lead to psychological distress and contribute to the development of anxiety, mood, and substance use disorders (e.g., Moghaddam and Jackson 2004; Aldao et al. 2010; Sheppes et al. 2015; Girotti et al. 2018). How one appraises and interprets the value of emotional stimuli is critical for controlling physiological responses to stressors and overall successful emotion regulation. Notably, stress has been shown to interfere with cognitive and neural mechanisms that support the most commonly used cognitive regulatory strategies (e.g., reappraisal), marked by dysfunction in the prefrontal cortex, which plays a key role in multiple effortful goal-directed strategies (Moghaddam and Jackson 2004; Arnsten 2009). Such dysfunction can lead to issues in regulation of negative emotions via cognitive strategies while experiencing acute stress (Raio et al. 2013; Kogler et al. 2015). Thus, understanding potential alternatives that can aid emotion regulatory processes under conditions of stress is an important topic of investigation.

Everyday occurrences of social processing and memory recall are two potential alternatives or supplements to more effortful and deliberate forms of emotion regulation. First, social others can play an important role in emotional regulatory processes (for review see Zaki and Williams 2013; Reeck et al. 2016), and might be uniquely suited to fill the regulatory gap under stress when cognitive resources are limited. Second, recall of personally salient emotional experiences (i.e., autobiographical memory) may support cognitive emotional regulation by expanding one's current perspective and allowing multiple opportunities to regulate retrospectively through memory recall (Samide and Ritchey 2021). Although laboratory-based episodic memory paradigms (e.g., affective word lists, standard image sets) provide excellent experimental control, they may not be representative of the real-world functions or processes of memory recall (McDermott et al. 2009; Zlomuzica et al. 2014). Autobiographical memory recall paradigms are particularly useful in examining the functions of emotional memory because they typically capture a fuller range of features of naturalistic emotional responses (Kensinger and Ford 2020; Maguire 2022). Past emotional experiences may help in guiding us toward appropriate neurophysiological responses in the present through overlapping circuitry of stress (Ulrich-Lai and Herman 2009) and reward (Shohamy and Adcock 2010) systems. Social processes and autobiographical memory are also closely connected. Even in the physical absence of others, reminders of social relationships and past life experiences buffer stress in response to distressing stimuli (Speer and Delgado 2020; Pei et al. 2023), drawing parallels to social buffering effects during early development. We also must acknowledge that for some individuals, the emotional regulatory challenges brought on by stressful life events often persist long after the events end (Milad and Quirk 2012; de Quervain et al. 2017; Schwabe et al. 2022; Cohen and Kahana 2022). The extent to which recall of past idiosyncratic experiences helps or hinders one's ability to manage their emotions may depend not just on affective valence and social context, but regulatory strategy as well.

Though an entire volume could be dedicated to the many strategies and components of emotion regulation (for review see Braunstein et al. 2017), this chapter focuses on the emerging role of social relationships and autobiographical memory recall in regulating emotions under stress. In particular, we aim to highlight how social experiences may augment effortful emotion change processes such as cognitive reappraisal not just in the moment, but retrospectively as well. By considering contexts in which recall of autobiographical memory poses emotional regulatory opportunities and challenges for regulatory enhancement, the goal of this chapter is to illustrate that ongoing social and emotional events are often at the beginning of a cascade of both emotion regulation and memory processes.

2 Basic Processes of Emotion Regulation

Emotion regulation is a fundamental aspect of daily life, often used in stressful situations to decrease negative affect and control behavioral responses. Regulation plays a vital role in the ability for humans to manage one's emotional experiences and adapt to ever-changing environments, entailing the use of valuation systems that are dependent on changing environmental contexts to inform decision-making about emotional responses (Gross 2015). There are many ways that individuals regulate their emotions. Some of the most common strategies include cognitive reappraisal, suppression, and avoidance (Mihalca and Tarnavska 2013; Sheppes et al. 2014). Common strategies can be placed into categories contrasted by one's emotion regulation goals (e.g., implicit/explicit) and how deliberate or effortful the emotion change process is (e.g., automatic/controlled). Not only do these strategies have long-term effects on physical and mental health (DeSteno et al. 2013; Song et al. 2015), but they also play a role in social outcomes (Gross 2002; Gross and John 2003; van't Wout et al. 2010). This section highlights 1) the social and health outcomes of cognitive reappraisal, suppression, and avoidance, 2) identifies key brain regions associated with these regulation strategies, and 3) discusses the top-down effects of stress on emotion regulation.

2.1 Cognitive Reappraisal

Cognitive reappraisal often involves reevaluating negative stimuli or negative situations in a more positive way in order to reduce the impact of that specific event (Gross 2002). The use of cognitive reappraisal as an emotion regulation strategy typically leads to more positive physical and mental health outcomes, as these individuals tend to have an increased ability to cope with negative emotions (Gross and John 2003; Song et al. 2015; Wu et al. 2015). Cognitive reappraisal has been linked to a reduced risk for cardiovascular disease and better well-being (Gross and John 2003; DeSteno et al. 2013). Effective use of reappraisal strategies leads to decreases in subjective and physiological expressions of negative emotion (Ochsner et al. 2012), as well as marked changes in the neural circuitry involved in emotional processing. For instance, in a classic study of reappraisal use, Ochsner et al. (2002) showed that reappraising the emotional meaning of negative pictures (compared to attending to natural feelings upon viewing them) led to decreases in subjective emotional ratings of feelings and neural activity in the amygdala. Further, engagement of cognitive strategies (i.e., reappraisal) recruited prefrontal regions such as dorsolateral prefrontal cortex (dlPFC) and the ventrolateral prefrontal cortex (vlPFC) – the latter of which has been linked to reappraisal success (Wager et al. 2008). Based on a meta-analysis of reappraisal, the dlPFC, vlPFC, and the dorsomedial prefrontal cortex (dmPFC) comprise key regions that support effective reappraisal (Buhle et al. 2014).

2.2 Expressive Suppression

Expressive suppression is a type of response modulation described as inhibiting the expression of negative emotions (Gross 2002). The brain regions associated with this inhibition are the anterior cingulate cortex (ACC), paracingulate cortex (PCC), and the medial prefrontal cortex (mPFC), which have been linked to the regulation of behavioral emotional expressions (Hermann et al. 2014; Dixon et al. 2017). While suppression may seem similar to cognitive reappraisal in its goal of reducing negative affect, physiologically suppression does not confer reduction benefits as seen for cognitive reappraisal (Gross 2002). While suppression may reduce behavioral responses to negative events, this strategy leads to a decrease in positive emotions and an increase in feelings of negative emotions (Gross and John 2003; Cutuli 2014). Suppression is also negatively correlated with positive indicators of subjective well-being (Gross 2002; Ehring et al. 2010). Suppression has been shown to have negative social and health outcomes, with suppressors feeling less socially supported and having lower life satisfaction (Cutuli 2014; Dryman and Heimberg 2018). In fact, an overreliance on suppression has been linked to social anxiety, which may further affect social outcomes in individuals (Dryman and Heimberg 2018; Rozen and Aderka 2023). Suppression might cause a feeling of incongruence within one's self, leading others to think that suppressors are inauthentic thus distancing themselves from the suppressor (Gross and John 2003). Social distancing may lead to increased negative emotions toward the suppressor (DeSteno et al. 2013), promoting an aversion toward forming closer relationships (Gross and John 2003). The unfortunate consequence of a diminished social network or increased isolation are the implications to health outcomes including a higher mortality rate and poor well-being (Holt-Lunstad 2021; Hawkley 2022).

2.3 *Experiential Avoidance*

Experiential avoidance is attention control that can be described as an unwillingness to experience negative feelings, thoughts, or sensations (Sloan 2004; Mitmansgruber et al. 2009). Evidence from both expressive suppression and avoidant distraction has been shown to impair memory (Richards and Gross 2006; Dillon et al. 2007; Sheppes and Meiran 2007, 2008; Hayes et al. 2010). Experiential avoidance has been associated with decreased neural activation in frontal regions, the ACC, the amygdala, and the hippocampus, along with recruitment of a fronto-limbic-striatal network (Delgado et al. 2009; Aupperle et al. 2015; Pastor and Medina 2021). Similar to suppression, experiential avoidance tends to increase negative affect; however, individuals continue to use experiential avoidance as an emotion regulation strategy due to its seemingly positive outcomes, even if short-term (Sloan 2004; Fledderus et al. 2010). Experiential avoiders have increased emotional responses to negative stimuli when compared to non-experiential avoiders (Sloan 2004). Consistent with this, experiential avoiders have higher risks of developing psychopathology such as depression and anxiety (Fledderus et al. 2010; Berghoff et al. 2017). Experiential avoidance can also translate into social contexts by acting as a deterrent when it comes to pursuing new social interactions or maintaining existing social relationships (Kashdan et al. 2014).

2.4 *Potential Limitations of Cognitive Regulation Strategies*

Compared to suppression and avoidance, cognitive reappraisal appears to be the more adaptive emotion regulation strategy with more positive health and social outcomes (Mihalca and Tarnavska 2013). However, there are times when reappraisal may not be the most effective strategy. For instance, the use of reappraisal strategy may be difficult for some to implement or is not appropriate in some contexts (Troy et al. 2013), possibly due to the recruitment of more effortful cognitive control processes (Silvers et al. 2015; Strauss et al. 2016). Further, while reappraisal may be needed most under distress, the initial deployment of cognitive reappraisal may be impacted by the experience of stress. Using a cold pressor (i.e., cold water submersion) task to induce acute stress, Raio et al. (2013) showed that cognitive reappraisal of fear responses to conditioned aversive cues was less successful under stress relative to a non-stressed condition. Kogler et al. (2015) demonstrated similar effects of stress on reappraisal success using a psychosocial stress induction paradigm. Mechanistically, hyperactivity in the amygdala, which plays a role in processing negative emotions (Barrett et al. 2007; Zhang et al. 2018) may lead to increased emotional reactivity in stressful situations (Yamamoto et al. 2017). A hyperactive amygdala coupled with potential disruptions in prefrontal cortex activity (Moghaddam and Jackson 2004; Arnsten 2009) may lead to ineffective usage of cognitive regulatory strategies, allowing distressing events to snowball over time. Identifying additional strategies that can help individuals cope with the initial aspects of stress in the short-term would allow more adaptive cognitive regulation processes such as reappraisal to be used for long-term building of resilience processes.

Similar to the regulatory challenges imposed by stress on the prefrontal cortex, cognitive strategies like reappraisal are also difficult for children and adolescents to use effectively, primarily because their prefrontal cortical regions supporting such strategies are not yet fully developed (Casey et al. 2008). Nonetheless, stronger emotion regulation abilities in children and adolescents are associated with numerous positive socio-emotional (McLaughlin and Hatzenbuehler 2009; Raver et al. 2016) and academic achievement outcomes (Seibert et al. 2017). To understand how we can augment the regulatory toolkit and reinforce the cognitive resources necessary for more complex and effortful emotion regulation strategies, in the following section we will discuss early developmental mechanisms that support emotional regulation despite the underdeveloped neural architecture critical for effortful cognitive strategies.

3 Developmental Origins: Why Do We Need Emotion Regulation?

Developmentally, infants are initially reliant on others to resolve distress, but begin to deploy basic regulatory strategies, such as self-soothing and attention-seeking, requiring only minimal cognitive resources (Ratcliff et al. 2021). As infants age, they begin to engage in more distraction-based emotion regulation behavior, which is an early display of effortful emotional control (Ratcliff et al. 2021). Distraction is often categorized as a maladaptive avoidance strategy in response to aversive experiences and emotions, but it can be an effective tool under limited cognitive resources – particularly when used to shift attention temporarily, postponing engagement until the stressor can be managed in a more controlled state (Linehan 1993; Wolgast and Lundh 2017).

Children and adolescents do not yet have a fully developed and refined prefrontal cortex (Casey et al. 2008), which encompasses regions suggested to be critical for deploying effortful emotion regulation strategies like cognitive reappraisal (e.g., ventrolateral, dorsolateral, and dorsomedial PFC) (Kohn et al. 2014; Buhle et al. 2014; Silvers et al. 2017). In typically developing children, rapid development of reappraisal abilities begins around 10 years of age and continues to improve into late teen years (Silvers et al. 2017). Despite limited access to the full repertoire of more complex cognitive regulatory strategies, children and adolescents often face stressful life events. Differences in health and achievement outcomes associated with differing levels of regulatory success (McLaughlin and Hatzenbuehler 2009; Raver et al. 2016) demonstrate the critical role of emotion regulation in healthy development. For example, children who exhibit higher emotion regulation abilities tend to adjust more positively to school environments (Bailey et al. 2016), and children facing poverty-related stress who struggle to properly regulate their emotions show higher levels of internalizing behavioral problems (Raver et al. 2016). Emotion dysregulation (e.g., delayed, diminished, or inappropriate use of regulatory strategies), during childhood and adolescence in particular, is suggested to be a significant transdiagnostic factor that elevates the risk for psychopathological outcomes (Maughan and Cicchetti 2002; McLaughlin et al. 2011, 2015).

Children's parents also play an important role in their emotional development by providing social buffering. The idea of social buffering is rooted in neurophysiological research that demonstrates an individual's ability to regulate the stress response system of another in early development (i.e., infancy, pre-adolescence), which is typically a primary caregiver (Hostinar et al. 2014; Gee 2016; Gunnar 2017). Notably, children with strong mother-child relationships showed more mature "emotion regulation network" connectivity resembling that of adolescents (Gee et al. 2014). While most research examines the mother-child dynamic, research on father-child interactions found that strong father relationships were also related to less overall infant distress during play (Kerr et al. 2021).

An important consideration for social buffering is that not all children have emotionally stable caregivers, and these children are often in adverse environments. While lack of parental/caregiver buffering can result in greater long-term emotion regulation difficulties for children (Milojevich et al. 2020), social buffering from individuals outside of primary caregivers can serve as an alternative to achieve similar benefits. There is evidence that suggests high-quality friendships can buffer physiological reactivity to social rejection in adolescents who were previously institutionalized (Tang et al. 2022). Adolescents are thought to be particularly sensitive to peer perspectives relative to younger children (for review see Blakemore and Mills 2014). Disclosure or discussion of past or current personal problems with friends has been shown to enhance friendship quality, a protective factor for navigating certain socio-emotional challenges (Rose et al. 2007); however, excessive discussion of negative feelings or problems (i.e., "co-rumination") is often viewed as maladaptive in children and adults due to associations with anxiety and depression symptoms (Rose 2002). Thus, an important direction for future research is further neurophysiological investigation of peer relationships on emotion regulation across developmental stages.

4 Social Emotional Processes as a Regulatory Mechanism

Emotion regulation research in adults has traditionally centered on how individuals manage their own emotions (i.e., self-regulation). However, there has been an increasing acknowledgment that social others often play an important role in emotional regulatory processes. Experiences of social buffering during development build the framework for adult capabilities of this interpersonal emotion regulation (Eisenberg et al. 1998; Gunnar 2017). Parental responses to distress teach children that negative emotions can be alleviated by social support and allow for safe exploration and confrontation of external stressors (Rimé 2009). Disruptions in this social regulatory system may lead to decreased ability to modulate physiological stress mechanisms (Gunnar 2017) and missed critical periods for oxytocin sensitivity (Feldman 2017), presenting potential challenges for adulthood. Over time, parents become less powerful buffers as the buffering role shifts to a friend (Gunnar 2017) or a romantic partner (Stephens et al. 2013). Despite decreasing in effectiveness, parental buffering proves important to the foundation of these relationships. The extent to which a parent provides safety and support may influence how we seek social support during stress in adulthood (Ainsworth and Bell 1970). For example, connection to a maternal figure at age 16 predicts decreased stressor reaction when holding hands with a platonic friend at age 24 (Coan et al. 2013). This work highlights how members of our social network are integral to processing emotions.

Therefore, it's unsurprising that we turn to others in navigating affective experiences in adulthood (Zaki and Williams 2013). In fact, the mere presence of people dampens reactivity to stressors and improves negative affect (Beckes and Coan 2011). This effect is further modulated by context including social closeness (e.g., stranger vs. romantic partner) and relationship quality (Coan et al. 2006). While interacting with others, we almost automatically engage in behaviors that support emotional expression. Social sharing of emotion, for example, is ubiquitous in human communication (for review see Rimé 2009). Open communication of emotional reactions allows for "making sense" of an experience, convergence on shared understanding, and the opportunity to seek social support (Rimé et al. 2020), which helps foster social connections (Delgado et al. 2023). Even without an explicit goal for regulation, social sharing may attenuate emotional experiences (Zaki and Williams 2013). Importantly, how a listener reacts to sharing has important implications for regulating the shared emotion. While sharing a negative emotion could elicit support, it could also lead to negative experiences like co-rumination (Rimé et al. 2020). Therefore, it is critical to consider bidirectional relationships in social experiences of emotion.

Current definitions of interpersonal emotion regulation emphasize the role of the regulator and the target, accounting for their independent, personal goals (for review see Zaki and Williams 2013; Reeck et al. 2016). After a negative experience, a person (i.e., target) may select to share with a friend (i.e., regulator) to regulate. If this is met with supportive signals, the target may perceive an increase in emotional resources, increased capability to face a stressor, and attenuate the negative experience (Billings and Moos 1981). The decision to provide support depends on the regulator's goal, capability to identify the target's emotions, selection, and implementation of a regulation strategy (Reeck et al. 2016; Ochsner 2019). A regulator may act from pro-social (e.g., changing the target's emotion as an end goal) or self-serving motivations (e.g., generating a warm glow). How the target interprets the regulator's goal plays a large role in the efficacy of this interaction (Ochsner 2019). If successful, interpersonal emotion regulation has important implications for strengthening our social relationships.

Interpersonal emotion regulation plays a large role in building social connections, which further support these processes. According to the Tend-and-Befriend hypothesis, maintaining supportive social connections (i.e., tend) permits seeking social connection in stressful contexts to buffer the situation (i.e., befriend) (Taylor 2006). One way that we can build and maintain social connectedness is through sharing with others (Sprecher et al. 2013; Brudner et al. 2022). For example, disclosing personal information can create a reciprocal sharing within a dyad, increasing feelings of closeness in both strangers (Sprecher et al. 2013) and friends (Vijayakumar et al. 2020). Furthermore, we draw on the feedback given within social sharing to guide our closeness and trust decisions (Brudner et al. 2022). The cyclical relationship between sharing and relationship strengthening builds the foundation for increased interpersonal emotion regulation capacities. Overall, establishing social connections via social support helps us buffer against physiological and emotional ramifications of stress (Cohen and Wills 1985; Inagaki and Eisenberger 2016; Delgado et al. 2023), promoting longevity and well-being (Holt-Lunstad et al. 2010; Holt-Lunstad 2021). Importantly, however, research also suggests that individuals with anxiety disorders (e.g., social anxiety, phobias) may be at heightened risk of over relying on interpersonal emotion regulation, especially when it coincides with self-doubt about their ability to regulate emotions independently (Hofmann 2014). Interpersonal regulation strategies that promote perspective-taking, rather than passive soothing, may offer more adaptive benefits (Akkuş and Peker 2022). Thus, understanding the role of different interpersonal regulation strategies in the maintenance of anxiety disorders is an important area for additional research.

5 Social Memory as a Regulatory Opportunity

Social buffering and interpersonal emotional regulation are powerful tools helping to cope with the deleterious effects of stress. However, there are often times when we are not able to be physically close to those important to us. For example, the transient loss of ability to physically interact with others during periods of isolation resulting from the COVID-19 pandemic was felt by many globally (Tindle et al. 2022; King et al. 2023). Disruptions of social support networks can also occur more commonly. For example, during life transitions such as beginning at a new school, graduating college, or taking a job out of state, individuals – particularly emerging adults – often find themselves physically distant from their established social groups (Weiss et al. 2022; Hawkins et al. 2024).

Recent work has examined the potential for a form of interpersonal regulation that does not require live interactions with others to effectively support emotion regulation. Pei et al. (2023) demonstrated that just imagining a friend's perspective in response to hypothetically discussing a real negative autobiographical memory is sufficient to buffer negative emotion relative to passive processing of feelings after recall of the negative memory. Buffered negative emotion was marked by decreased negativity and emotional intensity as well as associations with activation of a distinct neural signature for this imagined social support. These findings expand upon prior work demonstrating the regulatory benefit of sharing negative memories with others conversationally (Pasupathi 2003) or disclosing negative emotional events like failures through expressive writing (Pennebaker and Chung 2011; DiMenichi et al. 2018). This intriguing line of research highlights not only the need for further exploration of naturalistic regulation of past negative experiences, but also the role of mentalizing and perspective-taking in retrospective emotion regulation – both of which are suggested to be key functions of autobiographical memory recall (Bluck 2003; Pasupathi 2003; Bluck et al. 2005).

Recalling a positive autobiographical memory also tends to recruit reward-related processes (Speer et al. 2014). Such recall increases positive affect and helps buffer against exposure to acute stress (Speer and Delgado 2017). One important factor that may drive this effect is the social context inherent in the memory recall process. That is, memories involving close social others (e.g., a graduation dinner with family) elicited positive feelings and were associated with greater reduction in cortisol response after acute stress exposure than less social memories that were equated in terms of positive feelings elicited (e.g., receiving a good grade on an important exam) (Speer and Delgado 2020). This growing body of research suggests that social memory context might confer an additive boost to the supportive function of positive memory recall for explicit effortful regulation through stress-buffering mechanisms, even in the physical absence of others. Social perspective-taking and social remembering (i.e., recalling memories in a social setting) might also support the process of retrospective positive reframing that is characteristic of negative event reappraisal.

In a similar yet distinct regulatory domain, reminders of social support figures (e.g., images of close others) have been shown to attenuate fear acquisition and either prevent recovery of fear or enhance fear extinction memory, potentially both (Hornstein and Eisenberger 2017; Hornstein et al. 2018). Fear extinction learning often involves a paradigm known as retrieval-extinction (or sometimes extinction-reconsolidation); retrieving or "reactivating" a conditioned fear memory to induce memory malleability before establishing new or modifying existing relationships between the distressing stimulus outcomes, updating expected outcomes over time (Monfils et al. 2009; Schiller et al. 2010). The emotion change processes involved are typically considered more automatic or passive (i.e., less cognitively effortful) and not inherently goal-directed in nature relative to cognitive reappraisal (for comparative review, see Silvers 2020). More work is needed to unpack the underlying mechanisms of this social reminder effect on extinction (for review see

Hornstein et al. 2022), but it is nonetheless intriguing that various forms of social memory enhance emotion regulation success for both automatic implicit regulatory processes and more cognitively effortful goal-oriented strategies. Implicit regulatory strategies may be critical for scaffolding healthy development of explicit regulatory strategies like cognitive reappraisal (for review see Silvers et al. 2017; Silvers 2020), and social experiences may play a supporting role. Given striking similarities to findings from more traditional social buffering effects, an intriguing opportunity for the future centers around using longitudinal approaches to understand more about the types of relationships necessary to confer these regulatory benefits across developmental stages.

It is important to recognize that not all relationships are created equal, and there are some relationships or social contexts that might up-regulate negative affect or contribute to the use of regulatory strategies that may be maladaptive in the long-term. For example, although the inherent social aspect of co-rumination can support navigation of socio-emotional challenges in adolescence by building closer social bonds (Rose et al. 2007), ruminative thought is often associated with incidental maintenance of negative emotions over time (Sumner 2012). To better understand other contexts in which memory differentially contributes to emotion regulation efficacy, in the following section we present current research on both the emotional regulatory opportunities and challenges generally brought forth by autobiographical memory processes.

6 Contextual Factors in the Role of Memory for Emotion Regulation

Research in the last decade has yielded significant insight into the neurophysiological processes underlying laboratory-based regulation of emotional memories, particularly fear memories (Nader et al. 2000; Schiller and Delgado 2010). From the involvement of cortical processes in aiding conditioned fear responses and decreasing amygdala activation via extinction (Milad and Quirk 2002; Phelps and LeDoux 2005; Schiller and Delgado 2010) to updating conditioned fear memories via reconsolidation processes (for review see Levy and Schiller 2021), the ability to update memory has been suggested as a potential way to change affective representations of negative experiences (Schiller et al. 2013; Lee et al. 2017). Yet, findings have been mixed for regulation of more naturalistic idiosyncratic emotional event memories (for review see LaBar 2015; Treanor et al. 2017), highlighting the need to bridge the gap between emotional memory modification paradigms using lab-conditioned threat memory and real-world emotional memory at both the conceptual and mechanistic level (for review see McDermott et al. 2009). In this section, we focus on key findings in the translation from conditioned fear memory to more naturalistic emotional memory recall and modification. In a recent review, Kensinger and Ford (2020) present a framework for conceptualizing the influence of memory recall on emotion. They emphasize that recall is not *just* "an end-point" where emotional content and retrieval goals determine which memories are accessible for recall, it is also "a starting point" at which emotions can be up or down regulated, influencing each subsequent memory recall experience. In applying this framework to emotion regulation more broadly, we highlight how memory recall may support emotion regulation success through mechanisms at multiple temporal stages while addressing potential limitations depending on the recall context.

6.1 Memory Recall as a Support for Emotion Regulation

Autobiographical memory recall is often an iterative reconstructive process involving both search and detail elaboration phases to successfully retrieve event details (Holland and Kensinger 2013a, b). These reconstructive phases may present modulation opportunities for emotional memory recall, paralleling retrieval-extinction paradigms for conditioned fear (Monfils et al. 2009; Schiller et al. 2010). Reinstantiation of neurophysiological arousal patterns associated with negative emotional memory may open the window to modify or incorporate new response patterns to the original memory (i.e., reconsolidation); however, instead of occurring primarily through passive extinction, the cognitive elaboration phase may allow incorporation of information from active reappraisal – the reframing of one's perspective to find a silver lining or emotionally distancing oneself in the current context relative to the original event memory. Kross et al. (2009) examined the effectiveness of regulatory strategies in response to negative autobiographical memory recall, finding that both an acceptance-based and reappraisal analog strategy led to significant group average drops in self-reported negative affect relative to a feeling-based strategy designed to mimic ruminative thought. Parikh et al. (2019) extended this work by combining an episodic memory reactivation and spatial distancing reappraisal paradigm, demonstrating that reactivation of emotional memories boosted subsequent reappraisal effectiveness by selectively reducing perceived arousal responses after a 48-h delayed recall relative to reappraisal without reactivation. Although this study used non-naturalistic laboratory-based stimuli (i.e., standardized IAPS image set) which may not rely on identical memory systems as recall of personally relevant memories (Cabeza et al. 2004; Chen et al. 2017), the findings are a promising step in a growing line of work demonstrating the role of memory recall in supporting cognitive emotion regulation strategies (for review see Engen and Anderson 2018; Samide and Ritchey 2021; Wardell and Palombo 2024).

Interestingly, in a related line of work examining the recall of specific autobiographical self-efficacy events (e.g., overcoming challenges), Paersch et al. (2022) demonstrate that even under emotional stress, individuals were able to use these memories to successfully reappraise past negative emotional experiences (e.g., failing an exam, death of a loved one), as evidenced by a reduction of perceived heart palpitations and self-reported distress. Notably, while both self-efficacy and positive memory inductions led to momentary increases in positive mood and reductions in negative valence appraisal, only self-efficacy memories resulted in the aforementioned reductions, which is in line with research showing that recall of autobiographical problem-solving and meaningful self-identity memories caused increased positive affect while similarly valenced control memories (i.e., any other autobiographical memory) did not (Hallford and Mellor 2016). These findings suggest that not all positive memories are created equal with regard to specific regulatory capabilities, which merits further investigation into more nuanced affective memory contexts.

In addition to more traditional reframing and distancing reappraisal strategies, positive emotion-focused reappraisal has been shown to reduce physiological and subjective emotional reactivity in response to distressing stimuli (Shiota and Levenson 2012). Positive affect induction during a fear extinction paradigm meant to resemble exposure therapy has also demonstrated enhanced extinction learning effectiveness (Zbozinek and Craske 2017). With regard to positive autobiographical memory, Speer et al. (2021) showed that finding positive meaning from *negative* memory recall (e.g., something valuable was learned, something positive happened as a result) boosts regulatory success. Regulatory success was demonstrated by increases in positive affect in subsequent recall and updating of memory content with more positive reflections from the elaboration phase. Strikingly, both positive affect and memory updating effects remained present during recall at a 2-month follow-up, suggesting that these effects might be relatively long-lasting. In line with the recent framework proposed by Samide and Ritchey (2021), these findings support viewing cognitive reappraisal as a "special case of memory modification."

Given limitations regarding the efficacy of cognitive reappraisal under stress, autobiographical memory recall could serve as a tool that allows multiple opportunities to regulate an associated stress response before helping drive cognitive reappraisal success. During a stressful event, thinking of a related positive memory might help to temporarily distract from the negative feelings, buffering the stress response before returning to engage in other cognitive regulatory strategies more effectively under less stress, such as finding a silver lining via reappraisal. Emerging support for retrospective memory updating and positive memory recall as tools to support future cognitive emotion regulation success appears promising. That said, it is unsurprising that for both social and positive memories, not all contexts appear equal in their regulatory effects, which merits a more fine-grained understanding of the mechanisms that support cognitive regulation beyond physiological stress buffering. Of equal importance is understanding contexts in which maladaptive memory processes contribute to failures in emotion regulation strategies.

6.2 Maladaptive Memory Processes and Emotional Dysregulation

Highly emotional events seem to effortlessly etch themselves into our memory (LaBar and Cabeza 2006; Yonelinas and Ritchey 2015) and have been suggested to do so through interactive neurophysiological mechanisms of the stress response

(McGaugh 2000; McIntyre et al. 2012; Clewett and Murty 2019). By allowing us to re-experience past moments, memory is a powerful tool to help guide our thought and behavior; however, recall can also constrain us to the past in ways that are no longer adaptive in the present. Certain failures in emotion regulation, and consequently various psychiatric disorders, might be best explained by maladaptive engagement (or lack thereof) with memory processes. For example, the adage "Don't cry because it's over; smile because it happened" might be easier said than done for some individuals. Although the precise mechanisms are not clear, there is evidence that while those experiencing major depression can access positive memories while sad, positive autobiographical memory recall does not repair mood and can even make it worse – opposite to the pattern observed in never-depressed individuals (Joormann et al. 2007). Other research suggests that healthy adults show a memory advantage for positive stimuli, and that depression symptoms reduce this advantage leading to positive memory deficits (Dillon 2015; Cataldo et al. 2023). In this section, we provide a few illustrative examples of memory processes that may contribute to emotional dysregulation.

Both emotional valence (i.e., positive/negative) (Bowen et al. 2018) and emotional arousal (i.e., intensity) (Clewett and Murty 2019) have been proposed as critical determinants of what types of memories can be recalled repeatedly over time. If an event is negative and particularly intense, such as a breakup with a romantic partner or the loss of a friend, one initially might perceive the situation as distressing and overwhelming, triggering heightened amygdala, norepinephrine, and cortisol system activity, contributing to a durable event memory (McIntyre et al. 2012; Clewett and Murty 2019). Upon recall of the breakup or loss, amygdala activation is reinstated or "recapitulated," meaning neural patterns that were active at the time of memory formation are re-engaged (Bowen et al. 2018), leading to re-experiencing negative feelings and likely a desire to change emotional states. For example, as previously mentioned, suppression can temporarily reduce negative emotions; however, it can actually increase negative affect in the long term via the increase of negative feelings upon subsequent recall for individuals who tend to experience clinical levels of negative affect (Dalgleish et al. 2009).

Rumination refers to the process of *perseveratively* recounting one's feelings, but not necessarily the detailed content of a memory (Nolen-Hoeksema et al. 2008; Colombo et al. 2021). Though the goal of rumination is typically to resolve negative feelings, this strategy maintains emotions over time and has been attributed to overgeneralized details of memory recall. Overgeneralization has been hypothesized to hinder regulation of negative affect due to disruption of detailed recall caused by fixation on the negative feelings when reinstated (Williams et al. 2007), and is suggested to contribute to the development of anxiety and fear disorders (Laufer et al. 2016). Consistent partial activation or overgeneralization such as in the case of rumination may not be sufficient to cause destabilization for modifying the emotion associated with the memory representation. In their recent review, Samide and Ritchey (2021) consider a similar hypothesis regarding the optimal extent of memory reactivation, but also raise the possibility that too "strong" reactivation of emotional memory might strengthen or entrench the memory representation based on a nonmonotonic plasticity hypothesis (i.e., U-shaped relationship) of memory retrieval (Ritvo et al. 2019). Indeed, in the context of PTSD and specific phobias, repeated involuntary recall of aversive memories often causes re-experiencing of symptoms, which can amplify fear responding and subsequent cognitive avoidance (Zlomuzica et al. 2014). Those with anxiety and fear-related disorders may also selectively retrieve threat-relevant information relative to threat-irrelevant information. Though the emotional and neural mechanisms require further investigation, if such negative memory representations are prone to both selective retrieval and subsequent strengthening, these biases present an obvious challenge for retrieval-based modification of the memory trace. However, reconsolidation-based approaches have shown promise (Monfils and Holmes 2018), and paradigms leveraging more naturalistic social and emotional memory may offer novel insights. Indeed, novel autobiographical memory-based interventions have shown preliminary evidence that these distortions can be reduced (Arditte Hall et al. 2018; Dalgleish and Hitchcock 2023).

It is also important to acknowledge that acute stress has been shown to impair certain memory retrieval processes (Roozendaal et al. 2009; Schwabe et al. 2012), and the effects are particularly strong for recall of emotionally arousing memories and information (Joëls et al. 2011). These stress effects present a potential challenge for memory-based support of reappraisal (e.g., perspective-taking, meaning finding) under stress that should be addressed. However, a few caveats should also be noted. First and foremost, it is not clear that all types of episodic memory, from word lists to personally relevant life events, rely on the same underlying mechanisms. For instance, there is evidence that neural processes involved in autobiographical memory retrieval are distinct, possibly related to differences in personal relevance, scene construction, or general complexity (Cabeza et al. 2004; McDermott et al. 2009; Chen et al. 2017). Additionally, autobiographical memory recall is not *always* effortful (i.e., involving a search and elaboration phase). Direct retrieval – often considered "automatic and effortless" – is also common in response to personal reminders or cues (e.g., familiar people or places) (Uzer et al. 2012; Uzer and Brown 2017). Given the previously mentioned research showing the effectiveness of social support reminders in more automatic regulatory processes like stress buffering and fear extinction learning, autobiographical memories appear to be able to serve a dual function; personally relevant cues or reminders, particularly of social support, may be able to trigger memories that buffer stress automatically, which may free up cognitive resources for more effortfully generative recall that supports cognitive regulation strategies like reappraisal.

7 Conclusion and Future Directions

Emotion regulation has long been a proposed function of autobiographical event memory (Robinson and Swanson 1990; Hyman and Faries 1992; Pasupathi 2003). A growing body of empirical evidence supports this hypothesized function of autobiographical memory for both effortful cognitive regulation and more automatic regulatory processes that may scaffold effortful strategies across development. Collectively, the reviewed research suggests that under stressful conditions when we are most likely to need emotion regulation, recall of positive – particularly social – personal memories might serve as an additional regulatory resource. Positive social memories buffer the physiological stress response by freeing up cognitive resources to reframe one's perspective and disrupt the emotional cascade initiated by a distressing life event and maintained via memory processes. Further examination of social-emotional contexts will be important for understanding the mechanisms underlying emotional memory for more ecologically valid stimuli than traditionally utilized in laboratory tasks.

Social memory recall as a regulatory tool is not without its potential limitations. Much of the current evidence supporting positive memory recall as an effective strategy has focused on either healthy or depressed populations, leaving open the question of its effectiveness for those experiencing fear and anxiety disorders. Given the high rate of comorbidity and shared underlying risk factors among mood and anxiety disorders (Saha et al. 2021), it is important to consider how these strategies might translate across these conditions. For example, individuals with depression and anxiety both show difficulty recalling specific positive autobiographical memories (Sumner 2012; Dalgleish and Hitchcock 2023), and both depression (Teo et al. 2013) and social anxiety (Rodebaugh et al. 2015) have been associated with lower quality social relationships. Memory biases, such as overgeneralization and selective recall, are suggested mechanisms present across clinical mood and anxiety disorders (Dalgleish and Hitchcock 2023). Dillon and Pizzagalli (2018) propose that disrupted memory processes during depression – such as poor pattern separation caused by stress – may interfere with encoding and retrieval of specific positive memories, contributing to overgeneralization. Reduced pattern separation for negative memory has also been implicated in overgeneralized fear responses in anxiety disorders (Kheirbek et al. 2012; Dunsmoor et al. 2017; Starita et al. 2019). Thus, at least initially, existing social and memory challenges could make recalling specific positive social memories more, rather than less, effortful.

Nevertheless, interventions aimed at improving positive memory recall and social connection merit further consideration. A recent meta-analysis suggests that poor family and social relationship quality have reciprocal relationships with social anxiety (Chen et al. 2024) – counter to assumptions that social anxiety is the sole cause – suggesting that improving relationships may decrease social anxiety symptoms. Further, various psychological interventions have been successful in increasing objective social contact, decreasing deficits in perceived quality of social connections, and decreasing symptoms of depression (Zagic et al. 2022). Preliminary evidence also shows that training in detailed positive autobiographical memory recall can increase post-recall mood and encourage spontaneous positive memory recall to alleviate low mood in individuals with remitted depression (Haag et al. 2024). Interestingly, Haag et al. (2024) noted that many of the positive memories involved other people or pets to whom individuals felt connected. Further, adolescents that recalled more specific positive memories over time were at a decreased risk for developing depression (Askelund et al. 2019). In the rodent literature, directly reactivating hippocampal cells previously active during a positive experience suppressed depression-related behavior (Ramirez et al. 2015), suggesting a potential neural mechanism underpinning the beneficial effects of positive memory retrieval. More empirical research is needed to elucidate the mechanisms and boundary conditions necessary for enhancing positive memory recall and leveraging its regulatory potential in humans.

Finally, while acknowledging the potential of emotional memory updating, Phelps and Hofmann (2019) raise an important cautionary note on the ethics of its use not just in clinical practice but everyday life as well. Although the opportunity to reduce pain or protect someone by updating memories might seem enticing, there can be unforeseen consequences of disrupting natural emotional experiences. Negative memories often serve a directive (i.e., motivational) function (Rasmussen and Berntsen 2009), and have been shown to be valued over and above the associated negative feelings for some individuals who report preferring to keep certain negative memories when given a hypothetical choice of removal (Burnell et al. 2020). Though it might take time, we can often find meaning in memories of negative life events (Speer et al. 2021; Wardell and Palombo 2024). Our personal memories are fundamental to who we are; positive *and* negative autobiographical memories have the potential to shape our interactions with the people and world around us in important ways.

Acknowledgments This manuscript was supported by funding from the National Institutes of Health (R01DA053311).

References

Ainsworth MD, Bell SM (1970) Attachment, exploration, and separation: illustrated by the behavior of one-year-olds in a strange situation. Child Dev 41:49–67. https://doi.org/10.1111/j.1467-8624.1970.tb00975.x

Akkuş K, Peker M (2022) Exploring the relationship between interpersonal emotion regulation and social anxiety symptoms: the mediating role of negative mood regulation expectancies. Cogn Ther Res 46:287–301. https://doi.org/10.1007/s10608-021-10262-0

Aldao A, Nolen-Hoeksema S, Schweizer S (2010) Emotion-regulation strategies across psychopathology: a meta-analytic review. Clin Psychol Rev 30:217–237. https://doi.org/10.1016/j.cpr.2009.11.004

Arditte Hall KA, De Raedt R, Timpano KR, Joormann J (2018) Positive memory enhancement training for individuals with major depressive disorder. Cogn Behav Ther 47:155–168. https://doi.org/10.1080/16506073.2017.1364291

Arnsten AFT (2009) Stress signalling pathways that impair prefrontal cortex structure and function. Nat Rev Neurosci 10:410–422. https://doi.org/10.1038/nrn2648

Askelund AD, Schweizer S, Goodyer IM, van Harmelen A-L (2019) Positive memory specificity is associated with reduced vulnerability to depression. Nat Hum Behav 3:265–273. https://doi.org/10.1038/s41562-018-0504-3

Aupperle RL, Melrose AJ, Francisco A et al (2015) Neural substrates of approach-avoidance conflict decision-making. Hum Brain Mapp 36:449–462. https://doi.org/10.1002/hbm.22639

Bailey CS, Denham SA, Curby TW, Bassett HH (2016) Emotional and organizational supports for preschoolers' emotion regulation: relations with school adjustment. Emotion 16:263–279. https://doi.org/10.1037/a0039772

Barrett LF, Bliss-Moreau E, Duncan SL (2007) The amygdala and the experience of affect. Soc Cogn Affect Neurosci 2:73

Beckes L, Coan JA (2011) Social baseline theory: the role of social proximity in emotion and economy of action. Soc Personal Psychol Compass 5:976–988. https://doi.org/10.1111/j.1751-9004.2011.00400.x

Berghoff CR, Tull MT, DiLillo D et al (2017) The role of experiential avoidance in the relation between anxiety disorder diagnoses and future physical health symptoms in a community sample of young adult women. J Contextual Behav Sci 6:29–34. https://doi.org/10.1016/j.jcbs.2016.11.002

Billings AG, Moos RH (1981) The role of coping responses and social resources in attenuating the stress of life events. J Behav Med 4:139–157. https://doi.org/10.1007/bf00844267

Blakemore S-J, Mills KL (2014) Is adolescence a sensitive period for sociocultural processing? Annu Rev Psychol 65:187–207. https://doi.org/10.1146/annurev-psych-010213-115202

Bluck S (2003) Autobiographical memory: exploring its functions in everyday life. Memory 11: 113–123. https://doi.org/10.1080/741938206

Bluck S, Alea N, Habermas T, Rubin DC (2005) A tale of three functions: the self-reported uses of autobiographical memory. Soc Cogn 23:91–117. https://doi.org/10.1521/soco.23.1.91.59198

Bowen HJ, Kark SM, Kensinger EA (2018) NEVER forget: negative emotional valence enhances recapitulation. Psychon Bull Rev 25:870–891. https://doi.org/10.3758/s13423-017-1313-9

Braunstein LM, Gross JJ, Ochsner KN (2017) Explicit and implicit emotion regulation: a multi-level framework. Soc Cogn Affect Neurosci 12:1545–1557. https://doi.org/10.1093/scan/nsx096

Brudner EG, Fareri DS, Shehata SG, Delgado MR (2022) Social feedback promotes positive social sharing, trust, and closeness. Emotion 23:1536. https://doi.org/10.1037/emo0001182

Buhle JT, Silvers JA, Wager TD et al (2014) Cognitive reappraisal of emotion: a meta-analysis of human neuroimaging studies. Cereb Cortex 24:2981–2990. https://doi.org/10.1093/cercor/bht154

Burnell R, Rasmussen AS, Garry M (2020) Negative memories serve functions in both adaptive and maladaptive ways. Memory 28:494–505. https://doi.org/10.1080/09658211.2020.1737133

Cabeza R, Prince SE, Daselaar SM et al (2004) Brain activity during episodic retrieval of autobiographical and laboratory events: an fMRI study using a novel photo paradigm. J Cogn Neurosci 16:1583–1594. https://doi.org/10.1162/0898929042568578

Casey BJ, Jones RM, Hare TA (2008) The adolescent brain. Ann N Y Acad Sci 1124:111–126. https://doi.org/10.1196/annals.1440.010

Cataldo AM, Scheuer L, Maksimovskiy AL et al (2023) Abnormal evidence accumulation underlies the positive memory deficit in depression. J Exp Psychol Gen 152:139–156. https://doi.org/10.1037/xge0001268

Chen H-Y, Gilmore AW, Nelson SM, McDermott KB (2017) Are there multiple kinds of episodic memory? An fMRI investigation comparing autobiographical and recognition memory tasks. J Neurosci 37:2764–2775. https://doi.org/10.1523/JNEUROSCI.1534-16.2017

Chen B, Sun X, Huang X, Yao L (2024) Examining the reciprocal link between social anxiety and social relationships spanning from childhood to adulthood: a meta-analysis of longitudinal studies. Dev Psychol 60:170–186. https://doi.org/10.1037/dev0001666

Clewett D, Murty VP (2019) Echoes of emotions past: how neuromodulators determine what we recollect. eNeuro 6:ENEURO.0108. https://doi.org/10.1523/ENEURO.0108-18.2019

Coan JA, Schaefer HS, Davidson RJ (2006) Lending a hand: social regulation of the neural response to threat. Psychol Sci 17:1032–1039. https://doi.org/10.1111/j.1467-9280.2006.01832.x

Coan JA, Beckes L, Allen JP (2013) Childhood maternal support and social capital moderate the regulatory impact of social relationships in adulthood. Int J Psychophysiol 88:224–231. https://doi.org/10.1016/j.ijpsycho.2013.04.006

Cohen RT, Kahana MJ (2022) A memory-based theory of emotional disorders. Psychol Rev 129: 742–776. https://doi.org/10.1037/rev0000334

Cohen S, Wills TA (1985) Stress, social support, and the buffering hypothesis. Psychol Bull 98: 310–357. https://doi.org/10.1037/0033-2909.98.2.310

Colombo D, Serino S, Suso-Ribera C et al (2021) The moderating role of emotion regulation in the recall of negative autobiographical memories. Int J Environ Res Public Health 18(13):7122. https://doi.org/10.3390/ijerph18137122

Cutuli D (2014) Cognitive reappraisal and expressive suppression strategies role in the emotion regulation: an overview on their modulatory effects and neural correlates. Front Syst Neurosci 8: 175. https://doi.org/10.3389/fnsys.2014.00175

Dalgleish T, Hitchcock C (2023) Transdiagnostic distortions in autobiographical memory recollection. Nat Rev Psychol 2:166–182. https://doi.org/10.1038/s44159-023-00148-1

Dalgleish T, Yiend J, Schweizer S, Dunn BD (2009) Ironic effects of emotion suppression when recounting distressing memories. Emotion 9:744–749. https://doi.org/10.1037/a0017290

de Quervain D, Schwabe L, Roozendaal B (2017) Stress, glucocorticoids and memory: implications for treating fear-related disorders. Nat Rev Neurosci 18:7–19. https://doi.org/10.1038/nrn.2016.155

Delgado MR, Jou RL, Ledoux JE, Phelps EA (2009) Avoiding negative outcomes: tracking the mechanisms of avoidance learning in humans during fear conditioning. Front Behav Neurosci 3: 33. https://doi.org/10.3389/neuro.08.033.2009

Delgado MR, Fareri DS, Chang LJ (2023) Characterizing the mechanisms of social connection. Neuron 111:3911–3925. https://doi.org/10.1016/j.neuron.2023.09.012

DeSteno D, Gross JJ, Kubzansky L (2013) Affective science and health: the importance of emotion and emotion regulation. Health Psychol 32:474–486. https://doi.org/10.1037/a0030259

Dillon DG (2015) The neuroscience of positive memory deficits in depression. Front Psychol 6: 1295. https://doi.org/10.3389/fpsyg.2015.01295

Dillon DG, Pizzagalli DA (2018) Mechanisms of memory disruption in depression. Trends Neurosci 41:137–149. https://doi.org/10.1016/j.tins.2017.12.006

Dillon DG, Ritchey M, Johnson BD, LaBar KS (2007) Dissociable effects of conscious emotion regulation strategies on explicit and implicit memory. Emotion 7:354–365. https://doi.org/10.1037/1528-3542.7.2.354

DiMenichi BC, Lempert KM, Bejjani C, Tricomi E (2018) Writing about past failures attenuates cortisol responses and sustained attention deficits following psychosocial stress. Front Behav Neurosci 12:45. https://doi.org/10.3389/fnbeh.2018.00045

Dixon ML, Thiruchselvam R, Todd R, Christoff K (2017) Emotion and the prefrontal cortex: an integrative review. Psychol Bull 143:1033–1081. https://doi.org/10.1037/bul0000096

Dryman MT, Heimberg RG (2018) Emotion regulation in social anxiety and depression: a systematic review of expressive suppression and cognitive reappraisal. Clin Psychol Rev 65:17–42. https://doi.org/10.1016/j.cpr.2018.07.004

Dunsmoor JE, Otto AR, Phelps EA (2017) Stress promotes generalization of older but not recent threat memories. Proc Natl Acad Sci U S A 114:9218–9223. https://doi.org/10.1073/pnas.1704428114

Ehring T, Tuschen-Caffier B, Schnülle J et al (2010) Emotion regulation and vulnerability to depression: spontaneous versus instructed use of emotion suppression and reappraisal. Emotion 10:563–572. https://doi.org/10.1037/a0019010

Eisenberg N, Cumberland A, Spinrad TL (1998) Parental socialization of emotion. Psychol Inq 9: 241–273. https://doi.org/10.1207/s15327965pli0904_1

Engen HG, Anderson MC (2018) Memory control: a fundamental mechanism of emotion regulation. Trends Cogn Sci 22:982–995. https://doi.org/10.1016/j.tics.2018.07.015

Feldman R (2017) The neurobiology of human attachments. Trends Cogn Sci 21:80–99. https://doi.org/10.1016/j.tics.2016.11.007

Fledderus M, Bohlmeijer ET, Pieterse ME (2010) Does experiential avoidance mediate the effects of maladaptive coping styles on psychopathology and mental health? Behav Modif 34:503–519. https://doi.org/10.1177/0145445510378379

Gee DG (2016) Sensitive periods of emotion regulation: influences of parental care on frontoamygdala circuitry and plasticity. New Dir Child Adolesc Dev 2016:87–110. https://doi.org/10.1002/cad.20166

Gee DG, Gabard-Durnam L, Telzer EH et al (2014) Maternal buffering of human amygdala-prefrontal circuitry during childhood but not during adolescence. Psychol Sci 25:2067–2078. https://doi.org/10.1177/0956797614550878

Girotti M, Adler SM, Bulin SE et al (2018) Prefrontal cortex executive processes affected by stress in health and disease. Prog Neuro-Psychopharmacol Biol Psychiatry 85:161–179. https://doi.org/10.1016/j.pnpbp.2017.07.004

Gross JJ (2002) Emotion regulation: affective, cognitive, and social consequences. Psychophysiology 39:281–291. https://doi.org/10.1017/s0048577201393198

Gross JJ (2015) The extended process model of emotion regulation: elaborations, applications, and future directions. Psychol Inq 26:130–137. https://doi.org/10.1080/1047840X.2015.989751

Gross JJ, John OP (2003) Individual differences in two emotion regulation processes: implications for affect, relationships, and well-being. J Pers Soc Psychol 85:348–362. https://doi.org/10.1037/0022-3514.85.2.348

Gunnar MR (2017) Social buffering of stress in development: a career perspective. Perspect Psychol Sci 12:355–373. https://doi.org/10.1177/1745691616680612

Haag C, So M, Vainre M et al (2024) Positive autobiographical memories to counteract low mood in remitted depression: a longitudinal daily-life investigation. Emotion 24:1709–1720. https://doi.org/10.1037/emo0001330

Hallford DJ, Mellor D (2016) Brief reminiscence activities improve state well-being and self-concept in young adults: a randomised controlled experiment. Memory 24:1311–1320. https://doi.org/10.1080/09658211.2015.1103875

Hawkins J, D'Aniello-Heyda C, Fife S et al (2024) Emerging adults' experiences of managing social group disruptions. Emerg Adulthood 12:346–357. https://doi.org/10.1177/21676968241233137

Hawkley LC (2022) Loneliness and health. Nat Rev Dis Primers 8:22. https://doi.org/10.1038/s41572-022-00355-9

Hayes JP, Morey RA, Petty CM et al (2010) Staying cool when things get hot: emotion regulation modulates neural mechanisms of memory encoding. Front Hum Neurosci 4:230. https://doi.org/10.3389/fnhum.2010.00230

Hermann A, Bieber A, Keck T et al (2014) Brain structural basis of cognitive reappraisal and expressive suppression. Soc Cogn Affect Neurosci 9:1435–1442. https://doi.org/10.1093/scan/nst130

Hofmann SG (2014) Interpersonal emotion regulation model of mood and anxiety disorders. Cogn Ther Res 38:483–492. https://doi.org/10.1007/s10608-014-9620-1

Holland AC, Kensinger EA (2013a) An fMRI investigation of the cognitive reappraisal of negative memories. Neuropsychologia 51:2389–2400. https://doi.org/10.1016/j.neuropsychologia.2013.02.012

Holland AC, Kensinger EA (2013b) The neural correlates of cognitive reappraisal during emotional autobiographical memory recall. J Cogn Neurosci 25:87–108. https://doi.org/10.1162/jocn_a_00289

Holt-Lunstad J (2021) The major health implications of social connection. Curr Dir Psychol Sci 30:251–259. https://doi.org/10.1177/0963721421999630

Holt-Lunstad J, Smith TB, Layton JB (2010) Social relationships and mortality risk: a meta-analytic review. PLoS Med 7:e1000316. https://doi.org/10.1371/journal.pmed.1000316

Hornstein EA, Eisenberger NI (2017) Unpacking the buffering effect of social support figures: social support attenuates fear acquisition. PLoS One 12:e0175891. https://doi.org/10.1371/journal.pone.0175891

Hornstein EA, Haltom KEB, Shirole K (2018) A unique safety signal: social-support figures enhance rather than protect from fear extinction. Clin Psychol Sci 6(3):407–415. https://doi.org/10.1177/2167702617743002

Hornstein EA, Craske MG, Fanselow MS, Eisenberger NI (2022) Reclassifying the unique inhibitory properties of social support figures: a roadmap for exploring prepared fear suppression. Biol Psychiatry 91:778–785. https://doi.org/10.1016/j.biopsych.2021.11.017

Hostinar CE, Sullivan RM, Gunnar MR (2014) Psychobiological mechanisms underlying the social buffering of the hypothalamic-pituitary-adrenocortical axis: a review of animal models and human studies across development. Psychol Bull 140:256–282. https://doi.org/10.1037/a0032671

Hyman IE, Faries JM (1992) The functions of autobiographical memory. In: Conway MA, Rubin DC, Spinnler H, Wagenaar WA (eds) Theoretical perspectives on autobiographical memory. Springer, Dordrecht, pp 207–221

Inagaki TK, Eisenberger NI (2016) Giving support to others reduces sympathetic nervous system-related responses to stress. Psychophysiology 53:427–435. https://doi.org/10.1111/psyp.12578

Joëls M, Fernandez G, Roozendaal B (2011) Stress and emotional memory: a matter of timing. Trends Cogn Sci 15:280–288. https://doi.org/10.1016/j.tics.2011.04.004

Joormann J, Siemer M, Gotlib IH (2007) Mood regulation in depression: differential effects of distraction and recall of happy memories on sad mood. J Abnorm Psychol 116:484–490. https://doi.org/10.1037/0021-843X.116.3.484

Kashdan TB, Goodman FR, Machell KA et al (2014) A contextual approach to experiential avoidance and social anxiety: evidence from an experimental interaction and daily interactions of people with social anxiety disorder. Emotion 14:769–781. https://doi.org/10.1037/a0035935

Kensinger EA, Ford JH (2020) Retrieval of emotional events from memory. Annu Rev Psychol 71: 251–272. https://doi.org/10.1146/annurev-psych-010419-051123

Kerr ML, Rasmussen HF, Smiley PA et al (2021) The development of toddlers' emotion regulation within the family system: associations with observed parent-child synchrony and interparental relationship satisfaction. Early Child Res Q 57:215–227. https://doi.org/10.1016/j.ecresq.2021.06.004

Kheirbek MA, Klemenhagen KC, Sahay A, Hen R (2012) Neurogenesis and generalization: a new approach to stratify and treat anxiety disorders. Nat Neurosci 15:1613–1620. https://doi.org/10.1038/nn.3262

King B, Adepoju OE, Woodard L et al (2023) The effects of COVID-19 lockdown on social connectedness and psychological distress in U.S. adults with chronic diseases. Int J Environ Res Public Health 20(13):6218. https://doi.org/10.3390/ijerph20136218

Kogler L, Gur RC, Derntl B (2015) Sex differences in cognitive regulation of psychosocial achievement stress: brain and behavior. Hum Brain Mapp 36:1028–1042. https://doi.org/10.1002/hbm.22683

Kohn N, Eickhoff SB, Scheller M et al (2014) Neural network of cognitive emotion regulation--an ALE meta-analysis and MACM analysis. NeuroImage 87:345–355. https://doi.org/10.1016/j.neuroimage.2013.11.001

Koolhaas JM, Bartolomucci A, Buwalda B et al (2011) Stress revisited: a critical evaluation of the stress concept. Neurosci Biobehav Rev 35:1291–1301. https://doi.org/10.1016/j.neubiorev.2011.02.003

Kross E, Davidson M, Weber J, Ochsner K (2009) Coping with emotions past: the neural bases of regulating affect associated with negative autobiographical memories. Biol Psychiatry 65:361–366. https://doi.org/10.1016/j.biopsych.2008.10.019

LaBar KS (2015) Therapeutic affect reduction, emotion regulation, and emotional memory reconsolidation: a neuroscientific quandary. Behav Brain Sci 38:e10

LaBar KS, Cabeza R (2006) Cognitive neuroscience of emotional memory. Nat Rev Neurosci 7:54–64. https://doi.org/10.1038/nrn1825

Lang PJ, Bradley MM (2010) Emotion and the motivational brain. Biol Psychol 84:437–450. https://doi.org/10.1016/j.biopsycho.2009.10.007

Laufer O, Israeli D, Paz R (2016) Behavioral and neural mechanisms of overgeneralization in anxiety. Curr Biol 26:713–722. https://doi.org/10.1016/j.cub.2016.01.023

Lee JLC, Nader K, Schiller D (2017) An update on memory reconsolidation updating. Trends Cogn Sci 21:531–545. https://doi.org/10.1016/j.tics.2017.04.006

Levy I, Schiller D (2021) Neural computations of threat. Trends Cogn Sci 25:151–171. https://doi.org/10.1016/j.tics.2020.11.007

Linehan M (1993) Cognitive-behavioral treatment of borderline personality disorder. Guilford Press

Maguire EA (2022) Does memory research have a realistic future? Trends Cogn Sci 26:1043–1046. https://doi.org/10.1016/j.tics.2022.07.006

Maughan A, Cicchetti D (2002) Impact of child maltreatment and interadult violence on children's emotion regulation abilities and socioemotional adjustment. Child Dev 73:1525–1542. https://doi.org/10.1111/1467-8624.00488

McDermott KB, Szpunar KK, Christ SE (2009) Laboratory-based and autobiographical retrieval tasks differ substantially in their neural substrates. Neuropsychologia 47:2290–2298. https://doi.org/10.1016/j.neuropsychologia.2008.12.025

McEwen BS, Akil H (2020) Revisiting the stress concept: implications for affective disorders. J Neurosci 40:12–21. https://doi.org/10.1523/JNEUROSCI.0733-19.2019

McGaugh JL (2000) Memory-a century of consolidation. Science 287:248–251. https://doi.org/10.1126/science.287.5451.248

McIntyre CK, McGaugh JL, Williams CL (2012) Interacting brain systems modulate memory consolidation. Neurosci Biobehav Rev 36:1750–1762. https://doi.org/10.1016/j.neubiorev.2011.11.001

McLaughlin KA, Hatzenbuehler ML (2009) Mechanisms linking stressful life events and mental health problems in a prospective, community-based sample of adolescents. J Adolesc Health 44:153–160. https://doi.org/10.1016/j.jadohealth.2008.06.019

McLaughlin KA, Hatzenbuehler ML, Mennin DS, Nolen-Hoeksema S (2011) Emotion dysregulation and adolescent psychopathology: a prospective study. Behav Res Ther 49:544–554. https://doi.org/10.1016/j.brat.2011.06.003

McLaughlin KA, Peverill M, Gold AL et al (2015) Child maltreatment and neural systems underlying emotion regulation. J Am Acad Child Adolesc Psychiatry 54:753–762. https://doi.org/10.1016/j.jaac.2015.06.010

Mihalca AM, Tarnavska Y (2013) Cognitive emotion regulation strategies and social functioning in adolescents. Procedia Soc Behav Sci 82:574–579. https://doi.org/10.1016/j.sbspro.2013.06.312

Milad MR, Quirk GJ (2002) Neurons in medial prefrontal cortex signal memory for fear extinction. Nature 420:70–74. https://doi.org/10.1038/nature01138

Milad MR, Quirk GJ (2012) Fear extinction as a model for translational neuroscience: ten years of progress. Annu Rev Psychol 63:129–151. https://doi.org/10.1146/annurev.psych.121208.131631

Milojevich HM, Machlin L, Sheridan MA (2020) Early adversity and children's emotion regulation: differential roles of parent emotion regulation and adversity exposure. Dev Psychopathol 32:1788–1798. https://doi.org/10.1017/S0954579420001273

Mitmansgruber H, Beck TN, Höfer S, Schüßler G (2009) When you don't like what you feel: experiential avoidance, mindfulness and meta-emotion in emotion regulation. Pers Individ Differ 46:448–453. https://doi.org/10.1016/j.paid.2008.11.013

Moghaddam B, Jackson M (2004) Effect of stress on prefrontal cortex function. Neurotox Res 6:73–78. https://doi.org/10.1007/BF03033299

Monfils MH, Holmes EA (2018) Memory boundaries: opening a window inspired by reconsolidation to treat anxiety, trauma-related, and addiction disorders. Lancet Psychiatry 5: 1032–1042. https://doi.org/10.1016/S2215-0366(18)30270-0

Monfils CKK, Klann E, LeDoux JE (2009) Extinction-reconsolidation boundaries: key to persistent attenuation of fear memories. Science 324:951–955. https://doi.org/10.1126/science.1167975

Nader K, Schafe GE, LeDoux JE (2000) The labile nature of consolidation theory. Nat Rev Neurosci 1:216–219. https://doi.org/10.1038/35044580

Nolen-Hoeksema S, Wisco BE, Lyubomirsky S (2008) Rethinking rumination. Perspect Psychol Sci 3:400–424. https://doi.org/10.1111/j.1745-6924.2008.00088.x

Ochsner KN (2019) From the self to the social regulation of emotion: an evolving psychological and neural model. In: Emotion in the mind and body, vol 272. Springer, Cham, pp 43–75. https://doi.org/10.1007/978-3-030-27473-3_3

Ochsner KN, Bunge SA, Gross JJ, Gabrieli JDE (2002) Rethinking feelings: an FMRI study of the cognitive regulation of emotion. J Cogn Neurosci 14:1215–1229. https://doi.org/10.1162/089892902760807212

Ochsner KN, Silvers JA, Buhle JT (2012) Functional imaging studies of emotion regulation: a synthetic review and evolving model of the cognitive control of emotion. Ann N Y Acad Sci 1251:E1–E24. https://doi.org/10.1111/j.1749-6632.2012.06751.x

Paersch C, Schulz A, Wilhelm FH et al (2022) Recalling autobiographical self-efficacy episodes boosts reappraisal-effects on negative emotional memories. Emotion 22:1148–1158. https://doi.org/10.1037/emo0000949

Parikh N, McGovern B, LaBar KS (2019) Spatial distancing reduces emotional arousal to reactivated memories. Psychon Bull Rev 26:1967–1973. https://doi.org/10.3758/s13423-019-01648-z

Pastor V, Medina JH (2021) Medial prefrontal cortical control of reward- and aversion-based behavioral output: bottom-up modulation. Eur J Neurosci 53:3039–3062. https://doi.org/10.1111/ejn.15168

Pasupathi M (2003) Emotion regulation during social remembering: differences between emotions elicited during an event and emotions elicited when talking about it. Memory 11:151–163. https://doi.org/10.1080/741938212

Pei R, Courtney AL, Ferguson I et al (2023) A neural signature of social support mitigates negative emotion. Sci Rep 13:17293. https://doi.org/10.1038/s41598-023-43273-w

Pennebaker JW, Chung CK (2011) Expressive writing: connections to physical and mental health. In: The Oxford handbook of health psychology. Oxford University Press

Phelps EA, Hofmann SG (2019) Memory editing from science fiction to clinical practice. Nature 572:43–50. https://doi.org/10.1038/s41586-019-1433-7

Phelps EA, LeDoux JE (2005) Contributions of the amygdala to emotion processing: from animal models to human behavior. Neuron 48:175–187. https://doi.org/10.1016/j.neuron.2005.09.025

Raio CM, Orederu TA, Palazzolo L et al (2013) Cognitive emotion regulation fails the stress test. Proc Natl Acad Sci U S A 110:15139–15144. https://doi.org/10.1073/pnas.1305706110

Ramirez S, Liu X, Mac Donald CJ et al (2015) Activating positive memory engrams suppresses depression-like behaviour. Nature 522:335–339. https://doi.org/10.1038/nature14514

Rasmussen AS, Berntsen D (2009) Emotional valence and the functions of autobiographical memories: positive and negative memories serve different functions. Mem Cogn 37:477–492. https://doi.org/10.3758/MC.37.4.477

Ratcliff KA, Vazquez LC, Lunkenheimer ES, Cole PM (2021) Longitudinal changes in young children's strategy use for emotion regulation. Dev Psychol 57:1471–1486. https://doi.org/10.1037/dev0001235

Raver CC, Roy AL, Pressler E et al (2016) Poverty-related adversity and emotion regulation predict internalizing behavior problems among low-income children ages 8-11. Behav Sci 7(1):2. https://doi.org/10.3390/bs7010002

Reeck C, Ames DR, Ochsner KN (2016) The social regulation of emotion: an integrative, cross-disciplinary model. Trends Cogn Sci 20:47–63. https://doi.org/10.1016/j.tics.2015.09.003

Richards JM, Gross JJ (2006) Personality and emotional memory: how regulating emotion impairs memory for emotional events. J Res Pers 40:631–651. https://doi.org/10.1016/j.jrp.2005.07.002

Rimé B (2009) Emotion elicits the social sharing of emotion: theory and empirical review. Emot Rev 1:60–85. https://doi.org/10.1177/1754073908097189

Rimé B, Bouchat P, Paquot L, Giglio L (2020) Intrapersonal, interpersonal, and social outcomes of the social sharing of emotion. Curr Opin Psychol 31:127–134. https://doi.org/10.1016/j.copsyc.2019.08.024

Ritvo VJH, Turk-Browne NB, Norman KA (2019) Nonmonotonic plasticity: how memory retrieval drives learning. Trends Cogn Sci 23:726–742. https://doi.org/10.1016/j.tics.2019.06.007

Robinson JA, Swanson KL (1990) Autobiographical memory: the next phase. Appl Cogn Psychol 4:321–335. https://doi.org/10.1002/acp.2350040407

Rodebaugh TL, Lim MH, Shumaker EA et al (2015) Social anxiety and friendship quality over time. Cogn Behav Ther 44:502–511. https://doi.org/10.1080/16506073.2015.1062043

Roozendaal B, McEwen BS, Chattarji S (2009) Stress, memory and the amygdala. Nat Rev Neurosci 10:423–433. https://doi.org/10.1038/nrn2651

Rose AJ (2002) Co-rumination in the friendships of girls and boys. Child Dev 73:1830–1843. https://doi.org/10.1111/1467-8624.00509

Rose AJ, Carlson W, Waller EM (2007) Prospective associations of co-rumination with friendship and emotional adjustment: considering the socioemotional trade-offs of co-rumination. Dev Psychol 43:1019–1031. https://doi.org/10.1037/0012-1649.43.4.1019

Rozen N, Aderka IM (2023) Emotions in social anxiety disorder: a review. J Anxiety Disord 95:102696. https://doi.org/10.1016/j.janxdis.2023.102696

Saha S, Lim CCW, Cannon DL et al (2021) Co-morbidity between mood and anxiety disorders: a systematic review and meta-analysis. Depress Anxiety 38:286–306. https://doi.org/10.1002/da.23113

Samide R, Ritchey M (2021) Reframing the past: role of memory processes in emotion regulation. Cogn Ther Res 45:848–857. https://doi.org/10.1007/s10608-020-10166-5

Schiller D, Delgado MR (2010) Overlapping neural systems mediating extinction, reversal and regulation of fear. Trends Cogn Sci 14:268–276. https://doi.org/10.1016/j.tics.2010.04.002

Schiller D, Monfils M-H, Raio CM et al (2010) Preventing the return of fear in humans using reconsolidation update mechanisms. Nature 463:49–53. https://doi.org/10.1038/nature08637

Schiller D, Kanen JW, LeDoux JE et al (2013) Extinction during reconsolidation of threat memory diminishes prefrontal cortex involvement. Proc Natl Acad Sci U S A 110:20040–20045. https://doi.org/10.1073/pnas.1320322110

Schwabe L, Joëls M, Roozendaal B et al (2012) Stress effects on memory: an update and integration. Neurosci Biobehav Rev 36:1740–1749. https://doi.org/10.1016/j.neubiorev.2011.07.002

Schwabe L, Hermans EJ, Joëls M, Roozendaal B (2022) Mechanisms of memory under stress. Neuron 110:1450–1467. https://doi.org/10.1016/j.neuron.2022.02.020

Seibert GS, Bauer KN, May RW, Fincham FD (2017) Emotion regulation and academic underperformance: the role of school burnout. Learn Individ Differ 60:1–9. https://doi.org/10.1016/j.lindif.2017.10.001

Sheppes G, Meiran N (2007) Better late than never? On the dynamics of online regulation of sadness using distraction and cognitive reappraisal. Personal Soc Psychol Bull 33:1518–1532. https://doi.org/10.1177/0146167207305537

Sheppes G, Meiran N (2008) Divergent cognitive costs for online forms of reappraisal and distraction. Emotion 8:870–874. https://doi.org/10.1037/a0013711

Sheppes G, Scheibe S, Suri G et al (2014) Emotion regulation choice: a conceptual framework and supporting evidence. J Exp Psychol Gen 143:163–181. https://doi.org/10.1037/a0030831

Sheppes G, Suri G, Gross JJ (2015) Emotion regulation and psychopathology. Annu Rev Clin Psychol 11:379–405. https://doi.org/10.1146/annurev-clinpsy-032814-112739

Shiota MN, Levenson RW (2012) Turn down the volume or change the channel? Emotional effects of detached versus positive reappraisal. J Pers Soc Psychol 103:416–429. https://doi.org/10.1037/a0029208

Shohamy D, Adcock RA (2010) Dopamine and adaptive memory. Trends Cogn Sci 14:464–472. https://doi.org/10.1016/j.tics.2010.08.002

Silvers JA (2020) Extinction learning and cognitive reappraisal: windows into the neurodevelopment of emotion regulation. Child Dev Perspect 14:178–184. https://doi.org/10.1111/cdep.12372

Silvers JA, Weber J, Wager TD, Ochsner KN (2015) Bad and worse: neural systems underlying reappraisal of high- and low-intensity negative emotions. Soc Cogn Affect Neurosci 10:172–179. https://doi.org/10.1093/scan/nsu043

Silvers JA, Insel C, Powers A et al (2017) vlPFC–vmPFC–amygdala interactions underlie age-related differences in cognitive regulation of emotion. Cereb Cortex 27:3502

Sloan DM (2004) Emotion regulation in action: emotional reactivity in experiential avoidance. Behav Res Ther 42:1257–1270. https://doi.org/10.1016/j.brat.2003.08.006

Song Y, Lu H, Hu S et al (2015) Regulating emotion to improve physical health through the amygdala. Soc Cogn Affect Neurosci 10:523–530. https://doi.org/10.1093/scan/nsu083

Speer ME, Delgado MR (2017) Reminiscing about positive memories buffers acute stress responses. Nat Hum Behav 1:0093. https://doi.org/10.1038/s41562-017-0093

Speer ME, Delgado MR (2020) The social value of positive autobiographical memory retrieval. J Exp Psychol Gen 149:790–799. https://doi.org/10.1037/xge0000671

Speer ME, Bhanji JP, Delgado MR (2014) Savoring the past: positive memories evoke value representations in the striatum. Neuron 84:847–856. https://doi.org/10.1016/j.neuron.2014.09.028

Speer ME, Ibrahim S, Schiller D, Delgado MR (2021) Finding positive meaning in memories of negative events adaptively updates memory. Nat Commun 12:6601. https://doi.org/10.1038/s41467-021-26906-4

Sprecher S, Treger S, Wondra JD et al (2013) Taking turns: reciprocal self-disclosure promotes liking in initial interactions. J Exp Soc Psychol 49:860–866. https://doi.org/10.1016/j.jesp.2013.03.017

Starita F, Kroes MCW, Davachi L et al (2019) Threat learning promotes generalization of episodic memory. J Exp Psychol Gen 148:1426–1434. https://doi.org/10.1037/xge0000551

Stephens JE, Hittner EF, Haase CM (2013) Emotion regulation in couples across the life span. In: Gross JJ (ed) The Oxford handbook of emotional development. The Guilford Press, pp 267–283

Strauss GP, Ossenfort KL, Whearty KM (2016) Reappraisal and distraction emotion regulation strategies are associated with distinct patterns of visual attention and differing levels of cognitive demand. PLoS One 11:e0162290. https://doi.org/10.1371/journal.pone.0162290

Sumner JA (2012) The mechanisms underlying overgeneral autobiographical memory: an evaluative review of evidence for the CaR-FA-X model. Clin Psychol Rev 32:34–48. https://doi.org/10.1016/j.cpr.2011.10.003

Tang A, McLaughlin KA, Sheridan MA et al (2022) Autonomic reactivity to social rejection, peer difficulties, and the buffering effects of adolescent friendships following early psychosocial deprivation. Emotion 22:318–330. https://doi.org/10.1037/emo0001016

Taylor SE (2006) Tend and befriend: biobehavioral bases of affiliation under stress. Curr Dir Psychol Sci 15:273–277

Teo AR, Choi H, Valenstein M (2013) Social relationships and depression: ten-year follow-up from a nationally representative study. PLoS One 8:e62396. https://doi.org/10.1371/journal.pone.0062396

Tindle R, Hemi A, Moustafa AA (2022) Social support, psychological flexibility and coping mediate the association between COVID-19 related stress exposure and psychological distress. Sci Rep 12:8688. https://doi.org/10.1038/s41598-022-12262-w

Treanor M, Brown LA, Rissman J, Craske MG (2017) Can memories of traumatic experiences or addiction be erased or modified? A critical review of research on the disruption of memory

reconsolidation and its applications. Perspect Psychol Sci 12:290–305. https://doi.org/10.1177/1745691616664725
Troy AS, Shallcross AJ, Mauss IB (2013) A person-by-situation approach to emotion regulation: cognitive reappraisal can either help or hurt, depending on the context. Psychol Sci 24:2505–2514. https://doi.org/10.1177/0956797613496434
Ulrich-Lai YM, Herman JP (2009) Neural regulation of endocrine and autonomic stress responses. Nat Rev Neurosci 10:397–409. https://doi.org/10.1038/nrn2647
Uzer T, Brown NR (2017) The effect of cue content on retrieval from autobiographical memory. Acta Psychol 172:84–91. https://doi.org/10.1016/j.actpsy.2016.11.012
Uzer T, Lee PJ, Brown NR (2012) On the prevalence of directly retrieved autobiographical memories. J Exp Psychol Learn Mem Cogn 38:1296–1308. https://doi.org/10.1037/a0028142
van't Wout M, Chang LJ, Sanfey AG (2010) The influence of emotion regulation on social interactive decision-making. Emotion 10:815–821. https://doi.org/10.1037/a0020069
Vijayakumar N, Flournoy JC, Mills KL et al (2020) Getting to know me better: an fMRI study of intimate and superficial self-disclosure to friends during adolescence. J Pers Soc Psychol 118: 885–899. https://doi.org/10.1037/pspa0000182
Wager TD, Davidson ML, Hughes BL et al (2008) Prefrontal-subcortical pathways mediating successful emotion regulation. Neuron 59:1037–1050. https://doi.org/10.1016/j.neuron.2008.09.006
Wardell V, Palombo DJ (2024) Stability and malleability of emotional autobiographical memories. Nat Rev Psychol 3:1–14. https://doi.org/10.1038/s44159-024-00312-1
Weiss J, Lawton LE, Fischer CS (2022) Life course transitions and changes in network ties among younger and older adults. Adv Life Course Res 52:100478. https://doi.org/10.1016/j.alcr.2022.100478
Williams JMG, Barnhofer T, Crane C et al (2007) Autobiographical memory specificity and emotional disorder. Psychol Bull 133:122–148. https://doi.org/10.1037/0033-2909.133.1.122
Wolgast M, Lundh L-G (2017) Is distraction an adaptive or maladaptive strategy for emotion regulation? A person-oriented approach. J Psychopathol Behav Assess 39:117–127. https://doi.org/10.1007/s10862-016-9570-x
Wu L, Winkler MH, Wieser MJ et al (2015) Emotion regulation in heavy smokers: experiential, expressive and physiological consequences of cognitive reappraisal. Front Psychol 6:1555. https://doi.org/10.3389/fpsyg.2015.01555
Yamamoto T, Toki S, Siegle GJ et al (2017) Increased amygdala reactivity following early life stress: a potential resilience enhancer role. BMC Psychiatry 17:27. https://doi.org/10.1186/s12888-017-1201-x
Yonelinas AP, Ritchey M (2015) The slow forgetting of emotional episodic memories: an emotional binding account. Trends Cogn Sci 19:259–267. https://doi.org/10.1016/j.tics.2015.02.009
Zagic D, Wuthrich VM, Rapee RM, Wolters N (2022) Interventions to improve social connections: a systematic review and meta-analysis. Soc Psychiatry Psychiatr Epidemiol 57:885–906. https://doi.org/10.1007/s00127-021-02191-w
Zaki J, Williams WC (2013) Interpersonal emotion regulation. Emotion 13:803–810. https://doi.org/10.1037/a0033839
Zbozinek TD, Craske MG (2017) The role of positive affect in enhancing extinction learning and exposure therapy for anxiety disorders. J Exp Psychopathol 8:13–39. https://doi.org/10.5127/jep.052615
Zhang X, Ge TT, Yin G et al (2018) Stress-induced functional alterations in amygdala: implications for neuropsychiatric diseases. Front Neurosci 12:367. https://doi.org/10.3389/fnins.2018.00367
Zlomuzica A, Dere D, Machulska A et al (2014) Episodic memories in anxiety disorders: clinical implications. Front Behav Neurosci 8:131. https://doi.org/10.3389/fnbeh.2014.00131

Part II
Risk Factors and Moderating Factors

Animal models

Developmental Shifts in Amygdala Function

J. Amiel Rosenkranz

Contents

Abstract Mammals have evolved with strategies to optimize survival and thrive in their native environment. This includes both physical and behavioral adaptations, and extends to their social environment. However, within a social context, the roles of an animal change across development, and their behavior and biology must update to match these changes. The amygdala has a key role in social and emotional processing and expression, and displays developmental changes in early juvenile, adolescent, and adult transitions. Furthermore, the amygdala is highly sensitive to the social environment. This chapter will describe the primary amygdala developmental changes, how this maps onto major changes in social and emotional domains, and propose a framework where developmental stage of intra-amygdala circuits and its regulation by cortical inputs biases the animal toward developmentally appropriate social and emotional behavior. This developmental plasticity also presents an opportunity for retuning the developmental trajectory in the presence of ongoing challenges during maturation, such as constant threat or resource scarcity, so there can be realignment of behavior to match environmental demands.

Keywords Amygdala · Fear · Maturation · Social

J. A. Rosenkranz (✉)
Center for Neurobiology of Stress Resilience and Psychiatric Disorders, Brain Science Institute, Rosalind Franklin University of Medicine and Science, North Chicago, IL, USA
e-mail: jeremy.rosenkranz@rosalindfranklin.edu

© The Author(s), under exclusive license to Springer Nature Switzerland AG 2024
Curr Topics Behav Neurosci (2025) 73: 207–229
https://doi.org/10.1007/7854_2024_538
Published Online: 16 November 2024

There is a typical socio-emotional developmental trajectory seen across most mammals. At its core, this includes transition from reliance on others for protection and care toward independent navigation of the social environment and assessment of threats (Broad et al. 2006; Scheiber et al. 2017). In the transition between these stages, young mammals have opportunity to explore, practice, and refine social and defensive behaviors (Ellis et al. 2012; Romer et al. 2017). As they mature, this gives way to a stage where individual survival can depend on social and emotional fitness for their environment, but this is highly dependent on the social milieu in which they live (MacLeod et al. 2023). This contributed to the proposal that sensitive periods during development are adaptive, and allow the animal to adjust behavior and developmental trajectory based on the environment (Stamps and Krishnan 2014; Fawcett and Frankenhuis 2015). The amygdala has a key role in social and emotional processing and expression, and displays developmental changes in early juvenile, adolescent, and adult transitions. Furthermore, the amygdala is highly sensitive to the social environment. This chapter will describe the primary amygdala developmental changes, how this maps onto major changes in social and emotional domains, and propose a framework where developmental stage of intra-amygdala circuits and its regulation by cortical inputs biases the animal toward developmentally appropriate social and emotional behavior.

The amygdala and extended amygdala can be portioned into many regions based on neural origin, biochemical markers, and functional roles. The primary focus here is on two regions, the medial amygdala (MeA) and the basolateral amygdala (BLA), whose role in social and emotional development has been examined in some depth in rodents. This will be complemented with information on central amygdala (CeA) and bed nucleus of the stria terminalis (BNST). This chapter will describe amygdala maturation during developmental periods defined by behavioral and hormonal indices (Spear 2000), including early postnatal, juvenile, adolescent, and adult transitions. Although very important, this chapter will not describe sex differences in depth (discussed in Romero et al. Sex differences in neural circuits underlying fear processing).

1 Social Development

Most mammals have evolved within a complex and dynamic social ecosystem. Their survival is tied to their ability to remain within a social structure that provides opportunity for reproduction, protection, and resources (Alexander 1974). A substantial amount of behavior is adapted toward optimization of social opportunities, and not surprisingly, a large network of connected neural structures supports social behaviors (Adolphs 2009). The amygdala is a crucial node in this network. However, individual roles within a social environment change with age. Development of the amygdala can be viewed from the perspective of the social context, with the premise that the amygdala develops and refines at a pace that supports emergence and maturation of age-appropriate social behaviors (Ferrara and Opendak 2023).

Humans, nonhuman primates, and rodents display social maturation in parallel with amygdala maturation across development (Chareyron et al. 2012a, b; Gabard-Durnam et al. 2014). Both of these developmental processes are sensitive to the social environment. Severe neglect and socially impoverished settings during development can produce social deficits (Curtiss et al. 1974; Bos et al. 2011). Even extensive attempts at treatment can be inadequate to reverse social deficits (Gillies et al. 2016). Therefore, it is important to understand the maturation of social behavior within the social environment.

Most mammals, including most rodents and primates, require substantial early postnatal care from adults (König and Markl 1987; Ošt'ádalová and Babický 2012). These caregivers provide the initial primary social environment for infants (Galef and Clark 1972; Einon et al. 1978), and have a large impact on socio-emotional development, ability to handle later challenges, and amygdala maturation (discussed in Niu et al. The impact of early life experiences on stress neurobiology and the development of anxiety). Because of reliance on a caregiver, among the most important elements of their survival is bonding and attachment to the caregiver, and early postnatal life is partly specialized for this (Harlow and Zimmermann 1959; Hess 1959). But at this early stage, attachment develops even to harmful caregivers perhaps because the amygdala does not yet support aversive learning (Sullivan et al. 2000; Moriceau et al. 2006) and because of a co-occurring stress hyporesponsive period (Collier and Bolles 1980). Furthermore, the early life hyporesponsiveness to stress is partly attributable to maternal presence that buffers stress (Stanton et al. 1988; Rosenfeld et al. 1991). But the quality of this care strongly influences the subsequent maturational trajectory of the amygdala (Caldji et al. 1998; Sánchez et al. 2001).

After early reliance on caregivers and attachment, different motivated social behaviors emerge as rodents mature. Prepubertal rodents display predominantly affiliative behaviors, while social hierarchy and territorial behaviors emerge after puberty (Pellis and Pellis 1990; Takahashi 1992). This overlaps with a shift from peak social interaction during peripubertal ages to lower levels in adults (Panksepp 1981; Varlinskaya et al. 1999). There are hints toward greater social drive in prepubertal rodents, including greater initiation of social exploration and heightened play behavior (Panksepp 1981; Varlinskaya et al. 1999), as well as greater seeking of social interaction (Calcagnetti and Schechter 1992; Douglas et al. 2004). Social interactions during this period are necessary for appropriate maturation toward adult social and emotional phenotypes. Rodent social behavior becomes more selective after postpubertal emergence of mating and reproductive behaviors alongside territorial and social defensive behaviors. Upon reaching adult social maturity, social behavioral patterns are more rigid and more resilient to disruption. Although behavioral patterns and preferences in adults become less flexible, there is also emergence of social hierarchies (Berdoy et al. 1995; Chase et al. 2002), which can shift, requiring ability to adapt to social environment even as adults (Chase 1982; Williamson et al. 2016). Furthermore, many adults display dramatic shifts in social behaviors, such as emergence of maternal behaviors, which relies on amygdala plasticity (Fleming and Korsmit 1996; Chen et al. 2019).

2 Development of Fear and Anxiety

Parts of the neural networks for social behavior overlap with another key component of survival, ability to identify and respond to threats. The amygdala is a critical component of this too, and amygdala development parallels a shift in threat processing and defensive behaviors. The development of defensive behaviors is thought to align with likelihood of encountering predator, which occurs as young rodents become less reliant on adults for food, and begin to explore further from the caregiver and even leave the burrow (Stansfield et al. 2004; Lloyd et al. 2023).

Emergence of defensive behaviors is most often examined using classical fear conditioning, where a discrete neutral conditioned stimulus (CS) is paired with an aversive unconditioned stimulus (US). During pairings, the CS begins to elicit a fear response, often measured as conditioned freezing. This basic form of learning, where the CS is presented for several seconds and the US is presented to overlap with the end of the CS, is called delay conditioning (delay in the onset of US). Classical fear conditioning relies on the basolateral amygdala (BLA) (Fanselow and LeDoux 1999; Maren 2003). In early life, fear conditioning is absent. There is a slow emergence of fear and other aversive conditioning over the first few weeks of life (Moye and Rudy 1985; Stanton et al. 1992) alongside the emergence of avoidance of unfamiliar adult male conspecifics (a source of threat in some species; Takahashi 1992; Wiedenmayer and Barr 1998). By weaning age (>postnatal day (PND) 21), basic acquisition and expression of delay fear conditioning is similar to adults (Stanton et al. 1992; Rudy 1993), with perhaps a peak of conditioned freezing in adolescence (Barnet and Hunt 2005; Den and Richardson 2013). Forms of fear conditioning that are cognitively more challenging have a different developmental trajectory, even though they also rely on the BLA. A prototypical example is trace fear conditioning, which is very similar to delay fear conditioning, except there is a short time gap (seconds) between the CS end and the beginning of the US (a memory trace for the CS must be maintained over the time gap). Trace fear conditioning has a slower maturational progression (Moye and Rudy 1987; Barnet and Hunt 2005; Den and Richardson 2013), which is not surprising because, in addition to BLA, it relies on later developing prefrontal cortical (PFC) and hippocampal function (Raybuck and Lattal 2014). However, there is a surprising peak of trace fear conditioning at early adolescence, compared to younger and older rats (Moye and Rudy 1987; Barnet and Hunt 2005; Den and Richardson 2013). Contextual fear (fear for the context where aversive events occur), which relies on hippocampus and BLA, is weaker in juveniles (Rudy 1993), but is stable by PND23 in rats, and maybe earlier in mice (Rudy and Morledge 1994; Akers et al. 2012). However, in conditions where rats are trained in delay fear conditioning (classical CS–US pairings), and fear for the training context is tested, this contextual fear expression is also present by >PND 24 (Rudy 1993; Akers et al. 2012), but there may be a transient strong reduction of contextual fear expression in early adolescence (~PND 29–33; Pattwell et al. 2011). This is in contrast to other forms of hippocampal-dependent fear learning, such as trace fear conditioning, that peaks at a similar age (Moye and Rudy 1987; Barnet and Hunt 2005; Den and Richardson 2013).

Extinction of conditioned fear (repeated presentation of the CS without the US, and parallel reduced conditioned response to CS) requires additional learning. Adolescence also emerges as a unique period for fear extinction. Acquisition of fear extinction emerges by PND 24, at which point it is reliant on BLA (Kim and Richardson 2008), and is similar into adulthood (Baker and Richardson 2015). However, retention and expression of extinction learning, which relies on PFC and BLA, is lower in adolescence (Kim et al. 2009; Baker and Richardson 2015; Gerhard and Meyer 2021). Generally, rats at >PND 23 (but not <PND 18) display context dependency of extinction and renewal, but reinstatement and spontaneous recovery of extinguished fear may be greater in adolescents than adults (Kim and Richardson 2007; Gerhard and Meyer 2021). These deficits are consistent with immature PFC function (e.g., Kim et al. 2009; Baker and Richardson 2015). However, their emergence may be sex dependent. In juvenile rats, males show a full absence of renewal and spontaneous recovery of extinguished fear, while females show modest renewal and spontaneous recovery (PND 18; Park et al. 2017, 2020). One interpretation is that extinction leads to full overwrite of the CS memory at early ages in both sexes (so no renewal/recovery), while extinction leads to the formation of a competing CS memory in adolescent rats in both sexes. However, in the juvenile transition period, females can already form a competing CS memory (and do not overwrite the old memory) while males cannot. This ability to form a competing CS representation during extinction relies on development of hippocampal and PFC connections with BLA, and may therefore suggest that these circuits arise earlier in females.

Anxiety and fear are phenotypically different, but share some overlapping circuitry. The development of anxiety in rodents is more difficult to determine because many assays rely on exploratory behaviors, which are also very sensitive to age. So, it is not surprising that there are mixed reports of differences in anxiety-like behavior across development. There is a peak of exploratory behavior and sensitivity to novelty in adolescence (Spear and Brake 1983; Lynn and Brown 2009). However, it appears that innate defensive behaviors emerge by PND 14–21 and are stable into adulthood (Bronstein and Hirsch 1976; Wiedenmayer and Barr 1998), albeit with differences in how defensive behaviors are expressed across development (for review, see Boulanger Bertolus et al. 2016). There may be age differences in risk assessment behaviors, with mixed reports of younger rodents, particularly adolescents, displaying less risk assessment behaviors in conditions that may be anxiogenic (Macrì et al. 2002; Loh et al. 2022).

There is a balance between risk-taking, anxiety, and fear that needs to be within limitations that are appropriate for the environment (Orsini et al. 2015; Heinz et al. 2021). Social motivation drives social affiliation, but this drive is balanced by normal caution that is part of an adaptive anxiety continuum (Woody and Szechtman 2011). Internal and external factors that promote caution, anxiety, and fear can oppose social drive (Kudryavtseva et al. 1991; Takahashi 1992). What emerges from the combined literature on the development of social, anxiety, and fear behaviors is high social motivation during adolescence that is not yet restricted based on hierarchy or sex. This is coupled with a high exploratory drive during adolescence and novelty-seeking. While this can help hone social and exploration skills, it could also expose adolescents to heightened risk for predation. However, there is also mature cued fear conditioning, with weaker retention of extinction and stronger trace fear conditioning. This may ensure that specific dangers learned during adolescence can be retained strongly. In contrast, there is weaker contextual fear conditioning during adolescent exploration, which may help ensure that developing rodents do not overly fear exploring independently.

3 Amygdala Development

A large number of studies have probed the anatomical substrates of social and emotional development. Amygdala maturation occurs in a context of development of related neural networks. Among the most relevant to amygdala, there is an established developmental trajectory of prefrontal cortex, hippocampus, and hypothalamic systems. Interactions between those systems and amygdala contribute to the maturation of a wide range of social, sexual, defensive, and cognitive functions. As the field of coordinated development across multiple amygdala regions is still growing, there is much still to learn. Even within regions of the amygdala, there is much unknown, as most of the developmental studies have focused on the MeA and BLA.

3.1 Medial Amygdala

MeA is typically divided into two primary regions, anterior and posterior, that are further subdivided into dorsal and ventral (Fig. 1). Anterior MeA may play a role in

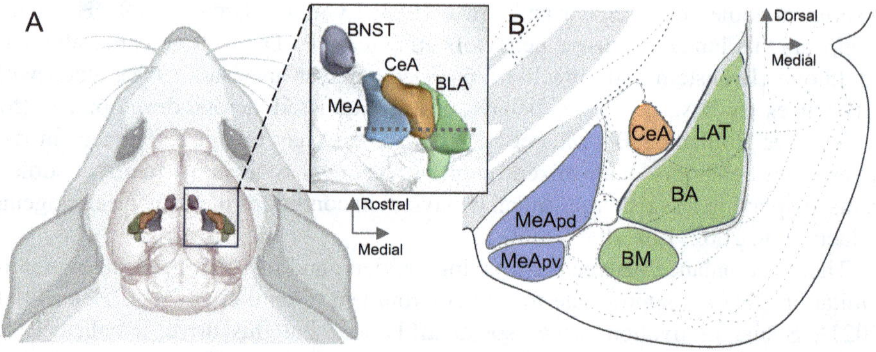

Fig. 1 Schematic of three main amygdala divisions. (**a**) Top-down view of basolateral (BLA; green), central (CeA; orange), and medial amygdala (MeA; blue), as well as BNST (gray). (**b**) Coronal view of MeA (MeApd and MeApv), CeA, and BLA (LAT, BA, BM). Figure modified from Swanson Brain Map 4.0 (plate 27; Swanson 2018 under Creative Commons License) and the free open resource Allen Mouse Brain (Allan Brain Science 2017; mouse.brain-map.org and atlas.brain-map.org), accessed via Scalable Mouse Atlas (Bakker et al. 2015)

initial filtering of signals to parse conspecific and heterospecific stimuli (Meredith and Westberry 2004; Petrulis 2009), with posterior MeA (MeAp) playing a critical role in conspecific defensive and mating behaviors, and projects heavily to medial BNST (Gomez and Newman 1992; Canteras et al. 1995). MeA neuronal composition varies across these regions, with primarily glutamatergic principal neurons in the ventral MeAp (MeApv) (Choi et al. 2005; Keshavarzi et al. 2014) and mixed glutamatergic and GABAergic principal neurons in dorsal MeAp (MeApd) (Johnson et al. 2021).

Immediate early gene mapping studies have provided strong evidence that the MeApd is activated during reproductive behaviors in rodents (Tetel et al. 1993) and has an important role in preference for same-sex or opposite-sex cues (Wu et al. 2009; Yao et al. 2017). However, aggressive conspecific encounter has occasionally been noted to activate MeApd (Veening et al. 2005). MeApv, in contrast, is consistently activated by conspecific aggression and predator cues (Kollack-Walker and Newman 1995; Dielenberg et al. 2001), but is also activated by same-sex or opposite-sex nonaggressive conspecific interactions (Kollack-Walker and Newman 1995; Weathington et al. 2012) and chemosensory cues (Been and Petrulis 2011; Kikusui et al. 2017). Furthermore, there is overlap in the MeApd and MeApv response to conspecific and predator odors (Bergan et al. 2014). Recent studies have begun to reexamine MeA organization based on neuron subpopulations (Yamaguchi et al. 2020; Lischinsky et al. 2023) and whether they project the ventromedial hypothalamus and medial preoptic hypothalamus, regions that participate in reproductive, aggressive, and defensive behaviors. MeApd neuron subpopulations (e.g., aromatase-containing, Lhx6-positive) have defined roles in these behaviors (Choi et al. 2005; Hong et al. 2014; Unger et al. 2015). Further, separate populations of MeApv neurons are engaged by cues that elicit defensive or sexual behaviors (Choi et al. 2005; Ishii et al. 2017). This suggests that MeApv is a point at which a specific social stimulus is directed toward driving either reproductive or defensive behaviors.

The MeA also has a role in suppression of social behaviors, not just in driving them. MeApv is sensitive to anxiogenic manipulations that reduce social behaviors (Kellogg et al. 1998; van-Hover and Li 2015), MeA stimulation suppresses normal sexual and maternal behavior (Stark et al. 1998; Morgan et al. 1999), and MeA lesions increase social interaction in a novel environment (Wang et al. 2014). A population of MeA neurons project to the BNST (Ishii et al. 2017). Lesions and inactivation of BNST regions reduce anxiety and avoidance responses (Walker and Davis 1997; Fendt et al. 2003). Social conditions that elicit social caution, such as intruders, unfamiliar conspecifics, social defeat, and alarm pheromone, increase cFos in BNST and MeA (Wang et al. 1997; Martinez et al. 1998; Kim et al. 2015), including within the MeApv specifically (Weathington et al. 2012; Kim et al. 2015) and within MeApv neurons that project to BNST (Been and Petrulis 2011). BNST activation by unfamiliar conspecifics requires MeA (Wang et al. 2013) and MeA or BNST inactivation reduces social anxiety caused by social defeat (Markham and Huhman 2008; Markham et al. 2009). These findings suggest that MeApv → hypothalamus drives affiliative, agonistic, and active defensive behaviors and

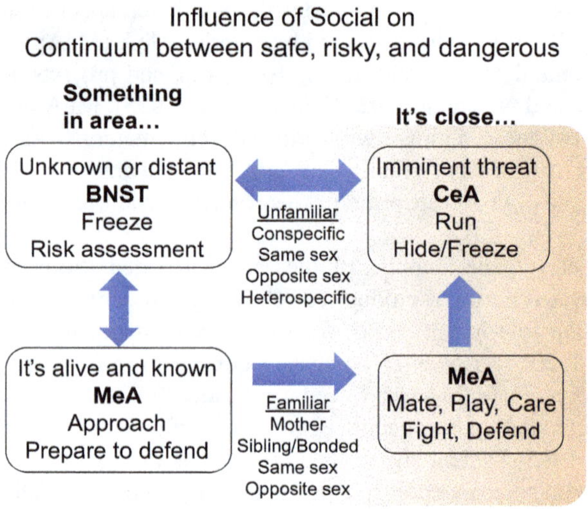

Fig. 2 A general framework for the interaction between MeA, CeA, and BNST. CeA and BNST participate in evaluation of threats. There is a postulated balance between CeA and BNST that contributes an appropriate response that depends on the proximity of the threat (Moscarello and Penzo 2022). The MeA is proposed to feed into this balance via a social buffering signal that decreases the level of danger based on the presence of social protection (e.g., caregiver). However, MeA also contributes by identifying whether a conspecific is a likely source of threat

MeApv → BNST may suppress these behaviors. However, there are MeAp projections to BNST that have a known role in initiation of some social behaviors, such as aggression (Nordman et al. 2020). Furthermore, MeA and BNST undergo coordinated changes during other social behaviors, including transition from pup attack to maternal care (Komisaruk et al. 2000). Indeed, newer perspectives propose a more nuanced role for inputs to BNST in behavioral response selection (Flanigan and Kash 2022).

There is substantial postnatal growth of the MeA (Chareyron et al. 2012b). MeAp dendritic complexity and synaptic input shows peripubertal maturation (Zehr et al. 2006; Cooke 2011; Adams and Rosenkranz 2016), along with increased androgen receptor positive neurons (Kashon et al. 1995), and is responsive to pubertal hormones (Gomez and Newman 1991; Rasia-Filho et al. 2004). There are substantial sex differences in MeA, and full mature sex differences require peri-adolescent sex hormones (Cooke 2006; Sano et al. 2016). However, it is not only neurons in the MeA. Microglia in MeA contributes to regulation of proliferation of multiple cell types in the MeA (VanRyzin et al. 2019) and subsequent development of sex differences in play behavior (Krebs-Kraft et al. 2010; Argue et al. 2017).

Overall, these studies point to a prominent role of MeA, and both its connections to extended amygdala and hypothalamus, in social behavior in young and adult rodents. However, its morphology and function undergoes early life and peripubertal maturation. These maturational processes contribute to sex-specific selection of social, reproductive, and defensive behaviors (Fig. 2).

3.2 Basolateral Amygdala

The basolateral amygdala complex includes the lateral, basal, and basomedial nuclei (Fig. 1). It is comprised of predominantly glutamatergic projection neurons, many with pyramidal-like features, and populations of inhibitory interneurons that potently regulate BLA neurons (Veres et al. 2014; McDonald 2020). These inhibitory interneurons are activated by excitatory inputs or by recurrent collaterals. In the mature BLA, afferents from thalamus and sensory association cortex provide a strong excitatory drive, and afferents from PFC can exert excitatory and feedforward inhibitory effects by activation of BLA inhibitory interneurons. In addition, lateral intercalated inhibitory neurons and some medial intercalated inhibitory neurons can provide inhibition into BLA (Mańko et al. 2011; Asede et al. 2015).

The BLA has a well-studied role in fear conditioning (Walker et al. 2009; Paré et al. 2004). However, that is only a portion of its role. Early studies identified that the BLA responds to a range of stimuli, including both appetitive and aversive stimuli, particularly those with biological importance (for review, see Urban and Rosenkranz 2020). This has expanded to a role of BLA in identifying novelty and appropriate habituation (Wilson and Rolls 1993; Burns et al. 1996; Herry et al. 2007), valence and salience encoding (Beyeler et al. 2016; Sengupta et al. 2018), and subsequent conditioned and unconditioned approach and avoidance (e.g., Ambroggi et al. 2008; Stuber et al. 2011; Tye et al. 2011), and in conditions where there is a balance of approach and avoidance, such as social encoding and motivation (Fustiñana et al. 2021; Ferrara et al. 2023).

The lateral and basal nuclei of the BLA have complementary roles in fear acquisition and extinction, and likely contribute differently to social behaviors. However, there is very little known about differences in their developmental trajectory, and are therefore considered together here. There is substantial postnatal growth of the BLA (Berdel et al. 1997). There are morphological changes of BLA neurons, increased dendritic complexity, and spine density between PND 20 and 35 (Koss et al. 2014; Ryan et al. 2016), although this may be sex-dependent (Guily et al. 2022), then increased branching but reduced neuron number between PND 35 and 90 (Rubinow and Juraska 2009; Koss et al. 2014). There is also maturation of membrane excitability early and up to adulthood (Ehrlich et al. 2012; Guily et al. 2022). By early adolescence, GABAergic circuits are intact within BLA (Dávila et al. 2008; Ehrlich et al. 2013). However, activation of these GABAergic circuits by extrinsic inputs to BLA still develops during adolescence (Arruda-Carvalho et al. 2017; Selleck et al. 2018).

Thalamic inputs emerge by PND 7–9 (Bouwmeester et al. 2002) and cortical inputs by PND 10 (Ryazantseva et al. 2020). There is strengthening of cortical inputs during the juvenile period (Arruda-Carvalho et al. 2017; Lugarà et al. 2019), and these cortical inputs have unique strong excitatory features in the juvenile–adolescent stage (Gambino et al. 2010). PFC inputs to BLA emerge in a similar timeframe as other cortical inputs expand (PND 7–13; Bouwmeester et al. 2002). However, PFC inputs are then pruned through adolescence (Cressman et al. 2010). The net

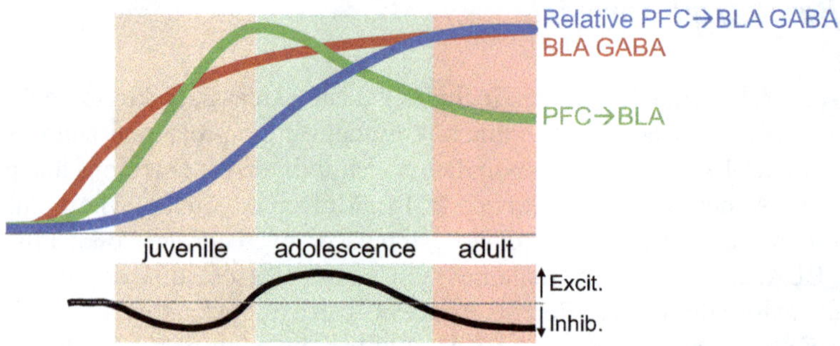

Fig. 3 Maturation of the excitatory–inhibitory balance in BLA. There is substantial early postnatal and juvenile growth of excitatory inputs to BLA alongside maturation of GABAergic inhibitory interneurons. The result is somewhat balanced BLA excitation and inhibition during the juvenile period. Additional cortical inputs continue to innervate the BLA, particularly from PFC. There is a period of peak excitatory input to BLA in late juvenile–early adolescence, comprised of thalamic and cortical inputs that have an excitatory influence and that drive BLA activity. However, these inputs do not yet exert the mature efficiency in activating inhibitory interneurons in the BLA, so there is more relative excitation compared to inhibition. During adolescence, PFC inputs to BLA are pruned, and there is a hypothesized honing of PFC excitation onto BLA inhibitory interneurons. This helps stabilize the balance of excitation–inhibition into adulthood, with the ability of PFC to regulate BLA function

result of the staggered development of these excitatory inputs, along with changes in synaptic inputs to BLA inhibitory interneurons, is substantial cortical and thalamic excitatory drive in adolescent BLA, but PFC inputs do not yet exert their mature actions that balance excitation and inhibition (Arruda-Carvalho et al. 2017; Selleck et al. 2018) (Fig. 3). This contributes to a BLA that may robustly respond to stimuli with less regulation and therefore less specificity during the juvenile and adolescent periods.

3.3 *CeA and BNST Development*

The rodent CeA is usually divided into three primary regions (Fig. 1), the medial (CeM) and capsular (CeC) with extensive projections to brainstem and hypothalamic regions, and lateral (CeL) with an important role in regulation of CeM. CeA has a critical role in fear, anxiety, and defensive responses (Moscarello and Penzo 2022). The BNST can be divided into more than ten regions, with functional divisions often based on anterior and posterior location, whose roles have been most characterized in stress responses and reproductive behaviors, respectively. BNST has a prominent role in assessment of risk, anticipation of threat, and unpredicted or ambiguous threat (Asok et al. 2018; Ressler et al. 2020).

There is relatively less known about the development of CeA and BNST. Similar to many other regions, there is developmental neuron loss (Chung et al. 2000; Gotsiridze et al. 2007), but increased size due to proliferation of oligodendrocytes

(Chareyron et al. 2012b). Principal GABAergic neurons in CeA develop early (Avila et al. 2011) with some continued proliferation of neurons in the first 2 weeks postnatal (Cox et al. 2012). Spine density stabilizes by around PND 14, as does BLA input to CeA (Ryazantseva et al. 2020). There is still some maturation of CeA neuronal excitability between adolescence and adult (Rouzer and Diaz 2021). Similar to MeA, there are sex differences in BNST that emerge due to organizational effects of hormones and later peripubertal hormones (Guillamón et al. 1988; Tsukahara et al. 2011).

There are also changes in peptide systems in the CeA and BNST that impact social and emotion function. There is a developmental shift in oxytocin (OT) and vasopressin (VP) receptors, with decreased V1aR in adults (Snijdewint et al. 1989; Smith et al. 2017) and increased OTR (Snijdewint et al. 1989; Prounis et al. 2018). A similar pattern is seen in BNST (Shapiro and Insel 1989; Smith et al. 2017; Prounis et al. 2018). These receptors in CeA and BNST play a role in balancing social affiliation and territorial aggression (Lubin et al. 2003; Veenema et al. 2010; Bosch and Neumann 2012; Paul et al. 2014), as well as balancing fear and anxiety (Roozendaal et al. 1992; Viviani et al. 2011; Duque-Wilckens et al. 2016; Martinon et al. 2019). What emerges from this literature is similar to MeA. CeA and BNST mature relatively early, but become more specialized at puberty.

4 Framework for Coordinated Development of Amygdala Function

CeA, MeA, and BNST modulate behaviors in response to external challenges. CeA is most well-known for rapid fear responses in the presence of an immediate threat, but it also

processes and promotes appetitive functions (Fadok et al. 2018; Warlow and Berridge 2021). BNST is associated with preparatory responses to an expected threat, ongoing stress, and some aspects of social function (Goode and Maren 2017; Flanigan and Kash 2022; Chudoba and Dabrowska 2023). MeA is involved in identification of conspecifics and heterospecifics, and guiding an appropriate social, aggressive, or defensive response (Masini et al. 2009; Nordman et al. 2020; Yamaguchi et al. 2020). While this generality is helpful as a framework for understanding their functions, in practice these regions are involved in a coordinated response to many appetitive and aversive conditions. The full roles of these regions and their coordination emerges when a balance between possible behavioral responses is required (Jennings et al. 2013; Hong et al. 2014; Fadok et al. 2017).

CeA, BNST, and MeA regions are highly interconnected (Sun et al. 1991; Canteras et al. 1995; Jolkkonen and Pitkänen 1998; Dong et al. 2001; Bienkowski et al. 2013). Both the GABAergic phenotype of primary connections and functional studies suggest that these regions can suppress each other. This is demonstrated most clearly for the antagonist balance between rapid fear responses mediated by CeA and prolonged anxiety mediated by BNST (Walker et al. 2009; Asok et al. 2018; Yamauchi et al. 2018; Ressler et al. 2020). Despite sharing many similar downstream targets, CeA- and BNST-mediated outputs are associated with temporally distinct and unique autonomic and behavioral responses. What emerges is a competition between these regions for the optimal physiological and behavioral response to different conditions. When MeA is layered onto this, an additional element of competition between social and defensive drives emerges (Fig. 2).

Although not studied comprehensively across all three regions, connections between MeA, CeA, and BNST appear mature by PND 10–15, and their outputs to other regions appear to be mature by PND 15–20 (Card et al. 2005; Cooke and Simerly 2005; Lanzillo et al. 2021). In younger animals, the balance between CeA, BNST, and MeA biases behaviors toward nonspecific affiliation between conspecifics and less caution of conspecifics. In addition, high drive for exploration and novelty is sufficient to override the contribution of these systems to avoidance of unknown contexts. However, as described above, there is a peri-pubertal rebalancing of these regions. This leads to discrimination between potential mates and competition or territorial threats, establishment of social hierarchies, and subsequent changes in response to social opportunities, rewards, and threats (Bell et al. 2013; Bergan et al. 2014; Ahern et al. 2016; Cooper et al. 2023).

BLA sends projections to CeA, BNST, and MeA regions (Pitkänen et al. 1995; Dong et al. 2001; Bienkowski and Rinaman 2013) and these paths are mostly separate (Bienkowski and Rinaman 2013; Beyeler et al. 2018; but see Reichard et al. 2017). This positions the BLA to guide appropriate behavioral responses via different effects on these regions. BLA can guide this based on prior experience and valence of stimuli (Lang and Davis 2006; O'Neill et al. 2018). During early amygdala development, maternal presence can suppress BLA activation (Perry et al. 2016; Opendak et al. 2019) and completely suppress fear learning (Moriceau and Sullivan 2006; Upton and Sullivan 2010). As pups gain independence, maternal presence buffers, but does not block, fear learning (PND 18; Upton and Sullivan 2010; Opendak et al. 2019), and this is associated with shifted impact of maternal presence on BLA and MeA, as well as networks of regions connected with amygdala (Perry et al. 2016; Opendak et al. 2019). At this stage, immaturity of contextual fear and context-dependency of extinction still points to less developed BLA guidance of these regions in younger animals. However, earlier developing fear conditioning provides protective effects even to juvenile animals. As they enter the adolescent period, BLA responsiveness peaks due to the high degree of excitatory input with relatively less developed recruitment of BLA inhibitory circuits. This is expected to make the BLA less selective, and perhaps increases BLA impact on these other regions. This begins to refine as PFC inputs to BLA are pruned and adult maturity is reached, with PFC guidance and regulation of BLA, and subsequent biasing of activity between CeA, BNST, and MeA (Fig. 4).

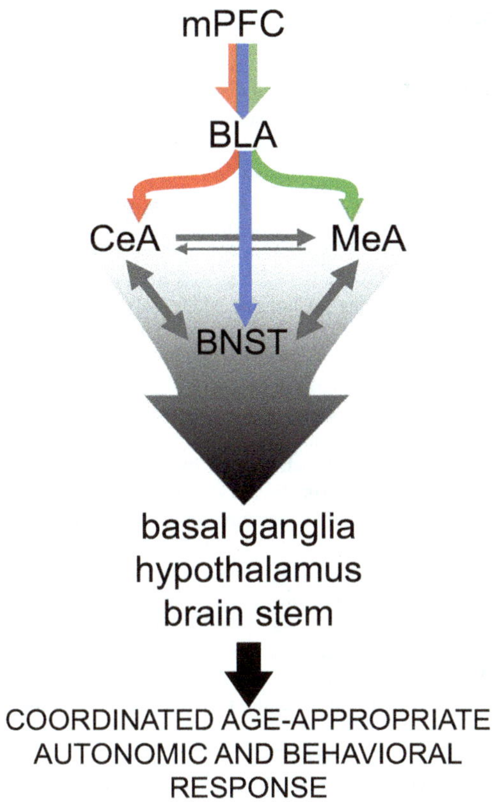

Fig. 4 Overall scheme for the mature amygdala. CeA and BNST have a balance that integrates a range of internal and external information to bias behavioral engagement. MeA can influence this balance based on social factors. BLA inputs to these regions can help tip the CeA–BNST balance and select appropriate behavioral outcome based on experience and valence. PFC inputs guide BLA, and can bias BLA inputs to CeA–BNST, particularly when there are competing choices that require a decision

5 Conclusion

Experiences during development complement ongoing maturational processes, but can also alter development. Periods of maturation allow revision of the developmental trajectory based on the environment. The amygdala is highly sensitive to the environment both early postnatally and during adolescence. This provides opportunity for optimization of social and emotional function by tuning the amygdala developmental trajectory. A huge amount has been learned in the last decades about the impact of the early and adolescent environment, particularly stressors, on amygdala and socio-emotional behaviors. A major challenge for the future is building an integrated concept of how these challenges during development exert a coordinated action across the amygdala to optimize behavioral adaptation to the environment, and how the development of the amygdala aligns with other neural regions to contribute to the developmental trajectory of social and emotional behaviors.

Acknowledgments The funding for the study is provided by the NIH (R01MH118237). I thank Dr. Nicole Ferrara for discussions and suggestions on this chapter.

References

Adams T, Rosenkranz JA (2016) Social isolation during postweaning development causes hypoactivity of neurons in the medial nucleus of the male rat amygdala. Neuropsychopharmacology 41:1929–1940

Adolphs R (2009) The social brain: neural basis of social knowledge. Annu Rev Psychol 60:693–716

Ahern M, Goodell DJ, Adams J, Bland ST (2016) Brain regional differences in social encounter-induced Fos expression in male and female rats after post-weaning social isolation. Brain Res 1630:120–133

Akers KG, Arruda-Carvalho M, Josselyn SA, Frankland PW (2012) Ontogeny of contextual fear memory formation, specificity, and persistence in mice. Learn Mem 19:598–604

Alexander R (1974) The evolution of social behavior. Annu Rev Ecol Syst 5:325–383

Allen Institute for Brain Science (2017) Allen mouse brain atlas [Common Coordinate Framework version 3]. Available from mouse.brain–map.org

Ambroggi F, Ishikawa A, Fields HL, Nicola SM (2008) Basolateral amygdala neurons facilitate reward-seeking behavior by exciting nucleus accumbens neurons. Neuron 59:648–661

Argue KJ, VanRyzin JW, Falvo DJ, Whitaker AR, Yu SJ, McCarthy MM (2017) Activation of both CB1 and CB2 endocannabinoid receptors is critical for masculinization of the developing medial Amygdala and Juvenile social play behavior. eNeuro 4. ENEURO.0344

Arruda-Carvalho M, Wu WC, Cummings KA, Clem RL (2017) Optogenetic examination of prefrontal-amygdala synaptic development. J Neurosci 37:2976–2985

Asede D, Bosch D, Lüthi A, Ferraguti F, Ehrlich I (2015) Sensory inputs to intercalated cells provide fear-learning modulated inhibition to the basolateral amygdala. Neuron 86:541–554

Asok A, Draper A, Hoffman AF, Schulkin J, Lupica CR, Rosen JB (2018) Optogenetic silencing of a corticotropin-releasing factor pathway from the central amygdala to the bed nucleus of the stria terminalis disrupts sustained fear. Mol Psychiatry 23:914–922

Avila MA, Real MA, Guirado S (2011) Patterns of GABA and GABA Transporter-1 immunoreactivities in the developing and adult mouse brain amygdala. Brain Res 1388:1–11

Baker KD, Richardson R (2015) Forming competing fear learning and extinction memories in adolescence makes fear difficult to inhibit. Learn Mem 22:537–543

Bakker R, Tiesinga P, Kötter R (2015) The scalable brain atlas: instant web-based access to public brain atlases and related content. Neuroinformatics 13:353–366

Barnet RC, Hunt PS (2005) Trace and long-delay fear conditioning in the developing rat. Learn Behav 33:437–443

Been LE, Petrulis A (2011) Chemosensory and hormone information are relayed directly between the medial amygdala, posterior bed nucleus of the stria terminalis, and medial preoptic area in male Syrian hamsters. Horm Behav 59:536–548

Bell MR, De Lorme KC, Figueira RJ, Kashy DA, Sisk CL (2013) Adolescent gain in positive valence of a socially relevant stimulus: engagement of the mesocorticolimbic reward circuitry. Eur J Neurosci 37:457–468

Berdel B, Moryś J, Maciejewska B (1997) Neuronal changes in the basolateral complex during development of the amygdala of the rat. Int J Dev Neurosci 15:755–765

Berdoy M, Smith P, Macdonald DW (1995) Stability of social status in wild rats: age and the role of settled dominance. Behaviour 132:193–212

Bergan JF, Ben-Shaul Y, Dulac C (2014) Sex-specific processing of social cues in the medial amygdala. eLife 3:e02743

Beyeler A, Namburi P, Glober GF, Simonnet C, Calhoon GG, Conyers GF, Luck R, Wildes CP, Tye KM (2016) Divergent routing of positive and negative information from the amygdala during memory retrieval. Neuron 90:348–361

Beyeler A, Chang CJ, Silvestre M, Lévêque C, Namburi P, Wildes CP, Tye KM (2018) Organization of valence-encoding and projection-defined neurons in the basolateral amygdala. Cell Rep 22:905–918

Bienkowski MS, Rinaman L (2013) Common and distinct neural inputs to the medial central nucleus of the amygdala and anterior ventrolateral bed nucleus of stria terminalis in rats. Brain Struct Funct 218:187–208

Bienkowski MS, Wendel ES, Rinaman L (2013) Organization of multisynaptic circuits within and between the medial and the central extended amygdala. J Comp Neurol 521:3406–3431

Bos K, Zeanah CH, Fox NA, Drury SS, McLaughlin KA, Nelson CA (2011) Psychiatric outcomes in young children with a history of institutionalization. Harv Rev Psychiatry 19:15–24

Bosch OJ, Neumann ID (2012) Both oxytocin and vasopressin are mediators of maternal care and aggression in rodents: from central release to sites of action. Horm Behav 61:293–303

Boulanger Bertolus J, Mouly AM, Sullivan RM (2016) Ecologically relevant neurobehavioral assessment of the development of threat learning. Learn Mem 23:556–566

Bouwmeester H, Smits K, Van Ree JM (2002) Neonatal development of projections to the basolateral amygdala from prefrontal and thalamic structures in rat. J Comp Neurol 450:241–255

Broad KD, Curley JP, Keverne EB (2006) Mother-infant bonding and the evolution of mammalian social relationships. Philos Trans R Soc Lond B Biol Sci 361:2199–2214

Bronstein PM, Hirsch SM (1976) Ontogeny of defensive reactions in Norway rats. J Comp Physiol Psychol 90:620–629

Burns LH, Annett L, Kelley AE, Everitt BJ, Robbins TW (1996) Effects of lesions to amygdala, ventral subiculum, medial prefrontal cortex, and nucleus accumbens on the reaction to novelty: implication for limbic-striatal interactions. Behav Neurosci 110:60–73

Calcagnetti DJ, Schechter MD (1992) Place conditioning reveals the rewarding aspect of social interaction in juvenile rats. Physiol Behav 51:667–672

Caldji C, Tannenbaum B, Sharma S, Francis D, Plotsky PM, Meaney MJ (1998) Maternal care during infancy regulates the development of neural systems mediating the expression of fearfulness in the rat. Proc Natl Acad Sci U S A 95:5335–5340

Canteras NS, Simerly RB, Swanson LW (1995) Organization of projections from the medial nucleus of the amygdala: a PHAL study in the rat. J Comp Neurol 360:213–245

Card JP, Levitt P, Gluhovsky M, Rinaman L (2005) Early experience modifies the postnatal assembly of autonomic emotional motor circuits in rats. J Neurosci 25:9102–9111

Chareyron LJ, Lavenex PB, Amaral DG, Lavenex P (2012a) Postnatal development of the amygdala: a stereological study in macaque monkeys. J Comp Neurol 520:1965–1984

Chareyron LJ, Lavenex PB, Lavenex P (2012b) Postnatal development of the amygdala: a stereological study in rats. J Comp Neurol 520:3745–3763

Chase ID (1982) Dynamics of hierarchy formation: the sequential development of dominance relationships. Behaviour 80:218–240

Chase ID, Tovey C, Spangler-Martin D, Manfredonia M (2002) Individual differences versus social dynamics in the formation of animal dominance hierarchies. Proc Natl Acad Sci U S A 99:5744–5749

Chen PB, Hu RK, Wu YE, Pan L, Huang S, Micevych PE, Hong W (2019) Sexually dimorphic control of parenting behavior by the medial amygdala. Cell 176:1206–1221.e18

Choi GB, Dong HW, Murphy AJ, Valenzuela DM, Yancopoulos GD, Swanson LW, Anderson DJ (2005) Lhx6 delineates a pathway mediating innate reproductive behaviors from the amygdala to the hypothalamus. Neuron 46:647–660

Chudoba R, Dabrowska J (2023) Distinct populations of corticotropin-releasing factor (CRF) neurons mediate divergent yet complementary defensive behaviors in response to a threat. Neuropharmacology 228:109461

Chung WC, Swaab DF, De Vries GJ (2000) Apoptosis during sexual differentiation of the bed nucleus of the stria terminalis in the rat brain. J Neurobiol 43:234–243

Collier AC, Bolles RC (1980) The ontogenesis of defensive reactions to shock in preweanling rats. Dev Psychobiol 13:141–150

Cooke BM (2006) Steroid-dependent plasticity in the medial amygdala. Neuroscience 138:997–1005

Cooke BM (2011) Synaptic reorganisation of the medial amygdala during puberty. J Neuroendocrinol 23:65–73

Cooke BM, Simerly RB (2005) Ontogeny of bidirectional connections between the medial nucleus of the amygdala and the principal bed nucleus of the stria terminalis in the rat. J Comp Neurol 489:42–58

Cooper MA, Grizzell JA, Whitten CJ, Burghardt GM (2023) Comparing the ontogeny, neurobiology, and function of social play in hamsters and rats. Neurosci Biobehav Rev 147:105102

Cox ET, Hodge CW, Sheikh MJ, Abramowitz AC, Jones GF, Jamieson-Drake AW, Makam PR, Zeskind PS, Johns JM (2012) Delayed developmental changes in neonatal vocalizations correlates with variations in ventral medial hypothalamus and central amygdala development in the rodent infant: effects of prenatal cocaine. Behav Brain Res 235:166–175

Cressman VL, Balaban J, Steinfeld S, Shemyakin A, Graham P, Parisot N, Moore H (2010) Prefrontal cortical inputs to the basal amygdala undergo pruning during late adolescence in the rat. J Comp Neurol 518:2693–2709

Curtiss S, Fromkin V, Krashen S, Rigler D, Rigler M (1974) The linguistic development of genie. Language 50:528–554

Dávila JC, Olmos L, Legaz I, Medina L, Guirado S, Real MA (2008) Dynamic patterns of colocalization of calbindin, parvalbumin and GABA in subpopulations of mouse basolateral amygdalar cells during development. J Chem Neuroanat 35:67–76

Den ML, Richardson R (2013) Enhanced sensitivity to learning fearful associations during adolescence. Neurobiol Learn Mem 104:92–102

Dielenberg RA, Hunt GE, McGregor IS (2001) "When a rat smells a cat": the distribution of Fos immunoreactivity in rat brain following exposure to a predatory odor. Neuroscience 104:1085–1097

Dong HW, Petrovich GD, Swanson LW (2001) Topography of projections from amygdala to bed nuclei of the stria terminalis. Brain Res Brain Res Rev 38:192–246

Douglas LA, Varlinskaya EI, Spear LP (2004) Rewarding properties of social interactions in adolescent and adult male and female rats: impact of social versus isolate housing of subjects and partners. Dev Psychobiol 45:153–162

Duque-Wilckens N, Steinman MQ, Laredo SA, Hao R, Perkeybile AM, Bales KL, Trainor BC (2016) Inhibition of vasopressin V1a receptors in the medioventral bed nucleus of the stria terminalis has sex- and context-specific anxiogenic effects. Neuropharmacology 110:59–68

Ehrlich DE, Ryan SJ, Rainnie DG (2012) Postnatal development of electrophysiological properties of principal neurons in the rat basolateral amygdala. J Physiol 590:4819–4838

Ehrlich DE, Ryan SJ, Hazra R, Guo JD, Rainnie DG (2013) Postnatal maturation of GABAergic transmission in the rat basolateral amygdala. J Neurophysiol 110:926–941

Einon DF, Morgan MJ, Kibbler CC (1978) Brief periods of socialization and later behavior in the rat. Dev Psychobiol 11:213–225

Ellis BJ, Del Giudice M, Dishion TJ, Figueredo AJ, Gray P, Griskevicius V, Hawley PH, Jacobs WJ, James J, Volk AA, Wilson DS (2012) The evolutionary basis of risky adolescent behavior: implications for science, policy, and practice. Dev Psychol 48:598–623

Fadok JP, Krabbe S, Markovic M, Courtin J, Xu C, Massi L, Botta P, Bylund K, Müller C, Kovacevic A, Tovote P, Lüthi A (2017) A competitive inhibitory circuit for selection of active and passive fear responses. Nature 542:96–100

Fadok JP, Markovic M, Tovote P, Lüthi A (2018) New perspectives on central amygdala function. Curr Opin Neurobiol 49:141–147

Fanselow MS, LeDoux JE (1999) Why we think plasticity underlying Pavlovian fear conditioning occurs in the basolateral amygdala. Neuron 23:229–232

Fawcett TW, Frankenhuis WE (2015) Adaptive explanations for sensitive windows in development. Front Zool 12(Suppl 1):S3

Fendt M, Endres T, Apfelbach R (2003) Temporary inactivation of the bed nucleus of the stria terminalis but not of the amygdala blocks freezing induced by trimethylthiazoline, a component of fox feces. J Neurosci 23:23–28

Ferrara NC, Opendak M (2023) Amygdala circuit transitions supporting developmentally-appropriate social behavior. Neurobiol Learn Mem 201:107762

Ferrara NC, Trask S, Padival M, Rosenkranz JA (2023) Maturation of a cortical-amygdala circuit limits sociability in male rats. Cereb Cortex

Flanigan ME, Kash TL (2022) Coordination of social behaviors by the bed nucleus of the stria terminalis. Eur J Neurosci 55:2404–2420

Fleming AS, Korsmit M (1996) Plasticity in the maternal circuit: effects of maternal experience on Fos-Lir in hypothalamic, limbic, and cortical structures in the postpartum rat. Behav Neurosci 110:567–582

Fustiñana MS, Eichlisberger T, Bouwmeester T, Bitterman Y, Lüthi A (2021) State-dependent encoding of exploratory behaviour in the amygdala. Nature 592:267–271

Gabard-Durnam LJ, Flannery J, Goff B, Gee DG, Humphreys KL, Telzer E, Hare T, Tottenham N (2014) The development of human amygdala functional connectivity at rest from 4 to 23 years: a cross-sectional study. Neuroimage 95:193–207

Galef BG, Clark MM (1972) Mother's milk and adult presence: two factors determining initial dietary selection by weanling rats. J Comp Physiol Psychol 78:220–225

Gambino F, Khelfaoui M, Poulain B, Bienvenu T, Chelly J, Humeau Y (2010) Synaptic maturation at cortical projections to the lateral amygdala in a mouse model of Rett syndrome. PLoS One 5: e11399

Gerhard DM, Meyer HC (2021) Extinction trial spacing across days differentially impacts fear regulation in adult and adolescent male mice. Neurobiol Learn Mem 186:107543

Gillies D, Maiocchi L, Bhandari AP, Taylor F, Gray C, O'Brien L (2016) Psychological therapies for children and adolescents exposed to trauma. Cochrane Database Syst Rev 10:CD012371

Gomez DM, Newman SW (1991) Medial nucleus of the amygdala in the adult Syrian hamster: a quantitative Golgi analysis of gonadal hormonal regulation of neuronal morphology. Anat Rec 231:498–509

Gomez DM, Newman SW (1992) Differential projections of the anterior and posterior regions of the medial amygdaloid nucleus in the Syrian hamster. J Comp Neurol 317:195–218

Goode TD, Maren S (2017) Role of the bed nucleus of the stria terminalis in aversive learning and memory. Learn Mem 24:480–491

Gotsiridze T, Kang N, Jacob D, Forger NG (2007) Development of sex differences in the principal nucleus of the bed nucleus of the stria terminalis of mice: role of Bax-dependent cell death. Dev Neurobiol 67:355–362

Guillamón A, Segovia S, del Abril A (1988) Early effects of gonadal steroids on the neuron number in the medial posterior region and the lateral division of the bed nucleus of the stria terminalis in the rat. Brain Res Dev Brain Res 44:281–290

Guily P, Lassalle O, Chavis P, Manzoni OJ (2022) Sex-specific divergent maturational trajectories in the postnatal rat basolateral amygdala. iScience 25:103815

Harlow HF, Zimmermann RR (1959) Affectional responses in the infant monkey; orphaned baby monkeys develop a strong and persistent attachment to inanimate surrogate mothers. Science 130:421–432

Heinz DE, Schöttle VA, Nemcova P, Binder FP, Ebert T, Domschke K, Wotjak CT (2021) Exploratory drive, fear, and anxiety are dissociable and independent components in foraging mice. Transl Psychiatry 11:318

Herry C, Bach DR, Esposito F, Di Salle F, Perrig WJ, Scheffler K, Lüthi A, Seifritz E (2007) Processing of temporal unpredictability in human and animal amygdala. J Neurosci 27:5958–5966

Hess EH (1959) Imprinting, an effect of early experience, imprinting determines later social behavior in animals. Science 130:133–141

Hong W, Kim DW, Anderson DJ (2014) Antagonistic control of social versus repetitive self-grooming behaviors by separable amygdala neuronal subsets. Cell 158:1348–1361

Ishii KK, Osakada T, Mori H, Miyasaka N, Yoshihara Y, Miyamichi K, Touhara K (2017) A labeled-line neural circuit for pheromone-mediated sexual behaviors in mice. Neuron 95:123–137.e8

Jennings JH, Sparta DR, Stamatakis AM, Ung RL, Pleil KE, Kash TL, Stuber GD (2013) Distinct extended amygdala circuits for divergent motivational states. Nature 496:224–228

Johnson CS, Hong W, Micevych PE (2021) Posterodorsal medial amygdala regulation of female social behavior: GABA versus glutamate projections. J Neurosci 41:8790–8800

Jolkkonen E, Pitkänen A (1998) Intrinsic connections of the rat amygdaloid complex: projections originating in the central nucleus. J Comp Neurol 395:53–72

Kashon ML, Hayes MJ, Shek PP, Sisk CL (1995) Regulation of brain androgen receptor immunoreactivity by androgen in prepubertal male ferrets. Biol Reprod 52:1198–1205

Kellogg CK, Awatramani GB, Piekut DT (1998) Adolescent development alters stressor-induced Fos immunoreactivity in rat brain. Neuroscience 83:681–689

Keshavarzi S, Sullivan RK, Ianno DJ, Sah P (2014) Functional properties and projections of neurons in the medial amygdala. J Neurosci 34:8699–8715

Kikusui T, Kajita M, Otsuka N, Hattori T, Kumazawa K, Watarai A, Nagasawa M, Inutsuka A, Yamanaka A, Matsuo N, Covington HE, Mogi K (2017) Sex differences in olfactory-induced neural activation of the amygdala. Behav Brain Res

Kim JH, Richardson R (2007) A developmental dissociation in reinstatement of an extinguished fear response in rats. Neurobiol Learn Mem 88:48–57

Kim JH, Richardson R (2008) The effect of temporary amygdala inactivation on extinction and reextinction of fear in the developing rat: unlearning as a potential mechanism for extinction early in development. J Neurosci 28:1282–1290

Kim JH, Hamlin AS, Richardson R (2009) Fear extinction across development: the involvement of the medial prefrontal cortex as assessed by temporary inactivation and immunohistochemistry. J Neurosci 29:10802–10808

Kim Y, Venkataraju KU, Pradhan K, Mende C, Taranda J, Turaga SC, Arganda-Carreras I, Ng L, Hawrylycz MJ, Rockland KS, Seung HS, Osten P (2015) Mapping social behavior-induced brain activation at cellular resolution in the mouse. Cell Rep 10:292–305

Kollack-Walker S, Newman SW (1995) Mating and agonistic behavior produce different patterns of Fos immunolabeling in the male Syrian hamster brain. Neuroscience 66:721–736

Komisaruk BR, Rosenblatt JS, Barona ML, Chinapen S, Nissanov J, O'Bannon RT, Johnson BM, Del Cerro MC (2000) Combined c-fos and 14C-2-deoxyglucose method to differentiate site-specific excitation from disinhibition: analysis of maternal behavior in the rat. Brain Res 859:262–272

König B, Markl H (1987) Maternal care in house mice: I. The weaning strategy as a means for parental manipulation of offspring quality. Behav Ecol Sociobiol

Koss WA, Belden CE, Hristov AD, Juraska JM (2014) Dendritic remodeling in the adolescent medial prefrontal cortex and the basolateral amygdala of male and female rats. Synapse 68:61–72

Krebs-Kraft DL, Hill MN, Hillard CJ, McCarthy MM (2010) Sex difference in cell proliferation in developing rat amygdala mediated by endocannabinoids has implications for social behavior. Proc Natl Acad Sci U S A 107:20535–20540

Kudryavtseva NN, Bakshtanovskaya IV, Koryakina LA (1991) Social model of depression in mice of C57BL/6J strain. Pharmacol Biochem Behav 38:315–320

Lang PJ, Davis M (2006) Emotion, motivation, and the brain: reflex foundations in animal and human research. Prog Brain Res 156:3–29

Lanzillo M, Gervais M, Croizier S (2021) Ontogeny of the projections from the dorsomedial division of the anterior bed nucleus of the Stria terminalis to hypothalamic nuclei. Front Neurosci 15:748186

Lischinsky JE, Yin L, Shi C, Prakash N, Burke J, Shekaran G, Grba M, Corbin JG, Lin D (2023) Transcriptionally defined amygdala subpopulations play distinct roles in innate social behaviors. Nat Neurosci 26:2131–2146

Lloyd A, Viding E, McKay R, Furl N (2023) Understanding patch foraging strategies across development. Trends Cogn Sci 27:1085–1098

Loh MK, Ferrara NC, Torres JM, Rosenkranz JA (2022) Medial orbitofrontal cortex and nucleus accumbens mediation in risk assessment behaviors in adolescents and adults. Neuropsychopharmacology

Lubin DA, Elliott JC, Black MC, Johns JM (2003) An oxytocin antagonist infused into the central nucleus of the amygdala increases maternal aggressive behavior. Behav Neurosci 117:195–201

Lugarà E, De Fusco A, Lignani G, Benfenati F, Humeau Y (2019) Synapsin I controls synaptic maturation of long-range projections in the lateral amygdala in a targeted selective fashion. Front Cell Neurosci 13:220

Lynn DA, Brown GR (2009) The ontogeny of exploratory behavior in male and female adolescent rats (Rattus norvegicus). Dev Psychobiol 51:513–520

MacLeod KJ, English S, Ruuskanen SK, Taborsky B (2023) Stress in the social context: a behavioural and eco-evolutionary perspective. J Exp Biol 226

Macrì S, Adriani W, Chiarotti F, Laviola G (2002) Risk taking during exploration of a plus-maze is greater in adolescent than in juvenile or adult mice. Anim Behav 64:541–546

Mańko M, Geracitano R, Capogna M (2011) Functional connectivity of the main intercalated nucleus of the mouse amygdala. J Physiol 589:1911–1925

Maren S (2003) The amygdala, synaptic plasticity, and fear memory. Ann N Y Acad Sci 985:106–113

Markham CM, Huhman KL (2008) Is the medial amygdala part of the neural circuit modulating conditioned defeat in Syrian hamsters. Learn Mem 15:6–12

Markham CM, Norvelle A, Huhman KL (2009) Role of the bed nucleus of the stria terminalis in the acquisition and expression of conditioned defeat in Syrian hamsters. Behav Brain Res 198:69–73

Martinez M, Phillips PJ, Herbert J (1998) Adaptation in patterns of c-fos expression in the brain associated with exposure to either single or repeated social stress in male rats. Eur J Neurosci 10: 20–33

Martinon D, Lis P, Roman AN, Tornesi P, Applebey SV, Buechner G, Olivera V, Dabrowska J (2019) Oxytocin receptors in the dorsolateral bed nucleus of the stria terminalis (BNST) bias fear learning toward temporally predictable cued fear. Transl Psychiatry 9:140

Masini CV, Sasse SK, Garcia RJ, Nyhuis TJ, Day HE, Campeau S (2009) Disruption of neuroendocrine stress responses to acute ferret odor by medial, but not central amygdala lesions in rats. Brain Res 1288:79–87

McDonald AJ (2020) Functional neuroanatomy of the basolateral amygdala: neurons, neurotransmitters, and circuits. Handb Behav Neurosci 26:1–38

Meredith M, Westberry JM (2004) Distinctive responses in the medial amygdala to same-species and different-species pheromones. J Neurosci 24:5719–5725

Morgan HD, Watchus JA, Milgram NW, Fleming AS (1999) The long lasting effects of electrical simulation of the medial preoptic area and medial amygdala on maternal behavior in female rats. Behav Brain Res 99:61–73

Moriceau S, Sullivan RM (2006) Maternal presence serves as a switch between learning fear and attraction in infancy. Nat Neurosci 9:1004–1006

Moriceau S, Wilson DA, Levine S, Sullivan RM (2006) Dual circuitry for odor-shock conditioning during infancy: corticosterone switches between fear and attraction via amygdala. J Neurosci 26:6737–6748

Moscarello JM, Penzo MA (2022) The central nucleus of the amygdala and the construction of defensive modes across the threat-imminence continuum. Nat Neurosci 25:999–1008

Moye TB, Rudy JW (1985) Ontogenesis of learning: VI. Learned and unlearned responses to visual stimulation in the infant hooded rat. Dev Psychobiol 18:395–409

Moye TB, Rudy JW (1987) Ontogenesis of trace conditioning in young rats: dissociation of associative and memory processes. Dev Psychobiol 20:405–414

Nordman JC, Ma X, Gu Q, Potegal M, Li H, Kravitz AV, Li Z (2020) Potentiation of divergent medial amygdala pathways drives experience-dependent aggression escalation. J Neurosci 40: 4858–4880

O'Neill PK, Gore F, Salzman CD (2018) Basolateral amygdala circuitry in positive and negative valence. Curr Opin Neurobiol 49:175–183

Opendak M, Robinson-Drummer P, Blomkvist A, Zanca RM, Wood K, Jacobs L, Chan S, Tan S, Woo J, Venkataraman G, Kirschner E, Lundström JN, Wilson DA, Serrano PA, Sullivan RM (2019) Neurobiology of maternal regulation of infant fear: the role of mesolimbic dopamine and its disruption by maltreatment. Neuropsychopharmacology 44:1247–1257

Orsini CA, Moorman DE, Young JW, Setlow B, Floresco SB (2015) Neural mechanisms regulating different forms of risk-related decision-making: insights from animal models. Neurosci Biobehav Rev 58:147–167

Ošt'ádalová I, Babický A (2012) Periodization of the early postnatal development in the rat with particular attention to the weaning period. Physiol Res 61:S1–S7

Panksepp J (1981) The ontogeny of play in rats. Dev Psychobiol 14:327–332

Paré D, Quirk GJ, Ledoux JE (2004) New vistas on amygdala networks in conditioned fear. J Neurophysiol 92:1–9

Park CHJ, Ganella DE, Kim JH (2017) Juvenile female rats, but not male rats, show renewal, reinstatement, and spontaneous recovery following extinction of conditioned fear. Learn Mem 24:630–636

Park CHJ, Ganella DE, Kim JH (2020) Context fear learning and renewal of extinguished fear are dissociated in juvenile female rats. Dev Psychobiol 62:123–129

Pattwell SS, Bath KG, Casey BJ, Ninan I, Lee FS (2011) Selective early-acquired fear memories undergo temporary suppression during adolescence. Proc Natl Acad Sci U S A 108:1182–1187

Paul MJ, Terranova JI, Probst CK, Murray EK, Ismail NI, de Vries GJ (2014) Sexually dimorphic role for vasopressin in the development of social play. Front Behav Neurosci 8:58

Pellis SM, Pellis VC (1990) Differential rates of attack, defense, and counterattack during the developmental decrease in play fighting by male and female rats. Dev Psychobiol 23:215–231

Perry RE, Al Aïn S, Raineki C, Sullivan RM, Wilson DA (2016) Development of odor hedonics: experience-dependent ontogeny of circuits supporting maternal and predator odor responses in rats. J Neurosci 36:6634–6650

Petrulis A (2009) Neural mechanisms of individual and sexual recognition in Syrian hamsters (Mesocricetus auratus). Behav Brain Res 200:260–267

Pitkänen A, Stefanacci L, Farb CR, Go GG, LeDoux JE, Amaral DG (1995) Intrinsic connections of the rat amygdaloid complex: projections originating in the lateral nucleus. J Comp Neurol 356: 288–310

Prounis GS, Thomas K, Ophir AG (2018) Developmental trajectories and influences of environmental complexity on oxytocin receptor and vasopressin 1A receptor expression in male and female prairie voles. J Comp Neurol 526:1820–1842

Rasia-Filho AA, Fabian C, Rigoti KM, Achaval M (2004) Influence of sex, estrous cycle and motherhood on dendritic spine density in the rat medial amygdala revealed by the Golgi method. Neuroscience 126:839–847

Raybuck JD, Lattal KM (2014) Bridging the interval: theory and neurobiology of trace conditioning. Behav Processes 101:103–111

Reichard RA, Subramanian S, Desta MT, Sura T, Becker ML, Ghobadi CW, Parsley KP, Zahm DS (2017) Abundant collateralization of temporal lobe projections to the accumbens, bed nucleus of stria terminalis, central amygdala and lateral septum. Brain Struct Funct 222:1971–1988

Ressler RL, Goode TD, Evemy C, Maren S (2020) NMDA receptors in the CeA and BNST differentially regulate fear conditioning to predictable and unpredictable threats. Neurobiol Learn Mem 174:107281

Romer D, Reyna VF, Satterthwaite TD (2017) Beyond stereotypes of adolescent risk taking: placing the adolescent brain in developmental context. Dev Cogn Neurosci 27:19–34

Roozendaal B, Wiersma A, Driscoll P, Koolhaas JM, Bohus B (1992) Vasopressinergic modulation of stress responses in the central amygdala of the Roman high-avoidance and low-avoidance rat. Brain Res 596:35–40

Rosenfeld P, Gutierrez YA, Martin AM, Mallett HA, Alleva E, Levine S (1991) Maternal regulation of the adrenocortical response in preweanling rats. Physiol Behav 50:661–671

Rouzer SK, Diaz MR (2021) Factors of sex and age dictate the regulation of GABAergic activity by corticotropin-releasing factor receptor 1 in the medial sub-nucleus of the central amygdala. Neuropharmacology 189:108530

Rubinow MJ, Juraska JM (2009) Neuron and glia numbers in the basolateral nucleus of the amygdala from preweaning through old age in male and female rats: a stereological study. J Comp Neurol 512:717–725

Rudy JW (1993) Contextual conditioning and auditory cue conditioning dissociate during development. Behav Neurosci 107:887–891

Rudy JW, Morledge P (1994) Ontogeny of contextual fear conditioning in rats: implications for consolidation, infantile amnesia, and hippocampal system function. Behav Neurosci 108:227–234

Ryan SJ, Ehrlich DE, Rainnie DG (2016) Morphology and dendritic maturation of developing principal neurons in the rat basolateral amygdala. Brain Struct Funct 221:839–854

Ryazantseva M, Englund J, Shintyapina A, Huupponen J, Shteinikov V, Pitkänen A, Partanen JM, Lauri SE (2020) Kainate receptors regulate development of glutamatergic synaptic circuitry in the rodent amygdala. eLife 9

Sánchez MM, Ladd CO, Plotsky PM (2001) Early adverse experience as a developmental risk factor for later psychopathology: evidence from rodent and primate models. Dev Psychopathol 13:419–449

Sano K, Nakata M, Musatov S, Morishita M, Sakamoto T, Tsukahara S, Ogawa S (2016) Pubertal activation of estrogen receptor α in the medial amygdala is essential for the full expression of male social behavior in mice. Proc Natl Acad Sci U S A 113:7632–7637

Scheiber IBR, Weiß BM, Kingma SA, Komdeur J (2017) The importance of the altricial – precocial spectrum for social complexity in mammals and birds – a review. Front Zool 14:3

Selleck RA, Zhang W, Mercier HD, Padival M, Rosenkranz JA (2018) Limited prefrontal cortical regulation over the basolateral amygdala in adolescent rats. Sci Rep 8:17171

Sengupta A, Yau JOY, Jean-Richard-Dit-Bressel P, Liu Y, Millan EZ, Power JM, McNally GP (2018) Basolateral amygdala neurons maintain aversive emotional salience. J Neurosci 38: 3001–3012

Shapiro LE, Insel TR (1989) Ontogeny of oxytocin receptors in rat forebrain: a quantitative study. Synapse 4:259–266

Smith CJW, Poehlmann ML, Li S, Ratnaseelan AM, Bredewold R, Veenema AH (2017) Age and sex differences in oxytocin and vasopressin V1a receptor binding densities in the rat brain: focus on the social decision-making network. Brain Struct Funct 222:981–1006

Snijdewint FG, Van Leeuwen FW, Boer GJ (1989) Ontogeny of vasopressin and oxytocin binding sites in the brain of Wistar and Brattleboro rats as demonstrated by lightmicroscopical autoradiography. J Chem Neuroanat 2:3–17

Spear LP (2000) The adolescent brain and age-related behavioral manifestations. Neurosci Biobehav Rev 24:417–463

Spear LP, Brake SC (1983) Periadolescence: age-dependent behavior and psychopharmacological responsivity in rats. Dev Psychobiol 16:83–109

Stamps JA, Krishnan VV (2014) Combining information from ancestors and personal experiences to predict individual differences in developmental trajectories. Am Nat 184:647–657

Stansfield KH, Philpot RM, Kirstein CL (2004) An animal model of sensation seeking: the adolescent rat. Ann N Y Acad Sci 1021:453–458

Stanton ME, Gutierrez YR, Levine S (1988) Maternal deprivation potentiates pituitary-adrenal stress responses in infant rats. Behav Neurosci 102:692–700

Stanton ME, Freeman JH, Skelton RW (1992) Eyeblink conditioning in the developing rat. Behav Neurosci 106:657–665

Stark CP, Alpern HP, Fuhrer J, Trowbridge MG, Wimbish H, Smock T (1998) The medial amygdaloid nucleus modifies social behavior in male rats. Physiol Behav 63:253–259

Stuber GD, Sparta DR, Stamatakis AM, van Leeuwen WA, Hardjoprajitno JE, Cho S, Tye KM, Kempadoo KA, Zhang F, Deisseroth K, Bonci A (2011) Excitatory transmission from the amygdala to nucleus accumbens facilitates reward seeking. Nature 475:377–380

Sullivan RM, Landers M, Yeaman B, Wilson DA (2000) Good memories of bad events in infancy. Nature 407:38–39

Swanson LW (2018) Brain maps 4.0-structure of the rat brain: an open access atlas with global nervous system nomenclature ontology and flatmaps. J Comp Neurol 526:935–943

Takahashi LK (1992) Developmental expression of defensive responses during exposure to conspecific adults in preweanling rats (Rattus norvegicus). J Comp Psychol 106:69–77

Tetel MJ, Getzinger MJ, Blaustein JD (1993) Fos expression in the rat brain following vaginal-cervical stimulation by mating and manual probing. J Neuroendocrinol 5:397–404

Tsukahara S, Tsuda MC, Kurihara R, Kato Y, Kuroda Y, Nakata M, Xiao K, Nagata K, Toda K, Ogawa S (2011) Effects of aromatase or estrogen receptor gene deletion on masculinization of the principal nucleus of the bed nucleus of the stria terminalis of mice. Neuroendocrinology 94: 137–147

Tye KM, Prakash R, Kim SY, Fenno LE, Grosenick L, Zarabi H, Thompson KR, Gradinaru V, Ramakrishnan C, Deisseroth K (2011) Amygdala circuitry mediating reversible and bidirectional control of anxiety. Nature 471:358–362

Unger EK, Burke KJ, Yang CF, Bender KJ, Fuller PM, Shah NM (2015) Medial amygdalar aromatase neurons regulate aggression in both sexes. Cell Rep 10:453–462

Upton KJ, Sullivan RM (2010) Defining age limits of the sensitive period for attachment learning in rat pups. Dev Psychobiol 52:453–464

Urban JH, Rosenkranz JA (2020) Handbook of amygdala structure and function. Academic Press

van-Hover C, Li C (2015) Stress-activated afferent inputs into the anterior parvicellular part of the paraventricular nucleus of the hypothalamus: insights into urocortin 3 neuron activation. Brain Res 1611:29–43

VanRyzin JW, Marquardt AE, Argue KJ, Vecchiarelli HA, Ashton SE, Arambula SE, Hill MN, McCarthy MM (2019) Microglial phagocytosis of newborn cells is induced by endocannabinoids and sculpts sex differences in juvenile rat social play. Neuron 102:435–449.e6

Varlinskaya EI, Spear LP, Spear NE (1999) Social behavior and social motivation in adolescent rats: role of housing conditions and partner's activity. Physiol Behav 67:475–482

Veening JG, Coolen LM, de Jong TR, Joosten HW, de Boer SF, Koolhaas JM, Olivier B (2005) Do similar neural systems subserve aggressive and sexual behaviour in male rats? Insights from c-Fos and pharmacological studies. Eur J Pharmacol 526:226–239

Veres JM, Nagy GA, Vereczki VK, Andrási T, Hájos N (2014) Strategically positioned inhibitory synapses of axo-axonic cells potently control principal neuron spiking in the basolateral amygdala. J Neurosci 34:16194–16206

Viviani D, Charlet A, van den Burg E, Robinet C, Hurni N, Abatis M, Magara F, Stoop R (2011) Oxytocin selectively gates fear responses through distinct outputs from the central amygdala. Science 333:104–107

Walker DL, Davis M (1997) Double dissociation between the involvement of the bed nucleus of the stria terminalis and the central nucleus of the amygdala in startle increases produced by conditioned versus unconditioned fear. J Neurosci 17:9375–9383

Walker DL, Miles LA, Davis M (2009) Selective participation of the bed nucleus of the stria terminalis and CRF in sustained anxiety-like versus phasic fear-like responses. Prog Neuropsychopharmacol Biol Psychiatry 33:1291–1308

Wang Z, Hulihan TJ, Insel TR (1997) Sexual and social experience is associated with different patterns of behavior and neural activation in male prairie voles. Brain Res 767:321–332

Wang Y, He Z, Zhao C, Li L (2013) Medial amygdala lesions modify aggressive behavior and immediate early gene expression in oxytocin and vasopressin neurons during intermale exposure. Behav Brain Res 245:42–49

Wang Y, Zhao S, Liu X, Fu Q (2014) Effects of the medial or basolateral amygdala upon social anxiety and social recognition in mice. Turk J Med Sci 44:353–359

Warlow SM, Berridge KC (2021) Incentive motivation: 'wanting' roles of central amygdala circuitry. Behav Brain Res 411:113376

Weathington JM, Strahan JA, Cooke BM (2012) Social experience induces sex-specific fos expression in the amygdala of the juvenile rat. Horm Behav 62:154–161

Wiedenmayer CP, Barr GA (1998) Ontogeny of defensive behavior and analgesia in rat pups exposed to an adult male rat. Physiol Behav 63:261–269

Williamson CM, Lee W, Curley JP (2016) Temporal dynamics of social hierarchy formation and maintenance in male mice. Anim Behav 115:259–272

Wilson FA, Rolls ET (1993) The effects of stimulus novelty and familiarity on neuronal activity in the amygdala of monkeys performing recognition memory tasks. Exp Brain Res 93:367–382

Woody EZ, Szechtman H (2011) Adaptation to potential threat: the evolution, neurobiology, and psychopathology of the security motivation system. Neurosci Biobehav Rev 35:1019–1033

Wu MV, Manoli DS, Fraser EJ, Coats JK, Tollkuhn J, Honda S, Harada N, Shah NM (2009) Estrogen masculinizes neural pathways and sex-specific behaviors. Cell 139:61–72

Yamaguchi T, Wei D, Song SC, Lim B, Tritsch NX, Lin D (2020) Posterior amygdala regulates sexual and aggressive behaviors in male mice. Nat Neurosci 23:1111–1124

Yamauchi N, Takahashi D, Sugimura YK, Kato F, Amano T, Minami M (2018) Activation of the neural pathway from the dorsolateral bed nucleus of the stria terminalis to the central amygdala induces anxiety-like behaviors. Eur J Neurosci 48:3052–3061

Yao S, Bergan J, Lanjuin A, Dulac C (2017) Oxytocin signaling in the medial amygdala is required for sex discrimination of social cues. eLife 6

Zehr JL, Todd BJ, Schulz KM, McCarthy MM, Sisk CL (2006) Dendritic pruning of the medial amygdala during pubertal development of the male Syrian hamster. J Neurobiol 66:578–590

Intergenerational Effects of Stress – A Focus on Learning and Memory

L. S. Vasquez, S. Stack, W. W. Taylor, and B. G. Dias

Contents

L. S. Vasquez, S. Stack, and W. W. Taylor
Neuroscience Graduate Program, University of Southern California, Los Angeles, CA, USA

Developmental Neuroscience and Neurogenetics Program, The Saban Research Institute, Los Angeles, CA, USA

B. G. Dias (✉)
Developmental Neuroscience and Neurogenetics Program, The Saban Research Institute, Los Angeles, CA, USA

Division of Endocrinology, Diabetes and Metabolism, Children's Hospital Los Angeles, Los Angeles, CA, USA

Department of Pediatrics, Keck School of Medicine of USC, Los Angeles, CA, USA
e-mail: bdias@chla.usc.edu

© The Author(s), under exclusive license to Springer Nature Switzerland AG 2025
Curr Topics Behav Neurosci (2025) 73: 231–251
https://doi.org/10.1007/7854_2025_578
Published Online: 22 March 2025

1 Introduction

Stress is a ubiquitous facet of life. Ranging in form (e.g., psychosocial, physical, nutritional, economic) and longevity (e.g., acute, chronic), stressors affect the biology of those directly in their line of attack. As is becoming increasingly appreciated, the pernicious effects of stress echo across generations (Dias et al. 2015; Yehuda and Lehrner 2018; Jawaid et al. 2021; Dion et al. 2022; Zhou and Ryan 2023; Dias 2024). With a focus on learning and memory, this chapter addresses how stressors derail learning and memory in the generation directly exposed to them and in future generations. To do so, with a specific emphasis on associative fear conditioning in humans and rodents, we touch upon the relevance of extinction training in the aftermath of such conditioning and the recall of such extinction training as windows into normative and disrupted learning. Next, we briefly discuss underlying neuroanatomical substrates mediating these processes. We then draw attention to influences of postnatal, in utero, and pre-conceptional stress on learning and memory across generations. Finally, we briefly outline biological factors that underlie how learning and memory is derailed by these stressors.

2 Extinction of Fear and Its Maintenance (Recall) (Fig. 1)

Achieved through the process of extinction learning and extinction recall, both humans and animals extinguish their fear toward stimuli previously associated with threat via newly formed associative learning that signals these stimuli are no longer dangerous. Deficits in extinction learning, and the recall of such learning, can be highly maladaptive and contribute to anxiety and trauma-related disorders (Fullana et al. 2020; Sangha et al. 2020; Cooper and Dunsmoor 2021). The amygdala, hippocampus, and medial prefrontal cortex play key roles in extinguishing fear and subsequent consolidation of extinction training and recall. Recent research has expanded the neuroanatomical framework underlying these processes to include noncanonical brain regions such as the zona incerta, nucleus reuniens, and paraventricular nucleus of the thalamus (Venkataraman and Dias 2023).

3 Intergenerational Legacies of Stress on Learning and Memory (Fig. 2)

Stress is experienced throughout all stages of life, from pre-conception to adulthood. Depending on their magnitude, duration, and developmental time point, stressors have the potential to influence not only those who directly experience them but also the future generations (McEwen and Gianaros 2011; Gunnar and Quevedo 2007;

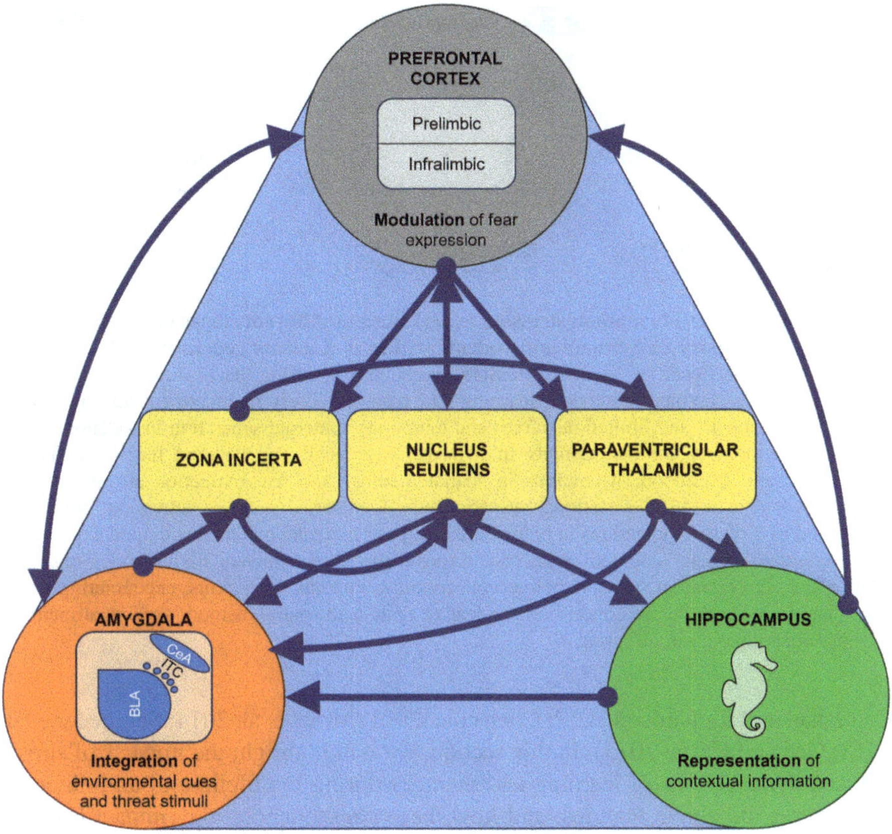

Fig. 1 Integration of canonical, thalamic, and subthalamic neurocircuitry underlying associative fear conditioning, extinction, and extinction recall. General connectivity diagram that includes the canonical tripartite circuit (circles in corners) and thalamic/subthalamic regions (yellow boxes). Within the canonical tripartite circuit, the basolateral amygdala (BLA) acts as a threat detector, pairing incoming noxious environmental stimuli with sensory cues. During the extinction of fear, excitatory projections from the infralimbic prefrontal cortex (IL-PFC) to the BLA become more active, engaging inhibitory circuitry within the amygdala (e.g., ITC cluster activity and their inhibition of the central amygdala - CeA) to reduce fear expression. The hippocampus (HPC) innervates both the amygdala and the PFC, providing contextual information. Inhibitory projections from the HPC to the IL and PL act to inhibit extinction and fear expression, respectively. Thalamic and subthalamic regions, including the zona incerta (ZI), nucleus reuniens (RE), and paraventricular nucleus of the thalamus (PVT), sit as additional nodes through which information is gated and processed within the tripartite fear circuit, adding an additional layer of complexity and computational power. The ZI sits within this circuitry, receiving inputs from the CeA that are necessary for fear memories and increasing activity in concert with the PFC to mediate extinction learning. Additionally, ZI projections to the RE are critical for fear generalization as well as fear extinction The RE sits between the PFC and HPC, acting as a mediator of cortical control of the HPC. Similarly, the RE sits between the PFC and BLA, mediating the expression of extinction at remote timepoints. The PVT is broadly interconnected within this circuitry. The PFC-PVT-amygdala pathway has been shown to support normative fear conditioning, fear recall, and extinction recall

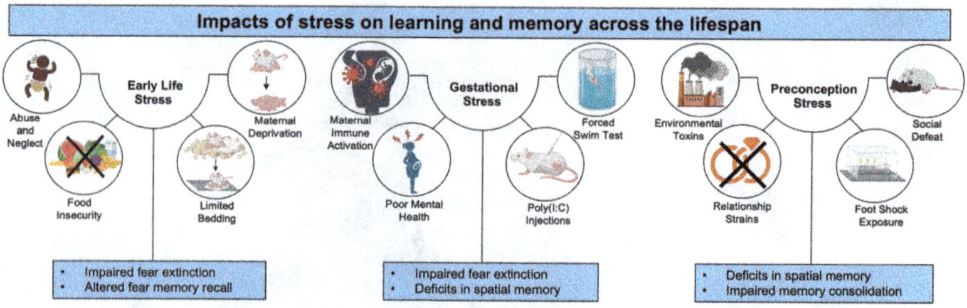

Fig. 2 Consequences of common stressors experienced at different stages of development on learning and memory in humans and rodents within and across generations. A conceptual diagram of the effects of typical stressors encountered during early life, gestation, and pre-conception on human and rodent learning and memory. Early maltreatment and nutritional insecurity in humans, and limited resources and temporary maternal separation in rodents, impact learning and memory through deficits in fear memory recall and delayed fear extinction in adolescence and adulthood. Alterations in spatial memory and fear extinction are observed in offspring when exposed to gestational stressors such as maternal immune activation, which can be induced by Poly(I:C) injections in pregnant dams, poor maternal mental health, and exposure to acute stressors such as a forced swim test. Exposure to environmental toxins and tumultuous relationships are common stressors before conception in humans. In rodents, encountering social defeat and footshocks before conception can lead to deficits in spatial memory and impairments in memory consolidation in offspring

Callaghan and Richardson 2012; Chen et al. 2021; Molet et al. 2014; Schroeder et al. 2018; Gee and Casey 2015). In this section, we will highlight the impacts of stress-induced perturbations of learning and memory during the lifetime of an individual considering the above features, and how these consequences may impact learning and memory within and across generations. To do so, we focus on three epochs of time at which stress may be experienced – early in life, during gestation, and prior to conceiving. More specifically, we discuss exposures to stress during these epochs and their effects on learning and memory within the generation exposed (G0 generation) and offspring (G1 generation).

3.1 *Early-Life Stress*

Stress experienced in infancy or adolescence significantly increases the likelihood of developing PTSD after a traumatic event (Teicher et al. 2002; Pratchett and Yehuda 2011). Common early-life stressors (ELS) in humans include physical and sexual abuse, nutritional insecurity, and neglect (Pechtel and Pizzagalli 2011; Moraes et al. 2018; Strzelewicz et al. 2019). Consequences of these adverse early experiences can predispose individuals to emotional dysregulation along with severe disruptions in learning and memory-related processes such as fear learning, extinction learning and recall, and well-established characteristics of PTSD (Haapanen et al. 2024; Lippard and Nemeroff 2023).

Fragmented care and unpredictable living situations have been shown to be associated with emotional and learning and memory deficits, specifically involving fear-related behaviors exhibited in childhood and beyond (Merz et al. 2013; Hodel et al. 2015; Bick et al. 2017; Sousa et al. 2023). Studies have shown impairments in extinction of fear, even though a threat is no longer present, and an increase in fear generalization to both threatening and non-threatening cues, in previously institutionalized children compared to children who have not been institutionalized (Waters et al. 2009; Cohen et al. 2013). Additionally, deprivation experienced through early-institutionalized care has also been linked to deficits in hippocampal-dependent tasks such as recognition memory and attention which may also contribute to deficits in fear-related behavior (Pollak et al. 2010; McLaughlin et al. 2017; Gunnar and Bowen 2021).

In the laboratory, rodent models provide researchers robust insights into determining how different early-life stressors impact later fear learning, extinction learning, and recall in both adolescence and adulthood (Green and McCormick 2013; Murthy and Gould 2018; Dutcher et al. 2020). Given the significance of maternal-infant interactions, manipulation of maternal care is commonly used to model early-life neglect, deprivation, and maltreatment (Murthy and Gould 2018; Dutcher et al. 2020). In rodent models, deprivation from maternal care via prolonged separation and fragmented maternal care bestowed via provision of limited nesting and bedding have been used to induce ELS (Chen and Baram 2016; Sheridan and McLaughlin 2014; Manzano-Nieves et al. 2018; Mishra et al. 2019; Guijarro et al. 2007). After pups are exposed to the above-mentioned conditions, animals typically undergo fear conditioning protocols at different ages to probe fear learning and memory-related behaviors. Maternal separation for 3 h from postnatal day 2–14 (PND2–14), deemed a chronic early-life stressor, resulted in deficits in fear extinction recall in infant (PND17) rats (Callaghan and Richardson 2011). To further demonstrate the importance of ontogeny to ELS-induced deficits in learning and memory, maternal separation and isolation for 6 h daily over 10 days (PND4–14) reduced fear extinction in adolescent rats but not in adults, highlighting the vulnerability of young adulthood to fear memory and extinction processes (Mishra et al. 2019). In pre-adolescent mice, exposure to limited bedding for 7 consecutive days (PND4–11) impacted fear memory recall after cued-fear conditioning, while maternal separation for 3 h daily (PND1–14) resulted in deficits in extinction recall in adolescent rats (Chocyk et al. 2014; Manzano Nieves et al. 2020). These findings collectively indicate that early life stress can affect learning and memory, which may endure long after the initial stressor has occurred.

3.2 *Gestational Stress*

Fetal development is highly vulnerable to environmental insults such as parental stress exposure. Common stressors experienced during pregnancy include maternal anxiety and depression, socioeconomic challenges, drug and alcohol abuse, maternal immune activation, and domestic violence (Graignic-Philippe et al. 2014; Henderson et al. 2007). These adverse experiences during gestation not only diminish parental well-being but can also heighten vulnerability in offspring to developing anxiety-like disorders such as PTSD after experiencing a traumatic event (Davis and Sandman 2012; Leen-Feldner et al. 2013).

To investigate the enduring influence of different gestational stressors on deficits in extinction, extinction recall, and fear generalization, we turn to rodent models. Male and female offspring of dams exposed to ethanol during pregnancy exhibited prolonged fear, as measured by sustained high levels of freezing, to a previously conditioned stimulus (tone) even in the absence of the aversive stimulus (footshock), after auditory fear conditioning (Yao et al. 2023; Plaza et al. 2020). Difficulties in extinction acquisition have also been shown to induce fear generalization to a neutral auditory stimulus reminiscent of the previously fear-conditioned auditory stimulus in rats born to dams that received random subcutaneous saline injections throughout gestation (Salm et al. 2015).

Recent attention has focused on how maternal immune activation and related immunological stressors may contribute to behaviors relevant to anxiety-like susceptibility in offspring (Conway and Brown 2019; Quagliato et al. 2021; Zhang et al. 2024). Preeclampsia, which leads to maternal immune activation, was induced in pregnant dams at the beginning of gestational day (GD) 12 via intraperitoneal infusions of an antiangiogenic drug, Fms-like tyrosine kinase-1 (sFlt-1). At PND 45, offspring from pre-conception-stressed dams underwent behavioral tests to assess spatial memory (Barnes Maze) and anxiety-like behavior (marble-burying test). Offspring exhibited both spatial learning deficits and increased anxiety-like behavior in their respective tasks suggesting deleterious effects of gestational maternal immune activation (Griffin et al. 2024). Another common procedure used to induce maternal immune activation is through intraperitoneal injections of polyriboinosinic-polyribocytidylic acid (poly(I:C)). Studies have demonstrated that poly(I:C) injections in mice dams at GD9.5 or GD12 resulted in both learning and memory deficits during novel object recognition tasks and increased anxiety-like behaviors during the elevated plus maze (EPM) in adult offspring, respectively (Gumusoglu et al. 2017; Perez-Palomar et al. 2023). Anxiety-like behavior was also exhibited during EPM in adult rat progeny of dams injected with poly(I:C) at GD10 (Hollins et al. 2018). Collectively, these experimental findings offer compelling proof that maternal gestational stress has enduring intergenerational effects.

3.3 *Pre-conception Stress*

The influence of the pre-conceptional period is typically ignored as an epoch of time in discussions of multi-generational legacies of stress. However, evidence in humans suggests that paternal exposure to stress, nutritional insults, and environmental toxins derails mental, physical, and metabolic health in offspring (Klaric et al. 2008; McPherson et al. 2014; Kaati et al. 2007; Suren et al. 2014; Pembrey 2010;

Yen et al. 2016; Kretschmer et al. 2023). Understanding the consequences from stress experienced during the pre-conception period both within a generation and across generations is paramount to shedding light on the etiology of anxiety-like disorders such as PTSD. Common pre-conception stressors such as poor physical and mental health, financial insecurity, and relationship strains, encompass a broad spectrum of challenges individuals or couples may encounter before attempting to conceive (Keenan et al. 2018; Rinne et al. 2023). Currently, there is limited human and animal literature examining the direct impacts of pre-conception stress on fear extinction and recall in offspring. However, a wealth of literature using rodent models and different forms of pre-conception stress have uncovered other processes that are perturbed in offspring. These behaviors range from increased anxiety-like behaviors to altered startle response, behavioral sensitivity, and spatial memory deficits (Bhattacharya et al. 2023; Shachar-Dadon et al. 2009; Dias and Ressler 2014; Aoued et al. 2019, 2020; Scarborough et al. 2021), all of which may have translational importance in contributing to increased susceptibility in developing anxiety-like disorders in future generations.

Adolescent G1 offspring born to male and female G0 mice previously exposed to psychosocial stress, by way of exposure to a rat, exhibited increased anxiety-like behaviors in commonly used assays to measure anxiety including the elevated plus maze, open field, and acoustic startle test. When the same G1 animals were exposed to a mild stressor (rat predator for 2 min) in adulthood, the animals not only did exhibit heightened sensitivity to the stressor but also displayed persistent anxiety-like behavior in the same behavioral tests from adolescence, compared to offspring from non-pre-conception stressed parents (Bhattacharya et al. 2023). Similar results were found in adult G1 animals of female mice that experienced 10 weeks of social isolation prior to breeding; however, notable sex differences emerged during the Y-maze test, used to measure spatial recognition memory, where only F1 males exhibited memory deficits (Scarborough et al. 2021). Exposing female rats to various unpredictable stressors for 7 consecutive days, one week prior to conception resulted in G1 males showing increased shock avoidance, a measure of exaggerated fear response, while G1 females displayed decreased shock avoidance (Shachar-Dadon et al. 2009). Evidence for behavioral sensitivity to a previously conditioned stimulus in the G0 generation persisting in offspring has also been demonstrated using chemosensation, namely, olfaction. Male mice that underwent olfactory fear conditioning (Odor A + shock) before conception sired male and female G1 offspring that exhibited heightened behavioral sensitivity to Odor A despite never having any prior exposure to Odor A (Dias and Ressler 2014; Aoued et al. 2019). Interestingly, the behavioral sensitivity in the G1 offspring to Odor A afforded female but not male G1 mice the ability to consolidate information about single trial Odor A conditioning demonstrating that some intergenerational legacies of stress could be to the advantage of descendant generations (Aoued et al. 2020). Altogether, these experimental results provide compelling evidence that maternal and paternal stress exposure prior to conception exerts lasting intergenerational consequences.

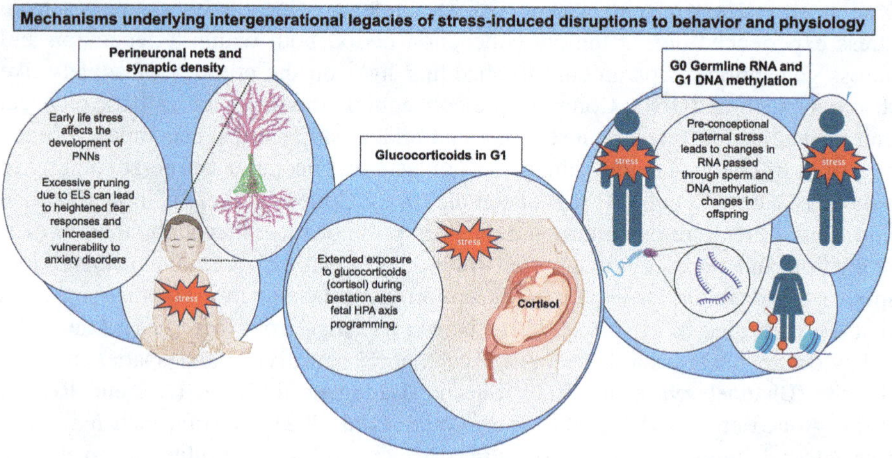

Fig. 3 Mechanisms underlying intergenerational legacies of stress-induced disruptions to behavior and physiology. A representative illustration demonstrating how stressors at different epochs of time influence intergenerational disruptions of learning and memory. Early-life stress (ELS) leads to neuroanatomical changes in the brain that impact normative age-related development of perineuronal nets (PNNs) and synaptic density. PNN accumulation is an age-dependent neuroanatomical change that can be altered by ELS in brain regions important for learning and memory. ELS can alter normative age-dependent synaptic pruning leading to changes in learning and memory. Gestational stress causes an influx of glucocorticoids across the placental barrier overriding the protective mechanisms the placenta has to limit cortisol exposure. Fetal exposure to cortisol causes alterations to fetal HPA axis programming leading to a blunted HPA axis response in G1 offspring. Pre-conceptional stress leaves an impression on G1 offspring via G0 germline RNA and DNA methylation in G1 cells. Taken together, these mechanisms contribute in part to intergenerational stress-induced disruptions of learning and memory and the development of trauma- and anxiety-related disorders

4 Mechanistic Underpinnings of Stress-Induced Intergenerational Disruptions of Learning and Memory (Fig. 3)

In this section, we review some factors and proximate mechanisms that ought to be considered in the discussion of each of the stress-induced intergenerational disruptions of learning and memory.

4.1 Early-Life Stress – Ontogeny of Learning, Brain Development, and Gonadal Steroids

Learning is a dynamic process that dovetails with the equally dynamic phenomena, that is, development. The age of an organism influences the expression (or not) of extinction learning and recall. There is remarkable conservation in developmental trajectories of extinction learning and recall across species (King et al. 2013; Spear 2004; Pattwell et al. 2012). Pre-adolescent children (age 6–11 years) exhibit within session extinction (extinction learning) but poor extinction recall (Marusak et al. 2018). Likewise, pre-adolescent rodents exhibit the same behavior in extinction learning (Pattwell et al. 2012) and recall (Hefner and Holmes 2007). Adolescent humans showed impaired extinction learning (Pattwell et al. 2012) and impaired extinction recall (Ganella et al. 2018). Likewise, adolescent rats (PND47-P48) demonstrated a deficit in extinction recall (Baker et al. 2013; Kim et al. 2011; McCallum et al. 2010). More specifically, adolescent rats showed extinction acquisition but failure to maintain that extinction when tested the following day (McCallum et al. 2010). Complementing the cue-based findings on developmental trajectories of learning and memory mentioned above, recently published data demonstrated that pre-adolescent mice (PND16-P20) exhibit imprecise contextual fear memories (Ramsaran et al. 2023). These mice showed high freezing in both Context A and B, after being initially conditioned in Context A. In contrast, older mice (PND24+) are able to distinguish between contexts (Ramsaran et al. 2023). One prevailing hypothesis of how early-life stress might derail learning and memory is by altering the cadence of maturation of brain circuitry and physiology (Hanson et al. 2021; Peña et al. 2019; Fan et al. 2014; Toya et al. 2014; Xu et al. 2020; Shin et al. 2023). Among the many ways ELS may disrupt the cadence of brain development and physiology are alterations in normative age-related development of neuroanatomy.

The end-points in the discussion of neuroanatomy are plenty. These include but are not limited to brain thickness, synaptic density, synaptic strength, and excitatory-inhibitory (E/I) balance. For the purposes of this chapter, we will focus our discussion of age-dependent changes in neuroanatomy to perineuronal nets constraining windows of developmental plasticity and synaptic density. For discussion on other dimensions of age-dependent changes in neuroanatomy, we refer the reader to the following reviews (Houston et al. 2014; Casey et al. 2005; Kolb and Gibb 2011; Tau and Peterson 2010).

Perineuronal nets (PNNs) are lattice-like structures that surround multiple cell types in the mammalian brain (Fawcett et al. 2019). Observed in the cortex, hippocampus, thalamus, and brainstem and spinal cord, they mostly envelope fast spiking inhibitory interneurons in the cortex (van't Spijker and Kwok 2017; Miyata et al. 2018; Fawcett et al. 2022). Of relevance to our discussion of age-dependent changes in neuroanatomy, PNNs accumulate across developmental time and serve to modulate E/I balance (Gogolla et al. 2009; Mauney et al. 2013; Rogers et al. 2018; Yamada and Jinno 2013). The formation of PNNs indicates a transition in plasticity resulting in the termination of a critical period (Pizzorusso et al. 2002; Carulli and Verhaagen 2021). Functionally, roles for PNNs in extinction learning and recall have been determined by dissolving PNNs using a substance like chondroitinase-ABC. Using this method, it has been shown that PNNs stabilize fear memories and are important in fear learning (Hylin et al. 2013; Banerjee et al. 2017). Specifically, the lack of PNNs in the amygdala leads to elimination of fear memories through extinction, providing evidence that PNNs protect fear memories (Gogolla et al. 2009). Recent data has also shown that deleting a critical regulator for PNN aggregation, and not PNNs directly, has the same effect in disrupting fear memory recovery (Lavertu-Jolin et al. 2023). PNNs also play a role in fear memory retrieval (Tsien 2013; Poli et al. 2023). Thompson et al. (2018) showed that recall of remote visual fear memories depends on the presence of PNNs in the visual cortex and that dissolving PNNs disrupts this recall. This evidence indicates that alterations to PNNs alter neuronal communication affecting memory engram reactivation. Taken together, this data elucidates the importance of PNNs in fear memory learning and recall. Rodent data shows that ELS alters PNN densities in regions essential for fear learning and recall (Gildawie et al. 2020; Santiago et al. 2018; Guadagno et al. 2020; Catale et al. 2022). Alterations to PNNs caused by ELS can alter memory engram circuits important for fear learning giving insight into ways in which ELS affects fear behavior.

Synaptic density is another dimension of neurobiology that tracks developmental time and is associated with normative extinction learning and recall. Synaptic density increases after birth, peaks around adolescence, and then gradually decreases (Kolb and Gibb 2011; Navlakha et al. 2015; Power and Schlaggar 2017; Sakai 2020). While this process of synaptic pruning follows a developmental cadence, it is extremely sensitive to experiences. Elimination of synapses is experience-driven and different areas of the brain experience pruning at different times (Tierney and Nelson 3rd 2009). To measure synaptic density, changes in the dendritic spine and spine densities are analyzed (Serrano et al. 2022) and manipulating spine densities can affect learning (Leuner and Shors 2004; Albarran et al. 2021; Guo et al. 2023). In the amygdala (Yang et al. 2016; Lai et al. 2018) and auditory cortex (Heinrichs et al. 2013; Moczulska et al. 2013; Keifer et al. 2015), fear conditioning increased dendritic spines while fear extinction decreased dendritic spines. In the amygdala, excessive pruning due to ELS can lead to heightened fear responses and increased vulnerability to anxiety disorders (Teicher et al. 2002; Gee et al. 2013). Similarly, the prefrontal cortex and the hippocampus are also susceptible to perturbations in synaptic pruning induced by ELS (Chocyk et al. 2013). Dayananda et al. (2023) discovered that pups exposed to limited bedding and impoverished maternal care had disruptions in microglial function. Microglia play an important role in synaptic pruning as they engulf excessive synaptic material. Downstream effects of ELS-induced deficits in synaptic pruning during the early postnatal period can thus lead to long-term changes in learning and memory-related circuitry important for mediating fear behavior in both adolescence and adulthood.

4.2 Gestational Stress – Glucocorticoids

To elucidate the biological mechanisms potentially responsible for the influences of gestational stress on learning and memory, significant focus has been placed on examining the interplay between the maternal-fetal hypothalamic-pituitary-adrenal

(HPA) axis and cortisol and its enduring impact on the stress response of future generations.

The release of glucocorticoids, cortisol in humans and corticosterone in rodents, is the end product of the hypothalamic-pituitary-adrenal axis (HPA axis), one of the primary biological responses to stress. Glucocorticoids bind mineralocorticoid and glucocorticoid receptors throughout the brain profoundly influencing learning and memory processes (Lupien et al. 2009). Mineralocorticoid receptors have a tenfold great affinity for glucocorticoids; thus, they are largely bound at baseline conditioning, leaving the glucocorticoid receptor as the primary actor in the stress response system (Finsterwald and Alberini 2014). Glucocorticoid receptors are highly expressed in brain regions like the amygdala and hippocampus that underlie fear conditioning, extinction, and extinction recall (Finsterwald and Alberini 2014). A wealth of literature has demonstrated that glucocorticoids enhance memory consolidation, showing that systemic administration of glucocorticoids following learning enhances future recall (Roozendaal and McGaugh 2011). In the context of associative fear extinction, glucocorticoid administration, including direct injection into the mPFC and amygdala, has been shown to enhance extinction learning and consolidation (de Quervain et al. 2011, 2019; Yang et al. 2006; Omoumi et al. 2023). Furthermore, disruption of glucocorticoid receptor expression within canonical fear circuitry has been shown to impair normative extinction and extinction recall (Scheimann et al. 2019; Taylor et al. 2021; Tertil et al. 2018; Wiktorowska et al. 2021). Conversely, it has been extensively demonstrated that acute glucocorticoid activity impairs recall of previously learned information, suggesting that glucocorticoids act during extinction to prioritize new learning while diminishing fear recall (de Quervain and Margraf 2008; de Quervain et al. 2009; Meir Drexler et al. 2019). Through this simultaneous enhancement of learning/consolidation and attenuation of recall, stress primes the brain through glucocorticoid action to encode extinction learning while limiting fear recall.

Throughout pregnancy, the body undergoes multiple adaptive processes to enhance the conditions within the uterus, promoting optimal development for the fetus. Among these processes is protecting the fetus from prolonged exposure to maternal cortisol which can be achieved through several mechanisms. Within the placenta, the enzyme 11β-hydroxysteroid dehydrogenase type 2 (11β-HSD2) converts cortisol to its inactive metabolite, cortisone, regulating the amount of cortisol penetrating the fetus (Konstantakou et al. 2017). The same is found in rodents where 11β-HSD2 converts corticosterone to its inactive form dehydrocorticosterone (Hamden et al. 2019). Additionally, maternal cortisol-binding globulin (CBG) binds to circulating cortisol, thus reducing the amount of available cortisol in the bloodstream (Meyer et al. 2016). Together, these mechanisms maintain a stable environment to ensure healthy fetal development, yet these processes can become derailed as a result of stress. In addition to maternal adverse experiences during gestation increasing the release of cortisol, these experiences can also lead to epigenetic changes affecting 11β-HSD2 activity. Previous work has shown gestational stress promotes DNA methylation of the 11β-HSD2 gene promoter and downregulates 11β-HSD2 activity. This ultimately reduces circulating GBG resulting in enhanced fetal cortisol exposure (Rinne et al. 2023; Konstantakou et al. 2017; Meyer et al. 2016). While increased fetal exposure to cortisol has been shown to influence brain development, fetal HPA axis programming by altering later response sensitivity to stressful stimuli is an especially salient phenomenon in the discussion of gestational stress-induced disruptions in G1 learning and memory (Glover et al. 2010; Sandman et al. 2012; Moisiadis and Matthews 2014). For example, O'Connor et al. (2013), investigated whether higher amniotic fluid cortisol levels during gestation would influence cortisol levels in infants when presented with an acute stressor (20 min of maternal separation). Indeed, infants had overall higher baseline levels of cortisol pre-stressor and elicited an overall blunted stress response to acute maternal separation (O'Connor et al. 2013). Similar results were found in infants of mothers diagnosed with major depressive disorder (MDD) during pregnancy. Interestingly, newborn daughters of prenatally diagnosed MDD mothers had an approximate 50% to 75% increase in baseline and stress-responsive cortisol levels when exposed to a mild stressor (being handled while awake, asleep, and crying) compared to daughters of control mothers, which was not observed in sons (Stroud et al. 2016). Given that glucocorticoid-glucocorticoid receptor action has been shown to affect extinction learning and recall, this altered programming of the fetal HPA axis that extends well beyond the fetal epoch is a crucial contributor to gestational stress-induced deficits in learning and memory in G1 offspring.

In summary, the gestational period is critical for neurodevelopment and is consequently highly sensitive to external factors, especially parental stress exposure. These stressors can exert significant influence on offspring stress responses and may potentially contribute to later emergence of anxiety-like disorders.

4.3 *Pre-conceptional Stress – Germline RNA and DNA Methylation*

To begin unraveling the biological mechanisms contributing to the effects of G0 pre-conception stress on G1 offspring, special focus has been placed on studying epigenetic changes in gene expression patterns in G1 biology (Dias et al. 2015; Yehuda and Bierer 2009; Bowers and Yehuda 2016). Alterations in DNA methylation provide an important biological marker that links stress experienced by an ancestral generation to altered biology in offspring. Adult offspring of Holocaust survivors diagnosed with PTSD had altered DNA methylation for the *nr3c1* gene which encodes the glucocorticoid receptor (GR) responsible for mediating responsiveness to the cortisol (Yehuda et al. 2014). Interestingly, results suggested a difference in methylation that was dependent on whether both parents were Holocaust survivors with PTSD (less methylation) or if it was only the father (higher methylation) (Yehuda et al. 2014). Building upon these results, methylation of *fkbp5,* another important molecular component in GR signaling, was also susceptible to epigenetic alterations in Holocaust survivors, particularly through the maternal lineage, and their offspring. Although increased maternal *fkbp5* methylation was observed, offspring *fkbp5* methylation levels were decreased (Yehuda et al. 2016; Bierer et al. 2020). These differences may suggest a differential epigenetic influence on subsequent interpersonal interactions between mother and child.

One of the more exciting biological concepts that has arisen from work on legacies of pre-conceptional paternal stress has been the idea that the epigenome of the sperm carries the echoes of stress from one generation to the next. In this arena, studies in mice and humans have demonstrated changes in DNA methylation, histone modifications and RNA in sperm in response to dietary insults, psychosocial stress, and chemosensory experiences (Dias and Ressler 2014; Aoued et al. 2020; Rodgers et al. 2013; Sharma and Rando 2014; Carone et al. 2010; Bohacek et al. 2015; Dickson et al. 2018; Gapp et al. 2020; Duffy et al. 2021; Lismer and Kimmins 2023; Maitin-Shepard et al. 2024). In a departure from studies that have correlated these changes with the legacy of paternal stress being handed to the next generation, intra-zygotic injections of RNA from sperm of male mice exposed to a variety of perturbations have recapitulated the effects normally passed down to offspring via traditional mating (Aoued et al. 2020; Rodgers et al. 2013; Sharma and Rando 2014; Gapp et al. 2020; Sharma et al. 2013; Chen 2022) and made a compelling case for sperm RNA being a key contributor to intergenerational legacies of paternal stress. Studies investigating germline legacies of maternal stress are more limited. Yet, there is work that has shown that prolonged unpredictable stress before conception (involving physical restraint, exposure to predator odors, and forced swim tests over a span of 10 days) elevates corticotropin-releasing factor type 1 (*crf1*) messenger RNA levels, an important mediator in the stress response, in mature rat oocytes (Zaidan et al. 2013). Furthermore, offspring of these pre-conceptionally stressed dams also show increased CRF1 expression in the brain, suggesting potential stress response alterations (Zaidan et al. 2013). To further characterize these findings, Zaidan et al. (2021) investigated whether mouse oocyte microRNAs targeting corticotropin-releasing hormone receptor type 1 (*crhr1*) are altered in response to pre-conception stress and whether these alterations persist in F1 and F2 neonatal brains. Indeed, not only were there increased *crhr1*-targeting microRNAs in the oocytes, but similar microRNA changes were found in the F1 and F2 neonate prefrontal cortex, confirming that stress effects can be transferred within and across generations.

In summary, parental stress before conception can significantly impact offspring, potentially extending its effects to subsequent generations. Epigenetic regulation of gene expression that begins in the germline and developing embryo have emerged as a pivotal mechanism facilitating this phenomenon, which may ultimately increase susceptibility to anxiety-related disorders such as PTSD.

5 Summary

With a focus on extinction learning and recall, processes that are disrupted in conditions like PTSD, this chapter began with a discussion of the neuroanatomy underlying extinction learning and recall. We then added to the perspective that early-life stress significantly increased the likelihood of developing trauma and anxiety-related disorders like PTSD by noting that gestational and pre-conceptional stress also allow for echoes of stress to reverberate across generations. In conclusion, we discussed a suite of mechanisms that allow for stress at these various epochs to exert their action on biology and, in so doing, disrupt learning and memory.

References

Albarran E et al (2021) Enhancing motor learning by increasing the stability of newly formed dendritic spines in the motor cortex. Neuron 109(20):3298–3311.e4

Aoued HS et al (2019) Reversing behavioral, neuroanatomical, and germline influences of intergenerational stress. Biol Psychiatry 85(3):248–256

Aoued HS et al (2020) Proximate causes and consequences of intergenerational influences of salient sensory experience. Genes Brain Behav 19(4):e12638

Baker KD, McNally GP, Richardson R (2013) Memory retrieval before or after extinction reduces recovery of fear in adolescent rats. Learn Mem 20(9):467–473

Banerjee SB et al (2017) Perineuronal nets in the adult sensory cortex are necessary for fear learning. Neuron 95(1):169–179.e3

Bhattacharya S et al (2023) A short pre-conception bout of predation risk affects both children and grandchildren. Sci Rep 13(1):10886

Bick J et al (2017) Effects of early institutionalization on emotion processing in 12-year-old youth. Dev Psychopathol 29(5):1749–1761

Bierer LM et al (2020) Intergenerational effects of maternal holocaust exposure on FKBP5 methylation. Am J Psychiatry 177(8):744–753

Bohacek J et al (2015) Pathological brain plasticity and cognition in the offspring of males subjected to postnatal traumatic stress. Mol Psychiatry 20(5):621–631

Bowers ME, Yehuda R (2016) Intergenerational transmission of stress in humans. Neuropsychopharmacology 41(1):232–244

Callaghan BL, Richardson R (2011) Maternal separation results in early emergence of adult-like fear and extinction learning in infant rats. Behav Neurosci 125(1):20–28

Callaghan BL, Richardson R (2012) Early-life stress affects extinction during critical periods of development: an analysis of the effects of maternal separation on extinction in adolescent rats. Stress 15(6):671–679

Carone BR et al (2010) Paternally induced transgenerational environmental reprogramming of metabolic gene expression in mammals. Cell 143(7):1084–1096

Carulli D, Verhaagen J (2021) An extracellular perspective on CNS maturation: perineuronal nets and the control of plasticity. Int J Mol Sci 22(5):2434

Casey BJ et al (2005) Imaging the developing brain: what have we learned about cognitive development? Trends Cogn Sci 9(3):104–110

Catale C et al (2022) Early-life social stress induces permanent alterations in plasticity and perineuronal nets in the mouse anterior cingulate cortex. Eur J Neurosci 56(10):5763–5783

Chen Q (2022) Sperm RNA-mediated epigenetic inheritance in mammals: challenges and opportunities. Reprod Fertil Dev 35(2):118–124
Chen Y, Baram TZ (2016) Toward understanding how early-life stress reprograms cognitive and emotional brain networks. Neuropsychopharmacology 41(1):197–206
Chen Y et al (2021) Early life stress induces different behaviors in adolescence and adulthood may related with abnormal medial prefrontal cortex excitation/inhibition balance. Front Neurosci 15: 720286
Chocyk A et al (2013) Early-life stress affects the structural and functional plasticity of the medial prefrontal cortex in adolescent rats. Eur J Neurosci 38(1):2089–2107
Chocyk A et al (2014) The effects of early-life adversity on fear memories in adolescent rats and their persistence into adulthood. Behav Brain Res 264:161–172
Cohen MM et al (2013) Early-life stress has persistent effects on amygdala function and development in mice and humans. Proc Natl Acad Sci U S A 110(45):18274–18278
Conway F, Brown AS (2019) Maternal immune activation and related factors in the risk of offspring psychiatric disorders. Front Psych 10:430
Cooper SE, Dunsmoor JE (2021) Fear conditioning and extinction in obsessive-compulsive disorder: a systematic review. Neurosci Biobehav Rev 129:75–94
Davis EP, Sandman CA (2012) Prenatal psychobiological predictors of anxiety risk in preadolescent children. Psychoneuroendocrinology 37(8):1224–1233
Dayananda KK et al (2023) Early life stress impairs synaptic pruning in the developing hippocampus. Brain Behav Immun 107:16–31
de Quervain DJ, Margraf J (2008) Glucocorticoids for the treatment of post-traumatic stress disorder and phobias: a novel therapeutic approach. Eur J Pharmacol 583(2–3):365–371
de Quervain DJ et al (2009) Glucocorticoids and the regulation of memory in health and disease. Front Neuroendocrinol 30(3):358–370
de Quervain DJ et al (2011) Glucocorticoids enhance extinction-based psychotherapy. Proc Natl Acad Sci USA 108(16):6621–6625
de Quervain D, Wolf OT, Roozendaal B (2019) Glucocorticoid-induced enhancement of extinction-from animal models to clinical trials. Psychopharmacology 236(1):183–199
Dias BG (2024) Legacies of salient environmental experiences-insights from chemosensation. Chem Senses 49:bjae002
Dias BG, Ressler KJ (2014) Parental olfactory experience influences behavior and neural structure in subsequent generations. Nat Neurosci 17(1):89–96
Dias BG et al (2015) Epigenetic mechanisms underlying learning and the inheritance of learned behaviors. Trends Neurosci 38(2):96–107
Dickson DA et al (2018) Reduced levels of miRNAs 449 and 34 in sperm of mice and men exposed to early life stress. Transl Psychiatry 8(1):101
Dion A, Munoz PT, Franklin TB (2022) Epigenetic mechanisms impacted by chronic stress across the rodent lifespan. Neurobiol Stress 17:100434
Duffy KA, Bale TL, Epperson CN (2021) Germ cell drivers: transmission of preconception stress across generations. Front Hum Neurosci 15:642762
Dutcher EG et al (2020) Early-life stress and inflammation: a systematic review of a key experimental approach in rodents. Brain Neurosci Adv 4:2398212820978049
Fan Y et al (2014) Early life stress modulates amygdala-prefrontal functional connectivity: implications for oxytocin effects. Hum Brain Mapp 35(10):5328–5339
Fawcett JW, Oohashi T, Pizzorusso T (2019) The roles of perineuronal nets and the perinodal extracellular matrix in neuronal function. Nat Rev Neurosci 20(8):451–465
Fawcett JW et al (2022) The extracellular matrix and perineuronal nets in memory. Mol Psychiatry 27(8):3192–3203
Finsterwald C, Alberini CM (2014) Stress and glucocorticoid receptor-dependent mechanisms in long-term memory: from adaptive responses to psychopathologies. Neurobiol Learn Mem 112: 17–29

Fullana MA et al (2020) Human fear conditioning: from neuroscience to the clinic. Behav Res Ther 124:103528

Ganella DE et al (2018) Extinction of conditioned fear in adolescents and adults: a human fMRI study. Front Hum Neurosci 11:647

Gapp K et al (2020) Alterations in sperm long RNA contribute to the epigenetic inheritance of the effects of postnatal trauma. Mol Psychiatry 25(9):2162–2174

Gee DG, Casey BJ (2015) The impact of developmental timing for stress and recovery. Neurobiol Stress 1:184–194

Gee DG et al (2013) Early developmental emergence of human amygdala-prefrontal connectivity after maternal deprivation. Proc Natl Acad Sci USA 110(39):15638–15643

Gildawie KR, Honeycutt JA, Brenhouse HC (2020) Region-specific effects of maternal separation on perineuronal net and parvalbumin-expressing interneuron formation in male and female rats. Neuroscience 428:23–37

Glover V, O'Connor TG, O'Donnell K (2010) Prenatal stress and the programming of the HPA axis. Neurosci Biobehav Rev 35(1):17–22

Gogolla N et al (2009) Perineuronal nets protect fear memories from erasure. Science 325(5945): 1258–1261

Graignic-Philippe R et al (2014) Effects of prenatal stress on fetal and child development: a critical literature review. Neurosci Biobehav Rev 43:137–162

Green MR, McCormick CM (2013) Effects of stressors in adolescence on learning and memory in rodent models. Horm Behav 64(2):364–379

Griffin A et al (2024) Maternal immune suppression during pregnancy does not prevent abnormal behavior in offspring. Biol Sex Differ 15(1):27

Guadagno A et al (2020) It is all in the right amygdala: increased synaptic plasticity and perineuronal nets in male, but not female, juvenile rat pups after exposure to early-life stress. J Neurosci 40(43):8276–8291

Guijarro JZ et al (2007) Effects of brief and long maternal separations on the HPA axis activity and the performance of rats on context and tone fear conditioning. Behav Brain Res 184(2):101–108

Gumusoglu SB et al (2017) The role of IL-6 in neurodevelopment after prenatal stress. Brain Behav Immun 65:274–283

Gunnar MR, Bowen M (2021) What was learned from studying the effects of early institutional deprivation. Pharmacol Biochem Behav 210:173272

Gunnar M, Quevedo K (2007) The neurobiology of stress and development. Annu Rev Psychol 58: 145–173

Guo H et al (2023) Dendritic spine dynamics in associative memory: a comprehensive review. FASEB J 37(5):e22896

Haapanen MJ et al (2024) Early growth, stress, and socioeconomic factors as predictors of the rate of multimorbidity accumulation across the life course: a longitudinal birth cohort study. Lancet Healthy Longev 5(1):e56–e65

Hamden JE et al (2019) Measurement of 11-dehydrocorticosterone in mice, rats and songbirds: effects of age, sex and stress. Gen Comp Endocrinol 281:173–182

Hanson JL et al (2021) Impact of early life stress on reward circuit function and regulation. Front Psych 12:744690

Hefner K, Holmes A (2007) Ontogeny of fear-, anxiety- and depression-related behavior across adolescence in C57BL/6J mice. Behav Brain Res 176(2):210–215

Heinrichs SC et al (2013) Dendritic structural plasticity in the basolateral amygdala after fear conditioning and its extinction in mice. Behav Brain Res 248:80–84

Henderson J, Gray R, Brocklehurst P (2007) Systematic review of effects of low-moderate prenatal alcohol exposure on pregnancy outcome. BJOG 114(3):243–252

Hodel AS et al (2015) Duration of early adversity and structural brain development in post-institutionalized adolescents. NeuroImage 105:112–119

Hollins SL et al (2018) A rodent model of anxiety: the effect of perinatal immune challenges on gastrointestinal inflammation and integrity. Neuroimmunomodulation 25(3):163–175

Houston SM, Herting MM, Sowell ER (2014) The neurobiology of childhood structural brain development: conception through adulthood. Curr Top Behav Neurosci 16:3–17
Hylin MJ et al (2013) Disruption of the perineuronal net in the hippocampus or medial prefrontal cortex impairs fear conditioning. Learn Mem (Cold Spring Harbor, NY) 20(5):267–273
Jawaid A, Jehle KL, Mansuy IM (2021) Impact of parental exposure on offspring health in humans. Trends Genet 37(4):373–388
Kaati G et al (2007) Transgenerational response to nutrition, early life circumstances and longevity. Eur J Hum Genet 15(7):784–790
Keenan K et al (2018) Extending the developmental origins of disease model: impact of preconception stress exposure on offspring neurodevelopment. Dev Psychobiol 60(7):753–764
Keifer OP et al (2015) Voxel-based morphometry predicts shifts in dendritic spine density and morphology with auditory fear conditioning. Nat Commun 6:7582
Kim JH, Li S, Richardson R (2011) Immunohistochemical analyses of long-term extinction of conditioned fear in adolescent rats. Cereb Cortex 21(3):530–538
King EC et al (2013) Nonlinear developmental trajectory of fear learning and memory. Ann N Y Acad Sci 1304:62–69
Klaric M et al (2008) Psychological problems in children of war veterans with posttraumatic stress disorder in Bosnia and Herzegovina: cross-sectional study. Croat Med J 49(4):491–498
Kolb B, Gibb R (2011) Brain plasticity and behaviour in the developing brain. J Can Acad Child Adolesc Psychiatry 20(4):265–276
Konstantakou P et al (2017) Dysregulation of 11beta-hydroxysteroid dehydrogenases: implications during pregnancy and beyond. J Matern Fetal Neonatal Med 30(3):284–293
Kretschmer M, Fischer V, Gapp K (2023) When dad's stress gets under kid's skin-impacts of stress on germline cargo and embryonic development. Biomol Ther 13(12):1750
Lai CSW, Adler A, Gan W-B (2018) Fear extinction reverses dendritic spine formation induced by fear conditioning in the mouse auditory cortex. Proc Natl Acad Sci USA 115(37):9306–9311
Lavertu-Jolin M et al (2023) Acan downregulation in parvalbumin GABAergic cells reduces spontaneous recovery of fear memories. Mol Psychiatry 28(7):2946–2963
Leen-Feldner EW et al (2013) Offspring psychological and biological correlates of parental posttraumatic stress: review of the literature and research agenda. Clin Psychol Rev 33(8): 1106–1133
Leuner B, Shors TJ (2004) New spines, new memories. Mol Neurobiol 29(2):117–130
Lippard ETC, Nemeroff CB (2023) The devastating clinical consequences of child abuse and neglect: increased disease vulnerability and poor treatment response in mood disorders. Am J Psychiatry 180(8):548–564
Lismer A, Kimmins S (2023) Emerging evidence that the mammalian sperm epigenome serves as a template for embryo development. Nat Commun 14(1):2142
Lupien SJ et al (2009) Effects of stress throughout the lifespan on the brain, behaviour and cognition. Nat Rev Neurosci 10(6):434–445
Maitin-Shepard M et al (2024) Food, nutrition, and fertility: from soil to fork. Am J Clin Nutr 119(2):578–589
Manzano Nieves G et al (2020) Early life adversity decreases pre-adolescent fear expression by accelerating amygdala PV cell development. elife 9:e55263
Manzano-Nieves G et al (2018) Early life stress impairs contextual threat expression in female, but not male, mice. Behav Neurosci 132(4):247–257
Marusak HA et al (2018) Poor between-session recall of extinction learning and hippocampal activation and connectivity in children. Neurobiol Learn Mem 156:86–95
Mauney SA et al (2013) Developmental pattern of perineuronal nets in the human prefrontal cortex and their deficit in schizophrenia. Biol Psychiatry 74(6):427–435
McCallum J, Kim JH, Richardson R (2010) Impaired extinction retention in adolescent rats: effects of D-cycloserine. Neuropsychopharmacology 35(10):2134–2142
McEwen BS, Gianaros PJ (2011) Stress- and allostasis-induced brain plasticity. Annu Rev Med 62: 431–445

McLaughlin KA, Sheridan MA, Nelson CA (2017) Neglect as a violation of species-expectant experience: neurodevelopmental consequences. Biol Psychiatry 82(7):462–471

McPherson NO et al (2014) Paternal obesity, interventions, and mechanistic pathways to impaired health in offspring. Ann Nutr Metab 64(3–4):231–238

Meir Drexler S et al (2019) How stress and glucocorticoids timing-dependently affect extinction and relapse. Neurosci Biobehav Rev 98:145–153

Merz EC et al (2013) Inhibitory control and working memory in post-institutionalized children. J Abnorm Child Psychol 41(6):879–890

Meyer EJ et al (2016) Corticosteroid-binding globulin: a review of basic and clinical advances. Horm Metab Res 48(6):359–371

Mishra PK, Kutty BM, Laxmi TR (2019) The impact of maternal separation and isolation stress during stress hyporesponsive period on fear retention and extinction recall memory from 5-week- to 1-year-old rats. Exp Brain Res 237(1):181–190

Miyata S et al (2018) Structural variation of chondroitin sulfate chains contributes to the molecular heterogeneity of perineuronal nets. Front Integr Neurosci 12:3

Moczulska KE et al (2013) Dynamics of dendritic spines in the mouse auditory cortex during memory formation and memory recall. Proc Natl Acad Sci USA 110(45):18315–18320

Moisiadis VG, Matthews SG (2014) Glucocorticoids and fetal programming part 1: outcomes. Nat Rev Endocrinol 10(7):391–402

Molet J et al (2014) Naturalistic rodent models of chronic early-life stress. Dev Psychobiol 56(8): 1675–1688

Moraes JB et al (2018) In major affective disorders, early life trauma predict increased nitro-oxidative stress, lipid peroxidation and protein oxidation and recurrence of major affective disorders, suicidal behaviors and a lowered quality of life. Metab Brain Dis 33(4):1081–1096

Murthy S, Gould E (2018) Early life stress in rodents: animal models of illness or resilience? Front Behav Neurosci 12:157

Navlakha S, Barth AL, Bar-Joseph Z (2015) Decreasing-rate pruning optimizes the construction of efficient and robust distributed networks. PLoS Comput Biol 11(7):e1004347

O'Connor TG et al (2013) Prenatal cortisol exposure predicts infant cortisol response to acute stress. Dev Psychobiol 55(2):145–155

Omoumi S et al (2023) Corticosterone injection into the infralimbic prefrontal cortex enhances fear memory extinction: involvement of GABA receptors and the extracellular signal-regulated kinase. Physiol Behav 265:114156

Pattwell SS et al (2012) Altered fear learning across development in both mouse and human. Proc Natl Acad Sci USA 109(40):16318–16323

Pechtel P, Pizzagalli DA (2011) Effects of early life stress on cognitive and affective function: an integrated review of human literature. Psychopharmacology 214(1):55–70

Pembrey ME (2010) Male-line transgenerational responses in humans. Hum Fertil (Camb) 13(4): 268–271

Peña CJ et al (2019) Early life stress alters transcriptomic patterning across reward circuitry in male and female mice. Nat Commun 10(1):5098

Perez-Palomar B et al (2023) Maternal immune activation induces cortical Catecholaminergic hypofunction and cognitive impairments in offspring. J Neuroimmune Pharmacol 18(3): 348–365

Pizzorusso T et al (2002) Reactivation of ocular dominance plasticity in the adult visual cortex. Science (New York, NY) 298(5596):1248–1251

Plaza W et al (2020) Pre- and postnatal alcohol exposure delays, in female but not in male rats, the extinction of an auditory fear conditioned memory and increases alcohol consumption. Dev Psychobiol 62(4):519–531

Poli A et al (2023) Selective disruption of perineuronal nets in mice lacking Crtl1 is sufficient to make fear memories susceptible to erasure. Mol Neurobiol 60(7):4105–4119

Pollak SD et al (2010) Neurodevelopmental effects of early deprivation in postinstitutionalized children. Child Dev 81(1):224–236

Power JD, Schlaggar BL (2017) Neural plasticity across the lifespan. Wiley Interdiscip Rev Dev Biol 6(1):e216
Pratchett LC, Yehuda R (2011) Foundations of posttraumatic stress disorder: does early life trauma lead to adult posttraumatic stress disorder? Dev Psychopathol 23(2):477–491
Quagliato LA, de Matos U, Nardi AE (2021) Maternal immune activation generates anxiety in offspring: a translational meta-analysis. Transl Psychiatry 11(1):245
Ramsaran AI et al (2023) A shift in the mechanisms controlling hippocampal engram formation during brain maturation. Science 380(6644):543–551
Rinne GR et al (2023) Stress before conception and during pregnancy and maternal cortisol during pregnancy: a scoping review. Psychoneuroendocrinology 153:106115
Rodgers AB et al (2013) Paternal stress exposure alters sperm microRNA content and reprograms offspring HPA stress axis regulation. J Neurosci 33(21):9003–9012
Rogers SL et al (2018) Normal development of the perineuronal net in humans; in patients with and without epilepsy. Neuroscience 384:350–360
Roozendaal B, McGaugh JL (2011) Memory modulation. Behav Neurosci 125(6):797–824
Sakai J (2020) How synaptic pruning shapes neural wiring during development and, possibly, in disease. Proc Natl Acad Sci 117(28):16096–16099
Salm AK et al (2015) Analysis of extinction acquisition to attenuated tones in prenatally stressed and non-stressed offspring following auditory fear conditioning. Physiol Behav 139:157–166
Sandman CA et al (2012) Exposure to prenatal psychobiological stress exerts programming influences on the mother and her fetus. Neuroendocrinology 95(1):7–21
Sangha S et al (2020) Know safety, no fear. Neurosci Biobehav Rev 108:218–230
Santiago AN et al (2018) Early life trauma increases threat response of peri-weaning rats, reduction of axo-somatic synapses formed by parvalbumin cells and perineuronal net in the basolateral nucleus of amygdala. J Comp Neurol 526(16):2647–2664
Scarborough J et al (2021) A novel murine model to study the impact of maternal depression and antidepressant treatment on biobehavioral functions in the offspring. Mol Psychiatry 26(11): 6756–6772
Scheimann JR et al (2019) Conditional deletion of glucocorticoid receptors in rat brain results in sex-specific deficits in fear and coping behaviors. elife 8:e44672
Schroeder A et al (2018) On the developmental timing of stress: delineating sex-specific effects of stress across development on adult behavior. Brain Sci 8(7):121
Serrano ME et al (2022) Imaging synaptic density: the next holy grail of neuroscience? Front Neurosci 16:796129
Shachar-Dadon A, Schulkin J, Leshem M (2009) Adversity before conception will affect adult progeny in rats. Dev Psychol 45(1):9–16
Sharma U, Rando OJ (2014) Father-son chats: inheriting stress through sperm RNA. Cell Metab 19(6):894–895
Sharma S et al (2013) Maturation of the adolescent brain. In: Neuropsychiatric disease and treatment, p 449
Sheridan MA, McLaughlin KA (2014) Dimensions of early experience and neural development: deprivation and threat. Trends Cogn Sci 18(11):580–585
Shin HS et al (2023) A novel early life stress model affects brain development and behavior in mice. Int J Mol Sci 24(5):4688
Sousa M et al (2023) Academic performance in institutionalized and noninstitutionalized children: the role of cognitive ability and negative lability. Children (Basel) 10(8):1405
Spear LP (2004) Adolescent brain development and animal models. Ann N Y Acad Sci 1021:23–26
Stroud LR et al (2016) Prenatal major depressive disorder, placenta glucocorticoid and serotonergic signaling, and infant cortisol response. Psychosom Med 78(9):979–990
Strzelewicz AR et al (2019) Access to a high resource environment protects against accelerated maturation following early life stress: a translational animal model of high, medium and low security settings. Horm Behav 111:46–59

Suren P et al (2014) Parental obesity and risk of autism spectrum disorder. Pediatrics 133(5):e1128–e1138

Tau GZ, Peterson BS (2010) Normal development of brain circuits. Neuropsychopharmacology 35(1):147–168

Taylor WW et al (2021) Contributions of glucocorticoid receptors in cortical astrocytes to memory recall. Learn Mem 28(4):126–133

Teicher MH et al (2002) Developmental neurobiology of childhood stress and trauma. Psychiatr Clin North Am 25(2):397–426, vii-viii

Tertil M et al (2018) Glucocorticoid receptor signaling in astrocytes is required for aversive memory formation. Transl Psychiatry 8(1):255

Thompson EH et al (2018) Removal of perineuronal nets disrupts recall of a remote fear memory. Proc Natl Acad Sci USA 115(3):607–612

Tierney AL, Nelson CA 3rd (2009) Brain development and the role of experience in the early years. Zero Three 30(2):9–13

Toya S et al (2014) Early-life-stress affects the homeostasis of glutamatergic synapses. Eur J Neurosci 40(11):3627–3634

Tsien RY (2013) Very long-term memories may be stored in the pattern of holes in the perineuronal net. Proc Natl Acad Sci USA 110(30):12456–12461

van't Spijker HM, Kwok JCF (2017) A sweet talk: the molecular systems of perineuronal nets in controlling neuronal communication. Front Integr Neurosci 11:33

Venkataraman A, Dias BG (2023) Expanding the canon: an inclusive neurobiology of thalamic and subthalamic fear circuits. Neuropharmacology 226:109380

Waters AM, Henry J, Neumann DL (2009) Aversive Pavlovian conditioning in childhood anxiety disorders: impaired response inhibition and resistance to extinction. J Abnorm Psychol 118(2):311–321

Wiktorowska L et al (2021) Knockdown of the astrocytic glucocorticoid receptor in the central nucleus of the amygdala diminishes conditioned fear expression and anxiety. Behav Brain Res 402:113095

Xu J et al (2020) The effect of early life stress on memory is mediated by anterior hippocampal network. Neuroscience 451:137–148

Yamada J, Jinno S (2013) Spatio-temporal differences in perineuronal net expression in the mouse hippocampus, with reference to parvalbumin. Neuroscience 253:368–379

Yang YL, Chao PK, Lu KT (2006) Systemic and intra-amygdala administration of glucocorticoid agonist and antagonist modulate extinction of conditioned fear. Neuropsychopharmacology 31(5):912–924

Yang Y et al (2016) Selective synaptic remodeling of amygdalocortical connections associated with fear memory. Nat Neurosci 19(10):1348–1355

Yao H, Wang C, Xia Z (2023) Prenatal alcohol exposure enhanced alcohol preference and susceptibility to PTSD in a sex-dependent manner through the synaptic HCN1 channel. J Affect Disord 324:143–152

Yehuda R, Bierer LM (2009) The relevance of epigenetics to PTSD: implications for the DSM-V. J Trauma Stress 22(5):427–434

Yehuda R, Lehrner A (2018) Intergenerational transmission of trauma effects: putative role of epigenetic mechanisms. World Psychiatry 17(3):243–257

Yehuda R et al (2014) Influences of maternal and paternal PTSD on epigenetic regulation of the glucocorticoid receptor gene in Holocaust survivor offspring. Am J Psychiatry 171(8):872–880

Yehuda R et al (2016) Holocaust exposure induced intergenerational effects on FKBP5 methylation. Biol Psychiatry 80(5):372–380

Yen AM et al (2016) Longer duration and earlier age of onset of paternal betel chewing and smoking increase metabolic syndrome risk in human offspring, independently, in a community-based screening program in Taiwan. Circulation 134(5):392–404

Zaidan H, Leshem M, Gaisler-Salomon I (2013) Prereproductive stress to female rats alters corticotropin releasing factor type 1 expression in ova and behavior and brain corticotropin releasing factor type 1 expression in offspring. Biol Psychiatry 74(9):680–687

Zaidan H, Galiani D, Gaisler-Salomon I (2021) Pre-reproductive stress in adolescent female rats alters oocyte microRNA expression and offspring phenotypes: pharmacological interventions and putative mechanisms. Transl Psychiatry 11(1):113

Zhang HL et al (2024) Inhibition of NKCC1 ameliorates anxiety and autistic behaviors induced by maternal immune activation in mice. Curr Issues Mol Biol 46(3):1851–1864

Zhou A, Ryan J (2023) Biological embedding of early-life adversity and a scoping review of the evidence for intergenerational epigenetic transmission of stress and trauma in humans. Genes 14(8):1639

Insights into the Neurobiology of Behavioral Inhibition from Nonhuman Primate Models

Lillian J. Campos, Carly M. Drzewiecki, and Andrew S. Fox

Contents

Abstract Children with extreme behavioral inhibition (BI) are at a significantly greater risk to develop anxiety disorders later in life. We and others have identified similar early-life temperamental BI in nonhuman primates (NHPs), including rhesus monkeys. NHP models of BI provide a unique opportunity to study the neurobiology of BI in a species that shares biological, developmental, and socioemotional similarities with humans. Rhesus monkey models have identified a distributed brain circuit that includes the central extended amygdala (EAc) as being critical for the genesis of BI. By leveraging multimodal neuroimaging, brain lesions, RNA-sequencing, and viral vector manipulations in rhesus monkeys, these studies have identified specific brain regions, genetic factors, and molecular mechanisms that causally contribute to BI. Here, we discuss these findings from NHPs and how they

L. J. Campos and A. S. Fox (✉)
Department of Psychology, University of California, Davis, CA, USA

California National Primate Research Center, University of California, Davis, CA, USA
e-mail: dfox@ucdavis.edu

C. M. Drzewiecki
California National Primate Research Center, University of California, Davis, CA, USA

© The Author(s), under exclusive license to Springer Nature Switzerland AG 2024

Curr Topics Behav Neurosci (2025) 73: 253–275
https://doi.org/10.1007/7854_2024_561
Published Online: 31 December 2024

fit into a translational framework that can contribute to our understanding of the neural circuits that give rise to the risk to develop anxiety and depressive disorders.

Keywords Amygdala · Anxiety · Anxious temperament · Bed nucleus of stria terminalis · BST · BNST · Ce · Development · DTI · Extended amygdala · Freezing · Heritability · Macaque · MRI · Neuroimaging · PET · Threat · Viral vectors

1 Introduction

Imagine you're in a room and suddenly, a tiger enters, however, it is unaware of your presence. What would you do? If your instinct is to stay quiet and avoid drawing attention to yourself, you're likely increasing your chances of surviving the encounter. Although the scenario of a tiger strolling into a room is unlikely for most of us, we intuitively understand that inhibiting our behavior in this context can increase our chances of survival. Throughout evolution, animals that adaptively respond to their environment are more likely to survive (Fox 2018). Now, let's shift the scenario slightly. Imagine that you're in a room filled with unfamiliar people. Although experiencing some amount of anxiety can be adaptive, extreme levels of anxiety and behavioral inhibition in this context can be maladaptive, leading to psychological distress and rob people of the opportunity to form new relationships (Clauss and Blackford 2012; Fox et al. 2021a, b; Kagan et al. 1984, 1988; Schwartz et al. 2003). In short, although inhibiting behavior can be adaptive in some contexts and even increase the chances of survival, extreme responses in other contexts can be maladaptive.

Some children react with extreme behavioral inhibition (BI), tending to inhibit their behavior when faced with potentially threatening or novel stimuli (Clauss and Blackford 2012; Kagan et al. 1984, 1988). Extreme BI can be maladaptive, and decades of work have shown that children who show extreme levels of BI are at risk to develop anxiety and depressive disorders (Biederman et al. 2001; Clauss et al. 2014; Clauss and Blackford 2012; Fox et al. 2005a, b; Kagan et al. 1988). Childhood BI is stable over development, marked by heightened physiological reactivity, and partially heritable (Clauss et al. 2015; Fox et al. 2021a, b; Kagan et al. 1987; Schwartz et al. 2003). It is estimated that approximately 15–20% of children display BI (Clauss et al. 2015; Kagan et al. 1988). Understanding the neurobiology of BI provides a unique opportunity to guide the development of early-life behavioral, cognitive, and pharmacological interventions to prevent the onset of anxiety and depressive disorders.

Optimal progress toward elucidating the neurobiology of BI requires generating and testing novel hypotheses that are best suited for studies in animal models. In humans, BI is often accompanied by subjective feelings of anxiety and apprehension. Although subjective feelings themselves are only overtly measurable in humans, these feelings emerge from distributed anatomical and molecularly diverse neural circuits that are associated with BI (Etkin et al. 2011; Fox and Shackman 2019; Pine et al. 2021). Because BI can be important for survival, BI can be observed in many different species making it amenable to cross-species study (Cavigelli and McClintock 2003; Clauss et al. 2015; Kalin and Shelton 2003; Qi et al. 2010). In this section, we outline how nonhuman primates (NHPs) can be leveraged as a model organism for understanding human BI, and how we can leverage NHP models to perform both discovery-based and mechanistic studies to elucidate the neurobiology of BI and inform our understanding of the risk to develop stress-related psychopathology.

2 Modeling Behavioral Inhibition in NHPs

The neurobiology of BI in humans includes brain areas that are, in part, evolutionarily conserved and have been elaborated throughout evolution. The large evolutionary gap between rodents and humans often limits the direct clinical applicability of rodent findings (Azkona and Sanchez-Pernaute 2022; Belmonte et al. 2015). However, due to their phylogenetic proximity to humans, NHPs are a crucial animal model in neuroscience, helping bridge the gap between rodent and human studies. NHPs and humans share many biological, cognitive, and social similarities, enhancing the translatability of NHP findings to humans (Chang and Platt 2014; Dunbar and Shultz 2007; Kalin and Shelton 2003; Miller et al. 2016).

In general, NHPs can be classified into several simian species, including monkeys and apes, as well as prosimians such as lemurs (Welker 2017). Monkeys can be further categorized into two groups – Old World (Catarrhini) and New World (Platyrrhini) monkeys. Old World monkeys are native to Africa and Asia and include species such as macaques and baboons. New World monkeys are native to the Americas, and include species like marmosets and titi monkeys. Old World Monkeys diverged most recently from humans approximately 25 million years ago (MYA), whereas New World monkeys diverged approximately 35 MYA. Importantly, both Old and New World primate species diverged from humans much more recently than rodents, which diverged from humans ~75–90 MYA (Fig. 1).

This recent evolutionary divergence between humans and NHPs holds significance for neuroscience, as it is reflected in the composition and function of the molecules, cells, and circuits that give rise to complex behavior in humans. This is most notable in the prefrontal cortex (PFC), as NHPs and humans share highly elaborated PFC regions, which have been implicated in anxiety and emotion (Birn et al. 2014; Fox and Kalin 2014; Öngür and Price 2000; Petrides and Pandya 1999; Pine et al. 2021). However, similarities between NHPs and humans extend far beyond the PFC, as many brain regions which are generally considered to be evolutionarily conserved contain primate-specific cell-types, which may contribute to human behavior (Kamath et al. 2022; Krienen et al. 2020; Schmitz et al. 2022). The shared biology between NHPs and humans underlies similarities across developmental (e.g., Paus et al. 2001), social (e.g., Dunbar and Shultz 2007), affective

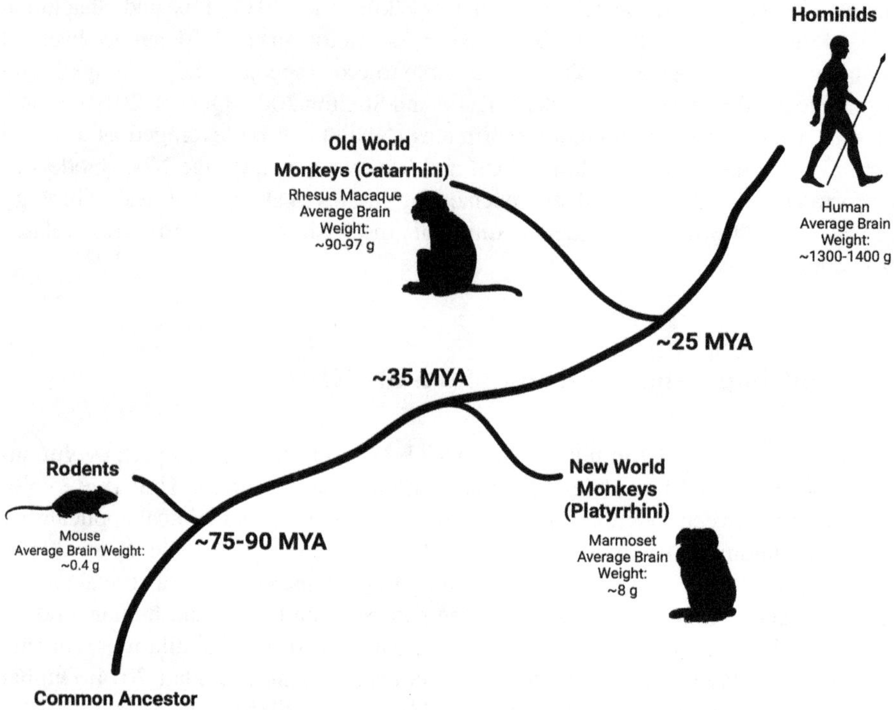

Fig. 1 Phylogenetic tree showing evolutionary divergence of humans from rodents and NHP species. Created with BioRender.com

(e.g., Fox and Kalin 2014), and cognitive (e.g., Rich and Wallis 2016) functions – including BI.

As might be expected from behaviors that increase the likelihood of survival, BI is not a distinctly human phenomenon; heightened reactivity to novel stimuli is widely observed across species (Fox et al. 2008; Qi et al. 2010). Over the years, researchers have developed various models to study the different facets of BI in animal species (Chun and Capitanio 2016; Clauss et al. 2015; Kalin and Shelton 2003). Due to the biological similarities in neurobiology alluded to above, NHPs are particularly well-suited for this purpose, as NHPs' brains reflect the integration between rodent-conserved emotion-related pathways with more recently expanded cortical areas, which can work in concert to increase the chances of survival (Fox 2018). As such, the shared biology between NHPs and humans can be leveraged to provide insight into the complex socio-emotional processes associated with BI. Here, we review rhesus monkey models of BI, which focus on heightened reactivity to novel or potentially threatening stimuli, and operationalized as trait-like tendencies toward increased timidity, freezing, or withdrawal responses. Just as in humans, rhesus monkey BI models require developmentally appropriate methods to quantify variation in BI. Below, we outline two such models of BI, focusing on the peri-adolescent and infant periods.

2.1 Peri-Adolescent Anxious Temperament (AT)

The human intruder paradigm (HIP) was developed in rhesus macaques by Kalin and Shelton to model earlier studies done in children (Kalin and Shelton 1989), but has since been adapted by many groups using other NHPs (Pryluk et al. 2020; Rahman et al. 2021). The HIP consists of various phases, each meant to elicit a specific behavioral response from the animals. During the *alone* phase, the animal is separated from conspecifics and placed alone in the testing environment. In this phase, animals display increased activity and cooing behaviors. These are believed to be affiliative behaviors, meant to recruit help from conspecifics in this unfamiliar environment. During the *stare* phase, an experimenter enters the room and makes direct eye contact with the animal, often inducing aggressive behaviors, including "barking" and teeth-bearing facial expressions.

The "no eye contact" (NEC) phase of the human intruder paradigm is specifically designed to elicit behaviors commonly used to assess BI. In this phase, an experimenter enters the room, presents their profile to the test subject while making no eye-contact, and remains motionless. During NEC, animals often freeze, inhibiting their movement and vocalizations. Like the previous example with a tiger, this behavior can function to reduce the chances of predator detection. Freezing behavior during NEC is often accompanied by decreases in coo vocalizations (Fox et al. 2005a, b, 2008), and increased cortisol levels (Kalin et al. 1998), consistent with broad-scale changes in behavior and physiology seen in childhood BI. This composite phenotype (increased freezing, decreased cooing, and increased cortisol) has been termed anxious temperament (AT); subjects that display BI are higher in AT than their less inhibited counterparts.

Decades of work have shown that the HIP can be used to model BI in young macaques, with extreme responders displaying freezing behaviors and a decrease in vocalizations, similar to childhood BI. Additionally, the context-specific behaviors during NEC begin to emerge around 3 to 4 months of age in rhesus monkeys (Kalin et al. 1991, 1998), a timeframe aligning with the onset of stranger anxiety in children (Brooker et al. 2013). The emergence of context-specific responses during infancy complicates the interpretation of behavior of infants exposed to the NEC context, making this paradigm ideally suited for childhood or adolescent assessment. To this end, Kalin and colleagues have assessed AT in large groups of ~1–3-year-old animals ($n > 500$), which is consistent with the peri-adolescent period (Fox et al. 2015a, b, 2018; Shackman et al. 2017). Together these findings highlight the translational relevance of using NEC to assess behavioral inhibition and suggest that the human intruder paradigm is an appropriate model of early-life behavioral inhibition in NHPs.

2.2 Infant Inhibited Temperament (IT)

A complementary measure for reliably assessing BI in infants was developed at the California National Primate Research Center (CNPRC) as a part of the Biobehavioral Assessment (BBA). The BBA is a research program created in order to characterize animals' psychological and physiological systems, referred to as their "biobehavioral organization" (Capitanio 2021). This includes characteristics like an animal's coping abilities, stress reactivity, personality, and temperament. Briefly, the BBA consists of a standardized series of assessments conducted over a 25-h period in 3–4-month-old rhesus monkeys (Capitanio 2021). The animals' young age enables researchers to evaluate subjects with minimal social exposure, facilitating the assessment of innate, temperamental characteristics unaffected by experiences (Capitanio 2021). To date, more than 4,000 animals have undergone BBA testing.

For most of the animals, enrollment in the BBA involves experiencing changes to their daily routines for the first time. This offers a unique opportunity to identify an inhibited phenotype in the BBA program by seeing how each animal responds and adapts to this change across the 25-h period (Capitanio 2021). To this end, the BBA conducts behavioral observations on animals at the beginning (Day 1) and at the end (Day 2) of the 25-h testing period (Capitanio 2018). Exploratory and confirmation analyses of behavior during the testing period revealed two latent traits, *Activity* and *Emotionality*. *Activity* is measured as time the animal spent locomoting, time eating or drinking water, and time not hanging from the top or side of their cage, or was observed crouching in the cage. *Emotionality* includes the animal's rates of vocalizations (cooing; barking), as well as any instances of lip-smacking, displaying threats, or scratching. The BBA's measure of *inhibited temperament* (IT) is the inverse-average of these two latent variables (i.e., *Activity* and *Emotionality*) across both days. Subjects exhibiting below-average scores on these dimensions on both Day 1 and Day 2 are categorized as behaviorally inhibited, indicating elevated levels of IT. Specifically, animals characterized as having an *inhibited temperament* show lower levels of *Activity* and *Emotionality* on both days of BBA testing. Presumably, this is driven by an inability to adapt to their new surroundings (Capitanio 2021). Importantly, the BBA finds that ~18% of the total number of animals that have undergone BBA testing fit this measure of infant *inhibited temperament*, a number similar to the 15% of children reported by Kagan (1989) to be BI (Capitanio 2018). This developmentally appropriate measure of infant IT provides an opportunity to understand the genesis of BI that can complement peri-adolescent IT measurements and strengthen our inferences about the underlying BI phenotype.

2.3 NHP BI Is Stable and Heritable

Though peri-adolescent AT and infant IT are assessed using different behaviors and methodologies, both consistently serve as reliable experimental measures of behavioral inhibition in NHPs. As outlined below, both research programs have found that high-BI animals tend to display inhibition and increased physiological reactivity across potentially threatening or novel contexts across development, as compared to their low-BI conspecifics. Moreover, similar to human studies (Clauss et al. 2015), individual differences in BI have also been shown to be heritable, suggesting a biological basis for BI.

Like human BI, both peri-adolescent AT and infant IT are stable over time and context. Repeated assessment of peri-adolescent AT has revealed that it is stable over time (Fox et al. 2008, 2019; Shackman et al. 2017). Since BI can often manifest as a latency for rewards, or withdrawals, we hypothesized that high-BI animals would forgo a reward in the presence of a potentially threatening stimulus. Indeed, we found that years after initial assessment, adult animals with high levels of IT during infancy were more likely to refuse treats from a potentially threatening human intruder (Fox et al. 2021a, b). This data suggests that infant IT is a trait-like measure that is stable across development. Similarly, peri-adolescent high-BI animals were less likely to reach for a treat in the presence of a live snake compared to their less inhibited counterparts (Shackman et al. 2017). These data demonstrate that peri-adolescent AT and infant IT are stable over time and context, indicative of trait-like inhibition and can be used to understand the early phenotypic risk to develop anxiety.

Both peri-adolescent AT and infant IT are significantly heritable, suggesting a genetic component for the development of BI. The first hint of the heritability of BI comes from studies in Wisconsin where infants and their mothers were shown to have similar behavioral and hormonal responses in response to exposure to the NEC context (Kalin et al. 1998). Both BI measures were collected in large groups of animals at National Primate Research Centers (NPRCs), which are breeding colonies for rhesus monkeys. As such, assessed animals are part of large multi-generational family pedigrees, which provide information about the estimated percentage of genetic similarity between animals. This enables well-powered heritability analyses of phenotypic measures, by determining how phenotypic measures, such as BI, fall through the family tree. Results have demonstrated that both peri-adolescent AT and infant IT are significantly heritable (Fox et al. 2015a, b, 2021a, b). Importantly, the findings from these studies in primates reveal heritability estimates for AT and IT that are similar to heritability estimates for anxiety disorders in humans (Clauss et al. 2015; Kagan et al. 1988). Overall, this data suggests that both peri-adolescent AT and infant IT are heritable endophenotypes that remain stable across development. As potentially continuous measures, this suggests that elucidating the biological underpinnings of BI is beneficial in identifying early-life interventions and better pharmacological interventions.

Finally, the consistent prediction of peri-adolescent AT and infant IT across various settings underscores their reliability as robust indicators of BI and BI-related traits. This consistent association between AT, IT, and other anxiety-related behaviors suggests that behaviors observed in tasks beyond the laboratory setting can effectively approximate composite BI measures.

3 The Neurobiology of BI

Early theories about the biological processes of BI proposed that the amygdala would be important in the emergence of BI because of the extensive body of work implicating it in the expression of fear (Blackford and Zald 2018; Kagan et al. 1988; Kagan and Snidman 1991). In support of this, the first neuroimaging study of BI found heightened amygdala reactivity in adult humans previously categorized as inhibited (Schwartz et al. 2003). In the decades since, converging evidence has not only supported the importance of the amygdala in BI, but has also revealed evidence of a distributed neural network involved in the expression of BI, trait-like anxiety, and anxiety disorders (Avery et al. 2016; Birn et al. 2014; Etkin et al. 2011; Kenwood et al. 2021; Oler et al. 2012).

A central element within the distributed BI-circuit is the central extended amygdala (EAc), strategically positioned to integrate environmental and interoceptive signals, facilitating adaptive behavioral responses to threatening stimuli (Holley and Fox 2022). In brief, the EAc is a functional unit composed of subcortical nuclei along the borders of the amygdala and the ventral striatum, including the central nucleus of the amygdala (Ce) and the bed nucleus of the stria terminalis (BST), as well as portions of the posterior shells of the nucleus accumbens and clusters of cells along the amygdalofugal pathway that connects these regions (often termed sublenticular extended amygdala, or SLEA) (Alheid 2003; Avery et al. 2016; Shackman and Fox 2016). The EAc receives multisensory information from different brain regions to coordinate behavioral responses via downstream projections to the periaqueductal gray (PAG), which mediate EAc-induced freezing behavior in mice (Dong and Swanson 2004; Tovote et al. 2016). In this section, we review the neural substrates identified by IT and AT in the generation and maintenance of BI.

3.1 *Neural Correlates of BI in NHPs*

To understand the neural correlates of BI in peri-adolescent monkeys, we have used positron emission tomography (PET) and magnetic resonance imaging (MRI). When used in combination with the NEC paradigm described above, [18F] fluorodeoxyglucose PET (FDG-PET) can be used to quantify individual differences in brain metabolism during exposure to the NEC context (i.e., a potential predator). Subjects are injected with FDG prior to being exposed to the NEC context where they display a range of defensive behaviors. Metabolically active cells take up FDG, which can be quantified in the post-NEC FDG-PET scan to assess circuits involved in BI-relevant behaviors.

NEC FDG-PET studies with peri-adolescent animals have rigorously identified AT-related metabolism during exposure to the NEC context across a network of regions including Ce, BST, anterior hippocampus, brainstem, and various PFC regions (Fox et al. 2015a, b; Oler et al. 2010) (Fig. 2). These findings have been

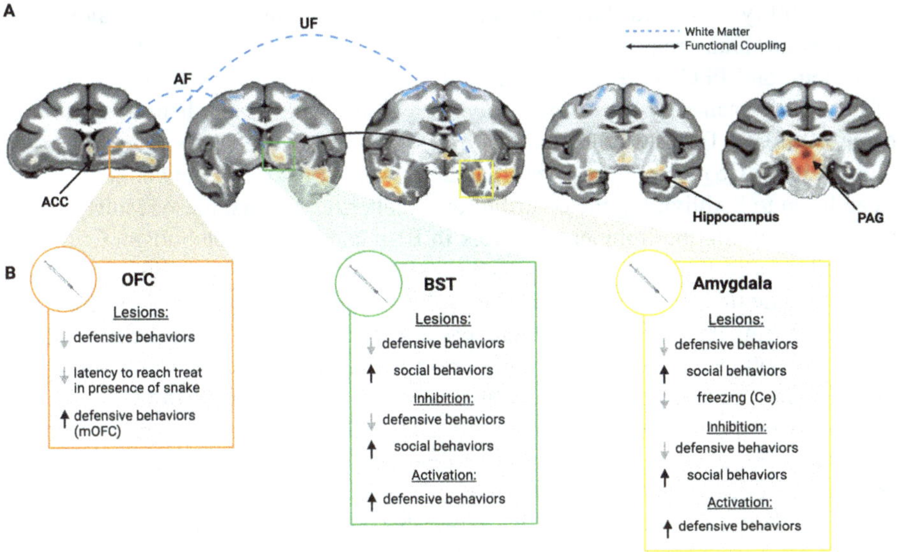

Fig. 2 Regions implicated to be associated with NHP BI through neuroimaging (**a**) and mechanistic studies including lesion and transient approaches (**b**). The direction of the effect is communicated through arrows. Created with BioRender.com

replicated across multiple cohorts and studies, ultimately encompassing a sample of over 600 animals (Fox et al. 2008, 2015a; Holley et al. 2024; Oler et al. 2010). These findings underscore a key insight from rhesus monkey FDG-PET studies – there is no singular brain region responsible for the inhibited phenotype; rather, a network of brain regions act together to induce behavioral inhibition (Fox et al. 2015a). In fact, metabolism in individual brain regions accounts for a small proportion of the variance in AT on their own (Fox et al. 2015a). In addition, FDG-PET studies have been used to assess brain metabolism in the absence of direct threats, enabling exploration of trait-like aspects of BI-related metabolism to determine whether BI is due to alterations in the brain's reactivity to the potential threat or baseline differences in brain metabolism. Several studies have shown that when animals are tested during the alone phase of the HIP (i.e., without the direct threat of the human intruder present), increased freezing and decreased coo vocalizations are associated with increased amygdala metabolism (Fox et al. 2005a, b, 2008, 2015a; Kalin et al. 2005; Oler et al. 2010). Moreover, peri-adolescent AT assessed during exposure to the NEC context is correlated with increased amygdala metabolism in non-stressful environments, as animals were allowed to freely go about life in their home-cage (Fox et al. 2008). Additionally, rhesus monkeys with high BI show sustained and increased BST metabolism and freezing during a 30-min recovery period following exposure to the human intruder in the NEC paradigm, when the threat is no longer present (Shackman et al. 2017). Together, these data suggest that BI is due to both baseline differences in brain regional metabolism and increased glucose metabolism in BI-relevant regions across a variety of threatening contexts.

Heritability analyses have revealed brain metabolism in the majority of AT-relevant brain regions to be significantly heritable, including the BST, Ce, brainstem, and PFC subregions (Fox et al. 2015a, b; Oler et al. 2010). The degree of heritability can vary significantly across regions, with metabolism in the hippocampus and BST showing heritability rates of 53% and 57% respectively, while heritability in Ce metabolism is less than 26% (Fox et al. 2015a, b; Oler et al. 2010). In addition to highlighting the contribution of inherited factors, these results emphasize a key role for environmental factors in BI-related brain metabolism. Critically, heritability of metabolism in AT-related regions does not necessarily implicate these regions in the intergenerational transfer of AT; different genes can both be heritable without sharing an overlapping genetic basis. For example, traits like height and eye color are both heritable, though they are not dependent on the same genes. To identify brain regions where metabolism mediates the intergenerational transfer of AT, we assessed the extent to which brain metabolism and AT fall together through the family tree using genetic correlation analyses. These analyses revealed significant genetic correlations with AT in a network of brain regions that include the OFC, BST, and PAG, suggesting that the genes that alter metabolism in these regions mediate the heritable components of AT (Fox et al. 2015a, b). It is worth noting that the identification of specific regions where metabolism is co-inherited with AT does not necessarily imply that these genes are influencing the function of those cells specifically, rather, it could be that genes influencing activity in one region are responsible for metabolic changes in another.

Functional MRI (fMRI) work can help shed light on the role that interactions between BI-relevant brain regions may play in AT. fMRI BOLD signals measured by "resting state" fMRI under light anesthesia can be used to identify significant correlations in fluctuations in activity across multiple brain regions, allowing researchers to infer connectivity between regions. Within the EAc, resting state connectivity analyses have revealed that the Ce and BST form a functional network with each other; BOLD signal in the Ce is significantly correlated with BOLD signal in the BST (Oler et al. 2012). This strong temporal coupling seems to influence the BI phenotype, as functional connectivity between the Ce and BST is associated with higher AT (Fox et al. 2018). Importantly, the Ce-BST functional connectivity is heritable and genetically correlated with AT, suggesting a shared genetic substrate (Fox et al. 2018). These data reinforce the importance of understanding the contribution of the EAc to BI.

Expanding on the network of BI-relevant brain regions identified using FDG-PET methods, fMRI research has also corroborated the role of multiple brain regions in influencing BI, primarily through their connectivity with the EAc. One region of interest is the expanded NHP cortex, where direct and indirect projections with the limbic system play important roles in complex aspects of threat processing. For example, the Ce shows intrinsic functional connectivity with AT-related subregions in the PFC, and increased functional connectivity between the dorsal lateral PFC (dlPFC) and Ce predicts lower levels of AT in young rhesus monkeys (Birn et al. 2014). The mechanism for this connectivity remains unknown and likely involves indirect projections, as there are no robust direct projections between Ce and dlPFC.

These data further support a distributed circuit engaging direct and indirect projections to influence AT and highlight a role for corticolimbic coupling in BI.

It is also possible that individual differences in the physical connections between BI-related regions can influence the BI phenotype. Recent studies in humans have suggested that alterations in the microstructure of the Uncinate Fasciculus (UF), a white matter tract that connects areas of the PFC to the amygdala, may be important in the development of anxiety disorders (Adluru et al. 2017; Baur et al. 2013; Phan et al. 2009; Tromp et al. 2012, 2019b). To further understand the contribution of fronto-limbic connections on BI, we used diffusion tensor imaging (DTI) to quantify changes in UF microstructure. In a study of over 500 young rhesus macaques, we found that changes in the microstructure of the UF were correlated with AT in male, but not female monkeys (Tromp et al. 2019a). This result complements human studies that found altered UF microstructure in boys with anxiety disorders (Aggarwal et al. 2022; Tromp et al. 2019a, b). Given the well-documented sex difference in the prevalence of anxiety disorders, identifying sex-specific mechanisms that underlie BI-related fronto-limbic circuitry is important. Interestingly, in a small preliminary sample, we found that changes in the microstructure of the Accumbofrontal Fasciculus (AF), a white matter tract that connects the PFC to areas of the limbic system, was significantly correlated with IT in a cohort of all female rhesus monkeys (unpublished data). Together, these data suggest the possibility of sexual dimorphism in the network dynamics of the BI phenotype.

Finally, results from the FDG-PET, fMRI, and DTI studies discussed above can complement each other, providing additional insights into how functional connectivity between two brain regions influences downstream brain metabolism. When leveraging these methods together, results show that Ce–BST functional connectivity, while not associated with Ce or BST metabolism, is associated with increased metabolism in shared downstream targets. Specifically, Ce-BST functional connectivity is associated with increased FDG metabolism in AT-related regions including the hypothalamus, SLEA, and PAG (A. S. Fox et al. 2018). These data further highlight the EAc's role in driving BI through a broader network of AT-relevant regions. Additionally, increased dlPFC-CeA connectivity, which is associated with AT, also predicts Ce metabolism (Birn et al. 2014). Together, these data underscore the importance of multimodal imaging techniques to further parse the influences of BI-relevant brain regions on the broader distributed network of BI-related regions.

In conclusion, collective evidence from a diverse array of imaging modalities underscores the intricate network dynamics underlying the expression of BI in peri-adolescent animals. Importantly, these studies consistently reveal that BI is not mediated by any singular brain region but rather by a network of interconnected regions, highlighting the distributed nature of the neural correlates of BI. Similarly, results from heritability analyses demonstrate the complex interplay between genetic predisposition and the environment, further highlighting the complexity of BI. Thus, integrating findings from multimodal imaging techniques will be necessary to provide a comprehensive understanding of how BI-relevant brain regions interact to drive behaviors.

3.2 Identification of BI-Related Molecules in NHPs

Having identified specific brain regions, researchers have begun to investigate the molecules that might give rise to BI using DNA and RNA sequencing. DNA sequencing is a powerful tool in NHPs, because it can be combined with the large pedigrees at National Primate Research Centers (NPRCs) to perform well-powered genome-wide association studies (GWAS) that leverage pedigree information, much like a large-scale family study (Glahn et al. 2019). Using this approach, we identified a single nucleotide polymorphism (SNP) in a primate-specific exon of the CRHR1 gene to be associated with individual differences in peri-adolescent AT (Rogers et al. 2013). Moreover, by sequencing the whole genome, we further identified a SNP near the CTNNA2 gene which was associated with infant IT (A. S. Fox et al. 2021b). These studies each identify a specific molecular process that is sufficient to give rise to BI, and are consistent with human research focusing on BI-related phenotypes and psychopathology (Gonda et al. 2021; Lesch et al. 2008; Smith et al. 2016). For example, genetic association studies of neuroticism and anxiety-related disorders in humans support our data implicating CRHR1 (Levey et al. 2020; Nagel et al. 2018; Stein et al. 2021). These data further support the NHP as a model organism for discovery, which can be further leveraged by the robust genetic variation in NHPs. Nevertheless, it is important to note that these studies are merely scratching the surface of the specific molecules that might give rise to the genesis of BI and more work needs to be done.

In addition to understanding the structural genetic changes that give rise to BI, another line of work has focused on RNA expression, searching for transcripts associated with individual differences in BI. This work has identified numerous molecules which are associated with variation in BI, including NTRK3, which was replicated in two independent samples of >20 animals (A. S. Fox et al. 2012, 2019). NTRK3 is a growth factor receptor, similar to the NTRK2 receptor, which binds BDNF, a gene that has been repeatedly implicated by rodent models of anxiety and depression (Amador-Arjona et al. 2010; Duman 2017; Farhang et al. 2014). In another gene expression study focusing on CeL neurons, we identified AT-related expression of additional genes including PRKCD, which was positively correlated with AT (Kovner et al. 2020). PRKCD-expressing neurons are particularly interesting because of the role that they play in rodent freezing behavior (Beyeler and Dabrowska 2020; Fadok et al. 2018; Haubensak et al. 2010). Interestingly, we found that PRKCD-expressing neurons represented the majority of BST-projecting CeL cells, highlighting a potential role in mediating Ce-BST functional connectivity outlined above. Similar studies have been undertaken in posterior OFC (consistent with the location of the AT-related metabolism in OFC), where we identified layer-specific AT-related expression in genes that included the gene that encodes the glucocorticoid receptor NR3C1. Each of these discovery-based studies begins to elucidate the molecular basis of BI, and the stage for mechanistic interventions that specifically target particular molecules to assess their causal role in altering BI.

3.3 Targeted Manipulations of BI-Related Regions

NHP lesion studies have begun to uncover the causal contributions of BI-related regions to threat processing and social behavior. For instance, lesions encompassing the whole amygdala are sufficient to decrease an animal's latency to retrieve a food-reward in the presence of a snake – the opposite of what we see in animals with high levels of peri-adolescent AT (Chudasama et al. 2009; Izquierdo et al. 2005; Kalin et al. 2001, 2004). Amygdala lesions also decrease freezing and other defensive responses, and are sufficient to decrease social reticence (Amaral 2002, though, c.f. Kalin et al. 2001). Moreover, Ce-specific lesions, which spare the majority of the amygdala including fibers of passage, are sufficient to decrease freezing and increase the frequency of coo vocalizations (Kalin et al. 2004). Again, these results are consistent with the neuroimaging data, and result in changes that are the opposite of what has been observed in high-BI animals. Critically, although lesions of the amygdala reduce BI, they do not abolish it. This is consistent with multiple brain circuits each contributing to an underlying inhibited temperament.

Another region that has been shown to causally contribute to BI is the OFC. Targeted lesions to the OFC have been shown to mediate aspects of threat processing. Importantly, this large and heterogeneous area of the cortex is structurally connected to the amygdala via the UF (Folloni et al. 2019; Price 2007; Wallis 2012). Similar to lesions of the amygdala, lesioning the OFC decreases freezing behaviors (Fox et al. 2010; Izquierdo et al. 2005; Kalin et al. 2007; Machado and Bachevalier 2008; Murray and Izquierdo 2007). Additionally, rhesus monkeys with OFC lesions or unilateral OFC and amygdala lesions show shorter latencies to reach for a preferred treat in the presence of a snake-like object (Kalin et al. 2007; Murray and Izquierdo 2007). However, monkeys with medial OFC lesions were also shown to have increased defensive behaviors to both threatening and non-threatening stimuli (Pujara et al. 2019). More recently, Kenwood et al. (2023) found that so-called "strip" lesions to the OFC, which spare most OFC gray-matter but damage fibers in the UF and AF, are sufficient to decrease freezing and BST metabolism during "no eye contact" (NEC). Overall, these studies demonstrate a causal role for OFC in threat-processing, and suggest that the OFC could contribute to the maladaptive regulation of anxiety and BI via projections to the BST and amygdala.

While lesion studies have supported the role of the amygdala and OFC in threat processing, results have been mixed due to inadvertent damage to nearby neuronal projection fibers or fibers passing through these regions, rather than the deficits being dependent on amygdala and OFC (Rudebeck et al. 2013). In support of this idea, excitotoxic lesions of the OFC, which spare fibers of passage, did not show changes in behavior during snake fear (Rudebeck et al. 2013). However, a follow-up study using excitotoxic lesions to the medial OFC (Walker's areas 14) saw increases in defensive behaviors (Pujara et al. 2019). Notably, this is in direct contrast to previous lesions to the OFC that found decreases in defensive responses (Fox et al. 2010; Izquierdo et al. 2005; Kalin et al. 2007; Machado and Bachevalier 2008). These findings underscore the importance of understanding precisely what regions and fibers are damaged when interpreting the results of lesion studies. Since both the OFC and amygdala are heterogeneous, the exact location of lesions may perturb different circuit-dynamics within these regions, as projections to and from these regions tend to be topographically organized (Price 2007; Saleem et al. 2008; Swanson and Petrovich 1998). Further research using more precise targeting strategies will provide additional insight into the role of these regions in anxiety-like behaviors and BI.

Another limitation of lesion experiments is their irreversibility, which allows for compensatory and indirect mechanisms to influence the observed effects. Efforts are ongoing to understand the specific circuit dynamics of these BI-related regions using more targeted and transient techniques. Consistent with the findings of lesion studies, transient activation and inactivation studies have demonstrated a causal role for amygdala and frontal regions in BI-related behaviors. For example, inhibition of the rhesus monkey amygdala using chemogenetics was also sufficient to decrease anxiety-like freezing during exposure to the NEC context (Roseboom et al. 2021). More recently, chemogenetic activation of the amygdala increased freezing, even in inappropriate contexts such as when a human intruder is staring at the animal (Mueller et al. 2023). Relatedly, transient inactivation of the BST (Jacobs et al. 2023) and Ce (Wellman et al. 2016) was sufficient to increase affiliative behaviors (e.g., grooming) toward a familiar conspecific in macaque monkeys. In marmosets, pharmacological overactivation of the ACC (Area 25) heightened anxiety-like behaviors in response to an unknown human (Roberts and Clarke 2019). In marmoset OFC, pharmacological inactivation increased anxiety-like behavior during a human intruder test (Alexander et al. 2020; Stawicka et al. 2022), mirroring the results of lesion studies in rhesus monkeys (Pujara et al. 2019). Together, these studies show that temporally specific manipulations of amygdala and OFC are sufficient to alter BI, opening the door to a more precise understanding of how these regions contribute to BI (Fig. 2).

Using similar techniques, researchers have begun to test the causality of some of the specific molecules (identified above) within these regions to alter BI. For example, using a viral vector to overexpress CRH, the ligand that binds to CRHR1, we were able to show that increasing CRH in the Ce-containing dorsal amygdala region is sufficient to increase peri-adolescent AT (Kalin et al. 2016). In contrast, overexpressing NTF3, the ligand that binds to NTRK3, was sufficient to decrease peri-adolescent AT (Fox et al. 2019). These studies demonstrate that altering a specific molecule in the dorsal amygdala is sufficient to produce large and meaningful alterations in BI. Critically, each of these studies leveraged neuroimaging to understand the impact of specific molecular interventions on the distributed AT circuit. Specifically, molecular manipulations in dorsal amygdala were sufficient to impact metabolism across the AT-network in regions including hippocampus, BST, and OFC (Fox et al. 2019). These results demonstrate that localized manipulation of a single molecule within a single region is sufficient to alter the function of a distributed neural circuit, underscoring the interconnections across the distributed network of brain regions that contribute to BI. Although this work is in its infancy and will require substantive development of molecular tools for optimal

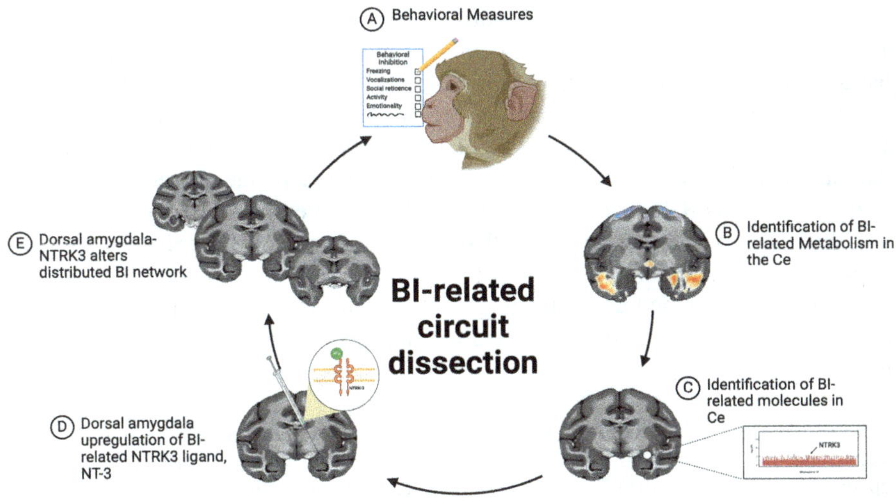

Fig. 3 Overview of Fox et al. (2019) highlights how NHP models can be used to dissect BI-related circuits using different levels of analysis. Specifically, NHPs can help (**a**) identify BI-related behaviors that are (**b**) correlated with specific circuits, (**c**) identify BI-related molecules in those circuits (**d**), which can then be targeted for viral-mediated manipulations (**e**), to examining their impact on distributed BI-related circuits (**e**), and behavior (**a**). Created with BioRender.com

manipulations of NHP brain circuits (e.g., Campos et al. 2023), these results highlight the promise of defining, monitoring, and manipulating specific regional molecular systems in NHPs.

4 Conclusion

Here we have introduced two rhesus monkey models of BI focused on distinct developmental periods, and outlined how these models can be leveraged to better understand the neurobiology of BI. As in humans, studies in rhesus monkeys have implicated a distributed brain network that includes the EAc in early-life BI. In addition to solidifying NHP models of BI, these studies complement human studies by implicating specific neural substrates in the intergenerational transmission of BI, suggesting that genes that affect metabolism in the BST and Ce-BST connectivity contribute to BI. Identifying the regions that contribute to BI has enabled researchers to begin to uncover the molecular underpinnings of BI using DNA and RNA-sequencing, such as *CRHR1* and *NTRK3*. Furthermore, the NHP model enables manipulation of brain regions and specific molecules to elucidate their causal contribution to BI. This work suggests that OFC may influence BI via projections to the BST, that the Ce causally contributes to BI, and demonstrated that manipulations of specific molecular pathways in the Ce, including CRH and

NTF-3, are sufficient to alter BI. Importantly, this work shows that manipulating a single molecule in a specific brain region is sufficient to alter function throughout the distributed network of BI-related brain regions. Despite this progress, there remain several challenges for NHP research to bridge the gap between rodent and human studies, including tools for viral-mediated transfection and cell-type-specific manipulations. This work is ongoing, and promises to aid in the dissection of these BI-related circuits (Chen et al. 2022, 2023; Chuapoco et al. 2023). Together, these data shed light on the neurobiology that underlies the risk to develop anxiety and depressive disorders, and underscore the utility of NHP models for understanding BI. Ultimately, a refined understanding of BI's neurobiology promises to guide the development of new strategies for treating and preventing debilitating psychopathology (Fig. 3).

Acknowledgments We thank Drs. Kalin and Capitanio for their efforts in creating NHP BI models, and the conversations that led to the thoughts outlined in this chapter. This work was supported by NIH grants R01MH121735 (PI: Fox), R21MH129851 (PI:Fox), and P51OD011107 (PI: Atkinson).

References

Adluru N, Luo Z, Van Hulle CA, Schoen AJ, Davidson RJ, Alexander AL, Goldsmith HH (2017) Anxiety-related experience-dependent white matter structural differences in adolescence: a monozygotic twin difference approach. Sci Rep 7. https://doi.org/10.1038/s41598-017-08107-6

Aggarwal N, Williams LE, Tromp DPM, Pine DS, Kalin NH (2022) A dynamic relation between whole-brain white matter microstructural integrity and anxiety symptoms in preadolescent females with pathological anxiety. Transl Psychiatry 12(1):57. https://doi.org/10.1038/s41398-022-01827-y

Alexander L, Wood CM, Gaskin PLR, Sawiak SJ, Fryer TD, Hong YT, McIver L, Clarke HF, Roberts AC (2020) Over-activation of primate subgenual cingulate cortex enhances the cardiovascular, behavioral and neural responses to threat. Nat Commun 11(1):5386. https://doi.org/10.1038/s41467-020-19167-0

Alheid GF (2003) Extended amygdala and basal forebrain. Ann N Y Acad Sci 985:185–205

Amador-Arjona A, Delgado-Morales R, Belda X, Gagliano H, Gallego X, Keck ME, Armario A, Dierssen M (2010) Susceptibility to stress in transgenic mice overexpressing TrkC, a model of panic disorder. J Psychiatr Res 44(3):157–167. https://doi.org/10.1016/j.jpsychires.2009.07.008

Amaral DG (2002) The primate amygdala and the neurobiology of social behavior: implications for understanding social anxiety. Biol Psychiatry 51:11–17

Avery SN, Clauss JA, Blackford JU (2016) The human BNST: functional role in anxiety and addiction. Neuropsychopharmacology 41(1):126–141. https://doi.org/10.1038/npp.2015.185

Azkona G, Sanchez-Pernaute R (2022) Mice in translational neuroscience: what R we doing? Prog Neurobiol 217:102330. https://doi.org/10.1016/j.pneurobio.2022.102330

Baur V, Brühl AB, Herwig U, Eberle T, Rufer M, Delsignore A, Jäncke L, Hänggi J (2013) Evidence of frontotemporal structural hypoconnectivity in social anxiety disorder: a quantitative fiber tractography study. Hum Brain Mapp 34(2):437–446. https://doi.org/10.1002/hbm.21447

Belmonte JCI, Callaway EM, Caddick SJ, Churchland P, Feng G, Homanics GE, Lee K-F, Leopold DA, Miller CT, Mitchell JF, Mitalipov S, Moutri AR, Movshon JA, Okano H, Reynolds JH, Ringach DL, Sejnowski TJ, Silva AC, Strick PL et al (2015) Brains, genes, and primates. Neuron 86(3):617–631. https://doi.org/10.1016/j.neuron.2015.03.021

Beyeler A, Dabrowska J (2020) Neuronal diversity of the amygdala and the bed nucleus of the stria terminalis. In: Handbook of behavioral neuroscience, vol 26. Elsevier, pp 63–100. https://doi.org/10.1016/B978-0-12-815134-1.00003-9

Biederman J, Hirshfeld-Becker DR, Rosenbaum JF, Hérot C, Friedman D, Snidman N, Kagan J, Faraone SV (2001) Further evidence of association between behavioral inhibition and social anxiety in children. Am J Psychiatry 158(10):1673–1679. https://doi.org/10.1176/appi.ajp.158.10.1673

Birn RM, Shackman AJ, Oler JA, Williams LE, McFarlin DR, Rogers GM, Shelton SE, Alexander AL, Pine DS, Slattery MJ, Davidson RJ, Fox AS, Kalin NH (2014) Evolutionarily conserved prefrontal-amygdalar dysfunction in early-life anxiety. Mol Psychiatry 19(8):915–922. https://doi.org/10.1038/mp.2014.46

Blackford JU, Zald DH (2018) Inhibited temperament and intrinsic versus extrinsic influences on fear circuits. In: The nature of emotion: fundamental questions, 2nd edn. Oxford University Press, pp 49–50

Brooker RJ, Buss KA, Lemery-Chalfant K, Aksan N, Davidson RJ, Goldsmith HH (2013) The development of stranger fear in infancy and toddlerhood: normative development, individual differences, antecedents, and outcomes. Dev Sci 16(6):864–878. https://doi.org/10.1111/desc.12058

Campos LJ, Arokiaraj CM, Chuapoco MR, Chen X, Goeden N, Gradinaru V, Fox AS (2023) Advances in AAV technology for delivering genetically encoded cargo to the nonhuman primate nervous system. Curr Res Neurobiol 4:100086. https://doi.org/10.1016/j.crneur.2023.100086

Capitanio JP (2018) Behavioral inhibition in nonhuman primates: the elephant in the room. In: Perez-Edgar K, Fox NA (eds) Behavioral inhibition: integrating theory, research, and clinical perspectives. Springer Nature Switzerland AG, pp 17–33

Capitanio JP (2021) Knowledge of biobehavioral organization can facilitate better science: a review of the BioBehavioral assessment program at the California National Primate Research Center. Animals 11(8):2445. https://doi.org/10.3390/ani11082445

Cavigelli SA, McClintock MK (2003) Fear of novelty in infant rats predicts adult corticosterone dynamics and an early death. Proc Natl Acad Sci 100(26):16131–16136. https://doi.org/10.1073/pnas.2535721100

Chang SWC, Platt ML (2014) Oxytocin and social cognition in rhesus macaques: implications for understanding and treating human psychopathology. Brain Res 1580:57–68. https://doi.org/10.1016/j.brainres.2013.11.006

Chen X, Ravindra Kumar S, Adams CD, Yang D, Wang T, Wolfe DA, Arokiaraj CM, Ngo V, Campos LJ, Griffiths JA, Ichiki T, Mazmanian SK, Osborne PB, Keast JR, Miller CT, Fox AS, Chiu IM, Gradinaru V (2022) Engineered AAVs for non-invasive gene delivery to rodent and non-human primate nervous systems. Neuron 110(14):2242–2257.e6. https://doi.org/10.1016/j.neuron.2022.05.003

Chen X, Wolfe DA, Bindu DS, Zhang M, Taskin N, Goertsen D, Shay TF, Sullivan EE, Huang S-F, Ravindra Kumar S, Arokiaraj CM, Plattner VM, Campos LJ, Mich JK, Monet D, Ngo V, Ding X, Omstead V, Weed N et al (2023) Functional gene delivery to and across brain vasculature of systemic AAVs with endothelial-specific tropism in rodents and broad tropism in primates. Nat Commun 14(1):3345. https://doi.org/10.1038/s41467-023-38582-7

Chuapoco MR, Flytzanis NC, Goeden N, Christopher Octeau J, Roxas KM, Chan KY, Scherrer J, Winchester J, Blackburn RJ, Campos LJ, Man KNM, Sun J, Chen X, Lefevre A, Singh VP, Arokiaraj CM, Shay TF, Vendemiatti J, Jang MJ et al (2023) Adeno-associated viral vectors for functional intravenous gene transfer throughout the non-human primate brain. Nat Nanotechnol. https://doi.org/10.1038/s41565-023-01419-x

Chudasama Y, Izquierdo A, Murray EA (2009) Distinct contributions of the amygdala and hippocampus to fear expression. Eur J Neurosci 30(12):2327–2337. https://doi.org/10.1111/j.1460-9568.2009.07012.x

Chun K, Capitanio JP (2016) Developmental consequences of behavioral inhibition: a model in rhesus monkeys (Macaca mulatta). Dev Sci 19(6):1035–1048. https://doi.org/10.1111/desc.12339

Clauss JA, Blackford JU (2012) Behavioral inhibition and risk for developing social anxiety disorder: a meta-analytic study. J Am Acad Child Adolesc Psychiatry 51(10):1066–1075.e1. https://doi.org/10.1016/j.jaac.2012.08.002

Clauss JA, Seay AL, VanDerKlok RM, Avery SN, Cao A, Cowan RL, Benningfield MM, Blackford JU (2014) Structural and functional bases of inhibited temperament. Soc Cogn Affect Neurosci 9(12):2049–2058. https://doi.org/10.1093/scan/nsu019

Clauss JA, Avery SN, Blackford JU (2015) The nature of individual differences in inhibited temperament and risk for psychiatric disease: a review and meta-analysis. Prog Neurobiol 127–128:23–45. https://doi.org/10.1016/j.pneurobio.2015.03.001

Dong H, Swanson LW (2004) Projections from bed nuclei of the stria terminalis, posterior division: implications for cerebral hemisphere regulation of defensive and reproductive behaviors. J Comp Neurol 471(4):396–433. https://doi.org/10.1002/cne.20002

Duman RS (2017) BDNF, 5-HT, and anxiety: identification of a critical periadolescent developmental period. Am J Psychiatry 174(12):1137–1139. https://doi.org/10.1176/appi.ajp.2017.17101084

Dunbar RIM, Shultz S (2007) Evolution in the social brain. Science 317(5843):1344–1347. https://doi.org/10.1126/science.1145463

Etkin A, Egner T, Kalisch R (2011) Emotional processing in anterior cingulate and medial prefrontal cortex. Trends Cogn Sci 15(2):85–93. https://doi.org/10.1016/j.tics.2010.11.004

Fadok JP, Markovic M, Tovote P, Lüthi A (2018) New perspectives on central amygdala function. Curr Opin Neurobiol 49:141–147. https://doi.org/10.1016/j.conb.2018.02.009

Farhang S, Barar J, Fakhari A, Mesgariabbasi M, Khani S, Omidi Y, Farnam A (2014) Asymmetrical expression of BDNF and NTRK3 genes in frontoparietal cortex of stress-resilient rats in an animal model of depression. Synapse 68(9):387–393. https://doi.org/10.1002/syn.21746

Folloni D, Sallet J, Khrapitchev AA, Sibson N, Verhagen L, Mars RB (2019) Dichotomous organization of amygdala/temporal-prefrontal bundles in both humans and monkeys. Elife 8:e47175. https://doi.org/10.7554/eLife.47175

Fox AS (2018) The brain is organized to emote. In: The nature of emotions: fundamental questions, 2nd edn. Oxford University Press, pp 118–121

Fox AS, Kalin NH (2014) A translational neuroscience approach to understanding the development of social anxiety disorder and its pathophysiology. Am J Psychiatry 171(11):1162–1173. https://doi.org/10.1176/appi.ajp.2014.14040449

Fox AS, Shackman AJ (2019) The central extended amygdala in fear and anxiety: closing the gap between mechanistic and neuroimaging research. Neurosci Lett 693:58–67. https://doi.org/10.1016/j.neulet.2017.11.056

Fox AS, Oakes TR, Shelton SE, Converse AK, Davidson RJ, Kalin NH (2005a) Calling for help is independently modulated by brain systems underlying goal-directed behavior and threat perception. Proc Natl Acad Sci U S A 102(11):4176–4179. https://doi.org/10.1073/pnas.0409470102

Fox NA, Henderson HA, Marshall PJ, Nichols KE, Ghera MM (2005b) Behavioral inhibition: linking biology and behavior within a developmental framework. Annu Rev Psychol 56(1):235–262. https://doi.org/10.1146/annurev.psych.55.090902.141532

Fox AS, Shelton SE, Oakes TR, Davidson RJ, Kalin NH (2008) Trait-like brain activity during adolescence predicts anxious temperament in primates. PLoS One 3(7):e2570. https://doi.org/10.1371/journal.pone.0002570

Fox A, Shelton S, Oakes T, Converse A, Davidson R, Kalin N (2010) Orbitofrontal cortex lesions alter anxiety-related activity in the primate bed nucleus of Stria Terminalis. J Neurosci 30(20):7023–7027. https://doi.org/10.1523/JNEUROSCI.5952-09.2010

Fox AS, Oler JA, Shelton SE, Nanda SA, Davidson RJ, Roseboom PH, Kalin NH (2012) Central amygdala nucleus (Ce) gene expression linked to increased trait-like Ce metabolism and anxious temperament in young primates. Proc Natl Acad Sci U S A 109(44):18108–18113. https://doi.org/10.1073/pnas.1206723109

Fox AS, Oler JA, Shackman AJ, Shelton SE, Raveendran M, McKay DR, Converse AK, Alexander A, Davidson RJ, Blangero J, Rogers J, Kalin NH (2015a) Intergenerational neural mediators of early-life anxious temperament. Proc Natl Acad Sci 112(29):9118–9122. https://doi.org/10.1073/pnas.1508593112

Fox AS, Oler JA, Tromp DPM, Fudge JL, Kalin NH (2015b) Extending the amygdala in theories of threat processing. Trends Neurosci 38(5):319–329. https://doi.org/10.1016/j.tins.2015.03.002

Fox AS, Oler JA, Birn RM, Shackman AJ, Alexander AL, Kalin NH (2018) Functional connectivity within the primate extended amygdala is heritable and associated with early-life anxious temperament. J Neurosci 38(35):7611–7621. https://doi.org/10.1523/JNEUROSCI.0102-18.2018

Fox AS, Souaiaia T, Oler JA, Kovner R, Kim JM (Hugo), Nguyen J, French DA, Riedel MK, Fekete EM, Rabska MR, Olsen ME, Brodsky EK, Alexander AL, Block WF, Roseboom PH, Knowles JA, Kalin NH (2019) Dorsal amygdala neurotrophin-3 decreases anxious temperament in primates. Biol Psychiatry 86(12):881–889. https://doi.org/10.1016/j.biopsych.2019.06.022

Fox AS, Harris RA, Rosso LD, Raveendran M, Kamboj S, Kinnally EL, Capitanio JP, Rogers J (2021a) Infant inhibited temperament in primates predicts adult behavior, is heritable, and is associated with anxiety-relevant genetic variation. Mol Psychiatry 26(11):6609–6618. https://doi.org/10.1038/s41380-021-01156-4

Fox NA, Buzzell GA, Morales S, Valadez EA, Wilson M, Henderson HA (2021b) Understanding the emergence of social anxiety in children with behavioral inhibition. Biol Psychiatry 89(7):681–689. https://doi.org/10.1016/j.biopsych.2020.10.004

Glahn DC, Nimgaonkar VL, Raventós H, Contreras J, McIntosh AM, Thomson PA, Jablensky A, McCarthy NS, Charlesworth JC, Blackburn NB, Peralta JM, Knowles EEM, Mathias SR, Ament SA, McMahon FJ, Gur RC, Bucan M, Curran JE, Almasy L et al (2019) Rediscovering the value of families for psychiatric genetics research. Mol Psychiatry 24(4):523–535. https://doi.org/10.1038/s41380-018-0073-x

Gonda X, Eszlari N, Torok D, Gal Z, Bokor J, Millinghoffer A, Baksa D, Petschner P, Antal P, Breen G, Juhasz G, Bagdy G (2021) Genetic underpinnings of affective temperaments: a pilot GWAS investigation identifies a new genome-wide significant SNP for anxious temperament in ADGRB3 gene. Transl Psychiatry 11(1):337. https://doi.org/10.1038/s41398-021-01436-1

Haubensak W, Kunwar PS, Cai H, Ciocchi S, Wall NR, Ponnusamy R, Biag J, Dong H-W, Deisseroth K, Callaway EM, Fanselow MS, Lüthi A, Anderson DJ (2010) Genetic dissection of an amygdala microcircuit that gates conditioned fear. Nature 468(7321):270–276. https://doi.org/10.1038/nature09553

Holley D, Fox AS (2022) The central extended amygdala guides survival-relevant tradeoffs: implications for understanding common psychiatric disorders. Neurosci Biobehav Rev 142:104879. https://doi.org/10.1016/j.neubiorev.2022.104879

Holley D, Campos LJ, Drzewiecki CM, Zhang Y, Capitanio JP, Fox AS (2024) Rhesus infant nervous temperament predicts peri-adolescent central amygdala metabolism & behavioral inhibition measured by a machine-learning approach. Transl Psychiatry 14(1):1–11. https://doi.org/10.1038/s41398-024-02858-3

Izquierdo A, Suda RK, Murray EA (2005) Comparison of the effects of bilateral orbital prefrontal cortex lesions and amygdala lesions on emotional responses in rhesus monkeys. J Neurosci 25(37):8534–8542. https://doi.org/10.1523/JNEUROSCI.1232-05.2005

Jacobs JT, Maior RS, Waguespack HF, Campos-Rodriguez C, Forcelli PA, Malkova L (2023) Pharmacological inactivation of the bed nucleus of the Stria Terminalis increases affiliative social behavior in rhesus macaques. J Neurosci 43(18):3331–3338. https://doi.org/10.1523/JNEUROSCI.2090-22.2023

Kagan J (1989) Temperamental contributions to social behavior. Am Psychol 44(4):668–674

Kagan J, Snidman N (1991) Temperamental factors in human development. Am Psychol 46(856–862)

Kagan J, Reznick JS, Clarke C, Snidman N (1984) Behavioral inhibition to the unfamiliar. Child Dev 55(6):2212–2225

Kagan J, Reznick JS, Snidman N (1987) The physiology and psychology of behavioral inhibition in children. Child Dev 58(6):1459–1473

Kagan J, Reznick JS, Snidman N (1988) Biological bases of childhood shyness. Science 240(4849): 167–171. https://doi.org/10.1126/science.3353713

Kalin NH, Shelton SE (1989) Defensive behaviors in infant rhesus monkeys: environmental cues and neurochemical regulation. Science 243(4899):1718–1721

Kalin NH, Shelton SE (2003) Nonhuman primate models to study anxiety, emotion regulation, and psychopathology. Ann N Y Acad Sci 1008(1):189–200. https://doi.org/10.1196/annals.1301.021

Kalin NH, Shelton SE, Takahashi LK (1991) Defensive behaviors in infant rhesus monkeys: ontogeny and context-dependent selective expression. Child Dev 62(5):1175–1183

Kalin NH, Larson C, Shelton SE (1998) Asymmetric frontal brain activity, cortisol, and behavior associated with fearful temperament in rhesus monkeys. Behav Neurosci 112(2):286–292

Kalin NH, Shelton SE, Davidson RJ, Kelley AE (2001) The primate amygdala mediates acute fear but not the behavioral and physiological components of anxious temperament. J Neurosci 21(6): 2067–2074. https://doi.org/10.1523/JNEUROSCI.21-06-02067.2001

Kalin NH, Shelton SE, Davidson RJ (2004) The role of the central nucleus of the amygdala in mediating fear and anxiety in the primate. J Neurosci Off J Soc Neurosci 24(24):5506–5515. https://doi.org/10.1523/JNEUROSCI.0292-04.2004

Kalin NH, Shelton SE, Fox AS, Oakes TR, Davidson RJ (2005) Brain regions associated with the expression and contextual regulation of anxiety in primates. Biol Psychiatry 58(10):796–804. https://doi.org/10.1016/j.biopsych.2005.05.021

Kalin NH, Shelton SE, Davidson RJ (2007) Role of the primate orbitofrontal cortex in mediating anxious temperament. Biol Psychiatry 62(10):1134–1139. https://doi.org/10.1016/j.biopsych.2007.04.004

Kalin NH, Fox AS, Kovner R, Riedel MK, Fekete EM, Roseboom PH, Tromp DPM, Grabow BP, Olsen ME, Brodsky EK, McFarlin DR, Alexander AL, Emborg ME, Block WF, Fudge JL, Oler JA (2016) Overexpressing corticotropin-releasing hormone (CRH) in the primate amygdala increases anxious temperament and alters its neural circuit. Biol Psychiatry 80(5):345–355. https://doi.org/10.1016/j.biopsych.2016.01.010

Kamath T, Abdulraouf A, Burris SJ, Langlieb J, Gazestani V, Nadaf NM, Balderrama K, Vanderburg C, Macosko EZ (2022) Single-cell genomic profiling of human dopamine neurons identifies a population that selectively degenerates in Parkinson's disease. Nat Neurosci 25(5): 588–595. https://doi.org/10.1038/s41593-022-01061-1

Kenwood MM, Kalin NH, Barbas H (2021) The prefrontal cortex, pathological anxiety, and anxiety disorders. Neuropsychopharmacology. https://doi.org/10.1038/s41386-021-01109-z

Kenwood MM, Oler JA, Tromp DPM, Fox AS, Riedel MK, Roseboom PH, Brunner KG, Aggarwal N, Murray EA, Kalin NH (2023) Prefrontal influences on the function of the neural circuitry underlying anxious temperament in primates. Oxford Open Neurosci 2:kvac016. https://doi.org/10.1093/oons/kvac016

Kovner R, Souaiaia T, Fox AS, French DA, Goss CE, Roseboom PH, Oler JA, Riedel MK, Fekete EM, Fudge JL, Knowles JA, Kalin NH (2020) Transcriptional profiling of primate central nucleus of the amygdala neurons to understand the molecular underpinnings of early-life anxious temperament. Biol Psychiatry 88(8):638–648. https://doi.org/10.1016/j.biopsych.2020.05.009

Krienen, F. M., Goldman, M., Zhang, Q., C. H. del Rosario, Florio, M., Machold, R., Saunders, A., Levandowski, K., Zaniewski, H., Schuman, B., Wu, C., Lutservitz, A., Mullally, C. D., Reed, N., Bien, E., Bortolin, L., Fernandez-Otero, M., Lin, J. D., Wysoker, A., . . . McCarroll, S. A. (2020). Innovations present in the primate interneuron repertoire. Nature 586(7828), 262–269. https://doi.org/10.1038/s41586-020-2781-z

Lesch K-P, Timmesfeld N, Renner TJ, Halperin R, Röser C, Nguyen TT, Craig DW, Romanos J, Heine M, Meyer J, Freitag C, Warnke A, Romanos M, Schäfer H, Walitza S, Reif A, Stephan DA, Jacob C (2008) Molecular genetics of adult ADHD: converging evidence from genome-wide association and extended pedigree linkage studies. J Neural Transm 115(11):1573–1585. https://doi.org/10.1007/s00702-008-0119-3

Levey DF, Gelernter J, Polimanti R, Zhou H, Cheng Z, Aslan M, Quaden R, Concato J, Radhakrishnan K, Bryois J, Sullivan PF, The Million Veteran Program, Stein MB (2020) Reproducible genetic risk loci for anxiety: results from ~200,000 participants in the Million Veteran Program. Am J Psychiatry 177(3):223–232. https://doi.org/10.1176/appi.ajp.2019.19030256

Machado CJ, Bachevalier J (2008) Behavioral and hormonal reactivity to threat: effects of selective amygdala, hippocampal or orbital frontal lesions in monkeys. Psychoneuroendocrinology 33(7):926–941. https://doi.org/10.1016/j.psyneuen.2008.04.012

Miller CT, Freiwald WA, Leopold DA, Mitchell JF, Silva AC, Wang X (2016) Marmosets: a neuroscientific model of human social behavior. Neuron 90(2):219–233. https://doi.org/10.1016/j.neuron.2016.03.018

Mueller SAL, Oler JA, Roseboom PH, Aggarwal N, Kenwood MM, Riedel MK, Elam VR, Olsen ME, DiFilippo AH, Christian BT, Hu X, Galvan A, Boehm MA, Michaelides M, Kalin NH (2023) DREADD-mediated amygdala activation is sufficient to induce anxiety-like responses in young nonhuman primates. Curr Res Neurobiol 5:100111. https://doi.org/10.1016/j.crneur.2023.100111

Murray EA, Izquierdo A (2007) Orbitofrontal cortex and amygdala contributions to affect and action in primates. Ann N Y Acad Sci 1121:273–296. https://doi.org/10.1196/annals.1401.021

Nagel M, Jansen PR, Stringer S, Watanabe K, de Leeuw CA, Bryois J, Savage JE, Hammerschlag AR, Skene NG, Muñoz-Manchado AB, White T, Tiemeier H, Linnarsson S, Hjerling-Leffler J, Polderman TJC, Sullivan PF, van der Sluis S, Posthuma D (2018) Meta-analysis of genome-wide association studies for neuroticism in 449,484 individuals identifies novel genetic loci and pathways. Nat Genet 50(7):920–927. https://doi.org/10.1038/s41588-018-0151-7

Oler JA, Fox AS, Shelton SE, Rogers J, Dyer TD, Davidson RJ, Shelledy W, Oakes TR, Blangero J, Kalin NH (2010) Amygdalar and hippocampal substrates of anxious temperament differ in their heritability. Nature 466(7308):864–868. https://doi.org/10.1038/nature09282

Oler JA, Birn RM, Patriat R, Fox AS, Shelton SE, Burghy CA, Stodola DE, Essex MJ, Davidson RJ, Kalin NH (2012) Evidence for coordinated functional activity within the extended amygdala of non-human and human primates. Neuroimage 61(4):1059–1066. https://doi.org/10.1016/j.neuroimage.2012.03.045

Öngür D, Price JL (2000) The organization of networks within the orbital and medial prefrontal cortex of rats, monkeys and humans. Cereb Cortex 10(3):206–219. https://doi.org/10.1093/cercor/10.3.206

Paus T, Collins DL, Evans AC, Leonard G, Pike B, Zijdenbos A (2001) Maturation of white matter in the human brain: a review of magnetic resonance studies. Brain Res Bull 54(3):255–266

Petrides M, Pandya DN (1999) Dorsolateral prefrontal cortex: comparative cytoarchitectonic analysis in the human and the macaque brain and corticocortical connection patterns: dorsolateral prefrontal cortex in human and monkey. Eur J Neurosci 11(3):1011–1036. https://doi.org/10.1046/j.1460-9568.1999.00518.x

Phan KL, Orlichenko A, Boyd E, Angstadt M, Coccaro EF, Liberzon I, Arfanakis K (2009) Preliminary evidence of White matter abnormality in the uncinate fasciculus in generalized social anxiety disorder. Biol Psychiatry 66(7):691–694. https://doi.org/10.1016/j.biopsych.2009.02.028

Pine DS, Wise SP, Murray EA (2021) Evolution, emotion, and episodic engagement. Am J Psychiatry 178(8):701–714. https://doi.org/10.1176/appi.ajp.2020.20081187

Price JL (2007) Definition of the orbital cortex in relation to specific connections with limbic and visceral structures and other cortical regions. Ann N Y Acad Sci 1121(1):54–71. https://doi.org/10.1196/annals.1401.008

Pryluk R, Shohat Y, Morozov A, Friedman D, Taub AH, Paz R (2020) Shared yet dissociable neural codes across eye gaze, valence and expectation. Nature 586(7827):95–100. https://doi.org/10.1038/s41586-020-2740-8

Pujara MS, Rudebeck PH, Ciesinski NK, Murray EA (2019) Heightened defensive responses following subtotal lesions of macaque orbitofrontal cortex. J Neurosci 39(21):4133–4141. https://doi.org/10.1523/JNEUROSCI.2812-18.2019

Qi C, Roseboom PH, Nanda SA, Lane JC, Speers JM, Kalin NH (2010) Anxiety-related behavioral inhibition in rats: a model to examine mechanisms underlying the risk to develop stress-related psychopathology. Genes Brain Behav 9(8):974–984. https://doi.org/10.1111/j.1601-183X.2010.00636.x

Rahman SS, Mulvihill K, Wood CM, Quah SKL, Horst NK, Clarke HF, Cockcroft GJ, Santangelo AM, Roberts AC (2021) Differential contribution of anterior and posterior midcingulate subregions to distal and proximal threat reactivity in marmosets. Cereb Cortex 31(10):4765–4780. https://doi.org/10.1093/cercor/bhab121

Rich EL, Wallis JD (2016) Decoding subjective decisions from orbitofrontal cortex. Nat Neurosci 19(7):973–980. https://doi.org/10.1038/nn.4320

Roberts AC, Clarke HF (2019) Why we need nonhuman primates to study the role of ventromedial prefrontal cortex in the regulation of threat- and reward-elicited responses. Proc Natl Acad Sci 116(52):26297–26304. https://doi.org/10.1073/pnas.1902288116

Rogers J, Raveendran M, Fawcett GL, Fox AS, Shelton SE, Oler JA, Cheverud J, Muzny DM, Gibbs RA, Davidson RJ, Kalin NH (2013) CRHR1 genotypes, neural circuits and the diathesis for anxiety and depression. Mol Psychiatry 18(6):700–707. https://doi.org/10.1038/mp.2012.152

Roseboom PH, Mueller SAL, Oler JA, Fox AS, Riedel MK, Elam VR, Olsen ME, Gomez JL, Boehm MA, DiFilippo AH, Christian BT, Michaelides M, Kalin NH (2021) Evidence in primates supporting the use of chemogenetics for the treatment of human refractory neuropsychiatric disorders. Mol Ther 29(12):3484–3497. https://doi.org/10.1016/j.ymthe.2021.04.021

Rudebeck PH, Saunders RC, Prescott AT, Chau LS, Murray EA (2013) Prefrontal mechanisms of behavioral flexibility, emotion regulation and value updating. Nat Neurosci 16(8):1140–1145. https://doi.org/10.1038/nn.3440

Saleem KS, Kondo H, Price JL (2008) Complementary circuits connecting the orbital and medial prefrontal networks with the temporal, insular, and opercular cortex in the macaque monkey. J Comp Neurol 506(4):659–693. https://doi.org/10.1002/cne.21577

Schmitz MT, Sandoval K, Chen CP, Mostajo-Radji MA, Seeley WW, Nowakowski TJ, Ye CJ, Paredes MF, Pollen AA (2022) The development and evolution of inhibitory neurons in primate cerebrum. Nature 603(7903):871–877. https://doi.org/10.1038/s41586-022-04510-w

Schwartz CE, Wright CI, Shin LM, Kagan J, Rauch SL (2003) Inhibited and uninhibited infants "grown up": adult amygdalar response to novelty. Science 300(5627):1952–1953. https://doi.org/10.1126/science.1083703

Shackman AJ, Fox AS (2016) Contributions of the central extended amygdala to fear and anxiety. J Neurosci 36(31):8050–8063. https://doi.org/10.1523/JNEUROSCI.0982-16.2016

Shackman AJ, Fox AS, Oler JA, Shelton SE, Oakes TR, Davidson RJ, Kalin NH (2017) Heightened extended amygdala metabolism following threat characterizes the early phenotypic risk to develop anxiety-related psychopathology. Mol Psychiatry 22(5):724–732. https://doi.org/10.1038/mp.2016.132

Smith DJ, Escott-Price V, Davies G, Bailey MES, Colodro-Conde L, Ward J, Vedernikov A, Marioni R, Cullen B, Lyall D, Hagenaars SP, Liewald DCM, Luciano M, Gale CR, Ritchie SJ, Hayward C, Nicholl B, Bulik-Sullivan B, Adams M et al (2016) Genome-wide analysis of over 106 000 individuals identifies 9 neuroticism-associated loci. Mol Psychiatry 21(6):749–757. https://doi.org/10.1038/mp.2016.49

Stawicka ZM, Massoudi R, Oikonomidis L, McIver L, Mulvihill K, Quah SKL, Cockcroft GJ, Clarke HF, Horst NK, Wood CM, Roberts AC (2022) Differential effects of the inactivation of anterior and posterior orbitofrontal cortex on affective responses to proximal and distal threat, and reward anticipation in the common marmoset. Cereb Cortex 32(7):1319–1336. https://doi.org/10.1093/cercor/bhab240

Stein MB, Levey DF, Cheng Z, Wendt FR, Harrington K, Pathak GA, Cho K, Quaden R, Radhakrishnan K, Girgenti MJ, Ho Y-LA, Posner D, Aslan M, Duman RS, Zhao H, Department of Veterans Affairs Cooperative Studies Program (no. 575B), Stein MB, Levey DF, Cheng Z et al (2021) Genome-wide association analyses of post-traumatic stress disorder and its symptom subdomains in the Million Veteran Program. Nat Genet 53(2):174–184. https://doi.org/10.1038/s41588-020-00767-x

Swanson LW, Petrovich GD (1998) What is the amygdala? Trends Neurosci 21(8):323–331

Tovote P, Esposito MS, Botta P, Chaudun F, Fadok JP, Markovic M, Wolff SBE, Ramakrishnan C, Fenno L, Deisseroth K, Herry C, Arber S, Lüthi A (2016) Midbrain circuits for defensive behaviour. Nature 534(7606):206–212. https://doi.org/10.1038/nature17996

Tromp DPM, Grupe DW, Oathes DJ, McFarlin DR, Hernandez PJ, Kral TRA, Lee JE, Adams M, Alexander AL, Nitschke JB (2012) Reduced structural connectivity of a major frontolimbic pathway in generalized anxiety disorder. Arch Gen Psychiatry 69(9):925. https://doi.org/10.1001/archgenpsychiatry.2011.2178

Tromp DPM, Fox AS, Oler JA, Alexander AL, Kalin NH (2019a) The relationship between the uncinate fasciculus and anxious temperament is evolutionarily conserved and sexually dimorphic. Biol Psychiatry 86(12):890–898. https://doi.org/10.1016/j.biopsych.2019.07.022

Tromp DPM, Williams LE, Fox AS, Oler JA, Roseboom PH, Rogers GM, Benson BE, Alexander AL, Pine DS, Kalin NH (2019b) Altered uncinate fasciculus microstructure in childhood anxiety disorders in boys but not girls. Am J Psychiatry:appiajp201818040425. https://doi.org/10.1176/appi.ajp.2018.18040425

Wallis JD (2012) Cross-species studies of orbitofrontal cortex and value-based decision-making. Nat Neurosci 15(1):13–19. https://doi.org/10.1038/nn.2956

Welker BJ (2017) The history of our tribe, Hominini. Open SUNY Textbooks

Wellman LL, Forcelli PA, Aguilar BL, Malkova L (2016) Bidirectional control of social behavior by activity within basolateral and central amygdala of primates. J Neurosci 36(33):8746–8756. https://doi.org/10.1523/JNEUROSCI.0333-16.2016

Humans

The Impact of Early Life Experiences on Stress Neurobiology and the Development of Anxiety

Yanbin Niu, M. Catalina Camacho, Shuang Wu, and Kathryn L. Humphreys

Contents

Abstract We examine the association between stress exposure during early development (i.e., the prenatal period through the first two postnatal years) and variation in brain structure and function relevant to anxiety. Evidence of stress-related effects occurring in regions essential for emotional processing and regulation may increase susceptibility to anxiety.

Keywords Anxiety · Brain Development · Stress

K. L. Humphreys (✉)
Vanderbilt University, Nashville, TN, USA
e-mail: k.humphreys@vanderbilt.edu

© The Author(s), under exclusive license to Springer Nature Switzerland AG 2024
Curr Topics Behav Neurosci (2025) 73: 279–300
https://doi.org/10.1007/7854_2024_542
Published Online: 13 November 2024

1 Introduction

Anxiety disorders are among the most prevalent mental health conditions globally (Kessler et al. 2005). Research points to stress as one etiological factor in the onset and maintenance of anxiety (Lezak et al. 2017; Pêgo et al. 2010). The developing brain undergoes major construction during the gestation period and the first few years of life (Johnston et al. 2001). Stressful experiences during periods of dramatic brain changes (periods of heightened plasticity) are likely to shape neural circuits in ways that increase the risk of anxiety across the lifespan (Andersen 2003; Pechtel and Pizzagalli 2011). This chapter focuses on stress experienced in early life, from the prenatal period until age 2 years, and links to neurobiological pathways hypothesized to predispose individuals to anxiety. We begin by reviewing the association between early life stress and the risk of anxiety, followed by the neurobiology of anxiety and the potential vulnerability of anxiety-related neural regions to early stress. We then explore the neural mechanisms that may underpin the link between early life stress and anxiety. Finally, we discuss the current state of knowledge and future research directions. While this chapter primarily centers on the potential neurological mechanisms underlying the link between early stress and anxiety, it is important to acknowledge that other mechanisms, including genetics, epigenetics, and other environmental contributors (e.g., teratogens, variation in caregiver behaviors), as well as the interplay between stress and these factors contribute to the development of anxiety (McEwen et al. 2015).

In this chapter, we consider stress as a subjective experience that results from adverse or very demanding circumstances. We intentionally avoid equating stress with biomarkers associated with stress (e.g., cortisol levels) given the low mapping of objective assessments of stress to these measures (Malanchini et al. 2021). In addition, elevated cortisol levels during pregnancy may not necessarily indicate stress but are a normal aspect of the pregnancy (De Weerth and Buitelaar 2005). Thus, for prenatal stress, we include 1) stressful life events experienced by the mother during pregnancy (e.g., illnesses or deaths in the close family, financial and relationship problems) and 2) exposure to a disaster (maternal hardship due to a natural disaster). For postnatal stress, we include a spectrum of severe and traumatic experiences, such as physical, emotional, and or sexual abuse, neglect, parental loss, and exposure to natural disaster, encountered up to 2 years old (Agorastos et al. 2019). Given that this overall volume includes perspectives from both research in animal models and humans, here we aim to clarify and align our definitions to facilitate translational understanding of stress impacts across species. Our definition of prenatal stress is intended to apply to pregnant mammals exposed to stressors such as predator scent or environmental challenges (St-Cyr et al. 2017). Similarly, our review of postnatal stress aligns with nonhuman animal research that involves maternal separation or environmental instability (Vivinetto et al. 2013). Note that pre and postnatal parental psychological *distress* (e.g., anxiety and depressive symptoms) is not included in this chapter, but this topic can be found elsewhere (Lautarescu et al. 2020). While the primary focus of this chapter is on how stress alters the brain in ways that may increase the risk of anxiety disorders, it is important to acknowledge that stress is a nonspecific risk factor implicated in a wide array of developmental disorders and health issues (Cowan et al. 2016; Lupien et al. 2009), consistent with the concept of multifinality. In terms of our approach to examining outcomes relevant to anxiety, we consider anxiety in a broad sense, encompassing anxiety-related behaviors that may manifest as heightened physiological arousal or excessive fear responses (Spielberger 2010). This perspective allows for a more inclusive discussion of anxiety manifestations, particularly in infants and toddlers, where traditional clinical characterizations may not be fully applicable. Moreover, given our focus on the subsequent risk of anxiety, we include anxiety-related behaviors assessed at any point – concurrently with early life stressors or at later stages.

2 Prenatal Stress and Anxiety

Prenatal stress is associated with an increased risk of psychopathology in children, including anxiety disorders (Van den Bergh et al. 2020). One pathway may be through infant temperament – individual differences in reactivity and self-regulation (Rothbart and Bates 2007) – that is associated with a heightened risk for later anxiety (Huizink 2008; Kagan and Snidman 1999). For example, prenatal stress has been linked to anxiety-related temperaments, including negative affect (Campbell et al. 2020; Lin et al. 2014), difficult temperament (Thiel et al. 2021), and fearful temperament in infants (Bergman et al. 2007). A prospective pregnancy cohort study of 445 mother–infant dyads from two urban US settings used a weighted prenatal stress index including lifetime, current life events, and the potential psychological sequelae from those events in the form of depressive and PTSD symptoms (Campbell et al. 2020). This composite containing both objective stress and distress was associated with greater infant negative affect ($\beta = 0.40$). In a longitudinal study of 125 mother–infant pairs, prenatal stress, assessed by the stressful life events questionnaire, was correlated with observed fearfulness in toddlerhood ($r = 0.33$), explaining 10% of the variance in fearfulness (Bergman et al. 2007). However, another recent systematic review examining the link between prenatal stress from potentially traumatic events and child temperament across 20 studies presented mixed outcomes (Rodríguez-Soto et al. 2021). Specifically, objective hardships due to disasters were correlated with greater child negative affectivity in 40% of studies (4 out of 10), lower levels of effortful control/regulation in 43% (3 out of 7) studies, and less positive affectivity in 11% (1 out of 9) studies. Additionally, experiences of intimate partner violence during pregnancy were associated with heightened child negative affectivity in 80% of studies (4 out of 5), lower levels of effortful control/regulation in 67% of studies (4 out of 6), and less positive affectivity in 50% studies (1 out of 2). This suggests that the effect of prenatal stress related to intimate partner experiences may be more consistently observed, while findings related to objective hardships due to disasters show greater variation in outcomes, with some studies not identifying statistically significant associations.

Additionally, prenatal stress is associated with infant fussiness (Wurmser et al. 2006) and emotional regulation difficulties (Lafortune et al. 2021; Thiel et al. 2021). In a longitudinal study, 86 pregnant participants were divided into high and low prenatal stress groups based on the occurrence of negative life events during pregnancy (Wurmser et al. 2006). Infants born to mothers with high prenatal stress displayed more crying and fussing during the first 6 months compared to infants of mothers from the low prenatal stress group. A longitudinal study involving 3,752 women found an association between prenatal life stress, as measured by self-reported difficult life events during pregnancy, and difficult infant temperament at 8 weeks, while accounting for only 1% of the variance (Thiel et al. 2021). Furthermore, after adjusting for confounders, perinatal life stress explained an additional 1% of the variance in social-emotional development outcomes, such as emotion regulation difficulties, when children were 2 years of age. A recent meta-analysis by Lafortune et al. (2021), which included data from 37 studies on natural disaster-related prenatal stress, found that such prenatal stress is significantly associated with a wide range of child development areas, including socio-emotional outcomes (although the magnitude of the overall effect size was very small: $r = 0.06$). One advantage of a focus on disaster-related stressful events is that these experiences are considered to be more "fated" than other types of stress (Hammen 2006), making estimates from these studies less likely to be influenced by gene–environment correlations.

Prenatal stress has been examined in relation to infants' reactivity to novelty; however, results are mixed – with higher prenatal stress being associated with lower levels of responses to novelty, or no relations (Möhler et al. 2006; Rothenberger et al. 2011). In a longitudinal study involving 104 mother–infant dyads, perceived stress and objective stress events were assessed three times during pregnancy (Rothenberger et al. 2011). Results indicated that higher perceived prenatal stress correlated with lower infant affective reactivity observed at 5 months. Additionally, in the extreme group comparison, infants in the high affectively reactive group had mothers with lower perceived stress and fewer stressful events at the end of pregnancy, compared to infants in the low affectively reactive group. Another study with 96 mother–infant pairs found no associations between self-report stressful events during pregnancy and either crying or motor responses to novelty observed at 4 months (Möhler et al. 2006). The observed negative or absent links suggest that while early heightened reactivity is often linked to later anxiety (Crockenberg and Leerkes 2006), the relation between prenatal stress and anxiety may not be explained by increased reactivity to novelty in infants.

While much of the research on prenatal stress focuses on early life temperament assessment due to its temporal proximity to the prenatal period, the longitudinal impacts of prenatal stress on child later anxiety have also been documented. In a study of 178 mother–child dyads, higher levels of perceived stress during pregnancy were associated with greater anxiety ($\beta = 0.20$) in children aged 6–9 years (Davis and Sandman 2012). These associations remain robust after adjusting for a range of obstetric, sociodemographic, and postnatal variables.

3 Early Postnatal Stress and Anxiety

Similar to prenatal stress, exposure to stress during the first 2 years of the postnatal period is also associated with a heightened risk of anxiety later in life (Juruena et al. 2020). This link is evidenced by increased negative affect (Davis et al. 2023; Kochukhova et al. 2016) and early emergence of anxiety symptoms (Burkholder et al. 2016; Ellis et al. 2004; Lewis and Olsson 2011). A nationally representative study in Australia followed a cohort of 3,425 children in the first 5 years of life and revealed that early life stress reported by parents in the first year of infant life was prospectively associated with higher composite anxiety and depressive problems (odd ratio = 1.31) when children were preschool age (Lewis and Olsson 2011). In another study involving a representative sample of 917 preschoolers (mean age = 3.84), a higher total count of adverse experiences faced by children was associated with greater negative emotionality ($\beta = 0.11$) reported by parents (Davis et al. 2023).

Some evidence on the effects of early life stress comes from investigations following children who experienced institutional rearing – an environment characterized by high child-to-caregiver ratios, rotating staff, and a lack of responsive care. In a sample of 51 children aged 2–6 years residing in institutional care in Romania at the time of the study, authors reported high rates of anxiety/affective symptoms (Ellis et al. 2004). Additionally, those children who experienced longer durations of institutionalization had higher levels of anxiety/affective symptoms ($\beta = 0.455$). In another sample of 96 toddlers, children reared in institutions exhibited less surgency/extraversion ($\eta^2 = 0.56$) and higher negative affectivity ($\eta^2 = 0.23$) compared to family-reared children (Kochukhova et al. 2016). Again, a longer duration spent in institutional care was associated with greater negative affectivity ($r = 0.41$). In another study of 161 older children, half of whom were internationally adopted from institutional care ($N = 79$) and half were age-matched control peers ($N = 82$) (Burkholder et al. 2016), youths with early institutional experiences demonstrated significantly more anxiety during laboratory tasks than their non-adopted peers ($\eta^2 = 0.12$). These studies establish important correlations between early stressful environments, such as institutional care, and anxiety. The Bucharest Early Intervention Project (BEIP), a landmark study in the field of developmental psychology, further offers causal evidence by the randomized controlled trial design that assigned children in institutions to be placed in either foster care settings or to care as usual (typically resulting in a longer stay in institutionalized care settings). Findings from the BEIP indicate the detrimental effects of, and recovery from, early institutionalization, including the prevalence of internalizing disorders such as anxiety (Zeanah et al. 2009, 2019). At 54 months of age, children with any history of institutional care were an estimated 2.8 times more likely to meet criteria for any anxiety disorder than those who were never institutionalized (30.6% vs. 13.6%), and children in the usual care group were 2.9 times more likely to meet criteria for any anxiety disorder compared to those in the foster care group (42.3% vs. 20.3%) (Zeanah et al. 2009). The latter findings comparing randomized arms from the intervention trial, which used a conservative intent-to-treat approach, provide causal evidence of the role of the caregiving environment (foster care vs. care as usual) in reducing risk for anxiety in early childhood.

In summary, experiences of early institutional care, as a form of early life stress, place children at an elevated risk of developing anxiety. Note that the effects of early institutional care are not limited to anxiety alone; it predisposes children to psychopathology across domains (Humphreys et al. 2017), with anxiety being one of the many potential outcomes.

4 Neurobiology of Anxiety

In adult humans, the neurobiology of anxiety involves a complex network of brain regions and pathways that broadly fall into either systems of salience and arousal or systems of self-regulation. Anxiety has been linked to hyperactivity in salience and arousal systems and/or a hypoactivity of regulation systems (Sylvester et al. 2012).

4.1 Salience and Arousal Systems

Human and nonhuman primate research has found that the amygdala, anterior cingulate cortex (ACC), and anterior insula are regions of salience and arousal systems that are closely linked to chronic arousal, hypersensitivity to external stimuli, and negative emotion states. In animals and humans, the amygdala has been shown to act as a salience detector to initiate physiological responses to external stimuli (Anderson and Phelps 2001; Davis and Whalen 2001; Duvarci and Pare 2014). Amygdala activation triggers a cascade of autonomic and endocrine responses (e.g., fight-or-flight reaction, elevating heart rate, blood pressure, and energy mobilization), and hyperactivation in the amygdala has been observed in anxiety disorders (Etkin and Wager 2007; Meaney et al. 2020). The insula has bidirectional connections to both the amygdala and nucleus accumbens (Reynolds and Zahm 2005), as well as the orbitofrontal cortex (Öngür and Price 2000). It is positioned to receive information about the salience (both appetitive and aversive) and relative value of the stimulus and integrate this information with body state information, likely contributing to bodily experience of tension and negative affect common in anxiety (Craig 2002). The ACC is shown to compute an error signal between a predicted and observed outcome for the allocation of attentional resources (Botvinick et al. 2004). In this context, the ACC may contribute to the enhanced processing of salient emotional stimuli (e.g., threats or errors), which may underlie the heightened vigilance, increased sensitivity to negative stimuli, and difficulty of disengaging from perceived threats that are characteristic of anxiety. Together, a lower threshold or increased sensitivity in those key regions within the salience network may contribute to heightened arousal and a bias toward stimulus-driven attention, making anxious individuals more responsive to environmental cues that they perceive as threatening or significant.

4.2 *Top-Down Regulatory Systems*

Aside from a possibly exaggerated or sensitive detection of stimulus-driven attention systems, individuals with anxiety may also exhibit compromised functionality in their top-down regulatory systems that are essential for modulating affective responses (Chavanne and Robinson 2021). The dorsolateral prefrontal cortex (dlPFC) is a key node in regulatory systems that is active during cognitive control and goal-directed behaviors including intentional emotion regulation (Etkin et al. 2015). When the amygdala detects and encodes salient stimuli, the dlPFC can modulate amygdala activity, thereby reducing the heightened intensity of affective responses (Eippert et al. 2007; Zilverstand et al. 2017). Anxiety disorders are often linked to dlPFC hypoactivity, indicating a diminished ability to exert regulation over emotional responses triggered by the salience network (Fitzgerald et al. 2013; Zilverstand et al. 2017). Moreover, dlPFC is an effective primary target in brain stimulation studies designed to alleviate anxiety in adults (Berlim and Van den Eynde 2014; Diefenbach et al. 2016). Additionally, the ventromedial prefrontal cortex (vmPFC) is involved in decision-making, automatic emotional regulation, and the processing of risk and fear, acting as a mediator for assigning emotional value to stimuli and guiding behavior based on these evaluations (Hiser and Koenigs 2018). Rodent studies provide compelling evidence for the vmPFC's regulation role in suppressing modulating amygdala activation (Adhikari et al. 2015; Quirk et al. 2000). Human functional imaging studies also support the vmPFC's regulatory functioning, showing inverse relations between vmPFC and amygdala activity during aversive stimuli (Kim et al. 2003; Motzkin et al. 2015), suppression/regulation of negative emotions (Fitzgerald et al. 2018; Urry et al. 2006), and the extinction of conditioned fear responses (Delgado et al. 2008; Phelps et al. 2004). However, there is also evidence suggesting the vmPFC's involvement in the generation of negative emotions (Hiser and Koenigs 2018). Individuals with vmPFC lesions have a general dampening of emotional experience and poorly modulated emotional reactions, indicating the complex role of vmPFC in both stimulus-driven responses and top-down regulatory systems (Barrash et al. 2000). Collectively, individuals with anxiety may have impaired functionality in their top-down regulatory systems, notably in the dlPFC and vmPFC, which is crucial for higher-order executive functioning and modulating emotional responses.

5 Development of Anxiety-Relevant Brain Circuitry

The human brain undergoes remarkable growth and development from the prenatal phase through the first 2 years of life, marking one of the most dynamic periods in the entire lifespan (Knickmeyer et al. 2008). Arousal/salience and regulation brain systems develop at uneven and asynchronous rates during this period. Overall, the fetal brain undergoes significant development, including intense proliferation and migration, neuronal aggregation, axonal growth, dendritic differentiation, and pathway formation, with all major brain subdivisions becoming clearly visible by the end of the embryonic period (Kostović et al. 2019). Following birth, the brain continues its rapid growth. During the first postnatal year, brain volume increases by 101% (Knickmeyer et al. 2008). Synaptogenesis, the formation of synapses, begins explosively at term, and myelination begins before birth but accelerates in the neonatal period (Dubois et al. 2014). As synaptogenesis and myelination progress, the neural networks become increasingly connected, facilitating efficient communication and coordination between brain regions. Due to the rapid changes occurring across all levels of brain functional and structural development during this period, the fetal period and infant period are the most plastic and susceptible to external influences such as prenatal and postnatal stress.

The salience and arousal systems, fundamental for detecting and responding to environmental stimuli, begin their development and maturation earlier than regulation systems to support processing external stimuli (Young et al. 2017). The amygdala, for example, undergoes significant development during prenatal and early postnatal life. The amygdala begins to form early in the embryonic period at the boundary between the pallium and subpallidum. By the term period, the amygdala demonstrates relatively few changes, indicating that most of the structural development is complete (Mulc et al. 2024). Throughout the first years of life, the amygdala continues to grow, forming extensive connections with other brain regions, such as the prefrontal cortex, thalamus, and hippocampus (Uematsu et al. 2012).

Additionally, the regulatory systems, crucial for modulating responses to stimulus-driven arousal and supporting higher-order cognitive functions, undergo a protracted developmental trajectory. Despite the extended period of maturation, these systems also experience a significant phase of rapid development early in life (Kolk and Rakic 2022). The prefrontal cortex develops rapidly between 0 and 2 years of age with the expansion of frontal lobe gray matter volume and dramatic changes in surface complexity (Hodel 2018). At birth, regions of the medial prefrontal cortex are among the thickest of all cortices in the brain (Li et al. 2015; Lyall et al. 2015). Relatively thicker prefrontal regions also show the fastest rate of growth in the first year of life, along with disproportionately rapid thickening in other frontal subregions (inferior frontal operculum, superior frontal gyrus) during the second year. By the age of two, the PFC's surface area is approximately 70% of its adult size, and cortical thickness approaches adult-like levels (Li et al. 2015; Lyall et al. 2015). Long-range frontal lobe connections are present from the second trimester of gestation and continue to develop postnatally. MRI studies have found that the onset of measurable myelination is slower in the frontal lobe compared to other brain lobes after birth, but displays a higher rate of myelin accrual over the first years of life compared to regions that started earlier (Deoni et al. 2011).

Moreover, regions within the salience/arousal and regulatory systems, like the amygdala and PFC, are rich in glucocorticoid receptors that bind to stress hormones (McEwen et al. 2016). Stress exposure prompts the release of cortisol, which can result in persistent binding of these hormones to receptors and lasting structural and functional changes. For example, in rats, stress results in an increase in dendritic length within the amygdala (Vyas et al. 2002). This response can be mimicked by administering a large bolus of corticosterone, which also induces anxiety-like behavior (Mitra and Sapolsky 2008). Traumatic stress can acutely increase spine density on amygdala neurons without initially altering dendritic length, a change that becomes apparent after a delay (Mitra et al. 2005). In the PFC, chronic stress leads to structural remodeling, including a reduction in dendritic length and spine density of neurons, particularly in the medial PFC (Cerqueira et al. 2005). Thus, the amygdala and PFC are sensitive to stress exposure particularly during early life when they experience significant growth (Malter Cohen et al. 2013). Stress during this formative period may have lasting impacts on the functioning of the salience/arousal and regulatory systems, potentially affecting emotional processing, stress reactivity, and cognitive functions.

6 Prenatal Stress, Neural Development, and Risks of Anxiety

The links between prenatal stress exposure and alterations in both brain structure and functioning have been well-documented (Lautarescu et al. 2020). Of particular relevance to this chapter, prenatal stress has been associated with changes in brain regions involved in stress response and regulation and implicated in anxiety (Favaro et al. 2015; Jones et al. 2019; Marečková et al. 2019; Mareckova et al. 2022). In this section, we review studies that examine these structural and functional changes, highlighting the potential pathways through which prenatal stress may link to the development of anxiety outcomes later in life.

Prenatal stress has been linked to structural changes in the amygdala. A study, investigating 68 children whose mothers were pregnant during the January 1998 Quebec Ice Storm, revealed that greater objective hardship was linked to larger amygdala volumes ($\beta s = 0.318–0.376$) in girls averaging 11.5 years old at the time of the MRI assessment (Jones et al. 2019). Furthermore, in a sample of 35 healthy women (mean age = 25.6 years), higher prenatal stress, evaluated by semi-structured interviews, was associated with smaller gray matter volume in the left medial temporal lobe and both amygdalae, while no significant changes were observed in the hippocampus. Amygdala volume was significantly predicted by prenatal stress (left amygdala: $\beta = -0.55$; right amygdala: $\beta = -0.52$), after accounting for age, postnatal stress, maternal smoking, and obstetric complications (Favaro et al. 2015). A more recent study with 129 participants (mean age = 23.9 years) reported no significant associations between overall amygdala volume in offspring and prenatal stress after adjusting for various confounders, such as maternal age at birth and birth weight (Mareckova et al. 2022). It is noteworthy that the direction of the association appears to depend on the age at which amygdala volume is assessed. In younger children, for example, prenatal stress is associated with larger amygdala volumes, while in older children, the opposite association is observed, potentially reflecting accelerated structural maturation related to stress (Callaghan and Tottenham 2016).

In addition to the amygdala, maternal stress during pregnancy has been linked to variations in other brain structures (Marečková et al. 2019). For example, a study with 93 participants found that higher prenatal stress predicted less volume within the ACC ($\beta = -0.21$), dlPFC ($\beta = -0.29$), and precuneus ($\beta = -0.20$) by young adulthood (Marečková et al. 2019). As previously mentioned, the ACC and dlPFC are crucial structures within the salience and arousal systems and the top-down regulation systems, respectively.

Furthermore, prenatal maternal stress has been linked to altered structural and functional connectivity between the amygdala and prefrontal cortical areas, a pathway critical for top-down regulation. The uncinate fasciculus is a long-range association tract connecting regions of the frontal lobe, specifically the orbitofrontal cortex, to the anterior temporal lobe, including the amygdala and anterior parahippocampal gyrus (Olson et al. 2015). One important function of the uncinate fasciculus is to transmit salience-laden stimulus representations stored in anterior-medial aspects of the temporal lobe to the lateral orbital frontal cortex (Von Der Heide et al. 2013). Compromised integrity of the uncinate fasciculus tract is associated with neurodevelopmental and psychiatric disorders, including anxiety disorders (Costanzo et al. 2016; Koch et al. 2017; Phan et al. 2009). Maternal prenatal stress exposure has been associated with changes in the uncinate fasciculus white matter microstructure (Lautarescu et al. 2020; Lean et al. 2022; Sarkar et al. 2014). Specifically, in 22 healthy children aged 6–9 years, self-reported prenatal stress exposure was correlated with lower fractional anisotropy (FA) in the right uncinate fasciculus ($r = 0.48$) of the child, but not in a control tract (Sarkar et al. 2014). Given the small sample size, the large effect size reported in one study might warrant cautious interpretation. Nonetheless, among 251 preterm infants assessed at term equivalence, higher stressful life event scores were associated with higher axial diffusivity ($\beta = 0.177$), radial diffusivity ($\beta = 0.133$), and mean diffusivity ($\beta = 0.149$) in the left uncinate fasciculus, and higher axial diffusivity ($\beta = 0.142$) in the right uncinate fasciculus (Lautarescu et al. 2020). Additionally, stress during pregnancy has also been associated with altered amygdala–prefrontal functional connectivity (Humphreys et al. 2020). A longitudinal study in a small sample of pregnant people and their infant offspring ($n = 33$ dyads) found that the total number of stressors experienced during pregnancy was associated with lower functional connectivity and greater structural connections between seeds of the amygdala and medial PFC ($\beta = -0.26$ for functional and $\beta = 0.29$ for structural connectivity, respectively) when infants were approximately 5 weeks (Humphreys et al. 2020). Altogether, these findings highlight that prenatal stress is linked to potentially compromised amygdala–prefrontal regulation connectivity, which may predispose infants to a higher risk of anxiety later in life.

7 Early Postnatal Stress, Neural Development, and Risks of Anxiety

Early postnatal stress experienced within the first 2 years of life is linked with changes in neural circuits associated with anxiety. As previously mentioned, the bulk of evidence on the effects of early life stress, particularly within the first 2 years, comes from studies on children who experienced early institutionalization followed by adoption, which allows researchers to have a better assessment of the timing of stress exposure. Early institutional experiences have been linked to the structure of amygdala. In a study of 78 children (mean age = 8.3 years) with histories of orphanage care and a control group, no significant group differences were observed in the volumes of the amygdala, hippocampus, and caudate (Tottenham et al. 2010). However, children adopted after 15 months of age exhibited significantly larger amygdala volumes compared to both early-adopted children and those in the comparison groups, highlighting the potential specific long-term associations between stress and the amygdala evident even years after the termination of adversity. Additionally, a longer duration of institutional care was associated with higher levels of anxiety and less effective emotion regulation abilities when exposed to negative emotional stimuli during an emotional go/no-go task. In another sample of 36 international adopted children (mean age at adoption = 38.08 months, with an average duration of institutional care = 29.52 months), on average, these children exhibited significantly smaller amygdala volume compared to the control participants, although no differences were observed in hippocampal volume (Hanson et al. 2015).

Beyond structural differences in the amygdala, early institutional experiences are associated with amygdala hyperactivity in response to emotional stimuli and heightened anxiety symptoms. In a study of 55 children (mean age = 10 years), those with a history of institutional care exhibited greater anxiety symptoms and heightened amygdala activity to fearful faces and distracter stimuli compared to a control group, with no group difference observed for neutral and target faces (Tottenham et al. 2011). Additionally, the emotion of the face stimuli (neural vs. fearful) interacted with the group in several other regions, including the vmPFC and ACC. For example, while both groups showed increased activity in the anterior cingulate and inferior/middle gyrus for fear relative to baseline, the early institution group exhibited a greater increase for neutral stimuli. These findings of the heightened amygdala activity and anxiety symptoms in children with early orphanage care were replicated in another sample (Gee et al. 2013). More intriguingly, building on those studies on institutional care, an innovative animal study was conducted to closely parallel the stress experienced in early orphanage life while being able to control for genetic and environmental confounds inherent in human studies (Malter Cohen et al. 2013). The study simulated the disruption of maternal care followed by adoption analogous to orphanage rearing. Results revealed early and persistent differences in anxiety-related behaviors and amygdala functionality that did not diminish even after the stressor is removed. Taken together, these findings from animal models to human conditions illustrate the connection between amygdala hyperactivity to emotionally relevant stimuli and stressful early postnatal environments like institutional care.

Other than heightened amygdala activity in response to emotional stimuli, prior exposure to institutional care has also been linked to impaired social threat discrimination. Specifically, children with prior exposure to institutional care exhibited less differentiation in amygdala response to trustworthy versus untrustworthy faces, compared to never-institutionalized children (Green et al. 2016). Among those with an institutional care history, their age of adoption was negatively correlated with amygdala differentiation ($r = -0.76$ and $r = -0.35$ for left and right amygdala difference, respectively), suggesting that greater exposure to depriving settings may result in the observed effect (Green et al. 2016). Further, the extent of amygdala differential response was longitudinally associated with separation anxiety symptoms over 2 years among both groups, suggesting that a lack of amygdala differential responding might be a potential mechanism underlying the associations between early postnatal stress and individual differences in anxiety – a compromised ability to evaluate and distinguish actual danger from non-threatening situations (Richards et al. 2014).

Institutional care experience has been associated with both structural and functional connectivity between the amygdala and prefrontal cortical regions. A study of 10 children (mean age = 8.83 years) adopted from Romania orphanages after residing in the orphanage for an average of 38 months before adoption found significantly lower glucose metabolism, an index of the brain's functional activity, in limbic and paralimbic structures, including the orbital frontal gyrus, infralimbic prefrontal cortex, medial temporal structures (amygdala and hippocampus head), lateral temporal cortex, and brain stem compared to never-institutionalized individuals. A follow-up study conducted by the same research team examined white matter tract integrity in tracts connecting these regions of reduced glucose metabolism, and found lower FA, specifically in the left uncinate fasciculus, among 7 institutionalized compared to 7 comparison children (Eluvathingal et al. 2006). Another study with 17 children with histories of institutional rearing (mean age = 10.9 years) who were subsequently adopted and 15 age-matched comparison children found lower FA in frontal, temporal, and parietal white matter including components of uncinate and superior longitudinal fasciculi among the children with histories of institutional rearing (Govindan et al. 2010). Further, investigation into children with a history of institutionalization observed that comparison children showed an immature pattern of amygdala–mPFC connectivity (positively coupled amygdala–mPFC activity), whereas children with a history of institutionalization displayed a mature pattern of connectivity (negatively coupled amygdala–mPFC activity), such that they resembled the adolescent phenotype (Gee et al. 2013). Despite being age-atypical, negative amygdala–mPFC coupling conferred some degree of reduced anxiety in previously institutionalized children, although anxiety still remained significantly higher, on average, among children with histories of institutional care. However, these findings should be interpreted with caution given the small sample sizes. Callaghan and Tottenham (2016) proposed that early stressful environments may prompt an accelerated development of emotion circuits, including the amygdala's precocious maturation as mentioned in the previous section, and amygdala–prefrontal connectivity. This is consistent with findings from another study of 23 infants aged 6–12 months, in which heightened posterior cingulate cortex (PCC)–mPFC connectivity, indicative of a more "mature" pattern, mediated the relationship between higher stress levels and greater negative infant emotionality (Graham et al. 2015). This accelerated development may confer immediate benefits (e.g., mitigating anxiety) as an adaptive mechanism. However, these short-term advantages may increase the risk of developing psychopathology, including anxiety, in the long run. Overall, studies of early deprivation in the form of institutional care provide evidence for the relation between early postnatal stress and neurobiological variation relevant to anxiety.

8 Current State of Knowledge, Level of Confidence, and Gaps

The prenatal period and the first 2 years of postnatal life constitute a critical window for human brain development, marked by remarkable malleability and plasticity, making it highly susceptible to environmental influences (Gee and Casey 2015). Previously, we reviewed evidence on how prenatal and early postnatal stress links to brain differences in regions/networks involved in salience/arousal and top-down regulation systems which might increase the risk of developing anxiety in children (see Fig. 1). The body of evidence acquired to date provides important insights into the link between early life stress and anxiety, as well as the potential neural mechanisms involved. However, limitations in this work warrant caution, including (1) consideration of the potential influence of confounding factors, (2) generalizability and small sample size constraints, (3) a narrowed regional brain focus, and (4) a general lack of specificity in types of stress.

The first limitation involves the presence of a range of confounding factors, including genetic factors, the subsequent postnatal stress environment (or prenatal stress), socioeconomic status, and the parents' own mental health and stress levels. For example, the temperament of an infant, often reported by parents, could be influenced by the parents' perceptions and their own psychological state, thus introducing bias. Adjusting for these factors is challenging. While prenatal stress studies often attempt to account for these potential confounders through statistical controls, the interactions of these variables may still obscure the conclusions. For

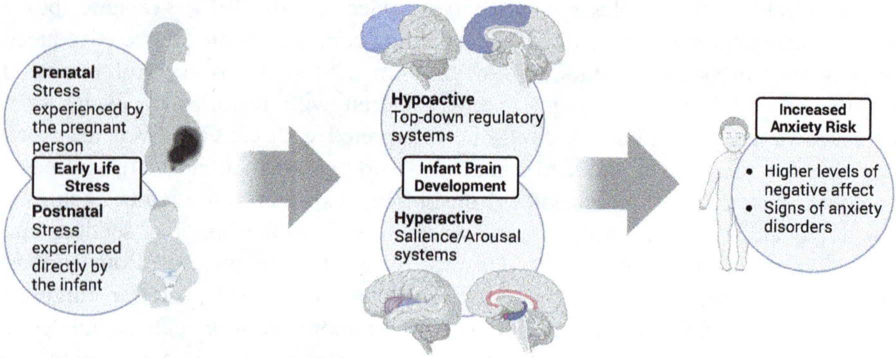

Fig. 1 Schematic diagram of the associations between prenatal and early postnatal stress, neurodevelopmental outcomes, and risk for anxiety phenotypes. This illustration represents the manner in which exposure to early life stress is thought to influence the development of neural structures and functions that, in turn, predispose children to a heightened risk of anxiety

instance, the negative effects of high prenatal stress may be greater under the condition of suboptimal nutritional status. It is important to note that while researchers typically use two distinct terms – prenatal stress and postnatal stress – stress during these periods is not likely to be orthogonal. Instead, it is more likely to represent a continuum of stress exposure throughout early development and some studies take advantage of natural experiments to try to determine the impact of the timing of stress to better identify sensitive periods. Understanding when stress has the most impact during development can help elucidate critical windows of vulnerability in the developing brain. Furthermore, cross-sectional studies, which predominate in existing literature, are subject to confounders related to individual neural differences at baseline (e.g., from the earliest stages of brain development).

Most studies examining the links between early stress exposure and brain development are based on a single type of stress (i.e., psychosocially depriving institutional care settings). While designs following children who experienced this type of stress have advantages in terms of more clearly defined timelines for the onset/offset of exposure to deprivation, there may be limited generalizability to other forms of stress. In addition, such studies typically involve small sample sizes, limiting their statistical power to detect subtle, but potentially important, effects.

One point to consider in interpreting the available literature is the predominant focus on specific brain regions – primarily the amygdala and prefrontal cortex – which may lead to the potential oversight of other critical areas and large-scale network organizations of the brain involved in stress response and anxiety. One such region is the bed nucleus of the stria terminalis (BNST), another key player in animal models of anxiety that has recently received attention in research on humans (Avery et al. 2016). BNST is tightly coupled with the central amygdala and is particularly implicated in the response to sustained, unpredictable, or diffuse threats (Lebow and Chen 2016). Furthermore, the BNST is also rich in glucocorticoid receptors, making it susceptible to the effects of stress hormones released during early life stress. The

BNST also has extensive connections to the hypothalamus and is likely to regulate the hypothalamic–pituitary–adrenal (HPA) axis to mediate sustained fear states, fear learning, and the stress response (Maita et al. 2021). Despite the BNST's importance to anxiety and its potential vulnerability to early life stress, no studies to date have examined this structure earlier than late childhood (Feola et al. 2021). This notable gap may be due to the small size of the BNST, especially in infant brains. Addressing this gap requires more sophisticated imaging methods with greater resolution to accurately trace and measure its structure and function, especially in the developing brains of young children.

Lastly, given challenges in the assessment of stress, prior work typically has not examined potential specificity in terms of the influences of different types of stress on brain and behavioral outcomes. For example, neglect versus physical abuse may affect distinct brain systems and pathways (Teicher and Samson 2016). Moving away from cumulative risk models, especially for postnatal stress exposure, will be important for creating datasets that allow for exploration of shared and unique associations of stress on development.

9 Future Directions

We have several recommendations for next steps in examining the potential role of stress exposures and developmental consequences. First, large longitudinal study designs are especially valuable. The Healthy Brain and Child Development (HBCD) Study (Volkow et al. 2021) is estimated to include 7,500 children with repeated assessments of stress exposure, brain structure and function, and social, emotional, and cognitive development, from prenatal development through the first decade of life. This unique dataset represents a powerful design for future studies on the influence of stress on brain and behavioral development. Frequent assessments of stress reduce the risk for memory biases related to retrospective report and better allow for determining stress onset/offset, chronicity, and intensity, which may be relevant for understanding risk and resilience, as well as identifications of critical and sensitive periods for stress (Gabard-Durnam and McLaughlin 2020; Nelson and Gabard-Durnam 2020). Moreover, large samples that encompass a wide range of early life experiences (e.g., abuse, neglect, violence exposure, and unpredictability) provide an opportunity for considering the potential for shared versus unique vulnerabilities relevant to dimensions of adversity (McLaughlin et al. 2021). Lastly, longitudinal models allow for better control over individual neurobiological differences affecting stress responses, allowing for isolation of stress effects from inherent neurobiological variances. Longitudinal data would also offer richer insights into the evolving impact of stress on developmental trajectories and anxiety risk.

An additional valuable approach is to utilize nonhuman animal research to gain evidence regarding the intensity and nature of adverse conditions. Such designs (e.g., Malter Cohen et al. 2013) offer a unique opportunity to control for genetic and environmental confounds that are inherent to human studies. By simulating human-like stressors in a controlled setting, researchers could make causal inferences on stress exposure and specific neurobiological outcomes, elucidating underlying mechanisms that could be examined in correlational human data. However, the interpretation of findings requires caution due to the challenges inherent in comparing animal models to human experiences (Brett et al. 2015).

Lastly, well-designed intervention studies are instrumental for inferring causality and identifying effective strategies to mitigate the impacts of early life stress. Such approach helps establish the causal links between components in early intervention and neurodevelopmental outcomes, enhancing our mechanic understanding of how early stress influences neurodevelopment and the onset of anxiety. Moreover, intervention studies lay the groundwork for evidence-based prevention and therapeutic practices. Incorporating a range of diverse interventions – from behavioral strategies targeting parents and caregivers to foster stable, nurturing environments, to interventions directly targeting children to improve emotional regulation and stress coping – offers potential to promote resilience in those exposed to early stress and at increased risk for anxiety disorders, ultimately improving overall well-being.

10 Conclusion

In conclusion, this chapter has explored the connection between early life stress (prenatal and up to 2 years old), the increased risk of developing anxiety disorders, and potential neural pathways involved. To date, we have substantial evidence linking early life stress exposures with increased vulnerability to anxiety. Further, there is some evidence suggesting that changes in brain regions associated with salience/arousal and top-down regulation systems, particularly the amygdala and prefrontal cortex, may be responsible for this association. However, given limitations posed by small samples, cross-sectional designs, and risk for confounding factors, more work is needed to be confident about causal pathways from early stress exposure to anxiety.

Acknowledgment of Funding This work was supported in part by funding from the National Science Foundation [CAREER Award 2042285].

References

Adhikari A, Lerner TN, Finkelstein J, Pak S, Jennings JH, Davidson TJ et al (2015) Basomedial amygdala mediates top-down control of anxiety and fear. Nature 527:179–185

Agorastos A, Pervanidou P, Chrousos GP, Baker DG (2019) Developmental trajectories of early life stress and trauma: a narrative review on neurobiological aspects beyond stress system dysregulation. Front Psych 10

Andersen SL (2003) Trajectories of brain development: point of vulnerability or window of opportunity? Neurosci Biobehav Rev 27:3–18

Anderson AK, Phelps EA (2001) Lesions of the human amygdala impair enhanced perception of emotionally salient events. Nature 411:305–309

Avery SN, Clauss JA, Blackford JU (2016) The human BNST: functional role in anxiety and addiction. Neuropsychopharmacology 41:126–141

Barrash J, Tranel D, Anderson SW (2000) Acquired personality disturbances associated with bilateral damage to the ventromedial prefrontal region. Dev Neuropsychol 18:355–381

Bergman K, Sarkar P, O'Connor TG, Modi N, Glover V (2007) Maternal stress during pregnancy predicts cognitive ability and fearfulness in infancy. J Am Acad Child Adolesc Psychiatry 46: 1454–1463

Berlim MT, Van den Eynde F (2014) Repetitive transcranial magnetic stimulation over the dorsolateral prefrontal cortex for treating posttraumatic stress disorder: an exploratory meta-analysis of randomized, double-blind and sham-controlled trials. Can J Psychiatry 59:487–496

Botvinick MM, Cohen JD, Carter CS (2004) Conflict monitoring and anterior cingulate cortex: an update. Trends Cogn Sci 8:539–546

Brett ZH, Humphreys KL, Fleming AS, Kraemer GW, Drury SS (2015) Using cross-species comparisons and a neurobiological framework to understand early social deprivation effects on behavioral development. Dev Psychopathol 27:347–367

Burkholder AR, Koss KJ, Hostinar CE, Johnson AE, Gunnar MR (2016) Early life stress: effects on the regulation of anxiety expression in children and adolescents. Soc Dev 25:777–793

Callaghan BL, Tottenham N (2016) The stress acceleration hypothesis: effects of early-life adversity on emotion circuits and behavior. Curr Opin Behav Sci 7:76–81

Campbell RK, Curtin P, Bosquet Enlow M, Brunst KJ, Wright RO, Wright RJ (2020) Disentangling associations among maternal lifetime and prenatal stress, psychological functioning during pregnancy, maternal race/ethnicity, and infant negative affectivity at age 6 months: a mixtures approach. Health Equity 4:489–499

Cerqueira JJ, Pêgo JM, Taipa R, Bessa JM, Almeida OFX, Sousa N (2005) Morphological correlates of corticosteroid-induced changes in prefrontal cortex-dependent behaviors. J Neurosci 25:7792–7800

Chavanne AV, Robinson OJ (2021) The overlapping neurobiology of induced and pathological anxiety: a meta-analysis of functional neural activation. Am J Psychiatry 178:156–164

Costanzo ME, Jovanovic T, Pham D, Leaman S, Highland KB, Norrholm SD, Roy MJ (2016) White matter microstructure of the uncinate fasciculus is associated with subthreshold posttraumatic stress disorder symptoms and fear potentiated startle during early extinction in recently deployed service members. Neurosci Lett 618:66–71

Cowan CSM, Callaghan BL, Kan JM, Richardson R (2016) The lasting impact of early-life adversity on individuals and their descendants: potential mechanisms and hope for intervention. Genes Brain Behav 15:155–168

Craig AD (2002) How do you feel? Interoception: the sense of the physiological condition of the body. Nat Rev Neurosci 3:655–666

Crockenberg SC, Leerkes EM (2006) Infant and maternal behavior moderate reactivity to novelty to predict anxious behavior at 2.5 years. Dev Psychopathol 18:17–34

Davis EP, Sandman CA (2012) Prenatal psychobiological predictors of anxiety risk in preadolescent children. Psychoneuroendocrinology 37:1224–1233

Davis M, Whalen PJ (2001) The amygdala: vigilance and emotion. Mol Psychiatry 6:13–34

Davis MM, Navarro E, Carpenter KLH, Copeland WE, Small B, Egger HL, Sheridan MA (2023) Early life adversity and psychopathology in preschoolers: mechanisms and moderators. Eur Child Adolesc Psychiatry

De Weerth C, Buitelaar JK (2005) Physiological stress reactivity in human pregnancy – a review. Neurosci Biobehav Rev 29:295–312

Delgado MR, Nearing KI, LeDoux JE, Phelps EA (2008) Neural circuitry underlying the regulation of conditioned fear and its relation to extinction. Neuron 59:829–838

Deoni SCL, Mercure E, Blasi A, Gasston D, Thomson A, Johnson M et al (2011) Mapping infant brain myelination with magnetic resonance imaging. J Neurosci 31:784–791

Diefenbach GJ, Bragdon LB, Zertuche L, Hyatt CJ, Hallion LS, Tolin DF et al (2016) Repetitive transcranial magnetic stimulation for generalised anxiety disorder: a pilot randomised, double-blind, sham-controlled trial. Br J Psychiatry 209:222–228

Dubois J, Dehaene-Lambertz G, Kulikova S, Poupon C, Hüppi PS, Hertz-Pannier L (2014) The early development of brain white matter: a review of imaging studies in fetuses, newborns and infants. Neuroscience 276:48–71

Duvarci S, Pare D (2014) Amygdala microcircuits controlling learned fear. Neuron 82:966–980

Eippert F, Veit R, Weiskopf N, Erb M, Birbaumer N, Anders S (2007) Regulation of emotional responses elicited by threat-related stimuli. Hum Brain Mapp 28:409–423

Ellis BH, Fisher PA, Zaharie S (2004) Predictors of disruptive behavior, developmental delays, anxiety, and affective symptomatology among institutionally reared Romanian children. J Am Acad Child Adolesc Psychiatry 43:1283–1292

Eluvathingal TJ, Chugani HT, Behen ME, Juhász C, Muzik O, Maqbool M et al (2006) Abnormal brain connectivity in children after early severe socioemotional deprivation: a diffusion tensor imaging study. Pediatrics 117:2093–2100

Etkin A, Wager TD (2007) Functional neuroimaging of anxiety: a meta-analysis of emotional processing in PTSD, social anxiety disorder, and specific phobia. Am J Psychiatry 164:1476–1488

Etkin A, Büchel C, Gross JJ (2015) The neural bases of emotion regulation. Nat Rev Neurosci 16: 693–700

Favaro A, Tenconi E, Degortes D, Manara R, Santonastaso P (2015) Neural correlates of prenatal stress in young women. Psychol Med 45:2533–2543

Feola B, Melancon SNT, Clauss JA, Noall MP, Mgboh A, Flook EA et al (2021) Bed nucleus of the stria terminalis and amygdala responses to unpredictable threat in children. Dev Psychobiol 63 (8):e22206

Fitzgerald KD, Liu Y, Stern ER, Welsh RC, Hanna GL, Monk CS et al (2013) Reduced error-related activation of dorsolateral prefrontal cortex across pediatric anxiety disorders. J Am Acad Child Adolesc Psychiatry 52:1183–1191.e1

Fitzgerald JM, DiGangi JA, Phan KL (2018) Functional neuroanatomy of emotion and its regulation in PTSD. Harv Rev Psychiatry 26:116

Gabard-Durnam L, McLaughlin KA (2020) Sensitive periods in human development: charting a course for the future. Curr Opin Behav Sci 36:120–128

Gee DG, Casey BJ (2015) The impact of developmental timing for stress and recovery. Neurobiol Stress 1:184–194

Gee DG, Gabard-Durnam LJ, Flannery J, Goff B, Humphreys KL, Telzer EH et al (2013) Early developmental emergence of human amygdala–prefrontal connectivity after maternal deprivation. Proc Natl Acad Sci 110:15638–15643

Govindan RM, Behen ME, Helder E, Makki MI, Chugani HT (2010) Altered water diffusivity in cortical association tracts in children with early deprivation identified with tract-based spatial statistics (TBSS). Cereb Cortex 20:561–569

Graham AM, Pfeifer JH, Fisher PA, Carpenter S, Fair DA (2015) Early life stress is associated with default system integrity and emotionality during infancy. J Child Psychol Psychiatry 56:1212–1222

Green SA, Goff B, Gee DG, Gabard-Durnam L, Flannery J, Telzer EH et al (2016) Discrimination of amygdala response predicts future separation anxiety in youth with early deprivation. J Child Psychol Psychiatry 57:1135–1144

Hammen C (2006) Stress generation in depression: reflections on origins, research, and future directions. J Clin Psychol 62:1065–1082

Hanson JL, Nacewicz BM, Sutterer MJ, Cayo AA, Schaefer SM, Rudolph KD et al (2015) Behavioral problems after early life stress: contributions of the hippocampus and amygdala. Biol Psychiatry 77:314–323

Hiser J, Koenigs M (2018) The multifaceted role of the ventromedial prefrontal cortex in emotion, decision making, social cognition, and psychopathology. Biol Psychiatry 83:638–647

Hodel AS (2018) Rapid infant prefrontal cortex development and sensitivity to early environmental experience. Dev Rev 48:113–144

Huizink AC (2008) Prenatal stress exposure and temperament: A review. Int J Dev Sci 2:77–99

Humphreys KL, Fox NA, Nelson CA, Zeanah CH (2017) Psychopathology following severe deprivation: history, research, and implications of the Bucharest early intervention project. Child maltreatment in residential care: history, research, and current practice, pp 129–148

Humphreys KL, Camacho MC, Roth MC, Estes EC (2020) Prenatal stress exposure and multimodal assessment of amygdala–medial prefrontal cortex connectivity in infants. Dev Cogn Neurosci 46:100877

Johnston MV, Nishimura A, Harum K, Pekar J, Blue ME (2001) CHAPTER 1 – sculpting the developing brain. Adv Pediatr 48:1–40

Jones SL, Dufoix R, Laplante DP, Elgbeili G, Patel R, Chakravarty MM et al (2019) Larger amygdala volume mediates the association between prenatal maternal stress and higher levels of externalizing behaviors: sex specific effects in project Ice Storm. Front Hum Neurosci 13. Retrieved from https://www.frontiersin.org. https://doi.org/10.3389/fnhum.2019.00144

Juruena MF, Eror F, Cleare AJ, Young AH (2020) The role of early life stress in HPA axis and anxiety. In: Kim Y-K (ed) Anxiety disorders: rethinking and understanding recent discoveries. Springer, Singapore, pp 141–153

Kagan J, Snidman N (1999) Early childhood predictors of adult anxiety disorders. Biol Psychiatry 46:1536–1541

Kessler RC, Berglund P, Demler O, Jin R, Merikangas KR, Walters EE (2005) Lifetime prevalence and age-of-onset distributions of DSM-IV disorders in the National Comorbidity Survey Replication. Arch Gen Psychiatry 62:593–602

Kim H, Somerville LH, Johnstone T, Alexander AL, Whalen PJ (2003) Inverse amygdala and medial prefrontal cortex responses to surprised faces. Neuroreport 14:2317

Knickmeyer RC, Gouttard S, Kang C, Evans D, Wilber K, Smith JK et al (2008) A structural MRI study of human brain development from birth to 2 years. J Neurosci 28:12176–12182

Koch SBJ, van Zuiden M, Nawijn L, Frijling JL, Veltman DJ, Olff M (2017) Decreased uncinate fasciculus tract integrity in male and female patients with PTSD: A diffusion tensor imaging study. J Psychiatry Neurosci 42:331–342

Kochukhova O, Mikhailova AA, Dyagileva JO, Makhin SA, Pavlenko VB (2016) Temperament differences between institution- and family-reared toddlers. Infant Behav Dev 45:91–97

Kolk SM, Rakic P (2022) Development of prefrontal cortex. Neuropsychopharmacology 47 (1):41–57

Kostović I, Sedmak G, Judaš M (2019) Neural histology and neurogenesis of the human fetal and infant brain. Neuroimage 188:743–773

Lafortune S, Laplante DP, Elgbeili G, Li X, Lebel S, Dagenais C, King S (2021) Effect of natural disaster-related prenatal maternal stress on child development and health: a meta-analytic review. Int J Environ Res Public Health 18:8332

Lautarescu A, Craig MC, Glover V (2020) Prenatal stress: effects on fetal and child brain development. Int Rev Neurobiol 150:17–40

Lean RE, Smyser CD, Brady RG, Triplett RL, Kaplan S, Kenley JK et al (2022) Prenatal exposure to maternal social disadvantage and psychosocial stress and neonatal white matter connectivity at birth. Proc Natl Acad Sci 119:e2204135119

Lebow MA, Chen A (2016) Overshadowed by the amygdala: the bed nucleus of the stria terminalis emerges as key to psychiatric disorders. Mol Psychiatry 21:450–463

Lewis AJ, Olsson CA (2011) Early life stress and child temperament style as predictors of childhood anxiety and depressive symptoms: findings from the longitudinal study of Australian children. Depress Res Treat 2011:e296026

Lezak KR, Missig G, Carlezon WA Jr (2017) Behavioral methods to study anxiety in rodents. Dialogues Clin Neurosci 19:181–191

Li G, Lin W, Gilmore JH, Shen D (2015) Spatial patterns, longitudinal development, and hemispheric asymmetries of cortical thickness in infants from birth to 2 years of age. J Neurosci 35: 9150–9162

Lin B, Crnic KA, Luecken LJ, Gonzales NA (2014) Maternal prenatal stress and infant regulatory capacity in Mexican Americans. Infant Behav Dev 37:571–582

Lupien SJ, McEwen BS, Gunnar MR, Heim C (2009) Effects of stress throughout the lifespan on the brain, behaviour and cognition. Nat Rev Neurosci 10:434–445

Lyall AE, Shi F, Geng X, Woolson S, Li G, Wang L et al (2015) Dynamic development of regional cortical thickness and surface area in early childhood. Cereb Cortex 25:2204–2212

Maita I, Bazer A, Blackford JU, Samuels BA (2021) Functional anatomy of the bed nucleus of the stria terminalis–hypothalamus neural circuitry: implications for valence surveillance, addiction, feeding, and social behaviors. Handb Clin Neurol 179:403–418

Malanchini M, Engelhardt LE, Raffington LA, Sabhlok A, Grotzinger AD, Briley DA et al (2021) Weak and uneven associations of home, neighborhood, and school environments with stress hormone output across multiple timescales. Mol Psychiatry 26:4823–4838

Malter Cohen M, Jing D, Yang RR, Tottenham N, Lee FS, Casey BJ (2013) Early-life stress has persistent effects on amygdala function and development in mice and humans. Proc Natl Acad Sci 110:18274–18278

Marečková K, Klasnja A, Bencurova P, Andrýsková L, Brázdil M, Paus T (2019) Prenatal stress, mood, and cray matter volume in young adulthood. Cereb Cortex 29:1244–1250

Mareckova K, Marecek R, Andryskova L, Brazdil M, Nikolova YS (2022) Impact of prenatal stress on amygdala anatomy in young adulthood: timing and location matter. Biol Psychiatry Cogn Neurosci Neuroimaging 7:231–238

McEwen BS, Bowles NP, Gray JD, Hill MN, Hunter RG, Karatsoreos IN, Nasca C (2015) Mechanisms of stress in the brain. Nat Neurosci 18:1353–1363

McEwen BS, Nasca C, Gray JD (2016) Stress effects on neuronal structure: hippocampus, amygdala, and prefrontal cortex. Neuropsychopharmacology 41:3–23

McLaughlin KA, Sheridan MA, Humphreys KL, Belsky J, Ellis BJ (2021) The value of dimensional models of early experience: thinking clearly about concepts and categories. Perspect Psychol Sci 16:1463–1472

Meaney MJ, Liebowitz MR, David Leonardo E (2020) Neurobiology of anxiety disorders. In: Tasman A, Riba MB, Alarcón RD, Alfonso CA, Kanba S, Ndetei DM et al (eds) Tasman's psychiatry. Springer, Cham, pp 1–43

Mitra R, Sapolsky RM (2008) Acute corticosterone treatment is sufficient to induce anxiety and amygdaloid dendritic hypertrophy. Proc Natl Acad Sci 105:5573–5578

Mitra R, Jadhav S, McEwen BS, Vyas A, Chattarji S (2005) Stress duration modulates the spatiotemporal patterns of spine formation in the basolateral amygdala. Proc Natl Acad Sci 102:9371–9376

Möhler E, Parzer P, Brunner R, Wiebel A, Resch F (2006) Emotional stress in pregnancy predicts human infant reactivity. Early Hum Dev 82:731–737

Motzkin JC, Philippi CL, Wolf RC, Baskaya MK, Koenigs M (2015) Ventromedial prefrontal cortex is critical for the regulation of amygdala activity in humans. Biol Psychiatry 77:276–284

Mulc D, Smilović D, Krsnik Ž, Junaković-Munjas A, Kopić J, Kostović I et al (2024) Fetal development of the human amygdala. J Comp Neurol 532:e25580

Nelson CA, Gabard-Durnam LJ (2020) Early adversity and critical periods: neurodevelopmental consequences of violating the expectable environment. Trends Neurosci 43:133–143

Olson IR, Heide RJVD, Alm KH, Vyas G (2015) Development of the uncinate fasciculus: implications for theory and developmental disorders. Dev Cogn Neurosci 14:50–61

Öngür D, Price JL (2000) The organization of networks within the orbital and medial prefrontal cortex of rats, monkeys and humans. Cereb Cortex 10:206–219

Pechtel P, Pizzagalli DA (2011) Effects of early life stress on cognitive and affective function: an integrated review of human literature. Psychopharmacology (Berl) 214:55–70

Pêgo JM, Sousa JC, Almeida O, Sousa N (2010) Stress and the neuroendocrinology of anxiety disorders. In: Stein MB, Steckler T (eds) Behavioral neurobiology of anxiety and its treatment. Springer, Berlin, pp 97–118

Phan KL, Orlichenko A, Boyd E, Angstadt M, Coccaro EF, Liberzon I, Arfanakis K (2009) Preliminary evidence of white matter abnormality in the uncinate fasciculus in generalized social anxiety disorder. Biol Psychiatry 66:691–694

Phelps EA, Delgado MR, Nearing KI, LeDoux JE (2004) Extinction learning in humans: role of the amygdala and vmPFC. Neuron 43:897–905

Quirk GJ, Russo GK, Barron JL, Lebron K (2000) The role of ventromedial prefrontal cortex in the recovery of extinguished fear. J Neurosci 20:6225–6231

Reynolds SM, Zahm DS (2005) Specificity in the projections of prefrontal and insular cortex to ventral striatopallidum and the extended amygdala. J Neurosci 25:11757–11767

Richards HJ, Benson V, Donnelly N, Hadwin JA (2014) Exploring the function of selective attention and hypervigilance for threat in anxiety. Clin Psychol Rev 34:1–13

Rodríguez-Soto NC, Buxó CJ, Morou-Bermudez E, Pérez-Edgar K, Ocasio-Quiñones IT, Surillo-González MB, Martinez KG (2021) The impact of prenatal maternal stress due to potentially traumatic events on child temperament: a systematic review. Dev Psychobiol 63:e22195

Rothbart MK, Bates JE (2007) Temperament. Handbook of child psychology, vol 3

Rothenberger SE, Resch F, Doszpod N, Moehler E (2011) Prenatal stress and infant affective reactivity at five months of age. Early Hum Dev 87:129–136

Sarkar S, Craig MC, Dell'Acqua F, O'Connor TG, Catani M, Deeley Q et al (2014) Prenatal stress and limbic-prefrontal white matter microstructure in children aged 6–9 years: a preliminary diffusion tensor imaging study. World J Biol Psychiatry 15:346–352

Spielberger CD (2010) State-trait anxiety inventory. In: The Corsini encyclopedia of psychology. Wiley, pp 1–1

St-Cyr S, Abuaish S, Sivanathan S, McGowan PO (2017) Maternal programming of sex-specific responses to predator odor stress in adult rats. Horm Behav 94:1–12

Sylvester CM, Corbetta M, Raichle ME, Rodebaugh TL, Schlaggar BL, Sheline YI et al (2012) Functional network dysfunction in anxiety and anxiety disorders. Trends Neurosci 35:527–535

Teicher MH, Samson JA (2016) Annual research review: enduring neurobiological effects of childhood abuse and neglect. J Child Psychol Psychiatry 57:241–266

Thiel F, Eberhard-Gran M, Garthus-Niegel S (2021) The impact of perinatal life stress on infant temperament and child development: a 2-year follow-up cohort study. J Dev Behav Pediatr 42:299

Tottenham N, Hare TA, Quinn BT, McCarry TW, Nurse M, Gilhooly T et al (2010) Prolonged institutional rearing is associated with atypically large amygdala volume and difficulties in emotion regulation. Dev Sci 13:46–61

Tottenham N, Hare TA, Millner A, Gilhooly T, Zevin JD, Casey BJ (2011) Elevated amygdala response to faces following early deprivation. Dev Sci 14:190–204

Uematsu A, Matsui M, Tanaka C, Takahashi T, Noguchi K, Suzuki M, Nishijo H (2012) Developmental trajectories of amygdala and hippocampus from infancy to early adulthood in healthy individuals. PLoS One 7:e46970

Urry HL, van Reekum CM, Johnstone T, Kalin NH, Thurow ME, Schaefer HS et al (2006) Amygdala and ventromedial prefrontal cortex are inversely coupled during regulation of negative affect and predict the diurnal pattern of cortisol secretion among older adults. J Neurosci 26:4415–4425

Van den Bergh BRH, van den Heuvel MI, Lahti M, Braeken M, de Rooij SR, Entringer S et al (2020) Prenatal developmental origins of behavior and mental health: the influence of maternal stress in pregnancy. Neurosci Biobehav Rev 117:26–64

Vivinetto AL, Suárez MM, Rivarola MA (2013) Neurobiological effects of neonatal maternal separation and post-weaning environmental enrichment. Behav Brain Res 240:110–118

Volkow ND, Gordon JA, Freund MP (2021) The healthy brain and child development study – shedding light on opioid exposure, COVID-19, and health disparities. JAMA Psychiatry 78: 471–472

Von Der Heide RJ, Skipper LM, Klobusicky E, Olson IR (2013) Dissecting the uncinate fasciculus: disorders, controversies and a hypothesis. Brain 136:1692–1707

Vyas A, Mitra R, Rao BSS, Chattarji S (2002) Chronic stress induces contrasting patterns of dendritic remodeling in hippocampal and amygdaloid neurons. J Neurosci 22:6810–6818

Wurmser H, Rieger M, Domogalla C, Kahnt A, Buchwald J, Kowatsch M et al (2006) Association between life stress during pregnancy and infant crying in the first six months postpartum: a prospective longitudinal study. Early Hum Dev 82:341–349

Young CB, Raz G, Everaerd D, Beckmann CF, Tendolkar I, Hendler T et al (2017) Dynamic shifts in large-scale brain network balance as a function of arousal. J Neurosci 37:281–290

Zeanah CH, Egger HL, Smyke AT, Nelson CA, Fox NA, Marshall PJ, Guthrie D (2009) Institutional rearing and psychiatric disorders in Romanian preschool children. Am J Psychiatry 166: 777–785

Zeanah CH, Fox NA, Nelson CA (2019) Recovery from severe deprivation: the Bucharest early intervention project. J Am Acad Child Psychiatry 58:S300–S301

Zilverstand A, Parvaz MA, Goldstein RZ (2017) Neuroimaging cognitive reappraisal in clinical populations to define neural targets for enhancing emotion regulation. A systematic review. Neuroimage 151:105–116

Genetic Vulnerability to Social Anxiety Disorder

Janna Marie Bas-Hoogendam

Contents

Abstract Most anxiety disorders 'run within families': people suffering from an anxiety disorder often have family members who are highly anxious as well. In this chapter, we explore recent work devoted to unraveling the complex interplay between genes and environment in the development of anxiety. We review studies focusing on the genetic vulnerability to develop social anxiety disorder (SAD), as SAD is one of the most prevalent anxiety disorders, with an early onset, a chronic

J. M. Bas-Hoogendam (✉)
Leiden University, Leiden, The Netherlands

Leiden University Medical Center, Leiden, The Netherlands

Leiden Institute for Brain and Cognition, Leiden, The Netherlands
e-mail: j.m.hoogendam@fsw.leidenuniv.nl

© The Author(s), under exclusive license to Springer Nature Switzerland AG 2024

Curr Topics Behav Neurosci (2025) 73: 301–328
https://doi.org/10.1007/7854_2024_544
Published Online: 14 November 2024

course, and associated with significant life-long impairments. More insight into the development of SAD is thus of uttermost importance.

First, we will discuss family studies, twin studies, and large-sized population-based registry studies and explain what these studies can reveal about the genetic vulnerability to develop anxiety. Next, we describe the endophenotype approach; in this context, we will summarize results from the Leiden Family Lab study on Social Anxiety Disorder. Subsequently, we review the relationship between the heritable trait 'behavioral inhibition' and the development of SAD, and highlight the relevance of this work for the development and improvement of preventative and therapeutic interventions for socially anxious youth.

Keywords Endophenotypes · Family-study · Genetics · Neuroimaging · Social anxiety disorder

1 Introduction

Most anxiety disorders 'run within families': people suffering from an anxiety disorder often have family members who are highly anxious as well. Here, we explore recent work devoted to unraveling the complex interplay between genes and environment in the development of anxiety. We review studies on the genetic vulnerability to develop social anxiety disorder (SAD), which is one of the most prevalent anxiety disorders, with an early onset, a chronic course, and associated with significant life-long impairments (Stein and Stein 2008; Kieling et al. 2024; Seidl et al. 2021; Aderka et al. 2012). More insight into the development of SAD is thus of uttermost importance.

First, we will discuss family studies, twin studies, and large-sized population-based registry studies and explore what these studies reveal about the genetic vulnerability to anxiety. Next, we describe the endophenotype approach, a method relating the genetic vulnerability to anxiety to neurobiological and psychological characteristics; here, we summarize results from the Leiden Family Lab study on Social Anxiety Disorder. Subsequently, we review the relationship between the heritable trait 'behavioral inhibition' and the development of SAD, and highlight the relevance of this work for the development and improvement of preventative and therapeutic interventions for socially anxious youth.

2 Anxiety Runs in Families

Almost 25 years ago, a meta-analysis investigated available evidence for familial aggregation of panic disorder, generalized anxiety disorder, and phobias, including SAD (Hettema et al. 2001). At that time, twin and family studies on anxiety were scarce and only a limited number of studies met inclusion criteria. Therefore, the results need to be interpreted with caution. However, this meta-analysis provided initial evidence that anxiety disorders have significant familial aggregation. An integrative review of intergenerational family studies on childhood anxiety subsequently confirmed that children whose parents have an anxiety disorder have an increased risk to develop anxiety themselves; moreover, children with anxiety disorders more often have a parent with anxiety (Murray et al. 2009).

A recent systematic review explored in more detail the association between parental anxiety and emotional development of their offspring (Sweeney and Wilson 2023), revealing that 'the presence of parental anxiety increases the risk of internalizing and externalizing problems in their offspring and is associated with mental health problems in later childhood and adolescence'. Furthermore, although the association between parental anxiety and negative developmental trajectories in their offspring was present across the whole developmental lifespan, *maternal* anxiety was more influential than *paternal* anxiety during adolescence, pointing at distinct roles for fathers and mothers during these developmental trajectories. Inconsistencies between study methods may explain why studies report different (even sometimes non-significant) associations on the relationship between anxiety in parents and emotional development of their children (Sweeney and Wilson 2023).

A large population-based multi-generational family study on the genetic vulnerability to social anxiety (data on 18,399 individuals with a diagnosis of SAD) confirmed that SAD runs in families and revealed that the risk for SAD in family members was positively related to genetic relatedness: first-degree relatives of individuals with SAD (50% genetic similarity) had a significantly higher risk of having SAD compared to second-degree relatives (25% genetic similarity) and third-degree relatives (12.5% genetic similarity) (Isomura et al. 2015). Additionally, this study revealed an increased risk of SAD in partners of people with SAD: these non-biological relatives (defined as 'individuals who have at least one child together with the patient') were four times more likely to have SAD compared to non-biological relatives of controls. Given the early age of onset of SAD, usually during childhood or adolescence (cf. (Bas-Hoogendam et al. 2020a; Haller et al. 2015) and chapter 'Temperamental Fear to Anxiety' by Dr. Perez-Edgar, this volume), the authors assume that these findings reflect 'assortative mating': they consider it most likely that socially anxious individuals seek partners with similar phenotypes, rather than to attribute the higher SAD prevalence in partners to the observation that partners who cohabit could become more similar in certain respects (Isomura et al. 2015).

An interesting question is: does parental anxiety always lead to anxiety in offspring? Uher and colleagues (Uher et al. 2023) recently summarized evidence from original family high-risk and registry studies (211 independent studies, >3 million offspring of parents with psychopathology and >20 million offspring of parents without psychopathology) and explored the transdiagnostic risk of developing a mental disorder in offspring of affected parents. This meta-analysis revealed that the relative risk (RR) of developing an anxiety disorder was 2.2 for offspring of parents with anxiety disorders, with a lifetime absolute risk (defined as 'up to age of assessment') of 31%. Risk for developing *any* mental disorder was even higher: the RR to develop a mental disorder appeared to be 3.0 for offspring of parents with anxiety disorders, with a lifetime risk of no less than 55% (Uher et al. 2023). Across all mental conditions studied, the risks for the development of any mental disorder were estimated to be among the highest for offspring of parents with anxiety disorders (Uher et al. 2023) stressing the need to develop effective preventative interventions for these children at risk.

3 Twin Studies

Complementing family studies, twin studies increase insights into the extent to which genes contribute to a specific phenotype, providing estimates of heritability, defined as 'the proportion of disease risk that is due to variation in genes in a population' (Smoller et al. 2009). A meta-analysis by Scaini and colleagues (Scaini et al. 2014) revealed that both genetic (meta-analytic estimate of proportion of total variance: 0.41) and non-shared environmental factors (meta-analytic estimate: 0.54) contribute to individual differences in social anxiety; shared environmental factors appeared less relevant for SAD.[1] Furthermore, the estimates of heritability were higher for original studies measuring social anxiety on a continuous scale (i.e., at the level of symptoms), compared to studies using a categorical approach (i.e., a DSM-based diagnosis of SAD) (Scaini et al. 2014). This increased sensitivity to detect the genetic contribution to the manifestation of SAD within individuals is in line with the notion that social anxiety is expressed on a continuum of severity, ranging from non-clinical ('discomfort in some situations') to clinical levels ('intense and incapacitating fear in most situations' (Miskovic and Schmidt 2012)) and emphasizes the added value of exploring the pathogenesis and pathophysiology of social anxiety on a continuous scale. In addition, a multiple-rater twin study, in which anxiety was rated by adolescent twins, their mothers and their fathers ($n = 1{,}394$), revealed rater-specific genetic effects: analyses based on parent-reported anxiety yielded higher heritability estimates compared to analyses on adolescent self-report data (Ask et al. 2014); cf. a recent study exploring why parents' survey reports of adolescent social anxiety fail to predict the anxiety that adolescents self-report (Keeley et al. 2024). Thus, multiple methodological choices can influence estimates of heritability of anxiety.

4 Complex Interplay Between Genes and Environment

Most family members share both genes as well as environment. Thus, family studies revealing aggregation of anxiety within families do not necessarily imply a genetic contribution to the risk to develop an anxiety disorder (Smoller et al. 2009).

[1]The influence of environmental factors on the development of SAD will not be extensively discussed here; for a recent overview on the role of environmental factors in the aetiology of SAD we refer to Norton and Abbott (2017).

Likewise, twin studies usually examine phenotypic variation within one generation and are often unable to exclude environmental interaction effects between parents and children (Bartels 2021).

In a systematic review and meta-analysis of genetically informed research, Ahmadzadeh and colleagues focused on gene-environment interactions between parents and offspring (Ahmadzadeh et al. 2021). The authors indicate that observational studies reporting on relationships between parent anxiety and internalizing symptoms in their offspring are often biased by unobserved variables, including genetics, hindering distinguishing genetic effects from environmental effects that are also present in the parent-child relationship (like social learning and parental modeling; cf. (Ahmadzadeh et al. 2021; Lieb et al. 2000)). Furthermore, they point at passive gene-environment correlations, encompassing the observation that 'the environment shared by parents and their offspring is passively correlated with the genes that they share' (Ahmadzadeh et al. 2021). Therefore, Ahmadzadeh et al. (2021) meta-analyzed studies which explored the relationship between parent anxiety and offspring internalizing outcomes while accounting for familial genetic resemblance, based on specific quasi-experimental study designs. Examples of such designs are adoption, sibling-comparison, children-of-twins, or in vitro fertilization studies. Eight publications (describing data from four independent cohorts) met the inclusion criteria for this meta-analysis (Ahmadzadeh et al. 2021). Based on these studies, the authors cautiously conclude that postnatal anxiety in at least one of the parents may be causally related to concurrent internalizing problems in offspring through environmental, nongenetic pathways. Although this observation might seem to contradict the findings of the meta-analysis discussed earlier (Scaini et al. 2014), it is worth noting that these meta-analyses differ in certain respects. Ahmadzadeh et al. (2021) included studies with quasi-experimental designs accounting for genetic confounding and explored associations between any anxiety disorder (parents) and broad internalizing problems (offspring), while Scaini et al. (2014) focused on SAD and included only studies with a classic twin design. It is possible that, compared to other anxiety disorders, the development of SAD is more influenced by a genetic vulnerability (e.g., temperamental traits like behavioral inhibition, which will be discussed in more detail later) than shared parent-child environmental influences. However, as the meta-analyses differed in multiple methodological aspects, this hypothesis needs to be tested in studies in which the contribution of genetic and environmental factors can be directly compared between several anxiety phenotypes, in both parents and children. Furthermore, the work by Ahmadzadeh and colleagues (2021) stresses the need for more research in this field, as their results were based on only four unique datasets. To enable identifying reproducible and reliable causal pathways between parent and offspring anxiety phenotypes, further studies are needed, preferably with a longitudinal design.

Kendler and colleagues also aimed to clarify the sources of parent-offspring transmission of anxiety disorders (Kendler et al. 2022). Using data from Swedish national samples, the authors determined six types of families in which offspring were raised, and they combined these household data with diagnostic data on anxiety disorders from national medical registries. This approach enabled them to explore three sources of parent-child resemblance: genes plus rearing, genes only, and rearing only (Kendler et al. 2022). Analyses on this extended dataset of over two million offspring revealed that anxiety disorders were transmitted from parents to offspring (intact families $r = 0.16$). By comparing data from the six family types, the authors distinguished to which extent genetical influences and rearing effects played a role in the intergenerational transmission of anxiety: genetics explained 70% of the parent-offspring resemblance, while environmental factors (i.e., rearing effects) accounted for the remaining 30%. These data thereby confirm that both genetic and environmental effects play a role in the intergenerational transmission of anxiety disorders (Kendler et al. 2022).

5 The Search for SAD-Genes

Given the evidence summarized above, pointing at genetic influences on the development of anxiety disorders in general, and SAD in particular, the question arises: which particular genes are implicated in this risk? Previous research showed that the exploration of 'SAD-genes' is hindered by multiple factors, including the heterogeneity of the disorder and the fact that the diagnosis is based on clinical assessments and not on biologically based measurements (Glahn et al. 2007; Gottesman and Gould 2003). Furthermore, accumulating evidence underscores that multiple genetic variants, each with a relatively small effect, interact and increase the risk to develop complex psychiatric disorders, including anxiety, referred to as 'polygenicity' (Binder 2012; Fox and Kalin 2014; Levey et al. 2020; O'Donovan and Owen 2016). These genetic variants are in turn influenced by environmental factors (Gottschalk and Domschke 2016; Morneau-Vaillancourt et al. 2020). Other articles in this volume will describe in more detail how studies have increased insights into genotypic alterations related to anxiety. Here, we continue with a discussion of an alternative approach to obtain insight into the genetic vulnerability to develop SAD: the endophenotype approach.

6 The Endophenotype Approach[2]

The endophenotype approach has been proposed to facilitate the investigation of genetic factors in psychiatric disorders (Glahn et al. 2014; Bas-Hoogendam et al. 2016). Endophenotypes are measurable characteristics on the causal pathway from genotype to phenotype and are thus thought to reflect underlying disease liability (Lenzenweger 2013a). The following criteria are most consistently mentioned to

[2]This part of the chapter is based on Bas-Hoogendam et al. (2016) and has been updated for the present work; excerpt with permission of Elsevier and Copyright Clearance Center.

define endophenotypes (Gottesman and Gould 2003; Lenzenweger 2013a): first an endophenotype is *associated with the disorder of interest*; second an endophenotype is a *stable, state-independent trait*, which is already present in a preclinical state; third an endophenotype is *heritable*; fourth an endophenotype *co-segregates with the disorder within a family*, with nonaffected family members showing altered levels of the endophenotype when compared to the general population. In addition, an endophenotype is ideally more strongly associated with a specific disorder of interest in comparison to other psychiatric conditions (Lenzenweger 2013b), but it is also possible that a certain endophenotype affects more than one disorder (Cannon and Keller 2006).

Originally, the usefulness of endophenotypes was supposed to lie in discovering the genes predisposing for complex disorders, based on the assumption that endophenotypes have a simpler genetic architecture than the disorders themselves (Glahn et al. 2007; Gottesman and Gould 2003). This idea was challenged by a meta-analytic review (Flint and Munafo 2007), which compared the effect sizes of genetic loci contributing to phenotypes (psychiatric disorders) and loci contributing to endophenotypes. Results showed comparable effect sizes, so the assumption that endophenotypes have a simpler genetic architecture than phenotypes was not supported (Flint and Munafo 2007). These findings were empirically confirmed by comparing GWAS studies investigating the genetic effects related to endophenotypes of schizophrenia (e.g., variation in brain structure and measures of cognitive performance) to studies aimed to identify risk genes for the disorder itself (Flint et al. 2014). Again, similar effect sizes were found. So, it is not necessarily true that the genetic architecture of endophenotypes is less complex than that of the disorders themselves.

This does not mean, however, that endophenotypes are of limited value (Roffman 2019). Their usefulness lies in understanding disease mechanisms: based on the assumption that complex disorders could be divided into simpler and more biologically coherent units, endophenotypes provide insight into the pathways leading to pathology and could help in discerning the origins of mental disorders (Flint et al. 2014). Furthermore, endophenotypes support a transdiagnostic perspective on mental disorders, given the fact that endophenotypes could cross traditional diagnostic boundaries (Beauchaine and Constantino 2017). Here, the endophenotype approach fits within the NIMH Research Domain Criteria (RDoC) initiative, a research framework in which not the clinical diagnoses are starting point for investigation, but core features of psychopathology, falling within five research domains (Morris et al. 2022; Auerbach 2022). The RDoC initiative explicitly acknowledges that these core features could be present in multiple psychiatric disorders and promotes the integration of data from several levels, from genes to neural systems to behavior (Insel 2014; Beauchaine and Hinshaw 2020). Endophenotypes fit perfectly within this framework, providing a bridge between genetic variations on the one hand and psychiatric disorders on the other.

Endophenotypes could also aid in the development of improved animal models for psychopathology (Gould and Gottesman 2006). A recent example is the study by Fox and colleagues, demonstrating that infant-inhibited temperament in rhesus monkeys is a heritable trait, which is stable over time and related to anxious behavior in adulthood (Fox et al. 2021). A preliminary genome-wide association analysis in these monkeys revealed a promising genetic variant on chromosome 13 (nearby the *CTNNA2* gene) related to inhibited temperament. Interestingly, this *CTNNA2* gene has been previously found in association studies on anxiety in humans, underscoring the value of translational research to explore molecular and genetic underpinnings of anxious behavior based on endophenotypes (Fox et al. 2021).

Furthermore, endophenotypes can be used to identify individuals at risk, as they are present prior to disease onset. This is of uttermost importance, because early detection of psychopathology and subsequent preventive interventions can improve long-term prognosis, reduce the substantial burden and cost of SAD, and lower the risk of developing co-morbid psychopathology (Beauchaine et al. 2008). Moreover, endophenotypes could provide clues for improvement of treatments and guide in the selection of appropriate pharmacological interventions (Garner et al. 2009).

7 Neurobiological Endophenotypes of SAD

The Leiden Family Lab study on Social Anxiety disorder (LFLSAD) was especially designed to investigate endophenotypes of SAD, using a two-generation design (Bas-Hoogendam et al. 2018a). Data acquisition for this study took place between 2013 and 2015 and electroencephalography (EEG) and magnetic resonance imaging (MRI) data revealed multiple promising candidate endophenotypes of SAD. Here, we describe the design of the LFLSAD and discuss several neurobiological endophenotypes based on MRI.

To the best of our knowledge, the LFLSAD is the first and only multiplex (i.e., multiple cases of the disorder within one family), multigenerational family study designed to unravel neurobiological SAD-endophenotypes, by exploring two endophenotype criteria. First, the family design enabled investigating whether a presumed endophenotype *co-segregated with the disorder* within the families. Furthermore, the *heritability* of candidate endophenotypes was estimated, using a joint mixed model taking the ascertainment process and familiar relationships into account (Tissier et al. 2017).

As described in a dedicated design paper (Bas-Hoogendam et al. 2018a), the multiplex, multigenerational design maximized statistical power to detect genetic and environmental influences on SAD-related characteristics. Families were considered eligible for inclusion when they contained at least one adult with a primary diagnosis of SAD ('proband'), of whom at least one child showed SAD symptoms at a clinical or subclinical level ('proband's SA-child', age 8–21 years). In addition to the proband and its SA-child, the proband's spouse, other children (age $\geq$ 8 years), and the proband's sibling(s) and their spouse(s) with their child(ren) (age $\geq$ 8 years) were invited to participate. Families with at least eight family members were included to enable reliable estimations of the relation between endophenotype and SAD. This multiplex design enriched the sample for a heritable basis of SAD and facilitated the detection of genetic factors. Furthermore, a sample consisting of large families, composed of several related nuclear families (parents with their children), is likely to share more heritable factors than a same-sized sample of unrelated nuclear families, hence more statistical power to distinguish shared environmental effects from genetic effects (Williams and Blangero 1999).

Based on an extensive literature review (Bas-Hoogendam et al. 2016), we identified several neurobiological measurements as SAD-endophenotypes, namely first the function and functional connectivity of the amygdala, second the function of the medial prefrontal cortex (PFC), third whole-brain functional connectivity, and fourth structural-anatomical brain changes. These candidate endophenotypes were explored in a one-hour MRI session (neuroimaging data from 8 families, total $n = 110$, age range 9.2–61.5 years). Results are summarized below, partly based on a discussion previously published in the thesis 'Extremely Shy and Genetically Close' (Bas-Hoogendam 2020a).

8 Functional Brain Characteristics as SAD-Endophenotypes

Previous functional (f)MRI studies indicated an *association* between SAD and hyperactivation of subcortical, frontal, parietal, and occipital brain areas, evoked by paradigms addressing specific SAD-related fears (Brühl et al. 2014a; Cremers and Roelofs 2016). In the LFLSAD, we employed two functional paradigms: a neutral faces paradigm and a social norm processing task. Each paradigm targeted different neurocognitive components of SAD. We used these paradigms to examine evidence for brain activation as candidate endophenotype of SAD, by exploring the *co-segregation within families* and *heritability* criteria.

8.1 Amygdala and Hippocampus Activation as Candidate SAD-Endophenotypes

The first fMRI paradigm concerned the processing of faces with a neutral expression, being strong social stimuli with an ambiguous meaning. The Neutral Faces Paradigm (NFP) was created to explore brain activation related to two different aspects of processing such faces. In the habituation phase (HP), we tested whether impaired habituation to neutral faces (i.e., the adaptive decline in brain activation to a stimulus which is presented multiple times without meaningful consequences (Ramaswami 2014; Rankin et al. 2009)) could be considered a candidate SAD-endophenotype. This hypothesis was based on previous research considering individuals with inhibited temperament, which is an important risk factor for SAD, and in participants with high levels of social fearfulness (Avery and Blackford 2016; Blackford et al.

2013); both studies reported failed habituation. We found that the neural habituation response in the right hippocampus and amygdala was impaired in family members with high levels of social anxiety, providing support for the endophenotype criterion of *co-segregation within the families*. Subsequent *heritability* analyses revealed that the neural habituation response within the right hippocampus was at least moderately heritable, in line with work in monkeys (Oler et al. 2010). These findings indicate that altered neural habituation in the hippocampus is a putative SAD-endophenotype (Bas-Hoogendam et al. 2019).

The second phase of the NFP concerned the social-evaluative conditioning of the faces (Bas-Hoogendam et al. 2020b). By consistently pairing three neutral faces with social-evaluative sentences with a positive ('He says you are smart'), negative ('He says you are stupid'), or neutral ('He says you are in Leiden') content, participants learned the social-evaluative value of each face. Previous work indicated amygdala engagement during this learning process (Davis et al. 2010), but the relation between amygdala activation related to this social-evaluative conditioning paradigm (SCP) and social anxiety has not been investigated, let alone within families genetically enriched for SAD. Data from the LFLSAD indicated bilateral amygdala hyperactivation to faces conditioned with a social-evaluative meaning; levels of this amygdala activation *co-segregated* with social anxiety and displayed at least moderate *heritability*. Interestingly, this amygdala hyperreactivity was present for all conditions of the SCP, indicating that being directly addressed ('He says you are. . .') strongly activates the amygdala in socially anxious family members, independent from the content of the evaluation (i.e., positive, negative, or neutral). These results provide evidence for amygdala activation in response to faces with a learned social-evaluative meaning as a neurobiological candidate endophenotype of SAD (Bas-Hoogendam et al. 2020b).

8.2 Medial PFC Activation as Candidate SAD-Endophenotype

The second fMRI paradigm employed in the LFLSAD, the Social Norm Processing Task (SNPT), taps into the fear of socially anxious individuals that they will *unintentionally* break a social norm in the presence of others. In this task, three different types of stories on social situations are presented: stories on situations in which *no social norm* (SN) was violated, stories describing *unintentional SN violations* and stories outlining *intentional SN violations*. These conditions enable investigating the behavioral and neurobiological correlates of processing intentional and unintentional social norm violations (Berthoz et al. 2002). Building upon previous versions of the SNPT (Berthoz et al. 2002; Blair et al. 2010), we created a revised version of the paradigm (SNPT-R) which facilitates using the paradigm in participants of different ages; furthermore, methodological improvements were made and we ensured that the SNPT-R is now publicly available (Bas-Hoogendam et al. 2017a). A validation study in healthy individuals demonstrated that participants rated the stories differently, depending on the intention underlying the social norm violation: *intentional* social norm violations were considered as more inappropriate and more embarrassing when compared to *unintentional* social norm violations. fMRI data revealed both overlapping as well as differential brain activation patterns for reading intentional and unintentional social norm violations (Bas-Hoogendam et al. 2017b). Next, we demonstrated a positive relationship between social anxiety and the ratings, with the most pronounced effect for the embarrassment ratings of *unintentional* social norm violations. While individuals with low-to-intermediate social anxiety levels rated the unintentional social norm transgressions as less embarrassing when compared to intentional social norm transgressions (i.e., these individuals make a distinction between breaking conventional rules, by *intention*, and committing a blunder, *unintentionally*, when they rate the stories on embarrassment), individuals with high social anxiety levels consider unintentional social norm violations as equally embarrassing as intentional social norm transgressions (Bas-Hoogendam et al. 2018b).

Consequently, we explored within the LFLSAD whether the neurobiological and behavioral correlates of processing unintentional social norm transgressions could serve as endophenotypes of SAD. Brain responses to unintentional social norm violations were positively *related to levels of social anxiety* within the LFLSAD families, in the medial PFC and in a cluster encompassing the medial temporal gyrus, superior temporal gyrus, and superior temporal sulcus; these brain activation levels were at least moderately *heritable* (Bas-Hoogendam et al. 2020c). Our hypothesis with respect to the ratings of embarrassment was partly supported: data revealed a positive correlation between social anxiety and embarrassment, but this effect was not specific for the unintentional condition and heritability estimates of these ratings were low or even absent. Thereby, this study provided evidence for hyperactivation in the medial PFC and temporal brain regions as putative SAD-endophenotypes (Bas-Hoogendam et al. 2020c).

8.3 Brain Connectivity as SAD-Endophenotype

Brain regions do not function in isolation but are part of large-scale networks (Bassett et al. 2018), and changes in brain connectivity have been associated with psychopathology in studies on individuals with SAD and children at risk for developing social anxiety (Zugman et al. 2023; Mizzi et al. 2021, 2024; Taber-Thomas et al. 2016; Pannekoek et al. 2013; Geiger et al. 2016). Within the LFLSAD sample, both intrinsic functional connectivity and structural brain connectivity were examined.

Functional brain networks were derived from fMRI data acquired during the resting-state, and analyses were performed within the default mode network, dorsal attention network, executive control network, frontoparietal network, limbic network, and salience network. Alterations in intrinsic functional connectivity in the dorsal attention network and the frontoparietal network met both endophenotype criteria (*co-segregation* and *heritability*) (Bas-Hoogendam et al. 2021), making them promising SAD-endophenotypes.

Structural white matter (WM) tracts were investigated using diffusion tensor imaging scans. We explored WM-microstructure in the uncinate fasciculus, superior longitudinal fasciculus, and inferior longitudinal fasciculus, building on prior work (Baur et al. 2011, 2013; Liao et al. 2011). Increased fractional anisotropy in the bilateral superior longitudinal fasciculus co-segregated with social anxiety within families enriched for SAD, and, in line with previous work, all WM characteristics were estimated to be at least moderately heritable (Kochunov et al. 2015). Thus, altered WM-microstructure could be a candidate SAD-endophenotype (Roelofs et al. 2020), but more research is needed to determine which WM-alterations are reliably and reproducible related to SAD (cf. this recent meta-analysis (Parsaei et al. 2024)).

8.4 Structural Brain Characteristics as SAD-Endophenotypes

Previous work provided evidence for associations between SAD and gray matter brain characteristics, although results are often inconsistent between studies, probably due to small sample sizes and heterogenous methodology (Bas-Hoogendam 2020b). Building on these studies, data from the LFLSAD allowed for exploring evidence for the *co-segregation* of social anxiety with gray matter characteristics within families. Analyses were restricted to regions of interest (exploratory results on all areas are available online (Bas-Hoogendam et al. 2018c)), revealing a positive association between pallidum volume and social anxiety, supporting the *co-segregation* endophenotype criterion. Furthermore, pallidum volume was moderately *heritable*, making pallidum volume a promising candidate SAD-endophenotype. Moreover, several cortical gray matter characteristics, extracted from frontal, parietal, and temporal regions, *co-segregated* with social anxiety within the families and had moderate to high *heritability*. It should be noted that these association results did not survive correction for the number of statistical tests, but these findings provide preliminary evidence that gray matter characteristics of various brain regions are candidate SAD-endophenotypes (Bas-Hoogendam et al. 2018d; Frick and Månsson 2018).

8.5 Insights into the Genetic Vulnerability to SAD from the LFLSAD

When considering these results from the LFLSAD together with other recent neuroimaging work on SAD, several interesting patterns emerge which provide novel insights into the genetic vulnerability to SAD.

First of all, results illustrate that brain alterations underlying the genetic vulnerability to SAD are diffusely spread over the brain, as they involve subcortical, prefrontal, parietal, and temporal regions (Fig. 1). These regions, whose function and/or structure qualifies as candidate SAD-endophenotypes, are to a great extent in line with the regions summarized in a neurobiological model of SAD proposed by Brühl and colleagues a decade ago (Brühl et al. 2014a). This model, based on a qualitative review and meta-analysis of 76 neuroimaging studies on adults with SAD, described SAD-related changes in brain function in subcortical, frontal, parietal, and occipital areas, as well as alterations in the connections between these regions. Brühl et al. (2014a) extended an older neurobiological model outlined by Etkin and Wager, describing functional alterations in the so-called 'fear circuit' (amygdala, parahippocampal gyrus, pallidum, insula, inferior frontal gyrus) as well as in the fusiform gyrus and superior temporal gyrus (Etkin and Wager 2007). The results from the LFLSAD in turn further extend the Brühl-model (Brühl et al. 2014a), because the LFLSAD provided insights into SAD-related brain characteristics which are not just *biomarkers* (i.e., associated with the disorder, but not necessarily located on the causal pathway from genotype to phenotype), but qualify as candidate *endophenotypes* and are as such most likely part of the neural circuitry that translates genetic effects into disorder phenotypes (Meyer-Lindenberg and Weinberger 2006). This distinction is important, as it implies that these brain alterations likely reflect the genetic vulnerability to *develop* SAD and are not the *result* of the (often chronic) course of the disorder, nor could they be attributed to the effects of psychological treatment, pharmacological medication, or comorbid psychopathology (Lenzenweger 2013b; Beauchaine and Constantino 2017). As such, the LFLSAD findings indicate that SAD is a multi-encompassing brain disorder already at the level of the endophenotype.

Second, the LFLSAD results stress the importance of considering the dorsal striatum, including the pallidum and putamen, in research on SAD. These subcortical brain areas have increasingly received attention in anxiety research (Guyer et al. 2006, 2012; Tang et al. 2022), but were not yet part of the neurobiological model by Brühl et al. (2014a). In two separate studies, being a mega-analysis on a large international dataset of individuals with SAD and healthy participants (Bas-Hoogendam et al. 2017c) and the endophenotype LFLSAD study (Bas-Hoogendam et al. 2018d), we found associations between social anxiety and gray matter volume of this region; furthermore, these alterations co-segregated with social anxiety within families and were moderately heritable. These findings were supported by other studies with relevance for SAD. Research on healthy participants demonstrated a positive correlation between the concept 'intolerance of uncertainty' and striatal volume (Kim et al. 2017), while a study on healthy women demonstrated that socially anxious tendencies were associated with an enlarged striatum (Günther et al. 2018). A recent mega-analysis within the ENIGMA (*E*nhancing *N*euro*I*maging *G*enetics by *M*eta-*A*nalysis)-Anxiety working group (Bas-Hoogendam et al. 2020d) on data from 37 international samples revealed alterations in putamen and pallidum volumes in adults with SAD (Groenewold et al. 2023). Furthermore, a meta-analysis of fMRI studies on pathological and induced anxiety revealed increased activation of, among others, the putamen (Chavanne and Robinson 2020). Interestingly, a transdiagnostic study on the common underlying structural brain alterations across multiple psychiatric disorders reported strong evidence for putamen enlargement as

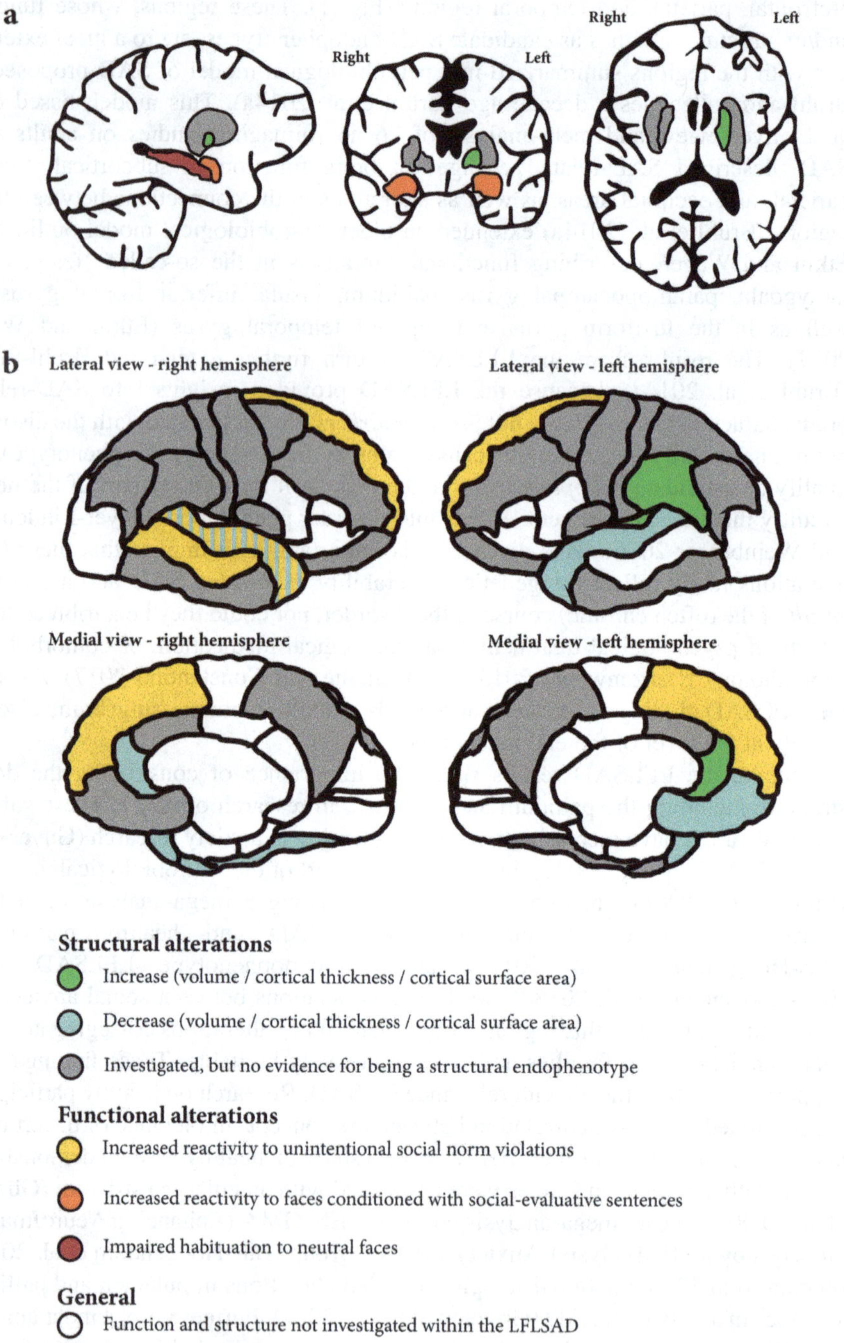

Fig. 1 Summary of candidate SAD-endophenotypes based on findings from the Leiden Family Lab study on Social Anxiety Disorder (LFLSAD). (**a**) Subcortical brain regions. (**b**) Cortical brain regions. (Adapted from Bas-Hoogendam (2020a))

a transdiagnostic marker of the familial vulnerability to psychopathology (Gong et al. 2019). The findings from the LFLSAD concur with this result, not only with respect to the involvement of the dorsal striatum in psychopathology but also in light of the genetic susceptibility to develop psychopathology. However, these observations also question the specificity of striatal enlargement as an endophenotype for SAD, suggesting that the putamen is a transdiagnostic marker of the vulnerability to develop psychopathology. Nevertheless, specificity is not a prerequisite for an endophenotype (Cannon and Keller 2006): endophenotypes could predispose for multiple anxiety and mood disorders. In line with this reasoning, we propose that striatal enlargement reflects the shared genetic background of anxiety disorders, depression, and related phenotypes (Shimada-Sugimoto et al. 2015); cf. work of the Brainstorm consortium showing a high degree of genetic correlation among psychiatric disorders (Anttila et al. 2018). The idea of striatal alterations as a transdiagnostic feature is reinforced by a review by Lago and colleagues, highlighting the important role of the striatum in multiple behavioral processes that are very relevant in psychopathology (Lago et al. 2017). Furthermore, a large (>30,000 MRI scans) genome-wide association study revealed several genetic variants influencing variation in putamen volume; these genetic variants were thought to affect developmental pathways such as apoptosis, axon guidance, and vesicle transport, and could aid in determining mechanisms of neuropsychiatric disorders (Hibar et al. 2015). Taken together, the role of the dorsal striatum in anxiety, both with respect to its structure as well as its function, deserves attention in future research on the genetic vulnerability to psychopathology, and SAD in particular.

While case-control studies provide insights into the *association with the disorder*, and LFLSAD data revealed initial evidence for *co-segregation within families* and *heritability*, these insights into the genetic susceptibility to SAD need to be complemented by further work focusing on the endophenotype criterion of *state-stability* (Gottesman and Gould 2003; Lenzenweger 2013a; Bearden and Freimer 2006). As the LFLSAD was a cross-sectional study (Bas-Hoogendam et al. 2018a), data from this project did not allow for investigating this criterion. To examine the stability of the candidate endophenotypes over time, longitudinal neuroimaging family-studies with a substantial sample size, involving adults with SAD as well as their offspring at risk, are important (Haller et al. 2018). To the best of our knowledge, no such studies have been undertaken, probably due to the high costs involved and the labor-intensity of such work.

9 Inhibited Temperament as a Marker of the Genetic Vulnerability to SAD

An alternative way to investigate stable neurobiological characteristics associated with the heritable risk to develop SAD is to focus on *inhibited temperament*. Temperament involves the 'innate individual differences in behavioral and emotional tendencies that appear in infancy and are relatively stable across context and time' (Clauss et al. 2015). One of the most commonly used constructs distinguishes individuals based on their tendency to *approach* or *avoid* novelty. The propensity to withdraw from new persons, situations, or objects is referred to as 'behavioral inhibition' or 'inhibited temperament' (IT) (Clauss et al. 2015; Pérez-Edgar and Guyer 2014; Kagan et al. 1987). IT is a heritable trait, constant across time and also measurable in animals, offering opportunities for translational research (Blackford et al. 2018; Caspi 2000; Kalin 2017). Variations in IT are present at a continuous scale within the population, and around 20% of young children present with high IT (Sylvester and Pine 2018). Essentially, having an IT at a young age has often long-term consequences: a study with a prospective longitudinal design over three decades revealed that infants with IT (determined at 14 months of age) became introverted adults, with poorer outcomes when it comes to social relationships; furthermore, infant IT was a major risk factor for internalizing psychopathology (Tang et al. 2020). Especially stably elevated childhood IT is associated with an increased risk of developing social anxiety (Schwartz et al. 1999; Biederman et al. 2001; Goldsmith et al. 2022). A meta-analysis quantified that almost half of all children with elevated IT will develop SAD, compared with only 12% of non-inhibited children (Clauss and Blackford 2012), and social anxiety shares a substantial proportion of genetic and environmental variance with IT, with tentative evidence for causality (Bourdon et al. 2019). Taken together, these findings indicate that IT is a strong innate risk factor for the later development of SAD, although it should be noted that not all children starting with similar levels of IT become anxious adults ('multifinality') (Liu and Pérez-Edgar 2019; Klein and Mumper 2018).

It has been proposed that brain characteristics provide the foundation for IT and the associated anxiety vulnerability (Blackford et al. 2018). However, our knowledge with respect to this 'neural risk signature' is still limited, although several neuroimaging studies have examined neurobiological characteristics associated with IT (Liu and Pérez-Edgar 2019). As also summarized in Bas-Hoogendam et al. (2022), some studies used a *cross-sectional* approach, examining children and early adolescents with high IT (Auday and Pérez-Edgar 2019) or including young adults who displayed inhibited behavior as a child (determined retrospectively) and at present (Clauss et al. 2014a, b). Other studies have *longitudinal* designs, in which infant temperament was established during early childhood, while neuroimaging was subsequently performed at various ages (Schwartz et al. 2003, 2012; Hill et al. 2010; Pérez-Edgar et al. 2014; Sylvester et al. 2015; Filippi et al. 2020; Abend et al. 2020). These studies suggest that IT is associated with structure and function of brain networks involved in processing fear, reward, and emotion regulation; key regions are the anterior cingulate cortex, insula, amygdala, PFC, and striatum (cf. (Clauss et al. 2015; Blackford et al. 2018)). However, such studies into the neurobiological characteristics associated with the temperamental vulnerability to develop SAD are scarce and the majority of them are based on a single longitudinal sample; furthermore, analyses are often restricted to a limited number of regions, and findings lack replication (Blackford et al. 2018). Moreover, sample sizes of current studies are relatively small (between 22 and 135 participants, including subjects with various levels of IT).

In a recently pre-registered ENIGMA-Anxiety mega-analysis (Bas-Hoogendam et al. 2022), combining structural MRI data and data on temperament assessments from almost 5,000 young participants (age 6–25 years), we aim to unravel which structural brain characteristics are associated with the temperamental risk for SAD. As described in more detail in the pre-registration (Bas-Hoogendam et al. 2022), we expect that structural alterations in multiple brain regions, in particular gray matter volumes of several subcortical structures (amygdala, hippocampus, striatum including caudate and putamen), and characteristics of frontal and temporal cortical areas are related to inhibited temperament during childhood.

10 Moving Forward: Neurobiological Endophenotypes as Targets for Interventions[3]

As argued above, unraveling neurobiological endophenotypes of SAD is important to gain insights into the genetic vulnerability to develop anxiety. Another avenue for future research concerns the use of candidate SAD-endophenotypes as targets for therapeutic or even preventive interventions. Nowadays, there are several cutting-edge techniques which enable altering brain function of specific regions or brain networks. Here, we highlight the possibilities of real-time fMRI-based neurofeedback and non-invasive brain stimulation, and take amygdala hyperreactivity (Bas-Hoogendam et al. 2020b) as an example of a promising treatment target.

10.1 Real-Time fMRI-Based Neurofeedback

In the last decade, real-time fMRI-based (rtfMRI) neurofeedback has increasingly attracted attention in the neuroimaging field. RtfMRI neurofeedback uses the latest developments of data processing and pattern analysis, making it possible to provide immediate information about brain activation or brain connectivity to participants, who, in turn, can learn to control the level of these indices (Sitaram et al. 2017; Paret and Hendler 2020). There are several important questions that need to be resolved, for example with respect to the sustainability, transferability, and feasibility of the technique (Thibault et al. 2018; Barreiros et al. 2019; Goldway et al. 2022), and the impact of psychological factors like attention, motivation, and mood influence on the outcome of rtfMRI neurofeedback experiments (Cohen Kadosh and Staunton 2019).

[3]This section is partly based on the discussion of the thesis 'Extremely Shy and Genetically Close' (Bas-Hoogendam 2020a) and updated for the present chapter.

However, multiple studies reported promising findings with respect to the possibility of rtfMRI neurofeedback to alter amygdala activation or amygdala connectivity. Two studies indicated that participants were able to downregulate activation of the amygdala during viewing faces with a negative expression (proof-of-principle study and replication (Brühl et al. 2014b; Herwig et al. 2019)). Furthermore, decreased amygdala activation was found during cognitive reappraisal in participants who received rtfMRI neurofeedback based on lateral PFC activation (Sarkheil et al. 2015), and two studies demonstrated that participants could regulate functional connectivity between regions of the PFC and the amygdala (Koush et al. 2015; Paret et al. 2016a). Furthermore, studies on small clinical samples, involving individuals with borderline personality disorder (Paret et al. 2016b), post-traumatic stress disorder (Nicholson et al. 2017; Lieberman et al. 2023; Zweerings et al. 2020), and depression (Young et al. 2017), yielded promising results with respect to their ability to regulate amygdala activation and connectivity. Lastly, work in adolescent girls described modulation of functional connectivity between the PFC and amygdala (Zich et al. 2020; Lisk et al. 2020).

To the best of our knowledge, rtfMRI neurofeedback has not been applied in participants with (sub)clinical social anxiety. Based on the findings summarized above, we speculate that rtfMRI neurofeedback could be a useful tool to downregulate the hyperresponsiveness of the amygdala, offering the possibility to alter the expression of this SAD-endophenotype (Bas-Hoogendam et al. 2020b). Furthermore, the effect of this downregulation on behavioral outcomes and levels of social anxiety is worthy of investigation. Confirmative results open the way for the use of rtfMRI neurofeedback in preventive and therapeutic interventions (Lipp and Cohen Kadosh 2020).

10.2 Brain Stimulation

Another method to non-invasively influence brain activation is brain stimulation, either by repetitive transcranial magnetic stimulation (rTMS) or transcranial direct current stimulation (tDCS) (Hoogendam et al. 2010; Hallett 2000; Nitsche et al. 2008). Research on non-invasive brain stimulation in the context of anxiety disorders, and SAD in particular, is still scarce (Sylvester et al. 2024; Vergallito et al. 2021; Vicario et al. 2019). Two case studies from the same research group demonstrated reductions in the level of social anxiety following rTMS over the right ventromedial PFC (Paes et al. 2013a, b), while a double-blind within-subject protocol in individuals with SAD revealed that tDCS over the left dorsolateral PFC reduced the attentional bias to threat, but not levels of social anxiety (Heeren et al. 2017). Unfortunately, these proof-of-principle studies did not include fMRI measurements to assess the effect of stimulation on amygdala activation.

Recently, a groundbreaking study combined rTMS with fMRI, in a considerable sample of 117 participants, exploring the neural circuit underlying emotion regulation (He et al. 2023). Active rTMS (versus sham) over the right ventrolateral PFC was associated with reduced amygdala activation during reappraisal of pictures depicting social exclusion; furthermore, active rTMS was related to lower negative feelings during reappraisal. Although these findings concern healthy participants, they highlight the potential of rTMS to effectively influence emotion regulation at the neurobiological and behavioral level. Furthermore, these findings align with the recommendations from a systematic review, in which the authors propose a treatment model in which stimulation of the dorsolateral PFC, via the corticolimbic pathway, results in down-regulation of the amygdala (Vicario et al. 2019). Although this model is speculative, a randomized tDCS study in women with high trait anxiety yielded promising results, by revealing that active tDCS (versus sham stimulation) of the dorsolateral PFC significantly reduced bilateral amygdala reactivity to threat (Ironside et al. 2019). Another study indicated that rTMS over the right dorsolateral PFC attenuated the amygdala response to visual stimuli with a strong negatively loaded emotional content (Baeken et al. 2010). Based on these encouraging results, we plead for double-blind placebo-controlled studies in healthy participants, individuals with SAD and young people with the genetic vulnerability to develop SAD, in order to characterize the neurobiological effects of non-invasive brain stimulation on amygdala reactivity, to determine the long-term outcomes of stimulation, and to explore the potential of these stimulation methods in the treatment and prevention of SAD (cf. the recommendations by (Sagliano et al. 2019) regarding the application of non-invasive brain stimulation in the treatment of generalized anxiety disorder).

To conclude, the work reviewed in this chapter highlights the impact of genetics on the development of SAD. Recent work on endophenotypes and inhibited temperament has further informed our understanding of this genetic vulnerability and state-of-the-art neuroimaging and neuromodulation techniques open the way to develop novel interventions to target this genetic vulnerability. More research, preferably in large samples is needed to further elucidate reliable and robust targets for these interventions.

References

Abend R, Swetlitz C, White LK et al (2020) Levels of early-childhood behavioral inhibition predict distinct neurodevelopmental pathways to pediatric anxiety. Psychol Med 50:96–106. https://doi.org/10.1017/S0033291718003999

Aderka IM, Hofmann SG, Nickerson A et al (2012) Functional impairment in social anxiety disorder. J Anxiety Disord 26:393–400. https://doi.org/10.1016/j.janxdis.2012.01.003

Ahmadzadeh YI, Schoeler T, Han M et al (2021) Systematic review and meta-analysis of genetically informed research: associations between parent anxiety and offspring internalizing problems. J Am Acad Child Adolesc Psychiatry. https://doi.org/10.1016/j.jaac.2020.12.037

Anttila V, Bulik-Sullivan B, Finucane HK et al (2018) Analysis of shared heritability in common disorders of the brain. Science 360:eaap8757. https://doi.org/10.1126/science.aap8757

Ask H, Torgersen S, Seglem KB, Waaktaar T (2014) Genetic and environmental causes of variation in adolescent anxiety symptoms: a multiple-rater twin study. J Anxiety Disord 28:363–371. https://doi.org/10.1016/j.janxdis.2014.04.003

Auday ES, Pérez-Edgar KE (2019) Limbic and prefrontal neural volume modulate social anxiety in children at temperamental risk. Depress Anxiety 36:690–700. https://doi.org/10.1002/da.22941

Auerbach RP (2022) RDoC and the developmental origins of psychiatric disorders: how did we get here and where are we going? J Child Psychol Psychiatry 63:377–380. https://doi.org/10.1111/jcpp.13582

Avery SN, Blackford JU (2016) Slow to warm up: the role of habituation in social fear. Soc Cogn Affect Neurosci 11:1832–1840. https://doi.org/10.1093/scan/nsw095

Baeken C, De Raedt R, Van Schuerbeek P et al (2010) Right prefrontal HF-rTMS attenuates right amygdala processing of negatively valenced emotional stimuli in healthy females. Behav Brain Res 214:450–455. https://doi.org/10.1016/j.bbr.2010.06.029

Barreiros AR, Almeida I, Baía BC, Castelo-Branco M (2019) Amygdala modulation during emotion regulation training with fMRI-based neurofeedback. Front Hum Neurosci 13:89

Bartels M (2021) Editorial: the value of genetically informative designs to understand pathways of intergenerational transmission and direction of causality. J Am Acad Child Adolesc Psychiatry. https://doi.org/10.1016/j.jaac.2021.02.017

Bas-Hoogendam JM (2020a) Extremely shy and genetically close. Investigating neurobiological endophenotypes of social anxiety disorder. Leiden University

Bas-Hoogendam JM (2020b) Gray matter matters: the structure of the socially-anxious brain. EBioMedicine 59:102937. https://doi.org/10.1016/j.ebiom.2020.102937

Bas-Hoogendam JM, Blackford JU, Brühl AB et al (2016) Neurobiological candidate endophenotypes of social anxiety disorder. Neurosci Biobehav Rev 71:362–378. https://doi.org/10.1016/j.neubiorev.2016.08.040

Bas-Hoogendam JM, van Steenbergen H, Kreuk T et al (2017a) Revised social norm processing task (SNPT-R). In: Database: open science framework. https://doi.org/10.17605/OSF.IO/M8R76

Bas-Hoogendam JM, van Steenbergen H, Kreuk T et al (2017b) How embarrassing! The behavioral and neural correlates of processing social norm violations. PLoS One 12:e0176326. https://doi.org/10.1371/journal.pone.0176326

Bas-Hoogendam JM, van Steenbergen H, Pannekoek JN et al (2017c) Voxel-based morphometry multi-center mega-analysis of brain structure in social anxiety disorder. Neuroimage Clin 16:678–688. https://doi.org/10.1016/j.nicl.2017.08.001

Bas-Hoogendam JM, Harrewijn A, Tissier RLM et al (2018a) The Leiden family lab study on social anxiety disorder: a multiplex, multigenerational family study on neurocognitive endophenotypes. Int J Methods Psychiatr Res 27:e1616. https://doi.org/10.1002/mpr.1616

Bas-Hoogendam JM, van Steenbergen H, van der Wee NJA, Westenberg PM (2018b) Not intended, still embarrassed: social anxiety is related to increased levels of embarrassment in response to unintentional social norm violations. Eur Psychiatry 52:15–21. https://doi.org/10.1016/j.eurpsy.2018.03.002

Bas-Hoogendam JM, van Steenbergen H, Tissier RLM et al (2018c) Gray matter characteristics as endophenotypes of social anxiety disorder. Database: Open Science Framework. https://doi.org/10.17605/OSF.IO/M8Q2Z

Bas-Hoogendam JM, van Steenbergen H, Tissier RLM et al (2018d) Subcortical brain volumes, cortical thickness and cortical surface area in families genetically enriched for social anxiety disorder – a multiplex multigenerational neuroimaging study. EBioMedicine:410–428. https://doi.org/10.1016/j.ebiom.2018.08.048

Bas-Hoogendam JM, van Steenbergen H, Blackford JU et al (2019) Impaired neural habituation to neutral faces in families genetically enriched for social anxiety disorder. Depress Anxiety 36:1143–1153. https://doi.org/10.1002/da.22962

Bas-Hoogendam JM, Roelofs EF, Westenberg PM, van der Wee NJA (2020a) Pathogenesis of SAD. In: Simon NM, Hollander E, Rothbaum BO, Stein DJ (eds) Textbook of anxiety, trauma and OCD-related disorders, 3rd edn. The American Psychiatric Association Publishing, Washington, pp 429–444

Bas-Hoogendam JM, van Steenbergen H, van der Wee NJA, Westenberg PM (2020b) Amygdala hyperreactivity to faces conditioned with a social-evaluative meaning - a multiplex, multigenerational fMRI study on social anxiety endophenotypes. Neuroimage Clin:102247. https://doi.org/10.1016/j.nicl.2020.102247

Bas-Hoogendam JM, van Steenbergen H, Tissier RLM et al (2020c) Altered neurobiological processing of unintentional social norm violations: a multiplex, multigenerational functional magnetic resonance imaging study on social anxiety Endophenotypes. Biol Psychiatry Cogn Neurosci Neuroimaging 5:981–990. https://doi.org/10.1016/j.bpsc.2019.03.003

Bas-Hoogendam JM, Groenewold NA, Aghajani M et al (2020d) ENIGMA-anxiety working group: rationale for and organization of large-scale neuroimaging studies of anxiety disorders. Hum Brain Mapp:1–30. https://doi.org/10.1002/hbm.25100

Bas-Hoogendam JM, van Steenbergen H, Cohen Kadosh K et al (2021) Intrinsic functional connectivity in families genetically enriched for social anxiety disorder – an endophenotype study. EBioMedicine 69:103445. https://doi.org/10.1016/j.ebiom.2021.103445

Bas-Hoogendam JM, Bernstein RA, Benson BE et al (2022) Structural brain correlates of childhood inhibited temperament: an ENIGMA-anxiety mega-analysis. J Am Acad Child Adolesc Psychiatry 61:1182–1188. https://doi.org/10.1016/j.jaac.2022.04.023

Bassett DS, Xia CH, Satterthwaite TD (2018) Understanding the emergence of neuropsychiatric disorders with network neuroscience. Biol Psychiatry Cogn Neurosci Neuroimaging 3:742–753. https://doi.org/10.1016/j.bpsc.2018.03.015

Baur V, Hänggi J, Rufer M et al (2011) White matter alterations in social anxiety disorder. J Psychiatr Res 45:1366–1372. https://doi.org/10.1016/j.jpsychires.2011.05.007

Baur V, Brühl AB, Herwig U et al (2013) Evidence of frontotemporal structural hypoconnectivity in social anxiety disorder: a quantitative fiber tractography study. Hum Brain Mapp 34:437–446. https://doi.org/10.1002/hbm.21447

Bearden CE, Freimer NB (2006) Endophenotypes for psychiatric disorders: ready for primetime? Trends Genet 22:306–313. https://doi.org/10.1016/j.tig.2006.04.004

Beauchaine TP, Constantino JN (2017) Redefining the endophenotype concept to accommodate transdiagnostic vulnerabilities and etiological complexity. Biomark Med 11:769–780. https://doi.org/10.2217/bmm-2017-0002

Beauchaine TP, Hinshaw SP (2020) RDoC and psychopathology among youth: misplaced assumptions and an agenda for future research. J Clin Child Adolesc Psychol 49:322–340. https://doi.org/10.1080/15374416.2020.1750022

Beauchaine TP, Neuhaus E, Brenner SL, Gatzke-Kopp L (2008) Ten good reasons to consider biological processes in prevention and intervention research. Dev Psychopathol 20:745–774. https://doi.org/10.1017/S0954579408000369

Berthoz S, Armony JL, Blair RJR, Dolan RJ (2002) An fMRI study of intentional and unintentional (embarrassing) violations of social norms. Brain 125:1696–1708. https://doi.org/10.1093/brain/awf190

Biederman J, Hirshfeld-Becker DR, Rosenbaum JF et al (2001) Further evidence of association between behavioral inhibition and social anxiety in children. Am J Psychiatry 158:1673–1679

Binder EB (2012) The genetic basis of mood and anxiety disorders – changing paradigms. Biol Mood Anxiety Disord 2:1–3. https://doi.org/10.1186/2045-5380-2-17

Blackford JU, Allen AH, Cowan RL, Avery SN (2013) Amygdala and hippocampus fail to habituate to faces in individuals with an inhibited temperament. Soc Cogn Affect Neurosci 8:143–150. https://doi.org/10.1093/scan/nsr078

Blackford JU, Clauss JA, Benningfield MM (2018) The neurobiology of behavioral inhibition as a developmental mechanism. In: Pérez-Edgar K, Fox NA (eds) Behavioral inhibition. Springer, Cham, pp 113–134

Blair KS, Geraci M, Hollon N et al (2010) Social norm processing in adult social phobia: atypically increased ventromedial frontal cortex responsiveness to unintentional (embarrassing) transgressions. Am J Psychiatry 167:1526–1532. https://doi.org/10.1176/appi.ajp.2010.09121797

Bourdon JL, Savage JE, Verhulst B et al (2019) The genetic and environmental relationship between childhood behavioral inhibition and preadolescent anxiety. Twin Res Hum Genet 22: 48–55. https://doi.org/10.1017/thg.2018.73

Brühl AB, Delsignore A, Komossa K, Weidt S (2014a) Neuroimaging in social anxiety disorder – a meta-analytic review resulting in a new neurofunctional model. Neurosci Biobehav Rev 47: 260–280. https://doi.org/10.1016/j.neubiorev.2014.08.003

Brühl AB, Scherpiet S, Sulzer J et al (2014b) Real-time neurofeedback using functional MRI could improve down-regulation of amygdala activity during emotional stimulation: a proof-of-concept study. Brain Topogr 27:138–148. https://doi.org/10.1007/s10548-013-0331-9

Cannon TD, Keller MC (2006) Endophenotypes in the genetic analyses of mental disorders. Annu Rev Clin Psychol 2:267–290. https://doi.org/10.1146/annurev.clinpsy.2.022305.095232

Caspi A (2000) The child is father of the man: personality continuities from childhood to adulthood. J Pers Soc Psychol 78:158–172. https://doi.org/10.1037/0022-3514.78.1.158

Chavanne AV, Robinson OJ (2020) The overlapping neurobiology of induced and pathological anxiety: a meta-analysis of functional neural activation. Am J Psychiatry 178:156–164. https://doi.org/10.1176/appi.ajp.2020.19111153

Clauss JA, Blackford JU (2012) Behavioral inhibition and risk for developing social anxiety disorder: a meta-analytic study. J Am Acad Child Adolesc Psychiatry 51:1066–1075. https://doi.org/10.1016/j.jaac.2012.08.002

Clauss JA, Seay AL, Vanderklok RM et al (2014a) Structural and functional bases of inhibited temperament. Soc Cogn Affect Neurosci 9:2049–2058. https://doi.org/10.1093/scan/nsu019

Clauss JA, Avery SN, Vanderklok RM et al (2014b) Neurocircuitry underlying risk and resilience to social anxiety disorder. Depress Anxiety 31:822–833. https://doi.org/10.1002/da.22265

Clauss JA, Avery SN, Blackford JU (2015) The nature of individual differences in inhibited temperament and risk for psychiatric disease: a review and meta-analysis. Prog Neurobiol 127–128:23–45. https://doi.org/10.1016/j.pneurobio.2015.03.001

Cohen Kadosh K, Staunton G (2019) A systematic review of the psychological factors that influence neurofeedback learning outcomes. Neuroimage 185:545–555. https://doi.org/10.1016/j.neuroimage.2018.10.021

Cremers HR, Roelofs K (2016) Social anxiety disorder: a critical overview of neurocognitive research. Wiley Interdiscip Rev Cogn Sci 7:218–232. https://doi.org/10.1002/wcs.1390

Davis FC, Johnstone T, Mazzulla EC et al (2010) Regional response differences across the human amygdaloid complex during social conditioning. Cereb Cortex 20:612–621. https://doi.org/10.1093/cercor/bhp126

Etkin A, Wager TD (2007) Functional neuroimaging of anxiety: a meta-analysis of emotional processing in PTSD, social anxiety disorder, and specific phobia. Am J Psychiatry 164:1476–1488. https://doi.org/10.1176/appi.ajp.2007.07030504

Filippi CA, Sachs JF, Phillips D et al (2020) Infant behavioral reactivity predicts change in amygdala volume 12 years later. Dev Cogn Neurosci 42:100776. https://doi.org/10.1016/j.dcn.2020.100776

Flint J, Munafo MR (2007) The endophenotype concept in psychiatric genetics. Psychol Med 37: 163–180. https://doi.org/10.1017/S0033291706008750

Flint J, Timpson N, Munafò M (2014) Assessing the utility of intermediate phenotypes for genetic mapping of psychiatric disease. Trends Neurosci 37:733–741. https://doi.org/10.1016/j.tins.2014.08.007

Fox AS, Kalin NH (2014) A translational neuroscience approach to understanding the development of social anxiety disorder and its pathophysiology. Am J Psychiatry 171:1162–1173. https://doi.org/10.1176/appi.ajp.2014.14040449

Fox AS, Harris RA, Del RL et al (2021) Infant inhibited temperament in primates predicts adult behavior, is heritable, and is associated with anxiety-relevant genetic variation. Mol Psychiatry. https://doi.org/10.1038/s41380-021-01156-4

Frick A, Månsson KNT (2018) Brain changes in social anxiety disorder run in the family. EBioMedicine. https://doi.org/10.1016/j.ebiom.2018.09.009

Garner M, Möhler H, Stein DJ et al (2009) Research in anxiety disorders: from the bench to the bedside. Eur Neuropsychopharmacol 19:381–390. https://doi.org/10.1016/j.euroneuro.2009.01.011

Geiger MJ, Domschke K, Ipser J et al (2016) Altered executive control network resting-state connectivity in social anxiety disorder. World J Biol Psychiatry 17:47–57. https://doi.org/10.3109/15622975.2015.1083613

Glahn DC, Thompson PM, Blangero J (2007) Neuroimaging endophenotypes: strategies for finding genes influencing brain structure and function. Hum Brain Mapp 28:488–501. https://doi.org/10.1002/hbm.20401

Glahn DC, Knowles EEM, McKay DR et al (2014) Arguments for the sake of endophenotypes: examining common misconceptions about the use of endophenotypes in psychiatric genetics. Am J Med Genet B Neuropsychiatr Genet 165B:122–130. https://doi.org/10.1002/ajmg.b.32221

Goldsmith HH, Hilton EC, Phan JM et al (2022) Childhood inhibition predicts adolescent social anxiety: findings from a longitudinal twin study. Dev Psychopathol:1–20. https://doi.org/10.1017/S0954579422000864

Goldway N, Jalon I, Keynan JN et al (2022) Feasibility and utility of amygdala NeuroFeedback. Neurosci Biobehav Rev:104694. https://doi.org/10.1016/j.neubiorev.2022.104694

Gong Q, Scarpazza C, Dai J et al (2019) A transdiagnostic neuroanatomical signature of psychiatric illness. Neuropsychopharmacology 44:869–875. https://doi.org/10.1038/s41386-018-0175-9

Gottesman II, Gould TD (2003) The endophenotype concept in psychiatry: etymology and strategic intentions. Am J Psychiatry 160:636–645. https://doi.org/10.1176/appi.ajp.160.4.636

Gottschalk MG, Domschke K (2016) Novel developments in genetic and epigenetic mechanisms of anxiety. Curr Opin Psychiatry 29(1):32–38. https://doi.org/10.1097/YCO.0000000000000219

Gould TD, Gottesman II (2006) Psychiatric endophenotypes and the development of valid animal models. Genes Brain Behav 5:113–119. https://doi.org/10.1111/j.1601-183X.2005.00186.x

Groenewold NA, Bas-Hoogendam JM, Amod AR et al (2023) Volume of subcortical brain regions in social anxiety disorder: mega-analytic results from 37 samples in the ENIGMA-Anxiety Working Group. Mol Psychiatry. https://doi.org/10.1038/s41380-022-01933-9

Günther V, Ihme K, Kersting A et al (2018) Volumetric associations between amygdala, nucleus Accumbens, and socially anxious tendencies in healthy women. Neuroscience 374:25–32. https://doi.org/10.1016/j.neuroscience.2018.01.034

Guyer AE, Nelson EE, Perez-Edgar K et al (2006) Striatal functional alteration in adolescents characterized by early childhood behavioral inhibition. J Neurosci 26:6399–6405. https://doi.org/10.1523/JNEUROSCI.0666-06.2006

Guyer AE, Choate VR, Detloff A et al (2012) Striatal functional alteration during incentive anticipation in pediatric anxiety disorders. Am J Psychiatry 169:205–212. https://doi.org/10.1176/appi.ajp.2011.11010006

Haller SPW, Cohen Kadosh K, Scerif G, Lau JYF (2015) Social anxiety disorder in adolescence: how developmental cognitive neuroscience findings may shape understanding and interventions for psychopathology. Dev Cogn Neurosci 13:11–20. https://doi.org/10.1016/j.dcn.2015.02.002

Haller SPW, Mills KL, Hartwright CE et al (2018) When change is the only constant: the promise of longitudinal neuroimaging in understanding social anxiety disorder. Dev Cogn Neurosci 33:73–82. https://doi.org/10.1016/j.dcn.2018.05.005

Hallett M (2000) Transcranial magnetic stimulation and the human brain. Nature 406:147–150. https://doi.org/10.1038/35018000

He Z, Li S, Mo L et al (2023) The VLPFC-engaged voluntary emotion regulation: combined TMS-fMRI evidence for the neural circuit of cognitive reappraisal. J Neurosci 43:6046–6060. https://doi.org/10.1523/JNEUROSCI.1337-22.2023

Heeren A, Billieux J, Philippot P et al (2017) Impact of transcranial direct current stimulation on attentional bias for threat: a proof-of-concept study among individuals with social anxiety disorder. Soc Cogn Affect Neurosci 12:251–260. https://doi.org/10.1093/scan/nsw119

Herwig U, Lutz J, Scherpiet S et al (2019) Training emotion regulation through real-time fMRI neurofeedback of amygdala activity. Neuroimage 184:687–696. https://doi.org/10.1016/j.neuroimage.2018.09.068

Hettema JM, Neale MC, Kendler KS (2001) A review and meta-analysis of the genetic epidemiology of anxiety disorders. Am J Psychiatry 158:1568–1578. https://doi.org/10.1176/appi.ajp.158.10.1568

Hibar DP, Stein JL, Renteria ME et al (2015) Common genetic variants influence human subcortical brain structures. Nature 520:224–229. https://doi.org/10.1038/nature14101

Hill SY, Tessner K, Wang S et al (2010) Temperament at 5 years of age predicts amygdala and orbitofrontal volume in the right hemisphere in adolescence. Psychiatry Res 182:14–21. https://doi.org/10.1016/j.pscychresns.2009.11.006

Hoogendam JM, Ramakers GMJ, Di Lazzaro V (2010) Physiology of repetitive transcranial magnetic stimulation of the human brain. Brain Stimul 3:95–118. https://doi.org/10.1016/j.brs.2009.10.005

Insel TR (2014) The NIMH research domain criteria (RDoC) project: precision medicine for psychiatry. Am J Psychiatry 171:395–397. https://doi.org/10.1176/appi.ajp.2014.14020138

Ironside M, Browning M, Ansari TL et al (2019) Effect of prefrontal cortex stimulation on regulation of amygdala response to threat in individuals with trait anxiety: a randomized clinical trial. JAMA Psychiatry 76:71–78. https://doi.org/10.1001/jamapsychiatry.2018.2172

Isomura K, Boman M, Rück C et al (2015) Population-based, multi-generational family clustering study of social anxiety disorder and avoidant personality disorder. Psychol Med 45:1581–1589. https://doi.org/10.1017/S0033291714002116

Kagan J, Reznick JS, Snidman N (1987) The physiology and psychology of behavioral inhibition in children. Child Dev 58:1459–1473

Kalin NH (2017) Mechanisms underlying the early risk to develop anxiety and depression: a translational approach. Eur Neuropsychopharmacol 27:543–553. https://doi.org/10.1016/j.euroneuro.2017.03.004

Keeley LM, Laird RD, Qasmieh N et al (2024) When parents and adolescents make discrepant reports about parental monitoring: links to adolescent social anxiety when interacting with unfamiliar peers. J Psychopathol Behav Assess. https://doi.org/10.1007/s10862-024-10132-5

Kendler KS, Abrahamsson L, Ohlsson H et al (2022) An extended Swedish adoption study of anxiety disorder and its cross-generational familial relationship with major depression. Am J Psychiatry 179:640–649. https://doi.org/10.1176/appi.ajp.21111110

Kieling C, Buchweitz C, Caye A et al (2024) Worldwide prevalence and disability from mental disorders across childhood and adolescence: evidence from the global burden of disease study. JAMA Psychiatry. https://doi.org/10.1001/jamapsychiatry.2023.5051

Kim MJ, Shin J, Taylor JM et al (2017) Intolerance of uncertainty predicts increased striatal volume. Emotion 17:895–899. https://doi.org/10.1037/emo0000331

Klein DN, Mumper EE (2018) Behavioral inhibition as a precursor to psychopathology BT – behavioral inhibition: integrating theory, research, and clinical perspectives. In: Pérez-Edgar K, Fox NA (eds) Behavioral inhibition. Springer, Cham, pp 283–307

Kochunov P, Jahanshad N, Marcus D et al (2015) Heritability of fractional anisotropy in human white matter: a comparison of human connectome project and ENIGMA-DTI data. Neuroimage 111:300–311. https://doi.org/10.1016/J.NEUROIMAGE.2015.02.050

Koush Y, Meskaldji D-E, Pichon S et al (2015) Learning control over emotion networks through connectivity-based neurofeedback. Cereb Cortex 27:1193–1202. https://doi.org/10.1093/cercor/bhv311

Lago T, Davis A, Grillon C, Ernst M (2017) Striatum on the anxiety map: small detours into adolescence. Brain Res 1654:177–184. https://doi.org/10.1016/j.brainres.2016.06.006

Lenzenweger MF (2013a) Thinking clearly about the endophenotype-intermediate phenotype-biomarker distinctions in developmental psychopathology research. Dev Psychopathol 25:1347–1357. https://doi.org/10.1017/S0954579413000655

Lenzenweger MF (2013b) Endophenotype, intermediate phenotype, biomarker: definitions, concept comparisons, clarifications. Depress Anxiety 30:185–189. https://doi.org/10.1002/da.22042

Levey DF, Gelernter J, Polimanti R et al (2020) Reproducible genetic risk loci for anxiety: results from ~200,000 participants in the million veteran program. Am J Psychiatry. appi.ajp.2019.19030256. https://doi.org/10.1176/appi.ajp.2019.19030256

Liao W, Xu Q, Mantini D et al (2011) Altered gray matter morphometry and resting-state functional and structural connectivity in social anxiety disorder. Brain Res 1388:167–177. https://doi.org/10.1016/j.brainres.2011.03.018

Lieb R, Wittchen H-U, Höfler M et al (2000) Parental psychopathology, parenting styles, and the risk of social phobia in offspring: a prospective-longitudinal community study. Arch Gen Psychiatry 57:859–866

Lieberman JM, Rabellino D, Densmore M et al (2023) Posterior cingulate cortex targeted real-time fMRI neurofeedback recalibrates functional connectivity with the amygdala, posterior insula, and default-mode network in PTSD. Brain Behav 13:e2883. https://doi.org/10.1002/brb3.2883

Lipp A, Cohen Kadosh K (2020) Training the anxious brain: using fMRI-based neurofeedback to change brain activity in adolescence. Dev Med Child Neurol 62:1239–1244. https://doi.org/10.1111/dmcn.14611

Lisk S, Kadosh KC, Zich C et al (2020) Training negative connectivity patterns between the dorsolateral prefrontal cortex and amygdala through fMRI-based neurofeedback to target adolescent socially-avoidant behaviour. Behav Res Ther 135:103760. https://doi.org/10.1016/j.brat.2020.103760

Liu P, Pérez-Edgar KE (2019) Developmental pathways from early behavioral inhibition to later anxiety: an integrative review of developmental psychopathology research and translational implications. Adolesc Res Rev 4:45–58. https://doi.org/10.1007/s40894-018-0092-5

Meyer-Lindenberg A, Weinberger DR (2006) Intermediate phenotypes and genetic mechanisms of psychiatric disorders. Nat Rev Neurosci 7:818–827. https://doi.org/10.1038/nrn1993

Miskovic V, Schmidt LA (2012) Social fearfulness in the human brain. Neurosci Biobehav Rev 36:459–478. https://doi.org/10.1016/j.neubiorev.2011.08.002

Mizzi S, Pedersen M, Lorenzetti V et al (2021) Resting-state neuroimaging in social anxiety disorder: a systematic review. Mol Psychiatry. https://doi.org/10.1038/s41380-021-01154-6

Mizzi S, Pedersen M, Rossell SL et al (2024) Resting-state amygdala subregion and precuneus connectivity provide evidence for a dimensional approach to studying social anxiety disorder. Transl Psychiatry 14:147. https://doi.org/10.1038/s41398-024-02844-9

Morneau-Vaillancourt G, Coleman JRI, Purves KL et al (2020) The genetic and environmental hierarchical structure of anxiety and depression in the UK Biobank. Depress Anxiety 37:512–520. https://doi.org/10.1002/da.22991

Morris SE, Sanislow CA, Pacheco J et al (2022) Revisiting the seven pillars of RDoC. BMC Med 20:220. https://doi.org/10.1186/s12916-022-02414-0

Murray L, Creswell C, Cooper PJ (2009) The development of anxiety disorders in childhood: an integrative review. Psychol Med 39:1413–1423. https://doi.org/10.1017/S0033291709005157

Nicholson AA, Rabellino D, Densmore M et al (2017) The neurobiology of emotion regulation in posttraumatic stress disorder: amygdala downregulation via real-time fMRI neurofeedback. Hum Brain Mapp 38:541–560. https://doi.org/10.1002/hbm.23402

Nitsche MA, Cohen LG, Wassermann EM et al (2008) Transcranial direct current stimulation: state of the art 2008. Brain Stimul 1:206–223. https://doi.org/10.1016/j.brs.2008.06.004

Norton AR, Abbott MJ (2017) The role of environmental factors in the aetiology of social anxiety disorder: a review of the theoretical and empirical literature. Behav Chang:1–22. https://doi.org/10.1017/bec.2017.7

O'Donovan MC, Owen MJ (2016) The implications of the shared genetics of psychiatric disorders. Nat Med 22:1214–1219. https://doi.org/10.1038/nm.4196

Oler JA, Fox AS, Shelton SE et al (2010) Amygdalar and hippocampal substrates of anxious temperament differ in their heritability. Nature 466:864–868. https://doi.org/10.1038/nature09282

Paes F, Baczynski T, Novaes F et al (2013a) Repetitive transcranial magnetic stimulation (rTMS) to treat social anxiety disorder: case reports and a review of the literature. Clin Pract Epidemiol Ment Health 9:180–188. https://doi.org/10.2174/1745017901309010180

Paes F, Machado S, Arias-Carrion O et al (2013b) rTMS to treat social anxiety disorder: a case report. Braz J Psychiatry 35:99–100. https://doi.org/10.1590/S1516-44462013000100020

Pannekoek JN, Veer IM, van Tol M-J et al (2013) Resting-state functional connectivity abnormalities in limbic and salience networks in social anxiety disorder without comorbidity. Eur Neuropsychopharmacol 23:186–195. https://doi.org/10.1016/j.euroneuro.2012.04.018

Paret C, Hendler T (2020) Live from the "regulating brain": harnessing the brain to change emotion. Emotion 20:126–131. https://doi.org/10.1037/emo0000674

Paret C, Ruf M, Gerchen MF et al (2016a) fMRI neurofeedback of amygdala response to aversive stimuli enhances prefrontal-limbic brain connectivity. Neuroimage 125:182–188. https://doi.org/10.1016/j.neuroimage.2015.10.027

Paret C, Kluetsch R, Zaehringer J et al (2016b) Alterations of amygdala-prefrontal connectivity with real-time fMRI neurofeedback in BPD patients. Soc Cogn Affect Neurosci 11:952–960. https://doi.org/10.1093/scan/nsw016

Parsaei M, Hasehmi SM, Seyedmirzaei H et al (2024) Microstructural white matter alterations associated with social anxiety disorders: a systematic review. J Affect Disord 350:78–88. https://doi.org/10.1016/j.jad.2024.01.118

Pérez-Edgar KE, Guyer AE (2014) Behavioral inhibition: temperament or Prodrome? Curr Behav Neurosci Rep. https://doi.org/10.1007/s40473-014-0019-9

Pérez-Edgar K, Hardee JE, Guyer AE et al (2014) DRD4 and striatal modulation of the link between childhood behavioral inhibition and adolescent anxiety. Soc Cogn Affect Neurosci 9:445–453. https://doi.org/10.1093/scan/nst001

Ramaswami M (2014) Network plasticity in adaptive filtering and behavioral habituation. Neuron 82:1216–1229. https://doi.org/10.1016/j.neuron.2014.04.035

Rankin CH, Abrams T, Barry RJ et al (2009) Habituation revisited: an updated and revised description of the behavioral characteristics of habituation. Neurobiol Learn Mem 92:135–138. https://doi.org/10.1016/j.nlm.2008.09.012

Roelofs EF, Bas-Hoogendam JM, van Ewijk H et al (2020) Investigating microstructure of white matter tracts as candidate endophenotypes of social anxiety disorder – findings from the Leiden family lab study on social anxiety disorder (LFLSAD). Neuroimage Clin:102493. https://doi.org/10.1016/j.nicl.2020.102493

Roffman JL (2019) Endophenotype research in psychiatry – the grasshopper grows up. JAMA Psychiatry 76:1230–1231. https://doi.org/10.1001/jamapsychiatry.2019.2194

Sagliano L, Atripaldi D, De Vita D et al (2019) Non-invasive brain stimulation in generalized anxiety disorder: a systematic review. Prog Neuropsychopharmacol Biol Psychiatry 93:31–38. https://doi.org/10.1016/j.pnpbp.2019.03.002

Sarkheil P, Zilverstand A, Kilian-Hütten N et al (2015) fMRI feedback enhances emotion regulation as evidenced by a reduced amygdala response. Behav Brain Res 281:326–332. https://doi.org/10.1016/j.bbr.2014.11.027

Scaini S, Belotti R, Ogliari A (2014) Genetic and environmental contributions to social anxiety across different ages: a meta-analytic approach to twin data. J Anxiety Disord 28:650–656. https://doi.org/10.1016/j.janxdis.2014.07.002

Schwartz CE, Snidman N, Kagan J (1999) Adolescent social anxiety as an outcome of inhibited temperament in childhood. J Am Acad Child Adolesc Psychiatry 38:1008–1015. https://doi.org/10.1097/00004583-199908000-00017

Schwartz CE, Wright CI, Shin LM et al (2003) Inhibited and uninhibited infants "grown up": adult amygdalar response to novelty. Science 300:1952–1953. https://doi.org/10.1126/science.1083703

Schwartz CE, Kunwar PS, Greve DN et al (2012) A phenotype of early infancy predicts reactivity of the amygdala in male adults. Mol Psychiatry 17:1042–1050. https://doi.org/10.1038/mp.2011.96

Seidl E, Venz J, Ollmann TM et al (2021) How current and past anxiety disorders affect daily life in adolescents and young adults from the general population – an epidemiological study with ecological momentary assessment. Depress Anxiety 38:272–285. https://doi.org/10.1002/da.23133

Shimada-Sugimoto M, Otowa T, Hettema JM (2015) Genetics of anxiety disorders: genetic epidemiological and molecular studies in humans. Psychiatry Clin Neurosci 69:388–401. https://doi.org/10.1111/pcn.12291

Sitaram R, Ros T, Stoeckel L et al (2017) Closed-loop brain training: the science of neurofeedback. Nat Rev Neurosci 18:86–100

Smoller JW, Block SR, Young MM (2009) Genetics of anxiety disorders: the complex road from DSM to DNA. Depress Anxiety 26:965–975. https://doi.org/10.1002/da.20623

Stein MB, Stein DJ (2008) Social anxiety disorder. Lancet 371:1115–1125. https://doi.org/10.1016/S0140-6736(08)60488-2

Sweeney S, Wilson C (2023) Parental anxiety and offspring development: a systematic review. J Affect Disord 327:64–78. https://doi.org/10.1016/j.jad.2023.01.128

Sylvester CM, Pine DS (2018) The biological bridge between behavioral inhibition and psychopathology BT – behavioral inhibition: integrating theory, research, and clinical perspectives. In: Pérez-Edgar K, Fox NA (eds) Behavioral inhibition. Springer, Cham, pp 309–335

Sylvester CM, Barch DM, Harms MP et al (2015) Early childhood behavioral inhibition predicts cortical thickness in adulthood. J Am Acad Child Adolesc Psychiatry. https://doi.org/10.1016/j.jaac.2015.11.007

Sylvester CM, Luby JL, Pine DS (2024) Novel mechanism-based treatments for pediatric anxiety and depressive disorders. Neuropsychopharmacology 49:262–275. https://doi.org/10.1038/s41386-023-01709-x

Taber-Thomas BC, Morales S, Hillary FG, Pérez-Edgar KE (2016) Altered topography of intrinsic functional connectivity in childhood risk for social anxiety. Depress Anxiety 33:995–1004. https://doi.org/10.1002/da.22508

Tang A, Crawford H, Morales S et al (2020) Infant behavioral inhibition predicts personality and social outcomes three decades later. Proc Natl Acad Sci:201917376. https://doi.org/10.1073/pnas.1917376117

Tang A, Harrewijn A, Benson B et al (2022) Striatal activity to reward anticipation as a moderator of the association between early behavioral inhibition and changes in anxiety and depressive symptoms from adolescence to adulthood. JAMA Psychiatry. https://doi.org/10.1001/jamapsychiatry.2022.3483

Thibault RT, MacPherson A, Lifshitz M et al (2018) Neurofeedback with fMRI: a critical systematic review. Neuroimage 172:786–807. https://doi.org/10.1016/j.neuroimage.2017.12.071

Tissier R, Tsonaka R, Mooijaart SP et al (2017) Secondary phenotype analysis in ascertained family designs: application to the Leiden longevity study. Stat Med 36:2288–2301. https://doi.org/10.1002/sim.7281

Uher R, Pavlova B, Radua J et al (2023) Transdiagnostic risk of mental disorders in offspring of affected parents: a meta-analysis of family high-risk and registry studies. World Psychiatry 22:433–448. https://doi.org/10.1002/wps.21147

Vergallito A, Gallucci A, Pisoni A et al (2021) Effectiveness of noninvasive brain stimulation in the treatment of anxiety disorders: a meta-analysis of sham or behaviour-controlled studies. J Psychiatry Neurosci 46:E592–E614. https://doi.org/10.1503/jpn.210050

Vicario CM, Salehinejad MA, Felmingham K et al (2019) A systematic review on the therapeutic effectiveness of non-invasive brain stimulation for the treatment of anxiety disorders. Neurosci Biobehav Rev 96:219–231. https://doi.org/10.1016/j.neubiorev.2018.12.012

Williams JT, Blangero J (1999) Power of variance component linkage analysis to detect quantitative trait loci. Ann Hum Genet 63:545–563. https://doi.org/10.1017/S0003480099007848

Young KD, Misaki M, Harmer CJ et al (2017) Real-time functional magnetic resonance imaging amygdala neurofeedback changes positive information processing in major depressive disorder. Biol Psychiatry 82:578–586. https://doi.org/10.1016/j.biopsych.2017.03.013

Zich C, Johnstone N, Lührs M et al (2020) Modulatory effects of dynamic fMRI-based neurofeedback on emotion regulation networks in adolescent females. Neuroimage 220: 117053. https://doi.org/10.1016/j.neuroimage.2020.117053

Zugman A, Jett L, Antonacci C et al (2023) A systematic review and meta-analysis of resting-state fMRI in anxiety disorders: need for data sharing to move the field forward. J Anxiety Disord 99: 102773. https://doi.org/10.1016/j.janxdis.2023.102773

Zweerings J, Sarkheil P, Keller M et al (2020) Rt-fMRI neurofeedback-guided cognitive reappraisal training modulates amygdala responsivity in posttraumatic stress disorder. Neuroimage Clin 28: 102483. https://doi.org/10.1016/j.nicl.2020.102483

Neurodevelopmental Pathways from Temperamental Fear to Anxiety

Eunkyung Shin and Koraly Pérez-Edgar

Contents

Abstract Early indicators of anxiety risk can appear as early as infancy, informing developmental pathways in which individual differences in temperament elevate the likelihood of future anxiety disorders. Clarifying the mechanisms that connect these early biological predispositions to later anxiety offers a foundation for designing targeted early intervention and prevention efforts. In this chapter, we aim to describe the association between fearful temperament and the development of anxiety disorders, highlighting how the interplay between biological and environmental factors shape vulnerability to anxiety from early in life. We describe (a) fearful temperament as a potential marker for vulnerability to anxiety, (b) neural mechanisms underlying fearful temperament and anxiety through detection and regulation processes, (c) internal and external factors that moderate the association between fearful temperament and anxiety, focusing on attentional bias and parental factors to understand distinct etiological process.

E. Shin (✉) and K. Pérez-Edgar
Department of Psychology, Pennsylvania State University, University Park, PA, USA
e-mail: eshin@psu.edu; kxp24@psu.edu

© The Author(s), under exclusive license to Springer Nature Switzerland AG 2025
Curr Topics Behav Neurosci (2025) 73: 329–353
https://doi.org/10.1007/7854_2024_571
Published Online: 12 March 2025

Keywords Anterior cingulate cortex · Anxiety · Amygdala · Behavioral inhibition · Delta-beta coupling · Dysregulated · Fear · Fearful temperament

Anxiety disorders are one of the most prevalent mental health problems in the United States (Szuhany and Simon 2022). More than 40 million adults over 18 years old experience anxiety symptoms (Booth 2023), leading to impairments in cognitive functioning, social relationships, and physical health. While the median age of diagnosis for anxiety disorders is around 6–10 years of age (Merikangas et al. 2010), early signs of anxiety risk can appear in infancy, including negative affect, hypervigilance, and withdrawal to novel stimuli (Fox et al. 2021). These precursors of anxiety symptoms suggest a developmental cascade, where individual differences in temperament can increase vulnerability to developing anxiety disorder later in life (Buss 2011; Buzzell et al. 2017; Shackman and Fox 2016; Taber-Thomas and Perez-Edgar 2015). Elucidating the link between early-appearing biological predispositions and anxiety offers a foundation for the development of early intervention and prevention strategies.

In this chapter, we aim to describe the association between fearful temperament and the development of anxiety disorders, highlighting how the interplay between biological and environmental factors shape vulnerability to anxiety from early in life. We describe (a) fearful temperament as a potential marker for vulnerability to anxiety, (b) neural mechanisms underlying fearful temperament and anxiety through detection and regulation processes, and (c) internal and external factors that moderate the association between fearful temperament and anxiety, focusing on attentional bias and parental factors to understand the distinct etiological process.

1 Fearful Temperament as a Potential Precursor to Anxiety

Temperament refers to consistent individual differences in behavioral and emotional tendencies, often observed in early infancy, that are thought to have biological underpinnings (Fox et al. 2021; Buss and Kiel 2013). These differences manifest in how infants interact with the world, including their emotional reactivity, self-regulation skills, attentional focus, and motor responses (Rothbart 2007). Fearful temperament, characterized by wariness, fear, and avoidance of novelty (Buss and Kiel 2013; Morales and Fox 2019), has long been investigated as a precursor to anxiety disorders in childhood and adolescence (Abend et al. 2020; Fox et al. 2021; Liu and Pérez-Edgar 2019). Traditionally, temperamental anxiety risk has been assessed through the lens of behavioral inhibition, primarily measured in toddlers (Fox et al. 2021). However, recent research has proposed dysregulated fear as a more nuanced predictor of social wariness across early childhood, potentially identifying a distinct subgroup of children with a fearful temperament at heightened risk for anxiety (Buss 2011; Kiel and Kalomiris 2019; Zhou et al. 2022). In this section, we review the literature on fearful temperament, focusing on both behavioral inhibition and dysregulated fear as potential markers for vulnerability to anxiety.

1.1 Behavioral Inhibition

Behavioral inhibition (BI) is a temperamental trait characterized by elevated negative reactivity and withdrawal in response to novelty, particularly in social contexts (Fox et al. 2021). This early-appearing biological predisposition manifests in negative reactivity (Kagan et al. 1984; Pérez-Edgar et al. 2011), attentional hypervigilance to potential threats (Filippi et al. 2022), and reticent behavior in unfamiliar settings or interactions (Pérez-Edgar and Guyer 2014). Kagan's pioneering observations of healthy (primarily middle-class and White) infants and toddlers interacting with novel stimuli identified qualitatively distinct behavioral and affective patterns across contexts in BI children compared to their less inhibited counterparts (Kagan et al. 1984). Notably, toddlers identified as BI displayed hypervigilance, high levels of negative affect, and increased motor activity in response to novelty (Fox et al. 2021). Longitudinal studies suggest moderate continuity of BI from toddlerhood through early adulthood, persisting across middle childhood and adolescence (Pérez-Edgar and Guyer 2014).

The neural underpinnings of behavioral inhibition were initially associated with amygdala function, particularly as seen in a hyper-reactive response to novel stimuli (see below). Evidence from animal research suggests that the amygdala plays a crucial role in fear acquisition and emotional reactivity (Kagan 1994). Kagan (1994) proposed that individuals with a lower threshold for amygdala responsivity tend to interpret novel stimuli as threats, leading to heightened fear and arousal (Kiel and Kalomiris 2019). This hyper-reactivity triggers distinct behavioral and affective profiles, serving as a robust individual difference predictor of risk for later anxiety (Liu and Pérez-Edgar 2019; Morales and Fox 2019; Roy et al. 2014).

1.2 Dysregulated Fear

While previous research examining the association between fearful temperament and anxiety has mainly relied on observations of BI, a new perspective focusing on dysregulated fear responses in low-threat situations offers a unique approach to identifying children with extreme fearfulness (Buss 2011). Dysregulated fear focuses on the context of emotional behavior and regulation, emphasizing the ability to adjust emotional responses flexibly to situational demands (e.g., Goldsmith and Davidson 2004). BI measures incorporate general fearfulness across various contexts by averaging fear responses across different situations. Although this approach captures the overall intensity and frequency of fear, this measure may miss children who only show extreme fear in specific contexts (Kiel and Kalomiris 2019). In contrast, dysregulated fear focuses on excessive fear in low-threat situations. For example, almost all children (regardless of temperament) will display fear in an acutely threatening situation. However, only a subset of children will continue to display signs of fear in putatively low-fear contexts, such as during routine activities where the contexts are predictable and familiar, and coping resources are accessible. Most children, for instance, did not display signs of fear during a "friendly" interaction with a puppet (Buss 2011; Kiel and Kalomiris 2019). Children who show a heightened fear response that does not align with the actual threat level (i.e., high fear in low-threat situations) may have difficulty regulating their fear, leading to maladaptive behavior, and increasing their risk for anxiety disorders (Buss 2011). Focusing on the context-sensitivity of fear offers a more nuanced understanding of the specific vulnerabilities associated with anxiety disorders and holds the potential for developing targeted interventions addressing context-inappropriate fear.

Dysregulated fear is an additional early risk factor for anxiety disorders, above and beyond BI (Zhou et al. 2022). Buss (2011) found that dysregulated fear at age 2 was associated with heightened maternal-reported anxiety and social withdrawal at ages 3 and 4, as well as teacher-reported anxiety during the kindergarten transition. The significant association between dysregulated fear and anxiety after controlling for BI suggests a distinct mechanism underlying this link. Furthermore, children with dysregulated fear at age 2 were nearly four times more likely to develop social anxiety disorder (SAD) by age 5, compared to children with less dysregulated fear and children displaying social reticence with unfamiliar peers (Buss et al. 2013). Longitudinal studies have also replicated the link between dysregulated fear and anxiety symptoms from toddlerhood to adolescence. Buss et al. (2021) found that dysregulated fear in toddlerhood uniquely predicted adolescent social anxiety symptoms, even after accounting for concurrent parenting and peer influences. Similarly, Zhou et al. (2022) reported that dysregulated fear at 24 months predicted social anxiety symptoms in early adolescence, after controlling for other potential risk factors, including social wariness and peer difficulties. These findings suggest a long-term and specific effect of dysregulated fear on anxiety development.

2 Neural Mechanisms Underlying Fearful Temperament and Anxiety Disorders

Children with fearful temperament have unique response patterns to their environment, potentially serving as a phenotypically observable and biologically linked marker for understanding the pathway to anxiety disorders (Filippi et al. 2021). Although not all children with a fearful temperament develop anxiety disorders (Buss and Kiel 2013), the overlap in biological factors linked to both fearful temperament and anxiety suggests a potential association (Shackman and Fox 2016). The continuity between early fearful temperament and increased risk for anxiety disorders across the lifespan suggests that they may both reflect early deviations from typical brain development patterns (Rogers et al. 2017). This section describes alterations in brain structure, function, and connectivity associated with fearful temperament and anxiety across development. In doing so, we focus on the literature detailing amygdala contributions, given the initial theoretical formulation for fearful temperament (Kagan 1994; Liu and Pérez-Edgar 2019).

2.1 Detection

2.1.1 Amygdala and the Fearful Brain

At the neural level, fearful temperament and anxiety disorders have shown distinct threat-sensitive circuitry within brain regions, including amygdala, bed nucleus of the stria terminalis (BNST), insula, hippocampus, and prefrontal cortex (PFC) (Kredlow et al. 2022; Peng et al. 2023). These regions are recruited for detecting and interpreting potential threats in the environment and alterations in these regions may contribute to heightened sensitivity to novel and salient stimuli, potentially leading individuals to perceive stimuli as threatening (Filippi et al. 2021). Robust evidence has demonstrated the role of amygdala in threat or novelty detection (Bas-Hoogendam et al. 2020; Mizzi et al. 2024), fear acquisition (Kredlow et al. 2022), fear conditioning (Keifer et al. 2015), and withdrawal behavior (Schwarzlose et al. 2023), all of which are key features of fearful temperament and anxiety.

As a limbic part of the "fear network" (Gorman et al. 2000), the amygdala detects threats and can motivate the individual to withdraw from potential harm and danger (Kragel and LaBar 2015). Neuroimaging studies have reported amygdala activation to fearful faces (Pérez-Edgar et al. 2007; Schwartz et al. 2012; Watve et al. 2024) and disturbing sounds (Lau et al. 2011; Tzovara et al. 2024). Increased activation in the amygdala, in turn, is associated with the physiological readiness and emotional arousal to respond to threats in the environment (Kragel and LaBar 2015). While the threat-detection ability heightened by the amygdala is an adaptive response to environmental threats, a lower-than-typical threshold for amygdala activation can contribute to hypervigilance and oversensitivity to novel and salient stimuli, triggering excessive fear and arousal (Morales and Fox 2019).

The function of the amygdala extends beyond simply detecting potential threats. The amygdala's involvement in fear learning allows it to not only respond to immediate dangers but also link current stimuli and fear-related memories, contributing to the development of conditioned fear responses (LeDoux 2003). Lesions or temporary inactivation of the amygdala can decrease reticent and freezing behavior, as well as normal fearfulness in novel social situations in both animals and humans (LeDoux 2003; Fox and Kalin 2014; Feinstein et al. 2011; Murray and Izquierdo 2007). As a critical hub for processing sensory input, the amygdala can sift information from the environment to identify, interpret, and consolidate fear-related memories (LeDoux 2003). Pavlovian fear conditioning is partially reliant on the amygdala in generalizing fear to non-threatening stimuli (Harnett et al. 2018).

2.1.2 Structural Difference

Amygdala volume in early childhood has been linked to the intensity of the vigilance and fear response (Cismaru et al. 2016; Lautarescu et al. 2020). Preterm infants who presented with larger amygdala volumes at birth exhibited more escape behavior to threat stimuli at 12 months, while smaller volumes predicted blunted reactions (Cismaru et al. 2016). In another longitudinal study with preschool children (Hill et al. 2010), children who were behaviorally inhibited had larger amygdala volume in adolescence versus participants who were less inhibited. Furthermore, larger amygdala volume in adults who were inhibited showed increased activation in response to neutral faces (Clauss et al. 2014).

However, other studies have found an inverse relationship between amygdala volume and BI. Infants characterized with negative reactivity at 4 months displayed a pattern of minimal growth in the left amygdala during middle childhood (Filippi et al. 2020). In a similar vein, recent studies found smaller volume and slower amygdala growth and individuals with anxiety (Burrows et al. 2024; Groenewold et al. 2023). These inconsistent findings can be attributed to factors including the non-linear development of the brain (Mills et al. 2016), limited sample sizes, and methodological differences. Large, multi-sites datasets employing standardized methodologies (e.g., HEALthy Brain and Child Development study, Volkow et al. 2021) will offer the potential to replicate existing findings and generate robust evidence regarding the link between amygdala maturation and behavioral inhibition, potentially enhancing our understanding of the development of anxiety.

2.1.3 Heightened and Sustained Amygdala Activation

While findings on amygdala volume remain inconsistent, amygdala hyperactivation is consistently linked to both fearful temperament and anxiety (Blackford et al. 2009; Pérez-Edgar et al. 2007; Schwartz et al. 2012). With advancements in functional magnetic resonance imaging (fMRI), researchers have been able to test Kagan's hypothesis that amygdala hypersensitivity to novel or salient stimuli contributes to BI. For instance, Schwartz et al. (2012) reported that early childhood fearful temperament predicted greater amygdala activation in adults' response to novel stimuli for male and unfamiliar neutral faces. Blackford et al. (2009) further demonstrated longer and greater amygdala reactivity to all faces in young adults who were inhibited. Similarly, Pérez-Edgar et al. (2007) found increased amygdala activation in response to fearful faces in young adolescents with a history of behavioral inhibition in childhood. Extensive evidence further supports amygdala hyperactivation in individuals with anxiety disorders (Bas-Hoogendam et al. 2020; Hahn et al. 2011; Mizzi et al. 2024). These findings suggest that fearful temperament and anxiety disorders may share a common neurobiological foundation.

Another neurobiological signature of children with a fearful temperament is a failure of amygdala habituation to repeated presentations of stimuli. Habituation, functioning as a sensory "firewall" (Poon and Young 2006), plays a crucial role in adaptively allocating attention and processing resources by filtering out repetitive and irrelevant sensory information (Avery et al. 2021; Blackford et al. 2018). When habituation fails to occur rapidly in response to neutral stimuli, it may lead to prolonged feelings of uncertainty or unfamiliarity, as these stimuli could be interpreted as novel, unpredictable, and potentially threatening (Avery and Blackford 2016). Studies have found that young adults characterized by inhibited temperament exhibited sustained amygdala activation to familiarized faces (Blackford et al. 2011, 2013). Additionally, adults who were classified as high-reactive in infancy demonstrate greater amygdala reactivity and slower habituation to neutral faces compared to low-reactive counterparts (Schwartz et al. 2012). Furthermore, individuals with higher levels of social fearfulness exhibit slower habituation in social brain regions, including the amygdala (Avery and Blackford 2016). These findings collectively suggest a link between individual differences in the rate of amygdala habituation and fearful temperament.

2.1.4 Functional Connectivity

The amygdala is not solely recruited for emotion processing (Sladky et al. 2022). Instead, it interacts with various cortical and subcortical regions, forming a complex "multi-road" network for emotion perception and experience (Pessoa and Adolphs 2011). While past research has focused on the amygdala's individual role, this approach might not capture its full functionality (Sylvester et al. 2020). Examining how brain regions connect and communicate with the amygdala offers a more comprehensive understanding of its influence. Studies investigating the amygdala's function within a distributed network of brain regions have identified distinct amygdala-related circuits in individuals with fearful temperament and anxiety disorders (Filippi et al. 2021; Fu et al. 2017; Rogers et al. 2017; Roy et al. 2014; Sladky et al. 2013, 2022; Schwarzlose et al. 2023; Watve et al. 2024). These findings suggest potential distributed differences in how individuals with fearful temperament and anxiety disorders perceive, attend to, and process emotional experiences.

Connectivity between fear-related regions, such as the amygdala and insula, and key attentional networks, including the dorsal attention network (DAN) and salience network (SN) has been consistently found in individuals with fearful temperament and anxiety. This connectivity pattern may integrate emotional meaning with self-relevant information and evaluate the salience of stimuli (Zhou et al. 2018) by routing information between the limbic systems and other regions for emotional processing (Sladky et al. 2013). The first study (Roy et al. 2014) exploring amygdala intrinsic functional connectivity in behavioral inhibition found a negative amygdala–insula functional connectivity in young adults with a history of childhood BI. Supporting this result, Thomas et al. (2019) found that amygdala-insular connectivity at 6 months predicted toddler fear at 24 months. Interestingly, this association is particularly related to fear and not sadness, implying specificity within negative emotionality. Recent studies with newborns suggest an early departure from neurobiological structure and function for fearful temperament and anxiety. Newborn's amygdala functional connectivity with specific regions predicted BI and generalized anxiety at age 2 years (Rogers et al. 2017). More specifically, amygdala–dorsal anterior cingulate connectivity predicted generalized anxiety symptom levels, and amygdala–medial prefrontal cortex (mPFC) connectivity predicted BI in infancy. Furthermore, amygdala–cingulate connectivity in 4-month-old infants was positively related to novelty-evoked distress, which is a precursor of BI (Filippi et al. 2021).

These findings imply that individual differences in functional connectivity between the amygdala and other brain regions might contribute to the continuity of anxiety disorders. Increased attention to and the salience of a threat facilitated by neural networks may potentially lead to freezing or avoidance behavior, ultimately triggering anxiety responses. Additionally, linking novel stimuli with heightened arousal early in development might establish and enhance fear circuits (Blackford et al. 2018), potentially laying the foundation for anxiety in later life.

Beyond systems directing attention and detecting potential threats, the amygdala's connections with brain regions involved in emotion regulation are altered in individuals with fearful temperament and anxiety across development (Fu et al. 2017; Rogers et al. 2017; Sladky et al. 2013, 2022; Schwarzlose et al. 2023; Watve et al. 2024). Notably, the PFC, including mPFC, dorsolateral prefrontal cortex (dlPFC), ventrolateral prefrontal cortex (vlPFC), and middle cingulate cortex (mCC), is implicated in top-down regulation. These regions are key for flexible responses to threats and contribute to dampening or amplifying emotional reactions (Sylvester and Pine 2018). Failure of top-down regulation of threat-related processing could lead to increased fear and shape the risk of anxiety disorders through altered amygdala connectivity with PFC. Indeed, a recent work leveraging dynamic causal modeling found evidence for the orbitofrontal cortex (OFC; Sladky et al. 2022) and medial OFC (mOFC; Watve et al. 2024) in the downregulation of amygdala activation during emotion processing and expression.

Disrupted amygdala-prefrontal circuitry has been shown in individuals with fearful temperament and anxiety across development. Newborn's neural response to novelty in PFC was related to BI at 13 months (Schwarzlose et al. 2023), and left amygdala-mPFC connectivity predicted BI at 2 years old (Rogers et al. 2017), implying that a disruption of the amygdala-prefrontal circuitry is one neurobiological mechanism contributing to early temperamental risk.

In a longitudinal study (Abend et al. 2020), children with a history of high early-childhood BI showed increased anxiety symptoms at age 13 if they exhibited weaker connectivity between the right amygdala and left dlPFC during threat processing tasks at age 10. Conversely, children with low-BI showed an increased positive relation between anxiety and amygdala-dlPFC connectivity with age. Similarly, Fu et al. (2017) found that dlPFC activation in children ages 9–12 years during an attention bias task was associated with anxiety levels, and this association was mediated by BI. These findings suggest that variations in communication between the amygdala and PFC may contribute to responses to threats and the regulation of emotional arousal, potentially linking to the development of anxiety symptoms.

Similar to BI, anxiety disorders are characterized by hyperactivation of the fear circuitry (Mao et al. 2020). Studies on anxiety offer robust evidence that a deficit in PFC's ability to downregulate the amygdala is associated with anxiety disorders (Bouras et al. 2023; Heitmann et al. 2017; Liu et al. 2020; Minkova et al. 2017; Sandman et al. 2020; Young et al. 2017). For example, people with SAD exhibit disrupted amygdala–PFC circuitry in response to harsh faces (Minkova et al. 2017), and the symptom severity in people with SAD has been positively associated with amygdala activation and negatively related to dlPFC (Heitmann et al. 2017) and vmPFC (Young et al. 2017). Furthermore, strengthened functional connectivity between the amygdala and dmPFC following cognitive behavioral therapy predicted greater social anxiety symptom reduction at 6–12-month follow-up (Sandman et al. 2020).

2.2 Regulation

While many children with a fearful temperament exhibit a biological predisposition for heightened threat detection in their environment, the developmental trajectory of anxiety disorders remains heterogeneous. Not all children classified as BI exhibit long-term persistence of this temperamental trait, nor do they inevitably develop anxiety disorders (Clauss and Blackford 2012a, b). This heterogeneity can be attributed, in part, to the interplay between threat detection and control processes (Fox et al. 2021) as children become more skilled in modulating their initial responses to the environment. While detection serves to register potentially salient information in the environment, control mechanisms help govern how this information is utilized. According to the dual-processing model of information processing (Henderson et al. 2015), individual differences in social and emotional outcomes reflect how automatic, stimulus-driven processing and controlled and goal-directed processing functionally interact to support adaptation. Compared to behaviors driven by the automatic mode, the control mode leverages a broader range of information, facilitating more deliberate, strategic, and goal-oriented behaviors (Henderson et al. 2015). Therefore, the control and regulatory processes that children with a fearful temperament deploy when responding to perceived threats differentially influence their susceptibility to developing anxiety disorders (Henderson et al. 2015). This section explores neural markers of emotion regulation, particularly their association with fearful temperament and anxiety symptoms. In addition, we discuss the developmental processes impacting the evolving balance between regulation and emotion. In doing so, we focus on biologically based markers that can help track the developmental trajectories of this balance, namely, delta-beta coupling, captured via electroencephalogram (EEG), and activity in the anterior cingulate cortex (ACC), captured via fMRI and building on the parallel discussion above on functional connectivity between prefrontal regions and the amygdala.

2.2.1 Delta-Beta Coupling

Delta-Beta Coupling (DBC) reflects the correlation between relative power in the delta and beta frequency bands of the EEG signal. Specifically, beta waves, associated with the cortex, are thought to reflect top-down cognitive processes whereas delta waves, linked to subcortical brain regions, are believed to reflect bottom-up emotional processes (Myruski et al. 2022). Research suggests that positive DBC is linked to communication between subcortical emotional centers and cortical regulatory systems (Knyazev and Slobodskaya 2003; Knyazev 2007), such that DBC may be a real-time indicator of brain processes related to emotion regulation (Anaya et al. 2021a).

While DBC might reflect the effective deployment of regulatory strategies in typical development, heightened coupling in clinical populations might reflect efforts to downregulate high levels of arousal from subcortical regions (Poole et al. 2020). Rigid and overcontrolled responses to threats or challenges may hinder flexible coping mechanisms, as observed in children with a fearful temperament (Liu et al. 2018). For example, children with a fearful temperament might employ excessive monitoring and control strategies in novel situations (Buss 2011). This overcontrol may limit their ability to adapt and cope with stressful situations by using more flexible emotion regulation strategies, increasing their vulnerability to anxiety disorders (Ramos et al. 2024).

Emerging evidence supports a link between individual differences in DBC and both fearful temperament and anxiety from infancy to adolescence (Brooker et al. 2016). Highly fearful toddlers exhibit greater DBC (Najjar and Brooker 2017) as do young children with high BI (Poole et al. 2020). Similarly, infants exhibiting high reactivity during a task eliciting fear (Brooker et al. 2016) also showed higher DBC compared to their peers. Furthermore, 10-year-olds with higher social anxiety exhibited a stronger and more persistent coupling pattern between delta and beta waves, even when the power of these individual waves deviates from normal levels, suggesting inflexibility in communication between subcortical emotional centers and cortical regulatory systems (Anaya et al. 2021b). Longitudinal studies provide additional evidence for a potential long-term association between DBC and social anxiety. Dysregulated fear in toddlerhood predicted increased DBC in preschool (Phelps et al. 2016). Children with increased cortisol levels and higher social anxiety exhibited stronger DBC 1 year later (Poole and Schmidt 2019). More recently, adolescents characterized as having high BI in childhood who also had greater DBC experienced more social anxiety during the COVID-19 pandemic (Ramos et al. 2024).

Taken together, these results suggest that neural markers associated with overcontrolled emotion regulation may reflect an increased readiness to perceive threats and fear, leading to hypervigilance and withdrawal behaviors. Repeated experiences of anxiety and fear may lead to hyperactivity in the limbic system which in turn could trigger overactive control mechanisms that ultimately impact patterns of DBC (Poole et al. 2020; Ramos et al. 2024). Alternatively, inflexibility in

DBC could stem from early developmental disruptions in the timing and communication between cortical and subcortical brain regions (Anaya et al. 2021a). Such disruptions might lead to abnormally high synchronization between these regions, potentially contributing to the development of an anxious and overcontrolled behavioral profile.

2.2.2 Anterior Cingulate Cortex

Extensive evidence suggests that reciprocal connections within the anterior limbic network (ALN) facilitate the integration of internal and external sensory information, contributing to emotional homeostasis (Beard et al. 2022; Blackford et al. 2014; Clauss et al. 2014). The regions of the ALN encompass the amygdala, mPFC, insula, vlPFC, dlPFC, and ACC (Strawn et al. 2014). These regions are mutually involved in processing salient information and modulating complex social and emotional behaviors, sharing a common neuroanatomical substrate and functional features (Morgane et al. 2005). Beyond salience detection, these regions are recruited for subprocesses of explicit emotion regulation (Phillips et al. 2008). Explicit emotion regulation involves cognitive strategies, such as cognitive reappraisal, and has been associated with a reduced risk of anxiety (Gyurak et al. 2011). Conversely, deficits in regulating emotions are implicated in the emergence of anxiety disorders, highlighting the importance of a balanced interplay between cognitive and emotional processing (Prater et al. 2013).

As noted earlier in this chapter, the amygdala has been central to research examining early emotion reactivity. Studies examining the neural mechanisms of emotion regulation, in comparison, often focus on prefrontal control of limbic systems, with the ACC emerging as a critical region in this network (Perlman and Pelphrey 2010). The ACC is anatomically positioned to integrate emotional and cognitive information since it is situated at the interface between the limbic system and the PFC (Stevens et al. 2011). This strategic location along with its unique neural connections facilitates communication with other brain regions, further supporting its role as a bridge between emotion and cognition (Shackman et al. 2011). Posner and Rothbart (1998) emphasized the role of the ACC in the development of emotion regulation during infancy and suggested that variations in ACC maturation might be related to emotion regulation abilities, potentially underlying developmental and temperamental differences in attentional control. Indeed, subsequent research reported that increased activation in the ACC, along with the dlPFC, is observed in healthy individuals during emotion regulation tasks, whereas people with SAD exhibit less activation (Ochsner and Gross 2005).

Aberrant activity in subregions of the ACC, like the dorsal ACC (dACC) and subgenual ACC (sgACC), are linked to fearful temperament and anxiety via their interplay with limbic system structures, such as the amygdala. The dACC is, in part, responsible for monitoring stimuli, recognizing and regulating emotions, cognitive processes, inhibitory control, and decision-making (Beard et al. 2022; Clauss et al. 2014). It is part of the intrinsic cortical network involved in salience, emotions, cognitive control, and default mode networks (Clauss et al. 2014). Indeed, the dACC is activated in response to conditioned fear and exhibits an inverse relationship with the amygdala response (Sarinopoulos et al. 2010). Increased amygdala activity and decreased PFC activity, including the dACC, tilt the fear network toward threat-related responses, acting as a contributing mechanism to the development of anxiety disorder, particularly in individuals at temperamental risk (Clauss et al. 2014).

Children with a fearful temperament recruit less dACC than children with a less fearful temperament (Perlman and Pelphrey 2010) while processing emotional faces. Similarly, adults who were inhibited showed decreased activity in the dACC, whereas adults who were uninhibited increased activity while observing unfamiliar faces (Clauss et al. 2011). Decreased dACC activity may make it more difficult for fearful individuals to downregulate heightened emotional arousal. This, in turn, could contribute to the increased, and extended, amygdala response (Clauss et al. 2011).

Neuroimaging research in adults illustrates the potential role of disrupted amygdala-dACC connectivity. Reduced connectivity between the amygdala and two regions of the ACC, dACC and rostral ACC (rACC), was found in individuals with higher levels of social inhibition (Blackford et al. 2014). In a similar vein, the reduced connectivity between the amygdala and dACC was observed in people with social anxiety (Clauss et al. 2014) and between the amygdala and the rACC in people with a generalized anxiety disorder (Prater et al. 2013). These data suggest that individual susceptibility to social inhibition and anxiety stems from alterations in brain connectivity (Blackford et al. 2014). These alterations may serve as potential biomarkers for the identification and intervention of SADs.

With development, the ability to engage dACC for emotion regulation matures alongside the development of the PFC (Casey et al. 2000). This shift reflects a transition from using the emotional areas of the ACC (e.g., ventral ACC) to the more cognitive regions (e.g., dACC) for emotional regulation (Perlman and Pelphrey 2010). For example, older children showed higher activation in the dACC during emotion regulation compared to younger children (Perlman and Pelphrey 2010). In a similar vein, brain regions essential for emotional regulation undergo significant structural and functional changes during childhood and adolescence (Blackford et al. 2018). Particularly, maturation of the PFC extends into young adulthood, with increased synaptogenesis, white matter volume, and myelination (Kolk and Rakic 2022). Age-related decreases in amygdala activation, particularly among adolescents, are linked to increased anatomical connectivity in the uncinate fasciculus, a white matter tract that connects the PFC and amygdala, as compared to younger children (Tan et al. 2020). Early maturing of the amygdala can lead to an imbalance with the PFC in children, leading to challenges in regulating emotions (Gee and Casey 2015). Balanced PFC development is especially crucial for children with a fearful temperament, as they tend to have hyperactive amygdala activity (Blackford et al. 2018). Insufficient PFC development can lead to difficulties in regulating emotions and behaviors effectively for these children.

3 Multiple Pathways from Fearful Temperament and Anxiety Disorders

The pathway from fearful temperament to anxiety is not linear and is influenced by a complex interplay of internal and external factors (Fox et al. 2021; Pérez-Edgar and Fox 2018). For example, a child with a fearful temperament may be more vulnerable to feeling anxious, when they experience overcontrolling parenting. However, other children with a similar level of fearfulness may rely on cognitive control mechanisms to moderate anxiety risk. In this section, we discuss internal and external factors that moderate the link between fearful temperament and anxiety, focusing on attentional bias, environmental exploration, and parental factors.

3.1 Attention Bias

Attention serves as a critical developmental gateway, selecting information and experiences from the environment. This filtered input shapes the development of knowledge and skills across various domains, fostering a coherent understanding of the world (Pérez-Edgar 2018). Additionally, attention actively selects specific elements in the environment for thorough exploration, allowing individuals a deeper understanding of information relevant to their short- and long-term goals (Amso and Scerif 2015). This allows individuals to focus on information relevant to their goals, filtering out distractions. Attention is a central developmental mechanism in that early forms of self-regulation in infancy, including disengaging from distractions and shifting focus, are rooted in attentional abilities (Rothbart et al. 1994). Through this ongoing process of selection and prioritization, unique learning patterns emerge for each individual (Wass et al. 2012), and these patterns underlie a child's working memory and problem-solving abilities. Ultimately, attention helps shape developmental trajectories in both cognitive and socio-emotional domains (Pérez-Edgar 2018).

Given the role of attention in information and emotional processing early in life, attention biases might interact with fearful temperament to increase vulnerability to anxiety (Liu and Bell 2020). In support of this possibility, Pérez-Edgar et al. (2010) first found that toddlers characterized with high BI tend to become socially withdrawn adolescents when they also present with biased attention toward threats. Other studies found the moderating effects of attention bias on the link between BI on both social withdrawal at age 5 (Pérez-Edgar et al. 2011) and anxiety at age 7 (White et al. 2017), suggesting that attention bias may exacerbate the early temperamental vulnerability for subsequent anxiety.

Studies using ERPs and neuroimaging advanced our understanding of attention processes at the neural level. Attention processing interacts dynamically with bottom-up, automatic processing and top-down, control processes, modulating the response from environmental inputs. Specifically, attention bias toward threat is thought to disrupt the frontolimbic system's response to salient stimuli in the environment (Taylor and Whalen 2015). Given that frontolimbic activity is influenced by the duration of threat exposure (Lipka et al. 2011), the balance of neural dominance between limbic and prefrontal systems shifts between brief and sustained threat presentations (Auday et al. 2018). Studies have highlighted the role of prefrontal functioning in regulating heightened limbic activity within a bidirectional system (Thai et al. 2016).

For instance, an ERP study with 9–12-year-old children (Thai et al. 2016) revealed that a larger P2 component, associated with top-down goal-directed perception and decision-making, weakened the link between behavioral inhibition and social anxiety symptoms. Furthermore, this study found that children who avoided threat cues exhibited a greater N2 component, another marker of top-down executive control. Conversely, children with an attention bias toward threats might fail to recruit these regulatory resources. In parallel, increased engagement of the PFC may serve as a protective factor against the development of anxiety disorders. In another EEG study with 9-year-olds (Liu and Bell 2023), children with a fearful temperament exhibited attention away from threat when they showed greater left frontal activation, suggesting left dlPFC-based cognitive control. Additionally, the attention bias modification (ABM) intervention resulted in heightened activation in the vlPFC, mediating the relations between ABM and reductions in anxiety symptoms (Liu et al. 2018).

In contrast, other studies reported conflicting results regarding attention bias and the PFC. Children aged 9–12 years old with higher BI exhibited hyperactivation in the right dorsolateral PFC, which subsequently predicted increased anxiety levels (Fu et al. 2017). These findings suggest that children with BI might have to engage more effortful control resources to shift attention away from threats. In a similar vein, greater P2 amplitude was observed for disgusted faces in children aged 8–12 years old with high social anxiety (Wauthia et al. 2023). Similarly, adults with low anxiety showed a reduced P2 amplitude during an attention task, while adults with high social anxiety did not show a significant reduction in P2 amplitude (Yuan et al. 2019). These inconsistent findings arise from methodological discrepancies between studies and variations in the type and severity of anxiety disorders (Liu and Bell 2023). Further research is needed to develop more reliable measures to enhance our understanding of the complex relations between attention bias, temperament, and the development of anxiety (Fu and Pérez-Edgar 2019).

3.2 *Limited Exploration*

Temperament reflects normative preferences for specific stimuli early in life (LoBue et al. 2010). These preferences can guide a child's interaction with the environment, shaping the information and experiences that contribute to cognitive mechanisms that are responsible for developing more controlled processes (Pérez-Edgar 2018). This combination of innate predispositions and environmental experiences contributes to distinct developmental trajectories. For a temperamentally fearful child, the social world may present a constant stream of unfamiliar objects, people, and situations. As children become more independent, they may seek out familiar environments because they feel overstimulated by novel stimuli (Blackford et al. 2018). For example, children with dysregulated fear experience excessive fear and anxiety in low-threat situations and their negative experiences can reinforce the perceptions of low-threat situations as threatening, creating a vicious cycle of fear and avoidance (Buss et al. 2013). Their tendency to rely on existing environments, and the skills associated with them, rather than exploring novel environments potentially limits their exposure to new experiences and opportunities for learning adaptive skills in diverse social settings (Ostlund and Pérez-Edgar 2023). The lack of social engagement can hinder the development of social competence and may exacerbate anxiety symptoms in these children. Furthermore, limited social interaction and exploration of novel environments may alter neural pathways associated with social engagement. This lack of fine-tuning of neural networks through experience can contribute to an increased risk of anxiety disorder (Jarcho and Guyer 2018), via alterations in brain regions and connectivity associated with fear processing (Burrows et al. 2024; Clauss et al. 2014; Fu et al. 2017) and emotion regulation (Abend et al. 2020; Roy et al. 2014; Sylvester et al. 2016).

3.3 Parental Factors

With birth, indeed starting even before birth, a parent is a crucial conduit for the genes and environments that the individual experiences and must adapt to. Extensive work suggests intergenerational transmission of psychopathology risk through fetal programming during pregnancy (Ostlund and Pérez-Edgar 2023; Seckl and Holmes 2007) and postnatal parent–child interaction (Perlman et al. 2022). The brain undergoes rapid development during gestation and early infancy, with the most significant growth occurring in utero and continuing until the fourth postnatal month (Thomas et al. 2019). This period is followed by a prolonged refinement of functional brain networks (Keller et al. 2023). The concept of fetal programming (Seckl and Holmes 2007) suggests that disruptions during fetal development can have lasting effects on a biological system. Disruptions in the intrauterine environment can influence the development and adaptability of biological systems, including the central nervous, autonomic nervous, and brain systems which increase susceptibility to pathology (Bale 2015). For example, prenatal stress is hypothesized to influence *in utero* glucocorticoid levels, potentially impacting amygdala development (Doyle et al. 2021).

Psychological challenges during pregnancy can lead to structural and functional alterations in brain regions related to reactivity and emotion regulation, increasing the risk for anxiety disorder later in life (Hendrix et al. 2022; Humphreys et al. 2020; Rajasilta et al. 2023). For example, newborns prenatally exposed to maternal stress showed decreased amygdala-mPFC functional connectivity (Humphreys et al. 2020) and decreased amygdala-subcortical connectivity (Scheinost et al. 2016). In addition, levels of prenatal maternal cortisol were associated with right amygdala volumes (Buss et al. 2012). Larger amygdala volumes, in turn, mediated the relation between prenatal maternal cortisol levels and internalizing problems in 7-year-olds.

In a similar vein, maternal anxiety during pregnancy contributes to mechanisms underlying fearful temperament and anxiety. Prenatal anxiety was associated with brain responses in infants, including the insula (Rifkin-Graboi et al. 2015), as well as functional connectivity between the amygdala and clusters in the bilateral parietal regions of preschool-aged children (Donnici et al. 2021). Moreover, a recent fMRI study (Turk et al. 2023) noted that adults prenatally exposed to higher maternal anxiety exhibited weaker functional connectivity between the mPFC and the left inferior frontal gyrus compared to adults exposed to low to medium maternal anxiety.

Another pathway for intergenerational anxiety transmission lies in direct parent–child interactions. Prior research has focused on the role of parental control behaviors, including overcontrol and overprotection (Fox et al. 2023). Overprotected infants are more likely to develop into inhibited toddlers (Mudra et al. 2022) and perceived maternal control behavior during childhood was associated with social anxiety among adolescents (Lewis-Morrarty et al. 2012). Some parents of children with a fearful temperament tend to exert excessive control or protection over their child, perceiving their child as vulnerable (Lewis-Morrarty et al. 2012). Other parents are protective and controlling to reduce the child's negative reactions to novel stimuli, which can hinder the child's opportunity to develop coping skills for unfamiliar or anxious-inducing situations on their own (Fox et al. 2023). The lack of independent experiences may influence the neural circuits involved in emotion regulation and problem solving, ultimately exacerbating vulnerabilities in children with a fearful temperament (Jarcho and Guyer 2018). Indeed, there is a positive association between maternal overprotection and increased amygdala response to angry facial expressions (Farber et al. 2019). Taken together, overprotective parenting can exacerbate existing vulnerabilities in children with BI, potentially hindering their development of healthy social skills and increasing social anxiety risk (Fox et al. 2023).

While harsh and hostile parenting behaviors are associated with increased amygdala reactivity and disrupted amygdala-vlPFC connectivity (Jiang et al. 2021; Pozzi et al. 2021), warm and supportive parenting behaviors may serve as a buffer, supporting children in coping with potential threats in their environment. Adolescents and children who experienced warm and supportive parenting show reduced amygdala activity when processing fearful faces (Romund et al. 2016a, b) and decreased connectivity between parietal regions and executive control networks, suggesting a more mature neural profile (Pozzi et al. 2021). Furthermore, adolescents who experienced more warmth and acceptance from their parents showed less brain activity in the regions engaged in salience detection and emotion regulation, such as left amygdala, bilateral insula, sgACC, right vlPFC, and ACC, when receiving critical feedback from their parents (Butterfield et al. 2021). When children feel loved, accepted, and safe with their parents, criticism becomes less significant and threatening to them.

4 Conclusions

Technological advancements in neuroscience have enabled researchers to explore the biological underpinnings of fearful temperament and anxiety disorder. Accumulating evidence suggests that fearful temperament and anxiety disorders may share neural processes and reflect early deviations from typical trajectories and profiles of brain development. Here, we have discussed unique response patterns associated with fearful temperament and their role in the development and continuity of anxiety across the lifespan. We have reviewed the literature examining the neural basis of temperamental vulnerability to anxiety, focusing on how children with a fearful temperament process information from the environment and regulate arousal. We have also discussed the interaction between these neural processes and environmental factors, illustrating how their interplay shapes socioemotional profiles. By understanding alterations in brain structure, function, and connectivity associated with fearful temperament and anxiety across development, we can gain insights into the underlying neural mechanisms shaping developmental outcomes.

While fearful temperament is a strong vulnerability marker for subsequent anxiety, we know that not all children with a fearful temperament develop into adults with anxiety. This heterogeneity in developmental trajectories suggests the interplay between internal and external factors that can exacerbate or buffer the risk associated with fearful temperament. The word *vulnerability* stems from the Latin word *vulnerare*, which means "to wound." In other words, children with a fearful temperament are "capable of being wounded" or "open to attack or damage" (Brown 2015). Thus, the temperamental foundation is probabilistic, rather than deterministic (Ostlund and Pérez-Edgar 2023). As a result, research examining the moderating forces within the child or their environment, such as warm and supportive parenting (Butterfield et al. 2021; Tan et al. 2020), can empower children to embrace their unique characteristics and effectively adapt to their environment.

References

Abend R, Swetlitz C, White LK et al (2020) Levels of early-childhood behavioral inhibition predict distinct neurodevelopmental pathways to pediatric anxiety. Psychol Med 50:96–106. https://doi.org/10.1017/S0033291718003999

Amso D, Scerif G (2015) The attentive brain: insights from developmental cognitive neuroscience. Nat Rev Neurosci 16:606–619. https://doi.org/10.1038/nrn4025

Anaya B, Vallorani A, Pérez-Edgar K (2021a) Dyadic behavioral synchrony between behaviorally inhibited and non-inhibited peers is associated with concordance in EEG frontal alpha asymmetry and Delta-Beta coupling. Biol Psychol 159:108018. https://doi.org/10.1016/j.biopsycho.2021.108018

Anaya B, Vallorani AM, Pérez-Edgar K (2021b) Individual dynamics of delta–beta coupling: using a multilevel framework to examine inter- and intraindividual differences in relation to social anxiety and behavioral inhibition. J Child Psychol Psychiatry 62:771–779. https://doi.org/10.1111/jcpp.13319

Auday ES, Taber-Thomas BC, Pérez-Edgar KE (2018) Neural correlates of attention bias to masked facial threat cues: examining children at-risk for social anxiety disorder. NeuroImage Clinical 19:202–212. https://doi.org/10.1016/j.nicl.2018.04.003

Avery SN, Blackford JU (2016) Slow to warm up: the role of habituation in social fear. Soc Cogn Affect Neurosci 11:1832–1840. https://doi.org/10.1093/scan/nsw095

Avery SN, McHugo M, Armstrong K et al (2021) Stable habituation deficits in the early stage of psychosis: a 2-year follow-up study. Transl Psychiatry 11:1–10. https://doi.org/10.1038/s41398-020-01167-9

Bale TL (2015) Epigenetic and transgenerational reprogramming of brain development. Nat Rev Neurosci 16:332–344. https://doi.org/10.1038/nrn3818

Bas-Hoogendam JM, van Steenbergen H, van der Wee NJA, Westenberg PM (2020) Amygdala hyperreactivity to faces conditioned with a social-evaluative meaning– a multiplex, multigenerational fMRI study on social anxiety endophenotypes. NeuroImage Clin 26: 102247. https://doi.org/10.1016/j.nicl.2020.102247

Beard SJ, Hastings PD, Ferrer E et al (2022) Neural response to social exclusion moderates the link between adolescent anxiety symptoms and substance use. Biol Psychiatry Cogn Neurosci Neuroimaging 7:180–191. https://doi.org/10.1016/j.bpsc.2021.06.006

Blackford JU, Avery SN, Shelton RC, Zald DH (2009) Amygdala temporal dynamics: temperamental differences in the timing of amygdala response to familiar and novel faces. BMC Neurosci 10:145. https://doi.org/10.1186/1471-2202-10-145

Blackford JU, Avery SN, Cowan RL et al (2011) Sustained amygdala response to both novel and newly familiar faces characterizes inhibited temperament. Soc Cogn Affect Neurosci 6:621–629. https://doi.org/10.1093/scan/nsq073

Blackford JU, Allen AH, Cowan RL, Avery SN (2013) Amygdala and hippocampus fail to habituate to faces in individuals with an inhibited temperament. Soc Cogn Affect Neurosci 8: 143–150. https://doi.org/10.1093/scan/nsr078

Blackford JU, Clauss JA, Avery SN et al (2014) Amygdala–cingulate intrinsic connectivity is associated with degree of social inhibition. Biol Psychol 99:15–25. https://doi.org/10.1016/j.biopsycho.2014.02.003

Blackford JU, Clauss JA, Benningfield MM (2018) The neurobiology of behavioral inhibition as a developmental mechanism. In: Pérez-Edgar K, Fox NA (eds) Behavioral inhibition: integrating theory, research, and clinical perspectives. Springer, Cham, pp 113–134

Booth J (2023) Anxiety statistics and facts. In: Forbes health. https://www.forbes.com/health/mind/anxiety-statistics/. Accessed 13 Apr 2024

Bouras NN, Mack NR, Gao W-J (2023) Prefrontal modulation of anxiety through a lens of noradrenergic signaling. Front Syst Neurosci 17. https://doi.org/10.3389/fnsys.2023.1173326

Brooker RJ, Phelps RA, Davidson RJ, Goldsmith HH (2016) Context differences in delta beta coupling are associated with neuroendocrine reactivity in infants. Dev Psychobiol 58:406–418. https://doi.org/10.1002/dev.21381

Brown B (2015) Daring greatly: how the courage to be vulnerable transforms the way we live, love, parent, and lead. Avery

Burrows CA, Lasch C, Gross J et al (2024) Associations between early trajectories of amygdala development and later school-age anxiety in two longitudinal samples. Dev Cogn Neurosci 65: 101333. https://doi.org/10.1016/j.dcn.2023.101333

Buss KA (2011) Which fearful toddlers should we worry about? Context, fear regulation, and anxiety risk. Dev Psychol 47:804–819. https://doi.org/10.1037/a0023227

Buss KA, Kiel EJ (2013) Temperamental risk factors for pediatric anxiety disorders. In: Vasa RA, Roy AK (eds) Pediatric anxiety disorders: a clinical guide. Springer, New York, pp 47–68

Buss C, Davis EP, Shahbaba B et al (2012) Maternal cortisol over the course of pregnancy and subsequent child amygdala and hippocampus volumes and affective problems. Proc Natl Acad Sci 109:E1312–E1319. https://doi.org/10.1073/pnas.1201295109

Buss KA, Davis EL, Kiel EJ et al (2013) Dysregulated fear predicts social wariness and social anxiety symptoms during kindergarten. J Clin Child Adolesc Psychol 42:603–616. https://doi.org/10.1080/15374416.2013.769170

Buss KA, Cho S, Morales S et al (2021) Toddler dysregulated fear predicts continued risk for social anxiety symptoms in early adolescence. Dev Psychopathol 33:252–263. https://doi.org/10.1017/S0954579419001743

Butterfield RD, Silk JS, Lee KH et al (2021) Parents still matter! Parental warmth predicts adolescent brain function and anxiety and depressive symptoms 2 years later. Dev Psychopathol 33:226–239. https://doi.org/10.1017/S0954579419001718

Buzzell GA, Troller-Renfree SV, Barker TV et al (2017) A neurobehavioral mechanism linking behaviorally inhibited temperament and later adolescent social anxiety. J Am Acad Child Adolesc Psychiatry 56:1097–1105. https://doi.org/10.1016/j.jaac.2017.10.007

Casey BJ, Giedd JN, Thomas KM (2000) Structural and functional brain development and its relation to cognitive development. Biol Psychol 54:241–257. https://doi.org/10.1016/S0301-0511(00)00058-2

Cismaru AL, Gui L, Vasung L et al (2016) Altered amygdala development and fear processing in prematurely born infants. Front Neuroanat 10. https://doi.org/10.3389/fnana.2016.00055

Clauss JA, Blackford JU (2012a) Behavioral inhibition and risk for developing social anxiety disorder: a meta-analytic study. J Am Acad Child Adolesc Psychiatry 51:1066–1075.e1. https://doi.org/10.1016/j.jaac.2012.08.002

Clauss JA, Blackford JU (2012b) Behavioral inhibition and risk for developing social anxiety disorder: a meta-analytic study. J Am Acad Child Adolesc Psychiatry 51:1066–1075.e1. https://doi.org/10.1016/j.jaac.2012.08.002.

Clauss JA, Cowan RL, Blackford JU (2011) Expectation and temperament moderate amygdala and dorsal anterior cingulate cortex responses to fear faces. Cogn Affect Behav Neurosci 11:13–21. https://doi.org/10.3758/s13415-010-0007-9

Clauss JA, Seay AL, VanDerKlok RM et al (2014) Structural and functional bases of inhibited temperament. Soc Cogn Affect Neurosci 9:2049–2058. https://doi.org/10.1093/scan/nsu019

Donnici C, Long X, Dewey D et al (2021) Prenatal and postnatal maternal anxiety and amygdala structure and function in young children. Sci Rep 11:4019. https://doi.org/10.1038/s41598-021-83249-2

Doyle CM, Lasch C, Elison JT (2021) Emerging evidence for putative neural networks and antecedents of pediatric anxiety in the fetal, neonatal, and infant periods. Biol Psychiatry 89: 672–680. https://doi.org/10.1016/j.biopsych.2020.11.020

Farber MJ, Kim MJ, Knodt AR, Hariri AR (2019) Maternal overprotection in childhood is associated with amygdala reactivity and structural connectivity in adulthood. Dev Cogn Neurosci 40:100711. https://doi.org/10.1016/j.dcn.2019.100711

Filippi CA, Sachs JF, Phillips D et al (2020) Infant behavioral reactivity predicts change in amygdala volume 12 years later. Dev Cogn Neurosci 42:100776. https://doi.org/10.1016/j.dcn.2020.100776

Filippi CA, Ravi S, Bracy M et al (2021) Amygdala functional connectivity and negative reactive temperament at age 4 months. J Am Acad Child Adolesc Psychiatry 60:1137–1146. https://doi.org/10.1016/j.jaac.2020.11.021

Filippi CA, Valadez EA, Fox NA, Pine DS (2022) Temperamental risk for anxiety: emerging work on the infant brain and later neurocognitive development. Curr Opin Behav Sci 44:101105. https://doi.org/10.1016/j.cobeha.2022.101105

Fox AS, Kalin NH (2014) A translational neuroscience approach to understanding the development of social anxiety disorder and its pathophysiology. Am J Psychiatry 171:1162–1173. https://doi.org/10.1176/appi.ajp.2014.14040449

Fox NA, Hane AA, Pérez-Edgar K (2015) Psychophysiological methods for the study of developmental psychopathology. In: Developmental psychopathology. Wiley, pp 381–426

Fox NA, Buzzell GA, Morales S et al (2021) Understanding the emergence of social anxiety in children with behavioral inhibition. Biol Psychiatry 89:681–689. https://doi.org/10.1016/j.biopsych.2020.10.004

Fox NA, Zeytinoglu S, Valadez EA et al (2023) Annual research review: developmental pathways linking early behavioral inhibition to later anxiety. J Child Psychol Psychiatry 64:537–561. https://doi.org/10.1111/jcpp.13702

Fu X, Pérez-Edgar K (2019) Threat-related attention bias in socioemotional development: a critical review and methodological considerations. Dev Rev 51:31–57. https://doi.org/10.1016/j.dr.2018.11.002

Fu X, Taber-Thomas BC, Pérez-Edgar K (2017) Frontolimbic functioning during threat-related attention: relations to early behavioral inhibition and anxiety in children. Biol Psychol 122:98–109. https://doi.org/10.1016/j.biopsycho.2015.08.010

Gee DG, Casey BJ (2015) The impact of developmental timing for stress and recovery. Neurobiol Stress 1:184–194. https://doi.org/10.1016/j.ynstr.2015.02.001

Goldsmith HH, Davidson RJ (2004) Disambiguating the components of emotion regulation. Child Dev 75:361–365. https://doi.org/10.1111/j.1467-8624.2004.00678.x

Gorman JM, Kent JM, Sullivan GM, Coplan JD (2000) Neuroanatomical hypothesis of panic disorder, revised. Am J Psychiatry 157:493–505. https://doi.org/10.1176/appi.ajp.157.4.493

Groenewold NA, Bas-Hoogendam JM, Amod AR et al (2023) Volume of subcortical brain regions in social anxiety disorder: mega-analytic results from 37 samples in the ENIGMA-Anxiety Working Group. Mol Psychiatry 28:1079–1089. https://doi.org/10.1038/s41380-022-01933-9

Gyurak A, Gross JJ, Etkin A (2011) Explicit and implicit emotion regulation: a dual-process framework. Cognit Emot 25:400–412. https://doi.org/10.1080/02699931.2010.544160

Hahn A, Stein P, Windischberger C et al (2011) Reduced resting-state functional connectivity between amygdala and orbitofrontal cortex in social anxiety disorder. Neuroimage 56:881–889. https://doi.org/10.1016/j.neuroimage.2011.02.064

Harnett NG, Ference EW, Wood KH et al (2018) Trauma exposure acutely alters neural function during Pavlovian fear conditioning. Cortex 109:1–13. https://doi.org/10.1016/j.cortex.2018.08.015

Heitmann CY, Feldker K, Neumeister P et al (2017) Brain activation to task-irrelevant disorder-related threat in social anxiety disorder: the impact of symptom severity. NeuroImage Clinical 14:323–333. https://doi.org/10.1016/j.nicl.2017.01.020

Henderson HA, Pine DS, Fox NA (2015) Behavioral inhibition and developmental risk: a dual-processing perspective. Neuropsychopharmacology 40:207–224. https://doi.org/10.1038/npp.2014.189

Hendrix CL, Srinivasan H, Feliciano I, Carré JM, Thomason ME (2022) Fetal hippocampal connectivity shows dissociable associations with maternal cortisol and self-reported distress during pregnancy. Life. 2022 Jun 23;12(7):943.

Hill SY, Tessner K, Wang S et al (2010) Temperament at 5years of age predicts amygdala and orbitofrontal volume in the right hemisphere in adolescence. Psychiatry Res Neuroimaging 182: 14–21. https://doi.org/10.1016/j.pscychresns.2009.11.006

Humphreys KL, Camacho MC, Roth MC, Estes EC (2020) Prenatal stress exposure and multimodal assessment of amygdala–medial prefrontal cortex connectivity in infants. Dev Cogn Neurosci 46:100877. https://doi.org/10.1016/j.dcn.2020.100877

Jarcho JM, Guyer AE (2018) The neural mechanisms of behavioral inhibition. In: Pérez-Edgar K, Fox NA (eds) Behavioral inhibition: integrating theory, research, and clinical perspectives. Springer, Cham, pp 59–90

Jiang N, Xu J, Li X, Wang Y, Zhuang L, Qin S (2021) Negative Parenting Affects Adolescent Internalizing Symptoms Through Alterations in Amygdala-Prefrontal Circuitry: A Longitudinal Twin Study. Biological Psychiatry 89(6):560–569. https://doi.org/10.1016/j.biopsych.2020.08.002

Kagan J, Reznick JS, Clarke C et al (1984) Behavioral inhibition to the unfamiliar. Child Dev 55: 2212–2225. https://doi.org/10.2307/1129793

Kagan J (1994) Galen's prophecy. Basic Books, New York
Keifer OP, Hurt RC, Ressler KJ, Marvar PJ (2015) The physiology of fear: reconceptualizing the role of the central amygdala in fear learning. Physiology 30:389–401. https://doi.org/10.1152/physiol.00058.2014
Keller AS, Sydnor VJ, Pines A et al (2023) Hierarchical functional system development supports executive function. Trends Cogn Sci 27:160–174. https://doi.org/10.1016/j.tics.2022.11.005
Kiel EJ, Kalomiris AE (2019) Emotional development and anxiety. In: LoBue V, Pérez-Edgar K, Buss KA (eds) Handbook of emotional development. Springer, Cham, pp 665–693
Knyazev GG (2007) Motivation, emotion, and their inhibitory control mirrored in brain oscillations. Neurosci Biobehav Rev 31:377–395. https://doi.org/10.1016/j.neubiorev.2006.10.004
Knyazev G, Slobodskaya H (2003) Personality trait of behavioral inhibition is associated with oscillatory systems reciprocal relationships. Int J Psychophysiol 48:247–261. https://doi.org/10.1016/S0167-8760(03)00072-2
Kolk SM, Rakic P (2022) Development of prefrontal cortex. Neuropsychopharmacology 47:41–57. https://doi.org/10.1038/s41386-021-01137-9
Kragel PA, LaBar KS (2015) Multivariate neural biomarkers of emotional states are categorically distinct. Soc Cogn Affect Neurosci 10:1437–1448. https://doi.org/10.1093/scan/nsv032
Kredlow MA, Fenster RJ, Laurent ES et al (2022) Prefrontal cortex, amygdala, and threat processing: implications for PTSD. Neuropsychopharmacology 47:247–259. https://doi.org/10.1038/s41386-021-01155-7
Lautarescu A, Pecheva D, Nosarti C et al (2020) Maternal prenatal stress is associated with altered uncinate fasciculus microstructure in premature neonates. Biol Psychiatry 87:559–569. https://doi.org/10.1016/j.biopsych.2019.08.010
LeDoux J (2003) The emotional brain, fear, and the amygdala. Cell Mol Neurobiol 23:727–738. https://doi.org/10.1023/A:1025048802629
Lewis-Morrarty E, Degnan KA, Chronis-Tuscano A et al (2012) Maternal over-control moderates association between early childhood behavioral inhibition and adolescent social anxiety symptoms. J Abnorm Child Psychol 40:1363–1373. https://doi.org/10.1007/s10802-012-9663-2
Lipka J, Miltner WHR, Straube T (2011) Vigilance for threat interacts with amygdala responses to subliminal threat cues in specific phobia. Biol Psychiatry 70:472–478. https://doi.org/10.1016/j.biopsych.2011.04.005
Liu R, Bell MA (2020) Fearful temperament and the risk for child and adolescent anxiety: the role of attention biases and effortful control. Clin Child Fam Psychol Rev 23:205–228. https://doi.org/10.1007/s10567-019-00306-z
Liu R, Bell MA (2023) Fearful temperament in middle childhood predicts adolescent attention bias and anxiety symptoms: the moderating role of frontal EEG asymmetry. Dev Psychopathol 35:1335–1345. https://doi.org/10.1017/S0954579421001231
Liu P, Pérez-Edgar KE (2019) Developmental pathways from early behavioral inhibition to later anxiety: an integrative review of developmental psychopathology research and translational implications. Adolescent Res Rev 4:45–58. https://doi.org/10.1007/s40894-018-0092-5
Liu P, Taber-Thomas BC, Fu X, Pérez-Edgar KE (2018) Biobehavioral markers of attention bias modification in temperamental risk for anxiety: a randomized control trial. J Am Acad Child Adolesc Psychiatry 57:103–110. https://doi.org/10.1016/j.jaac.2017.11.016
Liu W-Z, Zhang W-H, Zheng Z-H et al (2020) Identification of a prefrontal cortex-to-amygdala pathway for chronic stress-induced anxiety. Nat Commun 11:2221. https://doi.org/10.1038/s41467-020-15920-7
LoBue V, Rakison DH, DeLoache JS (2010) Threat perception across the life span: evidence for multiple converging pathways. Curr Dir Psychol Sci 19:375–379. https://doi.org/10.1177/0963721410388801
Mao Y, Zuo X-N, Ding C, Qiu J (2020) OFC and its connectivity with amygdala as predictors for future social anxiety in adolescents. Dev Cogn Neurosci 44:100804. https://doi.org/10.1016/j.dcn.2020.100804

Merikangas KR, He J, Burstein M, et al (2010) Lifetime Prevalence of Mental Disorders in U.S. Adolescents: Results from the National Comorbidity Survey Replication–Adolescent Supplement (NCS-A). Journal of the American Academy of Child & Adolescent Psychiatry 49:980–989. https://doi.org/10.1016/j.jaac.2010.05.017

Mills KL, Goddings A-L, Herting MM et al (2016) Structural brain development between childhood and adulthood: convergence across four longitudinal samples. Neuroimage 141:273–281. https://doi.org/10.1016/j.neuroimage.2016.07.044

Minkova L, Sladky R, Kranz GS, Woletz M, Geissberger N, Kraus C, Lanzenberger R, Windischberger C. (2017) Taskdependent modulation of amygdala connectivity in social anxiety disorder. Psychiatry Res Neuroimaging. 2017 Apr 30;262:39–46. https://doi.org/10.1016/j.pscychresns.2016.12.016. Epub 2017 Feb 9. PMID: 28226306.

Mizzi S, Pedersen M, Rossell SL et al (2024) Resting-state amygdala subregion and precuneus connectivity provide evidence for a dimensional approach to studying social anxiety disorder. Transl Psychiatry 14:1–10. https://doi.org/10.1038/s41398-024-02844-9

Morales S, Fox NA (2019) A neuroscience perspective on emotional development. In: LoBue V, Pérez-Edgar K, Buss KA (eds) Handbook of emotional development. Springer, Cham, pp 57–81

Morgane P, Galler J, Mokler D (2005) A review of systems and networks of the limbic forebrain/limbic midbrain. Prog Neurobiol 75:143–160. https://doi.org/10.1016/j.pneurobio.2005.01.001

Mudra S, Göbel A, Möhler E et al (2022) Behavioral inhibition in the second year of life is predicted by prenatal maternal anxiety, overprotective parenting and infant temperament in early infancy. Front Psych 13:844291. https://doi.org/10.3389/fpsyt.2022.844291

Murray EA, Izquierdo A (2007) Orbitofrontal cortex and amygdala contributions to affect and action in primates. Ann N Y Acad Sci 1121:273–296. https://doi.org/10.1196/annals.1401.021

Myruski S, Bagrodia R, Dennis-Tiwary T (2022) Delta-beta correlation predicts adaptive child emotion regulation concurrently and two years later. Biol Psychol 167:108225. https://doi.org/10.1016/j.biopsycho.2021.108225

Najjar R, Brooker RJ (2017) Delta-Beta coupling is associated with paternal caregiving behaviors during preschool. Int J Psychophysiol 112:31–39. https://doi.org/10.1016/j.ijpsycho.2016.11.014

Ochsner K, Gross J (2005) The cognitive control of emotion. Trends Cogn Sci 9:242–249. https://doi.org/10.1016/j.tics.2005.03.010

Ostlund B, Pérez-Edgar K (2023) Two-hit model of behavioral inhibition and anxiety. Annu Rev Dev Psychol 5:239–261. https://doi.org/10.1146/annurev-devpsych-120621-043722

Peng Y, Knotts JD, Young KS et al (2023) Threat neurocircuitry predicts the development of anxiety and depression symptoms in a longitudinal study. Biol Psychiatry Cogn Neurosci Neuroimaging 8:102–110. https://doi.org/10.1016/j.bpsc.2021.12.013

Pérez-Edgar K (2018) Attention mechanisms in behavioral inhibition: exploring and exploiting the environment. In: Pérez-Edgar K, Fox NA (eds) Behavioral inhibition: integrating theory, research, and clinical perspectives. Springer, Cham, pp 237–261

Pérez-Edgar K, Fox NA (2018) Next steps: behavioral inhibition as a model system. In: Pérez-Edgar K, Fox NA (eds) Behavioral inhibition: integrating theory, research, and clinical perspectives. Springer, Cham, pp 357–372

Pérez-Edgar KE, Guyer AE (2014) Behavioral inhibition: temperament or prodrome? Curr Behav Neurosci Rep 1:182–190. https://doi.org/10.1007/s40473-014-0019-9

Pérez-Edgar K, Roberson-Nay R, Hardin MG et al (2007) Attention alters neural responses to evocative faces in behaviorally inhibited adolescents. Neuroimage 35:1538–1546. https://doi.org/10.1016/j.neuroimage.2007.02.006

Pérez-Edgar K, Bar-Haim Y, McDermott JM et al (2010) Attention biases to threat and behavioral inhibition in early childhood shape adolescent social withdrawal. Emotion 10:349–357. https://doi.org/10.1037/a0018486

Pérez-Edgar K, Reeb-Sutherland BC, McDermott JM et al (2011) Attention biases to threat link behavioral inhibition to social withdrawal over time in very young children. J Abnorm Child Psychol 39:885–895. https://doi.org/10.1007/s10802-011-9495-5

Perlman SB, Pelphrey KA (2010) Regulatory brain development: balancing emotion and cognition. Soc Neurosci 5:533–542. https://doi.org/10.1080/17470911003683219

Perlman SB, Lunkenheimer E, Panlilio C, Pérez-Edgar K (2022) Parent-to-child anxiety transmission through dyadic social dynamics: a dynamic developmental model. Clin Child Fam Psychol Rev 25:110–129. https://doi.org/10.1007/s10567-022-00391-7

Pessoa L, Adolphs R (2011) Emotion and the brain: multiple roads are better than one. Nat Rev Neurosci 12:425–425. https://doi.org/10.1038/nrn2920-c2

Phelps RA, Brooker RJ, Buss KA (2016) Toddlers' dysregulated fear predicts delta–beta coupling during preschool. Dev Cogn Neurosci 17:28–34. https://doi.org/10.1016/j.dcn.2015.09.007

Phillips M, Ladouceur C, Drevets W (2008) A neural model of voluntary and automatic emotion regulation: implications for understanding the pathophysiology and neurodevelopment of bipolar disorder. Mol Psychiatry 13:829–857. https://doi.org/10.1038/mp.2008.65

Poole KL, Schmidt LA (2019) Frontal brain delta-beta correlation, salivary cortisol, and social anxiety in children. J Child Psychol Psychiatry 60:646–654. https://doi.org/10.1111/jcpp.13016

Poole KL, Anaya B, Pérez-Edgar KE (2020) Behavioral inhibition and EEG delta-beta correlation in early childhood: comparing a between-subjects and within-subjects approach. Biol Psychol 149:107785. https://doi.org/10.1016/j.biopsycho.2019.107785

Poon C-S, Young DL (2006) Nonassociative learning as gated neural integrator and differentiator in stimulus-response pathways. Behav Brain Funct 2:29. https://doi.org/10.1186/1744-9081-2-29

Posner MI, Rothbart MK (1998) Attention, self–regulation and consciousness. Philos Trans R Soc Lond B Biol Sci 353:1915–1927. https://doi.org/10.1098/rstb.1998.0344

Pozzi E, Vijayakumar N, Byrne ML et al (2021) Maternal parenting behavior and functional connectivity development in children: a longitudinal fMRI study. Dev Cogn Neurosci 48: 100946. https://doi.org/10.1016/j.dcn.2021.100946

Prater KE, Hosanagar A, Klumpp H et al (2013) Aberrant amygdala-frontal cortex connectivity during perception of fearful faces and at rest in generalized social anxiety disorder. Depress Anxiety 30:234–241. https://doi.org/10.1002/da.22014

Rajasilta O, Häkkinen S, Björnsdotter M et al (2023) Maternal psychological distress associates with alterations in resting-state low-frequency fluctuations and distal functional connectivity of the neonate medial prefrontal cortex. Eur J Neurosci 57:242–257. https://doi.org/10.1111/ejn.15882

Ramos ML, Zhou AM, Lytle MN et al (2024) Interactions among stress, behavioral inhibition, and delta–beta coupling predict adolescent anxiety during the COVID-19 pandemic. Dev Psychobiol 66:e22485. https://doi.org/10.1002/dev.22485

Rifkin-Graboi A, Meaney MJ, Chen H et al (2015) Antenatal maternal anxiety predicts variations in neural structures implicated in anxiety disorders in newborns. J Am Acad Child Adolesc Psychiatry 54:313–321.e2. https://doi.org/10.1016/j.jaac.2015.01.013

Rogers CE, Sylvester CM, Mintz C et al (2017) Neonatal amygdala functional connectivity at rest in healthy and preterm infants and early internalizing symptoms. J Am Acad Child Adolesc Psychiatry 56:157–166. https://doi.org/10.1016/j.jaac.2016.11.005

Romund L, Raufelder D, Flemming E et al (2016a) Maternal parenting behavior and emotion processing in adolescents – an fMRI study. Biol Psychol 120:120–125. https://doi.org/10.1016/j.biopsycho.2016.09.003

Romund L, Raufelder D, Flemming E et al (2016b) Maternal parenting behavior and emotion processing in adolescents – an fMRI study. Biol Psychol 120:120–125. https://doi.org/10.1016/j.biopsycho.2016.09.003.

Rothbart MK (2007) Temperament, development, and personality. Curr Dir Psychol Sci 16:207–212. https://doi.org/10.1111/j.1467-8721.2007.00505.x

Rothbart MK, Posner MI, Rosicky J (1994) Orienting in normal and pathological development. Dev Psychopathol 6:635–652. https://doi.org/10.1017/S0954579400004715

Roy AK, Benson BE, Degnan KA et al (2014) Alterations in amygdala functional connectivity reflect early temperament. Biol Psychol 103:248–254. https://doi.org/10.1016/j.biopsycho.2014.09.007

Sandman CF, Young KS, Burklund LJ et al (2020) Changes in functional connectivity with cognitive behavioral therapy for social anxiety disorder predict outcomes at follow-up. Behav Res Ther 129:103612. https://doi.org/10.1016/j.brat.2020.103612

Sarinopoulos I, Grupe DW, Mackiewicz KL et al (2010) Uncertainty during anticipation modulates neural responses to aversion in human insula and amygdala. Cereb Cortex 20:929–940. https://doi.org/10.1093/cercor/bhp155

Scheinost D, Kwon SH, Lacadie C et al (2016) Prenatal stress alters amygdala functional connectivity in preterm neonates. NeuroImage Clinical 12:381–388. https://doi.org/10.1016/j.nicl.2016.08.010

Schwartz CE, Kunwar PS, Greve DN et al (2012) A phenotype of early infancy predicts reactivity of the amygdala in male adults. Mol Psychiatry 17:1042–1050. https://doi.org/10.1038/mp.2011.96

Schwarzlose RF, Filippi CA, Myers MJ et al (2023) Neonatal neural responses to novelty related to behavioral inhibition at 1 year. Dev Psychol. https://doi.org/10.1037/dev0001654

Seckl JR, Holmes MC (2007) Mechanisms of disease: glucocorticoids, their placental metabolism and fetal "programming" of adult pathophysiology. Nat Rev Endocrinol 3:479–488. https://doi.org/10.1038/ncpendmet0515

Shackman AJ, Fox AS (2016) Contributions of the central extended amygdala to fear and anxiety contributions of the central extended amygdala to fear and anxiety. J Neurosci 36:8050–8063. https://doi.org/10.1523/JNEUROSCI.0982-16.2016

Shackman AJ, Salomons TV, Slagter HA et al (2011) The integration of negative affect, pain and cognitive control in the cingulate cortex. Nat Rev Neurosci 12:154–167. https://doi.org/10.1038/nrn2994

Sladky R, Baldinger P, Kranz GS et al (2013) High-resolution functional MRI of the human amygdala at 7T. Eur J Radiol 82:728–733. https://doi.org/10.1016/j.ejrad.2011.09.025

Sladky R, Hahn A, Karl I-L et al (2022) Dynamic causal modeling of the prefrontal/amygdala network during processing of emotional faces. Brain Connect 12:670–682. https://doi.org/10.1089/brain.2021.0073

Stevens FL, Hurley RA, Taber KH et al (2011) Anterior cingulate cortex: unique role in cognition and emotion. J Neuropsychiatry Clin Neurosci 23:121–125. https://doi.org/10.1176/jnp.23.2.jnp121

Strawn JR, Dominick KC, Patino LR et al (2014) Neurobiology of pediatric anxiety disorders. Curr Behav Neurosci Rep 1:154–160. https://doi.org/10.1007/s40473-014-0014-1

Sylvester CM, Pine DS (2018) The biological bridge between behavioral inhibition and psychopathology. In: Pérez-Edgar K, Fox NA (eds) Behavioral inhibition: integrating theory, research, and clinical perspectives. Springer, Cham, pp 309–335

Sylvester CM, Barch DM, Harms MP et al (2016) Early childhood behavioral inhibition predicts cortical thickness in adulthood. J Am Acad Child Adolesc Psychiatry 55:122–129.e1. https://doi.org/10.1016/j.jaac.2015.11.007

Sylvester CM, Yu Q, Srivastava AB et al (2020) Individual-specific functional connectivity of the amygdala: a substrate for precision psychiatry. Proc Natl Acad Sci U S A 117:3808–3818. https://doi.org/10.1073/pnas.1910842117

Szuhany KL, Simon NM (2022) Anxiety disorders: a review. JAMA 328:2431–2445. https://doi.org/10.1001/jama.2022.22744

Taber-Thomas B, Perez-Edgar K (2015) Emerging adulthood brain development. In: Arnett JJ (ed) Oxford handbook of emerging adulthood. Oxford University Press, pp 126–141

Taber-Thomas BC, Morales S, Hillary FG, Pérez-Edgar KE (2016) Altered topography of intrinsic functional connectivity in childhood risk for social anxiety. Depress Anxiety 33:995–1004. https://doi.org/10.1002/da.22508

Tan PZ, Oppenheimer CW, Ladouceur CD et al (2020) A review of associations between parental emotion socialization behaviors and the neural substrates of emotional reactivity and regulation in youth. Dev Psychol 56:516–527. https://doi.org/10.1037/dev0000893

Taylor JM, Whalen PJ (2015) Neuroimaging and anxiety: the neural substrates of pathological and non-pathological anxiety. Curr Psychiatry Rep 17:49. https://doi.org/10.1007/s11920-015-0586-9

Thai N, Taber-Thomas BC, Pérez-Edgar KE (2016) Neural correlates of attention biases, behavioral inhibition, and social anxiety in children: an ERP study. Dev Cogn Neurosci 19:200–210. https://doi.org/10.1016/j.dcn.2016.03.008

Thomas E, Buss C, Rasmussen JM et al (2019) Newborn amygdala connectivity and early emerging fear. Dev Cogn Neurosci 37:100604. https://doi.org/10.1016/j.dcn.2018.12.002

Turk E, van den Heuvel MI, Sleurs C et al (2023) Maternal anxiety during pregnancy is associated with weaker prefrontal functional connectivity in adult offspring. Brain Imaging Behav 17:595–607. https://doi.org/10.1007/s11682-023-00787-1

Tzovara A, Fedele T, Sarnthein J et al (2024) Predictable and unpredictable deviance detection in the human hippocampus and amygdala. Cereb Cortex 34:bhad532. https://doi.org/10.1093/cercor/bhad532

Volkow ND, Gordon JA, Freund MP (2021) The healthy brain and child development study – shedding light on opioid exposure, COVID-19, and health disparities. JAMA Psychiatry 78: 471–472. https://doi.org/10.1001/jamapsychiatry.2020.3803

Wass SV, Scerif G, Johnson MH (2012) Training attentional control and working memory – is younger, better? Dev Rev 32:360–387. https://doi.org/10.1016/j.dr.2012.07.001

Watve A, Haugg A, Frei N et al (2024) Facing emotions: real-time fMRI-based neurofeedback using dynamic emotional faces to modulate amygdala activity. Front Neurosci:17. https://doi.org/10.3389/fnins.2023.1286665

Wauthia E, Rossignol M, Blekic W et al (2023) Electrophysiological correlates of attentional bias towards threat in children with high levels of social anxiety. Cogn Affect Behav Neurosci 23: 190–202. https://doi.org/10.3758/s13415-022-01042-8

White LK, Degnan KA, Henderson HA et al (2017) Developmental relations between behavioral inhibition, anxiety, and attention biases to threat and positive information. Child Dev 88:141–155. https://doi.org/10.1111/cdev.12696

Young KS, Burklund LJ, Torre J et al (2017) Treatment for social anxiety disorder alters functional connectivity in emotion regulation neural circuitry. Psychiatry Res 261:44–51. https://doi.org/10.1016/j.pscychresns.2017.01.005

Yuan J, Mao N, Chen R et al (2019) Social anxiety and attentional bias variability: electrophysiological evidence of attentional control deficits. Neuroreport 30:887. https://doi.org/10.1097/WNR.0000000000001294

Zhou Y, Friston KJ, Zeidman P et al (2018) The hierarchical organization of the default, dorsal attention and salience networks in adolescents and young adults. Cereb Cortex 28:726–737. https://doi.org/10.1093/cercor/bhx307

Zhou AM, Trainer A, Vallorani A et al (2022) Are fearful boys at higher risk for anxiety? Person-centered profiles of toddler fearful behavior predict anxious behaviors at age 6. Front Psychol 13:911913. https://doi.org/10.3389/fpsyg.2022.911913

suggesting different roles for this area in both sexes (Allen and Gorski 1992; Hines et al. 1992; Uchida et al. 2019; Tsukahara and Morishita 2020). For instance, in a contextual fear conditioning paradigm, cFos expression was upregulated in the BNST of male rats but not females, consistent with prior research indicating that BNST lesions reduced freezing responses in male rats (Sullivan et al. 2004; Urien and Bauer 2022). Conversely, studies in females have found that BNST lesions impaired fear extinction specifically to long-duration cues (Waddell et al. 2006). Moreover, research has shown that BNST neurons projecting to the CeA exhibited upregulation of the immediate-early gene ARC only in males during contextual fear expression (Urien et al. 2021). These findings align with human studies reporting greater BNST-CeA fMRI connectivity in men compared to women (Hofmann and Straube 2019).

To summarize, females demonstrate increased activation in BLA during fear acquisition, extinction, and retrieval. Inhibition of the CeA showed sex opposite effects, increased fear memory in females but decreased in males. Lesions in BNST demonstrated its involvement in fear extinction to prolonged fear cues in females while reducing freezing behavior in males. BLA-BNST connectivity seems essential only for fear extinction in males. Moreover, BNST-CeA projections are active exclusively in males during contextual fear conditioning. These findings emphasize the effect of sex on complex neural mechanisms within the extended nuclei of the amygdala, crucial for understanding variability in fear and anxiety processing.

2.2 *Hippocampus*

The HPC is essential to integrate the contextual framework and spatial characteristics of the context during fear conditioning and fear extinction (Fanselow and Dong 2010; Moustafa et al. 2013; VanElzakker et al. 2014; Zemla and Basu 2017). This area is known to be involved in fear conditioning, as lesions in this region or administration of cholinergic antagonists, gamma-aminobutyric acid (GABA) antagonists, or N-methyl-D-aspartate (NMDA) antagonists impaired only the contextual fear conditioning, and not the cued fear conditioning (Kim and Fanselow 1992; Young et al. 1994; Bast et al. 2001; Wallenstein and Vago 2001).

A study showed that male rats retain contextual fear conditioning longer than female rats. The expression of the extracellular signal-regulated kinase (ERK) in the entral hippocampus (vHPC) was found to be higher in male mice than females. This x-dependent increase was specific to the formation of the contextual fear memory resack et al. 2009). This pathway is an important regulator of fear extinction, as wn by a study wherein increased ERK2 phosphorylation enhanced the extinction ar memory in male mice (Matsuda et al. 2010). Later, they showed that females etter at recovering fear memory than males, as they are more resistant to fear tion, requiring more sessions of extinction than males, increasing ERK2 horylation levels in the dorsal hippocampus (Matsuda et al. 2015).

The endocannabinoid (eCB) system, prevalent in the limbic system, is involved in the modulation of fear conditioning and fear extinction. A study showed that females classified as "darters", i.e., those that display fear as rapid movements called "darting", responded very differently than "non-darters" to the administration of CB1R antagonist. The "non-darter" females showed an increase in freezing during cued-fear conditioning. This is due to the potential role of the vHPC-BLA pathway in associative learning. The vHPC-BLA projections are two times more in female rats than males, and this is the first known evidence of sexual dimorphism in the efferent neuronal density of the hippocampal circuit. In females, loss of CB1R signaling caused contextual fear generalization, whereas increased freezing in cued fear conditioning. This can also be explained by the increased density of vHPC-BLA projections in females (Huckleberry et al. 2023).

In brief, the HPC is critical for integrating contextual and spatial elements during fear conditioning and extinction, with notable sex differences in response and involvement of the eCB system, where males show longer retention of contextual fear and higher ERK expression, while females exhibit greater resistance to fear extinction and distinct reactions to cannabinoid signaling, potentially due to a higher density of vHPC-BLA projections.

2.3 *Medial Prefrontal Cortex*

The mPFC is implicated in several neuropsychiatric disorders such as addiction, depression, and schizophrenia (Anastasiades and Carter 2021) and it is an important regulator of fear processing in humans and rodents (Milad and Quirk 2002). It collects both emotional and spatial information about the context during fear conditioning and fear extinction (Knapska et al. 2012; Zelikowsky et al., 2014).

The infralimbic cortex (IL) and the prelimbic cortex (PL) are part of the mPFC. The IL participates and is essential for the long-term storage of extinction memory and extinction retrieval (Do-Monte et al. 2015a). The latter is associated with fear acquisition and consolidation of fear memory (Choi et al. 2010). For instance, PL neurons increased firing during tone presentation in fear conditioning (Baeg et al. 2001; Burgos-Robles et al. 2009). In addition, these neurons are associated with fear memory retrieval (Corcoran and Quirk 2007).

Sex differences are present in mPFC, with lesions showing impaired fear extinction in male rats 24 h after cued-fear conditioning only after a 1-h delay extinction. In contrast, females with lesions showed increased freezing after 24 h regardless of the presence of 1-h delay extinction. Differences were also seen during fear conditioning, wherein females with lesions were conditioned faster. Moreover, the females also showed less freezing when compared to males in the last CS presentation. These sex differences reflect the role of mPFC in fear extinction recall in males and maintenance of fear extinction memory, and fear conditioning in females (Baran et al. 2010).

Disruption of the connection between mPFC and the AMY is believed to be the cause of fear-based disorders like PTSD, and inappropriate fear responses could be due to the inability of mPFC to regulate the activity of the AMY (Milad et al. 2009). In this regard, it has been shown that this pathway is responsible for fear extinction in male but not in female rats (Quirk et al. 2003). In rodents, a microcircuit in the BLA is activated by efferences from the IL, causing suppression of fear response post-extinction (Herry et al. 2008). Moreover, a study showed that these IL-BLA neurons undergo sex-specific dendritic remodeling in response to stress (Shansky et al. 2009). Structural differences were also observed in the IL-BLA projections between "high freezing" and "low freezing" males. "High-freezing" males showed significantly less dendritic length compared to "low-freezing" males. They also had less dendritic material at regions where most of the IL afferents from the BNST have their synaptic targets, suggesting a role of the BNST-IL-BLA in the high freezing. The lack of morphological differences in the "high freezing" vs "low freezing" females is predicted to be due to the involvement of an alternative circuit that drives freezing during fear extinction retrieval (Shansky et al. 2009). The IL is predicted to be involved in fear extinction via connections to the PL and not directly to the BLA (Gruene et al. 2015b).

Infusion of the brain-derived neurotrophic factor (BDNF), a synaptic plasticity mediator, into the IL cortex, decreased fear expression in male rats. Decreased BDNF was also observed in the inputs to the IL from the HPC in male rats failing to learn extinction (Peters et al. 2010). Similarly, another study showed that the activation of BDNF signaling blocked the return of fear in female mice subjected to extinction training (Baker-Andresen et al. 2013).

The PL cortex, and not the IL cortex is involved in fear expression after cued fear conditioning in female rats, and they showed persistent gamma activation (30-120 Hz) as determined by in vivo electrophysiology (Fenton et al. 2014, 2016; Vidal-Gonzalez et al. 2006). Interestingly, impaired extinction retrieval (cued fear conditioning) in female rats is also related to increased cfos expression in the PL (Gruene et al. 2014). Also, polymorphisms in the gene encoding the pituitary adenylate cyclase-activating polypeptide (PACAP) receptor type 1 (PAC1R) are associated with PTSD and anxiety symptom severity in women only (Ross et al. 2020; Ressler et al. 2011). In female mice, the administration of PAC1R antagonist in the PL cortex impaired cued fear learning after trace fear conditioning and showed higher PAC1R mRNA levels in the PL cortex (Kirry et al. 2018).

The mPFC is critical for fear processing and involved in neuropsychiatric disorders including PTSD, showing sex-specific roles in fear conditioning and extinction, particularly through its IL and PL regions, with disruptions linked to impaired fear extinction and differential memory consolidation between males and females.

2.4 *Thalamus*

The thalamus is constituted by several nuclei facilitating communication between the cortex and subcortical structures, actively contributing as an integral component in various functions typically attributed solely to the cortex (Vertes et al. 2015). One of the main thalamic nuclei related to fear learning is the paraventricular nucleus of the thalamus (PVT), which is mostly constituted by glutamatergic neurons, and therefore, is characterized by a purely excitatory nature (Zhou and Zhu 2019). These glutamatergic neurons are known to project to some of the main brain areas involved in fear learning, such as the mPFC, CeA, CeM, and the BNST (Spencer et al. 2004). Additionally, the thalamus is known to be involved in fear conditioning when auditory cues are used as CS (Bordi and LeDoux 1994).

The PVT has been reported to be involved in the modulation of CeA neuron activation in response to an acute emotional stressor since PVT lesions resulted in a significant increase in stress-induced CeA cFos expression (Spencer et al. 2004). This thalamic region not only contains fear-modulating connections with the AMY, but it also communicates with the PFC. It has been reported that connections between the IL and the PVT were activated during the late stages of cued fear retrieval. These connections are strictly necessary for fear retrieval, as inhibiting them impaired this response at late time points. Also, connections between the PVT and the CeA were seen to be necessary for retrieval at late stages (Do-Monte et al. 2015b). Additionally, during cued fear extinction retrieval, the PVT and the IL were strongly activated, as seen by an increased cFos expression. In addition, the silencing of PVT-IL projections inhibits extinction retrieval (Tao et al. 2021).

To the best of our knowledge, we are not aware of studies showing sex differences in the thalamus regarding fear processing. However, we encourage research to focus on this topic because it is worth exploring.

2.5 *Hypothalamus*

The hypothalamus serves as one of the main connections between the nervous system and various endocrine systems, such as the hypothalamic-pituitary-adrenal axis (HPA), which is the primary stress and fear-responsive neuroendocrine system (Fischer 2021). The hypothalamus releases the corticotropin-releasing factor (CRF) and arginine-vasopressin, which, in turn, stimulates the anterior pituitary gland triggering the release of the adrenocorticotropic hormone (ACTH). ACTH is transported by the blood to the adrenal cortex, where it rapidly stimulates the release of glucocorticoids. After this glucocorticoid rise, a negative feedback system brings them back to baseline levels (Goel et al. 2014).

Sex differences are found in the general functioning of this neuroendocrine axis, as female rats secreted higher concentrations of corticosterone than males when facing both physical and emotional stress, of a predictable and unpredictable nature (Heinsbroek et al. 1991; Rivier 1993; Mitsushima et al. 2003). Not only does corticosterone reach higher levels in females, but this surge happens faster than in males and the levels remain elevated for a longer period (Iwasaki-Sekino et al. 2009). In line with the findings, females showed greater ACTH and corticosterone response than males during both contextual fear conditioning and fear extinction. As for the behavioral outcome, the opposite phenomenon was seen, where females showed lower levels of freezing than males (Daviu et al. 2014). These opposite results could be explained by a potential anxiolytic effect of corticosterone (Albrecht et al. 2012).

Additionally, in female rats, optogenetic stimulation of fear-activated oxytocin neurons in the supraoptic nucleus (SON) of the hypothalamus, resulted in reduced freezing in a contextual fear paradigm (Hasan et al. 2019). The chemogenetic silencing of these neurons impaired context-specific fear extinction. Further exploration of SON-AMY projections showed that these cells displayed an enhanced glutamate release when rats were exposed to fear. Upon re-exposure, when the second fear conditioning was performed in the same environment, the same number of cells were activated, but when fear conditioning happened in a novel environment, a different and larger subset of neurons were recruited. These results showed that the involvement of the hypothalamus in fear processing is not only limited to the neuroendocrine stress response but is also directly involved in fear-modulating neural circuitry (Hasan et al. 2019).

Future studies should address sex differences in the neural circuitry within the hypothalamus and its connectivity with other brain areas.

3 Conclusion

Impairments in fear memory processing have a central role in the development of fear-related disorders such as PTSD, panic disorder, and specific phobias. Understanding the interconnectivity between the brain regions involved is crucial for developing targeted treatments for fear-based disorders. In this chapter, we have revised the scarce literature available on sex differences in fear-related neurocircuitry.

Among differences in cued fear conditioning, inhibition of CeA during fear conditioning results in opposite effects on fear memory, enhancing it in females and reducing it in males. Similarly, inhibition of BLA to BNST projections impaired fear memory only in males, as these connections were not as involved as other subpopulations in females. Nevertheless, females showed an increased cFos expression in BLA during fear conditioning. In addition, prefrontal areas showed high importance for males during fear extinction as IL-BLA and IL-PVT were related to fear extinction and expression retrieval, respectively. Besides, studies in females showed that proper fear extinction retrieval was related to increased cFos expression in IL, whereas impaired fear extinction retrieval was related to higher cFos levels in PL. Separately, PL activity was related to increased fear expression in females (Fig. 1).

Regarding contextual fear conditioning, HPC showed upregulated cFos and activation of the ERK in males. Meanwhile, females exhibited twice as many vHPC-BLA projections as males as well as upregulated cFos in BA. In males, BNST-CeA projections showed upregulated ARC during fear expression. Also,

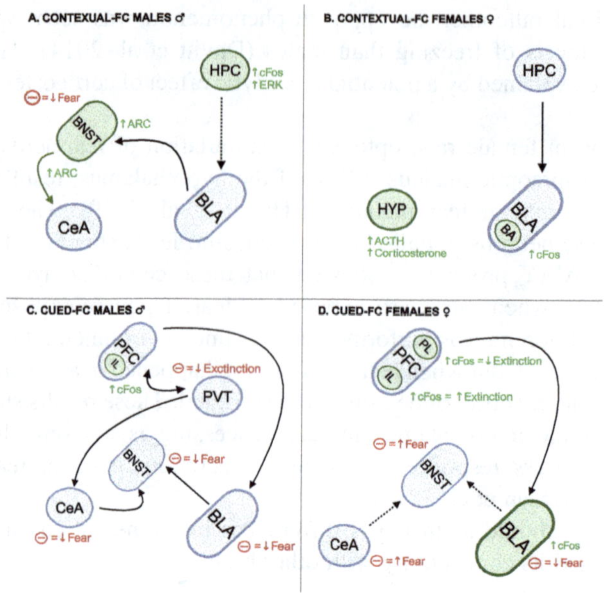

Fig. 1 Schematic representation of sex differences in neural circuits for contextual and cued fear conditioning. Shown contextual fear conditioning in males (**a**), contextual fear conditioning in females (**b**), cued fear conditioning in males (**c**), and cued fear conditioning in females (**d**). Green coloring denotes high activation of nuclei or projecting neurons. Dotted arrows denote fewer active projections in comparison with the opposite sex. Red denotes inhibition of nuclei or projections

lesions in BNST resulted in impaired fear expression. Lastly, females presented enhanced HPA neuroendocrine responses during both fear conditioning and fear extinction (Fig1).

Most neuroscience research has been conducted exclusively with male subjects, and when authors carry out experiments with both male and female animal models, analyses comparing these two sexes are rarely performed. Increasing research with both sexes and focusing on sex differences would not only provide a better understanding of neural, behavioral, and epidemiological sex-based variations, but it would also allow for the definition and adoption of personalized prevention measures. Additionally, it would facilitate the development of more precise therapeutic approaches for fear-based disorders, and therefore tackle the problem of the higher prevalence of these disorders in women.

References

Abbafati C, Abbas KM, Abbasi M et al (2020) Global burden of 369 diseases and injuries in 204 countries and territories, 1990-2019: a systematic analysis for the Global Burden of Disease Study 2019. Lancet 396:1204–1222. https://doi.org/10.1016/S0140-6736(20)30925-9

Albrecht A, Çalişkan G, Oitzl MS et al (2012) Long-lasting increase of corticosterone after fear memory reactivation: anxiolytic effects and network activity modulation in the ventral hippocampus. Neuropsychopharmacology 38:386–394. https://doi.org/10.1038/npp.2012.192

Allen LS, Gorski RA (1992) Sexual orientation and the size of the anterior commissure in the human brain. Proc Natl Acad Sci U S A 89:7199–7202. https://doi.org/10.1073/PNAS.89.15.7199

Amano T, Duvarci S, Popa D, Paré D (2011) The fear circuit revisited: contributions of the basal amygdala nuclei to conditioned fear. J Neurosci 31:15481–15489. https://doi.org/10.1523/JNEUROSCI.3410-11.2011

Anagnostaras SG, Gale GD, Fanselow MS (2001) Hippocampus and contextual fear conditioning: recent controversies and advances. Hippocampus 11:8–17

Anastasiades PG, Carter AG (2021) Circuit organization of the rodent medial prefrontal cortex. Trends Neurosci 44:550. https://doi.org/10.1016/J.TINS.2021.03.006

Arnegard ME, Whitten LA, Hunter C, Clayton JA (2020) Sex as a biological variable: a 5-year progress report and call to action. J Womens Health (Larchmt) 29:858–864. https://doi.org/10.1089/JWH.2019.8247

Arnold AP (2009) Mouse models for evaluating sex chromosome effects that cause sex differences in non-gonadal tissues. J Neuroendocrinol 21:377. https://doi.org/10.1111/J.1365-2826.2009.01831.X

Arnold AP (2017) A general theory of sexual differentiation. J Neurosci Res 95:291–300. https://doi.org/10.1002/JNR.23884

Baeg EH, Kim YB, Jang J et al (2001) Fast spiking and regular spiking neural correlates of fear conditioning in the medial prefrontal cortex of the rat. Cereb Cortex 11:441–451. https://doi.org/10.1093/cercor/11.5.441

Baker-Andresen D, Flavell CR, Li X, Bredy TW (2013) Activation of BDNF signaling prevents the return of fear in female mice. Learn Mem 20:237–240. https://doi.org/10.1101/LM.029520.112

Bandelow B, Michaelis S (2015) Epidemiology of anxiety disorders in the 21st century. Dialogues Clin Neurosci 17:327–335. https://doi.org/10.31887/DCNS.2015.17.3/BBANDELOW

Baran SE, Armstrong CE, Niren DC, Conrad CD (2010) Prefrontal cortex lesions and sex differences in fear extinction and perseveration. Learn Mem 17:267–278. https://doi.org/10.1101/lm.1778010

Bast T, Zhang WN, Feldon J (2001) The ventral hippocampus and fear conditioning in rats. Different anterograde amnesias of fear after tetrodotoxin inactivation and infusion of the GABA(A) agonist muscimol. Exp Brain Res 139:39–52. https://doi.org/10.1007/S002210100746

Beery AK, Zucker I (2011) Sex bias in neuroscience and biomedical research. Neurosci Biobehav Rev 35:565–572. https://doi.org/10.1016/J.NEUBIOREV.2010.07.002

Blume SR, Freedberg M, Vantrease JE et al (2017) Sex- and estrus-dependent differences in rat basolateral amygdala. J Neurosci 37:10567–10586. https://doi.org/10.1523/JNEUROSCI.0758-17.2017

Blume SR, Padival M, Urban JH, Rosenkranz JA (2019) Disruptive effects of repeated stress on basolateral amygdala neurons and fear behavior across the estrous cycle in rats. Sci Rep 9:1–18. https://doi.org/10.1038/s41598-019-48683-3

Bordi F, LeDoux JE (1994) Response properties of single units in areas of rat auditory thalamus that project to the amygdala – I. Acoustic discharge patterns and frequency receptive fields. Exp Brain Res 98:261–274. https://doi.org/10.1007/BF00228414/METRICS

Burgos-Robles A, Vidal-Gonzalez I, Quirk GJ (2009) Sustained conditioned responses in prelimbic prefrontal neurons are correlated with fear expression and extinction failure. J Neurosci 29:8474–8482. https://doi.org/10.1523/jneurosci.0378-09.2009

Calandreau L, Desmedt A, Decorte L, Jaffard R (2005) A different recruitment of the lateral and basolateral amygdala promotes contextual or elemental conditioned association in Pavlovian fear conditioning. Learn Mem 12:383–388. https://doi.org/10.1101/LM.92305

Chen LS, Tzeng WY, Chuang JY et al (2014) Roles of testosterone and amygdaloid LTP induction in determining sex differences in fear memory magnitude. Horm Behav 66:498–508. https://doi.org/10.1016/J.YHBEH.2014.07.008

Choi DC, Maguschak KA, Ye K et al (2010) Prelimbic cortical BDNF is required for memory of learned fear but not extinction or innate fear. Proc Natl Acad Sci U S A 107:2675–2680. https://doi.org/10.1073/pnas.0909359107

Ciocchi S, Herry C, Grenier F et al (2010) Encoding of conditioned fear in central amygdala inhibitory circuits. Nature 468:277–282. https://doi.org/10.1038/NATURE09559

Colon LM, Poulos AM (2020) Contextual processing elicits sex differences in dorsal hippocampus activation following footshock and context fear retrieval. Behav Brain Res 393:112771. https://doi.org/10.1016/J.BBR.2020.112771

Colon L, Odynocki N, Santarelli A, Poulos AM (2018) Sexual differentiation of contextual fear responses. Learn Mem 25:230–240. https://doi.org/10.1101/LM.047159.117

Corcoran KA, Quirk GJ (2007) Activity in prelimbic cortex is necessary for the expression of learned, but not innate, fears. J Neurosci 27:840–844. https://doi.org/10.1523/jneurosci.5327-06.2007

Crawford M, Masterson FA (1982) Species-specific defense reactions and avoidance learning. An evaluative review. Pavlov J Biol Sci 17:204–214. https://doi.org/10.1007/BF03001275

Davis P, Reijmers LG (2018) The dynamic nature of fear engrams in the basolateral amygdala. Brain Res Bull 141:44. https://doi.org/10.1016/J.BRAINRESBULL.2017.12.004

Daviu N, Andero R, Armario A, Nadal R (2014) Sex differences in the behavioural and hypothalamic-pituitary-adrenal response to contextual fear conditioning in rats. Horm Behav 66:713–723. https://doi.org/10.1016/J.YHBEH.2014.09.015

Day HLL, Reed MM, Stevenson CW (2016) Sex differences in discriminating between cues predicting threat and safety. Neurobiol Learn Mem 133:196. https://doi.org/10.1016/J.NLM.2016.07.014

Do Monte FH, Quirk GJ, Li B, Penzo MA (2016) Retrieving fear memories, as time goes by. Mol Psychiatry 21:1027–1036. https://doi.org/10.1038/mp.2016.78

Do-Monte FH, Manzano-Nieves G, Quiñones-Laracuente K et al (2015a) Revisiting the role of infralimbic cortex in fear extinction with optogenetics. J Neurosci 35:3607. https://doi.org/10.1523/JNEUROSCI.3137-14.2015

Do-Monte FH, Quinõnes-Laracuente K, Quirk GJ (2015b) A temporal shift in the circuits mediating retrieval of fear memory. Nature 519:460–463. https://doi.org/10.1038/nature14030

Dong HW, Petrovich GD, Watts AG, Swanson LW (2001) Basic organization of projections from the oval and fusiform nuclei of the bed nuclei of the stria terminalis in adult rat brain. J Comp Neurol 436:430–455. https://doi.org/10.1002/CNE.1079

Duvarci S, Pare D (2014) Amygdala microcircuits controlling learned fear. Neuron 82:966–980. https://doi.org/10.1016/J.NEURON.2014.04.042

Ehrlich I, Humeau Y, Grenier F et al (2009) Amygdala inhibitory circuits and the control of fear memory. Neuron 62:757–771. https://doi.org/10.1016/J.NEURON.2009.05.026

Fadok JP, Markovic M, Tovote P, Lüthi A (2018) New perspectives on central amygdala function. Curr Opin Neurobiol 49:141–147. https://doi.org/10.1016/J.CONB.2018.02.009

Fanselow MS, Dong HW (2010) Are the dorsal and ventral hippocampus functionally distinct structures? Neuron 65:7–19. https://doi.org/10.1016/J.NEURON.2009.11.031

Fenton GE, Pollard AK, Halliday DM et al (2014) Persistent prelimbic cortex activity contributes to enhanced learned fear expression in females. Learn Mem 21:55–60. https://doi.org/10.1101/lm.033514.113

Fenton GE, Halliday DM, Mason R et al (2016) Sex differences in learned fear expression and extinction involve altered gamma oscillations in medial prefrontal cortex. Neurobiol Learn Mem 135:66–72. https://doi.org/10.1016/j.nlm.2016.06.019

Fischer S (2021) The hypothalamus in anxiety disorders. Handb Clin Neurol 180:149–160. https://doi.org/10.1016/B978-0-12-820107-7.00009-4

Florido A, Perez-Caballero L, Velasco ER et al (2021) Direct and indirect measurements of sex hormones in rodents during fear conditioning. Curr Protoc 1. https://doi.org/10.1002/cpz1.102

Florido A, Velasco ER, Romero LR et al (2024) Sex differences in neural projections of fear memory processing in mice and humans. Sci Adv 10:3365

Fredrikson M, Annas P, Fischer H, Wik G (1996) Gender and age differences in the prevalence of specific fears and phobias. Behav Res Ther 34:33–39. https://doi.org/10.1016/0005-7967(95)00048-3

Giovanniello J, Yu K, Furlan A et al (2020) A central amygdala-globus pallidus circuit conveys unconditioned stimulus-related information and controls fear learning. J Neurosci 40:9043–9054. https://doi.org/10.1523/JNEUROSCI.2090-20.2020

Goel N, Workman JL, Lee TT et al (2014) Sex differences in the HPA axis. Compr Physiol 4:1121–1155. https://doi.org/10.1002/CPHY.C130054

Goode TD, Ressler RL, Acca GM et al (2019) Bed nucleus of the stria terminalis regulates fear to unpredictable threat signals. Elife 8. https://doi.org/10.7554/ELIFE.46525

Goosens KA, Maren S (2004) NMDA receptors are essential for the acquisition, but not expression, of conditional fear and associative spike firing in the lateral amygdala. Eur J Neurosci 20:537–548. https://doi.org/10.1111/J.1460-9568.2004.03513.X

Gresack JE, Schafe GE, Orr PT, Frick KM (2009) Sex differences in contextual fear conditioning are associated with differential ventral hippocampal extracellular signal-regulated kinase activation. Neuroscience 159:451–467. https://doi.org/10.1016/J.NEUROSCIENCE.2009.01.009

Gruene TM, Lipps J, Rey CD et al (2014) Heat exposure in female rats elicits abnormal fear expression and cellular changes in prefrontal cortex and hippocampus. Neurobiol Learn Mem 115:38–42

Gruene TM, Flick K, Stefano A et al (2015a) Sexually divergent expression of active and passive conditioned fear responses in rats. Elife 4. https://doi.org/10.7554/ELIFE.11352.001

Gruene TM, Roberts E, Thomas V et al (2015b) Sex-specific neuroanatomical correlates of fear expression in prefrontal-amygdala circuits. Biol Psychiatry 78:186–193. https://doi.org/10.1016/j.biopsych.2014.11.014

Hasan MT, Althammer F, Silva da Gouveia M et al (2019) A fear memory engram and its plasticity in the hypothalamic oxytocin system. Neuron 103:133–146.e8. https://doi.org/10.1016/j.neuron.2019.04.029

Haskell SG, Gordon KS, Mattocks K et al (2010) Gender differences in rates of depression, PTSD, pain, obesity, and military sexual Trauma among Connecticut War Veterans of Iraq and Afghanistan. J Womens Health (Larchmt). https://home.liebertpub.com/jwh 19:267–271. https://doi.org/10.1089/JWH.2008.1262

Haubensak W, Kunwar PS, Cai H et al (2010) Genetic dissection of an amygdala microcircuit that gates conditioned fear. Nature 468:270–276. https://doi.org/10.1038/NATURE09553

Heinsbroek RPW, van Haaren F, Feenstra MGP et al (1991) Controllable and uncontrollable footshock and monoaminergic activity in the frontal cortex of male and female rats. Brain Res 551:247–255. https://doi.org/10.1016/0006-8993(91)90939-S

Hermann C, Ziegler S, Birbaumer N, Flor H (2002) Psychophysiological and subjective indicators of aversive Pavlovian conditioning in generalized social phobia. Biol Psychiatry 52:328–337. https://doi.org/10.1016/S0006-3223(02)01385-9

Herry C, Ciocchi S, Senn V et al (2008) Switching on and off fear by distinct neuronal circuits. Nature 454:600–606

Hines M, Allen LS, Gorski RA (1992) Sex differences in subregions of the medial nucleus of the amygdala and the bed nucleus of the stria terminalis of the rat. Brain Res 579:321–326. https://doi.org/10.1016/0006-8993(92)90068-K

Hofmann D, Straube T (2019) Resting-state fMRI effective connectivity between the bed nucleus of the stria terminalis and amygdala nuclei. Hum Brain Mapp 40:2723–2735. https://doi.org/10.1002/HBM.24555

Hoover WB, Vertes RP (2007) Anatomical analysis of afferent projections to the medial prefrontal cortex in the rat. Brain Struct Funct 212:149–179. https://doi.org/10.1007/S00429-007-0150-4

Huckleberry KA, Calitri R, Li AJ et al (2023) CB1R blockade unmasks TRPV1-mediated contextual fear generalization in female, but not male rats. Neuropsychopharmacology 48:1500. https://doi.org/10.1038/S41386-023-01650-Z

Hulsman AM, Terburg D, Roelofs K, Klumpers F (2021) Roles of the bed nucleus of the stria terminalis and amygdala in fear reactions. Handb Clin Neurol 179:419–432. https://doi.org/10.1016/B978-0-12-819975-6.00027-3

Iwasaki-Sekino A, Mano-Otagiri A, Ohata H et al (2009) Gender differences in corticotropin and corticosterone secretion and corticotropin-releasing factor mRNA expression in the paraventricular nucleus of the hypothalamus and the central nucleus of the amygdala in response to footshock stress or psychological stress in rats. Psychoneuroendocrinology 34:226–237. https://doi.org/10.1016/J.PSYNEUEN.2008.09.003

Kaufman MR, Eschliman EL, Karver TS (2023) Differentiating sex and gender in health research to achieve gender equity. Bull World Health Organ 101:666. https://doi.org/10.2471/BLT.22.289310

Keiser AA, Turnbull LM, Darian MA et al (2017) Sex differences in context fear generalization and recruitment of hippocampus and amygdala during retrieval. Neuropsychopharmacology 42:397–407. https://doi.org/10.1038/NPP.2016.174

Kessler RC, McGonagle KA, Zhao S et al (1994) Lifetime and 12-month prevalence of DSM-III-R psychiatric disorders in the United States: results from the National Comorbidity Survey. Arch Gen Psychiatry 51:8–19. https://doi.org/10.1001/ARCHPSYC.1994.03950010008002

Kessler RC, Wai TC, Demler O, Walters EE (2005) Prevalence, severity, and comorbidity of 12-month DSM-IV disorders in the National Comorbidity Survey Replication. Arch Gen Psychiatry 62:617–627. https://doi.org/10.1001/ARCHPSYC.62.6.617

Kim JJ, Fanselow MS (1992) Modality-specific retrograde amnesia of fear. Science 256:675–677. https://doi.org/10.1126/SCIENCE.1585183

Kim JJ, Jung MW (2006) Neural circuits and mechanisms involved in Pavlovian fear conditioning: a critical review. Neurosci Biobehav Rev 30:188–202. https://doi.org/10.1016/J.NEUBIOREV.2005.06.005

Kim WB, Cho JH (2017) Encoding of Discriminative Fear Memory by Input-Specific LTP in the Amygdala. Neuron 95:1129–1146.e5. https://doi.org/10.1016/J.NEURON.2017.08.004

Kirry AJ, Herbst MR, Poirier SE et al (2018) Pituitary adenylate cyclase-activating polypeptide (PACAP) signaling in the prefrontal cortex modulates cued fear learning, but not spatial working memory, in female rats. Neuropharmacology 133:145–154. https://doi.org/10.1016/J.NEUROPHARM.2018.01.010

Klein SL, Schiebinger L, Stefanick ML et al (2015) Opinion: sex inclusion in basic research drives discovery. Proc Natl Acad Sci U S A 112:5257–5258. https://doi.org/10.1073/PNAS.1502843112/ASSET/BE0D977B-977D-49C0-8337-F4B6AC1B442E/ASSETS/GRAPHIC/PNAS.1502843112FIG01.JPEG

Knapska E, Macias M, Mikosz M et al (2012) Functional anatomy of neural circuits regulating fear and extinction. Proc Natl Acad Sci USA 109:17093–17098. https://doi.org/10.1073/PNAS.1202087109/SUPPL_FILE/PNAS.201202087SI.PDF

Lange MD, Daldrup T, Remmers F et al (2017) Cannabinoid CB1 receptors in distinct circuits of the extended amygdala determine fear responsiveness to unpredictable threat. Mol Psychiatry 22:1422–1430. https://doi.org/10.1038/MP.2016.156

LeDoux JE (2000) Emotion circuits in the brain. Annu Rev Neurosci 23:155–184. https://doi.org/10.1146/ANNUREV.NEURO.23.1.155

LeDoux J (2012) Rethinking the emotional BRAIN. Neuron 73:653. https://doi.org/10.1016/J.NEURON.2012.02.004

Lerner TN, Ye L, Deisseroth K (2016) Communication in neural circuits: tools, opportunities, and challenges. Cell 164:1136–1150. https://doi.org/10.1016/J.CELL.2016.02.027/ATTACHMENT/9F87FB3A-D7E9-4BBD-826D-64C7DB1FE54B/MMC1.PDF

Mahan AL, Ressler KJ (2012) Fear conditioning, synaptic plasticity and the amygdala: implications for posttraumatic stress disorder. Trends Neurosci 35:24–35. https://doi.org/10.1016/J.TINS.2011.06.007

Mamlouk GM, Dorris DM, Barrett LR, Meitzen J (2020) Sex bias and omission in neuroscience research is influenced by research model and journal, but not reported NIH funding. Front Neuroendocrinol 57. https://doi.org/10.1016/J.YFRNE.2020.100835

Maren S, Quirk GJ (2004) Neuronal signalling of fear memory. Nat Rev Neurosci 5:844–852. https://doi.org/10.1038/NRN1535

Matsuda S, Matsuzawa D, Nakazawa K et al (2010) d-serine enhances extinction of auditory cued fear conditioning via ERK1/2 phosphorylation in mice. Prog Neuropsychopharmacol Biol Psychiatry 34:895–902. https://doi.org/10.1016/J.PNPBP.2010.04.013

Matsuda S, Matsuzawa D, Ishii D et al (2015) Sex differences in fear extinction and involvements of extracellular signal-regulated kinase (ERK). Neurobiol Learn Mem 123:117–124. https://doi.org/10.1016/J.NLM.2015.05.009

Mazure CM, Jones DP (2015) Twenty years and still counting: including women as participants and studying sex and gender in biomedical research. BMC Womens Health 15. https://doi.org/10.1186/S12905-015-0251-9

Mcdonald AJ (1992) Cell types and intrinsic connections of amygdala. In: Aggleton JP (ed) The amygdala: neurobiological aspects of emotion, memory, and mental dysfunction, pp 67–96

Milad MR, Quirk GJ (2002) Neurons in medial prefrontal cortex signal memory for fear extinction. Nature 420:70–74

Milad MR, Quirk GJ (2012) Fear extinction as a model for translational neuroscience: ten years of progress. Annu Rev Psychol 63:129–151. https://doi.org/10.1146/ANNUREV.PSYCH.121208.131631/CITE/REFWORKS

Milad MR, Pitman RK, Ellis CB et al (2009) Neurobiological basis of failure to recall extinction memory in posttraumatic stress disorder. Biol Psychiatry 66:1075–1082. https://doi.org/10.1016/j.biopsych.2009.06.026

Mitsushima D, Masuda J, Kimura F (2003) Sex differences in the stress-induced release of acetylcholine in the hippocampus and corticosterone from the adrenal cortex in rats. Neuroendocrinology 78:234–240. https://doi.org/10.1159/000073707

Moustafa AA, Gilbertson MW, Orr SP et al (2013) A model of amygdala–hippocampal–prefrontal interaction in fear conditioning and extinction in animals. Brain Cogn 81:29–43. https://doi.org/10.1016/J.BANDC.2012.10.005

Nabavi S, Fox R, Proulx CD et al (2014) Engineering a memory with LTD and LTP. Nature 511:348–352. https://doi.org/10.1038/NATURE13294

Neueder D, Andreatta M, Pauli P (2019) Contextual fear conditioning and fear generalization in individuals with panic attacks. Front Behav Neurosci 13:452174. https://doi.org/10.3389/FNBEH.2019.00152/BIBTEX

Olff M (2017) Sex and gender differences in post-traumatic stress disorder: an update. Eur J Psychotraumatol 8:4. https://doi.org/10.1080/20008198.2017.1351204

Pape HC, Pare D (2010) Plastic synaptic networks of the amygdala for the acquisition, expression, and extinction of conditioned fear. Physiol Rev 90:419–463. https://doi.org/10.1152/PHYSREV.00037.2009

Paré D, Collins DR (2000) Neuronal correlates of fear in the lateral amygdala: multiple extracellular recordings in conscious cats. J Neurosci 20:2701–2710. https://doi.org/10.1523/JNEUROSCI.20-07-02701.2000

Peters J, Dieppa-Perea LM, Melendez LM, Quirk GJ (2010) Induction of fear extinction with hippocampal-infralimbic BDNF. Science 328:1288–1290. https://doi.org/10.1126/SCIENCE.1186909

Ploski JE, Pierre VJ, Smucny J et al (2008) The activity-regulated cytoskeletal-associated protein (Arc/Arg3.1) is required for memory consolidation of pavlovian fear conditioning in the lateral amygdala. J Neurosci 28:12383–12395. https://doi.org/10.1523/JNEUROSCI.1662-08.2008

Quirk GJ, Mueller D (2008) Neural mechanisms of extinction learning and retrieval. Neuropsychopharmacology 33:56–72

Quirk GJ, Likhtik E, Pelletier JG, Paré D (2003) Stimulation of medial prefrontal cortex decreases the responsiveness of central amygdala output neurons. J Neurosci 23:8800–8807. https://doi.org/10.1523/JNEUROSCI.23-25-08800.2003

Rainnie D, Mania I, Mascagni F, McDonald AJ (2006) Physiological and morphological characterization of parvalbumin-containing interneurons of the rat basolateral amygdala. J Comp Neurol 498:142–161. https://doi.org/10.1002/cne.21049

Ressler KJ, Mercer KB, Bradley B et al (2011) Post-traumatic stress disorder is associated with PACAP and the PAC1 receptor. Nature 477:120. https://doi.org/10.1038/nature10396

Rivier C (1993) Female rats release more corticosterone than males in response to alcohol: influence of circulating sex steroids and possible consequences for blood alcohol levels. Alcohol Clin Exp Res 17:854–859. https://doi.org/10.1111/J.1530-0277.1993.TB00853.X

Ross RA, Hoeppner SS, Hellberg SN et al (2020) Circulating PACAP peptide and PAC1R genotype as possible transdiagnostic biomarkers for anxiety disorders in women: a preliminary study. Neuropsychopharmacol 45(7):1125–1133. https://doi.org/10.1038/s41386-020-0604-4

Rubenstein DR, Hofmann HA (2015) Proximate pathways underlying social behavior. Curr Opin Behav Sci 6:154–159. https://doi.org/10.1016/J.COBEHA.2015.11.007

Russell JS, Trouche S, Reijmers LG (2020) Functional Characterization of the Basal Amygdala-Dorsal BNST Pathway during Contextual Fear Conditioning. eNeuro 7:1–12. https://doi.org/https://doi.org/10.1523/ENEURO.0163-20.2020

Sanchis-Segura C, Becker JB (2016) Why we should consider sex (and study sex differences) in addiction research. Addict Biol 21:995–1006. https://doi.org/10.1111/ADB.12382

Sears RM, Schiff HC, Ledoux JE (2014) Molecular mechanisms of threat learning in the lateral nucleus of the amygdala. Prog Mol Biol Transl Sci 122:263–304. https://doi.org/10.1016/B978-0-12-420170-5.00010-6

Senn V, Wolff SBE, Herry C et al (2014) Long-range connectivity defines behavioral specificity of amygdala neurons. Neuron 81:428–437. https://doi.org/10.1016/j.neuron.2013.11.006

Shansky RM, Murphy AZ (2021) Considering sex as a biological variable will require a global shift in science culture. Nature Neuroscience 2021 24:4 24:457–464. https://doi.org/10.1038/s41593-021-00806-8

Shansky RM, Hamo C, Hof PR et al (2009) Stress-induced dendritic remodeling in the prefrontal cortex is circuit specific. Cereb Cortex 19:2479–2484

Spencer SJ, Fox JC, Day TA (2004) Thalamic paraventricular nucleus lesions facilitate central amygdala neuronal responses to acute psychological stress. Brain Res 997:234–237. https://doi.org/10.1016/J.BRAINRES.2003.10.054

Stein DJ, Scott KM, de Jonge P, Kessler RC (2017) Epidemiology of anxiety disorders: from surveys to nosology and back. Dialogues Clin Neurosci 19:127. https://doi.org/10.31887/DCNS.2017.19.2/DSTEIN

Sullivan GM, Apergis J, Bush DEA et al (2004) Lesions in the bed nucleus of the stria terminalis disrupt corticosterone and freezing responses elicited by a contextual but not by a specific cue-conditioned fear stimulus. Neuroscience 128:7–14. https://doi.org/10.1016/j.neuroscience.2004.06.015

Sun N, Cassell MD (1993) Intrinsic GABAergic neurons in the rat central extended amygdala. J Comp Neurol 330:381–404. https://doi.org/10.1002/CNE.903300308

Tao Y, Cai CY, Xian JY et al (2021) Projections from Infralimbic cortex to paraventricular thalamus mediate fear extinction retrieval. Neurosci Bull 37:229–241. https://doi.org/10.1007/S12264-020-00603-6/METRICS

Tipps M, de Velasco EMF, Schaeffer A, Wickman K (2018) Inhibition of pyramidal neurons in the basal amygdala promotes fear learning. eNeuro 5:10.1523/ENEURO.0272-18.2018

Tolin DF, Foa EB (2006) Sex differences in trauma and posttraumatic stress disorder: a quantitative review of 25 years of research. Psychol Bull 132:959–992. https://doi.org/10.1037/0033-2909.132.6.959

Tovote P, Fadok JP, Lüthi A (2015) Neuronal circuits for fear and anxiety. Nat Rev Neurosci 16: 317–331. https://doi.org/10.1038/NRN3945

Trouche S, Sasaki JM, Tu T, Reijmers LG (2013) Fear extinction causes target-specific remodeling of perisomatic inhibitory synapses. Neuron 80:1054–1065. https://doi.org/10.1016/J.NEURON.2013.07.047

Tsukahara S, Morishita M (2020) Sexually dimorphic formation of the preoptic area and the bed nucleus of the Stria terminalis by neuroestrogens. Front Neurosci 14. https://doi.org/10.3389/FNINS.2020.00797

Uchida K, Otsuka H, Morishita M et al (2019) Female-biased sexual dimorphism of corticotropin-releasing factor neurons in the bed nucleus of the stria terminalis. Biol Sex Differ 10:2. https://doi.org/10.1186/S13293-019-0221-2

Urien L, Bauer EP (2022) Sex differences in BNST and amygdala activation by contextual, cued, and unpredictable threats. eNeuro 9:10.1523/ENEURO.0233-21.2021

Urien L, Stein N, Ryckman A et al (2021) Extended amygdala circuits are differentially activated by context fear conditioning in male and female rats. Neurobiol Learn Mem 180:107401. https://doi.org/10.1016/J.NLM.2021.107401

VanElzakker MB, Kathryn Dahlgren M, Caroline Davis F et al (2014) From Pavlov to PTSD: the extinction of conditioned fear in rodents, humans, and anxiety disorders. Neurobiol Learn Mem 113:3–18. https://doi.org/10.1016/J.NLM.2013.11.014

Vantrease JE, Avonts B, Padival M et al (2022) Sex differences in the activity of basolateral Amygdalar neurons that project to the bed nucleus of the Stria Terminalis and their role in anticipatory anxiety. J Neurosci 42:4488–4504. https://doi.org/10.1523/JNEUROSCI.1499-21.2022

Velasco ER, Florido A, Milad MR, Andero R (2019) Sex differences in fear extinction. Neurosci Biobehav Rev 103:81–108. https://doi.org/10.1016/J.NEUBIOREV.2019.05.020

Velasco ER, Florido A, Perez-Caballero L et al (2023) The impacts of sex differences and sex hormones on fear extinction. Curr Top Behav Neurosci 64:105–132. https://doi.org/10.1007/7854_2023_426

Vertes RP, Linley SB, Groenewegen HJ, Witter MP (2015) Thalamus. The rat nervous system: fourth edition, pp 335–390. https://doi.org/10.1016/B978-0-12-374245-2.00016-4

Vidal-Gonzalez I, Vidal-Gonzalez B, Rauch SL, Quirk GJ (2006) Microstimulation reveals opposing influences of prelimbic and infralimbic cortex on the expression of conditioned fear. Learn Mem 13:728–733. https://doi.org/10.1101/lm.306106

Waddell J, Morris RW, Bouton ME (2006) Effects of bed nucleus of the stria terminalis lesions on conditioned anxiety: aversive conditioning with long-duration conditional stimuli and reinstatement of extinguished fear. Behav Neurosci 120:324–336. https://doi.org/10.1037/0735-7044.120.2.324

Wallenstein GV, Vago DR (2001) Intrahippocampal scopolamine impairs both acquisition and consolidation of contextual fear conditioning. Neurobiol Learn Mem 75:245–252. https://doi.org/10.1006/NLME.2001.4005

Wiltgen BJ, Sanders MJ, Behne NS, Fanselow MS (2001) Sex differences, context preexposure, and the immediate shock deficit in Pavlovian context conditioning with mice. Behav Neurosci 115:26–32. https://doi.org/10.1037/0735-7044.115.1.26

Wiltgen BJ, Zhou M, Cai Y et al (2010) The hippocampus plays a selective role in the retrieval of detailed contextual memories. Curr Biol 20:1336–1344. https://doi.org/10.1016/J.CUB.2010.06.068

Young SL, Bohenek DL, Fanselow MS (1994) NMDA processes mediate anterograde amnesia of contextual fear conditioning induced by hippocampal damage: immunization against amnesia by context preexposure. Behav Neurosci 108:19–29. https://doi.org/10.1037/0735-7044.108.1.19

Zelikowsky M, Hersman S, Chawla MK et al (2014) Neuronal ensembles in amygdala, hippocampus, and prefrontal cortex track differential components of contextual fear. J Neurosci 34:8462. https://doi.org/10.1523/JNEUROSCI.3624-13.2014

Zemla R, Basu J (2017) Hippocampal function in rodents. Curr Opin Neurobiol 43:187–197. https://doi.org/10.1016/J.CONB.2017.04.005

Zhou K, Zhu Y (2019) The paraventricular thalamic nucleus: a key hub of neural circuits underlying drug addiction. Pharmacol Res 142:70–76. https://doi.org/10.1016/J.PHRS.2019.02.014

Sex Differences in the Neurobiology of Fear and Anxiety

Katelyn I. Oliver, Dasani DelRosario, and Jennifer S. Stevens

Contents

Abstract Although women are diagnosed with anxiety and stress-related disorders at twice the rate of men, there remains a lack of clarity around how to enhance treatment within each sex to reduce disparate rates of anxiety. However, in recent years, a growing literature has identified neural, cognitive, and physiological mechanisms that contribute to sex differences in fear and anxiety, with the promise of informing tailored treatment approaches. Here, we review recent findings, focusing on human studies among healthy populations as well as among patients with generalized anxiety, social anxiety disorder, post-traumatic stress disorder, and panic disorder. The literature reveals nuanced differences in the types of stimuli that preferentially evoke anxiety and stress responses in women and men, as well as sex differences in threat neurocircuitry that mediates the behavioral, physiological, and subjective components of fear and anxiety.

Keywords Anxiety · Brain · Fear · Sex differences · Stress

K. I. Oliver, D. DelRosario, and J. S. Stevens (✉)
Department of Psychiatry and Behavioral Sciences, Emory University School of Medicine, Atlanta, GA, USA
e-mail: jennifer.stevens@emory.edu

© The Author(s), under exclusive license to Springer Nature Switzerland AG 2024
Curr Topics Behav Neurosci (2025) 73: 377–404
https://doi.org/10.1007/7854_2024_536
Published Online: 08 November 2024

1 Introduction

In this chapter, we aim to present a review of relevant literature describing biological sex differences in human fear and anxiety neurobiology. We have broken down this topic into two major categories, focusing first on how sex impacts baseline fear and anxiety in healthy human populations using basic laboratory tasks, and then evaluating translational and clinical studies that look at sex differences related to symptoms of anxiety and stress-related disorders.

Here, we use the term "sex" to refer to the biological attributes such as reproductive anatomy and genetics that classifies individuals as either "male" or "female" assigned at birth. In contrast, "gender" refers to the social and cultural construct that makes up an individual's identity. Although these two terms are often conflated, generally the studies reviewed here measured sex assigned at birth or sex as a biological variable (SABV). We focus on sex differences in this chapter, but the contributions of sex and gender to risk for anxiety and stress-related disorders cannot be easily separated. Sex and gender are inextricably linked and interact with one another. Gender roles contribute to the presentation and prevalence of anxiety disorders. For example, in a cross-sectional study conducted in Ghana it was found that anxiety and depressive symptoms in adolescent girls were linked to social pressure to conform to prescribed gender roles in their households (Anyan and Hjemdal 2018). However, cultural factors can also influence sex-specific contributors to anxiety risk. Gendered societal experiences can contribute to chronic stress, which in extreme cases can have large effects on levels of ovarian hormones, such as in the case of hypothalamic amenorrhea, as well as in the levels of testosterone in men (Giles and Berga 1993; Marcus et al. 2001; Retana-Márquez et al. 2003; Michopoulos et al. 2013; Gray et al. 2020). Conversely, the levels of these hormones can influence subjective and physiological components of the stress response (Viau and Meaney 1996; Gordon et al. 2016). Moreover, some research suggests that evaluating the brain within a binary framework may be limiting to our understanding of human neurobiology (Joel 2021). These complex interactions mean that scientific efforts to separate effects of gender vs sex may be misplaced. We will proceed by emphasizing the strong body of literature identifying biological contributors to sex differences, while acknowledging that gender also plays a (largely unmodeled) role in contributing to these effects.

2 Healthy Human Populations and Fear-Related Sex Differences

Sex differences observed in human fear and anxiety processes can be partially attributed to multiple potentially interacting factors, including hormonal, sociocultural (Miller and Halpern 2014; Rippon et al. 2014), genetic (Domschke et al. 2012; Cerasa et al. 2014), neurostructural (Yamasue et al. 2008; Pöhlchen et al. 2023), and

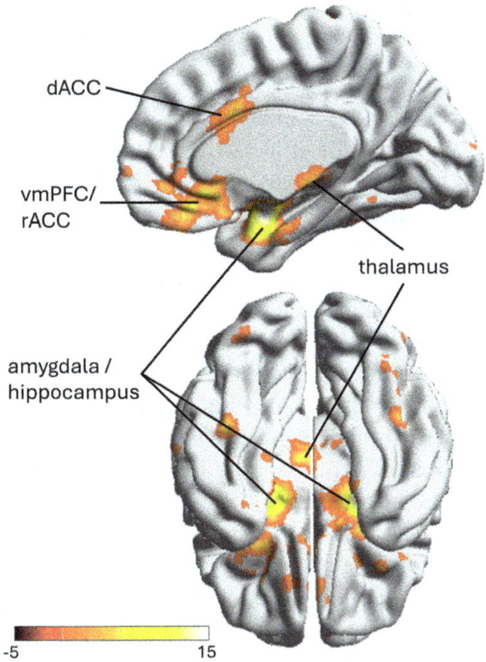

Fig. 1 Brain regions associated with fear and anxiety. A text-based meta-analysis of functional MRI findings in studies that contained the terms "fear" or "anxiety" was created using the automated meta-analysis tool at neurosynth.org. Results reflect fear or anxiety association tests, with coordinates extracted from >400 published studies. Maps for each search term were almost identical, with the exception that "anxiety" did not show dACC activation. The dorsal anterior cingulate cortex (**dACC**) is thought to be involved in conflict and salience detection, as well as fear expression (Shin and Liberzon 2010). The ventromedial prefrontal cortex (**vmPFC**) and rostral ACC (**rACC**) are involved in emotional regulation via top-down connections with the **amygdala** and **insula** (not pictured), emotional and fear response regions, as well as the **hippocampus**, which is involved with contextual learning of fear cues (Shin and Liberzon 2010). The **thalamus** synthesizes converging sensory information and helps relay relevant salient details to the rest of the cortex (Venkataraman and Dias 2023)

even metabolic (Hakamata et al. 2009) differences. A wide variety of experimental paradigms have been used to probe anxiety and fear, including those that evoke negative affective responses with threatening facial or bodily images and sounds, fear conditioning and extinction paradigms, and HPA-axis-activating tasks like those that use threat anticipation or psychosocial stress. In this section, we outline how sex impacts behavioral and functional neural outcomes in healthy human studies of fear and anxiety. We grouped sections together based on the cognitive or affective process of interest, in order to observe how biological sex might differentially impact various components of anxiety and fear. The brain regions illustrated in Fig. 1 are those that are most commonly implicated in fear and anxiety.

2.1 Unconditioned Fear and Anxiety Responses

Generally, women subjectively rate affective response or stress induction tasks as more anxiety inducing than men (Schienle et al. 2005; Kogler et al. 2015; Osinsky et al. 2017; Rattel et al. 2019; Li et al. 2020). In particular, a study that utilized analogue trauma films found that healthy women report more intrusions and distress as well as elevated state anxiety after watching aversive films compared to men (Rattel et al. 2019). Despite these sex differences in ratings, there have not been any reported differences between men and women in sympathetic arousal responses, assessed via the skin conductance response (SCR) during acoustic startle (Hubbard et al. 2011) or psychosocial stress (Kogler et al. 2015) tasks. However, women showed greater nociceptive flexion reflex across trials and higher threat-potentiated SCRs compared to men during noxious nerve stimulation tasks (Hubbard et al. 2011). This indicates that women may be responding more strongly to some forms of threat, such as the threat of pain, compared to men. This appears to be accompanied by the engagement of neural circuits that are important for the processing of threat cues. For example, in psychosocial and other stress inducing paradigms, generally women show more engagement of salience (Kogler et al. 2015; Qiu et al. 2022) or cognitive control (Osinsky et al. 2017) networks than men.

In neuroimaging studies that use visually threatening stimuli, some interesting patterns emerge. A meta-analysis of almost 2000 participants viewing a variety of emotionally evocative images (negative complex scene stimuli, fearful faces, shock or pain anticipation) found that men generally showed greater medial prefrontal cortex (mPFC), anterior cingulate cortex (ACC), frontal pole, insula, and thalamus responses, while women had greater bilateral amygdala, hippocampal, locus coeruleus, and dorsal midbrain activity such as periaqueductal gray/superior colliculus (Filkowski et al. 2017). This analysis combines study responses from both pleasant and unpleasant visual cues, but it is noteworthy that the hyperactive brain regions described in women are frequently implicated in trauma- and anxiety-related disorders such as post-traumatic stress disorder (PTSD) or generalized anxiety disorder (GAD), of which women are twice as likely to be diagnosed as men. Another meta-analysis likewise found that women elicit greater left amygdala activation in response to specifically negative emotional stimuli, whereas men showed greater amygdala activation than women during positive emotional stimuli presentation (Stevens and Hamann 2012), supporting the idea that women respond more strongly to negative emotional stimuli.

The amygdala is an especially important region of interest since hyperactivity in this region may indicate overexpression of the fear response. While one study did not find any sex differences in response to fearful or angry faces (Hornung et al. 2019), many others have noted that women may show greater amygdala fear responses to threatening faces. Women, but not men, with the anxiety disorder risk allele COMT 158al showed greater left amygdala activity in response to fearful or angry faces (Domschke et al. 2012). Additionally, women show more accuracy and greater left amygdala reactivity to both fearful and neutral faces with averted gaze compared to men (Im et al. 2018), which may further indicate overgeneralization of an amygdala fear response to human faces regardless of the emotional expression. In comparison, men showed greater bilateral amygdala activity than women when viewing attacks by humans or animals (Schienle et al. 2005). This may indicate that men have a greater fear response bias for threatening body positions. Supporting this idea, another study found that men showed greater neural response than women in the extrastriate body area, superior temporal sulcus, fusiform, and premotor areas when viewing threatening male body images (Kret et al. 2011). However, it is worth noting that this study had a smaller sample size of only 28 participants (14 women, 14 men), and therefore more neuroimaging studies are needed to further explore these neural correlates. Taken together, it may be the case that, overall, women attend to or respond more strongly to threatening facial cues but that men have a fear bias toward more broad physical threats such as body positioning.

In auditory emotion tasks, men and women seem to differentially respond to particular qualities of an auditory threat. During a stressful script-driven imagery task (with the script read aloud to participants), men had greater activity in threat processing regions such as the amygdala, rostral ACC, insula, and mPFC compared to women, and the men with greater laboratory-induced stress showed more activity in emotion-modulatory regions like the vmPFC (Seo et al. 2011). However, another task using magnetoencephalography (MEG) recordings during presentation of emotional music or video scenes found that women showed stronger beta–gamma frequency coupling in frontotemporal areas to fearful audiovisual stimuli (Yang and Lin 2017). The authors suggest that heightened frequency coupling of beta and gamma oscillatory responses reflects faster and more sensitive integration of sensory-related information during emotional processing, meaning that women may be more sensitively integrating auditory stimuli related to fearful moods. Likewise, women show higher accuracy as well as occipital N1 and N2 activity, which reflects greater attention, while listening to angry relative to neutral voices (Schirmer et al. 2019). This suggests that while men may have greater fear responses to specific words or fearful auditory imagery, women show greater attentional bias and threat response to more general threat-related qualities such as the emotion or mood of the sound. However, many of these auditory task studies used smaller sample sizes, so more work is needed to further expand on the relationship between sex and the response to auditory threats.

Section Summary

- *Women subjectively rate stress and threat tasks as being more anxiety inducing than men, but women only show greater physiological responses (i.e., SCR) during a few specific anxiety tasks such as threat of pain.*
- *Women typically show greater activity in fear expression-related regions such as the amygdala while viewing threatening faces, whereas men respond preferentially to threatening body postures.*
- *Men and women differ in their neural responses to auditory threat – while men show greater response to specific fearful words, women attend and respond more strongly to the emotional tone of auditory threats.*

2.2 Cortisol and Stress

The stress hormone cortisol is regulated by the hypothalamic–pituitary–adrenal axis (HPA axis) and is secreted during times of stress or triggered fear. However, cortisol also acts as a negative feedback control on the HPA axis when it reaches the hypothalamus, shutting down its own production by inhibiting hypothalamic and pituitary signals, such as corticotropin-releasing hormone (CRH) and adrenocorticotropic hormone (ACTH), which normally signal the production of cortisol. Experimental paradigms such as those using psychosocial or performative stress tasks are designed to evoke an HPA response and measure cortisol changes through salivatory or blood samples. Many of these stress studies report no sex differences in cortisol reactivity (Antov and Stockhorst 2014; Kogler et al. 2015; Poppelaars et al. 2019; Dong et al. 2022), even when controlling for menstrual cycle phase (Antov and Stockhorst 2014). However, a subset of this literature shows sex differences in the stress response, with one study showing higher increases in cortisol during stress tasks in women compared to men (Paris et al. 2010), and a number of other studies reporting that men demonstrate higher cortisol responses to stress (Kogler et al. 2017; Wheelock et al. 2021; Henze et al. 2021).

These mixed findings might be explained in part by potential sex differences in glucocorticoid receptor (GR) sensitivity. At baseline, men generally have higher levels of cortisol than women (Paris et al. 2010), a phenotype often associated with lower GR sensitivity (Pariante and Miller 2001), suggesting that men may have lower GR sensitivity and would therefore require more cortisol to shut down the HPA axis. A study by Kuhn and colleagues did not see any sex differences in the cortisol response to a stressor initially (Kuhn et al. 2023), but when the authors subdivided the groups into cortisol responders vs. nonresponders, they found that in the responders, men showed higher cortisol spikes directly after the stressor that remained significantly high up to 45 min post-task. In comparison, the women responders only showed a significant increase in cortisol 15 min after the stressor that was similar to the men's levels, but these levels soon returned to baseline (Kuhn et al. 2023). This study supports the idea that men may have innately lower GR sensitivity, resulting in a prolonged HPA response and high levels of cortisol for longer durations. In contrast, women, who experienced similarly high cortisol levels at the 15-min timepoint, may have more quickly shut down the HPA axis due to their greater GR sensitivity. These effects also appear to interact with age. Younger men show higher ACTH responses to stress than older men or women of any age, whereas older women or older men show greater total plasma cortisol or salivary cortisol (Kudielka et al. 2004). This may indicate initially higher HPA response to stressors in younger men, and then reduced GR sensitivity with age in both older men and women due to lifetime wear and tear on HPA feedback mechanisms. Further research is needed to investigate sex differences in the expression and regulation of the GR and to identify links to the stress response.

Among the neuroimaging studies that found increased cortisol responses in men, it was shown that men had positive, while women generally showed negative, correlations between cortisol and neural responses to stress in affective and sensory regions. Increasing cortisol in men only was associated with greater hippocampal, as well as precuneus, superior temporal gyrus, insula, parahippocampal, amygdala, cingulate cortex, and nucleus caudate and accumbens responses to stress tasks (Kogler et al. 2017; Henze et al. 2021). However, in women, increasing cortisol response to stress was associated with decreases in thalamus and nucleus accumbens activity during stress tasks (Henze et al. 2021). Similarly, men showed positive associations of cortisol with resting state functional connectivity (rsFC) between the amygdala and striatal regions, orbitofrontal gyrus, rostral anterior cingulate gyrus, middle and superior frontal gyrus, SMA, and parietal–occipital sulcus, compared to women who showed negative correlations of cortisol with rsFC between the amygdala and these same regions (Kogler et al. 2016). Several of these brain regions, including the amygdala and hippocampus, are key regions responsible for emotional response and memory formation. Further supporting these findings, generalized fractional anisotropy of the cingulum bundle, a white matter tract connecting the prefrontal cortex and various limbic regions, was shown to be higher in men with greater cortisol and stress reactivity (Wheelock et al. 2021). Perhaps in men where we hypothesize that GR sensitivity may be lower, increased activation and connectivity in these salience regions during stress tasks produces increased cortisol responses that cannot be as easily downregulated by negative feedback systems. However, in women with greater GR sensitivity, increased cortisol may lead to negative feedback and downregulation of the salience network response.

Threat conditioning studies using exogenous cortisol treatment find that boosting cortisol may have a dampening effect on fear conditioning and extinction learning in men while increasing this learning in women. One study found that men showed reduced SCR to the conditioned stimulus (CS+) as well as reduced CS+/CS− insula activity during fear conditioning with cortisol treatment, whereas women had increased CS+/CS− insula response with cortisol treatment (Merz et al. 2010). This suggests that cortisol increased differential learning of the CS+ vs. CS− in women, but reduced this distinction in men. Likewise, cortisol increased SCR and right amygdala activation during extinction recall in men, whereas women actually showed a decrease in amygdala activation associated with cortisol treatment (Kinner et al. 2018), again suggesting that women downregulate their amygdala fear response better with cortisol whereas men were still reactive. Together, these results may again suggest that women are more sensitive and show quicker negative feedback responses to cortisol. However, similar to the endogenous cortisol studies, there have been some contradictory findings, with one study noting no sex differences in BOLD activation associated with cortisol treatment during any phase of conditioning (Merz et al. 2012). Therefore, more studies may be needed to fully investigate the potential of exogenous cortisol to differentially enhance or inhibit fear associations among men and women.

Section Summary

- *Studies investigating reactivity of the stress hormone cortisol are split, either reporting no sex differences or suggesting that men may have higher cortisol responses to stress. However, sex differences and age effects on glucocorticoid (GR) sensitivity may explain some of these discrepancies.*
- *Increased salience network activation is associated with increasing cortisol in men, but in women greater cortisol is associated with lower neural responses in these regions.*
- *Additionally, treatments with exogenous cortisol may dampen fear conditioning and extinction learning in men whereas cortisol treatment seems to improve extinction learning in women.*

2.3 Fear Conditioning and Extinction in Healthy Human Subjects

Threat conditioning and extinction laboratory tasks have been widely used to study the behavioral and neural mechanisms of fear acquisition and extinction in both healthy human participants and those with anxiety or trauma-related disorders. During threat conditioning tasks, one type of neutral stimuli is repeatedly paired with an aversive stimulus (e.g., shock or loud sound) to produce a threat association with that neutral stimuli (CS+), whereas another type of neutral stimuli is not paired with any aversive experience (CS−). Responses to these two stimuli can be compared (CS+ vs. CS−) to determine how well an individual has learned to discriminate between threat (CS+) and safety (CS−) cues. During extinction phases, the aversive stimulus is no longer present and participants learn that the previously threatening cue (CS+) is safe again, leading to an extinction of the fear response. These experimental paradigms allow researchers to study differences in the strength or duration of an acquired fear response.

Studies examining sex differences in sympathetic arousal using the skin conductance response (SCR) have been very mixed. It seems that men and women show similar baseline skin conductance (Milad et al. 2006). However, while a few studies found that men had greater unconditioned responses to aversive stimuli (Wen et al. 2022) or larger SCR responses to either CS+ or CS− in general (Lonsdorf et al. 2015), a different study reported that women in their sample had greater SCR during presentations of the CS+ or CS− during conditioning (Rosenbaum et al. 2015). Of note, Rosenbaum and colleagues also noted that factors such as age or education had greater predictability on SCR than sex in their sample. Therefore, it is unclear if a sex bias exists in skin conductance responses to stimuli presentation during conditioning. Differential SCR for the CS+ vs. CS− indicates how well an individual has learned to distinguish between threat and safety cues. Interestingly, several studies report no sex differences in differential SCR response between the CS+/CS− during any phase of conditioning or extinction (Lebron-Milad et al. 2012; Pohlack et al.

2015; Wen et al. 2022), indicating equal fear acquisition and extinction learning abilities between men and women. However, a few other studies have reported that women show less distinction between the CS+ and CS− as evidenced by lower differential SCR response compared to men during conditioning and multiple extinction phases (Milad et al. 2006; Rosenbaum et al. 2015; Lonsdorf et al. 2015). A smaller differential SCR might indicate more generalization of the fear response in women but potentially better extinction learning as well. More research in this area is needed to determine if sex influences psychophysiological arousal during fear acquisition and extinction.

Interestingly, neuroimaging studies of fear conditioning report that women have greater amygdala activation in response to the CS+ (Wen et al. 2022) or higher amygdala, dACC, and rACC response for CS+ > CS− (Lebron-Milad et al. 2012). The amygdala and ACC are implicated in fear expression and therefore greater activation may indicate that women may show heightened levels of conditioned fear compared to men. Further supporting these ideas, a study by Merz and colleagues found that exposing participants to psychosocial stress enhanced late conditioning CS+/CS− amygdala and ACC activity in women but reduced activity in these regions in men, suggesting that psychosocial stress may impair fear acquisition in men but facilitate this learning in women (Merz et al. 2013). Slightly contradicting these ideas, another study found that men showed greater amygdala response to cued threats (Dark et al. 2022). However, Dark and colleagues' study specifically looked at adolescent populations between the ages of 11–19 years when the effects of puberty and dynamic hormone changes may have a larger impact on neural responses to threat. Additionally, this study found that women elicited greater amygdala responses to non-cued threats alone. Overall, these neuroimaging findings are especially interesting since they would suggest that women show hyperactivation in fear response regions despite potentially lower or similar psychophysiological response as men during conditioning.

Additionally, during both extinction (Wen et al. 2022) and extinction recall (Lebron-Milad et al. 2012), men showed greater rACC activity during CS + E > CS + U, interpreted as reflecting better top-down regulation of the fear response during differentiation between the CS+ that was extinguished (CS + E) compared to the unextinguished CS+ (CS + U). In comparison, women showed greater functional connectivity in various visual, ventral attention, and somatomotor regions as well as increased extinction recall CS + E > CS + U activity in the amygdala (Wen et al. 2022) and insula (Lebron-Milad et al. 2012) during extinction learning, indicating continued activation of emotional response regions. Taken together, these two studies indicate that women may still be experiencing overactivation in aspects of threat and salience neurocircuitry during times of safety, consistent with overgeneralization of fear, whereas men are showing better top-down neural regulation of their emotional responses. This seems to stand in contrast to the pattern observed for tasks that elicit an HPA axis response (discussed in Sect. 2.2), in which women showed quicker reduction of the stress response due to cortisol sensitivity. However, fear learning tasks often do not elicit a full HPA-axis response, instead requiring only a sympathetic or noradrenergic response to the US and CS+ cues. It may be that men show greater regulation of sympathetic arousal, but are slower to dampen cortisol stress responses.

Section Summary

- *Findings evaluating sympathetic arousal during fear conditioning paradigms are mixed, reporting either contradictory results or no sex differences in skin conductance. However, there may be more evidence that men show greater SCR and distinction between threat-paired and -unpaired stimuli during conditioning.*
- *By contrast, women seem to have greater neural responses in fear expression regions during fear conditioning.*
- *During extinction learning, it seems that women are still expressing fearful responses, whereas men show greater activity in emotion regulatory regions.*

2.4 Hormone Effects in Fear Conditioning

Hormonal changes throughout the menstrual cycle in women may be able to explain some of the mixed findings in the fear conditioning literature. Both men and women produce the same sex hormones (testosterone, progesterone, and estrogen) but at differing levels and patterns (Fig. 2). While men have more consistent levels of these gonadal hormones, women experience larger fluctuations in hormone production over a month-long cycle. Many of the brain regions frequently studied in fear conditioning tasks contain a high density of receptors for cycling sex hormones such as estradiol (E2) and progesterone (Barth et al. 2015). Therefore, the higher variability in these fluctuating hormone levels in women may have the potential to influence neural sex differences during fear conditioning experiments. It is worth mentioning that although there are several gonadal hormones that fluctuate with the female reproductive cycle, a majority of the prior preclinical and human literature has focused on estradiol and fear conditioning paradigms. Influences of other sex hormones such as progesterone and testosterone on fear and anxiety are discussed below, but not extensively due to a gap in the human literature.

Studies investigating estradiol level influences on sympathetic arousal during fear acquisition suggest that E2 may not differentially impact fear learning processes. Similar to the previous section investigating fear conditioning in women vs. men, a few studies have noted that men show higher conditioned responses (Milad et al. 2010) and differential SCR (Milad et al. 2006; Bierwirth et al. 2021) during conditioning compared to groups of women with either low or high E2. Similarly, most studies have reported no conditioning SCR differences among women with low vs. high E2 (Zeidan et al. 2011; Graham and Milad 2013; Graham et al. 2017). Additionally, a few samples showed no differences among men, women with high-E2, or women with low-E2 (Merz et al. 2012; Antov and Stockhorst 2014). Lonsdorf and colleagues' investigations are particularly interesting since this study used a large sample size ($n = 377$) and found that when all women were pooled together in

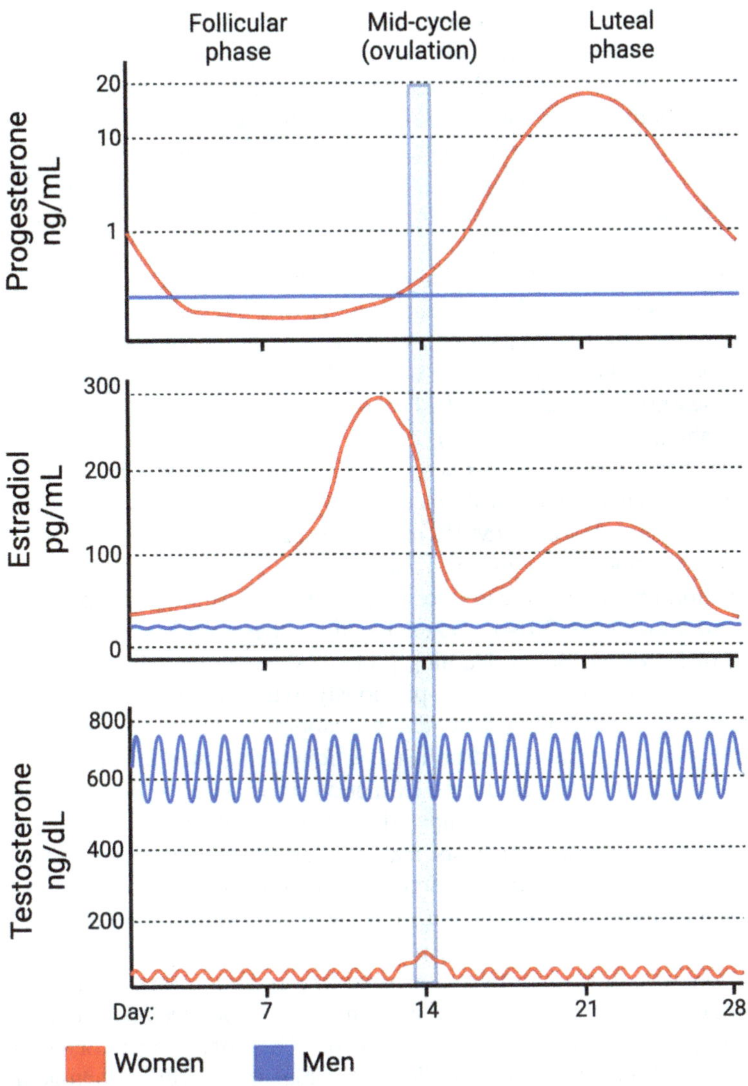

Fig. 2 Conceptual comparison of gonadal hormones between men and women. A schematic representation of fluctuations in progesterone, estradiol, and testosterone are illustrated for women and men. The *y*-axis represents day of the menstrual cycle in women, and shows diurnal variation in men. Women of reproductive age experience large fluctuations in progesterone and estradiol levels throughout the menstrual cycle (Stricker et al. 2006), whereas men typically have lower levels with fewer fluctuations of these two hormones (Oettel and Mukhopadhyay 2004; Brambilla et al. 2009; Parikh et al. 2018). Men experience a larger diurnal fluctuation and higher levels of daily testosterone (Bremner et al. 1983), while women experience much lower levels and smaller fluctuations that peak around ovulation (Stricker et al. 2006)

one group, men had greater conditioned responses during conditioning than women. However, when women were split to account for hormone influences (116 men, 89 free cycling women, and 172 women taking hormonal contraceptives – i.e., low E2 and progesterone), there were no significant differences among the three groups (Lonsdorf et al. 2015). Taken together, it is still somewhat unclear if men vs. women show different psychophysiological responses during conditioning, but E2 does not appear to be an influential factor during the acquisition phase. Similarly, a study evaluating time-of-day testosterone in men did not report any differences in fear acquisition, suggesting that daily fluctuating testosterone levels did not impact fear learning processes in men (Pace-Schott et al. 2013).

Interestingly, sex hormones seem to have a more robust impact on extinction learning and consolidation. Women with higher estradiol show lower SCR during extinction learning (Zeidan et al. 2011) and reduced fear recovery during extinction recall (Graham and Milad 2013) compared to women with lower estradiol, suggesting that higher E2 levels were linked with enhanced extinction of conditioned fear. When grouped with men, high-E2 women show greater extinction retention than low-E2 women (Milad et al. 2010). In groups exposed to psychosocial stress, women in mid-cycle (high-E2) and men showed less fear recovery and greater fear extinction retention compared to women in the early follicular phase (low-E2) (Antov and Stockhorst 2014). Taken together, these studies suggest a positive influence of E2 on extinction learning processes and show that hormones like E2 may influence the mixed findings previously reported in the fear conditioning literature when investigating potential psychophysiological sex differences. An interesting note – Lonsdorf and colleagues found that men and high-E2 women had greater SCR differentials during extinction recall, suggesting worse extinction retention in these two groups compared to women with low-E2 (Lonsdorf et al. 2015). However, this still suggests that men and high-E2 women show similar profiles of SCR activity during extinction recall compared to women with low-E2. These results may in part be different from the rest of the literature due to a slightly different experimental conditioning paradigm that employed a block design with multiple contexts. Also, this study compared free cycling women with those taking hormonal contraceptives but did not clarify the type of contraceptives, indicating that women might have been taking a variety of E2, progesterone, or combination treatments. Additionally, a few studies contradict these patterns altogether. One study found that men had greater differential SCRs than either women with low or high E2 during both extinction learning and recall (Bierwirth et al. 2021), and another study found that sex or hormone status did not influence SCR during extinction at all (Merz et al. 2012). Still, an overwhelming number of studies seem to point to a potential role of high estradiol supporting extinction learning and consolidation in women. Similar to E2, testosterone may likewise enhance fear extinction processes. Men with greater morning testosterone showed enhanced extinction recall and renewal via smaller differential SCRs and better general of the unextinguished CS+ with the extinguished CS+ (Pace-Schott et al. 2013). Therefore, gonadal hormones such as estradiol and testosterone may facilitate the consolidation of fear extinction memories.

Neuroimaging studies likewise support the idea that sex hormones like estradiol may mediate some of the behavioral differences seen in the psychophysiological data, but the findings are a bit mixed. Some studies report no group differences in engagement of brain regions responsible for regulating threat/arousal among women taking oral contraceptives (low E2 and progesterone), women in low- vs. high-E2 phases of the menstrual cycle, or men during conditioning (Zeidan et al. 2011; Merz et al. 2013), further supporting the idea that neither hormones nor sex majorly influences fear acquisition. Hwang and colleagues found that women with higher E2 levels had greater insula, amygdala, and hypothalamus responses during conditioning as well as greater insula and rACC activity during extinction and recall compared to men or women with low-E2 (Hwang et al. 2015). As discussed previously, the amygdala and insula are threat response regions, whereas the rACC helps to regulate amygdala threat responses. Therefore, Hwang et al.'s results suggest that E2 may upregulate activity in both threat response and regulatory regions, which may facilitate both conditioning and extinction learning. Similarly, when looking just within groups of women, higher-E2 individuals show greater vmPFC activation to CS+ > CS− during extinction learning and recall compared to lower-E2 participants (Zeidan et al. 2011). This suggests better top-down control of fear responses in high-E2 groups during extinction processes. Likewise, women with high E2 and men were shown to have reduced amygdala, thalamus, and dACC responses to CS+ > CS− during extinction learning, compared to low-E2 groups, indicating that women with lower E2 retain activity in regions associated with threat responses during safety learning. Taken together, these results point to an influential role of E2 on neural circuits that may help facilitate similar levels of extinction learning as seen in men.

A few studies have also investigated hormone influences on neural responses in nonconditioning, affective imagery tasks. Healthy women in the follicular (low E2 and progesterone) phase of their cycles were given either progesterone or placebo. During progesterone administration, women had increased amygdala responses to salient imagery, suggesting that increases in this hormone may have increased amygdala reactivity (Van Wingen et al. 2008). Another neuroimaging study that used negative affective pictures found that men and early follicular (low-E2) women showed very similar neural responses, while high-E2 women had greater OFC, medial PFC, hippocampal, and periaqueductal gray responses to these images (Goldstein et al. 2010). These findings slightly contradict the idea that men and high-E2 women may have similar neural and behavioral profiles as seen in the conditioning literature; however, it is worth noting that this study used a relatively small sample size ($N = 25$). Additionally, women who were administered testosterone showed reduced fear-potentiated startle during a threat-of-shock task (Hermans et al. 2006) and reduced SCR and startle response when viewing strongly aversive images (Hermans et al. 2007), suggesting a protective role for testosterone in preventing fear responses. However, this study only looked within women and did not evaluate the effects of exogenous testosterone on fear or anxiety processes in men. Together, these studies highlight an important gap that needs to be explored to fully understand how hormones may influence sex differences in fear and anxiety.

Section Summary

- *When taking estradiol levels into account, there may not be any sex differences in sympathetic arousal responses during fear conditioning.*
- *Women with higher estradiol and men seem to show similarly better safety learning patterns compared to women with low estradiol as indicated by extinction learning and recall SCR findings.*
- *Likewise, men seem to have better fear extinction recall in the mornings when testosterone is higher.*
- *Neuroimaging studies are mixed, but some studies find that E2 may increase activation in regions supporting emotional regulation during extinction and consolidation processes, helping women achieve similar levels of extinction as seen in men.*
- *More studies evaluating other sex hormones such as progesterone and testosterone are needed to explore the interactions of these hormones on fear and anxiety processes in women and men.*

3 Translational/Clinical: How Sex Differences are Related to Symptoms of Anxiety and Stress-Related Disorders

3.1 Prevalence of Anxiety and Stress-Related Disorders in Women and Men

Women exhibit a greater lifetime vulnerability to anxiety disorders than men, with the discrepancy emerging during adolescence and intensifying significantly between ages 14 and 25, while also reporting being more impacted by symptoms (Beesdo et al. 2009; Faravelli et al. 2013; Altemus et al. 2014). In the United States, generalized anxiety disorder (GAD) is nearly twice as prevalent in women than men, and has an earlier age of onset in women (McLean et al. 2011; Salminen et al. 2022). Several studies have reported a significantly increased prevalence of anxiety disorders in postmenopausal women (Alblooshi et al. 2023). This sex-based trend is common across all anxiety and stress-related disorders, including PTSD, with two notable exceptions being bipolar disorder and obsessive-compulsive disorder, which have relatively similar prevalence across women and men (Bangasser and Cuarenta 2021). Nevertheless, each anxiety disorder displays sex-based differences in its presentation. Women are more likely than men to be diagnosed with comorbid anxiety disorders and experience depressive symptoms, while men are more likely to be diagnosed with a comorbid substance use disorder or antisocial personality disorder and are predisposed to somatic symptoms (Jalnapurkar 2018; Boehlen et al. 2020).

3.1.1 Hormones

There is a strong body of evidence suggesting that gonadal hormone fluctuations are a prominent biological risk factor for anxiety disorders (Kundakovic and Rocks 2022). While women experience substantial fluctuations throughout their menstrual cycles, pregnancy, and menopause, men maintain relatively stable gonadal hormone levels. Male gonadal hormones, particularly testosterone, were not found to have the same robust impact on PTSD symptomology as estradiol and progesterone (Reijnen et al. 2015). Plasma testosterone levels in soldiers were measured before and after deployment, and there was no difference found between the testosterone levels of men reporting high and low PTSD symptoms; however, pre-deployment testosterone levels predicted PTSD symptoms at 1 and 2 years post-deployment, with high testosterone having a protective effect. This is in accordance with previous literature, suggesting that testosterone may influence HPA-axis functioning (Handa and Weiser 2014).

Gonadal hormone fluctuations in the late luteal phase have been reported to be associated with exacerbated anxiety symptoms and emotional instability, most significantly, the dropping of progesterone and estradiol levels in the premenstrual, or late luteal phase (Strine et al. 2005; Eisenlohr-Moul et al. 2015). During this phase, many women report experiencing premenstrual exacerbation (PME), which is a phenomenon reported by about 60% of women with mood disorders in which their symptoms are exacerbated in the period between ovulation and menstruation (Li and Graham 2017). Women with generalized anxiety disorder (GAD) were also found to be more likely to have comorbid premenstrual dysphoric disorder (PMDD), this association is mediated by mood symptoms and behavior inhibition (Yen et al. 2020). In addition to these effects, fluctuations in estradiol and progesterone levels may impact cognitive processes that contribute to PTSD such as fear extinction and memory retrieval and may mediate the HPA axis stress response (Nillni et al. 2015). Nillni and colleagues also found that post-trauma symptoms including those of anxiety and depression appeared to vary across menstrual cycle phases, with the greatest exacerbation being observed in the early follicular phase as compared to the luteal phase. In summary, the dropping of progesterone and estradiol levels in the late luteal phase has been found to be associated with exacerbated anxiety symptoms, while post-trauma symptoms in PTSD have been found to be exacerbated most notably in the early follicular phase. Both phases are marked by low circulating levels of estradiol and progesterone.

Section Summary

- *Sex hormones are the main biological factor for sex differences in anxiety disorder prevalence and symptom presentation.*
- *Fluctuations of female sex hormones can influence cognitive processes.*

3.2 *Neuroimaging Studies of Anxiety and Stress-Related Disorders in Women and Men*

Sex differences in brain structure and function have a dynamic relationship with gonadal sex hormones throughout development. Even when no anatomical differences are present, sex hormones may still exert an underlying influence on altering brain function. Studies have shown that brain structure and function fluctuate throughout the menstrual cycle (Altemus et al. 2014; Dubol et al. 2021; Kundakovic and Rocks 2022). Determining which sex differences are related to anxiety disorders and in what ways structural disparities are linked to brain function has proven to be quite challenging. The primary brain regions implicated in anxiety disorders are the amygdala, prefrontal cortex, and hippocampus.

3.2.1 Structural MRI Studies

Structural neuroimaging studies for GAD have yielded conflicting results in sex-based variations in the size of gray matter volume or white matter tracts. Within the body of literature, sample sizes are small, and few studies have specifically looked for sex differences in structure in adult populations, with existing literature primarily focusing on PTSD and childhood adversity in adolescents (Bellis et al. 2015). Overall, the literature is characterized by mixed findings with respect to sex differences in brain structure in the context of GAD. Moon and colleagues found that women with GAD had a significantly larger prefrontal cortex volume relative to men with GAD (Moon and Jeong 2017). In a study of adolescents by Liao and colleagues, dlPFC volume in girls with GAD was positively correlated with anxiety symptom severity. Tromp et al. (2012) used DTI and fMRI and found reduced structural connectivity in the frontolimbic pathway in GAD but no sex differences (Tromp et al. 2012). A recent large study used DTI to analyze sex differences in the microstructural integrity of white matter in a prefrontal tract, the uncinate fasciculus, which had previously been found to have reduced microstructural integrity in boys with anxiety disorders but not girls (Aggarwal et al. 2024). The study collected data from 165 adolescents who met the DSM-5 criteria for current generalized, separation, or social anxiety disorder and 132 healthy controls, and found a significant decrease in the microstructural integrity of the uncinate fasciculus white matter in boys but not girls, and also found that boys with anxiety disorders had several other regions with significantly altered white matter integrity. These findings provide some suggestion of sex-dependent prefrontal structural changes associated with GAD and anxiety symptoms. However, a meta-analysis by Harrewijn and colleagues found no main effect of GAD on brain structure or any significant interactions involving GAD age or sex (Harrewijn et al. 2021). Conflicting findings in structural studies may potentially be attributed to the regular fluctuation of gray matter volume in correspondence with increasing and decreasing estradiol levels throughout the menstrual cycle (Meeker et al. 2020), but results may also be restricted by sample size (Salminen et al. 2022) or heterogeneity across samples. This evidence cumulatively suggests that structural alterations do not largely influence the sexual dimorphic presentations of GAD.

As in GAD, few neuroimaging studies have looked at sex differences in panic disorder (PD), and existing studies were conducted with relatively small sample sizes. In a 2009 study, Asami and colleagues found that there was significant sexual dimorphism in the right amygdala and bilateral insular cortex, and left occipitotemporal gyrus in 24 people with PD vs 24 controls (15 women, 9 men), with the structures being larger in men (Asami et al. 2009). They also observed women-specific effects of PD in the dlPFC, thalamus, parietal cortex, and right cerebellar vermis, which were not present in the men. In contrast, Wang et al. (2021) found no sex differences in gray matter volume in adults with PD and social anxiety disorder (Wang et al. 2021). More recently, Kim et al. (2023) investigated gender differences in anxiety and depression symptoms in PD through network analysis (Kim et al. 2023). The group found significant sex differences in the cingulate gyrus white matter tracts and central symptoms of PD in women and men. They identified agoraphobia as the central symptom in men, and loss of pleasure as the central symptom in women.

Investigations into PTSD have yielded similar results with a similar trend of studies focusing on adolescents in gray matter sexual dimorphism. In an Emergency Department study where participants experienced a criterion-A-trauma as defined in the DSM-IV-TR in the past 24 h, MRI sessions within 3 weeks of their 1-month follow up visits revealed that greater right rACC thickness predicted the severity of PTSD symptoms at the 6-month mark after the traumatic incident in women, while in men, smaller right rACC thickness predicted the severity of depressive symptoms at the 6-month post-trauma visit (Roeckner et al. 2022). Similarly, Heyn et al. (2022) utilized T1-weighted MRI to investigate sex-based variations in brain structure in adolescents with PTSD (Heyn et al. 2022). Their analysis revealed gray matter differences in the vlPFC and frontal pole, with increased volume in girls with PTSD compared to boys. The vlPFC and frontal pole volume were negatively associated with PTSD symptoms in men 1 year later. De Bellis et al. (2015) investigated gray matter volume in adolescents with PTSD, and their findings suggest that adolescent boys with PTSD exhibit a deficit in gray matter volume in the PFC. Together the findings suggest that lower prefrontal gray matter may only predict PTSD symptoms in men, but not women.

Section Summary

- *While the conclusions have been mixed in structural studies, most studies find that men with PTSD have a lower gray matter volume than women with PTSD. However, a second common finding is that there are no differences between sexes in brain structure when comparing men with and without PTSD to women with and without PTSD.*
- *Findings indicate that brain structure is likely not a key factor in sexually dimorphic presentations of anxiety disorders.*

3.2.2 Functional Neuroimaging

There is a significant gap in the literature regarding sexual dimorphism in functional neuroimaging of anxiety disorders. While sexual dimorphism has been observed in the amygdala and prefrontal cortex in healthy populations, there has been limited investigation into sex differences in neural activation within anxiety disorders themselves. In one large-scale developmental study that included individuals diagnosed with PTSD, GAD, separation anxiety, and phobic disorders, a trait anxiety dimension measured by STAI responses was elevated in women, and this was mediated by left amygdala perfusion among $n = 477$ adolescent girls and $n = 398$ boys (Kaczkurkin et al. 2016). Another study found that women with panic disorder (PD) demonstrated significantly stronger activation than men in various brain regions including the amygdala, prefrontal cortex, temporal and occipital cortical areas, basal ganglia, and thalamus, while processing neutral or negative facial expressions (Ohrmann et al. 2010). The authors additionally found that functional connectivity between the amygdala, prefrontal cortical areas, and thalamus during the processing of angry facial expressions was notably stronger in women than in men. These results emphasize sex as a crucial variable influencing neural activation patterns in emotional processing in PD. However, the study's sample size was relatively small, consisting of 20 participants (12 women, 8 men).

Cognitive bias has been identified as a core aspect of GAD (Nejati et al. 2021). Cognitive bias can be defined as a subconscious cognitive inclination in thinking that goes against logical and reason and influences human decision-making. To investigate sex differences in cognitive bias, Kinney and colleagues utilized an exogenous spatial cueing task to investigate sex differences in attentional bias for either positive or negative word stimuli in people with GAD. They found that there was no difference in bias for negative words, but women displayed a significantly greater attentional bias for positive words than men (Kinney et al. 2017). While specific brain regions including the dlPFC and vmPFC have been linked to cognitive biases in GAD patients, there is not yet research looking into sex differences in these regional activation patterns (MacLeod and Mathews 2012; Nejati et al. 2021).

More extensive functional MRI (fMRI) investigation exists in the context of mechanisms that lead to increased vulnerability to PTSD in women. Milad and colleagues found no sex differences in extinction, which is typically impaired in PTSD, in their investigation of PTSD patients (Milad et al. 2009). However, during extinction recall, men with PTSD exhibited increased activation in the left rostral dACC compared with women with PTSD (Shvil et al. 2014). Findings from an Emergency Department study suggest that lower vmPFC activation during a response inhibition task administered 2 weeks post-trauma predicts PTSD symptoms 6 months post-trauma in women but not men (Borst et al. 2024). They also found that greater right inferior frontal gyrus (rIFG) activation during the response inhibition task predicted lower PTSD symptoms at the 6-month mark in men but not women. Felmingham and colleagues found that men with PTSD exhibited greater hippocampal activity in response to fearful face expression stimuli than women with

PTSD (Felmingham et al. 2010). The study design included women and men with PTSD, trauma-exposed controls, and non-trauma-exposed-controls with similar ratios of men to women. Women with and without PTSD exhibited enhanced brainstem activity associated with their exposure to trauma, and both sexes exhibited greater amygdala activity in the PTSD group. Helpman and colleagues investigated resting-state functional connectivity in trauma-exposed women and men with and without PTSD using resting state fMRI and found that women with PTSD exhibited weaker functional connectivity between the amygdala and posterior cingulate cortex and between the precuneus and hippocampus than both men with PTSD and trauma-exposed women (Helpman et al. 2021). Between the precuneus and anterior cingulate cortex, women with PTSD were found to exhibit stronger functional connectivity than men with PTSD and trauma-exposed women. These patterns concur with studies of women with PTSD, which find amygdala hyperreactivity and decreased functional connectivity between the amygdala and vmPFC (Fonzo et al. 2010; Stevens et al. 2013).

The current body of literature in brain metabolites in anxiety disorders is sparse, and few studies investigate sex differences within subjects. Magnetic resonance spectroscopy (MRS) is a noninvasive method to study brain chemistry and is the primary method used in clinical and research settings to measure brain metabolites. Investigation into hormonal fluctuations and brain metabolites emphasizes studies in only women. However, there is evidence of some sex differences in brain metabolites in patients with anxiety disorders. Pigoni et al. (2020) found an association between sex and brain metabolites, specifically glutamate, in an outpatient clinic study of patients with PD, GAD, and MDD (Pigoni et al. 2020). Participants underwent magnetic resonance imaging and spectroscopy focused on the dorsolateral cortex. Men with PD had reduced glutamate (Glu) and glycerophosphocholine plus phosphocholine (GPC + PC) concentrations in comparison to women in the same group, while healthy men had higher Glu and GPC + PC levels than women. The reduction in Glu and GPC + PC was still present when men and women with PD were combined and compared to the healthy control groups. These findings suggest that there may be sex-specific metabolic alterations in anxiety disorders. Hormonal fluctuations across the menstrual cycle can influence sex differences in brain metabolites (Gillies and McArthur 2010), but this has not been directly examined in the context of anxiety disorders.

Section Summary

- *Some evidence shows that regions of the brain related to fear circuitry including the amygdala and prefrontal cortex exhibit hyperactivity in women with anxiety disorders compared to men.*
- *In PTSD, several studies have found weaker functional connectivity between the amygdala and prefrontal cortical areas in women compared to men. Relative to men with PTSD and trauma-exposed women, women with PTSD were found to exhibit greater functional activity within the amygdala and posterior cingulate cortex, as well as between the precuneus and hippocampus.*

- *Men with panic disorder show reduced brain metabolites such as Glu and GPC + PC compared to women with PD and healthy control groups.*

4 Conclusions and Future Directions

Among studies that looked at anxiety and fear within healthy human populations, we see that women generally rate stress tasks as being more anxiety inducing than men. However, neural patterns suggest that men and women both show hyperresponsiveness in fear response regions such as the amygdala, albeit these responses are different for particular qualities or types of salient stimuli (i.e., threatening faces vs. physical threats; auditory words vs. tone). Men typically show increased responsiveness to more overt cues such as threatening body images or words whereas women respond more to intimate threat details such as facial cues or vocal mood. This may indicate that female neurobiology is more sensitive to the finer details of fear or anxiety inducing stimuli. Interestingly, tasks that activate the HPA-axis likewise suggest that women have higher glucocorticoid sensitivity, resulting in a quicker shutdown of the HPA-axis, lower basal cortisol levels, and heightened responsiveness during fear and extinction learning when cortisol is administered exogenously. In most conditioning and extinction tasks, men and women do not appear to show vast differences in measures of sympathetic arousal, but women tend to show heightened neural responses in regions related to threat response whereas men had greater responses in emotion regulatory regions. Some of these results may be explained by cycling sex hormones such as progesterone and estradiol, which fluctuate more in women. E2 may increase activation in regions supporting emotional regulation during extinction and consolidation processes, helping women in higher E2 phases of their cycle achieve similar levels of extinction as seen in men.

Gonadal hormones may also play a large role in sex differences observed among men and women with anxiety disorders. Women tend to be diagnosed with anxiety and stress-related disorders more than men, suggesting greater vulnerability in this population. Sex hormones such as progesterone, estradiol, and testosterone are thought to be protective. Men with higher testosterone predeployment were less likely to be diagnosed with PTSD, and women with anxiety disorders experience higher symptoms during early luteal and follicular phases of their cycles when estradiol and progesterone levels are low. Sex hormones may also influence cognitive processes and brain metabolites, which additionally impact anxiety disorder prevalence and symptoms. In structural brain studies, it has been shown in some cases that men with PTSD may have lower gray matter volume than women with PTSD. However, a majority of findings suggest that brain structure is not a key factor in sexually dimorphic presentations of anxiety disorders. Functional findings suggest that women with anxiety disorders exhibit hyperactivity in emotional response regions like the amygdala and show reduced functional connectivity between the

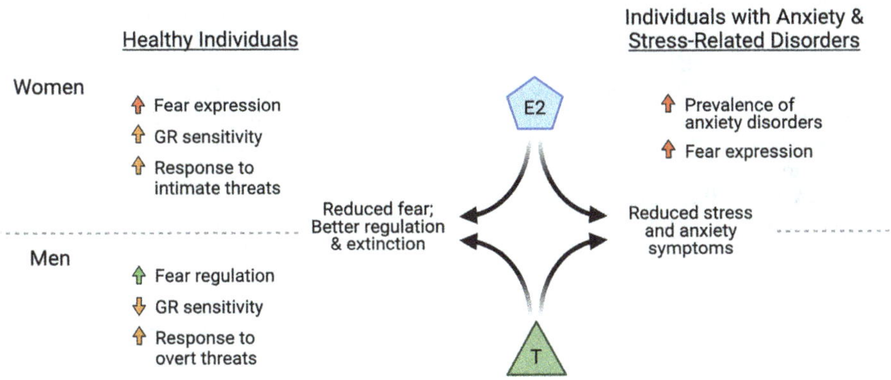

Fig. 3 Conceptual summary of sex differences in human fear and anxiety. E2 = estradiol, T = testosterone

amygdala and top-down regulatory regions in the prefrontal cortex as compared to men. See Fig. 3 for a summary of the findings.

While the literature thus far suggests that men and women exhibit interesting differences in the neurobiology of fear and anxiety, more work needs to be done to fully parse out what factors may influence these differences. First, neuroimaging studies evaluating healthy populations should consider further exploring more nuanced aspects of human threat in neuroimaging paradigms by using a wider range of stimuli such as combined visual and auditory tasks. Also, further work is needed to test whether men have lower GR sensitivity and dampened negative feedback on the HPA axis, relative to women. In addition to continuing to take into account cycle phase and measured estradiol levels, studies might evaluate other sex hormones such as progesterone or using stress tasks outside of fear conditioning to understand how cycling hormones may mechanistically impact brain function in women. It is likewise important that these future studies include men when possible to compare important hormonal sex differences in fear and anxiety. In clinical studies, it would be beneficial to investigate alternative sex hormones like testosterone when evaluating sex differences in brain metabolites. More structural neuroimaging studies are also needed to solve discrepancies in the literature evaluating structural brain differences among women and men with anxiety disorders. Lastly, functional neuroimaging studies in clinical populations need larger, more sexually and racially diverse sample sizes to fully understand what factors might impact the neurocircuitry of anxiety disorders.

References

Aggarwal N, Tromp DPM, Blackford JU et al (2024) Sex-specific distributed white matter microarchitectural alterations in preadolescent youths with anxiety disorders: a mega-analytic study. AJP 181:299–309. https://doi.org/10.1176/appi.ajp.20221048

Alblooshi S, Taylor M, Gill N (2023) Does menopause elevate the risk for developing depression and anxiety? Results from a systematic review. Australas Psychiatry 31:165–173. https://doi.org/10.1177/10398562231165439

Altemus M, Sarvaiya N, Epperson CN (2014) Sex differences in anxiety and depression clinical perspectives. Front Neuroendocrinol 35:320–330. https://doi.org/10.1016/j.yfrne.2014.05.004

Antov MI, Stockhorst U (2014) Stress exposure prior to fear acquisition interacts with estradiol status to alter recall of fear extinction in humans. Psychoneuroendocrinology 49:106–118. https://doi.org/10.1016/j.psyneuen.2014.06.022

Anyan F, Hjemdal O (2018) Stress of home life and gender role socializations, family cohesion, and symptoms of anxiety and depression. Women Health 58:548–564. https://doi.org/10.1080/03630242.2017.1316343

Asami T, Yamasue H, Hayano F et al (2009) Sexually dimorphic gray matter volume reduction in patients with panic disorder. Psychiatry Res Neuroimaging 173:128–134. https://doi.org/10.1016/j.pscychresns.2008.10.004

Bangasser DA, Cuarenta A (2021) Sex differences in anxiety and depression: circuits and mechanisms. Nat Rev Neurosci 22:674–684. https://doi.org/10.1038/s41583-021-00513-0

Barth C, Villringer A, Sacher J (2015) Sex hormones affect neurotransmitters and shape the adult female brain during hormonal transition periods. Front Neurosci 9. https://doi.org/10.3389/fnins.2015.00037

Beesdo K, Knappe S, Pine DS (2009) Anxiety and anxiety disorders in children and adolescents: developmental issues and implications for DSM-V. Psychiatr Clin North Am 32:483–524. https://doi.org/10.1016/j.psc.2009.06.002

Bellis MDD, Hooper SR, Chen SD et al (2015) Posterior structural brain volumes differ in maltreated youth with and without chronic posttraumatic stress disorder. Dev Psychopathol 27:1555–1576. https://doi.org/10.1017/S0954579415000942

Bierwirth P, Sperl MFJ, Antov MI, Stockhorst U (2021) Prefrontal theta oscillations are modulated by estradiol status during fear recall and extinction recall. Biol Psychiatry Cogn Neurosci Neuroimaging 6:1071–1080. https://doi.org/10.1016/j.bpsc.2021.02.011

Boehlen FH, Herzog W, Schellberg D et al (2020) Gender-specific predictors of generalized anxiety disorder symptoms in older adults: results of a large population-based study. J Affect Disord 262:174–181. https://doi.org/10.1016/j.jad.2019.10.025

Borst B, Jovanovic T, House SL et al (2024) Sex differences in response inhibition-related neural predictors of PTSD in recent trauma-exposed civilians. Biol Psychiatry Cogn Neurosci Neuroimaging:S2451902224000806. https://doi.org/10.1016/j.bpsc.2024.03.002

Brambilla DJ, Matsumoto AM, Araujo AB, McKinlay JB (2009) The effect of diurnal variation on clinical measurement of serum testosterone and other sex hormone levels in men. J Clin Endocrinol Metab 94(3):907–913. https://doi.org/10.1210/jc.2008-1902

Bremner WJ, Vitiello MV, Prinz PN (1983) Loss of circadian rhythmicity in blood testosterone levels with aging in normal men. J Clin Endocrinol Metab 56(6):1278–1281. https://doi.org/10.1210/jcem-56-6-1278

Cerasa A, Quattrone A, Piras F et al (2014) 5-HTTLPR, anxiety and gender interaction moderates right amygdala volume in healthy subjects. Soc Cogn Affect Neurosci 9:1537–1545. https://doi.org/10.1093/scan/nst144

Dark HE, Harnett NG, Hurst DR et al (2022) Sex-related differences in violence exposure, neural reactivity to threat, and mental health. Neuropsychopharmacology 47:2221–2229. https://doi.org/10.1038/s41386-022-01430-1

Domschke K, Baune BT, Havlik L et al (2012) Catechol-O-methyltransferase gene variation: impact on amygdala response to aversive stimuli. Neuroimage 60:2222–2229. https://doi.org/10.1016/j.neuroimage.2012.02.039

Dong D, Ironside M, Belleau EL et al (2022) Sex-specific neural responses to acute psychosocial stress in depression. Transl Psychiatry 12:2. https://doi.org/10.1038/s41398-021-01768-y

Dubol M, Epperson CN, Sacher J et al (2021) Neuroimaging the menstrual cycle: a multimodal systematic review. Front Neuroendocrinol 60:100878. https://doi.org/10.1016/j.yfrne.2020.100878

Eisenlohr-Moul TA, DeWall CN, Girdler SS, Segerstrom SC (2015) Ovarian hormones and borderline personality disorder features: preliminary evidence for interactive effects of estradiol and progesterone. Biol Psychol 109:37–52. https://doi.org/10.1016/j.biopsycho.2015.03.016

Faravelli C, Alessandra Scarpato M, Castellini G, Lo Sauro C (2013) Gender differences in depression and anxiety: the role of age. Psychiatry Res 210:1301–1303. https://doi.org/10.1016/j.psychres.2013.09.027

Felmingham K, Williams LM, Kemp AH et al (2010) Neural responses to masked fear faces: sex differences and trauma exposure in posttraumatic stress disorder. J Abnorm Psychol 119:241–247. https://doi.org/10.1037/a0017551

Filkowski MM, Olsen RM, Duda B et al (2017) Sex differences in emotional perception: meta analysis of divergent activation. Neuroimage 147:925–933. https://doi.org/10.1016/j.neuroimage.2016.12.016

Fonzo GA, Simmons AN, Thorp SR et al (2010) Exaggerated and disconnected insular-amygdalar blood oxygenation level-dependent response to threat-related emotional faces in women with intimate-partner violence posttraumatic stress disorder. Biol Psychiatry 68:433–441. https://doi.org/10.1016/j.biopsych.2010.04.028

Giles DE, Berga SL (1993) Cognitive and psychiatric correlates of functional hypothalamic amenorrhea: a controlled comparison*. Fertil Steril 60:486–492. https://doi.org/10.1016/S0015-0282(16)56165-2

Gillies GE, McArthur S (2010) Estrogen actions in the brain and the basis for differential action in men and women: a case for sex-specific medicines. Pharmacol Rev 62:155–198. https://doi.org/10.1124/pr.109.002071

Goldstein JM, Jerram M, Abbs B et al (2010) Sex differences in stress response circuitry activation dependent on female hormonal cycle. J Neurosci 30:431–438. https://doi.org/10.1523/JNEUROSCI.3021-09.2010

Gordon JL, Rubinow DR, Eisenlohr-Moul TA et al (2016) Estradiol variability, stressful life events and the emergence of depressive symptomatology during the menopause transition. Menopause 23:257. https://doi.org/10.1097/GME.0000000000000528

Graham BM, Ash C, Den ML (2017) High endogenous estradiol is associated with enhanced cognitive emotion regulation of physiological conditioned fear responses in women. Psychoneuroendocrinology 80:7–14. https://doi.org/10.1016/j.psyneuen.2017.02.023

Graham BM, Milad MR (2013) Blockade of estrogen by hormonal contraceptives impairs fear extinction in female rats and women. Biol Psychiatry 73:371–378. https://doi.org/10.1016/j.biopsych.2012.09.018

Gray PB, Straftis AA, Bird BM et al (2020) Human reproductive behavior, life history, and the challenge hypothesis: a 30-year review, retrospective and future directions. Horm Behav 123:104530. https://doi.org/10.1016/j.yhbeh.2019.04.017

Hakamata Y, Iwase M, Iwata H et al (2009) Gender difference in relationship between anxiety-related personality traits and cerebral brain glucose metabolism. Psychiatry Res Neuroimaging 173:206–211. https://doi.org/10.1016/j.pscychresns.2008.10.002

Handa RJ, Weiser MJ (2014) Gonadal steroid hormones and the hypothalamo–pituitary–adrenal axis. Front Neuroendocrinol 35:197–220. https://doi.org/10.1016/j.yfrne.2013.11.001

Harrewijn A, Cardinale EM, Groenewold NA et al (2021) Cortical and subcortical brain structure in generalized anxiety disorder: findings from 28 research sites in the ENIGMA-Anxiety Working Group. Transl Psychiatry 11:1–15. https://doi.org/10.1038/s41398-021-01622-1

Helpman L, Zhu X, Zilcha-Mano S et al (2021) Reversed patterns of resting state functional connectivity for females vs. males in posttraumatic stress disorder. Neurobiol Stress 15:100389. https://doi.org/10.1016/j.ynstr.2021.100389

Henze G-I, Konzok J, Kreuzpointner L et al (2021) Sex-specific interaction between cortisol and striato-limbic responses to psychosocial stress. Soc Cogn Affect Neurosci 16:972–984. https://doi.org/10.1093/scan/nsab062

Hermans EJ, Putman P, Baas JM et al (2006) A single administration of testosterone reduces fear-potentiated startle in humans. Biol Psychiatry 59:872–874. https://doi.org/10.1016/j.biopsych.2005.11.015

Hermans EJ, Putman P, Baas JM et al (2007) Exogenous testosterone attenuates the integrated central stress response in healthy young women. Psychoneuroendocrinology 32:1052–1061. https://doi.org/10.1016/j.psyneuen.2007.08.006

Heyn SA, Bailowitz S, Russell JD, Herringa RJ (2022) Sex-based variations of prefrontal structure and longitudinal symptoms in pediatric posttraumatic stress disorder. Depress Anxiety 39:902–912. https://doi.org/10.1002/da.23296

Hornung J, Noack H, Kogler L, Derntl B (2019) Exploring the fMRI based neural correlates of the dot probe task and its modulation by sex and body odor. Psychoneuroendocrinology 99:87–96. https://doi.org/10.1016/j.psyneuen.2018.08.036

Hubbard CS, Ornitz E, Gaspar JX et al (2011) Modulation of nociceptive and acoustic startle responses to an unpredictable threat in men and women. Pain 152:1632–1640. https://doi.org/10.1016/j.pain.2011.03.001

Hwang MJ, Zsido RG, Song H et al (2015) Contribution of estradiol levels and hormonal contraceptives to sex differences within the fear network during fear conditioning and extinction. BMC Psychiatry 15:295. https://doi.org/10.1186/s12888-015-0673-9

Im HY, Adams RB, Cushing CA et al (2018) S ex-related differences in behavioral and amygdalar responses to compound facial threat cues. Hum Brain Mapp 39:2725–2741. https://doi.org/10.1002/hbm.24035

Jalnapurkar I (2018) Sex differences in anxiety disorders: a review. PDA 4:1–9. https://doi.org/10.24966/PDA-0150/100011

Joel D (2021) Beyond the binary: rethinking sex and the brain. Neurosci Biobehav Rev 122:165–175. https://doi.org/10.1016/j.neubiorev.2020.11.018

Kaczkurkin AN, Moore TM, Ruparel K et al (2016) Elevated amygdala perfusion mediates developmental sex differences in trait anxiety. Biol Psychiatry 80:775–785. https://doi.org/10.1016/j.biopsych.2016.04.021

Kim H-J, Lee S-H, Pae C (2023) Gender differences in anxiety and depressive symptomatology determined by network analysis in panic disorder. J Affect Disord 337:94–103. https://doi.org/10.1016/j.jad.2023.05.087

Kinner VL, Wolf OT, Merz CJ (2018) Cortisol increases the return of fear by strengthening amygdala signaling in men. Psychoneuroendocrinology 91:79–85. https://doi.org/10.1016/j.psyneuen.2018.02.020

Kinney KL, Boffa JW, Amir N (2017) Gender difference in attentional bias toward negative and positive stimuli in generalized anxiety disorder. Behav Ther 48:277–284. https://doi.org/10.1016/j.beth.2016.06.002

Kogler L, Gur RC, Derntl B (2015) Sex differences in cognitive regulation of psychosocial achievement stress: brain and behavior. Hum Brain Mapp 36:1028–1042. https://doi.org/10.1002/hbm.22683

Kogler L, Müller VI, Seidel E-M et al (2016) Sex differences in the functional connectivity of the amygdalae in association with cortisol. Neuroimage 134:410–423. https://doi.org/10.1016/j.neuroimage.2016.03.064

Kogler L, Seidel E-M, Metzler H et al (2017) Impact of self-esteem and sex on stress reactions. Sci Rep 7:17210. https://doi.org/10.1038/s41598-017-17485-w

Kret ME, Pichon S, Grèzes J, Gelder BD (2011) Men fear other men most: gender specific brain activations in perceiving threat from dynamic faces and bodies – an fMRI study. Front Psychol 2. https://doi.org/10.3389/fpsyg.2011.00003

Kudielka BM, Buske-Kirschbaum A, Hellhammer DH, Kirschbaum C (2004) HPA axis responses to laboratory psychosocial stress in healthy elderly adults, younger adults, and children: impact of age and gender. Psychoneuroendocrinology 29:83–98. https://doi.org/10.1016/S0306-4530(02)00146-4

Kuhn L, Noack H, Wagels L et al (2023) Sex-dependent multimodal response profiles to psychosocial stress. Cereb Cortex 33:583–596. https://doi.org/10.1093/cercor/bhac086

Kundakovic M, Rocks D (2022) Sex hormone fluctuation and increased female risk for depression and anxiety disorders: from clinical evidence to molecular mechanisms. Front Neuroendocrinol 66:101010. https://doi.org/10.1016/j.yfrne.2022.101010

Lebron-Milad K, Abbs B, Milad MR et al (2012) Sex differences in the neurobiology of fear conditioning and extinction: a preliminary fMRI study of shared sex differences with stress-arousal circuitry. Biol Mood Anxiety Disord 2:7. https://doi.org/10.1186/2045-5380-2-7

Li G, Zhang S, Le TM et al (2020) Neural responses to negative facial emotions: sex differences in the correlates of individual anger and fear traits. Neuroimage 221:117171. https://doi.org/10.1016/j.neuroimage.2020.117171

Li SH, Graham BM (2017) Why are women so vulnerable to anxiety, trauma-related and stress-related disorders? The potential role of sex hormones. Lancet Psychiatry 4:73–82. https://doi.org/10.1016/S2215-0366(16)30358-3

Lonsdorf TB, Haaker J, Schümann D et al (2015) Sex differences in conditioned stimulus discrimination during context-dependent fear learning and its retrieval in humans: the role of biological sex, contraceptives and menstrual cycle phases. jpn 40:368–375. https://doi.org/10.1503/140336

MacLeod C, Mathews A (2012) Cognitive bias modification approaches to anxiety. Annu Rev Clin Psychol 8:189–217. https://doi.org/10.1146/annurev-clinpsy-032511-143052

Marcus MD, Loucks TL, Berga SL (2001) Psychological correlates of functional hypothalamic amenorrhea. Fertil Steril 76:310–316. https://doi.org/10.1016/S0015-0282(01)01921-5

McLean CP, Asnaani A, Litz BT, Hofmann SG (2011) Gender differences in anxiety disorders: prevalence, course of illness, comorbidity and burden of illness. J Psychiatr Res 45:1027–1035. https://doi.org/10.1016/j.jpsychires.2011.03.006

Meeker TJ, Veldhuijzen DS, Keaser ML et al (2020) Menstrual cycle variations in gray matter volume, white matter volume and functional connectivity: critical impact on parietal lobe. Front Neurosci 14

Merz CJ, Tabbert K, Schweckendiek J et al (2010) Investigating the impact of sex and cortisol on implicit fear conditioning with fMRI. Psychoneuroendocrinology 35:33–46. https://doi.org/10.1016/j.psyneuen.2009.07.009

Merz CJ, Tabbert K, Schweckendiek J et al (2012) Neuronal correlates of extinction learning are modulated by sex hormones. Soc Cogn Affect Neurosci 7:819–830. https://doi.org/10.1093/scan/nsr063

Merz CJ, Wolf OT, Schweckendiek J et al (2013) Stress differentially affects fear conditioning in men and women. Psychoneuroendocrinology 38:2529–2541. https://doi.org/10.1016/j.psyneuen.2013.05.015

Michopoulos V, Mancini F, Loucks TL, Berga SL (2013) Neuroendocrine recovery initiated by cognitive behavioral therapy in women with functional hypothalamic amenorrhea: a randomized, controlled trial. Fertil Steril 99:2084–2091.e1. https://doi.org/10.1016/j.fertnstert.2013.02.036

Milad MR, Goldstein JM, Orr SP et al (2006) Fear conditioning and extinction: influence of sex and menstrual cycle in healthy humans. Behav Neurosci 120:1196–1203. https://doi.org/10.1037/0735-7044.120.5.1196

Milad MR, Pitman RK, Ellis CB et al (2009) Neurobiological basis of failure to recall extinction memory in posttraumatic stress disorder. Biol Psychiatry 66:1075–1082. https://doi.org/10.1016/j.biopsych.2009.06.026

Milad MR, Zeidan MA, Contero A et al (2010) The influence of gonadal hormones on conditioned fear extinction in healthy humans. Neuroscience 168:652–658. https://doi.org/10.1016/j.neuroscience.2010.04.030

Miller DI, Halpern DF (2014) The new science of cognitive sex differences. Trends Cogn Sci 18: 37–45. https://doi.org/10.1016/j.tics.2013.10.011

Moon C-M, Jeong G-W (2017) Functional and morphological alterations associated with working memory dysfunction in patients with generalized anxiety disorder. Acta Radiol 58:344–352. https://doi.org/10.1177/0284185116649794

Nejati V, Khalaji S, Goodarzi H, Nitsche M (2021) The role of ventromedial and dorsolateral prefrontal cortex in attention and interpretation biases in individuals with general anxiety disorder (GAD): a tDCS study. J Psychiatr Res 144:269–277. https://doi.org/10.1016/j.jpsychires.2021.10.034

Nillni YI, Pineles SL, Patton SC et al (2015) Menstrual cycle effects on psychological symptoms in women with PTSD. J Trauma Stress 28:1–7. https://doi.org/10.1002/jts.21984

Oettel M, Mukhopadhyay AK (2004) Progesterone: the forgotten hormone in men? The aging male: the official journal of the International Society for the Study of the Aging Male 7(3):236–257. https://doi.org/10.1080/13685530400004199

Ohrmann P, Pedersen A, Braun M et al (2010) Effect of gender on processing threat-related stimuli in patients with panic disorder: sex does matter. Depress Anxiety 27:1034–1043. https://doi.org/10.1002/da.20721

Osinsky R, Karl C, Hewig J (2017) Dispositional anxiety and frontal-midline theta: on the modulatory influence of sex and situational threat. J Pers 85:300–312. https://doi.org/10.1111/jopy.12241

Pace-Schott EF, Spencer RMC, Vijayakumar S et al (2013) Extinction of conditioned fear is better learned and recalled in the morning than in the evening. J Psychiatr Res 47:1776–1784. https://doi.org/10.1016/j.jpsychires.2013.07.027

Pariante CM, Miller AH (2001) Glucocorticoid receptors in major depression: relevance to pathophysiology and treatment. Biol Psychiatry 49:391–404. https://doi.org/10.1016/S0006-3223(00)01088-X

Parikh TP, Stolze B, Ozarda Y, Jonklaas J, Welsh K, Masika L, Hill M, DeCherney A, Soldin SJ (2018) Diurnal variation of steroid hormones and their reference intervals using mass spectrometric analysis. Endocr connect 7(12):1354–1361. https://doi.org/10.1530/EC-18-0417

Paris JJ, Franco C, Sodano R et al (2010) Sex differences in salivary cortisol in response to acute stressors among healthy participants, in recreational or pathological gamblers, and in those with posttraumatic stress disorder. Horm Behav 57:35–45. https://doi.org/10.1016/j.yhbeh.2009.06.003

Pigoni A, Delvecchio G, Squarcina L et al (2020) Sex differences in brain metabolites in anxiety and mood disorders. Psychiatry Res Neuroimaging 305:111196. https://doi.org/10.1016/j.pscychresns.2020.111196

Pohlack ST, Nees F, Ruttorf M et al (2015) Neural mechanism of a sex-specific risk variant for posttraumatic stress disorder in the type I receptor of the pituitary adenylate cyclase activating polypeptide. Biol Psychiatry 78:840–847. https://doi.org/10.1016/j.biopsych.2014.12.018

Pöhlchen D, Fietz J, Czisch M et al (2023) Startle latency as a potential marker for amygdala-mediated hyperarousal. Biol Psychiatry Cogn Neurosci Neuroimaging 8:406–416. https://doi.org/10.1016/j.bpsc.2022.04.008

Poppelaars ES, Klackl J, Pletzer B et al (2019) Social-evaluative threat: stress response stages and influences of biological sex and neuroticism. Psychoneuroendocrinology 109:104378. https://doi.org/10.1016/j.psyneuen.2019.104378

Qiu Y, Fan Z, Zhong M et al (2022) Brain activation elicited by acute stress: an ALE meta-analysis. Neurosci Biobehav Rev 132:706–724. https://doi.org/10.1016/j.neubiorev.2021.11.020

Rattel JA, Wegerer M, Miedl SF et al (2019) Peritraumatic unconditioned and conditioned responding explains sex differences in intrusions after analogue trauma. Behav Res Ther 116: 19–29. https://doi.org/10.1016/j.brat.2019.01.009

Reijnen A, Geuze E, Vermetten E (2015) The effect of deployment to a combat zone on testosterone levels and the association with the development of posttraumatic stress symptoms: a longitudinal prospective Dutch military cohort study. Psychoneuroendocrinology 51:525–533. https://doi.org/10.1016/j.psyneuen.2014.07.017

Retana-Márquez S, Bonilla-Jaime H, Vázquez-Palacios G et al (2003) Changes in masculine sexual behavior, corticosterone and testosterone in response to acute and chronic stress in male rats. Horm Behav 44:327–337. https://doi.org/10.1016/j.yhbeh.2003.04.001

Rippon G, Jordan-Young R, Kaiser A, Fine C (2014) Recommendations for sex/gender neuroimaging research: key principles and implications for research design, analysis, and interpretation. Front Hum Neurosci 8. https://doi.org/10.3389/fnhum.2014.00650

Roeckner AR, Sogani S, Michopoulos V et al (2022) Sex-dependent risk factors for PTSD: a prospective structural MRI study. Neuropsychopharmacology 47:2213–2220. https://doi.org/10.1038/s41386-022-01452-9

Rosenbaum BL, Bui E, Marin M-F et al (2015) Demographic factors predict magnitude of conditioned fear. Int J Psychophysiol 98:59–64. https://doi.org/10.1016/j.ijpsycho.2015.06.010

Salminen LE, Tubi MA, Bright J et al (2022) Sex is a defining feature of neuroimaging phenotypes in major brain disorders. Hum Brain Mapp 43:500–542. https://doi.org/10.1002/hbm.25438

Schienle A, Schäfer A, Stark R et al (2005) Gender differences in the processing of disgust- and fear-inducing pictures: an fMRI study. Neuroreport 16:277–280. https://doi.org/10.1097/00001756-200502280-00015

Schirmer A, Wijaya M, Wu E, Penney TB (2019) Vocal threat enhances visual perception as a function of attention and sex. Soc Cogn Affect Neurosci 14:727–735. https://doi.org/10.1093/scan/nsz044

Seo D, Jia Z, Lacadie CM et al (2011) Sex differences in neural responses to stress and alcohol context cues. Hum Brain Mapp 32:1998–2013. https://doi.org/10.1002/hbm.21165

Shin LM, Liberzon I (2010) The neurocircuitry of fear, stress, and anxiety disorders. Neuropsychopharmacology: official publication of the American College of. Neuropsychopharmacology 35(1):169–191. https://doi.org/10.1038/npp.2009.83

Shvil E, Sullivan GM, Schafer S et al (2014) Sex differences in extinction recall in posttraumatic stress disorder: a pilot fMRI study. Neurobiol Learn Mem 113:101–108. https://doi.org/10.1016/j.nlm.2014.02.003

Stevens JS, Hamann S (2012) Sex differences in brain activation to emotional stimuli: a meta-analysis of neuroimaging studies. Neuropsychologia 50:1578–1593. https://doi.org/10.1016/j.neuropsychologia.2012.03.011

Stevens JS, Jovanovic T, Fani N et al (2013) Disrupted amygdala-prefrontal functional connectivity in civilian women with posttraumatic stress disorder. J Psychiatr Res 47:1469–1478. https://doi.org/10.1016/j.jpsychires.2013.05.031

Stricker R, Eberhart R, Chevailler MC, Quinn FA, Bischof P, Stricker R (2006) Establishment of detailed reference values for luteinizing hormone, follicle stimulating hormone, estradiol, and progesterone during different phases of the menstrual cycle on the Abbott ARCHITECT analyzer. Clin Chem Lab Med 44(7):883–887. https://doi.org/10.1515/CCLM.2006.160

Strine TW, Chapman DP, Kobau R, Balluz L (2005) Associations of self-reported anxiety symptoms with health-related quality of life and health behaviors. Soc Psychiatry Psychiatr Epidemiol 40:432–438. https://doi.org/10.1007/s00127-005-0914-1

Tromp DPM, Grupe DW, Oathes DJ et al (2012) Reduced structural connectivity of a major frontolimbic pathway in generalized anxiety disorder. Arch Gen Psychiatry 69:925–934. https://doi.org/10.1001/archgenpsychiatry.2011.2178

Van Wingen GA, Van Broekhoven F, Verkes RJ et al (2008) Progesterone selectively increases amygdala reactivity in women. Mol Psychiatry 13:325–333. https://doi.org/10.1038/sj.mp.4002030

Venkataraman A, Dias BG (2023) Expanding the canon: an inclusive neurobiology of thalamic and subthalamic fear circuits. Neuropharmacology 226:109380. https://doi.org/10.1016/j.neuropharm.2022.109380

Viau V, Meaney MJ (1996) The inhibitory effect of testosterone on hypothalamic-pituitary-adrenal responses to stress is mediated by the medial preoptic area. J Neurosci 16:1866–1876. https://doi.org/10.1523/JNEUROSCI.16-05-01866.1996

Wang X, Cheng B, Wang S et al (2021) Distinct grey matter volume alterations in adult patients with panic disorder and social anxiety disorder: a systematic review and voxel-based morphometry meta-analysis. J Affect Disord 281:805–823. https://doi.org/10.1016/j.jad.2020.11.057

Wen Z, Fried J, Pace-Schott EF et al (2022) Revisiting sex differences in the acquisition and extinction of threat conditioning in humans. Learn Mem 29:274–282. https://doi.org/10.1101/lm.053521.121

Wheelock MD, Goodman AM, Harnett NG et al (2021) Sex-related differences in stress reactivity and cingulum white matter. Neuroscience 459:118–128. https://doi.org/10.1016/j.neuroscience.2021.02.014

Yamasue H, Abe O, Suga M et al (2008) Gender-common and -specific neuroanatomical basis of human anxiety-related personality traits. Cereb Cortex 18:46–52. https://doi.org/10.1093/cercor/bhm030

Yang C, Lin C (2017) Magnetoencephalography study of different relationships among low- and high-frequency-band neural activities during the induction of peaceful and fearful audiovisual modalities among males and females. J Neurosci Res 95:176–188. https://doi.org/10.1002/jnr.23885

Yen J-Y, Lin P-C, Huang M-F et al (2020) Association between generalized anxiety disorder and premenstrual dysphoric disorder in a diagnostic interviewing study. Int J Environ Res Public Health 17:988. https://doi.org/10.3390/ijerph17030988

Zeidan MA, Igoe SA, Linnman C et al (2011) Estradiol modulates medial prefrontal cortex and amygdala activity during fear extinction in women and female rats. Biol Psychiatry 70:920–927. https://doi.org/10.1016/j.biopsych.2011.05.016

Part IV
Treatment

Neuromodulation of Fear and Anxiety Circuits

Joshua A. Brown, Kevin J. Clancy, and Wen Li

Contents

Abstract Anxiety is a complex and heterogeneous condition consisting of multiple component processes and neural underpinnings. Recent neural accounts for anxiety have expanded beyond the canonical anxiety and fear circuitry centered on the amygdala to a distributed network. A burgeoning form of clinical intervention with remarkable potential to directly engage the neural substrates of anxiety is non-invasive brain stimulation (NIBS), which is especially notable for the ability to modulate macro- and meso-scopic circuits and networks. NIBS is poised as a powerful research tool with the ability to extend basic research in animal models to humans, as well as develop novel mechanistic insights and therapeutic targets for underexplored elements of anxiety. This chapter provides an updated review of the neural anatomy of fear and anxiety and discusses the application of NIBS methodologies (primarily, transcranial magnetic, direct-current, and alternating-current

J. A. Brown (✉) and W. Li (✉)
Department of Psychiatry and Behavioral Sciences, University of Texas Health Center, Houston, TX, USA
e-mail: joshua.a.brown@uth.tmc.edu; wen.li.1@uth.tmc.edu

K. J. Clancy
Division of Depression and Anxiety Disorders, McLean Hospital, Belmont, MA, USA

Department of Psychiatry, Harvard Medical School, Boston, MA, USA

© The Author(s), under exclusive license to Springer Nature Switzerland AG 2025
Curr Topics Behav Neurosci (2025) 73: 407–431
https://doi.org/10.1007/7854_2024_573
Published Online: 12 March 2025

stimulation/transcranial magnetic stimulation, transcranial direct current stimulation, and transcranial alternating current stimulation) in testing, targeting, and normalizing the pathophysiology of fear and anxiety circuits. We conclude with a proposal of a unified approach for neuromodulation, promoting the synthesis of multiple neural circuits and systems involved in fear and anxiety.

Keywords Anxiety · Brain circuits · Brain stimulation (NIBS) · Fear · Large scale brain networks · Non-invasive · PTSD

1 Introduction

The growing prevalence of anxiety- and fear-related disorders has prompted an urgent need for the development of novel or improved clinical interventions. Existing evidence-based treatments, including pharmacotherapy and psychotherapy, demonstrate notable efficacy, and significant efforts are being made to address their limitations, such as increasing accessibility and scalability. Nonetheless, a significant portion of individuals do not benefit from existing treatments, and even in cases of clinical efficacy, there are high rates of relapse over time (Steenkamp et al. 2020). Such inconsistent response and elevated relapse rates thus call for new and innovative treatments to address the pressing clinical need. The mainline treatments for fear and anxiety – psychotherapy and pharmacotherapy – operate on behavioral and psychological processes and molecular and cellular functioning, respectively. Rapid advancements in non-invasive brain stimulation (NIBS) promise a new line of treatment for these disorders. Particularly, neuromodulatory interventions via transcranial magnetic stimulation (TMS) and transcranial electrical stimulation (tES) are filling a distinct and long-standing gap in clinical intervention by targeting meso- and macro-scopic circuitry and normalizing circuit and systems dysfunctions in these disorders (Fig. 1a; Leuchter et al. 2015; To et al. 2018; Deng et al. 2020). In this chapter, we will review the current use of neuromodulatory intervention (primarily via TMS and tES) in clinical and basic research of fear and anxiety. Given its mechanistic contribution to these disorders, we will particularly highlight the effect of neuromodulation on fear learning and related processes. We will end with a discussion of opportunities and challenges in this research and propose a unified approach for neuromodulation in future pursuits in both basic research and clinical intervention.

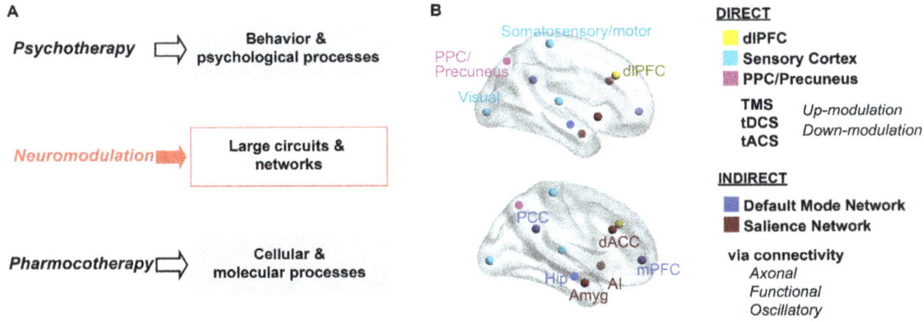

Fig. 1 Schematic of neuromodulatory intervention. (**a**) Neuromodulation targets large circuits and networks in the brain, bridging a long-standing gap between psychotherapy and pharmacotherapy that addresses behavior and psychological processes versus cellular and molecular processes, respectively. (**b**) An integrative approach for neuromodulation of fear and anxiety. TMS and tES can regulate the fear circuit by directly targeting dlPFC, sensory cortex, and the PPC/precuneus (Top). Also via these direct targets, TMS and tES can regulate the indirect targets, including hubs of the DMN and SN as well as the hippocampus (Bottom). These direct and indirect targets represent the multiple pathways involved in fear and anxiety. To date, these pathways have been modulated individually via NIBS. Future research is warranted to explore combinatory modulation of these interactive pathological pathways via multi-modal, multi-target NIBS, thereby achieving maximal treatment efficacy. Abbreviations: AI, anterior insula; Amyg, amygdala; dACC, dorsal anterior cingulate cortex; dlPFC, dorsolateral prefrontal cortex; Hip, hippocampus; PCC, posterior cingulate cortex; PPC, posterior parietal cortex; tACS, transcranial alternating current stimulation; tDCS, transcranial direct current stimulation; TMS; transcranial magnetic stimulation; vmPFC, ventromedial prefrontal cortex

2 Neural Circuits of Fear and Anxiety

Anxiety, characterized by heightened defensive responses and subjective feelings of fear and distress, is a heterogeneous construct and has multiple components that contribute to different phenotypes and clinical presentations. While each disorder has a unique constellation of observable phenotypes, including aberrant cognitive rumination and worry, exaggerated fear responses, avoidance, and pervasive generalization of learned threats, they all load heavily on two key constructs – fear and anxiety (LeDoux and Pine 2016).

The rich history of studies examining the neural basis of fear and anxiety is dominated by investigations into an amygdala–hippocampus–PFC circuit. This focus stems largely from fear conditioning research, which has presented the most influential experimental model for fear and anxiety (Maren 2001). While this canonical circuit remains essential in fear and anxiety, accruing, inter-disciplinary evidence has compelled an expansion to a distributed network to capture the multiple processes that are increasingly implicated in anxiety and fear (Pessoa and Adolphs 2010; Tovote et al. 2015). Moreover, with the discovery of large-scale intrinsic networks in the past decades, the conceptualization of fear and anxiety has also shifted to incorporate aberrations in major intrinsic networks (Menon 2011). These different circuitry systems have become targets of neuromodulatory intervention for fear and anxiety.

2.1 The Canonical Model: Amygdala–Hippocampus–PFC Circuit of Fear and Anxiety

Neurobiological models of threat processes have centered heavily on the amygdala, given its demonstrated role in acquiring conditioned (or learned) fear and initiating fear responses (LeDoux 2000; Maren 2001). In addition, the medial prefrontal cortex (mPFC) and hippocampus which are densely connected with the amygdala have also been included in this circuit. The (ventral) mPFC is known to be critical for fear extinction via "top-down" inhibitory regulation of amygdalar activity and the hippocampus is crucial for encoding the fear context. Accordingly, neurobiological models of anxiety and fear disorders have converged on the idea of deficient mPFC top-down inhibition of a hyperactive amygdala combined with impairments in the contextual-binding of learned fear due to dysfunction of the hippocampus (Milad et al. 2014; Liberzon and Abelson 2016; Fenster et al. 2018; Kessler et al. 2020). Notwithstanding its dominance in the conceptualization of anxiety and fear conceptualization and the tremendous research that it has thus motivated, this core amygdala–hippocampus–PFC circuit has been expanded upon substantially in recent years.

2.2 A Distributed Network of Fear and Anxiety

Though the amygdala has long been at the core of fear neurocircuitry, theories of a distributed fear network have gained increasing support (Headley et al. 2019; Josselyn and Tonegawa 2020; Li and Keil 2023). As demonstrated by recent meta- and mega-analyses of human neuroimaging studies, multiple brain regions contribute robustly to fear conditioning and extinction (Fullana et al. 2016, 2018; Visser et al. 2021; Wen et al. 2022). Within this distributed, complex fear network, the anterior insula (AI) and dorsal anterior cingulate cortex (dACC) are reliably identified in fear conditioning (Sehlmeyer et al. 2009). Accordingly, abnormal functioning of these areas, especially, hyperactivation and hyperconnectivity, has been observed in patients with fear and anxiety disorders, including posttraumatic stress disorder (PTSD) and social anxiety disorder (Hopper et al. 2007; Shin et al. 2011; Liberzon and Abelson 2016). For example, in PTSD, hyperactivation of the AI during anticipation of emotional stimuli is frequently identified (Lanius 2007; Lindauer et al. 2008; Aupperle et al. 2012). Furthermore, a recent meta-analysis of fear conditioning studies in healthy humans found that a wide network of cingulo-opercular regions form a generalization gradient, with greater activation being elicited with increasing similarity to the threat (Webler et al. 2021).

In addition, the cue-specific sensory cortex has received increasing recognition as a critical component of the network, particularly for its central role in the long-term storage of fear memory (Weinberger 2004; Concina et al. 2019; Li and Wilson 2024). Such fear memory traces in the sensory cortex (in the form of sensory cortical response pattern modification and response field adaptation, response disinhibition, response tuning shift, and so on) consequently serve as a neural representation of acquired fear, permitting fear encoding upon the encounter of the fear object (Li and Keil 2023). Accordingly, the sensory cortex would constitute a pathological substrate for anxiety and fear disorders as dysregulated bottom-up sensory afferents distort and inundate downstream regions including the amygdala and prefrontal cortex (PFC) (Clancy et al. 2017; Li 2019; Maron-Katz et al. 2020; Harnett et al. 2024). This conceptualization incorporating sensory–PFC–amygdala dysfunction can account for heightened or distorted sensory perception and higher rates of sensory complaints in fear and anxiety (McGaugh 2000; Olshausen and Field 2004; Hughes and Crunelli 2005; Keil et al. 2007; Sacco and Sacchetti 2010; Gdalyahu et al. 2012; Grosso et al. 2015; Cambiaghi et al. 2016; Meissner-Bernard et al. 2019; Ojala et al. 2022; Wen et al. 2022). Interestingly, a new meta-analysis has revealed pervasive, low-level sensory anomalies in multiple psychiatric disorders, in addition to fear and anxiety (van den Boogert et al. 2022), highlighting the importance of this hitherto overlooked sensory pathology in mental disorders.

2.3 Large-Scale Intrinsic Network Dysfunction in Fear and Anxiety

The conceptualization of fear and anxiety underwent another significant shift with the discovery of large-scale intrinsic networks in the 2010s (Menon 2011). Large-scale networks integrated across distributed structures have been reliably identified with resting-state functional magnetic resonance imaging (fMRI) and structural MRI in humans (Raichle 2015). Such intrinsic networks are recognized as the fundamental organizational architectures and operational units of the brain (Varela et al. 2001; Fox and Raichle 2007; Yuste 2015) and thus would play an important role in the pathophysiology of neuropsychiatric disorders. In line with that, a triple network model of psychopathology has gained strong traction by emphasizing the dysfunctions of three major higher-order large-scale networks that are centrally positioned in the human connectome – the default mode network (DMN), the salience network (SN), and the central executive network (CEN; Menon 2011; Margulies et al. 2016). Recently, the sensory network has also been recognized for its contribution to the maintenance of fear and anxiety disorders through interaction with those higher-order networks (Clancy et al. 2020a, b).

Salience Network

The most frequently implicated network in fear and anxiety is the SN, often interchangeably referred to as the cingulo-opercular network (Menon 2011). This network consists of two cortical hubs – AI (and adjacent operculum) and dACC (extending into the pre-supplementary motor area), in addition to the amygdala as a key subcortical node. The SN is believed to support bottom-up, stimulus-driven attentional capture and detection of biologically salient signals. It is also posited as a switchboard between other networks, particularly switching from internally-oriented, self-referential DMN to the externally-oriented, goal-directed CEN. Accordingly, recent work has recategorized the SN as an "action-mode" network (Dosenbach et al. 2024). Evidence to date suggests hyperactivity of the SN in fear and anxiety (Koch et al. 2016; Yin et al. 2018), which is linked to phenotypes of exaggerated (i.e., overgeneralized) detection of salience (i.e., threat) in the environment and resulting excessive arousal responses and defensive behaviors. As such, hyperactivity of the SN could disrupt the critical homeostatic balance in neural network dynamics and overwhelm the higher-order networks, ultimately contributing to their dysfunction and perpetuating a state of vigilant, anxious arousal that characterizes fear and anxiety disorders. These perturbations have been identified in PTSD even during rest, including hyperconnectivity within the SN and between the SN and DMN (Sripada et al. 2012; Nicholson et al. 2020).

Default Mode Network

The DMN is characteristic of introspection and waking rest, having been repeatedly identified in task-absent research (Raichle et al. 2001; Raichle 2015). Its primary hubs include the posterior cingulate cortex (PCC), mPFC, precuneus, and angular gyrus (AG). Aberrations in the DMN have been observed in several disorders but particularly PTSD, which has repeatedly been associated with hypoconnectivity and hypoactivity of the DMN (Bluhm et al. 2009; Lanius et al. 2010; Sripada et al. 2012; Koch et al. 2016; Miller et al. 2017a, b; Akiki et al. 2018; Clancy et al. 2020a, b). Clinically, this dysfunction contributes to an impaired ability to maintain internally-focused cognition and appropriately filter out external stimuli, fundamental symptoms of PTSD. Dysfunction of the DMN has also been identified in panic disorder (Lai and Wu 2014) and GAD (Yuan et al. 2023), although hyperconnectivity between DMN nodes during passive observation of images has been reported in GAD as well (Clancy et al. 2020a, b).

Sensory Network

Sensory networks (e.g., visual, somatomotor networks) are composed of basic (primary and secondary) sensory cortices, representing the peripheral components of the human connectome (Yeo et al. 2011; Margulies et al. 2016). It is posited that intrinsic disinhibition of the basic sensory cortices would underpin the pathology of fear and anxiety disorders (Clancy et al. 2017, 2020a, b). This disinhibition can manifest as impaired sensory gating and heightened sensory sensitivity at the neurophysiological (Morgan and Grillon 1999; Neylan et al. 1999; Skinner et al. 1999; Lewine et al. 2002; Javanbakht et al. 2011; Krusemark and Li 2012), behavioral (Hasson et al. 2013; Krusemark et al. 2013; Clancy et al. 2020a, b), and self-report levels (Stewart and White 2008; Engel-Yeger et al. 2013). Electrophysiologically, patients with PTSD demonstrate robust deficits in a key pattern of neural activity – alpha-frequency (8–12 Hz) oscillations (Kluetsch et al. 2014; Clancy et al. 2017, 2020a, b; du Bois et al. 2021). Similar reductions in alpha activity are seen in states of anxious arousal and fear-related disorders, albeit with no or opposite effects seen in generalized anxiety disorder (Eidelman-Rothman et al. 2016; Clancy et al. 2017). Furthermore, alpha deficit in PTSD can also impact feedforward inhibitory influence from the visual cortex to higher-order cortical networks – the DMN and SN (Clancy et al. 2020a, b), resulting in heightened connectivity of the sensory cortex with and thereby dysfunction of these networks (Breukelaar et al. 2021). Taken together, it is plausible that such sensory disinhibition sets the stage for bottom-up propagations of sensory signals across the cortical hierarchy, mirroring clinical accounts of patients with PTSD who report feeling "flooded" by everyday sensory stimuli (Stewart and White 2008; Engel-Yeger et al. 2013). The intrinsic nature of this disinhibition, present even in states of rest, further reflects the spontaneous sensory-based reliving of intrusive trauma memories – a core symptom of PTSD (Clancy et al. 2020a, b). Such disruptions are well-positioned to underpin the multifaceted features of anxiety and fear-related pathologies, presenting a promising, novel interventional target for circuit-based therapeutics.

3 Neuromodulation of Fear and Anxiety Circuits

The extant models of fear and anxiety, from the canonical fear circuit to the distributed network, and then the large-scale intrinsic networks, have identified key neural targets that could be attacked through neuromodulatory intervention. Recent advancements in non-invasive neuromodulation techniques are especially notable for progress in TMS and tES. Both methods are capable of tuning neural circuitry (Brown et al. 2024; Siddiqi and Fox 2024), enabling mechanistic investigation and manipulation of fear and anxiety circuits.

TMS applies magnetic pulses over the scalp to penetrate the skull with limited resistance or dispersion (Kim et al. 2009), inducing a relatively focal electrical current within a targeted brain region (Terao and Ugawa 2002). Magnetic pulses are often repeated, with high-frequency stimulation (>5 Hz) producing excitatory effects and low-frequency (e.g., 1 Hz) producing inhibitory effects (Klomjai et al. 2015). Its demonstrated clinical efficacy and advancing utility in mechanistic research make TMS a cornerstone in the neuromodulation of psychiatric disorders.

tES involves the administration of an electrical current via electrodes placed on the scalp to target brain regions or networks of interest. The most common forms of tES used in neuropsychiatric research consist of transcranial direct current stimulation (tDCS; with a constant current) and transcranial alternating current stimulation (tACS; with an alternating current) (Philip et al. 2017; Yavari et al. 2018). tDCS generally produces subthreshold membrane potential changes, specifically, long-term potentiation under an anode electrode and long-term depression under a cathode electrode, resulting in neural excitation and inhibition, respectively. tACS, a newer form of tES, applies a current in a sinusoidal pattern mimicking endogenous neural oscillations to entrain oscillations and synchronize neural networks. Notably, tACS uniquely tunes the phase of electrical oscillations, making it especially advantageous in modulating neural circuits.

3.1 Neuromodulation of the Fear Circuit: Fear Learning, Memory, and Extinction

Fear learning and memory are closely associated with fear and anxiety disorders (Davis 1992; Ressler 2020; Bienvenu et al. 2021). As discussed above, extensive research has been conducted on fear learning and memory, primarily using fear conditioning paradigms (where aversive unconditioned stimulus/US is paired with a neutral conditioned stimulus/CS). In recent years, neuromodulatory techniques (including TMS, tDCS, and tACS) have been used to modulate fear learning and memory. This literature, primarily from well-controlled laboratory settings, can provide models for neuromodulatory-based therapeutics for fear and anxiety disorders and their mechanisms of action. As the canonical circuit of fear and anxiety, the amygdala–hippocampus–PFC circuit has been the focus of neuromodulation. In addition, fear learning and memory involve several stages serving distinct functions, including fear learning (i.e., CS–US association acquisition), memory consolidation and reconsolidation, and fear extinction, all of which have been targeted by neuromodulation.

3.1.1 Targeting Fear Learning

Rodent studies have applied tDCS to the PFC during fear learning, but its effects are not straightforward to produce a useful animal model for this manipulation. Applied immediately prior to fear conditioning, with the effect likely persisting into the fear conditioning stage (Kasten et al. 2016), either anodal or cathodal stimulation of the lateral PFC mitigated fear recall (Abbasi et al. 2017). However, in other studies, when applied 24 h prior to learning, both anodal and cathodal tDCS of the lateral PFC enhanced fear recall (Manteghi et al. 2017; Nasehi et al. 2017a, b). These two sets of studies suggest that tDCS of PFC can cause immediate disruption but delayed enhancement in fear learning. These opposing patterns are difficult to interpret. Moreover, that anodal and cathodal tDCS, expected to produce opposite (excitatory or inhibitory, respectively) effects, caused similar changes in fear learning is also puzzling.

In human research, continuous theta-burst TMS (cTBS; presumably inducing neural inhibition) has been applied during the fear learning stage, and the results are somewhat consistent. One study applied cTBS to the somatosensory cortex (corresponding to the fingers exposed to the tactile CS) and found that it reduced conditioned fear response (indexed by skin conductance and pupil responses) during fear recall on the next day (Ojala et al. 2022). To note, this stimulation did not affect fear learning itself, implicating the effect of cTBS in subsequent processes. Another cTBS study was conducted among participants with PTSD symptoms (Webler et al. 2024). The researchers took a connectivity-based approach, where an fMRI scan was conducted to identify the stimulation site in the superficial posterior parietal cortex that has strong functional connectivity with the hippocampus (Hebscher et al. 2021). Immediately before visual fear conditioning, cTBS was applied to the target in the left posterior parietal cortex to inhibit this area and consequently the hippocampus. The study showed a reduction in fear generalization among patients with less fear sensitization but also failed to demonstrate clear effects on fear learning. Overall, extant findings in humans suggest that neuromodulation during fear learning can impact subsequent fear processes while preserving fear learning.

3.1.2 Non-Invasive Brain Stimulation Targeting Fear Memory Consolidation

Following fear learning, consolidation begins, consolidating the newly acquired, labile, and malleable association into a lasting one. As such, neuromodulation that disrupts this consolidation process would have important clinical implications by stifling the development of pathological fear and related symptoms. Indeed, several studies have applied neuromodulation of the PFC to disrupt fear memory consolidation.

An earlier study deployed cathodal (inhibitory), anodal (excitatory), or sham tDCS to the left dlPFC after fear learning (Asthana et al. 2013). The researchers found that at recall testing the next day, the anodal and sham stimulation groups showed conditioned fear responses, which nonetheless were absent in the cathodal stimulation group. This attenuation of fear memory by inhibitory neuromodulation of the dlPFC was reproduced in a new study (Battaglia et al. 2024). This study applied inhibitory low-frequency (1 Hz) repetitive TMS (rTMS) to the left dlPFC immediately after fear conditioning or after a delay of 6 h and showed that rTMS immediately after conditioning resulted in a reduction in fear recall when tested immediately, 1 h, or 1 day later. By contrast, the same stimulation administered not 6 h after learning did not produce any effects, suggesting the importance of targeting the early consolidation window. Another new study applied anodal tDCS targeting the vmPFC immediately after conditioning (Boehme et al. 2024). A similar disruption of fear memory was observed: conditioned fear response during fear recall on the next day was absent in the experimental group (while present in the sham group).

Therefore, research has produced consistent evidence to support the efficacy of PFC neuromodulation during the stage of fear memory consolidation. Notably, inhibitory stimulation of the dlPFC and excitatory stimulation of the vmPFC produced a similar reduction in fear recall, suggesting a function-structure double dissociation between the two stimulation protocols. Specifically, the dlPFC is known to figure importantly in memory retrieval (Moscovitch 1992; Cabeza and Nyberg 2000; Eichenbaum 2017) such that the inhibitory dlPFC stimulation may impede fear memory recall. The vmPFC plays a key role in downregulating the amygdala, the key structure for fear memory consolidation (Milad et al. 2004; Phelps and LeDoux 2005; Milad et al. 2006), and so excitatory stimulation of vmPFC can enhance amygdala downregulation, thereby impairing fear memory consolidation. The possibility of multiple processes at play here is encouraging, offering multiple pathways for NIBS to hinder the development of fear memory.

3.1.3 Non-Invasive Brain Stimulation Targeting Fear Memory Reconsolidation

A consolidated, stable memory may become labile and malleable when reactivated, as in memory recall, and then shortly thereafter begin to re-consolidate to restore its stability (Lee 2009). This reconsolidation stage presents a window during which a labile memory can be altered or erased, resulting in reduced or eliminated fear recall (Schiller et al. 2008; Schiller and Phelps 2011; Borgomaneri et al. 2021). Along this line, neuromodulation of the fear circuit during fear memory reconsolidation can steer fear memory to a desirable fate. Indeed, this reconsolidation stage has been leveraged by NIBS research: similar to neuromodulation during fear memory consolidation, TMS or tES is applied to the dlPFC (or, to some extent, the vmPFC) shortly after a CS reminder that reactivates fear memory.

One study applied low-frequency (inhibitory) rTMS to the dlPFC during reconsolidation (shortly after a CS reminder acquired the day before) and measured conditioned fear response (skin conductance response to CS+ vs. CS−) after extinction learning and after re-conditioning on the following day (Borgomaneri et al. 2020). The researchers found that this rTMS facilitated the reduction of conditioned fear response through extinction learning and prevented its return after reconditioning. By contrast, neither effect was observed for control rTMS. Importantly, these effects were not found when the recall test was conducted immediately after dlPFC-rTMS, namely, before memory reconsolidation had the time to transpire. Neither were they observed when participants received the dlPFC rTMS without the CS reminder, namely, without turning on the reconsolidation stage. These findings thus rigorously evince the efficacy of neuromodulation targeting fear reconsolidation. As an independent replication, a newer study applied inhibitory cTBS of the right dlPFC after a CS reminder and reduced fear recall and reinstatement (Su et al. 2022). Notably, this effect was demonstrated when the CS reminder and ensuing dlPFC cTBS were administered 2 weeks after fear conditioning. Furthermore, such effects were absent when the stimulation was applied 6 h after the reminder (i.e., beyond the reconsolidation window), highlighting the efficacy of neuromodulation of the fear circuit during the reconsolidation stage.

An earlier study applied tDCS in the opposite direction (i.e., enhancing the fear circuit during reconsolidation) and demonstrated the opposite effects accordingly. Specifically, researchers employed concurrent anodal tDCS of the right dlPFC with cathodal tDCS of the vPFC after fear reactivation a day after conditioning and demonstrated greater conditioned fear response in the active (vs. sham control) group at the fear recall the following day (Mungee et al. 2014). Therefore, the combination of dlPFC excitation and vPFC inhibition could intensify fear memory reconsolidation, thereby potentiating fear recall. By this token, the reversal of this montage would be expected to lead to the attenuation of fear recall, which was indeed tested in a follow-up study by the same group (Mungee et al. 2016). However, this stimulation protocol did not reduce fear recall, suggesting that the mechanism of action is not straightforward.

3.1.4 Non-Invasive Brain Stimulation Targeting Fear Extinction

Fear extinction involves the presentation of a CS without the pairing with a US, forming a new contingency that counteracts the CS-US association acquired from fear conditioning. Fear extinction thus presents an important opportunity to inhibit fear response and ultimately alleviate symptoms of fear and anxiety disorders. Accordingly, a good number of studies have applied NIBS immediately before or during extinction learning to modulate the fear circuit, thereby promoting fear extinction. The vPFC, a key structure underlying extinction learning, has been the main target of this research, through direct stimulation of this region or indirectly, via the stimulation of the dlPFC.

Clear evidence arose from three tDCS studies using anodal stimulation of the vmPFC during (Dittert et al. 2018; Vicario et al. 2020) or immediately before extinction learning (Boehme et al. 2024). The stimulation (vs. sham control) facilitated fear extinction, manifested as faster reduction (Dittert et al. 2018; Vicario et al. 2020) or elimination of conditioned fear response (Boehme et al. 2024). However, as mentioned earlier, given the deep location of the vmPFC, caution should be taken in interpreting the precise targeting of the region with these protocols.

Taking a connectivity-based approach to effectively target the vmPFC, a group of researchers identified a region of the dlPFC that was strongly connected to the vmPFC and applied excitatory 20 Hz rTMS during extinction learning, time-locked with the presentation of the CS (Raij et al. 2018). They found that the stimulation resulted in a strong reduction in conditioned fear response at fear recall the next day. In comparison, applying the 20 Hz rTMS at another dlPFC target that was weakly connected to the vmPFC, the researchers showed no effect on fear recall, highlighting the crucial role of vmPFC in fear extinction.

However, mixed effects have also been reported in this literature. One study applied anodal tDCS or 1 Hz tACS targeting the mPFC (with the anodal electrode placed at the middle of the forehead) and found that the tDCS (but not tACS) stimulation reduced fear response during fear recall next day (Abend et al. 2016). Notably, they also observed that both tDCS and tACS (vs. sham control) exacerbated subjective fear ratings of the CS+ during the fear recall, suggesting a dissociable effect of such neuromodulation on implicit fear memory versus explicit cognitive appraisal of the CS. In another set of studies, researchers applied anodal tDCS at the left frontal area to target the vmPFC during extinction and demonstrated no effect of stimulation on fear recall next day among healthy individuals (van't Wout et al. 2016) and a (non-significant) trend of fear reduction among patients with PTSD (Van't Wout et al. 2017). This set of findings does not negate the efficacy of neuromodulation of the fear circuit during extinction but rather cautions against the robustness of its effect.

Interestingly, some studies applied cathodal tDCS to the dlPFC during extinction learning, presumably inhibiting the dlPFC and thus weakening extinction. Results indeed indicated an increase in conditioned fear response at fear recall 1–3 months later (Ganho-Avila et al. 2019). While clinically undesirable, such stimulation may provide useful insight into the mechanism of action. Specifically, in a follow-up study with the same tDCS protocol (Ganho-Avila et al. 2022), fMRI recorded during extinction learning showed decreased activation of left PFC to the CS+ as well as increased connectivity between the dlPFC and a broad array of structures relevant to fear processing, including the amygdala and hippocampus (Ganho-Avila et al. 2022). These results evinced the modulation of the fear extinction circuit via dlPFC stimulation, laying the mechanistic foundation for such neuromodulation.

3.1.5 Interim Summary

In summary, investigations of neuromodulation of the fear circuit during distinct stages are by and large supportive of its efficacy. It is also worth noting that mixed or conflicting results exist in this literature. The precision in targeting, especially the deep structures, is a main challenge, and the lack thereof could contribute to the inconsistencies across studies. This issue has prompted a comparison of methodologies in recent reviews (Borgomaneri et al. 2021; Markovic et al. 2021) and warrants further research before turning such neuromodulatory techniques into therapeutics in clinical intervention. In that regard, neuroimaging acquisition that can validate target engagement (e.g., increased fMRI signal in the dlPFC consequent to anodal tDCS of the dlPFC) would be an important step forward in that direction. Finally, the literature is largely constrained to the canonical fear circuit, leaving a large territory and many opportunities unexplored. Notably, neuromodulation of the sensory cortex, a key structure for fear memory storage, has been attempted in only one study (Ojala et al. 2022). Nonetheless, the positive outcome of this study promises yield for future neuromodulation in this area.

Neuromodulation in Anxiety Interventions

The exciting potential for NIBS techniques to treat fear and anxiety disorders has prompted a significant number of studies and clinical trials. This literature, consisting largely of controlled, blinded studies with clinical populations, has focused on the PFC, including the mPFC (Herrmann et al. 2017; Van't Wout et al. 2017) and dlPFC (Deppermann et al. 2017; Ahmadizadeh and Rezaei 2018; Philip et al. 2019). By far, the dlPFC has been the primary target of neurostimulation in these clinical trials, paralleling interventions for depression (George et al. 2000). Clinical outcomes reported by the NIBS literature have been discussed in several recent meta-analyses, which broadly come to compatible conclusions: inhibition of the dlPFC over the course of several weeks can provide symptom relief for GAD and PTSD (Cirillo et al. 2019; Sagliano et al. 2019; Gay et al. 2022; Sa et al. 2023) as well as OCD (Sa et al. 2023). Such protocols are capable of reducing anxiety and depression symptoms, as measured by standardized questionnaires such as the Hamilton Anxiety Rating Scale (Diefenbach et al. 2016) as well as PTSD symptom profiles (Kozel et al. 2018). Other research has deployed excitation of the vmPFC as a conjunctive to virtual reality exposure therapy to promote safety learning in patients with PTSD (Van't Wout et al. 2017) or specific phobias (Herrmann et al. 2017), which successfully reduced PTSD symptoms following multiple visits as well as state anxiety during a single visit, respectively. Although the outcomes are promising, the mechanisms by which stimulation of the PFC, especially the dorsolateral portion, induces these benefits have been debated since the dawn of NIBS therapies and warrant further discussion.

3.1.6 Mechanisms of Neuromodulation Treatments

According to the theoretical conceptualization of anxiety that is centered on the hippocampus–amygdala–PFC fear circuit, clinical interventions have attempted to upregulate the mPFC to alleviate symptoms. Indeed, excitation of the mPFC may enhance exposure therapy by promoting fear extinction, allowing for a greater sense of calmness in the face of triggering stimuli (Herrmann et al. 2017; Van't Wout et al. 2017). Although grounded in a sound biological basis, stimulation of the vmPFC is difficult due to its deep locus and proximity to other fear regions, such as the dACC (Isserles et al. 2021). Accordingly, the more readily accessible dlPFC has been targeted by the large majority of NIBS clinical trials.

Additional models of anxiety (and depression) have characterized such disorders as rooted in a dysfunction of the dlPFC (Davidson and Irwin 1999; Etkin et al. 2009, Kolesar et al. 2019), often lateralized such that the left and right dlPFC contribute more to positive and negative emotional processing, respectively (Davidson and Irwin 1999; Sagliano et al. 2019). This lateralized model of anxiety, the hemispheric asymmetry model, has informed the neurostimulation literature, and many protocols apply excitatory stimulation to the left dlPFC (e.g., de Lima et al. 2019) or inhibitory stimulation to the right dlPFC (e.g., Kozel et al. 2018), in attempts to rectify an imbalance between the hemispheres and thereby reduce depression and anxiety. However, findings do not support significantly different effects of stimulation between hemispheres (Cirillo et al. 2019; Kan et al. 2020; Smits et al. 2020; Brown et al. 2024). In fact, excitatory stimulation of the right dlPFC has been associated with greater PTSD symptom improvement (Karsen et al. 2014; Philip et al. 2019). This body of work thus parallels inconsistent findings of both hyper- and hypo-activity of the right dlPFC in GAD (Kolesar et al. 2019), cautioning against a strictly lateralized model of dlPFC for anxiety.

Alternatively, stimulation of the dlPFC may cause symptom improvement via the structure's known role in cognitive and executive control functions. Extensive cognitive neuroscience research has investigated the dlPFC in cognitive functions, finding it robustly activated by tasks prompting decision-making (Krawczyk 2002), working memory (Curtis and D'Esposito 2003), and attention (Taylor and Fragopanagos 2005). Anodal tDCS targeting the dlPFC has produced cognitive improvements (Iyer et al. 2005; Wassermann and Grafman 2005; Fiori et al. 2011), prompting its extension to studies of anxiety, as impaired executive functioning is frequent in anxiety and depression (Snyder 2013). Accordingly, stimulation of the dlPFC may counteract executive impairments in clinical populations and thereby provide symptom relief. Within this cognitive framework, the role of the dlPFC in long-term memory may be responsible for symptom relief arising from its inhibition. As discussed in the above review of NIBS during fear reconsolidation, the dlPFC is activated when recalling fear memories and re-committing them to long-term storage. As PTSD is characterized by the inappropriate recall of fear memory, inhibition of the dlPFC may mitigate abnormal fear recall, presumably through interactions with the hippocampus.

Both inhibitory and excitatory TMS of the dlPFC appear to provide symptomatic relief, and an additional framework could explain this phenomenon. As discussed above, neurobiological models of fear and anxiety have increasingly incorporated intrinsic brain networks, whereby pathological fear and anxiety arise from disruption of the DMN and hyperactivation of the SN. The dlPFC is interconnected with many hubs of these large-scale networks, and symptomatic relief may potentially be mediated by large-scale networks (Opitz et al. 2016; Siddiqi et al. 2020; Siddiqi and Fox 2024). In fact, anterior and posterior divisions of the dlPFC may contribute differently to symptom relief based on their connectivity profiles with large-scale networks (Opitz et al. 2016; Siddiqi et al. 2020). For patients experiencing anxiety and depression symptoms, neuromodulatory inhibition of the dlPFC may help to quell dysphoric symptoms via indirect modulation of the SN and DMN (Siddiqi et al. 2020), in line with networks corresponding to symptom improvement in depression (Siddiqi and Fox 2024).

Although there are a significant number of studies using NIBS to modulate the dlPFC in clinical populations, few report neuroimaging data, leaving the underlying mechanisms obscure (Opitz et al. 2016). Moreover, the neurobiological motivations for targeting the dlPFC are not predominately centered on known fear and anxiety circuitry, often relying instead on clinical outcomes of previous trials. The dlPFC is a large, heterogeneous region with multiple functions (e.g., working memory and attention) and connections (e.g., CEN and AI). Therefore, work strictly within the frameworks of hemispheric asymmetry and cognitive control, without precise spatial targeting, may impede the advancement of NIBS treatments. New neural models of fear and anxiety have been developed, which present further opportunities for NIBS-based therapies.

3.2 *Opportunities in Neuromodulation Therapies for Anxiety*

3.2.1 Disrupting the Fear Circuit in Anxiety Disorders

A large body of work has identified specific impairments in fear learning in anxiety disorders. While neuromodulation of the fear circuit has been increasingly attempted in healthy populations during fear learning paradigms, there is a surprising lack of studies that target the fear circuit in clinical populations. The exception to this is the vmPFC, which is crucial for safety processing and amygdala regulation, and studies in clinical populations have shown promise as a conjunctive to exposure therapy. Most structures within the fear circuit are difficult to target in isolation, including the mPFC, due to the depth of these structures and the likelihood of inadvertent stimulation of nearby structures.

While the role of the sensory cortex in anxiety and fear-related processes has received growing recognition, the therapeutic targeting of the sensory cortex with NIBS techniques to treat anxiety and related disorders remains largely underexplored. Although limited evidence is currently available in humans, proof-of-concept studies have been conducted in healthy individuals with promising results that may extend to clinical populations. In particular, as discussed above, inhibition of the somatosensory cortex prior to fear learning impaired fear memory consolidation (Ojala et al. 2022). In addition, inhibition of the visual cortex was found to lessen the subjective impact of noxious stimuli (Clancy et al. 2018). Nevertheless, sensory regions remain absent from clinical NIBS trials, underscoring a gap in the literature between basic fear research and clinical applications.

3.2.2 Modulating Large-Scale Networks in Anxiety Disorders

As discussed above, neurobiological models of fear and anxiety have increasingly incorporated intrinsic brain networks, whereby pathological fear and anxiety arise from disruption of the DMN and hyperactivation of the SN. Targeting these networks with NIBS may treat the underlying causes of symptoms, and such interventions are attainable. Indeed, excitatory TMS of the dlPFC in PTSD has been capable of increasing within-network DMN connectivity and reducing DMN-SN and DMN-CEN coupling (Philip et al. 2019). Highlighting the utility of fear extinction research, this circuitry is consistent with a prior study that showed that dlPFC stimulation during the perception of learned threats promoted extinction learning, a process largely mediated by the vmPFC (Raij et al. 2018). As targeting of these networks becomes more precise, clinical presentation can inform TMS treatments, for anxiety symptom profiles will likely respond better to CEN suppression or DMN activation (Siddiqi et al. 2020; Brown et al. 2024; Siddiqi and Fox 2024). Upregulation of the DMN would be instrumental in developing new treatments for fear and anxiety disorders, and both TMS and tES have shown some capacity to achieve this.

In addition, the key hubs of SN, the ACC and the AI, have been successfully down-modulated via TMS (Gratton et al. 2013; Iwabuchi et al. 2017) and tDCS (Khan et al. 2020) although a recent meta-analysis also suggests up-modulation of the ACC via anodal tDCS of lPFC (Chan et al. 2021). These findings suggest that with careful consideration of stimulation polarity, TMS and tES can applied to mitigate SN hyperfunctioning in anxiety disorders.

3.2.3 Modulating Neural Oscillations via tACS in Anxiety Disorders

The importance of neural oscillations in normal and abnormal behavior has been increasingly recognized. As discussed above, tACS is a NIBS technique that is particularly effective at modulating neural oscillations. While still underutilized in anxiety and fear intervention, tACS has the potential to yield special benefits for patients exhibiting oscillatory dysrhythmia (also known as "oscillopathy"). Of note, as discussed above, oscillatory dysrhythmia, particularly in the alpha frequency, has been identified in PTSD, anxiety, and related disorders, such that upregulating alpha oscillations via tACS would be a promising treatment option.

Motivated by observed deficits in occipitoparietal alpha oscillations in PTSD (Clancy et al. 2017) and related states of anxious arousal, researchers (Clancy et al. 2018) targeted alpha oscillations in the occipitoparietal cortex to reduce anxiety in a proof-of-concept protocol among healthy individuals. Over a 4-day course with daily 30-min stimulation, participants who received active tACS demonstrated acute increases in alpha power over the visual cortex as well as sustained, lasting increases in posterior-to-frontal alpha-frequency connectivity. The connectivity increase further predicted subsequent reductions in anxious arousal over the 4 days.

The enhancement of interregional connectivity via tACS suggests that large-scale networks may be affected, a possibility that was examined in a follow-up study combining tACS with EEG and fMRI recordings in a healthy sample (Clancy et al. 2022). Once again, alpha-frequency tACS was targeted at the occipitoparietal cortex, especially the precuneus, which has strong connections with the posterior hub of the DMN – the PCC (Cavanna and Trimble 2006). Following stimulation, resting-state fMRI functional connectivity between the PCC and mPFC was increased alongside enhanced posterior-to-frontal alpha-frequency connectivity (Clancy et al. 2022), demonstrating the efficacy of alpha-tACS in upregulating DMN connectivity via the strengthening of the oscillatory communication between the two DMN hubs. By reifying the ability of alpha-frequency tACS to not only acutely modulate local cortical activity but also producing lasting impacts on long-range cortical connectivity, especially within the DMN, these data position tACS as a novel therapeutic intervention for the neural circuits and networks of anxiety and related disorders.

3.2.4 Sex Effects

Women and men have shown different responses to NIBS (Kedzior et al. 2014). These sex effects have likely contributed to inconsistent findings in the literature, and without careful consideration of these effects, the efficacy of NIBS treatments for fear and anxiety disorders can be undermined. To date, a nuanced understanding of these effects is still lacking. One potential source of the sex differences is physiological and hormonal disparities affecting the action of NIBS. Higher cortical bone density in women can undercut tES efficacy (Rudroff et al. 2020), while the larger distance between the scalp and the cortex and lower gray matter density and gyrification in men can limit TMS efficacy (Hanlon and McCalley 2022). Furthermore, hormonal fluctuations across the menstrual cycle can shift cortical excitability and excitation–inhibition balance, leading to variable responses to NIBS in women (Epperson et al. 2002). In addition, women and men also differ in symptom presentations and underlying pathology for fear and anxiety disorders (Maeng and Milad 2015; Jalnapurkar et al. 2018). NIBS likely interacts with the pathological processes differently for women and men, yielding sex-specific results. Research on this topic will provide highly valuable insights into and promote favorable therapeutic outcomes for NIBS intervention.

3.2.5 A Unified Approach for Targeting Fear Networks in Anxiety Disorders

The reviewed evidence supports NIBS neuromodulation as a valuable tool for novel clinical interventions of fear and anxiety disorders. Theory-driven but practical applications for NIBS will be critical for further refining existing protocols and capitalizing on underutilized processes. Integrating evidence from direct and indirect NIBS, we propose a unified approach modulating multiple pathways that can interact to treat fear and anxiety disorders (Fig. 1b).

Direct key targets would be proximal to the scalp (and thus accessible for NIBS) and play a crucial role in fear and anxiety (directly or through strong connectivity). Indirect key targets are deep substrates of fear and anxiety that are strongly connected, via axonal, functional, or oscillatory connectivity, to a surface structure accessible by NIBS. Qualifying key direct target regions are the dlPFC, the sensory cortex, and the posterior parietal/precuneus cortex (Fig. 1b Top). Excitatory TMS and tES of the dlPFC will facilitate extinction learning and fear reconsolidation, reducing fear recall. Inhibitory stimulation of the dlPFC and sensory cortex during fear memory processes will likely improve symptoms associated with fear memory. The stimulation of the posterior parietal/precuneus cortex would serve to attack the densely connected indirect targets.

Through strong connectivity, the direct targets can transmit the neuromodulatory effects to deep cortical and subcortical structures involved in fear and anxiety, that is, the qualifying key indirect target regions (Fig. 1b Bottom). (i) The vmPFC (critically involved in extinction learning) and the mPFC (the key anterior hub of the DMN) can be upregulated via stimulation of the dlPFC, resulting in enhanced extinction and strengthened DMN functioning, respectively. (ii) The PCC (the key posterior hub of the DMN) can be upregulated via stimulation of the precuneus, normalizing DMN functioning. (iii) The dACC, AI, and the amygdala, key components of both the fear circuit and the SN, can be downregulated via stimulation of the dlPFC and the sensory cortex, resulting in reduced threat response and anxious arousal. Finally, (iv) the hippocampus can be reached via stimulation of the posterior parietal cortex, resulting in normalized threat memory.

In summary, guided by this integrative model, researchers and clinicians can stimulate multiple pathways in combination, producing synergistic effects to normalize compounded neural dysfunctions and thus maximizing clinical improvement. For example, alpha tACS can be administered to promote local posterior alpha oscillations and long-range posterior-to-frontal connectivity, particularly in the DMN, rectifying alpha dysrhythmia and DMN hypofunctioning. Alternatively, the posterior dlPFC can be stimulated to up-modulate the vmPFC during extinction learning, thereby promoting extinction while additionally restoring DMN connectivity. Finally, future research may utilize advanced stimulation protocols. These may include combinatory applications that implement multi-modal, multi-target NIBS to achieve maximal effects, or closed-loop designs wherein optimal target stimulation is informed by real-time measurements of neural activity (Zrenner and Ziemann 2024).

4 Conclusion and Future Directions

In recent decades, great strides have been made in advancing the understanding of the neural basis of fear and anxiety disorders. The amygdala–hippocampus–PFC fear circuit has been expanded to include the sensory cortex, insula, and the dorsal ACC. Moreover, the large-scale networks, particularly the SN and DMN, have been incorporated into the conceptualization of fear and anxiety. Along this path, neuromodulation has helped to elucidate the mechanistic framework and presented exciting promise for treating these disorders. However, this literature is also littered with inconsistent and even contradictory findings, implicating the complexity and high dimensionality of the space of neuromodulation. Neuroimaging data are lacking, especially those that are synchronized with neuromodulation to elucidate the neural effects and the mechanism of action, leaving unresolved inconsistencies. Conversely, tremendous potentials exist for future neuromodulation of fear and anxiety that leverages insights into the neural mechanism of NIBS and precisely tackles the mechanism. Future research should also seek to attack the multiple pathways via the integrative application of NIBS, achieving multiplicative effects. Finally, technological innovations, such as individualized precision stimulation and closed-loop stimulation, have started to gain traction in research, poised to break new ground in the neuromodulation of fear and anxiety.

References

Abbasi S et al (2017) Effects of left prefrontal transcranial direct current stimulation on the acquisition of contextual and cued fear memory. Iran J Basic Med Sci 20(6):623–630

Abend R et al (2016) Modulation of fear extinction processes using transcranial electrical stimulation. Transl Psychiatry 6(10):e913

Ahmadizadeh MJ, Rezaei M (2018) Unilateral right and bilateral dorsolateral prefrontal cortex transcranial magnetic stimulation in treatment post-traumatic stress disorder: a randomized controlled study. Brain Res Bull 140:334–340

Akiki TJ et al (2018) Default mode network abnormalities in posttraumatic stress disorder: a novel network-restricted topology approach. NeuroImage 176:489–498

Asthana M et al (2013) Effects of transcranial direct current stimulation on consolidation of fear memory. Front Psych 4:107

Aupperle RL et al (2012) Dorsolateral prefrontal cortex activation during emotional anticipation and neuropsychological performance in posttraumatic stress disorder. Arch Gen Psychiatry 69(4):360–371

Battaglia S et al (2024) 'Nip it in the bud': low-frequency rTMS of the prefrontal cortex disrupts threat memory consolidation in humans. Behav Res Ther 178:104548

Bienvenu TCM et al (2021) The advent of fear conditioning as an animal model of post-traumatic stress disorder: learning from the past to shape the future of PTSD research. Neuron 109(15): 2380–2397

Bluhm RL et al (2009) Alterations in default network connectivity in posttraumatic stress disorder related to early-life trauma. J Psychiatry Neurosci 34(3):187–194

Boehme S et al (2024) Good moments to stimulate the brain - a randomized controlled double-blinded study on anodal transcranial direct current stimulation of the ventromedial prefrontal cortex on two different time points in a two-day fear conditioning paradigm. Behav Brain Res 460:114804

Borgomaneri S et al (2020) State-dependent TMS over prefrontal cortex disrupts fear-memory reconsolidation and prevents the return of fear. Curr Biol 30(18):3672–3679 e3674

Borgomaneri S et al (2021) Memories are not written in stone: re-writing fear memories by means of non-invasive brain stimulation and optogenetic manipulations. Neurosci Biobehav Rev 127: 334–352

Breukelaar IA et al (2021) The functional connectome in posttraumatic stress disorder. Neurobiol Stress 14:100321

Brown JC et al (2024) Critically assessing the unanswered questions of how, where, and when to induce plasticity in the posttraumatic stress disorder network with transcranial magnetic stimulation. Biol Psychiatry

Cabeza R, Nyberg L (2000) Neural bases of learning and memory: functional neuroimaging evidence. Curr Opin Neurol 13(4):415–421

Cambiaghi M et al (2016) Higher-order sensory cortex drives basolateral amygdala activity during the recall of remote, but not recently learned fearful memories. J Neurosci 36(5):1647–1659

Cavanna AE, Trimble MR (2006) The precuneus: a review of its functional anatomy and behavioural correlates. Brain 129(Pt 3):564–583

Chan MMY et al (2021) The neurobiology of prefrontal transcranial direct current stimulation (tDCS) in promoting brain plasticity: a systematic review and meta-analyses of human and rodent studies. Neurosci Biobehav Rev 125:392–416

Cirillo P et al (2019) Transcranial magnetic stimulation in anxiety and trauma-related disorders: a systematic review and meta-analysis. Brain Behav 9(6):e01284

Clancy K et al (2017) Restless 'rest': intrinsic sensory hyperactivity and disinhibition in post-traumatic stress disorder. Brain 140(7):2041–2050

Clancy KJ et al (2018) Lasting connectivity increase and anxiety reduction via transcranial alternating current stimulation. Soc Cogn Affect Neurosci 13(12):1305–1316

Clancy KJ et al (2020a) Intrinsic sensory disinhibition contributes to intrusive re-experiencing in combat veterans. Sci Rep 10(1):936

Clancy KJ et al (2020b) Posttraumatic stress disorder is associated with alpha dysrhythmia across the visual cortex and the default mode network. eNeuro 7(4)

Clancy KJ et al (2022) Transcranial stimulation of alpha oscillations up-regulates the default mode network. Proc Natl Acad Sci 119(1):e2110868119

Concina G et al (2019) The auditory cortex and the emotional valence of sounds. Neurosci Biobehav Rev 98:256–264

Curtis CE, D'Esposito M (2003) Persistent activity in the prefrontal cortex during working memory. Trends Cogn Sci 7(9):415–423

Davidson RJ, Irwin W (1999) The functional neuroanatomy of emotion and affective style. Trends Cogn Sci 3(1):11–21

Davis M (1992) The role of the amygdala in fear and anxiety. Annu Rev Neurosci 15:353–375

de Lima AL et al (2019) Transcranial direct current stimulation for the treatment of generalized anxiety disorder: a randomized clinical trial. J Affect Disord 259:31–37

Deng ZD et al (2020) Device-based modulation of neurocircuits as a therapeutic for psychiatric disorders. Annu Rev Pharmacol Toxicol 60:591–614

Deppermann S et al (2017) Neurobiological and clinical effects of fNIRS-controlled rTMS in patients with panic disorder/agoraphobia during cognitive-behavioural therapy. Neuroimage Clin 16:668–677

Diefenbach GJ et al (2016) Repetitive transcranial magnetic stimulation for generalised anxiety disorder: a pilot randomised, double-blind, sham-controlled trial. Br J Psychiatry 209(3): 222–228

Dittert N et al (2018) Augmentation of fear extinction by transcranial direct current stimulation (tDCS). Front Behav Neurosci 12:76

Dosenbach, N. U., et al. (2024). "The brain's cingulo-opercular action-mode network." preprint at PsyArXiv https://doi. org/10.31234/osf. io/2vt79

du Bois N et al (2021) Neurofeedback with low-cost, wearable electroencephalography (EEG) reduces symptoms in chronic post-traumatic stress disorder. J Affect Disord 295:1319–1334

Eichenbaum H (2017) Prefrontal-hippocampal interactions in episodic memory. Nat Rev Neurosci 18(9):547–558

Eidelman-Rothman M et al (2016) Alpha oscillations and their impairment in affective and post-traumatic stress disorders. Neurosci Biobehav Rev 68:794–815

Engel-Yeger B et al (2013) The sensory profile of people with post-traumatic stress symptoms. Occup Ther Ment Health 29(3):266–278

Epperson CN et al (2002) Cortical gamma-aminobutyric acid levels across the menstrual cycle in healthy women and those with premenstrual dysphoric disorder: a proton magnetic resonance spectroscopy study. Arch Gen Psychiatry 59(9):851–858

Etkin A et al (2009) Disrupted amygdalar subregion functional connectivity and evidence of a compensatory network in generalized anxiety disorder. Arch Gen Psychiatry 66(12):1361–1372

Fenster RJ et al (2018) Brain circuit dysfunction in post-traumatic stress disorder: from mouse to man. Nat Rev Neurosci 19(9):535–551

Fiori V et al (2011) Transcranial direct current stimulation improves word retrieval in healthy and nonfluent aphasic subjects. J Cogn Neurosci 23(9):2309–2323

Gay F et al (2022) Neuromodulation treatments of pathological anxiety in anxiety disorders, stressor-related disorders, and major depressive disorder: a dimensional systematic review and meta-analysis. Front Psych 13:910897

Fox MD, Raichle ME (2007) Spontaneous fluctuations in brain activity observed with functional magnetic resonance imaging. Nat Rev Neurosci 8(9):700–711

Fullana MA et al (2016) Neural signatures of human fear conditioning: an updated and extended meta-analysis of fMRI studies. Mol Psychiatry 21(4):500–508

Fullana MA et al (2018) Fear extinction in the human brain: a meta-analysis of fMRI studies in healthy participants. Neurosci Biobehav Rev 88:16–25

Ganho-Avila A et al (2019) The effect of cathodal tDCS on fear extinction: a cross-measures study. PLoS One 14(9):e0221282

Ganho-Avila A et al (2022) Offline tDCS modulates prefrontal-cortical-subcortical-cerebellar fear pathways in delayed fear extinction. Exp Brain Res 240(1):221–235

Gdalyahu A et al (2012) Associative fear learning enhances sparse network coding in primary sensory cortex. Neuron 75(1):121–132

George MS et al (2000) A controlled trial of daily left prefrontal cortex TMS for treating depression. Biol Psychiatry 48(10):962–970

Gratton C et al (2013) The effect of theta-burst TMS on cognitive control networks measured with resting state fMRI. Front Syst Neurosci 7:124

Grosso A et al (2015) Auditory cortex involvement in emotional learning and memory. Neuroscience 299:45–55

Hanlon CA, McCalley DM (2022) Sex/gender as a factor that influences transcranial magnetic stimulation treatment outcome: three potential biological explanations. Front Psych 13:869070

Harnett NG et al (2024) Affective visual circuit dysfunction in trauma and stress-related disorders. Biol Psychiatry

Hasson D et al (2013) Acute stress induces hyperacusis in women with high levels of emotional exhaustion. PLoS One 8(1):e52945

Headley DB et al (2019) Embracing complexity in defensive networks. Neuron 103(2):189–201

Hebscher M et al (2021) Enhanced reinstatement of naturalistic event memories due to hippocampal-network-targeted stimulation. Curr Biol 31(7):1428–1437 e1425

Herrmann MJ et al (2017) Medial prefrontal cortex stimulation accelerates therapy response of exposure therapy in acrophobia. Brain Stimul 10(2):291–297

Hopper JW et al (2007) Neural correlates of reexperiencing, avoidance, and dissociation in PTSD: symptom dimensions and emotion dysregulation in responses to script-driven trauma imagery. J Trauma Stress 20(5):713–725

Hughes SW, Crunelli V (2005) Thalamic mechanisms of EEG alpha rhythms and their pathological implications. Neuroscientist 11(4):357–372

Isserles M et al (2021) Deep transcranial magnetic stimulation combined with brief exposure for posttraumatic stress disorder: a prospective multisite randomized trial. Biol Psychiatry 90(10): 721–728

Iwabuchi SJ et al (2017) Targeted transcranial theta-burst stimulation alters fronto-insular network and prefrontal GABA. NeuroImage 146:395–403

Iyer MB et al (2005) Safety and cognitive effect of frontal DC brain polarization in healthy individuals. Neurology 64(5):872–875

Jalnapurkar I et al (2018) Sex differences in anxiety disorders: a review. J Psychiatry Depress Anxiety 4(12):3–16

Javanbakht A et al (2011) Event-related potential studies of post-traumatic stress disorder: a critical review and synthesis. Biol Mood Anxiety Disord 1(1):5

Josselyn SA, Tonegawa S (2020) Memory engrams: recalling the past and imagining the future. Science 367(6473)

Kan RLD et al (2020) Non-invasive brain stimulation for posttraumatic stress disorder: a systematic review and meta-analysis. Transl Psychiatry 10(1):168

Karsen EF et al (2014) Review of the effectiveness of transcranial magnetic stimulation for post-traumatic stress disorder. Brain Stimul 7(2):151–157

Kasten FH et al (2016) Sustained aftereffect of alpha-tACS lasts up to 70 min after stimulation. Front Hum Neurosci 10:245

Kedzior KK et al (2014) More female patients and fewer stimuli per session are associated with the short-term antidepressant properties of repetitive transcranial magnetic stimulation (rTMS): a meta-analysis of 54 sham-controlled studies published between 1997-2013. Neuropsychiatr Dis Treat 10:727–756

Keil A et al (2007) Adaptation in human visual cortex as a mechanism for rapid discrimination of aversive stimuli. NeuroImage 36(2):472–479

Kessler R et al (2020) Long-term neuroanatomical consequences of childhood maltreatment: reduced amygdala inhibition by medial prefrontal cortex. Front Syst Neurosci 14:28

Khan A et al (2020) Anodal transcranial direct current stimulation of anterior cingulate cortex modulates subcortical brain regions resulting in cognitive enhancement. Front Hum Neurosci 14:584136

Kim DR et al (2009) Transcranial magnetic stimulation in the treatment of psychiatric disorders. Curr Psychiatry Rep 11(6):447–452

Klomjai W et al (2015) Basic principles of transcranial magnetic stimulation (TMS) and repetitive TMS (rTMS). Ann Phys Rehabil Med 58(4):208–213

Kluetsch RC et al (2014) Plastic modulation of PTSD resting-state networks and subjective wellbeing by EEG neurofeedback. Acta Psychiatr Scand 130(2):123–136

Koch SB et al (2016) Aberrant resting-state brain activity in posttraumatic stress disorder: a meta-analysis and systematic review. Depress Anxiety 33(7):592–605

Kolesar TA et al (2019) Systematic review and meta-analyses of neural structural and functional differences in generalized anxiety disorder and healthy controls using magnetic resonance imaging. Neuroimage Clin 24:102016

Kozel FA et al (2018) Repetitive TMS to augment cognitive processing therapy in combat veterans of recent conflicts with PTSD: a randomized clinical trial. J Affect Disord 229:506–514

Krawczyk DC (2002) Contributions of the prefrontal cortex to the neural basis of human decision making. Neurosci Biobehav Rev 26(6):631–664

Krusemark EA, Li W (2012) Enhanced olfactory sensory perception of threat in anxiety: an event-related fMRI study. Chemosens Percept 5(1):37–45

Krusemark EA et al (2013) When the sense of smell meets emotion: anxiety-state-dependent olfactory processing and neural circuitry adaptation. J Neurosci 33(39):15324–15332

Lai CH, Wu YT (2014) The alterations in inter-hemispheric functional coordination of patients with panic disorder: the findings in the posterior sub-network of default mode network. J Affect Disord 166:279–284

Lanius R (2007) Complex adaptations to traumatic stress: from neurobiological to social and cultural aspects. Am J Psychiatry 164(11):1628–1630

Lanius RA et al (2010) Default mode network connectivity as a predictor of post-traumatic stress disorder symptom severity in acutely traumatized subjects. Acta Psychiatr Scand 121(1):33–40

LeDoux JE (2000) Emotion circuits in the brain. Annu Rev Neurosci 23:155–184

LeDoux JE, Pine DS (2016) Using neuroscience to help understand fear and anxiety: a two-system framework. Am J Psychiatry 173(11):1083–1093

Lee JL (2009) Reconsolidation: maintaining memory relevance. Trends Neurosci 32(8):413–420

Leuchter AF et al (2015) Rhythms and blues: modulation of oscillatory synchrony and the mechanism of action of antidepressant treatments. Ann N Y Acad Sci 1344(1):78–91

Lewine JD et al (2002) Abnormal stimulus-response intensity functions in posttraumatic stress disorder: an electrophysiological investigation. Am J Psychiatry 159(10):1689–1695

Li W (2019) Perceptual mechanisms of anxiety and its disorders. In: Olatunji B (ed) Cambridge handbook of anxiety and related disorders. Cambridge UP, Cambridge, UK

Li W, Keil A (2023) Sensing fear: fast and precise threat evaluation in human sensory cortex. Trends Cogn Sci 27(4):341–352

Li W, Wilson DA (2024) Threat memory in the sensory cortex: insights from olfaction. Neuroscientist 30(3):285–293

Liberzon I, Abelson JL (2016) Context processing and the neurobiology of post-traumatic stress disorder. Neuron 92(1):14–30

Lindauer RJ et al (2008) Effects of psychotherapy on regional cerebral blood flow during trauma imagery in patients with post-traumatic stress disorder: a randomized clinical trial. Psychol Med 38(4):543–554

Maeng LY, Milad MR (2015) Sex differences in anxiety disorders: interactions between fear, stress, and gonadal hormones. Horm Behav 76:106–117

Manteghi F et al (2017) Precondition of right frontal region with anodal tDCS can restore the fear memory impairment induced by ACPA in male mice. EXCLI J 16:1–13
Maren S (2001) Neurobiology of Pavlovian fear conditioning. Annu Rev Neurosci 24:897–931
Margulies DS et al (2016) Situating the default-mode network along a principal gradient of macroscale cortical organization. Proc Natl Acad Sci USA 113(44):12574–12579
Markovic V et al (2021) A systematic review on the effect of transcranial direct current and magnetic stimulation on fear memory and extinction. Front Hum Neurosci 15:655947
Maron-Katz A et al (2020) Individual patterns of abnormality in resting-state functional connectivity reveal two data-driven PTSD subgroups. Am J Psychiatry 177(3):244–253
McGaugh JL (2000) Memory – a century of consolidation. Science 287(5451):248–251
Meissner-Bernard C et al (2019) Encoding of odor fear memories in the mouse olfactory cortex. Curr Biol 29(3):367–380 e364
Menon V (2011) Large-scale brain networks and psychopathology: a unifying triple network model. Trends Cogn Sci 15(10):483–506
Milad MR et al (2004) Electrical stimulation of medial prefrontal cortex reduces conditioned fear in a temporally specific manner. Behav Neurosci 118(2):389–394
Milad MR et al (2006) Fear extinction in rats: implications for human brain imaging and anxiety disorders. Biol Psychol 73(1):61–71
Milad MR et al (2014) Neuroscience of fear extinction: implications for assessment and treatment of fear-based and anxiety related disorders. Behav Res Ther 62:17–23
Miller DR et al (2017a) Default mode network subsystems are differentially disrupted in posttraumatic stress disorder. Biol Psychiatry Cogn Neurosci Neuroimaging 2(4):363–371
Miller DR et al (2017b) Posttraumatic stress disorder symptom severity is associated with reduced default mode network connectivity in individuals with elevated genetic risk for psychopathology. Depress Anxiety 34(7):632–640
Morgan CA 3rd, Grillon C (1999) Abnormal mismatch negativity in women with sexual assault-related posttraumatic stress disorder. Biol Psychiatry 45(7):827–832
Moscovitch M (1992) Memory and working-with-memory: a component process model based on modules and central systems. J Cogn Neurosci 4(3):257–267
Mungee A et al (2014) Transcranial direct current stimulation of the prefrontal cortex: a means to modulate fear memories. Neuroreport 25(7):480–484
Mungee A et al (2016) No effect of cathodal transcranial direct current stimulation on fear memory in healthy human subjects. Brain Sci 6(4)
Nasehi M et al (2017a) The effect of left frontal transcranial direct-current stimulation on propranolol-induced fear memory acquisition and consolidation deficits. Behav Brain Res 331:76–83
Nasehi M et al (2017b) Interference effects of transcranial direct current stimulation over the right frontal cortex and adrenergic system on conditioned fear. Psychopharmacology 234(22): 3407–3416
Neylan TC et al (1999) Sensory gating in chronic posttraumatic stress disorder: reduced auditory P50 suppression in combat veterans. Biol Psychiatry 46(12):1656–1664
Nicholson AA et al (2020) Classifying heterogeneous presentations of PTSD via the default mode, central executive, and salience networks with machine learning. Neuroimage Clin 27:102262
Ojala KE et al (2022) Inhibiting human aversive memory by transcranial theta-burst stimulation to the primary sensory cortex. Biol Psychiatry 92(2):149–157
Olshausen BA, Field DJ (2004) Sparse coding of sensory inputs. Curr Opin Neurobiol 14(4): 481–487
Opitz A et al (2016) An integrated framework for targeting functional networks via transcranial magnetic stimulation. NeuroImage 127:86–96
Pessoa L, Adolphs R (2010) Emotion processing and the amygdala: from a 'low road' to 'many roads' of evaluating biological significance. Nat Rev Neurosci 11(11):773–783
Phelps EA, LeDoux JE (2005) Contributions of the amygdala to emotion processing: from animal models to human behavior. Neuron 48(2):175–187

Philip NS et al (2017) Low-intensity transcranial current stimulation in psychiatry. Am J Psychiatry 174(7):628–639

Philip NS et al (2019) Theta-burst transcranial magnetic stimulation for posttraumatic stress disorder. Am J Psychiatry 176(11):939–948

Raichle ME (2015) The brain's default mode network. Annu Rev Neurosci 38:433–447

Raichle ME et al (2001) A default mode of brain function. Proc Natl Acad Sci USA 98(2):676–682

Raij T et al (2018) Prefrontal cortex stimulation enhances fear extinction memory in humans. Biol Psychiatry 84(2):129–137

Ressler KJ (2020) Translating across circuits and genetics toward progress in fear- and anxiety-related disorders. Am J Psychiatry 177(3):214–222

Rudroff T et al (2020) Response variability in transcranial direct current stimulation: why sex matters. Front Psych 11:585

Sa KN et al (2023) Evidence-based umbrella review of non-invasive brain stimulation in anxiety disorders. The European Journal of Psychiatry 37(3):167–181

Sacco T, Sacchetti B (2010) Role of secondary sensory cortices in emotional memory storage and retrieval in rats. Science 329(5992):649–656

Sagliano L et al (2019) Non-invasive brain stimulation in generalized anxiety disorder: a systematic review. Prog Neuro-Psychopharmacol Biol Psychiatry 93:31–38

Schiller D, Phelps EA (2011) Does reconsolidation occur in humans? Front Behav Neurosci 5:24

Schiller D et al (2008) Evidence for recovery of fear following immediate extinction in rats and humans. Learn Mem 15(6):394–402

Sehlmeyer C et al (2009) Human fear conditioning and extinction in neuroimaging: a systematic review. PLoS One 4(6):e5865

Shin LM et al (2011) Exaggerated activation of dorsal anterior cingulate cortex during cognitive interference: a monozygotic twin study of posttraumatic stress disorder. Am J Psychiatry 168(9):979–985

Siddiqi SH, Fox MD (2024) Targeting symptom-specific networks with transcranial magnetic stimulation. Biol Psychiatry 95(6):502–509

Siddiqi SH et al (2020) Distinct symptom-specific treatment targets for circuit-based neuromodulation. Am J Psychiatry 177(5):435–446

Skinner RD et al (1999) Reduced sensory gating of the P1 potential in rape victims and combat veterans with posttraumatic stress disorder. Depress Anxiety 9(3):122–130

Smits FM et al (2020) Does non-invasive brain stimulation modulate emotional stress reactivity? Soc Cogn Affect Neurosci 15(1):23–51

Snyder HR (2013) Major depressive disorder is associated with broad impairments on neuropsychological measures of executive function: a meta-analysis and review. Psychol Bull 139(1): 81–132

Sripada RK et al (2012) Neural dysregulation in posttraumatic stress disorder: evidence for disrupted equilibrium between salience and default mode brain networks. Psychosom Med 74(9):904–911

Steenkamp MM et al (2020) First-line psychotherapies for military-related PTSD. JAMA 323(7): 656–657

Stewart LP, White PM (2008) Sensory filtering phenomenology in PTSD. Depress Anxiety 25(1): 38–45

Su S et al (2022) Continuous theta-burst stimulation over the right dorsolateral prefrontal cortex disrupts fear memory reconsolidation in humans. iScience 25(1):103614

Taylor JG, Fragopanagos NF (2005) The interaction of attention and emotion. Neural Netw 18(4): 353–369

Terao Y, Ugawa Y (2002) Basic mechanisms of TMS. J Clin Neurophysiol 19(4):322–343

To WT et al (2018) Changing brain networks through non-invasive neuromodulation. Front Hum Neurosci 12:128

Tovote P et al (2015) Neuronal circuits for fear and anxiety. Nat Rev Neurosci 16(6):317–331

van den Boogert F et al (2022) Sensory processing difficulties in psychiatric disorders: a meta-analysis. J Psychiatr Res 151:173–180
van't Wout M et al (2016) Can transcranial direct current stimulation augment extinction of conditioned fear? Brain Stimul 9(4):529–536
Van't Wout M et al (2017) Transcranial direct current stimulation may modulate extinction memory in posttraumatic stress disorder. Brain Behav 7(5):e00681
Varela F et al (2001) The brainweb: phase synchronization and large-scale integration. Nat Rev Neurosci 2(4):229–239
Vicario CM et al (2020) Anodal transcranial direct current stimulation over the ventromedial prefrontal cortex enhances fear extinction in healthy humans: a single blind sham-controlled study. Brain Stimul 13(2):489–491
Visser RM et al (2021) Robust BOLD responses to faces but not to conditioned threat: challenging the Amygdala's reputation in human fear and extinction learning. J Neurosci 41(50): 10278–10292
Wassermann EM, Grafman J (2005) Recharging cognition with DC brain polarization. Trends Cogn Sci 9(11):503–505
Webler RD et al (2021) The neurobiology of human fear generalization: meta-analysis and working neural model. Neurosci Biobehav Rev 128:421–436
Webler RD et al (2024) Causally probing the role of the hippocampus in fear discrimination: a precision functional mapping-guided, transcranial magnetic stimulation study in participants with posttraumatic stress symptoms. Biol Psychiatry Glob Open Sci 4(3):100309
Weinberger NM (2004) Specific long-term memory traces in primary auditory cortex. Nat Rev Neurosci 5(4):279–290
Wen Z et al (2022) Temporally and anatomically specific contributions of the human amygdala to threat and safety learning. Proc Natl Acad Sci USA 119(26):e2204066119
Yavari F et al (2018) Basic and functional effects of transcranial electrical stimulation (tES) – an introduction. Neurosci Biobehav Rev 85:81–92
Yeo BT et al (2011) The organization of the human cerebral cortex estimated by intrinsic functional connectivity. J Neurophysiol 106(3):1125–1165
Yin S et al (2018) Amygdala adaptation and temporal dynamics of the salience network in conditioned fear: a single-trial fMRI study. eNeuro 5(1)
Yuan M et al (2023) Dysfunction of default mode network characterizes generalized anxiety disorder relative to social anxiety disorder and post-traumatic stress disorder. J Affect Disord 334:35–42
Yuste R (2015) From the neuron doctrine to neural networks. Nat Rev Neurosci 16(8):487–497
Zrenner C, Ziemann U (2024) Closed-loop brain stimulation. Biol Psychiatry 95(6):545–552

Using Inhibitory Learning Theories to Optimise Treatment for Children with Anxiety Disorders

Wenting Chen, Melissa Aji, Chloe Y. S. Lim, Annabel Songco, and Jennifer L. Hudson

Contents

Abstract Anxiety disorders in children lead to substantial impairment in functioning and development. Even the most effective gold standard treatments for childhood anxiety have 50% remission rates, suggesting a critical need to improve current treatments. Optimising exposure, the key component of anxiety treatments,

W. Chen, M. Aji, C. Y. S. Lim, A. Songco, and J. L. Hudson (✉)
Black Dog Institute, University of New South Wales, Sydney, Australia
e-mail: jennie.hudson@unsw.edu.au

© The Author(s), under exclusive license to Springer Nature Switzerland AG 2025

Curr Topics Behav Neurosci (2025) 73: 433–472
https://doi.org/10.1007/7854_2024_553
Published Online: 16 January 2025

represents a promising way to do so. This chapter explains how to optimise exposure outcomes for childhood anxiety through inhibitory learning theory. This chapter describes the background of inhibitory learning, including its different components and the empirical evidence supporting it. We then discuss how to improve the formation of inhibitory associations through enhancing expectancy violation, the proposed mechanism underlying inhibitory learning. Strategies to enhance inhibitory learning for child anxiety treatment are provided. These include strategies to enhance the formation of inhibitory associations, such as psychoeducation, eliminating safety signals, deepened extinction, occasional reinforced extinction, and affect-based strategies. Additionally, strategies to enhance retrieval include variability, multiple contexts, and retrieval cues. Suggestions are made on how to adapt these strategies to child populations. Further, a clinical guide for using inhibitory learning strategies in child anxiety treatment is included as an appendix. This details how clinicians can utilise these strategies to enhance current treatments, including examples of case studies and scripts.

Keywords Anxiety disorders · Child · Inhibitory learning · Treatment

The recommended first-line treatment for anxiety disorders in children is Cognitive Behavioural Therapy (CBT; Higa-McMillan et al. 2016). CBT can improve anxiety symptoms with large pre- to post-effect sizes (In-Albon and Schneider 2007), and lead to approximately 50% remission in children (James et al. 2020). However, longitudinal studies indicate relapse in up to 50% of youth who initially respond to treatment (Kooiman et al. 2024). Thus, while CBT is beneficial, there is considerable room to improve efficacy. A core technique of CBT for anxiety is exposure, which involves encouraging children to face feared situations. Studies suggest that more in-session and more challenging exposures (e.g., deliberately making a mistake during a speech, compared to a less-challenging exposure such as practising a speech in front of trusted family members) can lead to greater effect sizes (Whiteside et al. 2016). Therefore, optimising this key component of exposure is one promising avenue for improving treatment outcomes.

We know exposure is extremely effective in reducing fears, but our understanding of how and why exposure works is still contested. One emerging theory is the inhibitory learning theory. We have devoted this review to describing inhibitory learning theory and its proposed mechanisms, which is intended to have utility in improving clinical practice. First, we will provide an overview of the background theory of inhibitory learning. As inhibitory learning originated in adult research, we will review how it has been or should be adapted for child anxiety. We will then further discuss inhibitory learning-based strategies, starting with expectancy violation, which has been proposed as a core mechanism underlying inhibitory learning (Gromer et al. 2023).

Subsequently, we will discuss strategies that maximise expectancy violation and inhibitory learning, which can be broadly grouped into two strategies: (1) Strategies to enhance the development of inhibitory associations, i.e., the child's nonthreat associations, or in simpler more child-friendly language: making "not scary" memories (Craske et al. 2014); and, (2) Strategies to enhance long-term fear extinction, i.e., the retrieval of inhibitory associations for sustained treatment gains: making the "not scary" memories stronger than the "scary" memories over time (Craske et al. 2014). Tables 2 and 3 present a summary description of these strategies alongside preliminary considerations for children. Given the limited research in child samples, considerations are based on available data and enhanced by clinical experience.

1 Strategies Based on Inhibitory Learning to Optimise Anxiety Outcomes

Inhibitory learning techniques in exposure were developed from principles of basic classical conditioning and extinction learning (Bouton 2004; Marks 1973). As early as 1924, Professor Mary Cover-Jones conducted seminal experimental research identifying that childhood phobias could be deconditioned through classical conditioning principles. For example, in the experiment "The Case of Peter", Cover-Jones (1924) found that childhood phobias could be reduced by conducting exposure to the stimuli and pairing it with a different association, in this case, with the child's favourite food, which would eventually reduce the child's fear to the stimuli. A summary of fundamental learning processes is provided in Table 1. Inhibitory learning refers to the process of developing new non-fearful associations, that

Table 1 Definitions for fundamental learning processes underlying exposure

Process	Description
Fear learning	In classical conditioning studies, this occurs by pairing a neutral conditioned stimulus (CS, e.g., a light) with a negative unconditioned stimulus (US, e.g., a shock), such that a feared association develops towards the neutral stimulus
Extinction	The process of presenting the neutral stimulus (CS) without the unconditioned stimulus (US), until the response of fear eventually subsides towards the conditioned stimulus. Extinction learning is the basis for exposure therapy
Spontaneous recovery	When the feared association suddenly reemerges over time, e.g., developing a fear of dogs again over time after exposure for no known reason
Context renewal	When the feared association returns because the neutral stimulus (CS) is presented in a context different from where extinction occurred, e.g., conducting exposure to dogs in a clinic but having the fear return after seeing the dog at a park
Reinstatement	When the feared association returns because of an un-signalled or unexpected presentation of the neutral stimulus (CS) with the feared outcome. For example, if a child has a fear of public speaking, and while giving a speech, another child gives them very negative feedback, this may reinstate their fear
Reacquisition	When the feared association returns because of repeated presentations of a neutral stimulus with the feared outcome. For example, after extinction learning, having the feared stimuli be repeatedly paired with a shock

compete with, or inhibit, established fearful associations or memories (Craske et al. 2008; Craske et al. 2014). For example, in traditional learning theories, a child fearful of dogs may have developed the fearful association between dogs (the conditioned stimulus (CS)) and a negative outcome of the dog biting the child (unconditioned stimulus (US)), i.e., "dogs are scary because they will bite me." Exposure involves presenting the dog (CS) to the child without the feared outcome (no-US). This new learned association is described as the CS-no US association: the child has learned a new non-threatening association of the feared outcome not occurring, i.e., "The dog did not bite me."

Inhibitory learning theory expands on traditional learning theories by suggesting that exposure works because the new non-threatening association (CS-no US) overshadows or inhibits the original feared association (CS-US), rather than replacing it (see Fig. 1). This theory has provided not only a framework for understanding how exposure works, but also why fears may relapse (because the original feared association is no longer inhibited). This is supported by neurobiological findings showing the neurocircuitry underlying fear extinction (amygdala, dorsal anterior cingulate cortex, and insular cortex) differs from that of fear inhibition (hippocampus and ventromedial prefrontal cortex) (Craske et al. 2018).

Inhibitory learning theory has resulted in the development of new exposure optimisation strategies in adult anxiety treatment (Craske et al. 2014; Tolin 2019; Weisman and Rodebaugh 2018). For example, inhibitory learning theory predicts exposure will be more effective when the outcome of the practice "violates" the client's expectations, that is, the person sees that nothing "bad" happened. The larger the violation between the predictions and outcomes, the better the learning. This contrasts with many exposure protocols that use a habituation model with the goal of

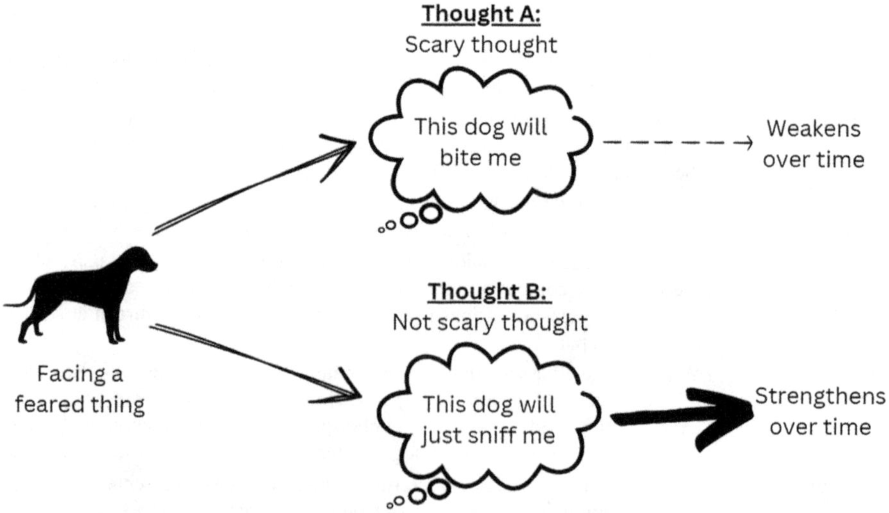

Fig. 1 Mechanism of inhibitory learning. Inhibitory learning theories suggest that exposure works by strengthening Thought B until it overshadows Thought A

reducing anxiety during exposure through graduated hierarchies, relaxation strategies, or combining other coping strategies.

Promising findings from the adult literature indicate that inhibitory learning is worth exploring for children and adolescents. In addition, there may be important modifications or unique considerations for this approach to children and young people. To date, there has been limited direct exploration of inhibitory learning theories applied to children (Teunisse et al. 2022). One review of the literature by McGuire et al. (2016) concluded that, compared to youth without anxiety disorders, children and youth with anxiety disorders demonstrated impairments in inhibitory learning, i.e., they found it harder to inhibit a feared response. Also, a study of 11 to 19-year-olds found that higher levels of reported fear prior to exposure practice were related to more favourable treatment outcomes (de Jong et al. 2024). Moreover, they found that if the young person's expectations of threat remained high after exposure (indicating that new learning did not occur), a less favourable treatment outcome was observed. In this study, exposures that led to greater expectancy violations (i.e., change in feared expectancy from pre- to post-exposure) did not predict better outcomes for adolescents, whereas greater expectancy change (i.e., differences between original and adjusted expectancies after exposure) did. This suggests that there are nuances in targeting expectancy violation effectively to improve treatment outcomes.

Perhaps the most exciting findings to date regarding the application of inhibitory learning theory in children are outcomes of a small pilot randomised controlled trial designed to assess the implementation and efficacy of using an inhibitory learning rationale to exposures in children aged 7 to 13 years (Kennedy and Hawks 2021). In this study, the acceptability and feasibility of the inhibitory-learning-based exposure program was like standard exposure. However, there was also a small effect in favour of an inhibitory learning-based exposure treatment across outcome measures, including anxiety symptoms, in comparison to a standard exposure. This is the first study to show that using an inhibitory learning approach may lead to more favourable outcomes. One review in adolescents (Plaisted et al. 2021) also suggested that exposure can be optimised in young people through inhibitory learning-based techniques, including dropping safety behaviours, discouraging avoidance, and use of homework-enhanced outcomes (Craske et al. 2014). Understanding the applicability of inhibitory learning to children could be a beneficial avenue to improve treatment outcomes.

2 Expectancy Violation

At the core of inhibitory learning is expectancy violation, as it is considered the proposed mechanism underlying inhibitory learning-based change in treatment (Gromer et al. 2023). By diagnostic definition, anxiety disorders are commonly associated with overprediction of aversive, negative outcomes (American Psychiatric Association and Association 2013). Maximising expectancy violation refers to designing exposures that maximally violate the child's biased expectancies about how frequent or intense their feared expectation would be (Craske et al. 2014). In clinical treatment, this may be observed when a child completes an exposure and has an "AHA!" moment because they learned something new and surprising. Exposures are created based on what the child needs to learn (regarding their feared outcomes), rather than staying in the situation until fears decline, as habituation models would indicate. Results from neurobiological studies support the claim that expectancy violation is the mechanism underlying exposure, including findings that dopaminergic neurons in the midbrain responsible for expectancy violation fire neural patterns during extinction, and extinction being prevented when these neurons are inhibited (Craske et al. 2022).

Experimental and clinical studies have also found that, in support of this theory, change in arousal across an exposure did not predict extinction retention in the long term, whereas violation of expectancies did (e.g., Baker et al. 2010; Meuret et al. 2012). A large-scale study examining over 8,000 exposures across 605 adult patients with anxiety disorders found that expectancy violation and expectancy change represented the underlying mechanism for successful exposures (Pittig et al. 2022). In this study, expectancy violation referred to the mismatch between expectancy towards threat and threat occurrence, whereas expectancy change was the difference between original expectancy and adjusted expectancy after exposure.

Importantly, Craske et al. (2022) recommend that more research is required to understand how expectancy violation can be further maximised. One experimental study by (Stemerding et al. 2023) found that greater expectancy violations and greater awareness of expectancy violations led to greater updating of threat expectancies. However, greater expectancy violation did not improve long-term retention of learning (as measured through physiological fear responses to a conditioned stimulus) in a subsequent test phase, whereas greater awareness of expectancy violation showed some reduction of fear at the test. Taken together, this suggests there is nuance in how to best ensure expectancy violation learning is consolidated in treatment. One difficulty with trying to maximise expectancy violation in children is that this technique requires children to reflect on and verbalise their expectancies, and some children demonstrate difficulties in future episodic thinking (McCormack et al. 2019). To manage this, a gentle and supportive approach should be used with developmentally appropriate questioning and ensuring expectancies are in the child's words.

3 Strategies Enhancing the Development of Inhibitory Associations Through Maximising Expectancy Violation

Table 2 provides a summary of the suggested strategies and developmental considerations for developing inhibitory associations.

Table 2 Strategies to enhance the formation of inhibitory associations ("not scary" memories)

Strategy	Brief description of strategy for children and families	Additional considerations for children and families
Expectancy violation	Maximise the surprise or violation of a child's initial expectancies, relating to how frequent or intense the feared outcome will be This strategy is considered the mechanism underlying inhibitory learning in exposure	Ideally requires child to identify their expectancies and reflect on them after the exposure. This can be difficult for some children to verbalise and harder to design optimised exposures. It can be helpful to provide examples of common expectancies for the child's diagnosis as a starting point, though answers are best in the child's own words
Inhibitory learning strategies aimed at maximising expectancy violation		
Appropriate initial psychoeducation	Provide a strong rationale to child and parent, based on inhibitory learning goals (focusing on forming new inhibitory associations and maximising expectancy violation)	Use developmentally appropriate language
Eliminating safety signals/ behaviours	Removing signals or behaviours that children could wrongfully attribute non-occurrence of threat to	Safety signals/behaviours may be useful to encourage mastery or reduce attrition in early exposures Parents may inadvertently maintain safety signals/behaviours too through accommodation behaviours
Deepened extinction	Conduct exposure to different feared stimuli separately, and then combine stimuli in later exposures	Combined stimuli can be good to plan during the weekend or school holidays when new stimuli can be more easily planned. Attending a camp can also often provide opportunities for deepened extinction
Occasional reinforced extinction	Allowing exposures where the threat expectation comes true (CS is paired with US)	Not appropriate for all expectations (e.g., death or serious injury). Consider activities like deliberately making a mistake on a school project, or deliberately getting a detention. This often reveals parental concerns about exaggerated feared consequences and can be a good opportunity for child and parental learning
Other strategies that complement inhibitory learning		
Affect labelling	Putting emotional responses into words during exposure (e.g., a child verbalising "I am scared" at the beginning of the practice)	Some children may find this difficult given their level of language or emotional development. The goal is to increase focus on the feared stimuli and so if the child must divert too much attention to completing this labelling task, it may serve as a distraction.

(continued)

Table 2 (continued)

Strategy	Brief description of strategy for children and families	Additional considerations for children and families
Enhancing positive affect	Inducing positive affect before or after exposure may enhance the formation of inhibitory associations	Previously used memory induction tasks (such as recalling a positive memory) could be cognitively demanding for younger children as it requires them to recall and savour the memory in the present tense. Other strategies (currently under investigation by our team) include watching positive video clips just before exposure practice

3.1 Treatment Rationale

Inhibitory learning-based psychoeducation would emphasise that the purpose of exposure is to maximise the violation of the threat association (the "scary" memory) and strengthen learning and recall of nonthreat associations ("not scary" memories) as this learning leads to better long-term remission of anxiety symptoms (McGuire and Storch 2019). This may be different from many CBT protocols for child anxiety that rely on habituation or coping models, providing parents and children with a rationale that the goal of exposure is to be in a situation until anxiety subsides or to develop coping skills.

For example, the Coping Cat program (Kendall and Hedtke 2006), a widely used and well-researched child-focused treatment, provides a rationale to children and parents that treatment involves learning new coping skills (cognitive restructuring, relaxation) and improving these skills by practising them in situations of increasing difficulty. This encourages learning of change in expectancy about a child's coping, and although it is in line with inhibitory learning, it could be supplemented by emphasising other new learning about the likelihood of severity of the feared stimuli as well. Many protocols also blend inhibitory learning and habituation approaches. The Cool Kids program (Rapee 2000), another widely implemented treatment protocol, explains to children and parents that exposure requires the child to stay in the situation until their anxiety has reduced and/or until the child has learned something new. The nuance between these treatment rationale explanations can be subtle and almost undetectable. The greatest difference in these approaches is the framework which can be seen in the way the treatment activities are designed, executed, and reflected upon.

3.2 Eliminating Safety Behaviours/Signals

Another core strategy to improve inhibitory learning during exposure is to remove safety behaviours and safety signals. For children, common safety signals may include items (e.g., toys, clothes), behaviours (e.g., over-preparation, delaying a task), and people (e.g., parents, clinicians). Safety behaviours or signals are suggested to impede exposure as the child maladaptively attributes the non-occurrence of the threat to the presence of the safety signal (Hedtke et al. 2009). For example, a child with social anxiety may think, "I did manage to go to this party today, but it was only because I brought my toy with me." Experimental studies in both humans and animals have shown that safety signals can reduce distress in the short term, but when the safety signal is removed, the fear returns (Lovibond et al. 2000; Pineño et al. 2008). From an inhibitory learning point of view (Blakey and Abramowitz 2019), safety behaviours may reduce attention towards the feared stimuli, reducing the possibility of the inhibitory association forming. Some traditional CBT programs use alongside exposure other techniques such as relaxation and breathing exercises (Kendall and Hedtke 2006), which could also inadvertently serve as a safety signal and reduce inhibitory learning (Craske et al. 2014).

To consider, some studies in adults have suggested that safety behaviours, such as wearing safety gear have been associated with more approach behaviours and acceptability during exposures such as with spider phobia or OCD (Levy and Radomsky 2014; Milosevic and Radomsky 2013). As motivation can be an issue for child treatment, safety behaviours aimed at improving the acceptability of exposure treatment may prove helpful in early exposures as the child is building mastery and may reduce attrition or resistance towards exposure. However, it is still recommended to reduce safety behaviours and signals over time or immediately if the child is willing as this will lead to a greater discrepancy between the child's expectations and the experience, and hence, better learning (Craske et al. 2014).

When providing treatment for child anxiety, it may also benefit to consider the role of parents in encouraging safety behaviours. Parental accommodation, or changes in parent behaviours to prevent or reduce distress in the child, could maintain a child's anxiety (Lebowitz et al. 2013). Parental accommodation could include allowing children to engage with their own safety behaviours, allowing them to avoid anxiety-provoking situations, or encouraging family-wide safety behaviours such as modifying family routines or providing excessive reassurance (Thompson-Hollands et al. 2014). Parental accommodation reduces a child's anxiety in the short term, but it also reduces a child's opportunities to form inhibitory associations. Contingency management strategies can help reduce parental accommodation and improve treatment outcomes (Manassis et al. 2014; Silverman et al. 2021). These strategies can supplement inhibitory learning by providing a rationale for parents to reduce accommodation behaviours as they diminish children's learning and recommending, and suggest for parents to positively reinforce approach behaviours to increase opportunities for new learning.

3.3 Deepened or Compound Extinction

Deepened or compound extinction refers to conducting exposures to extinguish multiple feared stimuli separately and then combining these stimuli in later exposures (Craske et al. 2018; Craske et al. 2014). For example, if a child has a social anxiety disorder, the child may complete separate exposures of reading to a small group, standing and reading in front of the group, and conducting a speech without a script. Then, they may combine exposures to do a speech while standing in front of a group without a script. When multiple fear stimuli are presented at the same time, this is expected to maximise the potential expectancy violation. There appears to be strong theoretical foundation for this technique and some support in human and animal studies (Culver et al. 2015; Rescorla 2006), though there is still limited evidence investigating it in clinical studies.

3.4 Occasional Reinforced Extinction

Occasionally reinforced extinction involves occasionally allowing exposure where the CS and US are paired, i.e., where the threat association comes true (Craske et al. 2018; Craske et al. 2014). The inhibitory learning model suggests that this technique may have multiple benefits on extinction learning. First, the participant may be less likely to expect the next exposure of the feared stimuli to predict the feared outcome because they have experienced both feared and inhibitory associations before (Bouton et al. 2004). Second, occasionally reinforcing the feared association during exposure could enhance the salience of the feared stimuli, which encourages new learning about it (Craske et al. 2014). Furthermore, this may reduce threat expectations, as the child may learn that even when the worst-case scenario occurs, it is not as catastrophic as they predicted (Tolin 2019). This strategy is thought to be especially helpful for reducing subsequent reacquisition of fear (Bouton et al. 2004). When feared stimuli are paired with the "scary" outcome again, this scary memory is less impactful. McGuire and Storch (2019) recommend that this strategy be utilised only after children have already demonstrated mastery in initial exposures.

4 Other Strategies to Support the Development of Inhibitory Associations

Beyond the strategies presented above, inhibitory learning models have included other strategies that enhance the development of inhibitory learning that are considered independent or complementary to extinction learning (Craske et al. 2014). Though the strategies below do not work by enhancing expectancy violation, they have been shown as promising strategies to promote inhibitory learning.

4.1 Affect-Related Strategies

Craske et al. (2018) proposed that affect labelling could enhance the formation of new inhibitory associations by increasing attention to the feared stimuli during exposure. Affect labelling, i.e., having the child identify an emotion such as "I feel scared", is considered a subtle method for downregulating emotion. It is suggested that through this technique, more attention can be focused on the feared stimuli during the exposure, which will allow for non-threat occurrences to be attributed to these stimuli (Craske et al. 2014). Although there has been a strong theoretical rationale for this technique, there has been some mixed evidence for the benefits of affect labelling on exposure in non-clinical samples. Affect labelling in adults was associated with reduced physiological reactivity after spider phobia exposures (Kircanski et al. 2012a) and public speaking anxiety exposures (Niles et al. 2015), compared to controls. However, these studies did not consistently find differences between groups on self-reported fear. Although affect labelling has not been examined in children, one study in 13- to 14-year-olds with public speaking anxiety also found that affect labelling did not enhance exposure at post-test or one-week follow-up (Plaisted et al. 2022). Importantly, the use of this strategy in child populations may require developmentally appropriate tailoring, dependent on aspects such as the child's language development, understanding of emotions, and articulation.

Another affect-based strategy that has emerged from inhibitory learning theories is enhancing positive affect. Experimental studies in non-clinical adults have found that greater trait positive affect assessed before or after exposure was associated with lower reacquisition of fear (Zbozinek and Craske 2017). Furthermore, Zbozinek et al. (2015) found that a positive mood induction reduced reinstatement of fear. This suggested that positive affect may benefit long-term treatment outcomes as it may reduce the return of fear after exposures. Neurobiological studies have also shown positive mood to increase activity in brain regions associated with semantic processing including the medial temporal lobe and prefrontal cortex (Kiefer et al. 2006; Nyberg 2002). Positive affect can also enhance the encoding, rehearsal, and retrieval of information, so it is theoretically expected that enhancing positive mood can improve the formation of inhibitory associations and their retrieval as well (Zbozinek and Craske 2017). Positive mood inductions are suggested to be beneficial before exposure to increase attention to the feared stimulus, or during counterconditioning (Craske et al. 2022). Counterconditioning refers to pairing the feared stimuli with a positive outcome, thus conditioning a positive outcome with the feared stimuli.

Hughes and Kendall (2008) found that children with an anxiety disorder who underwent a positive mood induction had reduced threat biases, i.e., tendencies to interpret ambiguous situations as threats, compared to those who completed a neutral mood induction. Furthermore, Newall et al. (2017) conducted a study in a non-clinical child sample, in which children underwent a fear acquisition of animals. They found that children had greater reductions of fear if they were in a counterconditioning condition (animal paired with a happy face), compared to a regular extinction phase (animal without the scared face) or a control condition (no extinction). Although there is still scant evidence supporting affect strategies in clinical samples, initial evidence suggests that enhancing positive mood may enhance inhibitory learning.

5 Strategies to Enhance Retrieval of Inhibitory Associations

To ensure treatment outcomes are maintained after treatment is over, inhibitory learning theories also suggest other strategies to reduce the return of fear and improve the long-term effectiveness of exposure. After a course of exposure treatment is complete, the original feared and inhibitory associations are still in competition. The feared association could be recovered through processes such as spontaneous recovery, context renewal, reinstatement, and reacquisition of fear, discussed in more detail in Table 1 (Craske et al. 2022):

Return of fear usually occurs because new nonthreat associations are formed in a specific context (usually in the therapist's office or clinic, or in the child's home). When the child encounters feared stimuli in another context or in another way, this fear may return. As such, having strategies to assist the child in retrieving new inhibitory associations across multiple contexts is essential for maintaining long-term outcomes. These strategies include increasing variability, conducting exposure in multiple contexts, and using retrieval cues (see Table 3). The first two strategies encourage greater expectancy violation by reducing the predictability of exposures. Retrieval cues are suggested to complement inhibitory learning, by providing tangible reminders of treatment learning. These strategies will be discussed more below.

5.1 Increased Variability of Exposures

Having variability in stimuli could include conducting exposures with different durations, different levels of intensity, and using different types of fear stimuli during exposures. In practice, this can also mean that exposure can be chosen at random, compared to traditional graded exposures where a child starts on a low-level exposure and moves up towards harder exposures in a progressive manner. As variable exposure is less predictable and more surprising than graded exposure, this is expected to encourage greater expectancy violation. Furthermore, as the post-treatment "real world" context is more likely to be unpredictable and variable, variability during exposure may be a more ecologically valid way of conducting exposures, which could offset context renewal effects as well (Bouton 1993; Craske et al. 2014).

In two clinical analogue studies, variable exposure was shown to lead to greater physical arousal and subjective anxiety than graded exposure at pre-exposure but

Table 3 Strategies to enhance retrieval of inhibitory associations (making "not scary" memories stronger)

Strategy	Brief description of strategy for children and families	Additional considerations for children
Variability of exposures	Varying the exposure duration, intensity, and stimulus. Therefore, exposures are conducted in "random" order or starting at a higher point, rather than in graded order from low to high	Starting children with exposures involving low fear may reduce dropout if children feel too overwhelmed yet it may slow progress. If random order is not possible or feasible, encourage the child to do the most difficult exposure on their list that they feel capable of that day (rather than gradually moving up a hierarchy)
Multiple contexts	Conducting exposures in various physical, internal (different levels of arousal or mood), and interpersonal contexts (with different people)	Conducting exposures at school and or exposures without parents can be challenging yet should not be avoided as practice only at home or in therapy may limit improvement. To begin with, use imaginal or virtual reality to access harder-to-coordinate contexts. Then include additional supports such as teachers or friends. Weekends and school holidays can also be a good time for planning multiple contexts for non-school-related fears
Retrieval cues	Using cues in extinction that can serve as a reminder of the nonthreat association in a new context	There is some evidence that internal cues could be more beneficial than external cues. External cues could be useful for children, especially in conjunction with internal cues

post-treatment outcomes for both exposures were similar (Kircanski et al. 2012b; Lang and Craske 2000). Despite that, variable exposures led to greater long-term benefits including better generalisation of fear learning and maintained reductions of fear at follow-up (Kircanski et al. 2012b). In a clinical context, variability is encouraged only after initial mastery of easier exposures to reduce patient dropout and issues with treatment acceptability. Other ways of allowing more variability may also include changing the time in which exposures are conducted (McGuire and Storch 2019) or varying emotional responses through conducting unpredictable lengths of exposure (Craske et al. 2014).

5.2 *Multiple Contexts*

Conducting exposure in multiple contexts is one of the most common ways of increasing variability. This could involve conducting imaginal, interoceptive, and in-vivo exposures in different contexts. These contexts could include conducting exposures in different external contexts (i.e., locations and surroundings), different internal contexts (i.e., physiological or physical states), or different interpersonal contexts (i.e., with or around different people).

Regarding external contexts, previous human and animal studies have found that conducting exposures in multiple locations reduced fear renewal, especially when exposures were conducted in vivo and locations where extinction learning environment is similar to that where the fear was acquired or those where renewal may occur (Balooch and Neumann 2011; de Jong et al. 2019). As it is sometimes impractical to know when a fear was acquired in clinical practice, it could be helpful to conduct exposures in multiple contexts where renewal may occur. For example, if a child has a fear of dogs, they could do exposures in the office, at home, and at the park. Internal states, such as arousal or mood, also represent a context that could become tied to the new inhibitory learning. Mystkowski et al. (2003) found that adult participants given caffeine at treatment and follow-up had reduced return of fear of spiders, compared to participants given caffeine only at follow-up. Some children may only wish to conduct exposure when they are feeling calm and ready (McGuire and Storch 2019), but encouraging exposure despite internal state (e.g., when feeling anxious, tired, alert, or calm) has benefits in reducing avoidance and in supporting inhibitory learning through reducing renewal. Lastly, varying the interpersonal context can be helpful, especially as children commonly experience anxiety with many different relationship groups. For example, a child with social anxiety may be able to do a speech with the therapist, parent, teacher, and peers.

5.3 Retrieval Cues

To support new learning being transferred to different contexts, children can also use retrieval cues during treatment. Retrieval cues refer to stimuli that could be to remind the child of their learning making inhibitory learning associations more accessible in multiple and diverse settings. For example, Culver et al. (2011) conducted a study in which adults with elevated public speaking fears conducted exposure with a pen and clipboard present or not. They found that having the pen/clipboard present during exposure reduced the renewal of fear in a follow-up test conducted in a novel setting. For children, this could include wearing a wristband during exposure that they carry with them or getting a certificate for completing an exposure that is placed at home on a wall. To note, some other studies on external retrieval cues have failed to find the benefits of context renewal (e.g., Dibbets et al. 2013; Laborda et al. 2016). However, internal retrieval cues could be another strategy to reduce renewal. This involves actively recalling the initial exposure context when encountering a new anxiety-provoking scenario after treatment. For example, this could involve asking the child, "Remember what you learned in our session last week?". This strategy has been shown to have a more robust effect on reducing renewal than external retrieval cues (Mystkowski et al. 2006). Blakey and Abramowitz (2019) suggested that internal cues may represent a psychological retrieval cue, which could be more practically useful than external cues (as it is always accessible). Internal cues could even serve as a reminder of previous successes, therefore enhancing self-efficacy for tolerating distress.

Importantly, retrieval cues can sometimes be risky if they become used as a safety signal (Dibbets et al. 2008). Whereas retrieval cues are aimed at retrieving the new inhibitory associations, safety signals become directly associated with the non-occurrence of the US. For example, a wristband could become a safety signal if the child believes that they only managed to complete the exposure because they had their wristband. On the other hand, the wristband could be a retrieval cue if the child thinks of their successes from exposure when seeing the wristband outside of the clinic. As such, retrieval cues are recommended as a relapse prevention skill (Craske et al. 2014). They should be used sparingly and only after the child has already formed inhibitory associations to mitigate its possibility of becoming a safety signal.

6 Suggestions for Further Research

More empirical evidence is required to examine how exposure can be optimised through inhibitory learning strategies in clinical trials with children who have anxiety diagnoses. Traditional methods, such as conducting multiple trials on different strategies separately would be a time-consuming and resource-heavy endeavour. Research efforts should go towards effective interventions that test multiple strategies together, which could allow for simultaneous examination of multiple strategies and their interaction effects. One study design that could achieve this is through factorial design optimisation techniques that test which combination of strategies is most effective to determine an optimised method of conducting exposure for children with anxiety. Our team is currently conducting a factorial-design RCT examining the potential benefit of five features based on inhibitory learning strategies (ANZCTR registration number: ACTRN12623001256662). This includes three features that we would expect to enhance outcomes, i.e., expectancy violation, parent training, and positive affect. We also will examine two features that are commonly utilised in current child treatments but have inconsistent effects on outcomes based on inhibitory learning, i.e., rewards and relaxation. These two features will allow us to understand if strategies in our current standard treatments may be reducing mechanisms for change.

7 Conclusion

This chapter provides an overview regarding the use of the inhibitory learning framework to maximise exposure outcomes in children with anxiety disorders. For further information on the clinical implications, and practical strategies to help guide clinicians in implementing an inhibitory learning-based model of exposure therapy for children with anxiety, please see Appendix A below.

Most of the research to date investigating inhibitory learning-based strategies to improve exposure has been conducted in adults with elevated anxiety or anxiety disorders. However, adapting these strategies for child populations has immense potential to improve treatment outcomes for children with anxiety and enhance new learning pathways. This could also reduce mental health burden early in life and prevent further exacerbation of difficulties at an older age. Given the differences in cognitive and emotional development in children, clinicians need to tailor and personalise strategies during therapy to optimise inhibitory learning for children with anxiety. Whilst there may be some challenges in shifting from traditional exposure-based models of therapy to an inhibitory learning framework, this modern learning theory is supported by a wealth of experimental research and promising clinical research and holds promising advantages for optimising treatment efficacy in children. Future translational research of a systematic nature is needed to better understand the optimal method for implementing this model in routine clinical care for children, as well as a more nuanced understanding of factors that may serve as moderators and mediators of these processes.

Appendix A: Clinical Implications

This section describes a clinical application that uses methods to optimise exposure through the lens of inhibitory learning theories. To assist in the application of these methods, we have supplemented the strategy descriptions with a case study of a 10-year-old boy presenting with a Specific Phobia (Natural environment type.) Ashwin and his mother Miriam (names and details changed for confidentiality and anonymity) received a blended care treatment involving both digital and person-to-person care as part of a research trial at the Black Dog Institute, Sydney Australia. Although we have chosen to use a concrete example of a specific fear, applications of inhibitory learning theories can be applied to a wide range of fears and worries, including abstract fears and worries often seen in children with generalised anxiety or obsessive-compulsive disorders.

Case Study Example

Ashwin, a 10-year-old boy, presents to the clinic with his mother, Miriam. An assessment indicates that Ashwin has a specific phobia of thunderstorms and rainstorms. Miriam tells their psychologist that Ashwin gets extremely upset when it storms, when there is heavy rain, or heavy clouds.

***Miriam**: "We could be driving in the car, and he'll see grey clouds in the distance, and he'll ask us if we're going in that direction. He'll keep asking us until we arrive. But if we are driving under grey clouds, he keeps asking us if we think it's going to storm and what happens if the lightning hits the car. He*

(continued)

keeps thinking it will storm and lightning is going to strike the house. There's been many times where we've had to cancel our plans because he won't leave the house. He misses out on a lot. He missed out on going with his friends to the new bike track that just opened because there was rain forecast. He was really disappointed to miss out. For the last two years, it feels like this is an issue that crops up at least once a day, even when the weather is good. He checks the weather. Sometimes we can distract him during a storm by watching TV or playing a game, but then he'll hear thunder and start getting worried again. If it's storming when it's his bedtime he will cry and beg us to stay with him. If he refuses to go to bed on his own, sometimes we let him sleep in our bed. He also gets really upset if there's a storm and he can't find Cody – his toy bear. We must turn the whole house upside down to find Cody, otherwise Ashwin is inconsolable."

Assessment

Few standardised assessment tools are available that adequately prepare clinicians, parents and children to design optimised exposure activities during treatment, with most tools focusing instead on the assessment of symptoms and diagnoses. These types of assessment methods (e.g., diagnostic interviews, rating scales, behavioural avoidance tasks) gather information from multiple informants (e.g., from home and school) and provide information (often generic) to help the clinician understand the extent and severity of the presenting problems. Although such assessment tools can provide some information about specific stimuli and expectancies, typically greater detail is required before designing and commencing personalised exposure activities. Although behavioural avoidance tasks are a valuable assessment method to quantify approach/avoidance behaviour and gain a more nuanced assessment of the child and parent's behaviour during feared situations, this method is rarely used in clinical practice and it also, like an adequately designed exposure activity requires the collection of more detailed information not typically available from standardised assessment tools.

In contrast, a robust assessment of fear stimuli (i.e., internal or external), cognitions (e.g., expectancies), and behaviours (e.g., safety-seeking, mental responses, avoidance, parental reinforcement, reassurance provision) will provide the kind of detail that would enable optimised personalised exposure activities. In addition, it is recommended that the clinician also conduct an exploration of factors that the child believes will increase or decrease the likelihood of the negative outcome occurring. This may include physical sensations, objects, thoughts or images, places, behaviours and external situations or settings. This level of detail will be key in developing the exposure activities and to ultimately elicit what the child needs to learn. Our team is in the process of piloting a new assessment tool to aid clinicians, parents, and young people in developing personalised exposure activities more rapidly from the assessment phase. In the future, it is hoped that such a tool will help to facilitate more rapid and extensive use of well-designed exposures.

Clarifying Expectancies

A particularly important component of assessment in the context of inhibitory learning theories is eliciting expectancies. The clinicians may ask questions to elicit expectancies towards the fear and related features including: (1) a probability rating, i.e., how likely the child believes the anticipated outcome will occur, (2) affect expectancy, i.e., how distressed they believe they would be if the outcome occurred, and (3) coping expectancy, i.e., how they believe they would cope if the outcome occurred. Rating these on a quantitative scale such as the scale in Fig. 2 can help when reviewing and comparing post-exposure affective expectancies with the child to highlight expectancy violation.

To clarify Ashwin's expectancies, the psychologist asks specific questions about Ashwin's fears. The psychologist provides multiple prompts to understand Ashwin's expectancies.

Psychologist: *"When you're in the car and you see grey clouds in the distance, what do you worry about?"*

Ashwin: *"I don't know... I just get worried."*

Psychologist: *"When you see the grey clouds, what do you think is going to happen?"*

Ashwin: *"I think it's going to storm."*

Psychologist: *"And what do you mean by storm? Like you're worried it will rain?"*

Ashwin: *"I guess so...".*

Psychologist: *"If it rains, what do you think is the worst thing that could happen?"*

Ashwin: *"Well if the rain gets heavy, it might end up storming."*

Psychologist: *"And what about the storm makes you feel worried?"*

Ashwin: *"I don't know... I just get worried."*

Psychologist: *"Is it that you don't like it because it's cold? Because there might be thunder and lightning? Because it gets a bit dark?*

Ashwin: *"Yeah, because there might be thunder and lightning."*

Psychologist: *"And what do you think will happen if there's thunder and lightning?"*

Ashwin: *"I'm worried the lightning will strike us."*

(continued)

***Psychologist**: "Okay, so when you're in the car and you see the grey clouds, you're worried it will end up storming and lightning will strike the car? Or just strike you?"*

***Ashwin**: "I'm worried the lightning will strike the car but then what if it passes through me?"*

***Psychologist**: "What do you think will happen if the lightning passes through you?"*

***Ashwin**: "It sounds like it'll be really scary and painful."*

***Psychologist**: "Okay, so when you're in the car, you're worried that the lightning will be scary and painful. And do you worry about that when you're at home too and it's storming?"*

***Ashwin**: "Yes."*

***Psychologist**: "What do you get worried about when you're at home and it's storming?"*

***Ashwin**: "I'm worried the lightning will hit our house."*

***Psychologist**: "How much do you believe that you, your car, or your house will be hit by lightning?"*

***Ashwin**: "A lot."*

***Psychologist**: "If you look on this piece of paper, there is a scale from 1 to 8, where 1 is it's not going to happen and 8 is it's definitely going to happen. Can you point to where on this scale you believe you will be hit by lightning?"*

***Ashwin**: "Maybe a 7 [points to scale]. Because it might not happen all the time, but I think it will probably happen at least once in my life."*

***Psychologist**: "How upset would you be if the lightning hit your car or your house, on this same scale?"*

***Ashwin**: "Definitely an 8 [points to scale], that would be horrible."*

***Psychologist**: "What do you think you would do to cope, if your house or car got hit by lightning?"*

***Ashwin**: "I would probably hug Mum or Dad."*

From this assessment, Ashwin's psychologist has established the stimuli (seeing grey clouds, hearing heavy rain or thunder) and his cognitions (he believes the lightning will strike his house, car, or himself; it will be painful and scary). Additional questions (not listed above) also revealed specific behaviours he engages in that will be important for developing exposure activities: (asks reassurance-seeking questions, avoids leaving the house when it is raining heavily, distracts himself by watching TV or playing games with family, avoids going to bed alone when storming, must have his toy bear Cody with him). A goal for treatment is for Ashwin to be able to ride his bike at the local outdoor bike track in the rain. Exposure activities should be created where Ashwin can learn to have "not scary" memories associated with grey clouds, heavy rain, and thunderstorms.

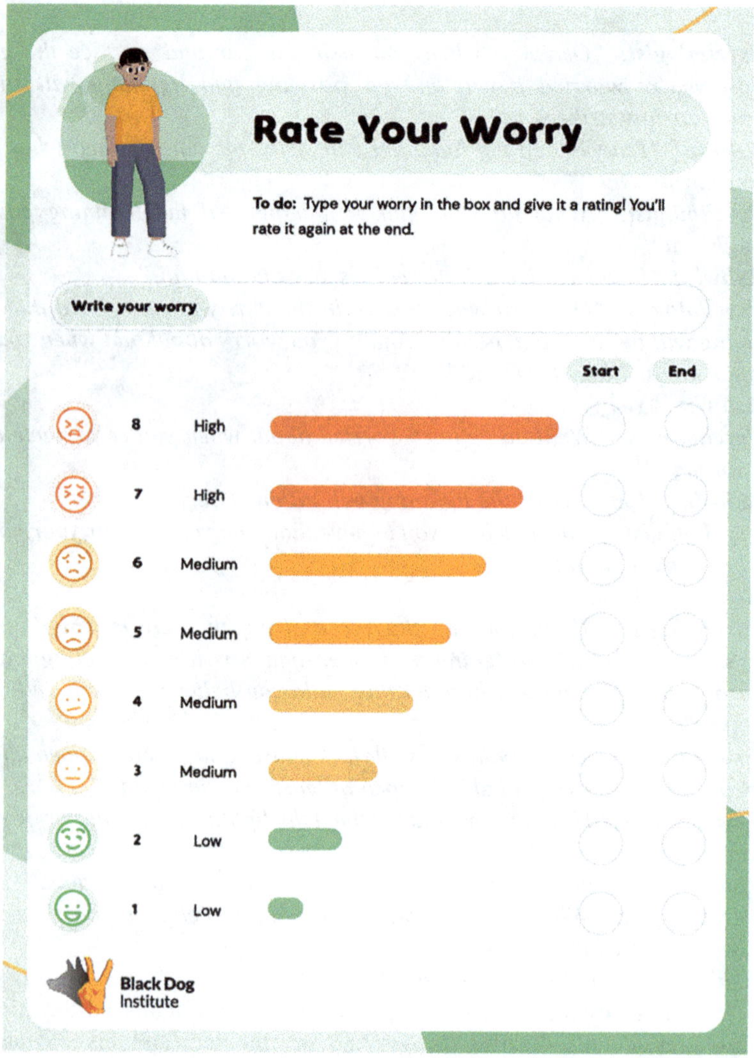

Fig. 2 Example worksheet for children to rate their feelings and beliefs pre- and post-exposure

How To Do Exposure Based on Inhibitory Learning

Comparison Between Exposure Approaches: Traditional and Inhibitory Learning Model

There are several differences between a traditional and an inhibitory learning model approach to exposure that impact both the set-up and administration of exposure. The Table 4 below summarises the different factors. The right-most column

Table 4 Varying features of exposure based on traditional versus inhibitory learning exposure approaches

Design and features of exposure	Traditional approach	Inhibitory learning approach
Treatment objective	Symptom reduction or skill practice	Facilitating the learning needed for the child to experience that the feared outcome did not occur or was not as bad as expected
Development of ladder steps	Developed and ordered based on increasing subjective units of distress (SUDS)	Developed through an assessment based on expectancies and learning needed, and SUDS In addition, incorporating as much variability between steps as possible (see Increased Variability of Exposures section)
Undertaking exposure steps	Gradual and linear fashion from low to high SUDS	Undertake exposure with variable SUDS levels. If this is too difficult and the child refuses to engage, start with a low-level exposure to increase confidence and move quickly to variable exposures or choose the most difficult the child feels capable of doing
Cessation of exposure activity (i.e., stop criterion during exposure)	A reduction in SUDS	Clinician and child decide on a temporal limit apriori that is based on the time required to allow for an adequate test of the expectancy
Incorporation of other CBT techniques	Frequently used in conjunction with psychoeducation, cognitive restructuring and relaxation with no specific instructions in use and combination	Rationale of psychoeducation tailored based on expectancy violation and learning. Strategies including cognitive restructuring and relaxation are not recommended to be used before or during exposure

summarises the key components of the model that can be applied clinically to optimise exposure.

Clinical Application of Strategies to Optimise Exposure

The section below provides further detail related to the clinical application of strategies based on the inhibitory learning model that can be applied in addition to usual exposure treatment protocols.

Psychoeducation

Psychoeducation based on the inhibitory model emphasises the importance of the learning process, i.e., "what does the child need to learn to eventually be less fearful or be able to function in daily life?" This compares to the rationale of "How do I get rid of fear?" (as per some traditional exposure models).

The following is an example of psychoeducation delivered to a parent which can be personalised based on the child's fears:

> *Exposure is a powerful strategy that involves your child engaging with and confronting their fears. It works by breaking the avoidance cycle. Through exposure, your child experiences opportunities to test out their scary beliefs. Depending on these beliefs, they may learn that perhaps the feared outcome doesn't happen. They might also learn that the feared outcome isn't that bad, or maybe that they were able to handle their fears even when they felt strong or unmanageable. In this way, your child will develop new, strong and safe memories. It is important to note that we aren't trying to get "rid" of your child's memories or their fears. Rather, our aim in exposure is to provide opportunities for learning, which can in turn help reduce your child's fear. We want your child to test it out for themselves. This means your child will be asked to temporarily tolerate their fear to allow for the learning opportunity. Over time, when your child is faced with their fear and their memories are activated, these new and safe memories that they have developed will compete with (and eventually outcompete) the old danger memories.*

Another way of explaining exposure, particularly for children, may involve the "scientist" metaphor:

> *We are going to pretend that we are scientists. It sounds like you have this big, scary idea in your head that you can't do something, like talk to new people. When scientists want to find out the truth about something, they use experiments. What if we were to use an experiment to test if your fear is true?*

Explaining Maximising of Expectancy Violation

Consistent with expectancy violation principles, the therapist empowers the parent to set up exposures in a way that maximises the discrepancy between the anticipated and real outcome for themselves. This may be explained as:

> *It is important for your child to really experience the difference between what they feared would happen compared to what happened in its fullest form. The bigger we can help them experience this "surprise" for themselves, the stronger the learning will be. For this reason, we want to encourage you to hold back from providing certain types of reassurance such as "nothing bad will happen" or "he looks like a nice dog, he won't bite" because this may reduce the surprise. Instead, we want to help you provide support around other things such as their willingness to approach their fear example, "I really like how you are giving this a go!" This can be really challenging as it is a different approach to what might feel natural. My job is to support you so that you can provide the best support you can to your child in these situations.*

Explaining Eliminating Safety Behaviours and Signals

The goal is to support the parent and child to do the exposure activities without the use of safety behaviours and signals. Psychoeducation to parents about safety behaviours and signals can also help to identify common safety behaviours the child has that they may be unaware of.

> *It is important that your child understands they can do these activities without relying on anything else. For example, your child might want to carry a specific toy when doing an exposure, or they might even want to just hold your hand. We call these safety behaviours or signals – things that make the child believe a situation is safe. If they always have safety behaviours or signals in exposures, they might only believe "Nothing went wrong because I had my toy," or "Nothing went wrong because my parent was there." We want them to learn that they can confront their fears without their safety behaviours. If it is hard to remove safety behaviours, we can have them at the start, but we will ideally work quickly towards your child not having any safety behaviours during exposure practice. This helps to maximise the surprise and the learning.*

Like safety behaviours, a discussion about the parents' accommodations around the child's anxiety and how this, while a natural response to seeing your child in distress, could maintain the anxiety as it hinders the learning process for children. It is also helpful to obtain a brief understanding of the parent's own anxieties and how they can model positive behaviour that could reinforce inhibitory learning.

> *A natural response for some parents is to allow your child to avoid the thing they're scared of. You might have changed your family's routines or provide lots of reassurance like "Everything will be okay." Just like safety behaviours, this helps to reinforce the "scary memories" about the feared situation and prevents new learning of "not scary" memories. Children also learn a lot from how their parents act, and if they see that you are making these accommodations, they will also learn to make accommodations too. It works the other way too. If you show them how important it is to face your fears, they will learn this as well. You could also think about your own fears, how you respond to them, if you make any accommodations, and what your child learns when they see you responding to your own fears.*

It would also be useful to provide psychoeducation to parents about the benefits of intrinsic rewards immediately after exposure.

> *It's often a good idea to reward your child after they do an exposure. Although material rewards like buying them a toy or chocolate can be motivating, we think it is also extremely helpful to reward them with non-material things like praising them or spending more time with them. Try praising them while being clear on what exactly you are praising them for, such as, "I am so proud of you for doing your exposure" or "Even though you were really scared to do that, you gave it a go. Well, done!" This helps them to learn that doing exposure practices can be associated with feeling good instead of feeling scared.*

The psychologist discusses the cycle of avoidance and expectancy violation as described above. Additionally, the psychologist highlights the safety behaviours and accommodations that could be maintaining Ashwin's anxiety:

(continued)

- Answering Ashwin's questions about whether it will storm
- Cancelling plans to let Ashwin stay at home and avoid the storm
- Distracting him when there is a storm
- Letting him sleep in his parents' bed
- Ashwin having a teddy bear (Cody) with him

During a discussion without Ashwin, the therapist encourages Miriam to consider how she responds to her own anxieties, and how she can model helpful behaviours when faced with her own worries.

Planning and Setting Up Exposure

Following a comprehensive assessment of the child's fears and worries (see Assessment section), the next step is the clinician working with the child and parents to identify a behavioural goal that the child would like to achieve. Once this has been established, the clinician and child can work backwards from this goal to identify the various related situations the child is fearful of. This can be developed in the form of a fear hierarchy (also known as a *step ladder*) which is a key feature of any exposure protocol. Items on the ladder can be developed based on their threat expectancies and subjective units of distress (SUDS) and ordered by subjective units of distress as per traditional protocols of exposure. Despite the hierarchal construction of the ladder, it is important to remember that the inhibitory model does not suggest attempting these activities in a graded fashion from low to high but instead in a way that will encourage the greatest expectancy violation (see Increased Variability in Exposure section).

It is important to remember that based on the inhibitory learning model, cognitive strategies such as restructuring should not be used before or during an exposure as it may prevent maximum violation. They can instead be introduced at post-exposure.

Increase Variability in Exposure Stimuli and Contexts

As discussed in the "Increased Variability of Exposures" section, the incorporation of variability has the potential to optimise exposure. Some examples of stimuli-level including variations in nature and types of stimuli, settings and environments the stimuli are experienced in (including unfamiliar places), interpersonal contexts, levels of intensity, duration, and number of fear cues present.

Specifying a Temporal Limit

In comparison to the habituation model where the exposure task is offset once fear subsides, the inhibitory model posits that the duration of exposure is determined by the time required for expectancies to be reduced. The clinician and child can predetermine at each level the time required that will allow for an adequate test of the expectancy (i.e., temporal hypothesis). For example, the child may suspect that a dog will bite them within 5 min of being at a dog park. The exposure task offset would then be 5 min as opposed to the previously used criteria of fear reduction. It is therefore important for the clinician to help the child specify the duration for each exposure task as this determines the offset criterion.

An exposure hierarchy is created for his goal of being able to go bike riding with his friends even if rain is forecast. The list includes inhibitory model features such as temporal limits, increased variability, etc. These changes are to ensure Ashwin does not rely on the presence of his Mum or Dad to feel safe, and he does not rely on being in his own home to feel safe. Inhibitory model features are underlined in the table below.

Distress	Steps	Exposure optimisation features introduced
Low	Watch a video about storms • *Short* video • *Long* documentary	• Temporal limit
	Look out the window for *5 min* when it is raining heavily • *At home* • *At school* • *At a friend's house*	• Temporal limit • Increasing variability in an external environment (familiar and unfamiliar places) and interpersonal contexts
	Be *visually* guided through storm imagery	• Changing the type of exposure (i.e., imaginal) (increasing variability)
	Listen to storm sounds online for *10 min*	• Temporal limit
Medium	Stand outside (under shelter) and hold Mum or Dad's hands when it is raining *for 5 min*, then stand *for 5 min* without holding their hands • *In front of the house* • *Outside the shops*	• Temporal limit • Increasing variability in an external environment (familiar and unfamiliar places)
	Stand outside (under shelter) and hold onto Cody when it is raining for *5 min*, then stand for *5 min* without holding Cody • *In front of the house* • *Outside the shops*	• Temporal limit • Increasing variability in an external environment (familiar and unfamiliar places)
	Only ask *one* reassurance-seeking question when seeing grey clouds; if I ask another then Mum and Dad won't answer me	• Limiting reassurance seeking
	Listen to the storm for *5 min* before distracting myself with TV or games	• Temporal limit
	When it is raining heavily, go to bed on my own for *10 min*, and then Mum and Dad can say goodnight to me	• Temporal limit

(continued)

Distress	Steps	Exposure optimisation features introduced
High	Stand outside (under shelter) and hold Mum or Dad's hands for *5 min*, then stand for *5 min* without holding their hands, when there is *both rain and thunder*	• Temporal limit • Increasing number of fear cues present (deepened extinction)
	Stand outside (under shelter) and hold onto Cody when there is *both rain and thunder* for *5 min*, then stand for *5 min* without holding Cody	• Increasing the number of fear cues present • Temporal limit • Removal of safety behaviour
	When there is both *rain and distant thunder*, leave the house, walk through the rain, and get in the car. Stay in the car for *5 min* before leaving the car, walking through the rain, and going back into the house	• Increasing the number of fear cues present • Temporal limit
	Stand outside on my own (without Mum, Dad, or Cody) when there is *both rain and thunder* for *5 min* then I can go indoors • *At home* • *In front of school* • *Outside the shops*	• Increasing the number of fear cues present • Temporal limit • Increasing variability in an external environment (familiar and unfamiliar places)
	Go to bed on my own when there is *both rain and thunder* for *5 min*, then Mum and Dad can say goodnight to me	• Increasing the number of fear cues present • Temporal limit
	Spend the afternoon at the new bike track with my friends when there is rain forecast and distant thunder	• Temporal limit • Increasing variability in an external environment (unfamiliar places) • Increasing number of fear cues

Conducting Exposure

Pre-exposure

Before undertaking the exposure, the clinician will facilitate the following key steps with the parent and child:

1. Selecting the exposure task to be undertaken from the hierarchy (see Exposure Task Selection and Progression section)
2. Eliciting and clarifying the expectancy related to the exposure task (see Clarifying Expectancies section)

Exposure Task Selection and Progression

It is important to balance factors that increase the efficacy of exposure with maintaining therapeutic alliance and engagement. Although varying the difficulty of exposure in a random fashion may increase efficacy, this needs to be balanced with the child's willingness to undertake each exposure. Similarly, the clinician may have an optimal exposure in mind that provides an opportunity for learning that directly targets a feared outcome. However, it is important to collaboratively explore what the child is willing to test out to maintain rapport and adherence.

While conducting exposures in session, the clinician may elicit affect labelling by asking children to say aloud what emotions they are feeling. An emotion wheel can be provided if children find it difficult to identify their emotions. Affect labelling can be elicited briefly multiple times (i.e., before, during, and after the exposure). As described above, the goal of this is to increase focus on the feared stimuli, so if this becomes a complex task that serves as a distraction or makes the child/parent focus on the wrong aspects of the tasks (e.g., reducing the anxiety rather than new learning) then it is not recommended.

With the exposure hierarchy developed, Ashwin and his psychologist started their exposure activities.

To increase the variability of the exposures, the psychologist asks Ashwin to pick the hardest activity he thinks he can do today. He selects:

"Stand outside the shops (under shelter) for 5 min and hold onto Cody when it is raining, then stand for 5 min without holding Cody" [Medium]

Due to the nature of Ashwin's fear, this exposure activity cannot be done immediately in-session or even at home, as Ashwin's family will have to wait for a day when it is raining. In this, case, the psychologist can set up an exposure where Ashwin and the psychologist close the curtains, turn the lights off, and play thunder sounds to help Ashwin's imagination.

Here is a scenario where it was already raining during Ashwin's in-person session and Ashwin has brought Cody along. The psychologist starts by first repeating the expectancy questions:

***Psychologist**: "Since it's already raining, we can stand outside the office and do this now. How do you feel about that?"*

***Ashwin**: "I feel worried."*

***Psychologist**: "What do you think might happen?"*

***Ashwin**: "I think it will start to storm and then I'll be struck by lightning."*

***Psychologist**: "Okay let's record it. Here is the scale of 1-8, how much do you think you will be struck by lightning when we go outside the office now?" [Ashwin points to 6]*

***Psychologist**: "And if does end up storming, and you hear thunder, how upset do you think you will be on this scale?" [Ashwin points to 7]*

***Psychologist**: "What do you think will happen once you stop holding Cody?"*

***Ashwin**: "I don't know... I just don't want to stop holding him."*

***Psychologist**: "What do you like holding onto Cody?"*

(continued)

Ashwin: *"I don't know... it feels nice to hold onto him."*
Psychologist: *"Does holding onto Cody make you feel safe?"*
Ashwin: *"Yeah, I guess so."*
Psychologist: *"Are you worried that if you stop holding onto Cody, you won't feel safe anymore?"*
Ashwin: *Yeah... and I'll probably get really scared again."*
Psychologist: *"Do you think you'll be able to stand in the rain for 5 min without Cody?"*
Ashwin: *"Probably not."*
Psychologist: *"Can you rate on the scale how much you think you can stand in the rain for 5 mins without Cody?" [Ashwin points to 1]*

They begin the exposure activity. When they reach the first 5-min mark, the psychologist elicits affect labelling, and ensures they reward Ashwin with verbal praises:

Psychologist: *"How are you feeling now?"*
Ashwin: *"I'm feeling better."*
Psychologist: *"What emotions are you feeling?"*
Ashwin: *"I'm feeling calm."*
Psychologist: *"And now the first 5 min are complete. Well, done, Ashwin! You managed to get through those 5 min standing outside when it's raining. Now, pass over Cody and we'll continue for another 5 min."*
Psychologist: *[After Ashwin passes Cody over] "How are you feeling now?"*
Ashwin: *"I'm feeling scared again because I can't hold onto Cody."*

Measuring Progress

In traditional applications of exposure, SUDS and other progress measures that track symptom reduction would be used as key indicators of exposure efficacy. For an inhibitory learning-based treatment, there would be less focus on the change of anxiety symptoms within the exposure. In other words, despite expectations that the worst-case scenario will happen, the aim of consistently facing the feared stimulus is to learn other potential scenarios, e.g., "This 'scary thing' is sometimes not scary," or "I coped really well in this situation." Given this, a more suitable progress measure is tracking the child's self-reported qualitative learnings and expectancy reappraisal, for instance the difference between pre- and post-ratings of expectancy occurrence. These perceived likelihood ratings (i.e., strength in belief of expectancy occurring) can be recorded on a scale such as that in Fig. 2 and used to track quantitative progress. See "Post Exposure Debrief" section for example questions to elicit this information.

SUDS ratings can be a useful complementary measure given that we expect to gradually see a reduction in distress as a by-product of the child's therapeutic learning. However, SUDS ratings should not be used as the key and sole marker of therapy progress. It is also possible that during an exposure or even between exposures, distress could fluctuate and is likely to even increase during initial exposures. Based on the inhibitory learning theory, exposure would still be considered effective if SUDS remained high but a reduction in threat reappraisal was observed. As per usual clinical practice, routine outcome monitoring of supplementary measures (e.g., fear, anxiety, life interference) can further help with assessing progress.

After the exposure activity, the psychologist measures Ashwin's progress by asking him to rate his distress. Again, the psychologist ensures they reward Ashwin with verbal praises.

Psychologist: *"That's time! Well, done Ashwin, you were able to stay for 5 min in the rain without Cody. I know you were scared before, but you were really brave giving it a go, and you completed the activity! How do you feel now?"*

Ashwin: *"I'm feeling better."*

Psychologist: *"What emotions are you feeling?"*

Ashwin: *"I'm feeling happy I guess."*

Psychologist: *"Happy? What do you feel happy about?"*

Ashwin: *"I guess I'm happy that I completed it, and kind of proud."*

Psychologist: *"Happy and proud, nice! If you were to do this activity again in the future, how much do you believe you will be struck by lightning on the scale of 1-8?" [Ashwin points to 5]*

Psychologist: *"And how much do you believe you can stand in the rain for 5 mins without Cody?" [Ashwin points to 7]*

Retrieval Cues

After conducting several exposures and once the child begins to form inhibitory associations, it may be helpful to introduce retrieval cues. One way to engage internal retrieval cues is to use coping mantras during the exposure, such as "ride the anxiety wave", which can be stated when encountering feared stimuli outside of treatment. It may be preferable to use internal retrieval cues, though this may depend on the child's developmental ability. Physical cues or reminders could also be used and may be preferred by younger children and can also be used in conjunction with internal reminders, for example, a picture on a laminated card or a bracelet that reminds them of the "not scary memory" the child can carry with them.

Post-exposure Debriefs

The post-exposure debrief is an important opportunity to highlight expectancy violation in the context of inhibitory learning and can be completed immediately after exposure practice. Post-exposure questioning aims to bring awareness to what the child expected would happen and reflect on what the child learnt.

The clinician can use a range of open- and close-ended questions to explore this. See examples below:

- What did you fear would happen?
- Did your greatest worry come true? How do you know?
  - If it didn't, then what actually happened?
  - If it did happen, how did you cope? How bad was it?
- How was that surprising?
- What did you learn? Imagine that you were going to repeat the same exposure task. How likely is it that your greatest fear will come true? (rated 1–8, using a worksheet such as in Fig. 2).

Using a quantitative scale will help the clinician refer to the pre-exposure rating to highlight these changes for the child.

For younger children, more directive and leading questions can be used to elicit learnings. For example, the clinician may say "You thought you weren't going to be able to cope with the scary thing. Did you notice that you managed to get through so well?". This process will facilitate attention to both the previously "scary thing" and the "not scary" memory.

At the end of the session, the clinician and child will discuss what exposure activities will be practised as a take-home task. The clinician will clearly set out the steps for the family to follow including: (1) clarifying expectancies pre-exposure, (2) practise exposure, and (3) post-exposure debriefs questions.

After the exposure activity, back in the office, the psychologist reviewed Ashwin's expectancies and whether the activity violated his expectancies.

***Psychologist**: "Let's put on our scientist hats and have a look at the data you gathered. Let's have a look at what you thought would happen before we started our practice. Do you remember what you told me?"*

***Ashwin**: "Um... I can't remember."*

***Psychologist**: "You can always look at what we wrote down if you can't remember."*

***Ashwin**: [Looks at the record] "I said I thought it would storm and I'd be struck by lightning."*

***Psychologist**: "Okay, so that's what you did before you did your practice. What do you think of that now?"*

***Ashwin**: "Well it didn't storm... but I still think that if it did storm, I might get struck by lightning."*

***Psychologist**: "That's okay, we haven't tested that out yet. Today our practice was about the rain, so let's focus on that. You said something*

(continued)

interesting just then. Before we did the activity, you thought that because it was raining, it would storm. And you just said that it actually didn't storm. What do you think about that?"

***Ashwin**: "I don't know. . ."*

***Psychologist**: "Let's put our scientist hats on and treat this like an experiment. We guessed that when it rains, it will storm. Is that what happened today?"*

***Ashwin**: "No. . ."*

***Psychologist**: "So maybe our guess wasn't correct? Maybe our conclusion is that it doesn't always storm whenever it rains?"*

***Ashwin**: "Yeah I guess so. . ."*

***Psychologist**: "And do you remember what you said when I asked if you thought you could stand in the rain for 5 min without Cody?"*

***Ashwin**: "I said I didn't think I could."*

***Psychologist**: "That's right. You pointed to 0 when you rated it on the scale, but you ended up lasting the whole 5 min without Cody! So, if you had your scientist hat on, what could your conclusion for this be?"*

***Ashwin**: "That I don't always need to have Cody to be able to stand outside when it's raining?"*

***Psychologist**: "Yeah, I think that sounds like a reasonable conclusion. And how do you feel about that?"*

***Ashwin**: "I guess I feel proud and brave. I didn't think I could do that, but I did!"*

***Psychologist**: "You did a wonderful job and I'm so glad you feel proud of yourself for it! Why don't we write down on your piece of paper that you DID end up standing in the rain without Cody for 5 min and that you felt happy and proud of yourself for doing so?"*

Towards the end of the session, Ashwin and his psychologist create his action plan for the week. This involves planning some exposure activities, with the psychologist encouraging variability of exposures.

***Psychologist**: "Our next appointment is next week. Let's set up an action plan for then. What do you think you can do before next week to face your fear of thunder and lightning?"*

***Ashwin**: "I could probably try some of the low fears."*

***Psychologist**: "Okay, let's have a look at your low fears again. Which one of these do you think you can do?"*

***Ashwin**: "I can probably watch a short video of storms, and also listen to storm sounds online for 10 min."*

***Psychologist**: "That's a great idea, Ashwin. What if it rains during the week, how do you feel about trying one of the other ones? We could pick it from the hat again to see what activity you could do. How do you feel about that?"*

(continued)

***Ashwin**: "I'm not sure... I'm a bit nervous about doing that."*

***Psychologist**: "What if we just picked one out of the hat now, and if it rains, you and Mum and Dad can decide if you'll do it or not?"*

***Ashwin**: "Okay I guess I can." [Picks "Go to bed on my own when it is raining heavily for 10 min, then Mum and Dad can say goodnight to me." [Medium]]*

***Psychologist**: "Okay so if it rains heavily, you can talk to Mum and Dad about trying this out. You can even try this out when it's not raining, if you get Mum and Dad to play rain sounds and you can pretend it's raining. But that's up to you to decide. How does that sound?"*

***Ashwin**: "I don't know... I don't think I can do that. It sounds scary."*

***Psychologist:** "That's totally understandable, Ashwin! Remember at the start of our session, we learnt that it's completely normal to not want to do things you're scared of? But remember when we looked at that cycle of avoidance and why it's important to act like scientists and test our fears?"*

***Ashwin:** "Yeah, I guess so..."*

***Psychologist:** "Do you remember how you felt today when you thought you couldn't stand in the rain without Cody, but you did? You told me you felt happy and proud! We even wrote that down. Do you think you can read that the next time you are worried about doing a practice, to remind yourself of how you felt today?"*

***Ashwin:** "Yes, I guess I can do that."*

Following the session, the psychologist spoke to Miriam to discuss the progress they made.

The psychologist encourages Miriam to reward Ashwin with verbal praises about his practice today, e.g., *"The psychologist told me you managed to stand outside while it was raining, and you even stood for 5 min without Cody! Well, done, I'm so proud of you!"* The psychologist shows Miriam what Ashwin wrote about being able to do the practice and feeling happy and proud.

The psychologist also tells Miriam about the action plan set for Ashwin. She shows Miriam the activities Ashwin picked, including the one if it rains. She shows Miriam the worksheet for Ashwin to fill out for each activity, and discusses potentially doing an imaginal exposure if it does not rain.

Challenging Issues

So far in this chapter, we have described an overview of inhibitory learning theory and identified the key mechanisms of change that could optimise exposure therapy. The following section explores common challenges that clinicians may face when implementing therapy sessions using an inhibitory learning-based model and further considerations for children with anxiety who may have specific needs. This is not intended to be an exhaustive list, but rather exemplary to demonstrate that an inhibitory learning-based model of exposure therapy can be tailored and adapted for anxious children.

Clinician Biases

Despite a strong and large evidence-based exposure, its use in clinical practice is relatively low with reported rates of usage as low as 5–19% (Reid et al. 2017; Whiteside et al. 2016). There is also evidence to suggest that exposure is particularly not utilised by clinicians in clients with anxiety (Meyer et al. 2014; Teunisse et al. 2022). Moreover, when exposure is used, it is often adapted in a conservative manner such as early termination of exposure or tasks that are less distressing (Stobie et al. 2007). Qualitative studies suggest various barriers including a lack of training (Wolitzky-Taylor et al. 2018), concerns about exacerbating and purposefully evoking distress (Olatunji et al. 2009) and clinicians' own anxiety (Waller and Turner 2016). These biases and challenging issues that clinicians encounter may also be experienced by parents, who play a key role in facilitating exposure training in between therapy sessions. Awareness of these biases and implementing additional strategies to overcome them may help to improve outcomes for inhibitory learning-based exposure for children.

Indeed, ensuring clinician training programs provide current and evidence-based knowledge will be a vital step to improving the science-practice gap. Existing training programs tend to provide only habituation-model-based exposure and the increasing support for the inhibitory learning model which provides an opportunity to optimise standard exposure practices warrants a review of outdated clinical programs. In addition to the growing evidence for inhibitory learning-based exposure with the potential to optimise exposure, it perhaps has an additional unintended benefit in providing a rationale that may ease clinician's concerns particularly given the shift from the client staying in the exposure until symptom reduction to a focus on threat appraisal.

Children with Comorbid Anxiety and Depression

Anxiety in youth often co-occurs alongside a comorbid diagnosis of internalising disorders, such as an additional anxiety disorder or depression (Mobach et al. 2019). Despite anxiety and depression commonly co-occurring in childhood and adolescence (Costello et al. 2003), there are few existing transdiagnostic treatments for comorbid anxiety and depression in youth, which often show high rates of relapse (Chu et al. 2016). Research shows that youth with depression display maladaptive information processing (Lau and Waters 2017), poor inhibitory control (Kertz et al. 2016), and deficits in processing positive affect and reward learning (Morris et al.

2015), which may hinder the efficacy of treatments that rely on these processes. Whilst cognitive impairments in children with depression may interfere with inhibitory learning (Craske et al. 2014), clinicians can implement additional strategies during therapy to help children better engage with exposure therapy (Craske et al. 2008; Craske et al. 2014).

Exposure therapy protocols for children with comorbid diagnoses can be tailored to optimise new learning and target impairments in children with comorbid anxiety and depression. One study conducted on youth found that reward learning was impaired in children with depression, social phobia, PTSD, and GAD, compared to children with no disorders, and those with ADHD (Hayes 2015). Integrating strategies that enhance reward learning during exposure therapy such as adapting rewards to focus on intrinsic motivations (e.g., verbal praise), rather than extrinsic motivations (e.g., materialistic rewards) may lead to optimal long-term treatment outcomes.

Children with Learning Disabilities

Exposure therapy using an inhibitory learning framework can be beneficial for children with learning disabilities. Research shows that anxiety and learning disabilities often co-occur (Nelson and Harwood 2011) with approximately one-third of children with a reading disability at an increased risk for developing anxiety problems (Francis et al. 2019). The various types of exposure therapy for children with learning disabilities can include imaginal exposure, where the child is instructed by the therapist to "imagine" reading to their peers, or "in vivo" exposure, where the child faces their fear in real-world scenarios such as completing exposure activities in school. However, current CBT exposure programs are not always accessible for children with learning disabilities as they often require complex reading, writing, and comprehension skills. Thus, to optimise inhibitory learning, tailoring CBT programs is critical to improving accessibility for children with disabilities.

Strategies to enhance inhibitory learning can be focused on improving accessibility by adapting design aspects of the exposure therapy intervention such as using animated videos to deliver psychoeducation, cognitive restructuring, and exposure modules; incorporating visual digital worksheets using images and limited text; and providing easy to understand language to encourage and optimise new learning (Francis et al. 2021). The inhibitory learning framework emphasises the importance of the learning process and provides children and parents with a strong rationale for forming new nonthreat associations during psychoeducation. Given the challenges anxious children with learning disabilities experience with the presentation format of traditional CBT programs, integrating novel features that enhance the inhibitory learning process and tailor exposure therapy for children can help maximise clinical outcomes.

Children with Diverse Backgrounds

Another consideration for implementing exposure therapy based on inhibitory learning theories is the role of culture. Culture permeates all aspects of a child's behaviour, beliefs, and attitudes, and influences the way cognitive processes, such as language, which are organised in the brain (Carter et al. 2005). Normative strategies for processing information differ in children and adolescents based on cultural biases and experiences in a child's physical and social environment. There is a growing recognition of the importance of developing and adapting appropriate treatments to meet the needs of diverse populations. Incorporating strategies that are sensitive to cultural differences in children when administering exposure therapy based on inhibitory learning theory, may help to optimise clinical outcomes during treatment. Awareness of cognitive cultural differences to optimise inhibitory learning during exposure therapy in children could help strengthen neural pathways associated with learning.

Furthermore, understanding the context and resources available to a child through the lens of their socio-economic background is an important challenging issue for clinicians to consider when conducting exposure therapy and implementing strategies to enhance inhibitory learning. A wealth of research has shown that low SES can impact the well-being and development of children and adolescents, internalising behaviours, externalising behaviours, as well as cognitive and language development (Mendelson et al. 2008). Studies have also shown that SES can affect child and adolescent development through parental mental health, parental resources, and parental behaviours (Kalil and DeLeire 2006). Parents play a vital role in influencing treatment outcomes for their child and parent involvement during exposure therapy could be adapted to optimise inhibitory learning processes. Strategies could involve clinicians providing extra guidance and additional psychoeducation sessions for parents to help prepare their child for practising exposures outside of therapy, ensuring parents are confident with expressing emotion labelling and have a good understanding of mental health literacy, and encouraging contingency management strategies for parents such as positive reinforcement to enhance treatment success for their child. Identifying pathways for how SES and cultural background may impact mental health and adapting strategies to improve current treatments for anxiety in children is crucial to integrating a wholistic approach to cater to children with specific needs.

References

American Psychiatric Association, D., and Association, A. P (2013) Diagnostic and statistical manual of mental disorders: DSM-5, vol 5. American Psychiatric Association Washington

Baker A, Mystkowski J, Culver N, Yi R, Mortazavi A, Craske MG (2010) Does habituation matter? Emotional processing theory and exposure therapy for acrophobia. Behav Res Ther 48(11): 1139–1143. https://doi.org/10.1016/j.brat.2010.07.009

Balooch SB, Neumann DL (2011) Effects of multiple contexts and context similarity on the renewal of extinguished conditioned behaviour in an ABA design with humans. Learn Motiv 42:53–63. https://doi.org/10.1016/j.lmot.2010.08.008

Blakey SM, Abramowitz JS (2019) Dropping safety aids and maximizing retrieval cues: two keys to optimizing inhibitory learning during exposure therapy. Cogn Behav Pract 26:166–175. https://doi.org/10.1016/j.cbpra.2018.03.001

Bouton ME (1993) Context, time, and memory retrieval in the interference paradigms of Pavlovian learning. Psychol Bull 114(1):80–99. https://doi.org/10.1037/0033-2909.114.1.80

Bouton ME (2004) Context and behavioral processes in extinction. Learn Mem 11(5):485–494

Bouton ME, Woods AM, Pineño O (2004) Occasional reinforced trials during extinction can slow the rate of rapid reacquisition. Learn Motiv 35(4):371–390. https://doi.org/10.1016/j.lmot.2004.05.001

Carter JA, Lees JA, Murira GM, Gona J, Neville BG, Newton CR (2005) Issues in the development of cross-cultural assessments of speech and language for children. Int J Lang Commun Disord 40(4):385–401. https://doi.org/10.1080/13682820500057301

Chu BC, Crocco ST, Esseling P, Areizaga MJ, Lindner AM, Skriner LC (2016) Transdiagnostic group behavioral activation and exposure therapy for youth anxiety and depression: initial randomized controlled trial. Behav Res Ther 76:65–75. https://doi.org/10.1016/j.brat.2015.11.005

Costello EJ, Mustillo S, Erkanli A, Keeler G, Angold A (2003) Prevalence and development of psychiatric disorders in childhood and adolescence. Arch Gen Psychiatry 60(8):837–844. https://doi.org/10.1001/archpsyc.60.8.837

Craske MG, Kircanski K, Zelikowsky M, Mystkowski J, Chowdhury N, Baker A (2008) Optimizing inhibitory learning during exposure therapy. Behav Res Ther 46(1):5–27. https://doi.org/10.1016/j.brat.2007.10.003

Craske MG, Treanor M, Conway CC, Zbozinek T, Vervliet B (2014) Maximizing exposure therapy: an inhibitory learning approach. Behav Res Ther 58:10–23. https://doi.org/10.1016/j.brat.2014.04.006

Craske MG, Hermans D, Vervliet B (2018) State-of-the-art and future directions for extinction as a translational model for fear and anxiety. Philos Trans R Soc B:373. https://doi.org/10.1098/rstb.2017.0025

Craske MG, Treanor M, Zbozinek TD, Vervliet B (2022) Optimizing exposure therapy with an inhibitory retrieval approach and the OptEx Nexus. Behav Res Ther 152:104069. https://doi.org/10.1016/j.brat.2022.104069

Culver NC, Stoyanova M, Craske MG (2011) Clinical relevance of retrieval cues for attenuating context renewal of fear. J Anxiety Disord 25(2):284–292. https://doi.org/10.1016/j.janxdis.2010.10.002

Culver NC, Vervliet B, Craske MG (2015) Compound extinction:using the Rescorla–Wagner model to maximize exposure therapy effects for anxiety disorders. Clin Psychol Sci 3(3): 335–348. https://doi.org/10.1177/2167702614542103

de Jong R, Lommen MJJ, de Jong PJ, Nauta MH (2019) Using multiple contexts and retrieval cues in exposure-based therapy to prevent relapse in anxiety disorders. Cogn Behav Pract 26:154–165. https://doi.org/10.1016/j.cbpra.2018.05.002

de Jong R, Lommen MJJ, de Jong PJ, van Hout W, Duin-van der Marel ACE, Nauta MH (2024) Effectiveness of exposure-based treatment for childhood anxiety disorders: an open clinical trial to test its relation with indices of emotional processing and inhibitory learning. J Behav Ther Exp Psychiatry 83:101942. https://doi.org/10.1016/j.jbtep.2024.101942

Dibbets P, Havermans R, Arntz A (2008) All we need is a cue to remember: the effect of an extinction cue on renewal. Behav Res Ther 46(9):1070–1077. https://doi.org/10.1016/j.brat.2008.05.007

Dibbets P, Moor C, Voncken MJ (2013) The effect of a retrieval cue on the return of spider fear. J Behav Ther Exp Psychiatry 44(4):361–367. https://doi.org/10.1016/j.jbtep.2013.03.005

Francis DA, Caruana N, Hudson JL, McArthur GM (2019) The association between poor reading and internalising problems: a systematic review and meta-analysis. Clin Psychol Rev 67:45–60. https://doi.org/10.1016/j.cpr.2018.09.002

Francis D, Hudson JL, Kohnen S, Mobach L, McArthur GM (2021) The effect of an integrated reading and anxiety intervention for poor readers with anxiety. PeerJ 9:e10987. https://doi.org/10.7717/peerj.10987

Gromer D, Hildebrandt LK, Stegmann Y (2023) The role of expectancy violation in extinction learning: a two-day online fear conditioning study. Clin Psychol Eur 5(2):e9627. https://doi.org/10.32872/cpe.9627

Hayes AM (2015) Facilitating emotional processing in depression: the application of exposure principles. Curr Opin Psychol 4:61–66. https://doi.org/10.1016/j.copsyc.2015.03.032

Hedtke KA, Kendall PC, Tiwari S (2009) Safety-seeking and coping behavior during exposure tasks with anxious youth. J Clin Child Adolesc Psychol 38(1):1–15. https://doi.org/10.1080/15374410802581055

Higa-McMillan CK, Francis SE, Rith-Najarian L, Chorpita BF (2016) Evidence base update: 50 years of research on treatment for child and adolescent anxiety. J Clin Child Adolesc Psychol 45(2):91–113. https://doi.org/10.1080/15374416.2015.1046177

Hughes AA, Kendall PC (2008) Effect of a positive emotional state on interpretation bias for threat in children with anxiety disorders. Emotion 8(3):414–418. https://doi.org/10.1037/1528-3542.8.3.414

In-Albon T, Schneider S (2007) Psychotherapy of childhood anxiety disorders: a meta-analysis. Psychother Psychosom 76(1):15–24. https://doi.org/10.1159/000096361

James AC, Reardon T, Soler A, James G, Creswell C (2020) Cognitive behavioural therapy for anxiety disorders in children and adolescents. Cochrane Database Syst Rev 11(11):CD013162. https://doi.org/10.1002/14651858.CD013162.pub2

Jones MC (1924) A laboratory study of fear: the case of Peter. Pedagogical Seminary 31:308–315

Kalil A, DeLeire T (2006) Family investments in children's potential: resources and parenting behaviors that promote success, vol 41. Libra Publishers Incorporated

Kendall PC, Hedtke KA (2006) Coping cat workbook. Workbook Publishing

Kennedy SM, Hawks JL (2021) Lessons learned from a randomized controlled pilot trial testing an inhibitory learning approach to exposure in anxious youth. Cogn Behav Pract 28:241–258

Kertz SJ, Belden AC, Tillman R, Luby J (2016) Cognitive control deficits in shifting and inhibition in preschool age children are associated with increased depression and anxiety over 7.5 years of development. J Abnorm Child Psychol 44(6):1185–1196. https://doi.org/10.1007/s10802-015-0101-0

Kiefer M, Schuch S, Schenck W, Fiedler K (2006) Mood states modulate activity in semantic brain areas during emotional word encoding. Cereb Cortex 17(7):1516–1530. https://doi.org/10.1093/cercor/bhl062

Kircanski K, Lieberman MD, Craske MG (2012a) Feelings into words: contributions of language to exposure therapy. Psychol Sci 23(10):1086–1091. https://doi.org/10.1177/0956797612443830

Kircanski K, Mortazavi A, Castriotta N, Baker AS, Mystkowski JL, Yi R, Craske MG (2012b) Challenges to the traditional exposure paradigm: variability in exposure therapy for contamination fears. J Behav Ther Exp Psychiatry 43(2):745–751. https://doi.org/10.1016/j.jbtep.2011.10.010

Kooiman BEAM, Hudson JL, Bertie L, Albers CJ, Nauta MH (2024) Predictors of relapse after successful CBT in anxiety disorders in children and young people. Under review

Laborda MA, Schofield CA, Johnson EM, Schubert JR, George-Denn D, Coles ME, Miller RR (2016) The extinction and return of fear of public speaking. Behav Modif 40(6):901–921. https://doi.org/10.1177/0145445516645766

Lang AJ, Craske MG (2000) Manipulations of exposure-based therapy to reduce return of fear: a replication. Behav Res Ther 38:1–12. https://doi.org/10.1016/S0005-7967(99)00031-5

Lau JY, Waters AM (2017) Annual research review: an expanded account of information-processing mechanisms in risk for child and adolescent anxiety and depression. J Child Psychol Psychiatry 58(4):387–407. https://doi.org/10.1111/jcpp.12653

Lebowitz ER, Woolston J, Bar-Haim Y, Calvocoressi L, Dauser C, Warnick E, Scahill L, Chakir AR, Shechner T, Hermes H, Vitulano LA, King RA, Leckman JF (2013) Family accommodation in pediatric anxiety disorders. Depress Anxiety 30(1):47–54. https://doi.org/10.1002/da.21998

Levy HC, Radomsky AS (2014) Safety behaviour enhances the acceptability of exposure. Cogn Behav Ther 43(1):83–92. https://doi.org/10.1080/16506073.2013.819376

Lovibond PF, Davis NR, O'Flaherty AS (2000) Protection from extinction in human fear conditioning. Behav Res Ther 38:967–983. https://doi.org/10.1016/S0005-7967(99)00121-7

Manassis, K., Lee, T. C., Bennett, K., Zhao, X. Y., Mendlowitz, S., Duda, S., Saini, M., Wilansky, P., Baer, S., Barrett, P., Bodden, D., Cobham, V. E., Dadds, M. R., Flannery-Schroeder, E., Ginsburg, G., Heyne, D., Hudson, J. L., Kendall, P. C., Liber, J., . . . Wood, J. J. (2014). Types of parental involvement in CBT with anxious youth: a preliminary meta-analysis. J Consult Clin Psychol, 82(6), 1163–1172. https://doi.org/10.1037/a0036969

Marks IM (1973) Reduction of fear: towards a unifying theory. Can Psychiatr Assoc J 18(1):9–12. https://doi.org/10.1177/070674377301800103

McCormack T, Burns P, O'Connor P, Jaroslawska A, Caruso EM (2019) Do children and adolescents have a future-oriented bias? A developmental study of spontaneous and cued past and future thinking. Psychol Res 83(4):774–787. https://doi.org/10.1007/s00426-018-1077-5

McGuire JF, Storch EA (2019) An inhibitory learning approach to cognitive-behavioral therapy for children and adolescents. Cogn Behav Pract 26(1):214–224. https://doi.org/10.1016/j.cbpra.2017.12.003

McGuire JF, Orr SP, Essoe JK, McCracken JT, Storch EA, Piacentini J (2016) Extinction learning in childhood anxiety disorders, obsessive compulsive disorder and post-traumatic stress disorder: implications for treatment. Expert Rev Neurother 16(10):1155–1174. https://doi.org/10.1080/14737175.2016.1199276

Mendelson T, Kubzansky LD, Datta GD, Buka SL (2008) Relation of female gender and low socioeconomic status to internalizing symptoms among adolescents: a case of double jeopardy? Soc Sci Med 66(6):1284–1296. https://doi.org/10.1016/j.socscimed.2007.11.033

Meuret AE, Seidel A, Rosenfield B, Hofmann SG, Rosenfield D (2012) Does fear reactivity during exposure predict panic symptom reduction? J Consult Clin Psychol 80(5):773–785. https://doi.org/10.1037/a0028032

Meyer JM, Farrell NR, Kemp JJ, Blakey SM, Deacon BJ (2014) Why do clinicians exclude anxious clients from exposure therapy? Behav Res Ther 54:49–53. https://doi.org/10.1016/j.brat.2014.01.004

Milosevic I, Radomsky AS (2013) Incorporating the judicious use of safety behavior into exposure-based treatments for anxiety disorders: a study of treatment acceptability. J Cogn Psychother 27(2):155–174. https://doi.org/10.1891/0889-8391.27.2.155

Mobach L, Gould K, Hudson JL (2019) Comorbidity in childhood anxiety disorders. In: Compton SN, Villabø MA, Kristensen H (eds) Pediatric anxiety disorders, London, pp 277–298. https://doi.org/10.1016/B978-0-12-813004-9.00013-X

Morris BH, Bylsma LM, Yaroslavsky I, Kovacs M, Rottenberg J (2015) Reward learning in pediatric depression and anxiety: preliminary findings in a high-risk sample. Depress Anxiety 32(5):373–381. https://doi.org/10.1002/da.22358

Mystkowski JL, Mineka S, Vernon LL, Zinbarg REJJ (2003) Changes in caffeine states enhance return of fear in spider phobia. J Consult Clin Psychol 71(2):243

Mystkowski JL, Craske MG, Echiverri AM, Labus JS (2006) Mental reinstatement of context and return of fear in spider-fearful participants. Behav Ther 37(1):49–60. https://doi.org/10.1016/j.beth.2005.04.001

Nelson JM, Harwood H (2011) Learning disabilities and anxiety: a meta-analysis. J Learn Disabil 44(1):3–17. https://doi.org/10.1177/0022219409359939

Newall C, Watson T, Grant KA, Richardson R (2017) The relative effectiveness of extinction and counter-conditioning in diminishing children's fear. Behav Res Ther 95:42–49. https://doi.org/10.1016/j.brat.2017.05.006

Niles AN, Craske MG, Lieberman MD, Hur C (2015) Affect labeling enhances exposure effectiveness for public speaking anxiety. Behav Res Ther 68:27–36. https://doi.org/10.1016/j.brat.2015.03.004

Nyberg L (2002) Levels of processing: a view from functional brain imaging. Memory 10(5–6): 345–348. https://doi.org/10.1080/09658210244000171

Olatunji BO, Deacon BJ, Abramowitz JS (2009) The cruelest cure? Ethical issues in the implementation of exposure-based treatments. Cogn Behav Pract 16:172–180. https://doi.org/10.1016/j.cbpra.2008.07.003

Pineño O, Zilski-Pineno JM, Miller RR (2008) Habituation of unconditioned fear can be attenuated by the presence of a safe stimulus: assessment using the neophobic response of the rat. Behav Processes 77(1):55–60. https://doi.org/10.1016/j.beproc.2007.06.003

Pittig, A., Heinig, I., Goerigk, S., Richter, J., Hollandt, M., Lueken, U., Pauli, P., Deckert, J., Kircher, T., Straube, B., Neudeck, P., Koelkebeck, K., Dannlowski, U., Arolt, V., Fydrich, T., Fehm, L., Ströhle, A., Totzeck, C., Margraf, J., . . . Wittchen, H.-U. (2022). Change of threat expectancy as mechanism of exposure-based psychotherapy for anxiety disorders: evidence from 8,484 exposure exercises of 605 patients. Clin Psychol Sci, 11(2), 199–217. https://doi.org/10.1177/21677026221101379

Plaisted H, Waite P, Gordon K, Creswell C (2021) Optimising exposure for children and adolescents with anxiety, OCD and PTSD: a systematic review. Clin Child Fam Psychol Rev 24(2): 348–369. https://doi.org/10.1007/s10567-020-00335-z

Plaisted H, Waite P, Creswell C (2022) Optimising exposure for adolescents with public speaking anxiety: affect labelling or positive coping statements? Behav Res Ther 148:103997. https://doi.org/10.1016/j.brat.2021.103997

Rapee RM (2000) Group treatment of children with anxiety disorders: outcome and predictors of treatment response. Aust J Psychol 52(3):125–129. https://doi.org/10.1080/00049530008255379

Reid AM, Bolshakova MI, Guzick AG, Fernandez AG, Striley CW, Geffken GR, McNamara JP (2017) Common barriers to the dissemination of exposure therapy for youth with anxiety disorders. Community Ment Health J 53(4):432–437. https://doi.org/10.1007/s10597-017-0108-9

Rescorla RA (2006) Deepened extinction from compound stimulus presentation. J Exp Psychol Anim Behav Process 32(2):135–144. https://doi.org/10.1037/0097-7403.32.2.135

Silverman WK, Rey Y, Marin CE, Jaccard J, Pettit JW (2021) Does training parents in reinforcement skills or relationship skills enhance individual youths' cognitive behavioral therapy for anxiety? Outcome, specificity, and mediation. Clin Psychol Sci

Stemerding LE, van Ast VA, Kindt M (2023) Manipulating expectancy violations to strengthen the efficacy of human fear extinction. Behav Res Ther 165:104319. https://doi.org/10.1016/j.brat.2023.104319

Stobie B, Taylor T, Quigley A, Ewing S, Salkovskis PM (2007) "Contents may vary": a pilot study of treatment histories of OCD patients. Behav Cogn Psychother *35*(3):273–282. https://doi.org/10.1017/S135246580700358X

Teunisse AK, Pembroke L, O'Gradey-Lee M, Sy M, Rapee RM, Wuthrich VM, Creswell C, Hudson JL (2022) A scoping review investigating the use of exposure for the treatment and targeted prevention of anxiety and related disorders in young people. JCPP Adv 2(2):e12080

Thompson-Hollands J, Kerns CE, Pincus DB, Comer JS (2014) Parental accommodation of child anxiety and related symptoms: range, impact, and correlates. J Anxiety Disord 28(8):765–773. https://doi.org/10.1016/j.janxdis.2014.09.007

Tolin DF (2019) Inhibitory learning for anxiety-related disorders. Cogn Behav Pract 26:225–236. https://doi.org/10.1016/j.cbpra.2018.07.008

Waller G, Turner H (2016) Therapist drift redux: why well-meaning clinicians fail to deliver evidence-based therapy, and how to get back on track. Behav Res Ther 77:128–137. https://doi.org/10.1016/j.brat.2015.12.005

Weisman JS, Rodebaugh TL (2018) Exposure therapy augmentation: a review and extension of techniques informed by an inhibitory learning approach. Clin Psychol Rev 59:41–51. https://doi.org/10.1016/j.cpr.2017.10.010

Whiteside SPH, Deacon BJ, Benito K, Stewart E (2016) Factors associated with practitioners'use of exposuretherapy for childhood anxiety disorders. J Anxiety Disord 40:29–36

Wolitzky-Taylor K, Fenwick K, Lengnick-Hall R, Grossman J, Bearman SK, Arch J, Miranda J, Chung B (2018) A preliminary exploration of the barriers to delivering (and receiving) exposure-based cognitive behavioral therapy for anxiety disorders in adult community mental health settings. Community Ment Health J 54(7):899–911. https://doi.org/10.1007/s10597-018-0252-x

Zbozinek TD, Craske MG (2017) The role of positive affect in enhancing extinction learning and exposure therapy for anxiety disorders. J Exp Psychopathol 8(1):13–39. https://doi.org/10.5127/jep.052615

Zbozinek TD, Holmes EA, Craske MG (2015) The effect of positive mood induction on reducing reinstatement fear: relevance for long term outcomes of exposure therapy. Behav Res Ther 71:65–75. https://doi.org/10.1016/j.brat.2015.05.016

Neural Markers of Treatment Response in Pediatric Anxiety and PTSD

Dana E. Díaz, Hannah C. Becker, and Kate D. Fitzgerald

Contents

Abstract Pediatric anxiety disorders and post-traumatic stress disorder (PTSD) are associated with elevated threat sensitivity and impaired emotion regulation, accompanied by dysfunction in the neural circuits involved in these processes. Despite established treatments like cognitive behavioral therapy (CBT) and selective serotonin reuptake inhibitors, many children do not achieve remission, underscoring the importance of understanding the neurobiological underpinnings of these disorders. This review synthesizes current research on the neural predictors of treatment response and the neurofunctional changes associated with treatment in pediatric

D. E. Díaz
Department of Psychiatry, Columbia University Irving Medical Center, New York, NY, USA

H. C. Becker
Department of Psychology, University of Michigan, Ann Arbor, MI, USA

K. D. Fitzgerald (✉)
Department of Psychiatry, Columbia University Irving Medical Center, New York, NY, USA

New York State Psychiatric Institute, New York, NY, USA
e-mail: kf2688@cumc.columbia.edu

© The Author(s), under exclusive license to Springer Nature Switzerland AG 2024
Curr Topics Behav Neurosci (2025) 73: 473–494
https://doi.org/10.1007/7854_2024_547
Published Online: 14 December 2024

anxiety and PTSD during threat and reward processing. Several key findings emerged. First, enhanced threat/safety discrimination in the amygdala predicted better outcomes of pediatric anxiety and PTSD treatments. Second, differences in pretreatment activation within the lateral prefrontal and dorsal anterior cingulate cortices predicted treatment response, likely reflecting baseline executive control differences. Third, post-CBT decreases in activation in default mode, visuo-attentional, and sensorimotor areas may support treatment-related increases in task engagement. Finally, functional connectivity between the amygdala and other limbic, prefrontal, and default mode network nodes predicts treatment response in anxiety and PTSD, highlighting its potential as a biomarker for therapeutic efficacy. Understanding these neurofunctional markers could lead to more targeted interventions, optimizing treatment planning and potentially leading to the development of "pretreatment" strategies to enhance the efficacy of existing treatments. This review highlights the necessity for future research to establish more direct links between neuroimaging findings and clinical outcomes to facilitate the translation of these findings into clinical practice.

Keywords Adolescents · Anxiety disorders · CBT · Children · Post-traumatic stress disorder · Neural biomarkers · SSRI · Reward processing · Threat processing

1 Introduction

Pediatric anxiety disorders are a prevalent mental health concern, affecting roughly 30% of the US population at some point during childhood or adolescence (Kessler et al. 2005; Merikangas et al. 2010). Anxiety disorders typically first emerge during childhood, with a median onset age of 6 years old (Beesdo et al. 2010; Kessler et al. 2012; Merikangas et al. 2010). Anxiety disorders that begin during youth are often chronic (Spence and Rapee 2016), impairing (Kendall et al. 2010; Ranøyen et al. 2018), and can increase risk for other disorders later in life (Kessler et al. 2011), including mood disorders, disruptive behaviors, substance use, and suicidality (Asselmann et al. 2018). Despite the availability of effective treatments, such as cognitive-behavioral therapy (CBT) and selective serotonin reuptake inhibitors (SSRIs), approximately 40–50% of people with anxiety disorders do not achieve remission (Ginsburg et al. 2011; Piacentini et al. 2014), highlighting the need to understand the mechanisms by which treatment can be successful. Pediatric anxiety is associated with biases in threat processing, emotion regulation, and executive control, as well as in the neural circuits that support these functions (Ashworth et al. 2021; Fitzgerald et al. 2021; Strawn et al. 2014). However, the mechanisms by which the neurobiology of pediatric anxiety disorders predicts or is modified by successful treatment are not well understood. This review aims to shed light on neural underpinnings and predictors of treatment efficacy in children and adolescents with anxiety and PTSD to inform future interventions targeted to individual neurobiological profiles.

2 Established Treatments for Pediatric Anxiety and PTSD

The most prevalent treatments for pediatric anxiety are CBT and SSRIs, which can be employed either separately or together, depending on the child's needs and the severity of their anxiety (Birmaher et al. 2003; Compton et al. 2004; Walkup et al. 2008). CBT uses cognitive restructuring to identify and modify negative thought patterns and beliefs fueling anxiety, equipping children with strategies to manage symptoms and minimize avoidance behaviors (Compton et al. 2004). These strategies are intended to promote confrontation and engagement with anxiety-inducing situations (i.e., gradual exposure) to learn that feared outcomes do not occur (Craske et al. 2022; McGuire and Storch 2019). Relaxation techniques are also usually taught to help reduce anxious distress. These components of CBT are typically packaged together for delivery over 10 to 16 weekly sessions (e.g., Nelson and Finch 2008; Rapee and Wignall 2002). Additionally, for children with PTSD, trauma-focused therapy can be integrated with CBT to address and process traumatic events (Cohen et al. 2017). SSRIs, such as fluoxetine, sertraline, and escitalopram, are also frequently prescribed due to their efficacy and safety in children and adolescents with anxiety disorders or PTSD (Strawn et al. 2015). These medications are believed to work by increasing serotonin levels in the brain to improve mood and reduce anxiety. Due to potential side effects of medications, use of SSRIs is typically considered after trying psychotherapy options like CBT and involves careful evaluation of benefits against risks, with monitoring for adverse effects. Other treatments for pediatric anxiety are available (e.g., mindfulness-based cognitive therapy), yet research on the neural mechanisms of their efficacy is limited, and therefore will not be discussed in this review.

In the following section, we will review studies that examine (1) pretreatment neural predictors of treatment response, and (2) pre-to-post-treatment changes in brain function that correlate with treatment response. These studies generally include youth with GAD, SAD, and/or separation anxiety disorders. Neural predictors of trauma-focused CBT have also been tested in adolescents with PTSD. Most studies on the neural predictors of treatment response in pediatric anxiety and other negative valence have focused on neural systems for threat responding (reviewed in La Buissonniere-Ariza et al. 2021; Strawn et al. 2021), based on prevailing theory that anxiety derives from hypersensitivity to threat and deficits of threat regulation. We will first review functional brain activation during threat processing paradigms as a marker (i.e., predictor and mechanism) of treatment response. Next, we will review several studies that have examined the neural correlates of reward responsiveness that predict treatment response. Finally, we will discuss the implications of this research for clinical application and outline potential future directions.

3 Neurofunctional Markers of Treatment Response During Threat Processing

3.1 Amygdala and Insula

The amygdala, a subcortical region responsible for threat detection and fear responding (Davis and Whalen 2001; Ohman 2005), has been implicated in the pathology of pediatric anxiety disorders (Ashworth et al. 2021) and PTSD (Alexandra Kredlow et al. 2022). Research in youth with GAD, SAD, panic disorder, separation anxiety, and post-traumatic stress symptoms have found that viewing (or anticipation of viewing) negative or threatening stimuli elicits amygdala hyperactivation compared to healthy controls (Beesdo et al. 2009; Garrett et al. 2012; Guyer et al. 2008; McClure et al. 2007b; Monk et al. 2008; Thomas et al. 2001), likely contributing to heightened threat perception and increased fear responses characteristic of these disorders (Ohman 2005).

Amygdala function during threat processing has emerged as a predictor of treatment response in pediatric anxiety and PTSD. For children and adolescents with GAD, greater left amygdala responses to fearful relative to happy faces prior to treatment with CBT or fluoxetine predicted greater posttreatment symptom improvement, regardless of treatment type (McClure et al. 2007a). Similarly, greater pretreatment amygdala response to fearful compared to neutral faces predicted greater symptom reduction in adolescent girls with PTSD, following trauma-focused CBT (Cisler et al. 2015). However, in this study, girls with greater amygdala activity to *both* fearful and neutral faces displayed poorer posttreatment symptom reduction, suggesting that greater threat-safety discrimination in the amygdala, rather than hyperactivity alone, may underlie treatment gains. Supporting this notion, PTSD symptom improvement following trauma-focused CBT was associated with decreases in amygdala and hippocampal activation to neutral, but not fearful or angry, faces (Garrett et al. 2019). This may further suggest that CBT reduces anxiety symptoms by improving the ability to discriminate between threat and safety, as evidenced by a reduced response to safe stimuli. Alternatively, amygdala hyperactivation to emotional stimuli may reflect greater symptom severity (Cisler et al. 2015), suggesting more to gain from treatment. It is also possible that greater amygdala activity could reflect greater sensitivity to exposure, widely held to be the most active "ingredient" in CBT (Bilek et al. 2022). Notably, not all studies observe amygdala function as a predictor of treatment response (Burkhouse et al. 2017) and amygdala activation may not change with treatment (Haller et al. 2024).

The insula is a limbic-related region implicated in integrating emotional responses with internal physiological states and the detection of salient stimuli (internal and external) to direct processing between internal and external demands as appropriate to context (Reynolds and Zahm 2005; Seeley 2019). Basic science has shown that the insula has strong reciprocal connections with the amygdala (Reynolds and Zahm 2005). Additionally, functional magnetic resonance imaging (fMRI) studies have used inter-regional correlations between low-frequency blood oxygenation level-dependent oscillations to characterize insula connectivity with dorsal anterior cingulate cortex to form the cingulo-opercular network which functions to detect salient stimuli from external and internal milieu (Seeley et al. 2007). The cingulo-opercular network may serve as a switch between frontoparietal network regions (e.g., dorsolateral prefrontal cortex, superior parietal cortex) for implementation of cognitive processes, and default mode network regions (e.g., ventromedial prefrontal cortex, posterior cingulate cortex) for internally directed, self-referential processes (Sridharan et al. 2008). In adults with anxiety and PTSD, insula hyperactivation during threat processing has been repeatedly demonstrated (Chavanne and Robinson 2021; Etkin and Wager 2007). This aligns with findings that clinically anxious youth display elevated amygdala-insula connectivity during threat processing compared to healthy controls (McClure et al. 2007b), suggesting that greater connectivity might underlie more severe anxiety, perhaps due to enhanced coupling of threat perception with bodily responses and/or increased influence of amygdala on the insula-based cingulo-opercular network for salience detection.

Functional connectivity between the amygdala and the insula is a marker of CBT outcomes in pediatric anxiety and PTSD. For example, greater pretreatment amygdala-insula connectivity on congruent trials of a threat attention task predicted poorer response to CBT in youth with GAD, SAD, and/or separation anxiety, controlling for pretreatment severity (White et al. 2017). Thus, greater amygdala-insula connectivity prior to CBT may identify a neural target for intervention. This hypothesis is supported by evidence that successful treatment can reduce amygdala-insula connectivity. Specifically, among adolescent girls with PTSD undergoing trauma-focused CBT, greater symptom reduction correlated with pre-to-post-treatment decreases in amygdala-insula connectivity when reappraising negative images (Cisler et al. 2016). Normalization of connectivity between neural pathways involved in threat response and interoception and relatedly, salience detection, may suggest a mechanism by which therapy reduces anxiety symptoms, possibly by attenuating threat sensitivity, including anxiety-induced bodily distress and/or attention to threat.

3.2 *Prefrontal and Anterior Cingulate Cortices*

Altered function of the prefrontal (PFC) and anterior cingulate cortices (ACC) during emotional tasks has also been implicated in pediatric anxiety (Britton et al. 2013; Burkhouse et al. 2018; McClure et al. 2007b; Monk et al. 2006; Strawn et al. 2012; Swartz et al. 2014) and PTSD (Killgore et al. 2014; Kim et al. 2008; Stevens et al. 2017). Broadly, these regions are involved in emotion regulation, cognitive appraisal, executive functioning, and attentional control processes, and play a pivotal role in managing fear and anxiety (Bishop et al. 2004; Etkin et al. 2006; Haber and Knutson 2010; Kenwood et al. 2022; Shackman et al. 2011). Specifically, the lateral PFC is more involved in cognitive processes (Koechlin et al. 2003), the ACC serves the spectrum of cognitive to emotional processes along the dorsal to rostral continuum (Bush et al. 2000), and the ventromedial PFC mediates the affective valuation of stimuli for both goal-directed and self-referential processing (D'Argembeau 2013). Altered function in ventral portions of the PFC and ACC during emotional tasks has also been associated with pediatric anxiety symptoms (Britton et al. 2013; Burkhouse et al. 2018; McClure et al. 2007b; Monk et al. 2006; Strawn et al. 2012; Swartz et al. 2014) and PTSD (Killgore et al. 2014; Kim et al. 2008; Stevens et al. 2017). Abnormalities in these regions (e.g., ventrolateral and ventromedial PFC, subgenual and rostral ACC) may contribute to threat biases, difficulties in regulating emotions, and avoidance tendencies that are common in anxiety and PTSD (Blackford and Pine 2012; Kenwood et al. 2022; Liberzon and Sripada 2007).

Activation of the PFC and ACC during threat processing also predicts treatment response in pediatric anxiety, providing insights into the regulatory and cognitive neural mechanisms underlying therapeutic success. For instance, elevated pretreatment activation in the ventrolateral and dorsolateral PFC during the evaluation of threatening emotional faces predicted greater symptom reduction following treatment with CBT or sertraline in youth with GAD, SAD, or separation anxiety disorder (Kujawa et al. 2016). Increased pretreatment activation in the ventrolateral and dorsolateral PFC may suggest greater capacity for cognitive control over emotions (e.g., reappraisal) and emotion regulation, which may enable youth to better engage with their fears and anxieties (i.e., exposure adherence) and thereby benefit from treatment. Moreover, pre-to-post-treatment increases in right ventrolateral PFC activation to angry faces were observed in a study of 14 youth with GAD undergoing CBT or fluoxetine (Maslowsky et al. 2010). Though these changes were not associated with changes in anxiety severity, they may reflect the development of more effective strategies for managing negative emotions, such as cognitive reappraisal, which is a core component of CBT response.

By contrast, *less* pretreatment activation in a cluster encompassing the dorsal ACC and dorsomedial PFC predicted greater symptom reduction following CBT or sertraline in youth with GAD or SAD (Burkhouse et al. 2017). In this study, pretreatment brain function was measured while children matched shapes that were flanked by emotional face distractors. Lower pretreatment dorsal ACC and dorsomedial PFC activation in CBT responders was suggested to identify children with the most to gain from treatment. Given that anxious youth show reduced rostral ACC activation during implicit emotion processing compared to healthy controls (Burkhouse et al. 2018; Swartz et al. 2014), lower pretreatment activation could indicate cognitive or attentional bias to irrelevant emotional stimuli when responding to cognitive demands with clinically significant anxiety, thereby reflecting a treatment target. Consistent with this possibility, increases in rostral ACC activation have been observed during implicit emotion processing following CBT or sertraline among youth with GAD, SAD, and/or separation anxiety (Burkhouse et al. 2018), with greater increases correlating with larger reductions in social anxiety and avoidance symptoms. The rostral ACC plays a role in conflict monitoring, cognitive appraisal, and attention (Bush et al. 2000; Etkin et al. 2011) and dorsomedial PFC is involved in appraising negative emotions (Etkin et al. 2011; Price and Drevets

2010). These abilities may be especially important when trying to exert cognitive control to perform tasks in the presence of negative emotional distractors – a process that is impaired in anxious youth (Cisler and Koster 2010) and targeted by CBT (i.e., cognitive restructuring and successful engagement with exposures).

Discrepancies in the direction of subregional activations in the PFC and ACC in predicting treatment response (e.g., elevated ventrolateral and dorsolateral PFC (Kujawa et al. 2016) versus less dorsal ACC and dorsomedial PFC (Burkhouse et al. 2017)) may stem from variations in task instructions. When participants are instructed to evaluate threat stimuli, as in Kujawa et al. (2016), it may encourage the use of regulatory strategies for engaging with these emotions, leading to greater lateral PFC activation. By contrast, when participants are tasked to ignore emotional stimuli to perform another task (e.g., shape-matching), as in Burkhouse et al. (2018), less activation in regions involved in emotional appraisal or conflict monitoring (e.g., dorsal ACC) may reflect better task engagement and/or less distraction by emotional stimuli. Future research that directly modulates task instructions is needed to elucidate how attentional mechanisms impact spatially localized neural activation and subsequent treatment outcomes.

Functional connectivity between the amygdala and prefrontal cortical structures has also been identified as a predictor of how well youth respond to anxiety treatment. In a sample of adolescents with GAD, more negative (i.e., anti-correlated) pretreatment functional connectivity between the amygdala and a cluster encompassing the bilateral ventromedial PFC (vmPFC) and subgenual ACC (sgACC) during implicit emotion processing predicted greater improvement in anxiety symptoms following treatment with escitalopram, but not for those who received a placebo (Lu et al. 2022). Moreover, participants who received escitalopram also exhibited more negative amygdala-vmPFC/sgACC connectivity at post-treatment compared to the placebo group. The vmPFC and sgACC are involved in emotion processing (Etkin et al. 2011; Hiser and Koenigs 2018) and regulating amygdala response via inhibitory connections (Etkin et al. 2010; Rosenkranz et al. 2003). Anxious youth have been found to display weaker negative coupling between the amygdala and ventrolateral PFC than healthy control subjects during threat processing (Monk et al. 2008), as well as reduced structural connectivity between limbic and prefrontal cortical regions (Aggarwal et al. 2024; Baur et al. 2013; Glenn et al. 2022), which may contribute to impaired prefrontal modulation of amygdala-driven responses to negative stimuli. Thus, larger amygdala-vmPFC/sgACC anti-correlations at baseline may suggest enhanced (i.e., more normalized) inhibitory control over the amygdala, which were further increased by treatment, leading to better treatment outcomes.

Alternatively, as the vmPFC plays a role in self-referential thinking, less positive amygdala-vmPFC/sgAGG connectivity of the amygdala may reflect greater separation of threat processing and introspective networks. This separation could result in less attention to internal distress elicited by threatening stimuli, allowing for better processing of cognitive task demands during implicit emotion processing. In line with this interpretation, among adolescent girls with PTSD undergoing trauma-focused CBT, less pretreatment functional connectivity between the amygdala and dorsal ACC to fearful faces predicted greater symptom reduction following treatment (Cisler et al. 2015). Whereas the vmPFC is associated with self-referential thinking and the integration of emotional and cognitive processes (Hiser and Koenigs 2018), the dorsal ACC plays a role in error detection, conflict monitoring, and adaptive response selection (Shenhav et al. 2016). These distinct functions may underlie the differential impact of amygdala-PFC/ACC connectivity on treatment outcomes in anxiety and PTSD. Further research is needed to elucidate the precise mechanisms by which amygdala connectivity with specific cortical regions influences treatment response across different psychiatric disorders.

3.3 *Frontoparietal and Cingulo-Opercular Networks*

Regions in the lateral PFC, posterior parietal cortex, dorsal ACC, and insula are nodes in the frontoparietal and cingulo-opercular networks, respectively, which are task control networks that are engaged when healthy participants perform cognitively demanding tasks. The cingulo-opercular network consists of the dorsal ACC, anterior insula, thalamus, and anterior PFC, and is involved in sustained performance monitoring and attending to salient information (Dosenbach et al. 2006, 2007). The frontoparietal network, which is engaged to adjust control when task demands change, consists of the dorsolateral PFC, posterior parietal cortex, and intraparietal sulcus (Dosenbach et al. 2007; Scolari et al. 2015). Baseline activation of task control networks during threat processing is relevant for treatment because it might indicate cognitive control and/or attentional mechanisms that may influence how youth with anxiety disorders or PTSD engage in CBT. For instance, post-treatment symptom reduction in anxious youth was predicted by elevated baseline cingulo-opercular network function (Burkhouse et al. 2017) but reduced frontoparietal network activity (Kujawa et al. 2016), possibly suggesting differential effects of continuous performance monitoring versus flexible control on one's ability to engage with treatment. Further, differences in functional connectivity between the amygdala and cingulo-opercular network predicted treatment response in anxiety and PTSD (Cisler et al. 2016), suggesting that the interaction between circuitry for threat processing and salience monitoring/cognitive control may be a biomarker for treatment outcomes.

3.4 *Default Mode Network*

Unlike the cingulo-opercular and frontoparietal networks, the default mode network is deactivated during tasks, and is instead engaged during introspection and self-referential thinking (Raichle et al. 2001; Shulman et al. 1997). Treatments for anxiety and PTSD have been shown to modulate the function of the default mode network, a network primarily consisting of the vmPFC/sgACC, posterior cingulate cortex, angular gyrus, and precuneus. Youth with psychopathology sometimes fail to suppress the default mode network during externally focused tasks (Han et al. 2016; Liddle et al. 2011; Yin et al. 2017), suggesting difficulties in task engagement and/or diverting attention away from introspection (Greicius and Menon 2004; Singh and Fawcett 2008). Among youth undergoing trauma-focused CBT for PTSD, symptom improvement pre-to-post-treatment was associated with reductions in activation of a cluster encompassing the posterior cingulate cortex and precuneus when viewing emotional faces (Garrett et al. 2019). Reduced activation in default mode network nodes following treatment (and at baseline) may indicate less internally focused and potentially maladaptive rumination on traumatic experiences or anxious thoughts, contributing to symptom improvement.

Treatment-related increases in functional connectivity between the amygdala and discrete nodes within the default mode network have also been observed in a study of medication for youth with pediatric anxiety disorders. Specifically, functional connectivity between the left amygdala and right angular gyrus increased following treatment with escitalopram in anxious youth during a task with emotional and neutral distractors (Lu et al. 2022). Further, increased amygdala-angular gyrus connectivity at the second week of treatment predicted greater posttreatment improvement in anxiety symptoms. The angular gyrus is involved in various aspects of cognition, including attention, spatial cognition, and theory of mind (Seghier 2013). Thus, enhanced connectivity between these regions suggests that escitalopram may influence the integration of emotional and cognitive processing. Speculatively, this may contribute to reductions in anxiety symptoms by regulating the amygdala and reducing maladaptive rumination. These findings also suggest that medications, as compared to CBT, may engage different neural mechanisms. Studies directly comparing medication and CBT treatment for pediatric anxiety disorders will be needed to directly assess this possibility.

3.5 *Vision/Attention Networks*

Neural activation in visuo-attention networks has also been shown to predict and change with treatment response in anxious youth, possibly due to input from the amygdala. When a threat is detected, the amygdala sends reentrant feedback signals to sensory regions (Vuilleumier 2005), leading to perceptual prioritization of anxiety-provoking stimuli (Anderson and Phelps 2001; Davidson 2002; Ohman 2005; Ohman et al. 2001). Anxious individuals display elevated amygdala activity relative to controls (Brühl et al. 2014; Etkin and Wager 2007), elevated visual cortical engagement (Brühl et al. 2014; Etkin and Wager 2007; Gentili et al. 2016; Straube et al. 2005), and increased functional connectivity between the amygdala and perceptual regions (Brühl et al. 2014) during threat processing.

A recent investigation assessed whole-brain activation in 69 youth with GAD, SAD, and/or separation anxiety disorder while they viewed neutral and angry faces, both before and after undergoing CBT (Haller et al. 2024). Following CBT, anxious youth displayed decreased activation in regions associated with visual perception and attention. For the superior parietal lobule, anxious youth exhibited elevated pretreatment activation compared to healthy controls, which normalized after therapy. In addition, anxious youth displayed pre-to-post-treatment reductions in function of the inferior parietal lobule, bilateral superior and inferior temporal gyri, and middle occipital gyri compared to healthy controls, despite there being no group differences at baseline. Together, these regions are broadly involved in visual perception, attention, and spatial awareness (Behrmann et al. 2004; Conway 2018; Shapiro et al. 2002), suggesting CBT may influence visual attention during threat processing, possibly by reducing hypervigilance and/or perceptual prioritization.

Similarly, in youth with PTSD, greater connectivity between the amygdala and visual cortex at baseline predicted faster reductions in symptoms during trauma-focused CBT (Cisler et al. 2015). This may suggest that amygdalar influence on visual attention networks identifies the patients that are most likely to respond and could "prime" the brain in ways that facilitate response to CBT. However, it is important to note that predictors do not necessarily equal mechanisms. FMRI studies that assess treatment mechanisms at pre- and post-treatment (i.e., brain changes with symptom changes) have been a challenge in the field and may require more sophisticated analytic methods sensitive to changes in individuals or patient subgroups (Becker et al. 2023).

Reduced cortical thickness in parietal and occipital cortical regions has also been found to predict poorer treatment outcomes following CBT in children and adolescents with GAD, SAD, and/or separation anxiety disorder. These regions included the fusiform gyrus and inferior parietal lobe, which have been shown to play roles in socio-emotional processing and are engaged by emotional face stimuli (Fusar-Poli et al. 2009; Igelström and Graziano 2017; Kanwisher and Yovel 2006). Occipital findings align with studies that observe function in visual cortical regions as a predictor of treatment response in adults with SAD (Doehrmann et al. 2013) and specific phobia (Hauner et al. 2012).

3.6 Sensorimotor Networks

Neurofunctional markers of treatment response have also been observed in the sensorimotor network, a large-scale brain network that processes sensory inputs and coordinates motor functions. For instance, greater pretreatment activation to threat faces in the pre- and postcentral gyri, regions responsible for voluntary movement and sensory processing, respectively (Buhle et al. 2014; Goldin et al. 2008; Kohn et al. 2014), predicted better treatment responses in anxious youth undergoing treatment with CBT or SSRIs (Kujawa et al. 2016). Thus, greater activation in these regions could represent more frequent or better task responding during threat processing, though this hypothesis was not tested and behavioral differences did not predict treatment response. In another sample, anxious youth showed elevated pretreatment activation in the bilateral supplementary motor area and superior parietal lobule compared to healthy controls during a dot probe paradigm, a task measuring attention to threat. In anxious youth, this hyperactivation normalized following CBT (Haller et al. 2024). The supplementary motor area is involved in planning and executing motor responses (Nachev et al. 2008) and interacts with the dACC, a cingulo-opercular node, during motor coordination (Diwadkar et al. 2017). The superior parietal lobule is a node in the dorsal attention network, which coordinates with the frontoparietal network to regulate attention allocation according to goals and task demands, and plays a role in perceptive, visuomotor, and somatosensory processes (Sulpizio et al. 2023; Wu et al. 2016). Thus, normalization of activation in these regions may have therefore contributed to improved behavioral performance (i.e., less slowing) observed in this sample at posttreatment versus pretreatment, relative to healthy controls (Haller et al. 2024). Alternatively, the pretreatment hyperactivation observed in these cingulo-opercular/frontoparietal-related regions in anxious youth may have indexed compensatory attempts to maintain higher levels of performance despite attention bias to more negative faces, although not entirely effective, as anxious youth responded more slowly than controls. Subsequently, when anxiety symptoms were reduced following treatment, this compensatory hyperactivation of task control networks might not have been necessary, leading to reduced engagement.

4 Neurofunctional Markers of Treatment Response During Reward Processing

While most pediatric anxiety research has focused on threat processing, emerging evidence points to aberrant reward circuitry functioning as a potential factor in anxiety pathophysiology (Bar-Haim et al. 2009; Guyer et al. 2006, 2012; Hardin et al. 2006). In addition to contributing to anxiety symptoms, disruptions in reward processing may affect the ability to engage with and benefit from positive reinforcement strategies commonly used in behavioral therapies (e.g., praise, stickers from therapists). These reinforcement techniques foster treatment engagement and build therapeutic alliance, both of which are linked to improved treatment outcomes (Shirk and Karver 2006). Aberrant reward processing may also reduce motivation to engage in situations that are typically experienced as rewarding (e.g., social interactions) in the face of anxiety about negative outcomes (e.g., embarrassment) in anxious youth. Given that exposure-based therapy for anxiety requires engagement with feared, but potentially rewarding stimuli, less sensitive neural response to reward could drive anxious behaviors and make affected youth less willing to engage with exposure, both naturalistically and in the context of treatment.

The striatum is a subcortical structure involved in motivation and reward processes (Daniel and Pollmann 2014). Emerging research finds that regions within the striatum, including the nucleus accumbens (NAcc), caudate, and putamen, are hyperactive in youth with anxiety (Guyer et al. 2012) and with temperamental risk factors for anxiety (Bar-Haim et al. 2009; Guyer et al. 2006). The ventral striatum, and NAcc in particular, has been linked to treatment outcomes. Higher pretreatment activation to reward in a cluster encompassing the NAcc and sgACC was observed in anxious youth who responded to treatment (>35% symptom reduction) compared to non-responders (Sequeira et al. 2021), regardless of the type of therapy they participated in. This study included youth with GAD, SAD, and/or, separation anxiety disorder who participated in CBT or a comparison, client-centered psychotherapy. Similarly, greater pretreatment activation to reward anticipation in the entire bilateral ventral striatum, including the NAcc and adjacent regions of the caudate, predicted reduced anxiety severity and a faster rate of improvement in anxiety symptoms following treatment with CBT monotherapy or CBT combined with SSRIs (citalopram or fluoxetine) among youth with depression and comorbid anxiety (Forbes et al. 2010). These findings suggest that ventral striatal responsiveness to reward underlies greater sensitivity to positive outcomes that derive from CBT (e.g., expectation of clinical improvement, in-session therapist approval, anticipation of rewarding outcomes in situations that also increase anxiety) and might facilitate engagement in therapy and thereby better treatment outcomes. For example, greater reward responsiveness could manifest as a more active participation in therapy sessions, a higher likelihood of completing therapeutic homework, or a greater capacity to apply coping strategies outside of the therapeutic setting.

Conversely, in the same study, lower reward-related activation in a region extending from dorsomedial PFC into pregenual ACC before treatment was related to faster improvement of anxiety symptoms (Forbes et al. 2010). The pregenual ACC was considered in relation to the ventromedial PFC node of the default mode network (Raichle 2015), which is involved in self-referential processing (Amodio and Frith 2006). Reduced activation was therefore interpreted to reflect less self-focused thoughts, enabling better task engagement and potentially better engagement with treatment. Alternatively, reduced dorsomedial PFC activation may reflect poorer emotion regulatory capacity (Etkin et al. 2011; Price and Drevets 2010) and therefore more room to benefit from treatment. However, as this finding involves youth with comorbid depression and anxiety, thus its applicability to anxiety disorders warrants further validation.

Relevant to these functional predictors, greater gray matter volume in the left NAcc predicted decreases in anxiety symptoms following treatment with CBT or sertraline, among children and adolescents with GAD or SAD (Burkhouse et al. 2020). As the NAcc is involved in reward processing (Daniel and Pollmann 2014), larger volumes may reflect increased reward sensitivity or reward seeking (Coveleskie et al. 2015; Damme et al. 2022; Thayer et al. 2012). Thus, greater NAcc volume might reflect more responsivity to the positive reinforcement and goal-oriented nature of therapies like CBT. SSRIs also increase serotonin levels in the brain, which can directly affect reward processing pathways, including the NAcc (Kitaichi et al. 2010). NAcc volume may be associated with differences in serotonergic response, potentially enhancing sensitivity to positive effects of SSRIs, though this would need to be tested directly.

5 Summary

We reviewed how the neural circuits responsive to threat and reward processing serve as predictors and/or mechanisms of therapeutic success in pediatric anxiety and PTSD. First, with respect to threat processing, enhanced pretreatment amygdala response to threatening versus safe stimuli predicts improved treatment outcomes in pediatric anxiety and PTSD (Cisler et al. 2015; McClure et al. 2007a), suggesting that accurate threat-safety discrimination may facilitate better engagement with CBT or enable better responses to SSRIs. Trauma-focused CBT led to reductions in amygdala responsiveness to benign stimuli (Garrett et al. 2019), marking a potential mechanism of symptom improvement in PTSD, which may also have implications for anxiety disorders. Second, neural activation in lateral prefrontal and more dorsal anterior cingulate cortices during threat processing likely reflects baseline differences in executive control processes and may be an important biomarker of treatment response for pediatric anxiety, though this may be moderated by attentional focus. Elevated pretreatment activation to threat in the dorsolateral and ventrolateral PFC predicts a better response to CBT/SSRIs (Kujawa et al. 2016), with posttreatment increases in ventrolateral PFC activation suggesting enhancements in cognitive control (Maslowsky et al. 2010). In contrast, during tasks that require ignoring threats, less activation in medial frontal cortex areas (e.g., dorsomedial PFC, dorsal ACC) predicts better responses to CBT and SSRIs (Burkhouse et al. 2017) with rostral ACC function increasing following treatment, corresponding with reductions in symptom severity (Burkhouse et al. 2018). Third, activation in other default mode network nodes (posterior cingulate cortex, precuneus) during threat processing was reduced following trauma-focused CBT for PTSD (Garrett et al. 2019), possibly indicating reduced self-focus or enhanced task engagement. Decreases in visuo-attentional and sensorimotor regions have also been observed following the treatment of anxiety with CBT (Haller et al. 2024), with elevated pretreatment sensorimotor activation predicting better responses to CBT and SSRIs (Kujawa et al. 2016). This may further support that treatment leads to improved task engagement, as was observed in one study. Finally, functional connectivity between the amygdala and other limbic, prefrontal, and default mode network nodes is also an important marker of treatment response in youth with anxiety and PTSD. Specifically, reduced pretreatment connectivity between the amygdala and the insula associated with better treatment outcomes for anxiety disorders (White et al. 2017) and reduced amygdala-dorsal ACC connectivity predicted better treatment outcomes for PTSD (Cisler et al. 2015). By contrast, greater anti-correlations between amygdala and vmPFC/sgACC activity are linked to greater symptom reduction for GAD following treatment with SSRIs (Lu et al. 2022). Lastly, increased amygdala connectivity with the default mode network (Lu et al. 2022) and visuospatial areas (Cisler et al. 2015) predicts better treatment responses for anxiety and PTSD, respectively, suggesting these connections could serve as potential biomarkers for predicting therapeutic efficacy.

Emerging evidence suggests that disruptions in reward circuitry, particularly within the ventral striatum, also play a key role in pediatric anxiety and may influence treatment outcomes. Higher pretreatment activation in the ventral striatum, and the NAcc in particular, has been linked to better responses to CBT and other treatments, likely indicating greater sensitivity to positive reinforcement (Forbes et al. 2010; Sequeira et al. 2021). Additionally, greater gray matter volume in the left NAcc predicted decreases in youth's anxiety symptoms following treatment with CBT or SSRIs (Burkhouse et al. 2020).

6 Treatment Implications

The findings of this review support the neurobiological basis of prediction of treatment response (specifically CBT and SSRIs) for youth with anxiety and PTSD. Importantly, the findings of region/network-specific prediction and change articulated in this review have implications for clinical translation. Better understanding of the function of brain regions that lead to better response to treatment (e.g., greater activation of the dorsolateral PFC during threat processing) may be helpful in determining if a given treatment for a child or adolescent with an anxiety disorder or PTSD will be worthwhile or effective. Having predictive information about baseline brain function, prior to the initiation of treatment, will help to identify which youth are more likely to respond to the treatment, helping to distribute scarce clinical resources effectively. Further, understanding the neurobiological mechanisms that help affected youth respond well to treatment could motivate the development of "pretreatment" interventions designed to strengthen the function of certain regions or networks prior to starting treatment, with the goal of improving efficacy of established interventions such as CBT.

The findings of neural change with treatment response also have important treatment implications. By demonstrating change in specific brain regions from pre- to post-treatment in youth with anxiety disorders, this work has potential to identify neurobiological mechanisms that are indirectly or directly related to symptom reduction. If these findings are replicated and a specific mechanism of change is established, novel interventions could be designed with the goal of enhancing this change. For instance, increases in right ventrolateral PFC activity observed with CBT in Maslowsky et al. (2010) indicate that a treatment designed to directly manipulate the right ventrolateral PFC, such as through repetitive transcranial magnetic stimulation, may be a successful and directed future treatment approach. Similarly, these brain regions could be targets of "booster" therapies that help prolong symptom reductions observed with CBT, without requiring affected youth to complete another full course of CBT. However, before these findings can be translated to the clinical space, there are several areas of future study that should be pursued.

7 Future Directions

As discussed in the previous sections of this review, there are some inconsistencies in the findings of prior work, specifically with regard to the direction (greater or weaker) and the specific neural location of brain regions associated with treatment response. For instance, results vary based on the type of task or even the instructions provided for similar task paradigms. This has led to mixed findings that make clinical translation more challenging. To address this, future studies should move toward a more consistent task-based neuroimaging paradigm, and specifically one that closely approximates real-world experiences of anxiety for youth (as discussed in La Buissonniere-Ariza et al. 2021). Alternatively, future studies could prioritize consistency and generalizability by using resting-state neuroimaging, a method that has been understudied in fMRI clinical trials of pediatric anxiety disorders, likely due to uncertainty about the clinical relevance of resting state function, along with challenges related to motion and compliance in youth. More individualized approaches, such as those that utilize repetitive precision scanning of a single individual, could also be considered as a method of better understanding the precise neural correlates of treatment response for a given child or adolescent with anxiety or PTSD. Second, few studies have utilized neuroimaging to examine how pre-to-post-CBT in brain function underpin changes in symptom severity. Even fewer studies have employed randomized control trial designs in which CBT can be directly compared to another intervention. Such work will be needed to isolate treatment-specific effects and determine whether neuroimaging could be used to optimize treatment selection or develop neurally targeted interventions for patients (or groups of patients) with different brain-based vulnerabilities or strengths. Third, a clinical model that relies on a neuroimaging scan prior to treatment will likely always impede feasibility of application. To avoid the need for a neuroimaging scan at the start of treatment, future studies could benefit from identifying symptom or behavior-based measures that associate with patterns of brain activity and that can serve as a proxy for the neural biomarkers of treatment response discussed here. Finally, future research that expands beyond imaging of threat processing is needed to understand the mechanisms of treatment with CBT and SSRIs in pediatric anxiety and PTSD. A promising area for exploration lies in the neural substrates of reward and motivation, which are relevant in youth with depression, but are underexplored in youth with anxiety and PTSD (Sequeira et al. 2021). Furthermore, the neural networks associated with cognitive control function (e.g., cingulo-opercular and frontoparietal networks) and the default mode network have been implicated in threat-processing tasks that involve attentional shifts, but have only recently begun to be examined in the context of anxiety disorders. Given the relevance of the default mode and cognitive control networks in anxiety disorders and CBT (Fitzgerald et al. 2021), investigating the interaction of these tripartite networks both at rest and during cognitive control tasks could yield valuable insights.

References

Aggarwal N, Tromp DPM, Blackford JU, Pine DS, Roseboom PH, Williams LE, Kalin NH (2024) Sex-specific distributed White matter microarchitectural alterations in preadolescent youths with anxiety disorders: a mega-analytic study. Am J Psychiatry., appiajp20221048:299

Alexandra Kredlow M, Fenster RJ, Laurent ES, Ressler KJ, Phelps EA (2022) Prefrontal cortex, amygdala, and threat processing: implications for PTSD. Neuropsychopharmacology 47(1): 247–259

Amodio DM, Frith CD (2006) Meeting of minds: the medial frontal cortex and social cognition. Nat Rev Neurosci 7(4):268–277

Anderson AK, Phelps EA (2001) Lesions of the human amygdala impair enhanced perception of emotionally salient events. Nature 411(6835):305–309

Ashworth E, Brooks SJ, Schiöth HB (2021) Neural activation of anxiety and depression in children and young people: a systematic meta-analysis of fMRI studies. Psychiatry Res Neuroimaging 311:111272

Asselmann E, Wittchen H-U, Lieb R, Beesdo-Baum K (2018) Sociodemographic, clinical, and functional long-term outcomes in adolescents and young adults with mental disorders. Acta Psychiatr Scand 137(1):6–17

Bar-Haim Y, Fox NA, Benson B, Guyer AE, Williams A, Nelson EE, Perez-Edgar K, Pine DS, Ernst M (2009) Neural correlates of reward processing in adolescents with a history of inhibited temperament. Psychol Sci 20(8):1009–1018

Baur V, Brühl AB, Herwig U, Eberle T, Rufer M, Delsignore A, Jäncke L, Hänggi J (2013) Evidence of frontotemporal structural hypoconnectivity in social anxiety disorder: a quantitative fiber tractography study. Hum Brain Mapp 34(2):437–446

Becker HC, Norman LJ, Yang H, Monk CS, Phan KL, Taylor SF, Liu Y, Mannella K, Fitzgerald KD (2023) Disorder-specific cingulo-opercular network hyperconnectivity in pediatric OCD relative to pediatric anxiety. Psychol Med 53(4):1468–1478

Beesdo K, Lau JYF, Guyer AE, McClure-Tone EB, Monk CS, Nelson EE, Fromm SJ, Goldwin MA, Wittchen H-U, Leibenluft E, Ernst M, Pine DS (2009) Common and distinct amygdala-function perturbations in depressed vs anxious adolescents. Arch Gen Psychiatry 66(3):275–285

Beesdo K, Pine DS, Lieb R, Wittchen H-U (2010) Incidence and risk patterns of anxiety and depressive disorders and categorization of generalized anxiety disorder. Arch Gen Psychiatry 67(1):47–57

Behrmann M, Geng JJ, Shomstein S (2004) Parietal cortex and attention. Curr Opin Neurobiol 14(2):212–217

Bilek E, Tomlinson RC, Whiteman AS, Johnson TD, Benedict C, Phan KL, Monk CS, Fitzgerald KD (2022) Exposure-focused CBT outperforms relaxation-based control in an RCT of treatment for child and adolescent anxiety. J Clin Child Adolesc Psychol 51(4):410–418

Birmaher B, Axelson DA, Monk K, Kalas C, Clark DB, Ehmann M, Bridge J, Heo J, Brent DA (2003) Fluoxetine for the treatment of childhood anxiety disorders. J Am Acad Child Adolesc Psychiatry 42(4):415–423

Bishop S, Duncan J, Brett M, Lawrence AD (2004) Prefrontal cortical function and anxiety: controlling attention to threat-related stimuli. Nat Neurosci 7(2):184–188

Blackford JU, Pine DS (2012) Neural substrates of childhood anxiety disorders: a review of neuroimaging findings. Child Adolesc Psychiatr Clin N Am 21(3):501–525

Britton JC, Grillon C, Lissek S, Norcross MA, Szuhany KL, Chen G, Ernst M, Nelson EE, Leibenluft E, Shechner T, Pine DS (2013) Response to learned threat: an FMRI study in adolescent and adult anxiety. Am J Psychiatry 170(10):1195–1204

Brühl AB, Delsignore A, Komossa K, Weidt S (2014) Neuroimaging in social anxiety disorder – a meta-analytic review resulting in a new neurofunctional model. Neurosci Biobehav Rev 47: 260–280

Buhle JT, Silvers JA, Wager TD, Lopez R, Onyemekwu C, Kober H, Weber J, Ochsner KN (2014) Cognitive reappraisal of emotion: a meta-analysis of human neuroimaging studies. Cereb Cortex 24(11):2981–2990

Burkhouse KL, Kujawa A, Klumpp H, Fitzgerald KD, Monk CS, Phan KL (2017) Neural correlates of explicit and implicit emotion processing in relation to treatment response in pediatric anxiety. J Child Psychol Psychiatry 58(5):546–554

Burkhouse KL, Kujawa A, Hosseini B, Klumpp H, Fitzgerald KD, Langenecker SA, Monk CS, Phan KL (2018) Anterior cingulate activation to implicit threat before and after treatment for pediatric anxiety disorders. Prog Neuropsychopharmacol Biol Psychiatry 84(Pt A):250–256

Burkhouse KL, Jimmy J, Defelice N, Klumpp H, Ajilore O, Hosseini B, Fitzgerald KD, Monk CS, Phan KL (2020) Nucleus accumbens volume as a predictor of anxiety symptom improvement following CBT and SSRI treatment in two independent samples. Neuropsychopharmacology 45(3):561–569

Bush G, Luu P, Posner MI (2000) Cognitive and emotional influences in anterior cingulate cortex. Trends Cogn Sci 4(6):215–222

Chavanne AV, Robinson OJ (2021) The overlapping neurobiology of induced and pathological anxiety: a meta-analysis of functional neural activation. Am J Psychiatry 178(2):156–164

Cisler JM, Koster EHW (2010) Mechanisms of attentional biases towards threat in anxiety disorders: an integrative review. Clin Psychol Rev 30(2):203–216

Cisler JM, Sigel BA, Kramer TL, Smitherman S, Vanderzee K, Pemberton J, Kilts CD (2015) Amygdala response predicts trajectory of symptom reduction during Trauma-focused cognitive-behavioral therapy among adolescent girls with PTSD. J Psychiatr Res 71:33–40

Cisler JM, Sigel BA, Steele JS, Smitherman S, Vanderzee K, Pemberton J et al (2016) Changes in functional connectivity of the amygdala during cognitive reappraisal predict symptom reduction duringtrauma-focused cognitive-behavioral therapy among adolescent girls with post-traumatic stress disorder. Psychol Med 46(14):3013–3023

Cohen JA, Mannarino AP, Deblinger E (2017) Treating Trauma and traumatic grief in children and adolescents, 2nd edn. Guilford Publications

Compton SN, March JS, Brent D, Albano AM, Weersing R, Curry J (2004) Cognitive-behavioral psychotherapy for anxiety and depressive disorders in children and adolescents: an evidence-based medicine review. J Am Acad Child Adolesc Psychiatry 43(8):930–959

Conway BR (2018) The organization and operation of inferior temporal cortex. Annu Rev Vision Sci 4:381–402

Coveleskie K, Gupta A, Kilpatrick LA, Mayer ED, Ashe-McNalley C, Stains J, Labus JS, Mayer EA (2015) Altered functional connectivity within the central reward network in overweight and obese women. Nutr Diabetes 5(1):e148

Craske MG, Treanor M, Zbozinek TD, Vervliet B (2022) Optimizing exposure therapy with an inhibitory retrieval approach and the OptEx Nexus. Behav Res Ther 152:104069

D'Argembeau A (2013) On the role of the ventromedial prefrontal cortex in self-processing: the valuation hypothesis. Front Hum Neurosci 7:372

Damme KSF, Alloy LB, Kelley NJ, Carroll A, Young CB, Chein J, Ng TH, Titone MK, Bart CP, Nusslock R (2022) Bipolar spectrum disorders are associated with increased gray matter volume in the medial orbitofrontal cortex and nucleus accumbens. JCPP Adv 2(1). https://doi.org/10.1002/jcv2.12068

Daniel R, Pollmann S (2014) A universal role of the ventral striatum in reward-based learning: evidence from human studies. Neurobiol Learn Mem 114:90–100

Davidson RJ (2002) Anxiety and affective style: role of prefrontal cortex and amygdala. Biol Psychiatry 51(1):68–80

Davis M, Whalen PJ (2001) The amygdala: vigilance and emotion. Mol Psychiatry 6(1):13–34

Diwadkar VA, Asemi A, Burgess A, Chowdury A, Bressler SL (2017) Potentiation of motor sub-networks for motor control but not working memory: interaction of dACC and SMA revealed by resting-state directed functional connectivity. PLoS One 12(3):e0172531

Doehrmann O, Ghosh SS, Polli FE, Reynolds GO, Horn F, Keshavan A, Triantafyllou C, Saygin ZM, Whitfield-Gabrieli S, Hofmann SG, Pollack M, Gabrieli JD (2013) Predicting treatment response in social anxiety disorder from functional magnetic resonance imaging. JAMA Psychiatry 70(1):87–97

Dosenbach NUF, Visscher KM, Palmer ED, Miezin FM, Wenger KK, Kang HC, Burgund ED, Grimes AL, Schlaggar BL, Petersen SE (2006) A core system for the implementation of task sets. Neuron 50(5):799–812

Dosenbach NUF, Fair DA, Miezin FM, Cohen AL, Wenger KK, Dosenbach RAT, Fox MD, Snyder AZ, Vincent JL, Raichle ME, Schlaggar BL, Petersen SE (2007) Distinct brain networks for adaptive and stable task control in humans. Proc Natl Acad Sci U S A 104(26):11073–11078

Etkin A, Wager TD (2007) Functional neuroimaging of anxiety: a meta-analysis of emotional processing in PTSD, social anxiety disorder, and specific phobia. Am J Psychiatry 164(10): 1476–1488

Etkin A, Egner T, Peraza DM, Kandel ER, Hirsch J (2006) Resolving emotional conflict: a role for the rostral anterior cingulate cortex in modulating activity in the amygdala. Neuron 51(6): 871–882

Etkin A, Prater KE, Hoeft F, Menon V, Schatzberg AF (2010) Failure of anterior cingulate activation and connectivity with the amygdala during implicit regulation of emotional processing in generalized anxiety disorder. Am J Psychiatry 167(5):545–554

Etkin A, Egner T, Kalisch R (2011) Emotional processing in anterior cingulate and medial prefrontal cortex. Trends Cogn Sci 15(2):85–93

Fitzgerald KD, Schroder HS, Marsh R (2021) Cognitive control in pediatric obsessive-compulsive and anxiety disorders: brain-behavioral targets for early intervention. Biol Psychiatry 89(7): 697–706

Forbes EE, Olino TM, Ryan ND, Birmaher B, Axelson D, Moyles DL, Dahl RE (2010) Reward-related brain function as a predictor of treatment response in adolescents with major depressive disorder. Cogn Affect Behav Neurosci 10(1):107–118

Fusar-Poli P, Placentino A, Carletti F, Landi P, Allen P, Surguladze S, Benedetti F, Abbamonte M, Gasparotti R, Barale F, Perez J, McGuire P, Politi P (2009) Functional atlas of emotional faces processing: a voxel-based meta-analysis of 105 functional magnetic resonance imaging studies. J Psychiatry Neurosci 34(6):418–432

Garrett AS, Carrion V, Kletter H, Karchemskiy A, Weems CF, Reiss A (2012) Brain activation to facial expressions in youth with PTSD symptoms. Depress Anxiety 29(5):449–459

Garrett A, Cohen JA, Zack S, Carrion V, Jo B, Blader J, Rodriguez A, Vanasse TJ, Reiss AL, Agras WS (2019) Longitudinal changes in brain function associated with symptom improvement in youth with PTSD. J Psychiatr Res 114:161–169

Gentili C, Cristea IA, Angstadt M, Klumpp H, Tozzi L, Phan KL, Pietrini P (2016) Beyond emotions: a meta-analysis of neural response within face processing system in social anxiety. Exp Biol Med 241(3):225–237

Ginsburg GS, Kendall PC, Sakolsky D, Compton SN, Piacentini J, Albano AM, Walkup JT, Sherrill J, Coffey KA, Rynn MA, Keeton CP, McCracken JT, Bergman L, Iyengar S, Birmaher B, March J (2011) Remission after acute treatment in children and adolescents with anxiety disorders: findings from the CAMS. J Consult Clin Psychol 79(6):806–813

Glenn DE, Merenstein JL, Bennett IJ, Michalska KJ (2022) Anxiety symptoms and puberty interactively predict lower cingulum microstructure in preadolescent Latina girls. Sci Rep 12(1):20755

Goldin PR, McRae K, Ramel W, Gross JJ (2008) The neural bases of emotion regulation: reappraisal and suppression of negative emotion. Biol Psychiatry 63(6):577–586

Greicius MD, Menon V (2004) Default-mode activity during a passive sensory task: uncoupled from deactivation but impacting activation. J Cogn Neurosci 16(9):1484–1492

Guyer AE, Nelson EE, Perez-Edgar K, Hardin MG, Roberson-Nay R, Monk CS, Bjork JM, Henderson HA, Pine DS, Fox NA, Ernst M (2006) Striatal functional alteration in adolescents characterized by early childhood behavioral inhibition. J Neurosci Off J Soc Neurosci 26(24): 6399–6405

Guyer AE, Lau JYF, McClure-Tone EB, Parrish J, Shiffrin ND, Reynolds RC, Chen G, Blair RJR, Leibenluft E, Fox NA, Ernst M, Pine DS, Nelson EE (2008) Amygdala and ventrolateral prefrontal cortex function during anticipated peer evaluation in pediatric social anxiety. Arch Gen Psychiatry 65(11):1303–1312

Guyer AE, Choate VR, Detloff A, Benson B, Nelson EE, Perez-Edgar K, Fox NA, Pine DS, Ernst M (2012) Striatal functional alteration during incentive anticipation in pediatric anxiety disorders. Am J Psychiatry 169(2):205–212

Haber SN, Knutson B (2010) The reward circuit: linking primate anatomy and human imaging. Neuropsychopharmacology 35(1):4–26

Haller SP, Linke JO, Grassie HL, Jones EL, Pagliaccio D, Harrewijn A, White LK, Naim R, Abend R, Mallidi A, Berman E, Lewis KM, Kircanski K, Fox NA, Silverman WK, Kalin NH, Bar-Haim Y, Brotman MA (2024) Normalization of Fronto-parietal activation by cognitive-behavioral therapy in unmedicated pediatric patients with anxiety disorders. Am J Psychiatry 181(3):201–212

Han DH, Kim SM, Bae S, Renshaw PF, Anderson JS (2016) A failure of suppression within the default mode network in depressed adolescents with compulsive internet game play. J Affect Disord 194:57–64

Hardin MG, Perez-Edgar K, Guyer AE, Pine DS, Fox NA, Ernst M (2006) Reward and punishment sensitivity in shy and non-shy adults: relations between social and motivated behavior. Personal Individ Differ 40(4):699–711

Hauner KK, Mineka S, Voss JL, Paller KA (2012) Exposure therapy triggers lasting reorganization of neural fear processing. Proc Natl Acad Sci U S A 109(23):9203–9208

Hiser J, Koenigs M (2018) The multifaceted role of the ventromedial prefrontal cortex in emotion, decision making, social cognition, and psychopathology. Biol Psychiatry 83(8):638–647

Igelström KM, Graziano MSA (2017) The inferior parietal lobule and temporoparietal junction: a network perspective. Neuropsychologia 105:70–83

Kanwisher N, Yovel G (2006) The fusiform face area: a cortical region specialized for the perception of faces. Philos Trans R Soc Lond B Biol Sci 361(1476):2109–2128

Kendall PC, Compton SN, Walkup JT, Birmaher B, Albano AM, Sherrill J, Ginsburg G, Rynn M, McCracken J, Gosch E, Keeton C, Bergman L, Sakolsky D, Suveg C, Iyengar S, March J, Piacentini J (2010) Clinical characteristics of anxiety disordered youth. J Anxiety Disord 24(3): 360–365

Kenwood MM, Kalin NH, Barbas H (2022) The prefrontal cortex, pathological anxiety, and anxiety disorders. Neuropsychopharmacology 47(1):260–275

Kessler RC, Berglund P, Demler O, Jin R, Merikangas KR, Walters EE (2005) Lifetime prevalence and age-of-onset distributions of DSM-IV disorders in the National Comorbidity Survey Replication. Arch Gen Psychiatry 62(6):593–602

Kessler, R. C., Ormel, J., Petukhova, M., McLaughlin, K. A., Green, J. G., Russo, L. J., Stein, D. J., Zaslavsky, A. M., Aguilar-Gaxiola, S., Alonso, J., Andrade, L., Benjet, C., de Girolamo, G., de Graaf, R., Demyttenaere, K., Fayyad, J., Haro, J. M., Hu, C., Karam, A., . . . Ustün, T. B. (2011). Development of lifetime comorbidity in the World Health Organization world mental health surveys. Arch Gen Psychiatry, 68(1), 90–100

Kessler RC, Avenevoli S, Costello EJ, Georgiades K, Green JG, Gruber MJ, He J-P, Koretz D, McLaughlin KA, Petukhova M, Sampson NA, Zaslavsky AM, Merikangas KR (2012) Prevalence, persistence, and sociodemographic correlates of DSM-IV disorders in the National Comorbidity Survey Replication Adolescent Supplement. Arch Gen Psychiatry 69(4):372–380

Killgore WDS, Britton JC, Schwab ZJ, Price LM, Weiner MR, Gold AL, Rosso IM, Simon NM, Pollack MH, Rauch SL (2014) Cortico-limbic responses to masked affective faces across ptsd, panic disorder, and specific phobia. Depress Anxiety 31(2):150–159

Kim MJ, Chey J, Chung A, Bae S, Khang H, Ham B, Yoon SJ, Jeong D-U, Lyoo IK (2008) Diminished rostral anterior cingulate activity in response to threat-related events in posttraumatic stress disorder. J Psychiatr Res 42(4):268–277

Kitaichi Y, Inoue T, Nakagawa S, Boku S, Kakuta A, Izumi T, Koyama T (2010) Sertraline increases extracellular levels not only of serotonin, but also of dopamine in the nucleus accumbens and striatum of rats. Eur J Pharmacol 647(1–3):90–96

Koechlin E, Ody C, Kouneiher F (2003) The architecture of cognitive control in the human prefrontal cortex. Science 302(5648):1181–1185

Kohn N, Eickhoff SB, Scheller M, Laird AR, Fox PT, Habel U (2014) Neural network of cognitive emotion regulation – an ALE meta-analysis and MACM analysis. Neuroimage 87:345–355

Kujawa A, Swain JE, Hanna GL, Koschmann E, Simpson D, Connolly S, Fitzgerald KD, Monk CS, Phan KL (2016) Prefrontal reactivity to social signals of threat as a predictor of treatment response in anxious youth. Neuropsychopharmacology 41(8):1983–1990

La Buissonniere-Ariza V, Fitzgerald K, Meoded A, Williams LL, Liu G, Goodman WK, Storch EA (2021) Neural correlates of cognitive behavioral therapy response in youth with negative valence disorders: a systematic review of the literature. J Affect Disord 282:1288–1307

Liberzon I, Sripada CS (2007) The functional neuroanatomy of PTSD: a critical review. In: De Kloet ER, Oitzl MS, Vermetten E (eds) Progress in brain research, vol 167. Elsevier, pp 151–169

Liddle EB, Hollis C, Batty MJ, Groom MJ, Totman JJ, Liotti M, Scerif G, Liddle PF (2011) Task-related default mode network modulation and inhibitory control in ADHD: effects of motivation and methylphenidate. J Child Psychol Psychiatry 52(7):761–771

Lu L, Li H, Baumel WT, Mills JA, Cecil KM, Schroeder HK, Mossman SA, Huang X, Gong Q, Sweeney JA, Strawn JR (2022) Acute neurofunctional effects of escitalopram during emotional processing in pediatric anxiety: a double-blind, placebo-controlled trial. Neuropsychopharmacology 47(5):1081–1087

Maslowsky J, Mogg K, Bradley BP, McClure-Tone E, Ernst M, Pine DS, Monk CS (2010) A preliminary investigation of neural correlates of treatment in adolescents with generalized anxiety disorder. J Child Adolesc Psychopharmacol 20(2):105–111

McClure EB, Adler A, Monk CS, Cameron J, Smith S, Nelson EE, Leibenluft E, Ernst M, Pine DS (2007a) fMRI predictors of treatment outcome in pediatric anxiety disorders. Psychopharmacology (Berl) 191(1):97–105

McClure EB, Monk CS, Nelson EE, Parrish JM, Adler A, Blair RJR, Fromm S, Charney DS, Leibenluft E, Ernst M, Pine DS (2007b) Abnormal attention modulation of fear circuit function in pediatric generalized anxiety disorder. Arch Gen Psychiatry 64(1):97–106

McGuire JF, Storch EA (2019) An inhibitory learning approach to cognitive-behavioral therapy for children and adolescents. Cogn Behav Pract 26(1):214–224

Merikangas KR, He J-P, Burstein M, Swanson SA, Avenevoli S, Cui L, Benjet C, Georgiades K, Swendsen J (2010) Lifetime prevalence of mental disorders in U.S. adolescents: results from the National Comorbidity Survey Replication – Adolescent Supplement (NCS-A). J Am Acad Child Adolesc Psychiatry 49(10):980–989

Monk CS, Nelson EE, McClure EB, Mogg K, Bradley BP, Leibenluft E, Blair RJR, Chen G, Charney DS, Ernst M, Pine DS (2006) Ventrolateral prefrontal cortex activation and attentional bias in response to angry faces in adolescents with generalized anxiety disorder. Am J Psychiatry 163(6):1091–1097

Monk CS, Telzer EH, Mogg K, Bradley BP, Mai X, Louro HMC, Chen G, McClure-Tone EB, Ernst M, Pine DS (2008) Amygdala and ventrolateral prefrontal cortex activation to masked angry faces in children and adolescents with generalized anxiety disorder. Arch Gen Psychiatry 65(5):568. https://doi.org/10.1001/archpsyc.65.5.568

Nachev P, Kennard C, Husain M (2008) Functional role of the supplementary and pre-supplementary motor areas. Nat Rev Neurosci 9(11):856–869

Nelson W, Finch AJ (2008) Cognitive-behavioral therapy for aggressive children [therapist manual for keeping your cool: the anger management workbook

Ohman A (2005) The role of the amygdala in human fear: automatic detection of threat. Psychoneuroendocrinology 30(10):953–958

Ohman A, Flykt A, Esteves F (2001) Emotion drives attention: detecting the snake in the grass. J Exp Psychol Gen 130(3):466–478

Piacentini J, Bennett S, Compton SN, Kendall PC, Birmaher B, Albano AM, March J, Sherrill J, Sakolsky D, Ginsburg G, Rynn M, Bergman RL, Gosch E, Waslick B, Iyengar S, McCracken J, Walkup J (2014) 24- and 36-week outcomes for the child/adolescent anxiety multimodal study (CAMS). J Am Acad Child Adolesc Psychiatry 53(3):297–310

Price JL, Drevets WC (2010) Neurocircuitry of mood disorders. Neuropsychopharmacology 35(1): 192–216

Raichle ME (2015) The brain's default mode network. Annu Rev Neurosci 38:433–447

Raichle ME, MacLeod AM, Snyder AZ, Powers WJ, Gusnard DA, Shulman GL (2001) A default mode of brain function. Proc Natl Acad Sci U S A 98(2):676–682

Ranøyen I, Lydersen S, Larose TL, Weidle B, Skokauskas N, Thomsen PH, Wallander J, Indredavik MS (2018) Developmental course of anxiety and depression from adolescence to young adulthood in a prospective Norwegian clinical cohort. Eur Child Adolesc Psychiatry 27(11):1413–1423

Rapee RM, Wignall A (2002) Cool kids program: family version session guide. Macquarie University Anxiety Research Unit, Sydney

Reynolds SM, Zahm DS (2005) Specificity in the projections of prefrontal and insular cortex to ventral striatopallidum and the extended amygdala. J Neurosci Off J Soc Neurosci 25(50): 11757–11767

Rosenkranz JA, Moore H, Grace AA (2003) The prefrontal cortex regulates lateral amygdala neuronal plasticity and responses to previously conditioned stimuli. J Neurosci Off J Soc Neurosci 23(35):11054–11064

Scolari M, Seidl-Rathkopf KN, Kastner S (2015) Functions of the human frontoparietal attention network: evidence from neuroimaging. Curr Opin Behav Sci 1:32–39

Seeley WW (2019) The salience network: a neural system for perceiving and responding to homeostatic demands. J Neurosci Off J Soc Neurosci 39(50):9878–9882

Seeley WW, Menon V, Schatzberg AF, Keller J, Glover GH, Kenna H, Reiss AL, Greicius MD (2007) Dissociable intrinsic connectivity networks for salience processing and executive control. J Neurosci Off J Soc Neurosci 27(9):2349–2356

Seghier ML (2013) The angular gyrus: multiple functions and multiple subdivisions. Neuroscientist 19(1):43–61

Sequeira SL, Silk JS, Ladouceur CD, Hanson JL, Ryan ND, Morgan JK, McMakin DL, Kendall PC, Dahl RE, Forbes EE (2021) Association of neural reward circuitry function with response to psychotherapy in youths with anxiety disorders. Am J Psychiatry 178(4):343–351

Shackman AJ, Salomons TV, Slagter HA, Fox AS, Winter JJ, Davidson RJ (2011) The integration of negative affect, pain and cognitive control in the cingulate cortex. Nat Rev Neurosci 12(3): 154–167

Shapiro K, Hillstrom AP, Husain M (2002) Control of visuotemporal attention by inferior parietal and superior temporal cortex. Curr Biol 12(15):1320–1325

Shenhav A, Cohen JD, Botvinick MM (2016) Dorsal anterior cingulate cortex and the value of control. Nat Neurosci 19(10):1286–1291

Shirk S, Karver M (2006) Process issues in cognitive-behavioral therapy for youth. In: Kendall P (ed) Child and adolescent therapy: cognitive-behavioral procedures. Guilford, New York, pp 465–491

Shulman GL, Corbetta M, Buckner RL, Fiez JA, Miezin FM, Raichle ME, Petersen SE (1997) Common blood flow changes across visual tasks: I. Increases in subcortical structures and cerebellum but not in nonvisual cortex. J Cogn Neurosci 9(5):624–647

Singh KD, Fawcett IP (2008) Transient and linearly graded deactivation of the human default-mode network by a visual detection task. Neuroimage 41(1):100–112

Spence SH, Rapee RM (2016) The etiology of social anxiety disorder: an evidence-based model. Behav Res Ther 86:50–67

Sridharan D, Levitin DJ, Menon V (2008) A critical role for the right fronto-insular cortex in switching between central-executive and default-mode networks. Proc Natl Acad Sci U S A 105(34):12569–12574

Stevens JS, Kim YJ, Galatzer-Levy IR, Reddy R, Ely TD, Nemeroff CB, Hudak LA, Jovanovic T, Rothbaum BO, Ressler KJ (2017) Amygdala reactivity and anterior cingulate habituation predict posttraumatic stress disorder symptom maintenance after acute civilian Trauma. Biol Psychiatry 81(12):1023–1029

Straube T, Mentzel H-J, Miltner WHR (2005) Common and distinct brain activation to threat and safety signals in social phobia. Neuropsychobiology 52(3):163–168. https://doi.org/10.1159/000087987

Strawn JR, Bitter SM, Weber WA, Chu W-J, Whitsel RM, Adler C, Cerullo MA, Eliassen J, Strakowski SM, DelBello MP (2012) Neurocircuitry of generalized anxiety disorder in adolescents: a pilot functional neuroimaging and functional connectivity study. Depress Anxiety 29(11):939–947

Strawn JR, Dominick KC, Patino LR, Doyle CD, Picard LS, Phan KL (2014) Neurobiology of pediatric anxiety disorders. Curr Behav Neurosci Rep 1(3):154–160

Strawn JR, Welge JA, Wehry AM, Keeshin B, Rynn MA (2015) Efficacy and tolerability of antidepressants in pediatric anxiety disorders: a systematic review and meta-analysis. Depress Anxiety 32(3):149–157

Strawn JR, Lu L, Peris TS, Levine A, Walkup JT (2021) Research review: pediatric anxiety disorders – what have we learnt in the last 10 years? J Child Psychol Psychiatry 62(2):114–139

Sulpizio V, Fattori P, Pitzalis S, Galletti C (2023) Functional organization of the caudal part of the human superior parietal lobule. Neurosci Biobehav Rev 153:105357

Swartz JR, Phan KL, Angstadt M, Klumpp H, Fitzgerald KD, Monk CS (2014) Altered activation of the rostral anterior cingulate cortex in the context of emotional face distractors in children and adolescents with anxiety disorders. Depress Anxiety 31(10):870–879

Thayer RE, Crotwell SM, Callahan TJ, Hutchison KE, Bryan AD (2012) Nucleus accumbens volume is associated with frequency of alcohol use among Juvenile Justice-involved adolescents. Brain Sci 2(4):605–618

Thomas KM, Drevets WC, Dahl RE, Ryan ND, Birmaher B, Eccard CH, Axelson D, Whalen PJ, Casey BJ (2001) Amygdala response to fearful faces in anxious and depressed children. Arch Gen Psychiatry 58(11):1057–1063

Vuilleumier P (2005) How brains beware: neural mechanisms of emotional attention. Trends Cogn Sci 9(12):585–594

Walkup JT, Albano AM, Piacentini J, Birmaher B, Compton SN, Sherrill JT, Ginsburg GS, Rynn MA, McCracken J, Waslick B, Iyengar S, March JS, Kendall PC (2008) Cognitive behavioral therapy, sertraline, or a combination in childhood anxiety. N Engl J Med 359(26):2753–2766

White LK, Sequeira S, Britton JC, Brotman MA, Gold AL, Berman E, Towbin K, Abend R, Fox NA, Bar-Haim Y, Leibenluft E, Pine DS (2017) Complementary features of attention bias modification therapy and cognitive-behavioral therapy in pediatric anxiety disorders. Am J Psychiatry 174(8):775–784

Wu Y, Wang J, Zhang Y, Zheng D, Zhang J, Rong M, Wu H, Wang Y, Zhou K, Jiang T (2016) The neuroanatomical basis for posterior superior parietal lobule control lateralization of visuospatial attention. Front Neuroanat 10:32

Yin D, Liu W, Zeljic K, Lv Q, Wang Z, You M, Men W, Fan M, Cheng W, Wang Z (2017) Failure in cognitive suppression of negative affect in adolescents with generalized anxiety disorder. Sci Rep 7(1):6583

Fears Worth Testing Out: A Systematic Review of the Neural Mechanisms of Treatment Outcome for Anxiety-Related Disorders

Robin Aupperle, Hannah Berg, and Jacob Armstrong

Contents

Abstract With the advent of human neuroimaging, researchers were drawn to the idea that by better understanding the human brain, more effective mental health interventions could be developed. It has been more than 20 years since the first functional magnetic resonance imaging (fMRI) studies were conducted to examine changes in brain activation with anxiety-related treatments and more than 60 studies have since been published in this vein. For the current review, we conduct a systematic review of this literature, focusing on adult studies using task-based fMRI to measure brain activation changes with pharmacologic or psychotherapy interventions for phobia, social anxiety disorder, panic disorder, generalized anxiety disorder, posttraumatic stress disorder, and obsessive-compulsive disorder. Neuroscientific theories of anxiety-related disorders and their treatment have focused on prefrontal-insula-amygdala networks. Treatment-related decreases in amygdala

R. Aupperle (✉)
Laureate Institute for Brain Research, Tulsa, OK, USA

School of Community Medicine, The University of Tulsa, Tulsa, OK, USA
e-mail: raupperle@laureateinstitute.org

H. Berg and J. Armstrong
Laureate Institute for Brain Research, Tulsa, OK, USA

© The Author(s), under exclusive license to Springer Nature Switzerland AG 2024
Curr Topics Behav Neurosci (2025) 73: 495–526
https://doi.org/10.1007/7854_2024_549
Published Online: 14 December 2024

and/or anterior insula activation were identified as the most consistent finding across disorders, with the most consistent results reported for specific phobia. Directionality of change and specific regions implicated in the prefrontal cortex were inconsistent across studies. The potential importance for probing other networks and processes as mechanisms of anxiety treatment was recognized, such as striatal regions underlying inhibitory learning or reward responsivity. Future treatment-fMRI research related to anxiety disorders would benefit from larger sample sizes, use of more nuanced computational approaches, and increased focus on replication. There is continued promise that fMRI research will enhance our understanding of how treatments work and inform the evolution of more effective or personalized mental health treatment.

Keywords Amygdala · Anxiety disorders · Cognitive-behavioral therapy · Exposure therapy · Functional MRI · Insula · OCD · Pharmacotherapy · Prefrontal cortex · PTSD · SSRI · Striatum

1 Introduction

Fear and anxiety are adaptive, cross-species processes supporting one's ability to respond to threat and/or anticipate potential threats in our environment, and thus crucial for supporting our functioning and survival (LeDoux and Pine 2016). However, excessive and chronic experiences of fear or anxiety can disrupt one's ability to function optimally and thrive. Anxiety-related disorders in the Diagnostic and Statistical Manual (DSM) include specific phobia, panic disorder, generalized anxiety disorder (GAD), and social anxiety disorder (SAD). Posttraumatic stress disorder (PTSD) and obsessive-compulsive disorder (OCD) were moved to their own categories in the transition to DSM-5 (American Psychiatric Association 2013), but still have overlapping symptoms with anxiety disorders (e.g., distress about perceived or potential threats, avoidance behavior) and similar treatment strategies (e.g., exposure-based interventions).

With the advent of human neuroimaging, researchers have been drawn to the idea that by better understanding the human brain, we could better understand emotion and mental health, and thus inform the development of more effective interventions. It has been over 20 years since the first functional magnetic resonance imaging (fMRI) studies were conducted to examine changes in brain activation from pre- to post-therapy (e.g., Hoehn-Saric et al. 2004; Paquette et al. 2003) and >60 studies have since been published. For the current review, we begin by briefly describing the first-line anxiety treatments and their underlying theoretical rationale. We then provide a summary of studies identified via systematic review of the literature, focusing on fMRI studies of brain activation changes in adult populations who engage in pharmacologic or psychotherapy interventions for specific phobia, panic disorder, GAD, SAD, OCD, and PTSD. We discuss this literature with the aim of clarifying insights gained, revealing gaps that remain, and inspiring future research to enhance translation toward more efficacious treatments.

2 Intervention Approaches for Anxiety- and Fear-Related Disorders

Treatment guidelines for anxiety disorders most often recommend selective serotonin reuptake inhibitors (SSRIs) and serotonin-norepinephrine reuptake inhibitors (SNRIs) as first-line pharmacologic agents and cognitive-behavioral interventions (including both exposure-based behavioral approaches and cognitive approaches) as first-line psychotherapy interventions (e.g. National Institute for Health and Care Excellence 2020; American Psychiatric Association 2010). Additional interventions (e.g., benzodiazepines, "third wave" psychological interventions) have also been found effective and are frequently utilized. Below, we summarize these intervention approaches and their basic underlying theoretical rationale.

Benzodiazepines are effective in the treatment of anxiety and are thus often prescribed, but have lower long-term acceptability of use compared to SSRIs, given risk of tolerance and abuse, adverse drug interactions, and withdrawal symptoms (Dubovsky and Marshall 2022). Benzodiazepines bind to the major inhibitory neurotransmitter, gamma-aminobutyric acid (GABA), and reach peak serum levels within 30 min to 2 h (Dubovsky and Marshall 2022). Benzodiazepine receptors are located throughout the brain, with high density in cortical regions, hippocampus, striatum, amygdala, cerebellum, and brainstem (Dubovsky and Marshall 2022; Zezula et al. 1988).

SSRIs were designed to specifically target serotonin reuptake while avoiding influencing other neuroreceptors and undesirable side effects, with SNRIs inhibiting serotonin and norepinephrine reuptake (Hillhouse and Porter 2015). SSRIs were developed as antidepressants but quickly recognized to also benefit anxiety (Lopez-Munoz and Alamo 2009). The clinical benefit of SSRIs are often observed 2 weeks or more after initiating treatment. Serotonin receptors are widespread throughout the brain, with high density in the substantia nigra, hippocampus, hypothalamus, amygdala, striatum, and frontal cortex (Charnay and Leger 2010).

Exposure therapy is related to the concept of fear extinction, with the idea that repeated exposure to a feared memory, image, stimulus, or situation – in the absence of the feared outcome occurring – will result in habituation or inhibitory learning (Craske et al. 2014). Cross-species fear extinction models involve pairing a previously neutral conditioned stimulus (CS+) with an aversive unconditioned stimulus (UCS) to elicit a conditional response (CR) of anticipatory fear. Fear extinction involves presentation of the CS+ repeatedly without occurrence of the UCS, resulting in habituation of the CR and learning of a new association between the CS and a safe outcome (inhibitory learning; Craske et al. 2014). Animal research highlights central roles of the basolateral amygdala, hippocampus, and ventromedial prefrontal cortex (vmPFC) in extinction learning (Bouton et al. 2021). Human neuroimaging research more consistently reports engagement of anterior insula, dorsal anterior cingulate cortex (dACC), and dorsolateral prefrontal cortex (dlPFC; Kirlic et al. 2022; Fullana et al. 2018; Kalisch in press).

Cognitive-therapy approaches to understanding psychiatric disorders were based on information processing models, which theorized that enduring cognitive schemas distort the processing of external and internal stimuli, resulting in maladaptive cognitive errors (e.g., overgeneralization, personalization, etc.) (Beck 2005; Clark and Beck 2011). Cognitive therapy focuses on identifying core schemas and exploring alternative ways of thinking (e.g., through cognitive restructuring). Human neuroimaging studies of emotional reappraisal consistently report involvement of a frontoparietal network, namely the dlPFC, dorsomedial PFC (dmPFC) or mid-cingulate cortex (MCC), ventrolateral PFC (vlPFC), and posterior parietal regions (Buhle et al. 2014).

"Third-wave" psychotherapy interventions include mind-body interventions such as mindfulness-based stress reduction (MBSR), mindfulness-based CBT (MBCT), and acceptance and commitment therapy (ACT) (Vollestad et al. 2012). Mindfulness can be operationalized as the awareness that emerges through non-judgmentally paying attention to the present moment. ACT aims to promote psychological flexibility through non-judgmental and non-reactive awareness of experience. Neuroimaging studies of mindfulness implicate default-mode, salience, and executive control networks, with anterior cingulate cortex (ACC) and dlPFC regions increasing in activation and mPFC and posterior cingulate cortex (PCC) decreasing (Ganesan et al. 2022).

Based on the theoretical rationales underlying anxiety treatments and human lesion and neuroimaging research related to emotional processing and anxiety disorders, there are a general set of brain regions hypothesized as potential mechanisms of anxiety treatment (McTeague et al. 2020; Feinstein 2013; Pico-Perez et al. 2017; Chavanne and Robinson 2021). This includes salience processing regions, such as the amygdala, considered central in coordinating the fear response and fear learning (Shin and Liberzon 2010; Milad et al. 2014), and the anterior insula, important for interoceptive processing (monitoring internal bodily states) (Paulus and Stein 2006, 2010). The PFC is also hypothesized to be central to intervention effects (Brooks and Stein 2015), including (a) mPFC/vmPFC and more ventral or rostral regions of the ACC (e.g., subgenual), important for safety learning, regulation, and inhibition of amygdala response (Duval et al. 2015; Etkin et al. 2011, 2015), (b) more dorsal regions of the ACC (particularly anterior dorsal regions), which are considered important for conflict/error monitoring and response inhibition (Etkin et al. 2011, 2015), and (c) dlPFC, which is part of the executive control network and implicated in cognitive control and explicit emotion regulation (Etkin et al. 2015). The hippocampus may also be important in anxiety treatment, given its role in emotion processing and contextual fear learning (Milad et al. 2014; Duval et al. 2015). Other regions are hypothesized as important for treatment of specific disorders, such as the basal ganglia in OCD (Rasgon et al. 2017). FMRI research examining the impact of anxiety treatment provides an experimental framework for better understanding how these networks relate to fear and anxiety, as well as informing our understanding and optimization of treatment.

3 Systematic Review

We conducted a systematic review to identify studies using fMRI to examine brain activation changes from pre- to post-treatment for adults with anxiety disorders. To sufficiently narrow the scope, we limited the review to studies reporting task-based activation, rather than connectivity. The systematic review was conducted according to PRISMA guidelines (Liberati et al. 2009); it was not pre-registered. An electronic search was performed on PubMed, Web of Science, and PsychInfo, before January 2024, using: the "name of the disorder" (i.e., PTSD, SAD, GAD, panic disorder, OCD, phobia, and their non-abbreviated names), and terms of pharmacologic (i.e., "pharmacotherapy", "medication", "SSRI", or "benzodiazepine") or psychological treatments (i.e., "psychotherapy", "behavior therapy", "CBT", "cognitive therapy", "exposure therapy", "EMDR", "acceptance and commitment therapy", "psychoanalysis", "psychodynamic", or "mindfulness"), and "functional neuroimaging" or "fMRI". Publications were included if they: (i) were written in English; (ii) used psychotherapy or medication approaches and not other methods (e.g., neuromodulation approaches); (iii) examined change in neural responses from before to after a single treatment (or in comparison to another treatment or wait-list control group), (iv) used fMRI and not other imaging modalities (e.g., structural MRI, diffusion techniques, EEG/ERP, etc.); (v) focused on adult populations. Reference lists of the articles obtained from the initial search were checked for further relevant articles. Identification of duplicate DOI numbers was conducted automatically in excel to exclude duplicate articles. Two independent raters (HB, JA) conducted an initial review of all articles identified in the search and indicated if they should be included or excluded. For articles in which the raters disagreed, a third rater (RA) made a final determination. Table 1 summarized results from each article reviewed.

3.1 Specific Phobia

The reviewed fMRI studies related to specific phobia primarily employed brief exposure-based CBT approaches, with the majority using single-session graded exposure (though some included up to 8 weeks of therapy). The 10 fMRI studies identified all utilized symptom provocation tasks, with patients being asked to passively observe phobic and neutral stimuli. Only a few studies employed comparison groups, including healthy individuals without specific phobia and/or comparison phobic subjects who were scanned twice with no intervention.

Evidence from the specific phobia literature strongly supports the notion that exposure therapy for specific phobia reduces salience network activation, with the

Table 1 Results from system review: list of studies, basic characteristics of study design, and list of primary findings

Study	Disorder	Treatment(s)	*n* (*n* per treatment arm)	Task	Changes with treatment
Paquette et al. (2003)	Spider phobia	Exposure (group, 4 weekly sessions)	12	Symptom provocation	↑ M occipital, sup parietal, L inf occipital, L FuG, R IFG ↓ **R dlPFC**, R PHG
Straube et al. (2006)	Spider phobia	CBT vs. waitlist (2 sessions)	25 (13 CBT; 12 WL)	Symptom provocation	↓ **dACC**, **L aIns**, L thal
Goossens et al. (2007a, b)	Spider phobia	CBT (1 session)	20 Phobia	Symptom provocation	↓ **L amyg**, **L aIns**, **dACC**
Schienle et al. (2007)[A]	Spider phobia	CBT vs. waitlist (1 session)	26 (14 CBT; 12 WL)	Symptom provocation	↑ **Medial OFC** ↑ Tx response assoc. w/ ↓ **R amyg**, **L aIns**
Schienle et al. (2009)[A]	Spider phobia	CBT (1 session)	10	Symptom provocation	↓ **R aIns**, **lat OFC** ↑ **Medial OFC**
Hauner et al. (2012)	Spider phobia	Exposure (1 session)	11	Symptom provocation	↓ PCC, r**ACC/dACC**, vmPFC, **aIns**, pIns, MTG, MFG, **R. amyg** ↑ **dlPFC** ↑ Tx response assoc. w/ ↓ ventral visual cortex
Lipka et al., (2014)	Spider phobia	CBT (2 sessions) vs. waitlist	28 (14 CBT; 14 WL)	Backward-masked symptom provocation	↓ **R amyg**, **dACC** ↑ Tx response assoc. w/ ↓ **R amyg**
Halsband and Wolf (2015)	Dental phobia	Hypnosis (1 session)	12	Symptom provocation	↓ **L amyg**, **aIns**, hipp
Alvarez-Perez et al. (2021)	Small animal phobia	CBT (8 sessions)	31 (14 VR; 17 non-VR)	Symptom provocation	↓ R thal, **amyg**, L occipital, **OFC**, **R dlPFC**, L vmPFC, **L ACC**, **aIns**, FuG, R precuneus

Nave et al. (2012)	Snake phobia	Exposure + DCS or placebo (1 session)	20	Symptom provocation	↓ **R dlPFC** ↑ **Medial OFC**, **sgACC**
Goldin et al. (2009)[B]	SAD	MBSR (8 sessions + retreat)	16	Self-referential processing (SRP)	↓ dmPFC, mPFC, L IFG (positive SRP) ↑ L inf parietal, medial precuneus (negative SRP)
Goldin and Gross (2010)[B]	SAD	MBSR (8 sessions + retreat)	14	Negative self-belief induction with attention regulation	↑ Inf/sup parietal, cuneus, precuneus, M occipital, PHG (breath vs. react) ↓ **R amyg** (react) ↑ Tx response assoc. w/ ↑ Medial cuneus, L cuneus, R M occipital
Goldin et al. (2012)[C]	SAD	MBSR (8 sessions + retreat) vs. Aerobic Exercise (8 weeks)	42 (24 MBSR; 18 AE)	Self-referential processing (SRP)	↓ dmPFC (AE, negative SRP) ↑ PCC (MBSR, negative SRP) ↑ Tx response assoc. w/ ↑ dmPFC (MBSR & AE, negative SRP) ↑ Mindfulness assoc. w/ ↑ dmPFC (MBSR, negative SRP) ↑ Mindfulness assoc. w/ ↓ dmPFC (MBSR, positive SRP)
Goldin et al. (2013)[C]	SAD	MBSR (8 sessions + retreat) vs. Aerobic Exercise (8 weeks)	42 (23 MBSR; 19 AE)	Meta-cognitive emotion regulation	↓ **R vlPFC** (MBSR, react) ↑ **R vlPFC** (AE, react) ↓ R superior temporal, lingual gyrus (MBSR, regulate vs. react) ↑ L IPL, R anterior IPL, R posterior IPL, R SPL (MBSR, regulate vs. react) ↓ R SPL, bilateral APL (AE, regulate vs. react) ↑ R lingual gyrus (AE, regulate vs. react)

(continued)

Table 1 (continued)

Study	Disorder	Treatment(s)	*n* (*n* per treatment arm)	Task	Changes with treatment
Goldin et al. (2013)[D]	SAD	CBT vs. waitlist (16 sessions)	60 (31 CBT; 29 WL)	Meta-cognitive emotion regulation	↑ mPFC (react) ↑ dmPFC and L **dlPFC** (reappraise)
Klumpp et al. (2013)	SAD	CBT (12 sessions)l	14	Emotional face processing	↓ **R Ins**, **R medial OFC**, R dmPFC (angry vs. happy)
Månsson et al. (2015)	SAD	iCBT (9 sessions) vs. attention training (ABM; 8 sessions)	22 (11 iCBT; 11 ABM)	Emotional face processing	↓ **L caudate**, cerebellum, **L dlPFC**, **putamen**, **R rACC** (iCBT) ↑ R postcentral gyrus, **putamen**, L SMA, L STG (ABM)
Goldin et al. (2014)[D]	SAD	CBT (16 sessions) vs. waitlist	60 (31 CBT; 29 WL)	Meta-cognitive emotion regulation	↑ R sup frontal, R inf parietal, R M occipital (CBT, react praise) ↑ R sup frontal, R IPL (CBT, react criticism) ↓ L sup temporal (CBT, react criticism) ↑ R sup/M frontal , R M occipital (CBT, reappraise criticism) ↓ L posterior temporal (CBT, reappraise criticism) ↑ Tx response assoc. w/ Δ in above stated activations to reappraise criticism, controlling for Δ neg. emotion, Δ activation to react criticism & react praise)
Young et al. (2017)[E]	SAD	CBT vs. ACT (combined treatments; 12 sessions)	42 (13 CBT; 16 ACT; 13 WL)	Self-referential processing (SRP)	↓ **Amyg**
Brown et al. (2019)[E]	SAD	CBT vs. ACT (combined treatments; 12 sessions)	37	Self-referential processing (SRP)	↓ **L Ins**, d**ACC**, several parietal/occipital regions (self and other)

Goldin et al. (2021)	SAD	CBT vs. MBSR (group, 12 sessions) vs. waitlist	95 (31 CBGT; 32 MBSR; 32 WL)	Meta-cognitive emotion regulation	↑ dmPFC, **dACC** (react) ↑ dmPFC, **dACC**, **L dlPFC**, **L vlPFC**, L SMG, L sup temporal, thal (reappraise) ↑ **L dlPFC**, dmPFC, **dACC**, thal (accept) ↑ Tx response assoc. w/ ↑ dmPFC, **dACC**, **L dlPFC**, **L vlPFC**, L SMG, L superior temporal, and thal (reappraise)
Hur et al. (2021)	SAD	VRET (6 session)	21	Self-referential processing (SRP)	↑ R PCC/precuneus, lingual gyrus, L inferior temporal, precentral/postcentral gyrus (positive) ↑ L M occipital, PHG, L superior frontal, **L caudate** (negative) ↑ Tx response assoc. w/ ↑ lingual gyrus
Gingnell et al. (2016)	SAD	iCBT + escitalopram or placebo (9 weeks)	48 (24 responders; 24 non-responders)	Emotional face-matching	↓ **Amyg** (iCBT+E > iCBT+P; responder > non-responder) ↓ Bil precentral gyrus, SMG, R sup. frontal, L rolandic operculum (iCBT+E > iCBT+P) ↑ Tx response assoc. w/ ↓ **R amyg**, precentral gyrus, R inf. parietal, angular gyrus, M temporal, precentral gyrus
Faria et al. (2017)	SAD	Overt vs. covert escitalopram (9 weeks)	46 (24 Overt; 22 Covert)	Emotional face-matching	↑ R PCC, L M temporal, L PCC, L IFG (overt SSRI > covert SSRI) ↑ Tx response assoc. w/ ↓ **R amyg** ↑ Tx response assoc. w/ ↑ PCC/precuneus

(continued)

Table 1 (continued)

Study	Disorder	Treatment(s)	*n* (*n* per treatment arm)	Task	Changes with treatment
Beutel et al. (2010)	PD	Inpatient psychodynamic (4 weeks intensive)	9	Emotional processing and response inhibition	↓ L hipp/**amyg** (NegativeGo - NeutralGo) ↓ **L amyg**, L hipp (NeuNogo - NeuGo) ↑ **vlPFC** (NegNogo - NegGo) ↑ **L vlPFC**, **lateral OFC** (PosNoGo - PosGo)
Kircher et al. (2013)[F]	PD/AG	CBT+/CBT– therapist support during exposure (12 sessions)	42 PD/A	Differential fear conditioning	↓ L IFG/**aIns** (CS+ anticipation) ↓ **L amyg**, **Ins**, basal ganglia, **ACC** (CS+ > CS-) ↑ Tx response assoc. w/ ↓ L IFG (CS+ > CS-)
Lueken et al. (2013)[F]	PD/AG	CBT+/CBT– therapist support during exposure (12 sessions)	42 (CBT+/CBT- combined; 24 responders vs.18 non-responders)	Differential fear conditioning	↓ R hipp, **R ACC** (non-responders, CS-) ↑ R hipp (responders, CS+ and CS-)
Straube et al. (2014)[F]	PD/AG	CBT+/CBT– therapist support during exposure (12 sessions)	42 (22 CBT+; 20 CBT-)	Differential fear conditioning	↓ L IFG ↑ Tx response assoc. w/ ↓ hipp
Wittmann et al. (2018)[F]	PD/AG	CBT+/CBT– therapist support during exposure (12 sessions)	51	Agoraphobic symptom provocation	↓ **R ventral striatum** (AA - NA) ↓ **R amyg** (AP - NP) ↑ Tx response assoc. w/ ↓ **R ventral striatum** (AA - NA)
Kunas et al. (2019)[F]	PD/AG	CBT+/CBT– therapist support during exposure (12 sessions)	42 (CBT+/CBT- combined)	Differential fear conditioning	↓ R M occipital, L sup occipital (PD/AG+DEP, CS+ and CS-) ↓ R angular gyrus (PD/AG +DEP, CS+) ↓ L IFG (PD/AG-DEP, acquisition CS+)

Neufang et al. (2019)	PD/AG	CBT (6 sessions)	40	Attention network task	↑ MFG, SPL
Bomyea et al. (2020)	PD and GAD	CBT (10 sessions)	30 (20 responders; 10 non-responder)	Emotion regulation	↓ FuG, PHG ↑ Tx response assoc. w/ ↓ PHG (reappraise)
Yang et al. (2020)	PD	CBT+/CBT− interoceptive fear augmentation (12 sessions)	39	Lexical processing of fear-related stimuli	↓ **ACC** ↑ Tx response assoc. w/ ↓ **ACC**
Liebscher et al. (2016)[F]	PD/AG	CBT+/CBT− therapist support during exposure (12 sessions); SSRI/SSNRI (7-8 weeks), waitlist	94 (29 CBT+; 22 CBT-; 28 SSRI/SSNRI; 15 WL)	Agoraphobic symptom provocation	↓ **R amyg** (CBT+) ↓ **L amyg** (CBT-) ↑ Tx response assoc. w/ ↓ **R amyg**, **L amyg**
Plag et al. (2018)[F]	PD/AG	SSRI/SSNRI (9 weeks)	28 (14 responders; 14 non-responders)	Agoraphobic symptom provocation	↓ **L amyg**, **R ACC**, **OFC** (Responders) ↓ **L ins**, **ACC** (Non-responders)
Holzel et al. (2013)	GAD	MBSR vs. Stress management education (8 sessions)	26 (15 MBSR; 11 SME)	Emotional faces	↓ **R amyg**, **R lateral PFC** (across treatments; neutral) ↑ **vlPFC** (MBSR > SME; neutral) ↑ **R vlPFC** to **Ins** and **R rostral M frontal** (MBSR > SME, angry) ↑ Tx response assoc. w/ greater ↑ **L. vlPFC**
Fonzo et al. (2014)	GAD	CBT (10 sessions)	21	Emotional faces	↓ **sgACC**, **L amyg** (ROI; angry and fearful) ↓ **R dlPFC** (whole-brain; angry and fearful) ↑ **R aIns**, **R pIns** (ROI; happy) ↑ Brainstem, visual cortex, cerebellum (whole-brain; happy)

(continued)

Table 1 (continued)

Study	Disorder	Treatment(s)	*n* (*n* per treatment arm)	Task	Changes with treatment
Bomyea et al. (2020)	GAD and PD	CBT (10 sessions)	30	Emotional reappraisal	↓ PHG, FuG ↑ Tx response assoc. w/ ↓ PHG (reappraise)
Hoehn-Saric et al. (2004)	GAD	Citalopram (7 weeks)	6	Worry induction	↓ R **M frontal**, bil pre/post-central gyri, **Ins**, **putamen**, **R sgACC, dACC** and PCC, M cingulate, R PHG, bilateral sup temporal, R M temporal, R IPL, R precuneus, L M occipital
Brown et al. (2015)	GAD	Alprazolam vs. placebo (28 days)	30 (19 alprazolam; 11 placebo)	Emotional faces	↑ PCC (T2 only, alprazolam > placebo)
Nabeyama et al. (2008)	OCD	Behavioral therapy/ERP (12 sessions)	11	Stroop	↑ R precuneus, cerebellum ↓ **R OFC**, **L M frontal**, L FuG, PHG, L precuneus
Freyer et al. (2011)	OCD	Intensive CBT (16 session)	10	Reinforcement learning	↑ **Caudate** ↑ Tx response assoc. w/ smaller ↑ palladial region
Schiepek et al. (2013)	OCD	CBT (8 weeks inpatient)	9	Symptom provocation	↓ dACC/SMA, dlPFC, **Ins**, parietal, cuneus ↑ Tx response assoc. w/ ↓ dACC/SMA, dlPFC, **Ins**, parietal, cuneus
Baioui et al. (2013)	OCD	CBT (32 sessions)	12	Symptom provocation	↓ **NAc**, L posterior SMG, **L OFC**, **R caudate**, **L vlPFC**, R posterior SMG

Morgieve et al. (2014)	OCD	CBT (15 sessions)	27	Symptom provocation	↓ dACC and **L OFC**
Viol et al. (2020)	OCD	Integrative psychotherapy (avg 86 days inpatient)	17	Symptom provocation	↓ **dACC/rACC**, **OFC**, Pcu, **Putamen**
Lee et al. (2023)	OCD	ACT vs. waitlist (8 sessions)	17	Symptom provocation for thought-action fusion	↑ **Ins**, STG ↑ Tx response assoc. w/ ↑ **Ins,** STG
Nakao et al. (2005)	OCD	Fluvoxamine (12 weeks) vs. behavioral therapy (12 sessions)	10 (4 FL; 6 BT)	Stroop	↑ **R dlPFC**, **putamen**, L temporal cortex, parietal, L occipital, cerebellum ↓ R frontal cortex, **cingulate**, R hipp, R temporal and parietal cortex
Farrow et al. (2005)	PTSD	CBT with forgiveness component (avg 7 sessions)	13	Theory of mind	↑ L M temporal (empathy) ↑ PCC (forgiveability)
Felmingham et al. (2007)	PTSD	Imaginal exposure & cognitive restructuring sessions)	8	Emotional faces	↑ **rACC**, L M temporal, R IFG, L parietotemporal, R hipp ↓ R postcentral gyrus, R M temporal, and L superior temporal ↑ Tx response assoc. w/ ↑ **R rACC** and ↓ **amyg**
Roy et al. (2010)[G]	PTSD	VRET vs. PE (12 sessions)	8	Emotional stroop	↑ d**ACC** (laterality unspecified) ↓ **Amyg**, subcallosal gyrus and **lateral PFC** (laterality unspecified)

(continued)

Table 1 (continued)

Study	Disorder	Treatment(s)	*n* (*n* per treatment arm)	Task	Changes with treatment
Thomaes et al. (2012)	PTSD	PE vs. Psychoeducation (20 sessions)	16 (9 EXP, 6 TAU)	Emotional stroop	↓ Premotor cortex/SMA and L IFG (incongruent) ↓ **dACC** and **L aIns**, more in EXP (incongruent) ↑ Tx response assoc. w/ ↓ **R dACC** (negative and trauma vs. neutral)
Simmons et al. (2013)	PTSD	PE (12 sessions)	24 (PE; 9 remitted vs. 15 non-remitted)	Emotional anticipation	↓ **L aIns** (remitted vs. non-remitted; neg. vs. pos. anticipation)
Aupperle et al. (2013)	PTSD	Cognitive trauma therapy for battered women (CTT-BW; 12 sessions)	14	Emotional anticipation	↑ **L rACC/dACC**, L PCC, cerebellum, R superior/M temporal (anticipation) ↓ **R aIns**, M frontal gyrus (anticipation) ↑ Precuneus, R IPL (presentation) ↓ **dlPFC**, **R amg** (presentation) ↑ Tx response assoc. w/ ↓ L **aIns**, PCC, precuneus, and ↑ R **pIns** (anticipation) ↑ Tx response assoc. w/ ↑ precuneus, R PCC, medial frontal, precentral gyrus, lingual gyrus, cerebellum, inferior temporal, **L pIns** (presentation) ↑ Tx response assoc. w/ ↓ R superior/transverse temporal, cerebellum, **caudate** (presentation)

Roy et al. (2014) [G]	PTSD	VRET vs. PE (12-20 sessions)	10 (combined tx groups)	Emotional stroop	↓ **Amyg** (unspecified laterality) ↑ vmPFC, **rACC** (unspecified laterality)
Van Rooij et al. (2015)	PTSD	Trauma-focused CBT w/ exposure and/or EMDR (avg 8-10 sessions)	41	Stop-signal anticipation	↓ L IPL (responders vs. non-responders; contextual cue processing; driven by pre-treatment group differences)
King et al. (2016)	PTSD	Mindfulness-based exposure group therapy (MBET) vs. Present-centered group therapy (PCGT) (16 sessions)	21 (13 MBET; 8 PCGT)	Emotional faces	↑ **L amyg**, R FuG, L PHG, precuneus, and PCC (MBET > PCGT; angry) ↑ L medial frontal cortex (MBET > PCGT; fearful) ↑ Bil FuG/lingual gyrus (MBET > PCGT; neutral) ↑ Tx response assoc. w/ ↑ L mPFC/**rACC**, L dmPFC, **L peri-amygdala area**
Helpman et al. (2016)	PTSD	PE (10 sessions)	16	Fear learning	↓ **R rACC** ↑ Tx response assoc. w/ ↓ **R sgACC**, L hipp/parahipp
Van Rooij et al. (2016)	PTSD	Trauma-focused CBT w/ exposure and/or EMDR (avg 7-11 sessions)	43 (21 remitted; 22 non-remitted)	Emotional processing	↓ **Amyg** (non-remitted > remitted; negative vs. neutral; driven by pre-treatment differences)
Fonzo et al. (2017)	PTSD	PE (9-12 sessions) vs. wait-list (10 weeks)	Emotional reactivity and conflict: 48 (24 PE; 24 WL); Emotional reappraisal: 50 (24 PE, 26 WL)	Emotional reactivity (faces); Emotional conflict; Emotional reappraisal	↑ **L lateral frontopolar cortex** (reappraise negative) ↑ Tx response assoc. w/ ↑ L frontopolar (reappraise negative)

(continued)

Table 1 (continued)

Study	Disorder	Treatment(s)	*n* (*n* per treatment arm)	Task	Changes with treatment
Rousseau et al. (2019a, b)	PTSD	EMDR (until trauma SUDS = 0) vs. supportive psychotherapy (SUP; unspecified duration)	24 (12 EMDR; 12 SUP)	Fear learning	Δ **Amyg**, L hipp, R frontal eye fields, R IFG and **insula**, LHeschl gyrus, L PCC (directionality not specified)
Rousseau et al. (2019b)	PTSD	EMDR (until trauma SUDS = 0)	16	Emotional faces	↓ L thalamus, **L caudate**, **L amyg**, L hippocampus and **Ins**, R inferior temporal, M/posterior cingulum, R parieto-central cortex, precuneus, vmPFC, **L dlPFC** ↑ Tx response assoc. w/ ↓ R thal (faces – shapes)
Harle et al. (2020)	PTSD	PE (12 sessions)	20	Stop signal task	↑ **R caudate** (P(stop) Success) ↓ **R caudate** (P(stop) Stop Error, SPE) ↑ Tx response assoc. w/ ↓ **R caudate** (SPE, P(Stop) Success)
Inoue et al. (2021)	PTSD	EMDR (until trauma SUDS = 0)	6	Symptom provocation: Trauma	↓ Bil hipp, auditory cortex (discrepancy in activations for trauma vs. tooth) ↑ Tx response assoc. w/ ↓ L hipp (discrepancy in activations for trauma vs. tooth)
MacNamara et al. (2016)	PTSD	Paroxetine (12 weeks)	17	Cognitive reappraisal	↑ **L dlPFC**, R SMA

Duval et al. (2020)	PTSD	PE + placebo vs. sertraline + enhanced med. management vs. PE + sertraline (≤13 weeks PE; 24 weeks sertraline)	36 (combined tx groups)	Shifted attention emotional appraisal	↓ **R aIns** (implicit processing) ↓ **R aIns** (appraisal)
Joshi et al. (2020)	PTSD	PE + placebo vs. sertraline + enhanced med. management vs. PE + sertraline (≤13 weeks PE; 24 weeks sertraline)	26 (combined tx groups)	Emotion regulation; Emotional faces	None

Note: Superscript letters denote studies using overlapping samples (e.g., those marked with "A" are from different publications but from overlapping samples); Brain regions are bilateral unless otherwise noted. List of results is not meant to be exhaustive; please see individual publications for more detailed description of findings

Abbreviations: CBT: cognitive-behavioral therapy; DCS: d-cycloserine; MBSR: mindfulness-based stress reduction; iCBT: internet CBT; ABM: internet attentional bias modification; ACT: acceptance and commitment therapy; VRET: virtual reality exposure therapy; SSRI: selective serotonin reuptake inhibitor; SSNRI: selective serotonin noradrenaline reuptake inhibitor; ERP: exposure and response prevention; PE: prolonged exposure; EMDR: eye movement desensitization and reprocessing; SUDS: subjective units of distress. M: middle; L: left; R: right; inf: inferior; sup: superior; FuG: fusiform gyrus; IFG: inferior frontal gyrus; mPFC: medial prefrontal cortex; dlPFC: dorsolateral prefrontal cortex; vlPFC: ventrolateral prefrontal cortex; dmPFC: dorsomedial prefrontal cortex; vmPFC: ventromedial prefrontal cortex; OFC: orbitofrontal cortex; ACC: anterior cingulate cortex, dACC: dorsal anterior cingulate cortex; sgACC: subgenual anterior cingulate cortex; rACC: rostral anterior cingulate cortex; PCC: posterior cingulate cortex; Ins: insula; aIns: anterior insula; pIns: posterior insula; SMG: supramarginal gyrus; PHG: parahippocampal gyrus; thal: thalamus; amyg: amygdala; hipp: hippocampus; MTG: medial temporal gyrus; IPL: inferior parietal lobule; SPL: superior parietal lobule; MFG: medial frontal gyrus; NAc: nucleus accumbens; STG: superior temporal gyrus most consistent findings being for anterior insula and amygdala (Alvarez-Perez et al. 2021; Goossens et al. 2007b; Halsband and Wolf 2015; Lipka et al. 2014; Hauner et al. 2012; Schienle et al. 2007; Straube et al. 2006), though a few studies also reported reduced ACC activation (Goossens et al. 2007a; Nave et al. 2012; Straube et al. 2006). There is some support for increases in medial OFC engagement (Nave et al. 2012; Schienle et al. 2007, 2009), which could relate to the proposed role of this region (and neighboring vmPFC) in fear suppression and stimulus valuation (Fettes et al. 2017). One strength of this literature is the consistent use of symptom provocation tasks and intervention approaches, allowing for replicability across studies. This likely contributes to the consistency in findings and provides an important framework for understanding potential mechanisms of treatment for the other anxiety disorders. However, future specific phobia studies using avoidance-based decision-making or fear learning tasks may be useful for assessing treatment-related changes that relate more directly to the treatment processes.

3.2 *Social Anxiety Disorder*

Almost all SAD articles reviewed focused on psychotherapy treatments (13/14 articles), of which the majority were exposure-based CBT, though four articles focused on MBSR. FMRI tasks included emotional face processing (4), self-referential processing (4), meta-cognitive emotion regulation (4), incidental emotion regulation (1), and attention regulation (1). Five studies included a healthy control or waitlist control condition, and eight included a comparison treatment condition (although four of these combined treatment arms for some or all analyses). Across studies, the amygdala and prefrontal regions were the most commonly identified regions of interest (ROIs).

Despite the centrality of amygdala to hypotheses concerning SAD treatment and the use of tasks thought to probe salience processing, only a few studies reported changes in activation within this region (Faria et al. 2017; Goldin and Gross 2010; Young et al. 2017). It is possible the stimuli used in fMRI tasks (e.g., emotional faces and other stimuli) are not as directly relevant to the symptoms of SAD (e.g., actual social interactions or public speaking) and thus not as sensitive to treatment-related changes in salience processing. One of the most consistent findings was that of increased PFC activation (medial and lateral PFC regions) that often related to symptom change (Goldin et al. 2009, 2012, 2013, 2021), potentially reflecting the importance of emotion regulation, cognitive control, and/or safety valuation in SAD treatment. Findings from the individual studies likely vary depending on task and valence of stimuli (e.g., when the valence of stimuli was negative rather than positive, symptom change related to decreased rather than increased dmPFC activity) (Goldin et al. 2012, 2021).

Notably, there are several symptom dimensions of SAD and different ways in which SAD is experienced across individuals. In one study, increased activation of the lingual gyrus was associated with decreased rumination on negative events, providing initial evidence for brain-behavior relationships for a specific symptom dimension with transdiagnostic relevance (Hur et al. 2021). To further identify consistent biomarkers of SAD treatment response, it may be useful for future studies to (a) attempt to replicate findings using the same tasks, treatments, and symptom measures as prior studies, (b) utilize symptom provocation tasks with more direct relevance to SAD feared situations, and/or (c) focus on specific symptom dimensions that may track more consistently with brain changes. Lastly, further research concerning the impact of pharmacologic interventions would be useful for better understanding which mechanisms may be common biomarkers of treatment response vs unique to specific treatment modalities.

3.3 Panic Disorder

The majority of panic disorder articles reviewed focused on psychotherapy (10/11 articles), of which the majority were exposure-based CBT, with one study of psychodynamic therapy. FMRI tasks varied considerably, but with fear conditioning (4) and agoraphobic symptom provocation (3) being the most common. All 11 studies included either no treatment comparator, or collapsed treatment arms for analysis, with six including either healthy control or waitlist control data for comparison. ROIs primarily included the amygdala, insula, hippocampus, ACC, and OFC.

Results from studies on panic disorder point to treatment-related decreases in salience processing regions, specifically amygdala and ACC, in response to a variety of tasks (Liebscher et al. 2016; Plag et al. 2018; Wittmann et al. 2018; Beutel et al. 2010; Kircher et al. 2013; Yang et al. 2020; Kunas et al. 2019; Lueken et al. 2013). Other regions, such as hippocampal and parahippocampal regions (Beutel et al. 2010; Straube et al. 2014; Lueken et al. 2013; Bomyea et al. 2020), anterior insula (Kircher et al. 2013; Plag et al. 2018), and IFG (Kircher et al. 2013; Kunas et al. 2019) were implicated in specific studies but findings across studies were inconsistent in the identification or directionality of effects. Results concerning hippocampus activity changes may be due to differences in tasks, with studies examining negative affective processing reporting decreased activation (Beutel et al. 2010; Straube et al. 2014), while a study examining fear learning reporting increased activation (Lueken et al. 2013).

Given that physiological and interoceptive symptoms are more central to the experience of panic than other anxiety disorders, it is interesting to note there were not consistent findings of decreased insula activation with treatment. This could potentially be due to limitations in the fMRI tasks utilized. The clinical presentation of panic disorder, which includes acute experiences of physiological anxiety to individually and context-specific triggers (with varying degrees of awareness for those triggers), presents unique obstacles to identifying the most relevant fMRI tasks (i.e., for symptom provocation). Additional research utilizing interoceptively focused tasks may be useful for specifically probing mechanisms of panic disorder treatment. Of note, seven of the publications arose from one, multi-site randomized clinical outcome study of panic disorder and agoraphobia (Gloster et al. 2009, 2011). This large study is unique and represents the type of work needed to increase sample sizes and identify mechanisms of treatment outcome. However, the fact that so many of the findings reported in the literature are from this same highly overlapping sample may also impact generalizability to independent samples. As in SAD, further replication of prior findings in larger and independent samples is needed, as well as further research on the impact of pharmacologic interventions.

3.4 Generalized Anxiety Disorder

Only six task-based fMRI studies examining treatment-related changes for GAD were identified, including three on psychotherapy (one MBSR, two CBT) and three pharmacotherapy (two citalopram; one alprazolam). Three studies utilized emotional face processing, while the others focused on emotional reappraisal, anticipation, and worry induction. In ROI approaches, the most common brain regions include the amygdala, insula, and anterior cingulate cortex (ACC).

Given the few number of studies, with varying tasks and interventions, it is difficult to identify consistent patterns of findings across studies. However, reports of decreased amygdala activity during emotional face processing were relatively consistent and reported for both psychotherapy and pharmacotherapy interventions (Holzel et al. 2013; Fonzo et al. 2014; Wise et al. 2020). Reports concerning changes in other salience processing regions, i.e., insula and ACC, have been less consistent in regard to significance or directionality (Fonzo et al. 2014; Hoehn-Saric et al. 2004; Brown et al. 2015; Andreescu et al. 2016); though two studies reported decreased PHG activation after CBT (Bomyea et al. 2020) and citalopram (Hoehn-Saric et al., 2004).

Directionality of change in PFC regions (e.g., lateral PFC) was also inconsistent across studies (Holzel et al. 2013; Brown et al. 2015; Fonzo et al. 2014; Hoehn-Saric et al. 2004), with no discernible pattern concerning reasons for the discrepancies (e.g., no identified task or intervention differences). Notably, theories of GAD have involved either *over-regulation* by PFC regions of emotional responses to anticipate and prepare for perceived future threats or a *lack of effective engagement* by prefrontal regions (Mochcovitch et al. 2014). One possibility is that there are individual differences for those with GAD in the over- vs under-engagement of PFC regions during emotional processing, leading to variability in the directionality of changes. It is also possible that an optimal outcome of GAD treatment is more effective modulation of PFC regions depending on the context and task-specific goal, which could be an interesting area for future research. It is clear that more fMRI research studies focused on GAD treatment are needed to inform a more holistic view of mechanisms and potential novel treatment targets.

3.5 Obsessive-Compulsive Disorder

OCD has been associated with hyperactivity of cortico-striatal-thalamic circuitry (CSTC) in the presence of triggering stimuli (e.g., a dirty surface), which is thought to contribute to the tendency for obsessive thoughts and compulsive behaviors. Existing fMRI treatment studies primarily employed CBT with a focus on exposure and response prevention (ERP), with one study on ACT and two on SSRIs. Of the 8 studies identified, only two included comparison conditions. The majority of fMRI tasks involved symptom provocation and the primary ROIs of interest included ACC, OFC, dlPFC, and striatum.

Perhaps the most consistent finding across OCD studies involved treatment-related decreases in ACC activation during symptom provocation (Morgieve et al. 2014; Schiepek et al. 2021; Viol et al. 2020; Nakao et al. 2005); several studies also reported decreases in OFC activation (Baioui et al. 2013; Morgieve et al. 2014; Viol et al. 2020; Nabeyama et al. 2008). Changes in dorsal striatal activation were dependent upon the cognitive process probed, with decreased activation for symptom provocation (Nakao et al. 2005; Viol et al. 2020; Baioui et al. 2013) and increased activation for inhibitory processes (Nakao et al. 2005; Freyer et al. 2011). This may reflect that treatment increases context-dependent striatal flexibility. Few of the OCD-related studies reviewed investigated the relationship with symptom improvement, limiting conclusions concerning treatment mechanisms. Most studies to date also did not include active comparator treatment, but there was some indication that ACT and CBT may have differential mechanisms (i.e., with ACT related to increases and CBT to decreases in insula activation) (Lee et al. 2023). These differences may be due to distinctions in emotion regulation strategies for the two interventions, with CBT focusing more on use of cognitive strategies to down-regulate one's emotional response vs ACT focusing more on noticing and accepting which could enhance interoceptive and emotional awareness. Future research is warranted to examine unique mechanisms for different treatment components or different types of OCD treatment. Most studies included passive viewing tasks with no changing contingencies, enabling conclusions about treatment effects on the brain's initial response to aversive stimuli. However, such tasks do not shed light on neural changes associated with behaviors or learning mechanisms likely important for psychotherapy approaches. Assessing changes in more dynamic tasks, e.g., tasks involving fear learning or compulsion-related decision-making (Cooper and Dunsmoor 2021) may provide further insights.

3.6 Posttraumatic Stress Disorder

PTSD is the most studied in regard to fMRI-based treatment changes, with the majority of studies focusing on psychotherapy (16 of 19 publications reviewed), and exposure-based interventions specifically (including prolonged exposure

[PE] and eye-movement desensitization and reprocessing [EMDR]). Most of the fMRI tasks focused on emotional face processing, emotional image anticipation or presentation, or fear learning. However, there have also been a few studies examining executive function (e.g., emotional Stroop, stop signal inhibition, etc.). The most common ROIs include amygdala, insula, ACC, mPFC and OFC, and lateral PFC.

The most consistent findings in PTSD appeared to be decreased anterior insula activation during emotional processing tasks (Thomaes et al. 2012; Aupperle et al. 2013; Simmons et al. 2013; Rousseau et al. 2019a; Duval et al. 2020; Fitzgerald et al. 2018). Given this, it may be important to integrate understanding of insula function and interoception into the predominant fear learning theories of PTSD (Joshi et al. 2020) and to further examine the potential for PTSD interventions targeting insula and interoceptive function, for example through interoceptive exposure, interoceptively focused mindfulness, or more novel interventions such as floatation therapy (Feinstein et al. 2018). Enhanced engagement of PFC/ACC regions was also reported rather consistently, including (1) dorsal or rostral ACC (Aupperle et al. 2013; Roy et al. 2014; Roy et al. 2010; Felmingham et al. 2007), (2) lateral PFC (Fonzo et al. 2017; MacNamara et al. 2016), and (3) vmPFC (King et al. 2016; Roy et al. 2014). While the intervention-related increases in dorsal/rostral ACC or vmPFC were observed during emotional face processing, emotional image anticipation, or emotional Stroop tasks, the increases in lateral PFC activation were observed specifically for emotional reappraisal tasks perhaps reflecting the more specific role of these regions in explicit regulation of emotion (Etkin et al. 2015). The few discrepancies observed in regard to directionality of PFC changes are likely due to differences in tasks, with studies reporting decreases in activation focusing primarily on fear extinction or emotional Stroop (Helpman et al. 2016; Rousseau et al. 2019b; Thomaes et al. 2012) while those reporting increases in activation focusing primarily on emotional faces or images (Aupperle et al. 2013; Fonzo et al. 2017; MacNamara et al. 2016; Felmingham et al. 2007). Surprisingly, reports concerning amygdala activation changes (Roy et al. 2010, 2014; Aupperle et al. 2013; Rousseau et al. 2019a; King et al. 2016) were not as consistent.

Similar to findings in panic disorder, studies reported PTSD treatment to relate to decreases in hippocampal engagement during emotional processing (e.g., emotional faces, trauma recall) (Felmingham et al. 2007; Rousseau et al. 2019a; Inoue et al. 2021) but increased hippocampal engagement during fear learning (Helpman et al. 2016). Striatal regions have also been implicated in a few studies, with reports of treatment-related decreases in caudate activation, the extent of which related to symptom improvement (Aupperle et al. 2013; Rousseau et al. 2019a). A more recent study utilized a computational approach to examine processes relevant for inhibitory learning, a presumed core component to exposure-based therapy. Harle et al. (2020) examined neural activation in relation to signal prediction error and computational parameters relating to probabilities of the need to stop during an inhibition task. They reported treatment-related decreases in caudate activation relating to the probability of encountering a stop trial, while behavioral performance increased – thus potentially reflecting a "fine-tuning of model-based behavior" and reduced discrepancies between expected and observed outcomes. This study offers a prime example for how future research can incorporate computational methods to more directly link changes in neural activation to changes in behavior.

Despite the reviewed PTSD literature being larger than that for the other anxiety disorders, conclusions were still limited by a lack of active comparator treatments in most studies as well as small sample sizes that limited power in detecting relationships with symptom change. PTSD in particular includes a number of different symptom dimensions and unique clinical presentations across individuals. Promising future avenues of research include probing dimensions of interest beyond negative valence processing and fear learning (e.g., reward or social processing) or examining how neural changes relate to specific symptom dimensions.

4 Conclusions

Human neuroimaging is a method that sits at the intersection of psychological and neurobiological research and holds the promise of helping to integrate the two into more cohesive and explanatory theories and inspire new treatment approaches. However, as time has passed and neuroimaging research has arguably not led to demonstrable changes in clinical care, this promise has begun to be questioned. It is important to recognize when a line of research is not fruitful so it can be abandoned to focus on other promising avenues. However, given the state of the literature, we argue that such abandonment of treatment-related fMRI research would be premature. Twenty years of research is not long when considering that each clinical outcome study takes several years to complete (with NIH-funded studies usually lasting ~5 years). In addition, the number of fMRI studies conducted is still relatively small and with relatively small sample sizes, ranging from 10–31 (M = 19.5) for phobia, 14–95 (M = 39.9) for SAD, 9–94 (M = 40.9) for SAD, 6–30 (M = 22.6) for GAD, 9–27 (M = 14.1) for OCD, and 6–48 (M = 21.42) for PTSD. The completion of multi-site studies with larger populations or the sharing of data across studies (e.g., through open-source data sharing archives) is crucial for moving the field toward identification of consistent, replicable, and generalizable results.

The most promising fMRI biomarkers of treatment response included decreases in the salience network, including amygdala (specific phobia, panic disorder, GAD), anterior insula (specific phobia, PTSD) or ACC (panic disorder, OCD). While these findings may reflect changes specifically in regions of a proposed "negative affect" circuit (Williams 2016), further work is needed (e.g., using tasks focused on processing of reward or positive affect) to determine if these findings are specific to negative valence vs more general to affective salience. It is possible that differences in regard to insula vs amygdala changes suggest differential importance for these regions depending on the disorder or treatment modality. However, it is also likely that different interventions, disorders, and even individuals, engage unique brain processes during the course of treatment that all lead to a common downstream effect of reducing amygdala and/or anterior insula reactivity. Increased engagement of medial and lateral PFC regions were also relatively consistent for PTSD and SAD, lending some support for the idea that increased engagement of PFC regions may be important for adaptive regulation. It may be useful for future research to examine specific learning mechanisms during the most active stages of treatment, when behavior or symptoms are just starting to shift (e.g., a few weeks after treatment initiation). The inclusion of early- to mid-treatment timepoints would also allow for potential early markers of treatment response.

The most consistent findings were identified for specific phobia, which also happened to be the most homogenous in regard to fMRI tasks and intervention approaches utilized. For many of the other disorders, it would be useful for future work to incorporate similar tasks as prior work in order to allow for testing replication. However, there is also benefit to considering additional cognitive and emotional processes that have not yet been examined but may be important treatment mechanisms. While most anxiety treatments focus on decreasing avoidance or threat processing, those who seek treatment often have other treatment goals – such as increasing their level of motivation, experiencing more joy or developing more meaningful social connections (e.g., Chevance et al. 2020). Utilizing tasks that probe effort and motivation, positive affect, or social processes may be important for identifying novel treatment targets. Further, computational tasks could be designed to specifically and uniquely probe the processes thought to underlie the different treatment modalities (Norbury et al. 2024).

The process of SSRI development may offer an example of how neuroimaging could contribute to future treatment development. While tricyclic antidepressants (TCAs) and monoamine oxidase inhibitors (MAOIs) were identified serendipitously as antidepressants, subsequent research identified serotonergic neurotransmission as an important mechanism of action for these agents (Lopez-Munoz and Alamo 2009). These findings helped inspire the development of SSRIs to more specifically target the serotonergic system, leading to one of the most effective pharmacological treatment approaches. Similarly, using fMRI to identify mechanisms of current interventions has the potential to inspire development of treatments that more specifically target these mechanisms. While fMRI research has not yet led to profound increases in treatment efficacy, it has arguably had more subtle impacts on clinical research. For example, helping inspire development of treatments targeting reward responsivity and positive affect (to impact striatal reward networks (Taylor et al. 2024)) or executive functions (to impact PFC networks (Clausen et al. 2019)), and the combining of neuromodulation approaches with therapy and medication (Gay et al. 2022). There is a possibility that fMRI research will never have a more profound effect on mental health treatment. However, in the spirit of inhibitory learning, we proposed this is a fear worth testing out. It may very well be that these endeavors inform the evolution of the mental health treatment landscape in ways that could never be predicted.

References

Alvarez-Perez Y, Rivero F, Herrero M, Vina C, Fumero A, Betancort M, Penate W (2021) Changes in brain activation through cognitive-behavioral therapy with exposure to virtual reality: a neuroimaging study of specific phobia. J Clin Med 10(16). https://doi.org/10.3390/jcm10163505

American Psychiatric Association (2010) Practice guidelines for the treatment of patients with acute stress disorder and posttraumatic stress disorder. https://psychiatryonline.org/pb/assets/raw/sitewide/practice_guidelines/guidelines/acutestressdisorderptsd-1410199309090.pdf

American Psychiatric Association (2013) Diagnostic and statistical manual of mental disorders (DSM-5®). American Psychiatric Pub

Andreescu C, Sheu LK, Tudorascu D, Gross JJ, Walker S, Banihashemi L, Aizenstein H (2016) Emotion reactivity and regulation in late-life generalized anxiety disorder: functional connectivity at baseline and post-treatment. Am J Geriatr Psychiatry 24(11):1040

Aupperle RL, Allard CB, Simmons AN, Flagan T, Thorp SR, Norman SB, Paulus MP, Stein MB (2013) Neural responses during emotional processing before and after cognitive trauma therapy for battered women. Psychiatry Res 214(1):48–55. https://doi.org/10.1016/j.pscychresns.2013.05.001

Baioui A, Pilgramm J, Kagerer S, Walter B, Vaitl D, Stark R (2013) Neural correlates of symptom reduction after CBT in obsessive-compulsive washers – an fMRI symptom provocation study. J Obs Compuls Relat Disord 2(3):322–330

Beck AT (2005) The current state of cognitive therapy: a 40-year retrospective. Arch Gen Psychiatry 62(9):953–959. https://doi.org/10.1001/archpsyc.62.9.953

Beutel ME, Stark R, Pan H, Silbersweig D, Dietrich S (2010) Changes of brain activation pre- post short-term psychodynamic inpatient psychotherapy: an fMRI study of panic disorder patients. Psychiatry Res 184(2):96–104. https://doi.org/10.1016/j.pscychresns.2010.06.005

Bomyea J, Ball TM, Simmons AN, Campbell-Sills L, Paulus MP, Stein MB (2020) Change in neural response during emotion regulation is associated with symptom reduction in cognitive behavioral therapy for anxiety disorders. J Affect Disord 271:207–214. https://doi.org/10.1016/j.jad.2020.04.001

Bouton ME, Maren S, McNally GP (2021) Behavioral and neurobiological mechanisms of Pavlovian and instrumental extinction learning. Physiol Rev 101(2):611–681. https://doi.org/10.1152/physrev.00016.2020

Brooks SJ, Stein DJ (2015) A systematic review of the neural bases of psychotherapy for anxiety and related disorders. Dialogues Clin Neurosci 17(3):261–279

Brown GG, Ostrowitzki S, Stein MB, von Kienlin M, Liu TT, Simmons A, Wierenga C, Stein OY, Bruns A, Bischoff-Grethe A, Paulus M (2015) Temporal profile of brain response to alprazolam in patients with generalized anxiety disorder. Psychiatry Res 233(3):394–401. https://doi.org/10.1016/j.pscychresns.2015.06.016

Brown LA, Young KS, Goldin PR, Torre JB, Burklund LJ, Davies CD, Niles AN, Lieberman MD, Saxbe DE, Craske MG (2019) Self-referential processing during observation of a speech performance task in social anxiety disorder from pre-to post-treatment: evidence of disrupted neural activation. Psychiatry Res Neuroimaging 284:13–20

Buhle JT, Silvers JA, Wager TD, Lopez R, Onyemekwu C, Kober H, Weber J, Ochsner KN (2014) Cognitive reappraisal of emotion: a meta-analysis of human neuroimaging studies. Cereb Cortex 24(11):2981–2990. https://doi.org/10.1093/cercor/bht154

Charnay Y, Leger L (2010) Brain serotonergic circuitries. Dialogues Clin Neurosci 12(4):471–487

Chavanne AV, Robinson OJ (2021) The overlapping neurobiology of induced and pathological anxiety: a meta-analysis of functional neural activation. Am J Psychiatry 178(2):156–164. https://doi.org/10.1176/appi.ajp.2020.19111153

Chevance A, Ravaud P, Tomlinson A, Le Berre C, Teufer B, Touboul S, Fried EI, Gartlehner G, Cipriani A, Tran VT (2020) Identifying outcomes for depression that matter to patients, informal caregivers, and health-care professionals: qualitative content analysis of a large international online survey. Lancet Psychiatry 7(8):692–702

Clark DA, Beck AT (2011) Cognitive therapy of anxiety disorders: science and practice. Guilford Press

Clausen AN, Thelen J, Francisco AJ, Bruce J, Martin L, McDowd J, Aupperle RL (2019) Computer-based executive function training for combat veterans with PTSD: a pilot clinical trial assessing feasibility and predictors of dropout. Front Psych 10:62. https://doi.org/10.3389/fpsyt.2019.00062

Cooper SE, Dunsmoor JE (2021) Fear conditioning and extinction in obsessive-compulsive disorder: a systematic review. Neurosci Biobehav Rev 129:75–94. https://doi.org/10.1016/j.neubiorev.2021.07.026

Craske MG, Treanor M, Conway CC, Zbozinek T, Vervliet B (2014) Maximizing exposure therapy: an inhibitory learning approach. Behav Res Ther 58:10–23. https://doi.org/10.1016/j.brat.2014.04.006

Dubovsky SL, Marshall D (2022) Benzodiazepines remain important therapeutic options in psychiatric practice. Psychother Psychosom 91(5):307–334. https://doi.org/10.1159/000524400

Duval ER, Javanbakht A, Liberzon I (2015) Neural circuits in anxiety and stress disorders: a focused review. Ther Clin Risk Manag 11:115–126. https://doi.org/10.2147/TCRM.S48528

Duval ER, Sheynin J, King AP, Phan KL, Simon NM, Martis B, Porter KE, Norman SB, Liberzon I, Rauch SAM (2020) Neural function during emotion processing and modulation associated with treatment response in a randomized clinical trial for posttraumatic stress disorder. Depress Anxiety 37(7):670–681. https://doi.org/10.1002/da.23022

Etkin A, Egner T, Kalisch R (2011) Emotional processing in anterior cingulate and medial prefrontal cortex. Trends Cogn Sci 15(2):85–93. https://doi.org/10.1016/j.tics.2010.11.004

Etkin A, Buchel C, Gross JJ (2015) The neural bases of emotion regulation. Nat Rev Neurosci 16(11):693–700. https://doi.org/10.1038/nrn4044

Faria V, Gingnell M, Hoppe JM, Hjorth O, Alaie I, Frick A, Hultberg S, Wahlstedt K, Engman J, Mansson KNT, Carlbring P, Andersson G, Reis M, Larsson EM, Fredrikson M, Furmark T (2017) Do you believe it? Verbal suggestions influence the clinical and neural effects of escitalopram in social anxiety disorder: a randomized trial. EBioMedicine 24:179–188. https://doi.org/10.1016/j.ebiom.2017.09.031

Farrow TF, Hunter MD, Wilkinson ID, Gouneea C, Fawbert D, Smith R, Lee KH, Mason S, Spence SA, Woodruff PW (2005) Quantifiable change in functional brain response to empathic and forgivability judgments with resolution of posttraumatic stress disorder. Psychiatry Res Neuroimaging 140(1):45–53

Feinstein JS (2013) Lesion studies of human emotion and feeling. Curr Opin Neurobiol 23(3): 304–309. https://doi.org/10.1016/j.conb.2012.12.007

Feinstein JS, Khalsa SS, Yeh H-w, Wohlrab C, Simmons WK, Stein MB, Paulus MP (2018) Examining the short-term anxiolytic and antidepressant effect of floatation-REST. PLoS One 13(2):e0190292

Felmingham K, Kemp A, Williams L, Das P, Hughes G, Peduto A, Bryant R (2007) Changes in anterior cingulate and amygdala after cognitive behavior therapy of posttraumatic stress disorder. Psychol Sci 18(2):127–129

Fettes P, Schulze L, Downar J (2017) Cortico-striatal-thalamic loop circuits of the orbitofrontal cortex: promising therapeutic targets in psychiatric illness. Front Syst Neurosci 11:25. https://doi.org/10.3389/fnsys.2017.00025

Fitzgerald JM, DiGangi JA, Phan KL (2018) Functional neuroanatomy of emotion and its regulation in PTSD. Harv Rev Psychiatry 26(3):116–128. https://doi.org/10.1097/HRP.0000000000000185

Fonzo GA, Ramsawh HJ, Flagan TM, Sullivan SG, Simmons AN, Paulus MP, Stein MB (2014) Cognitive-behavioral therapy for generalized anxiety disorder is associated with attenuation of limbic activation to threat-related facial emotions. J Affect Disord 169:76–85. https://doi.org/10.1016/j.jad.2014.07.031

Fonzo GA, Goodkind MS, Oathes DJ, Zaiko YV, Harvey M, Peng KK, Weiss ME, Thompson AL, Zack SE, Mills-Finnerty CE, Rosenberg BM, Edelstein R, Wright RN, Kole CA, Lindley SE, Arnow BA, Jo B, Gross JJ, Rothbaum BO, Etkin A (2017) Selective effects of psychotherapy on Frontopolar cortical function in PTSD. Am J Psychiatry 174(12):1175–1184. https://doi.org/10.1176/appi.ajp.2017.16091073

Freyer T, Kloppel S, Tuscher O, Kordon A, Zurowski B, Kuelz AK, Speck O, Glauche V, Voderholzer U (2011) Frontostriatal activation in patients with obsessive-compulsive disorder before and after cognitive behavioral therapy. Psychol Med 41(1):207–216. https://doi.org/10.1017/S0033291710000309

Fullana MA, Albajes-Eizagirre A, Soriano-Mas C, Vervliet B, Cardoner N, Benet O, Radua J, Harrison BJ (2018) Fear extinction in the human brain: a meta-analysis of fMRI studies in healthy participants. Neurosci Biobehav Rev 88:16–25. https://doi.org/10.1016/j.neubiorev.2018.03.002

Ganesan S, Beyer E, Moffat B, Van Dam NT, Lorenzetti V, Zalesky A (2022) Focused attention meditation in healthy adults: a systematic review and meta-analysis of cross-sectional functional MRI studies. Neurosci Biobehav Rev 141:104846. https://doi.org/10.1016/j.neubiorev.2022.104846

Gay F, Singier A, Aouizerate B, Salvo F, Bienvenu TC (2022) Neuromodulation treatments of pathological anxiety in anxiety disorders, stressor-related disorders, and major depressive disorder: a dimensional systematic review and meta-analysis. Front Psych 13:910897

Gingnell M, Frick A, Engman J, Alaie I, Björkstrand J, Faria V, Carlbring P, Andersson G, Reis M, Larsson EM, Wahlstedt K (2016) Combining escitalopram and cognitive–behavioural therapy for social anxiety disorder: randomised controlled fMRI trial. Br J Psychiatry 209(3):229–235

Gloster AT, Wittchen HU, Einsle F, Hofler M, Lang T, Helbig-Lang S, Fydrich T, Fehm L, Hamm AO, Richter J, Alpers GW, Gerlach AL, Strohle A, Kircher T, Deckert J, Zwanzger P, Arolt V (2009) Mechanism of action in CBT (MAC): methods of a multi-center randomized controlled trial in 369 patients with panic disorder and agoraphobia. Eur Arch Psychiatry Clin Neurosci 259(Suppl 2):S155–S166. https://doi.org/10.1007/s00406-009-0065-6

Gloster AT, Wittchen HU, Einsle F, Lang T, Helbig-Lang S, Fydrich T, Fehm L, Hamm AO, Richter J, Alpers GW, Gerlach AL, Strohle A, Kircher T, Deckert J, Zwanzger P, Hofler M, Arolt V (2011) Psychological treatment for panic disorder with agoraphobia: a randomized controlled trial to examine the role of therapist-guided exposure in situ in CBT. J Consult Clin Psychol 79(3):406–420. https://doi.org/10.1037/a0023584

Goldin PR, Gross JJ (2010) Effects of mindfulness-based stress reduction (MBSR) on emotion regulation in social anxiety disorder. Emotion 10(1):83–91. https://doi.org/10.1037/a0018441

Goldin P, Ramel W, Gross J (2009) Mindfulness meditation training and self-referential processing in social anxiety disorder: behavioral and neural effects. J Cogn Psychother 23(3):242–257. https://doi.org/10.1891/0889-8391.23.3.242

Goldin P, Ziv M, Jazaieri H, Gross JJ (2012) Randomized controlled trial of mindfulness-based stress reduction versus aerobic exercise: effects on the self-referential brain network in social anxiety disorder. Front Hum Neurosci 6:295. https://doi.org/10.3389/fnhum.2012.00295

Goldin P, Ziv M, Jazaieri H, Hahn K, Gross JJ (2013) MBSR vs aerobic exercise in social anxiety: fMRI of emotion regulation of negative self-beliefs. Soc Cogn Affect Neurosci 8(1):65–72. https://doi.org/10.1093/scan/nss054

Goldin PR, Ziv M, Jazaieri H, Weeks J, Heimberg RG, Gross JJ (2014) Impact of cognitive-behavioral therapy for social anxiety disorder on the neural bases of emotional reactivity to and regulation of social evaluation. Behav Res Ther 62:97–106

Goldin PR, Thurston M, Allende S, Moodie C, Dixon ML, Heimberg RG, Gross JJ (2021) Evaluation of cognitive behavioral therapy vs mindfulness meditation in brain changes during reappraisal and acceptance among patients with social anxiety disorder: a randomized clinical trial. JAMA Psychiatry 78(10):1134–1142. https://doi.org/10.1001/jamapsychiatry.2021.1862

Goossens L, Schruers K, Peeters R, Griez E, Sunaert S (2007a) Visual presentation of phobic stimuli: amygdala activation via an extrageniculostriate pathway? Psychiatry Res Neuroimaging 155(2):113–120. https://doi.org/10.1016/j.pscychresns.2006.12.005

Goossens L, Sunaert S, Peeters R, Griez EJ, Schruers KR (2007b) Amygdala hyperfunction in phobic fear normalizes after exposure. Biol Psychiatry 62(10):1119–1125. https://doi.org/10.1016/j.biopsych.2007.04.024

Halsband U, Wolf TG (2015) Functional changes in brain activity after hypnosis in patients with dental phobia. J Physiol Paris 109(4–6):131–142. https://doi.org/10.1016/j.jphysparis.2016.10.001

Harle KM, Simmons AN, Norman SB, Spadoni AD (2020) Neural affective mechanisms associated with treatment responsiveness in veterans with PTSD and comorbid alcohol use disorder. Psychiatry Res Neuroimaging 305:111172. https://doi.org/10.1016/j.pscychresns.2020.111172

Hauner KK, Mineka S, Voss JL, Paller KA (2012) Exposure therapy triggers lasting reorganization of neural fear processing. Proc Natl Acad Sci U S A 109(23):9203–9208. https://doi.org/10.1073/pnas.1205242109

Helpman L, Marin MF, Papini S, Zhu X, Sullivan GM, Schneier F, Neria M, Shvil E, Malaga Aragon MJ, Markowitz JC, Lindquist MA, Wager T, Milad M, Neria Y (2016) Neural changes in extinction recall following prolonged exposure treatment for PTSD: a longitudinal fMRI study. Neuroimage Clin 12:715–723. https://doi.org/10.1016/j.nicl.2016.10.007

Hillhouse TM, Porter JH (2015) A brief history of the development of antidepressant drugs: from monoamines to glutamate. Exp Clin Psychopharmacol 23(1):1–21. https://doi.org/10.1037/a0038550

Hoehn-Saric R, Schlund MW, Wong SH (2004) Effects of citalopram on worry and brain activation in patients with generalized anxiety disorder. Psychiatry Res 131(1):11–21. https://doi.org/10.1016/j.pscychresns.2004.02.003

Holzel BK, Hoge EA, Greve DN, Gard T, Creswell JD, Brown KW, Barrett LF, Schwartz C, Vaitl D, Lazar SW (2013) Neural mechanisms of symptom improvements in generalized anxiety disorder following mindfulness training. Neuroimage Clin 2:448–458. https://doi.org/10.1016/j.nicl.2013.03.011

Hur JW, Shin H, Jung D, Lee HJ, Lee S, Kim GJ, Cho CY, Choi S, Lee SM, Cho CH (2021) Virtual reality-based psychotherapy in social anxiety disorder: fMRI study using a self-referential task. JMIR Ment Health 8(4):e25731. https://doi.org/10.2196/25731

Inoue J, Matsuo K, Iwabuchi T, Takehara Y, Yamasue H (2021) How memory switches brain responses of patients with post-traumatic stress disorder. Cereb Cortex Commun 2(2):tgab021. https://doi.org/10.1093/texcom/tgab021

Joshi SA, Duval ER, Sheynin J, King AP, Phan KL, Martis B, Porter KE, Liberzon I, Rauch SAM (2020) Neural correlates of emotional reactivity and regulation associated with treatment response in a randomized clinical trial for posttraumatic stress disorder. Psychiatry Res Neuroimaging 299:111062. https://doi.org/10.1016/j.pscychresns.2020.111062

Kalisch R (in press) TBD. In: Blackford J, Milad M (eds) New discoveries in the brain sciences of fear and anxiety. Springer

King AP, Block SR, Sripada RK, Rauch SA, Porter KE, Favorite TK, Giardino N, Liberzon I (2016) A pilot study of mindfulness-based exposure therapy in OEF/OIF combat veterans with PTSD: altered medial frontal cortex and amygdala responses in social-emotional processing. Front Psych 7:154. https://doi.org/10.3389/fpsyt.2016.00154

Kircher T, Arolt V, Jansen A, Pyka M, Reinhardt I, Kellermann T, Konrad C, Lueken U, Gloster AT, Gerlach AL, Strohle A, Wittmann A, Pfleiderer B, Wittchen HU, Straube B (2013) Effect of cognitive-behavioral therapy on neural correlates of fear conditioning in panic disorder. Biol Psychiatry 73(1):93–101. https://doi.org/10.1016/j.biopsych.2012.07.026

Kirlic N, Kuplicki R, Touthang J, Cohen ZP, Stewart JL, Tulsa I, Paulus MP, Aupperle RL (2022) Behavioral and neural responses during fear conditioning and extinction in a large transdiagnostic sample. Neuroimage Clin 35:103060. https://doi.org/10.1016/j.nicl.2022.103060

Klumpp H, Fitzgerald DA, Phan KL (2013) Neural predictors and mechanisms of cognitive behavioral therapy on threat processing in social anxiety disorder. Prog Neuro-Psychopharmacol Biol Psychiatry 45:83–91

Kunas SL, Yang Y, Straube B, Kircher T, Gerlach AL, Pfleiderer B, Arolt V, Wittmann A, Stroehle A, Wittchen HU, Lueken U (2019) The impact of depressive comorbidity on neural plasticity following cognitive-behavioral therapy in panic disorder with agoraphobia. J Affect Disord 245:451–460. https://doi.org/10.1016/j.jad.2018.11.026

LeDoux JE, Pine DS (2016) Using neuroscience to help understand fear and anxiety: a two-system framework. Am J Psychiatry 173(11):1083–1093. https://doi.org/10.1176/appi.ajp.2016.16030353

Lee SW, Choi M, Lee SJ (2023) A randomized controlled trial of group-based acceptance and commitment therapy for obsessive-compulsive disorder. J Contextual Behav Sci 27:45–53

Liebscher C, Wittmann A, Gechter J, Schlagenhauf F, Lueken U, Plag J, Straube B, Pfleiderer B, Fehm L, Gerlach AL, Kircher T, Fydrich T, Deckert J, Wittchen HU, Heinz A, Arolt V, Strohle A (2016) Facing the fear – clinical and neural effects of cognitive behavioural and pharmacotherapy in panic disorder with agoraphobia. Eur Neuropsychopharmacol 26(3):431–444. https://doi.org/10.1016/j.euroneuro.2016.01.004

Liberati A, Altman DG, Tetzlaff J, Mulrow C, Gøtzsche PC, Ioannidis JP, Clarke M, Devereaux PJ, Kleijnen J, Moher D (2009) The PRISMA statement for reporting systematic reviews and meta-analyses of studies that evaluate health care interventions: explanation and elaboration. Ann Intern Med 151(4):W-65

Lipka J, Hoffmann M, Miltner WH, Straube T (2014) Effects of cognitive-behavioral therapy on brain responses to subliminal and supraliminal threat and their functional significance in specific phobia. Biol Psychiatry 76(11):869–877. https://doi.org/10.1016/j.biopsych.2013.11.008

Lopez-Munoz F, Alamo C (2009) Monoaminergic neurotransmission: the history of the discovery of antidepressants from 1950s until today. Curr Pharm Des 15(14):1563–1586. https://doi.org/10.2174/138161209788168001

Lueken U, Straube B, Konrad C, Wittchen HU, Strohle A, Wittmann A, Pfleiderer B, Uhlmann C, Arolt V, Jansen A, Kircher T (2013) Neural substrates of treatment response to cognitive-behavioral therapy in panic disorder with agoraphobia. Am J Psychiatry 170(11):1345–1355. https://doi.org/10.1176/appi.ajp.2013.12111484

MacNamara A, Rabinak CA, Kennedy AE, Fitzgerald DA, Liberzon I, Stein MB, Phan KL (2016) Emotion regulatory brain function and SSRI treatment in PTSD: neural correlates and predictors of change. Neuropsychopharmacology 41(2):611–618. https://doi.org/10.1038/npp.2015.190

Månsson KN, Frick A, Boraxbekk CJ, Marquand AF, Williams SCR, Carlbring P, Andersson G, Furmark T (2015) Predicting long-term outcome of Internet-delivered cognitive behavior therapy for social anxiety disorder using fMRI and support vector machine learning. Transl Psychiatry 5(3):e530–e530

McTeague LM, Rosenberg BM, Lopez JW, Carreon DM, Huemer J, Jiang Y, Chick CF, Eickhoff SB, Etkin A (2020) Identification of common neural circuit disruptions in emotional processing across psychiatric disorders. Am J Psychiatry 177(5):411–421. https://doi.org/10.1176/appi.ajp.2019.18111271

Milad MR, Rosenbaum BL, Simon NM (2014) Neuroscience of fear extinction: implications for assessment and treatment of fear-based and anxiety related disorders. Behav Res Ther 62:17–23. https://doi.org/10.1016/j.brat.2014.08.006

Mochcovitch MD, da Rocha Freire RC, Garcia RF, Nardi AE (2014) A systematic review of fMRI studies in generalized anxiety disorder: evaluating its neural and cognitive basis. J Affect Disord 167:336–342. https://doi.org/10.1016/j.jad.2014.06.041

Morgieve M, N'Diaye K, Haynes WI, Granger B, Clair AH, Pelissolo A, Mallet L (2014) Dynamics of psychotherapy-related cerebral haemodynamic changes in obsessive compulsive disorder using a personalized exposure task in functional magnetic resonance imaging. Psychol Med 44(7):1461–1473. https://doi.org/10.1017/S0033291713002237

Nabeyama M, Nakagawa A, Yoshiura T, Nakao T, Nakatani E, Togao O, Yoshizato C, Yoshioka K, Tomita M, Kanba S (2008) Functional MRI study of brain activation alterations in patients with obsessive-compulsive disorder after symptom improvement. Psychiatry Res 163(3):236–247. https://doi.org/10.1016/j.pscychresns.2007.11.001

Nakao T, Nakagawa A, Yoshiura T, Nakatani E, Nabeyama M, Yoshizato C, Kudoh A, Tada K, Yoshioka K, Kawamoto M, Togao O, Kanba S (2005) Brain activation of patients with obsessive-compulsive disorder during neuropsychological and symptom provocation tasks before and after symptom improvement: a functional magnetic resonance imaging study. Biol Psychiatry 57(8):901–910. https://doi.org/10.1016/j.biopsych.2004.12.039

National Institute for Health and Care Excellence (2020) Generalised anxiety disorder and panic disorder in adults: management. https://www.nice.org.uk/guidance/cg113

Nave AM, Tolin DF, Stevens MC (2012) Exposure therapy, D-cycloserine, and functional magnetic resonance imaging in patients with snake phobia: a randomized pilot study. J Clin Psychiatry 73(9):1179–1186. https://doi.org/10.4088/JCP.11m07564

Neufang S, Geiger MJ, Homola GA, Mahr M, Schiele MA, Gehrmann A, Schmidt B, Gajewska A, Nowak J, Meisenzahl-Lechner E, Pham M (2019) Cognitive-behavioral therapy effects on alerting network activity and effective connectivity in panic disorder. Eur Arch Psychiatry Clin Neurosci 269:587–598

Norbury A, Hauser TU, Fleming SM, Dolan RJ, Huys QJM (2024) Different components of cognitive-behavioral therapy affect specific cognitive mechanisms. Sci Adv 10(13):eadk3222. https://doi.org/10.1126/sciadv.adk3222

Paquette V, Levesque J, Mensour B, Leroux JM, Beaudoin G, Bourgouin P, Beauregard M (2003) “Change the mind and you change the brain”: effects of cognitive-behavioral therapy on the neural correlates of spider phobia. Neuroimage 18(2):401–409. https://doi.org/10.1016/s1053-8119(02)00030-7

Paulus MP, Stein MB (2006) An insular view of anxiety. Biol Psychiatry 60(4):383–387. https://doi.org/10.1016/j.biopsych.2006.03.042

Paulus MP, Stein MB (2010) Interoception in anxiety and depression. Brain Struct Funct 214(5–6): 451–463. https://doi.org/10.1007/s00429-010-0258-9

Pico-Perez M, Radua J, Steward T, Menchon JM, Soriano-Mas C (2017) Emotion regulation in mood and anxiety disorders: a meta-analysis of fMRI cognitive reappraisal studies. Prog Neuropsychopharmacol Biol Psychiatry 79(Pt B):96–104. https://doi.org/10.1016/j.pnpbp.2017.06.001

Plag J, Petzold MB, Gechter J, Liebscher C, Strohle A (2018) Patients' characteristics and their influence on course of fear during agoraphobic symptom provocation: may SS(N)RI treatment compensate unfavorable individual preconditions? Nord J Psychiatry 72(5):325–335. https://doi.org/10.1080/08039488.2018.1457178

Rasgon A, Lee WH, Leibu E, Laird A, Glahn D, Goodman W, Frangou S (2017) Neural correlates of affective and non-affective cognition in obsessive compulsive disorder: a meta-analysis of functional imaging studies. Eur Psychiatry 46:25–32. https://doi.org/10.1016/j.eurpsy.2017.08.001

Rousseau P, El Khoury-Malhame M, Reynaud E, Zendjidjian X, Samuelian J, Khalfa S (2019a) Neurobiological correlates of EMDR therapy effect in PTSD. Eur J Trauma Dissoc 3(2): 103–111

Rousseau PF, El Khoury-Malhame M, Reynaud E, Boukezzi S, Cancel A, Zendjidjian X, Guyon V, Samuelian JC, Guedj E, Chaminade T, Khalfa S (2019b) Fear extinction learning improvement in PTSD after EMDR therapy: an fMRI study. Eur J Psychotraumatol 10(1):1568132. https://doi.org/10.1080/20008198.2019.1568132

Roy MJ, Francis J, Friedlander J, Banks-Williams L, Lande RG, Taylor P, Blair J, McLellan J, Law W, Tarpley V, Patt I, Yu H, Mallinger A, Difede J, Rizzo A, Rothbaum B (2010) Improvement in cerebral function with treatment of posttraumatic stress disorder. Ann N Y Acad Sci 1208:142–149. https://doi.org/10.1111/j.1749-6632.2010.05689.x

Roy MJ, Costanzo ME, Blair JR, Rizzo AA (2014) Compelling evidence that exposure therapy for PTSD normalizes brain function. Stud Health Technol Inform 199:61–65

Schienle A, Schafer A, Hermann A, Rohrmann S, Vaitl D (2007) Symptom provocation and reduction in patients suffering from spider phobia: an fMRI study on exposure therapy. Eur Arch Psychiatry Clin Neurosci 257(8):486–493. https://doi.org/10.1007/s00406-007-0754-y

Schienle A, Schafer A, Stark R, Vaitl D (2009) Long-term effects of cognitive behavior therapy on brain activation in spider phobia. Psychiatry Res 172(2):99–102. https://doi.org/10.1016/j.pscychresns.2008.11.005

Schiepek G, Tominschek I, Heinzel S, Aigner M, Dold M, Unger A, Lenz G, Windischberger C, Moser E, Ploderl M, Lutz J, Meindl T, Zaudig M, Pogarell O, Karch S (2013) Discontinuous patterns of brain activation in the psychotherapy process of obsessive-compulsive disorder: converging results from repeated FMRI and daily self-reports. PLoS One 8(8):e71863. https://doi.org/10.1371/journal.pone.0071863

Schiepek G, Viol K, Aas B, Kastinger A, Kronbichler M, Scholler H, Reiter EM, Said-Yurekli S, Kronbichler L, Kravanja-Spannberger B, Stoger-Schmidinger B, Aichhorn W, Battaglia D, Jirsa V (2021) Pathologically reduced neural flexibility recovers during psychotherapy of OCD patients. Neuroimage Clin 32:102844. https://doi.org/10.1016/j.nicl.2021.102844

Shin LM, Liberzon I (2010) The neurocircuitry of fear, stress, and anxiety disorders. Neuropsychopharmacology 35(1):169–191. https://doi.org/10.1038/npp.2009.83

Simmons AN, Norman SB, Spadoni AD, Strigo IA (2013) Neurosubstrates of remission following prolonged exposure therapy in veterans with posttraumatic stress disorder. Psychother Psychosom 82(6):382–389. https://doi.org/10.1159/000348867

Straube T, Glauer M, Dilger S, Mentzel HJ, Miltner WH (2006) Effects of cognitive-behavioral therapy on brain activation in specific phobia. Neuroimage 29(1):125–135. https://doi.org/10.1016/j.neuroimage.2005.07.007

Straube B, Lueken U, Jansen A, Konrad C, Gloster AT, Gerlach AL, Strohle A, Wittmann A, Pfleiderer B, Gauggel S, Wittchen U, Arolt V, Kircher T (2014) Neural correlates of procedural variants in cognitive-behavioral therapy: a randomized, controlled multicenter FMRI study. Psychother Psychosom 83(4):222–233. https://doi.org/10.1159/000359955

Taylor CT, Stein MB, Simmons AN, He F, Oveis C, Shakya HB, Sieber WJ, Fowler JH, Jain S (2024) Amplification of positivity treatment for anxiety and depression: a randomized experimental therapeutics trial targeting social reward sensitivity to enhance social connectedness. Biol Psychiatry 95(5):434–443. https://doi.org/10.1016/j.biopsych.2023.07.024

Thomaes K, Dorrepaal E, Draijer N, de Ruiter MB, Elzinga BM, van Balkom AJ, Smit JH, Veltman DJ (2012) Treatment effects on insular and anterior cingulate cortex activation during classic and emotional Stroop interference in child abuse-related complex post-traumatic stress disorder. Psychol Med 42(11):2337–2349. https://doi.org/10.1017/S0033291712000499

Van Rooij SJ, Geuze E, Kennis M, Rademaker AR, Vink M (2015) Neural correlates of inhibition and contextual cue processing related to treatment response in PTSD. Neuropsychopharmacology 40(3):667–675

Van Rooij SJ, Kennis M, Vink M, Geuze E (2016) Predicting treatment outcome in PTSD: a longitudinal functional MRI study on trauma-unrelated emotional processing. Neuropsychopharmacology 41(4):1156–1165

Viol K, Schiepek G, Kronbichler M, Hartl A, Grafetstatter C, Strasser P, Kastinger A, Scholler H, Reiter EM, Said-Yurekli S, Kronbichler L, Kravanja-Spannberger B, Stoger-Schmidinger B, Hutt MT, Aichhorn W, Aas B (2020) Multi-level assessment of obsessive-compulsive disorder (OCD) reveals relations between neural and neurochemical levels. BMC Psychiatry 20(1):559. https://doi.org/10.1186/s12888-020-02913-5

Vollestad J, Nielsen MB, Nielsen GH (2012) Mindfulness- and acceptance-based interventions for anxiety disorders: a systematic review and meta-analysis. Br J Clin Psychol 51(3):239–260. https://doi.org/10.1111/j.2044-8260.2011.02024.x

Williams LM (2016) Precision psychiatry: a neural circuit taxonomy for depression and anxiety. Lancet Psychiatry 3(5):472–480. https://doi.org/10.1016/S2215-0366(15)00579-9

Wise T, Patrick F, Meyer N, Mazibuko N, Oates AE, van der Bijl AHM, Danjou P, O'Connor SM, Doolin E, Wooldridge C, Rathjen D, Macare C, Williams SCR, Perkins A, Young AH (2020) Cholinergic modulation of disorder-relevant neural circuits in generalized anxiety disorder. Biol Psychiatry 87(10):908–915. https://doi.org/10.1016/j.biopsych.2019.12.013

Wittmann A, Schlagenhauf F, Guhn A, Lueken U, Elle M, Stoy M, Liebscher C, Bermpohl F, Fydrich T, Pfleiderer B, Bruhn H, Gerlach AL, Straube B, Wittchen HU, Arolt V, Heinz A, Kircher T, Strohle A (2018) Effects of cognitive behavioral therapy on neural processing of agoraphobia-specific stimuli in panic disorder and agoraphobia. Psychother Psychosom 87(6): 350–365. https://doi.org/10.1159/000493146

Yang Y, Lueken U, Richter J, Hamm A, Wittmann A, Konrad C, Strohle A, Pfleiderer B, Herrmann MJ, Lang T, Lotze M, Deckert J, Arolt V, Wittchen HU, Straube B, Kircher T (2020) Effect of CBT on biased semantic network in panic disorder: a multicenter fMRI study using semantic priming. Am J Psychiatry 177(3):254–264. https://doi.org/10.1176/appi.ajp.2019.19020202

Young KS, Burklund LJ, Torre JB, Saxbe D, Lieberman MD, Craske MG (2017) Treatment for social anxiety disorder alters functional connectivity in emotion regulation neural circuitry. Psychiatry Res Neuroimaging 261:44–51. https://doi.org/10.1016/j.pscychresns.2017.01.005

Zezula J, Cortes R, Probst A, Palacios JM (1988) Benzodiazepine receptor sites in the human brain: autoradiographic mapping. Neuroscience 25(3):771–795. https://doi.org/10.1016/0306-4522(88)90036-x

Transcranial Magnetic Stimulation for Posttraumatic Stress Disorder and Generalized Anxiety Disorder

Samantha L. Cilli, Miriam A. Goldberg, Camila Cosmo, Amanda R. Arulpragasam, Amin Zand Vakili, Yosef A. Berlow, and Noah S. Philip

Contents

Abstract Posttraumatic stress disorder (PTSD) and generalized anxiety disorder (GAD) are debilitating psychiatric disorders. While treatments are often effective, many patients do not adequately respond or experience significant side effects. Transcranial magnetic stimulation (TMS) is an emerging approach for treating PTSD and GAD. Several randomized clinical trials have demonstrated that TMS over the dorsolateral prefrontal cortex may be efficacious in reducing psychiatric symptoms; however, results are inconsistent regarding whether any parameter or treatment paradigm is superior. Other RCTs have targeted novel brain regions using newer TMS modalities. Combining TMS with psychotherapy may augment response in patients with PTSD, yet results are inconclusive. Little research has been done on TMS in combination with psychotherapy for GAD, indicating a need

S. L. Cilli, M. A. Goldberg, C. Cosmo, A. R. Arulpragasam, A. Zand Vakili, Y. A. Berlow, and N. S. Philip (✉)
Center for Neurorestoration and Neurotechnology, VA Providence Healthcare System, Providence, RI, USA

Department of Psychiatry and Human Behavior, Alpert Medical School of Brown University, Providence, RI, USA
e-mail: Noah_Philip@Brown.edu

© The Author(s), under exclusive license to Springer Nature Switzerland AG 2024

Curr Topics Behav Neurosci (2025) 73: 527–554
https://doi.org/10.1007/7854_2024_540
Published Online: 07 November 2024

for further investigation. Future studies may assess TMS parameter optimization for enhancing effectiveness and improving therapeutic response duration. Identifying response biomarkers through functional magnetic resonance imaging and electroencephalography may offer a means to predict and monitor clinical response as precision methods to improve treatment response.

Keywords Generalized anxiety disorder · Neuromodulation · Non-invasive brain stimulation · Posttraumatic stress disorder · Transcranial magnetic stimulation

1 Introduction

PTSD and GAD pose significant burdens on patients and healthcare providers. PTSD is characterized by exposure to a traumatic event, followed by various symptoms and negative changes in mood and cognition (American Psychiatric Association 2013; Shalev et al. 2017). The lifetime prevalence of PTSD is ~6% (Goldstein et al. 2016). Some groups, particularly military veterans, experience rates up to 13% (Hoge et al. 2004). The societal burden of PTSD includes adverse health outcomes, disruption of social and occupational function, substance use, and suicide risk (Flory and Yehuda 2015; Pietrzak et al. 2011; Solomon and Davidson 1997). Up to one-third of patients with PTSD have GAD (Solomon and Davidson 1997). GAD is defined as excessive anxiety and worry, and is the most common anxiety disorder in older adults; it impacts social function and causes significant distress (American Psychiatric Association 2013; Flory and Yehuda 2015; Lee et al. 2016).

Despite PTSD and GAD's prevalence and impact, treatment options often do not meet patient needs. For PTSD, first-line treatments include trauma-focused cognitive behavioral therapies (CBT) and pharmacological interventions (Watts et al. 2013). However, recent studies indicate limitations regarding both approaches (Lee et al. 2016; Steenkamp et al. 2015; Imel et al. 2013; Watts et al. 2013). Similarly, many patients with GAD either do not respond sufficiently or have significant side effects to pharmacological interventions (Reinhold and Rickels 2015). Benzodiazepines are often contraindicated due to dependence and abuse concerns, and other medications are less often used (Garakani et al. 2020; Curtiss et al. 2017). Psychotherapies for GAD, such as CBT and exposure therapy, yield substantive improvement in half of the patients (Newman et al. 2020; Erickson and Newman 2005). Thus, novel interventions are needed.

Non-invasive brain stimulation has emerged as a promising new approach for PTSD and GAD. Various techniques, including transcranial magnetic stimulation (often abbreviated as rTMS, hereafter TMS) and transcranial direct current stimulation (tDCS), have been studied (Koek et al. 2019). TMS has been available for pharmacoresistant major depressive disorder (MDD) since 2008, demonstrating a strong safety and efficacy profile (George et al. 2010; O'Reardon et al. 2007; Cosmo et al. 2021). Since MDD is often comorbid with PTSD and GAD, there is increased interest in using TMS for these conditions. For example, Kan et al. reviewed articles from 2000 to 2020 that investigated the efficacy of TMS, tDCS, and theta burst stimulation (TBS) for PTSD (Kan et al. 2020). Recently, Saccenti et al. reviewed non-invasive brain stimulation and highlighted its ability to ameliorate PTSD and GAD (Saccenti et al. 2024). Of all modalities, TMS has been studied in the greatest detail and is thus the focus of this review.

This chapter will summarize the current literature on PTSD and GAD treatment, including an overview of issues for ongoing consideration. It will start by providing an overview of TMS, followed by discussions of TMS for PTSD and GAD, considering their use as monotherapy and in combination with psychotherapy, and conclude with considerations for future clinical and research use.

2 Principles of TMS

TMS is a form of non-invasive brain stimulation where a conductive coil is placed on the scalp to deliver an alternating current; this current generates a localized, fluctuating magnetic field, which induces an electrical field that depolarizes cortical neurons under the coil. Early studies stimulating the primary motor cortex (M1) informed the current practice of determining the motor threshold (MT), which is the least amount of energy necessary to elicit a motor response in the contralateral hand 50% of the time (Barker et al. 1986). The MT is used to calibrate the intensity for therapeutic TMS, expressed as a percentage (e.g., stimulating at 120% of MT). When applied clinically for psychiatric disorders, TMS is typically applied to the dorsolateral prefrontal cortex (DLPFC) (Perera et al. 2016) (Figs. 1 and 2).

When administered repetitively, these magnetic pulses form repetitive "trains," leading to the term repetitive TMS. Early physiology studies indicated that higher frequency protocols (>5 Hz) were excitatory and lower frequencies (≤1 Hz) were inhibitory, although this is now recognized as inconsistent within or between individuals (Pascual-Leone et al. 1994; Chen et al. 1997; Magnuson et al. 2023). When the FDA cleared TMS for pharmacoresistant MDD, patients received 10 Hz TMS at 120% of MT at the left DLPFC, with 3,000 pulses over ~37.5 min; patients received once-daily sessions per week for 6 weeks, followed by a six-session taper. Currently, this protocol is largely unchanged except for a reduction of the intertrain interval (Carpenter et al. 2021).

3 Early Formative TMS Studies for PTSD and GAD

In the 1990s, rapid advancement in functional imaging led to the conceptualization of PTSD as a disorder of abnormal neural circuitry involving trauma-recall-associated hyperactivity in limbic, paralimbic, and frontal cortical regions (Van Der Kolk

Fig. 1 Applying transcranial magnetic stimulation

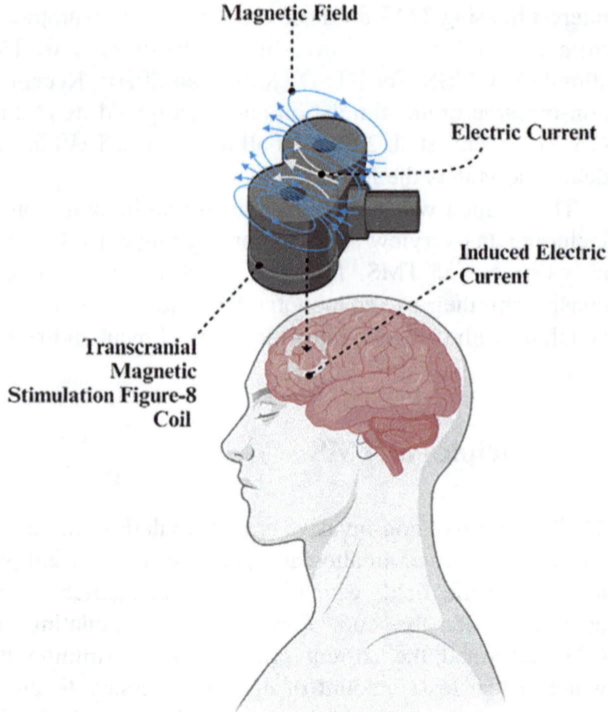

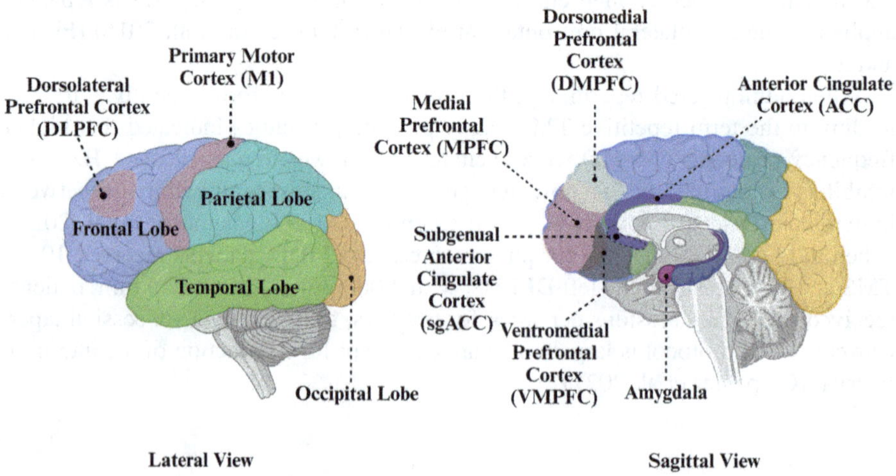

Fig. 2 Brain regions relevant to TMS targeting and related neural circuitry

et al. 1997). In 1995, TMS research showcased its ability to alter cortical activity and was successfully applied to treatment-resistant depression (George et al. 1995). TMS, therefore, was recognized as having the potential to alleviate PTSD by targeting involved regions. In 1998, two major studies were published. McCann et al. described PTSD improvement in two cases treated with multiple sessions of 1 Hz TMS to the right DLPFC, and improvements persisted for 1-month. Furthermore, positron emission tomography after TMS showed reduced right cerebral metabolism (McCann et al. 1998). Next, Grisaru et al. demonstrated transient improvement in PTSD symptoms with 0.3 Hz TMS sequentially applied to the left and right motor cortex (Grisaru et al. 1998). Four years later, Rosenberg et al. delivered 1 or 5 Hz TMS to the left DLPFC in patients with comorbid PTSD/MDD. Post-TMS, combat-related PTSD symptoms, anxiety, and anger significantly improved, though PTSD effects were smaller relative to MDD; no clear differences emerged between protocols. Although limited by its small sample size and unblinded design, this study was one of the first that indicated promise of TMS for comorbid PTSD and MDD (Rosenberg et al. 2002).

In 2004, the first randomized controlled trial (RCT) of TMS for PTSD was conducted; participants received either 10 Hz, 1 Hz, or sham TMS to the right DLPFC. Compared to sham and 1 Hz, 10 Hz TMS was superior in improving PTSD and anxiety symptoms (Cohen et al. 2004). At two weeks, participants reported reduced reexperiencing and avoidance symptoms. In both groups, depressive symptoms did not meaningfully improve. Within the active groups, however, total doses differed; the 1 Hz group received less pulses than the 10 Hz group, which may explain the latter group's superior outcomes.

In 2008, Bystritsky et al. investigated functional magnetic resonance imaging (fMRI)-guided low-frequency TMS for GAD (Bystritsky et al. 2008). Participants completed TMS over 3 weeks and measured outcomes via the HAM-A (Maier et al. 1988) and CGI-I (Guy 1976) (see Table 1 for Measure Acronyms Guide). Response was defined as a >50% reduction on the HAM-A and a CGI-I score of 1–2 (i.e., at least much improved). TMS significantly improved HAM-A scores ($p = 0.001$), and 60% of participants achieved clinical response and met remission criteria (HAM-A score < 8). The team collected HAM-A and CGI-I scores at 6-month follow-up, and results found continued improvement (Bystritsky et al. 2009). This data indicated the promise of TMS for GAD to yield short- and long-term benefits.

These studies demonstrated the safety and tolerability of TMS for PTSD and GAD and provided preliminary evidence for efficacy. Of note, parameters varied across studies, and total TMS doses were low compared to current protocols (Rosenberg et al. 2002; Cohen et al. 2004).

4 TMS Monotherapy

There have been several RCTs investigating the efficacy of TMS monotherapy for PTSD and GAD. Most used standard low- and high-frequency TMS, although several used newer approaches (Table 2). Several unblinded prospective studies and case series/registry studies have also been published and are described below. For further study details, including stimulation parameters, see Table 2.

Table 1 Measure acronyms guide

Acronym	Meaning
BAI	Beck Anxiety Inventory
BDI	Beck Depression Inventory
BNCE	Brief Neurobehavioral Cognitive Examination
CAPS	Clinician Administered PTSD Scale
CAS	Clinical Anxiety Scale
CGI	Clinical Global Impressions Scale
CGI-I	Clinical Global Impressions-Improvement Scale
DERS	Difficulties in Emotion Regulation Scale
GAD-7	Generalized Anxiety Disorder Assessment
HAM-A	Hamilton Anxiety Rating Scale
HAM-D	Hamilton Depression Rating Scale
HDRS-A/S	Hamilton Depression Rating Scale Anxiety/Somatization
IDS-SR	Inventory of Depressive Symptomatology Self-Report
IES	Impact of Event Scale
IPF	Inventory of Psychosocial Functioning
MADRS	Montgomery-Asberg Depression Rating Scale
MPSS	Modified PTSD Symptom Self-Report Scale
M-PTSD	Mississippi Scale for Combat-Related PTSD
NSI	Neurobehavioral Symptom Inventory
PCL (-M, -C, -5)	PTSD Checklist (Military Version, Civilian Version, DMS-5 Version)
PHQ-9	Patient Health Questionnaire 9-Item
PSS-SR	PTSD Symptom Scale-Self-Report
PTGI	Post-Traumatic Growth Inventory
QIDS-SR	Quick Inventory of Depressive Symptomatology-Self Report
QLESQ	Quality of Life Enjoyment and Satisfaction Questionnaire
SOFAS	Social and Occupational Functioning Assessment Scale
STAI	State Trait Anxiety Inventory

4.1 1 Versus 10 Hz Right-Sided TMS for PTSD and GAD

Following Cohen et al. (Cohen et al. 2004) several RCTs tested right-sided 1 Hz TMS versus 10 Hz TMS, sham, or both. Of note, different sham techniques have been used over time; while a discussion on sham TMS is outside the scope of this piece, we direct readers to *Rethinking the Role of Sham TMS* (Duecker and Sack 2015).

In 2012, a study compared 1 Hz TMS to the right DLPFC to sham. Participants reported significantly improved PTSD and MDD symptoms, with an attenuation of effect by 2 months and no change in STAI anxiety scores (Watts et al. 2012). In 2013, another study compared right-sided 1 Hz TMS to sham for PTSD (Nam et al. 2013). This study found low-frequency TMS improved PTSD symptoms, particularly re-experiencing symptoms on the CAPS (Blake et al. 1995). One year later, a case series was published documenting improvement in only hyperarousal with 1 Hz

Table 2 Randomized controlled trials of TMS monotherapy for PTSD and GAD

Study	Participants	N	TMS type	Groups	MT (%)	Total dose	Outcome measures	Main outcomes
Cohen et al. (2004)	PTSD	24	rTMS	1 Hz right DLPFC 10 Hz right DLPFC Sham	80%	10 sessions 1,000 pulses (1 Hz) 4,000 pulses (10 Hz)	PCL-5 PTSD scale HAM-A HAM-D CAPS	10 Hz had therapeutic effects; PTSD core symptoms (reexperiencing, avoidance) improved alongside anxiety symptoms
Boggio et al. (2010)	PTSD	30	rTMS	20 Hz right DLPFC 20 Hz left DLPFC Sham	80%	10 sessions 16,000 pulses	CAPS PCL-5 PTSD scale HAM-A HAM-D	Active conditions decreased PTSD symptoms; right rTMS induced a larger effect. Significant improvements in mood after left TMS and anxiety reduction following right TMS. Effects remained significant at 3 months
Watts et al. (2012)	PTSD	20	rTMS	1 Hz right DLPFC Sham	90%	10 sessions 4,000 pulses	CAPS PCL BDI STAI BNCE	1 Hz to the right DLPFC improved PTSD and depressive symptoms vs. sham. The effectiveness showed some degradation after 2 months
Nam et al. (2013)	PTSD	18	rTMS	1 Hz right DLPFC Sham	100%	15 sessions Total 18,000 pulses	CAPS	CAPS scores improved, with significant differences in the re-experiencing and total CAPS score between 1 Hz versus sham
Diefenbach et al. (2016)	GAD	25	rTMS	1 Hz right DLPFC Sham	90%	30 sessions 27,000 total pulses	CGI-I DERS	Self-reported emotion regulation improved in active group only. Improvements in goal-directed behaviors and impulse control were associated with a global clinician rating of improvement
Dilkov et al. (2017)	GAD	40	rTMS	20 Hz right DLPFC Sham	110%	25 sessions 20 trains, 9 s per train, 51 s intertrain intervals	HAM-A CGI HAM-D	Active stimulation improved anxiety. At 2 and 4-week follow-ups, HAM-A scores of the active group remained stable or slightly improved

(continued)

Table 2 (continued)

Study	Participants	N	TMS type	Groups	MT (%)	Total dose	Outcome measures	Main outcomes
Ahmadizadeh and Rezaei (2018)	PTSD	65	rTMS	20 Hz bilateral DLPFC 20 Hz right DLPFC Sham	100%	10 sessions 24,000 pulses	PCL-M	The bilateral group demonstrated significant PTSD symptom reductions earlier than sham; no differences between active groups at end-point. The unilateral group vs. sham showed greater overall symptom improvement
Kozel et al. (2019)	PTSD	44	rTMS	1 Hz right DLPFC 10 Hz right DLPFC	110%	36 sessions 86,400 pulses	CAPS-5 PCL-5 IPF MADRS QIDS-SR	Both groups had significant improvement in PTSD and depression. Only the 10 Hz group demonstrated significant improvement in function
Philip et al. (2019a)	PTSD/ MDD	22	sTMS	sTMS at IAF Sham	N/A	20–40 sessions	PCL-5 QIDS	Feasibility and tolerability of sTMS, with some indication favoring active sTMS
Philip et al. (2019b)	PTSD	50	TBS	TBS right DLPFC Sham	80%	10–20 sessions (10 blinded +10 unblinded) 18,000–36,000 pulses (1,800 pulses per 9.5 min sessions)	CAPS-5 SOFAS QLESQ PCL IDS-SR	TBS improved PTSD, depression, and social and occupational function; improvements were greater at 1 month. Pretreatment resting state functional connectivity predicted improvement
Leong et al. (2020)	PTSD	31	rTMS	1 Hz right DLPFC 10 Hz right DLPFC Sham 1 Hz Sham 10 Hz	120%	10 sessions 22,500 pulses (1 Hz) 30,000 pulses (10 Hz)	CAPS-IV	Only 1 Hz TMS significantly improved PTSD symptoms on the CAPS

Yuan et al. (2023)	PTSD	75	rTMS iTBS	10 Hz rTMS right DLPFC iTBS right DLPFC Sham iTBS right DLPFC	80%	15 sessions 27,000 total pulses	PCL-C PTGI	PCL-C and PTGI scores in iTBS and rTMS groups significantly improved post-treatment. No differences between the two active treatment groups

DLPFC dorsolateral prefrontal cortex, iTBS intermittent theta burst stimulation, MDD major depressive disorder, MT motor threshold, PTSD posttraumatic stress disorder, rTMS repetitive transcranial magnetic stimulation, sTMS synchronized transcranial magnetic stimulation, TBS theta-burst transcranial magnetic stimulation right-sided TMS (Oznur et al. 2014). However, this study was unblinded, and used a different scale (IES) (Weiss and Marmar 1997) and participant sample (military veterans). Regardless, these studies indicated that 1 Hz TMS at the right DLPFC for PTSD can be efficacious and indicated that a higher number of cumulative pulses may be required for superior outcomes.

In 2016, another study compared 1 Hz right-sided active TMS versus sham in patients with GAD (Diefenbach et al. 2016). Results indicated significant improvements in self-reported emotion regulation (assessed via DERS) at endpoint and 3-month follow-up; effects were only observed for active TMS. Improvements were found in goal-directed behaviors and impulse control, and were associated with global improvements (via CGI-I); anxiety symptoms were not directly assessed. These preliminary results provide some support regarding TMS for GAD and suggest a mechanism of change through improved emotion regulation.

A larger study in veterans compared right-sided 1 Hz with 10 Hz to understand the impact of TMS frequency (Kozel et al. 2019). Unlike the studies above, this protocol's parameters more closely approximated modern clinical practice. At endpoint, both groups had similar improvements in PTSD and MDD symptoms. For the 1 Hz group, response rates on the CAPS, PCL-5 (Wortmann et al. 2016), MADRS (Montgomery and Asberg 1979), and QIDS-SR (Rush et al. 2003) were 29%, 43%, 36%, and 29%, respectively. Comparatively, 10 Hz response rates were 31%, 31%, 25%, and 46%, respectively. There were no significant differences between groups other than improved psychosocial functioning scores in the 10 Hz group; no difference was found in self-reported functioning scores. Effects were largely maintained at 3 months. As such, this study indicated that either 1 or 10 Hz TMS could be efficacious for PTSD, in contrast with earlier work.

In 2020, a study randomized patients with PTSD into right-sided 1 Hz, 10 Hz, or sham (Leong et al. 2020). 1 Hz was superior to 10 Hz and sham for PTSD symptoms as determined by the CAPS at endpoint. There were no differences in anxiety outcomes (BAI and the GAD-7) between groups. Due to the sham group's high attrition, 3-month follow-up data analysis could not be conducted. Compared to the older RCTs, total doses in both groups were higher. Interestingly, the 1 Hz stimulation group was superior to 10 Hz despite receiving a lower TMS dose.

Collectively, these studies demonstrate that 1 and 10 Hz TMS targeting the right DLPFC can be effective for PTSD, with mixed evidence favoring right-sided low-frequency stimulation. Results were variable in terms of which symptoms of PTSD, GAD, and/or MDD showed improvements; no consistent picture emerged regarding specific subscales or anxiety. For example, while one study showcased CAPS improvement in the re-experiencing domain, another observed contradictory findings (Nam et al. 2013; Oznur et al. 2014). Overarchingly, both 1 Hz versus sham and 1 Hz versus 10 Hz showed nonsignificant improvement in anxiety scores; there were no statistically significant differences in anxiety scores (BAI and GAD-7) in both Kozel and Leong's studies (Kozel et al. 2019; Leong et al. 2020; Oznur et al. 2014). Several important questions arise from this work. Which, if any, parameters are important in optimizing treatment? At this point, there does appear to be a clear dose-response relationship, as studies with high cumulative TMS doses demonstrated less variable effects (Cohen et al. 2004). There also appears to be a dose-dependent effect on durability, with higher doses yielding better outcomes. Furthermore, since many of the studies found changes in components or subdomains of standardized rating scales, is it possible that negative outcomes may be related to reduced dimensionality? These remain important questions as the field moves forward and can be answered – at least in part – by some of the larger studies described subsequently.

4.2 Left-Sided Versus Right-Sided TMS for PTSD and GAD

When examining the efficacy of TMS for PTSD and GAD, laterality is another consideration. While the left DLPFC is the primary target for MDD, there is functional imaging evidence of right DLPFC hypermetabolism with trauma/fear and anxiety processing in patients with PTSD and GAD (Rauch et al. 1996). As part of the executive control network (ECN), the DLPFC is an important region for emotional regulation. Therefore, stimulating the right DLPFC may enhance top-down regulation of affective circuits, improving PTSD and anxiety symptoms (Koek et al. 2019). Comparatively, White et al. proposed a model of DLPFC-mediated emotional regulation to working memory (White et al. 2023). The DLPFC allows for retention and suppression of affective information and overall processing of mood. Said processing is dependent on whether the information is positive or negative and verbal or non-verbal. In the proposed framework, the left DLPFC is sensitive to verbal content, while the right DLPFC is sensitive to non-verbal or affective content. If accurate, the framework predicts that efficacy of DLPFC-targeted neuromodulation may depend on the affective state in which a patient receives TMS. Additionally, high-frequency stimulation at the left DLPFC could facilitate verbal cognitive emotion regulation strategies, while low-frequency TMS at the right DLPFC would affect anxiety by weakening the processes that maintain negative affect, decreasing anxiety.

To investigate the importance of laterality, two RCTs examined left-sided versus right-sided or bilateral TMS. One study randomized patients with PTSD to right, left, or sham 20 Hz TMS and found active TMS superior for PTSD (Boggio et al. 2010). The right-sided group had the greatest effect (36.9% versus 23.1% reduction in total PCL scores) and found superior outcomes for hyperarousal and avoidance symptom subscales. In this study, right-sided TMS improved anxiety scores (HAM-A) while left-sided TMS improved depression; both effects were sustained at 3 months. In comparison, another RCT analyzed right-sided 20 Hz, bilateral 20 Hz, and sham TMS at the DLPFC (Ahmadizadeh and Rezaei 2018). Both active groups were superior to sham, with no significant differences between right-sided and bilateral TMS. Notably, there was a quicker response in the bilateral group; at midpoint, the bilateral group showed significant improvement in PCL scores over the right-sided group.

Similarly, another study compared 20 Hz TMS at the right DLPFC to sham in patients with GAD (Dilkov et al. 2017). HAM-A scores assessed anxiety. At endpoint, the active group showed a clinically significant reduction in HAM-A scores compared to sham. At 2- and 4-week follow-up, anxiety symptoms in the active group remained stable and even slightly improved, demonstrating a sustained effect of TMS.

Following these studies was a case report of a veteran with PTSD, who received left versus right-sided 1 Hz TMS to reduce hyperarousal (Tillman et al. 2011). Physiological arousal was measured using electroencephalography (EEG)-measured P3a amplitude in response to combat-related threat stimuli, before and after TMS. Right-sided TMS reduced P3a amplitude, which correlated with self-reported symptom improvement (reduced total CAPS score); this did not occur with left-sided TMS. By using the P3a as a marker of hyperarousal, the researchers argued that TMS interfered with the right frontal lobe function, leading to symptom improvement. However, these findings have yet to be replicated.

To date, the largest source of efficacy data comes from investigating TMS outcomes in patients with comorbid MDD/PTSD. This is because TMS is FDA cleared for MDD, a diagnosis highly comorbid with PTSD (Rytwinski et al. 2013). These studies consistently demonstrate that TMS, regardless of application, improved PTSD symptoms with comorbid MDD. One early case report applied a standard 10 Hz protocol to an individual with comorbid PTSD and MDD, who then remitted on the BDI (Beck et al. 1961) and the PCL (Nakama et al. 2014). Because prior work showcased 5 Hz TMS for both MDD and anxiety symptoms, two studies found significant improvement in depression and PTSD with left-sided 5 Hz TMS (Carpenter et al. 2018; Philip et al. 2016). Additionally, they found a mean PCL reduction of 35.5% ($d = 1.04$) and a response rate of 48.6%, with similar reductions in IDS-SR scores (Carpenter et al. 2018).

Several studies investigated the neurobiological basis of TMS in PTSD, finding that fMRI and EEG identified biomarkers of change (Philip et al. 2018; Zandvakili et al. 2019). Philip et al. evaluated TMS-associated changes in brain connectivity using fMRI in patients with comorbid PTSD and MDD who received 5 Hz TMS to the left DLPFC (Philip et al. 2018). Analysis revealed that more negative pretreatment connectivity between the subgenual anterior cingulate cortex (sgACC) and the default mode network (DMN) predicted clinical improvement, as did more positive amygdala-to-ventromedial PFC (VMPFC) connectivity. After TMS, PTSD and MDD symptom reductions, measured by PCL-5 and IDS-SR respectively, correlated with reduced connectivity between the sgACC and the DMN, left dorsolateral prefrontal cortex, and insula, alongside reduced connectivity between the hippocampus and the salience network.

Zand Vakili et al. collected EEGs before and after 5 Hz TMS at the left DLPFC in participants with PTSD and MDD (Zandvakili et al. 2019). Using an 8-channel EEG device (ENOBIO8, Neuroelectrics, Cambridge, MA, USA), dry electrodes were placed over the TMS target using frontal cortex and midline sites. Twelve minutes of eyes closed data was analyzed. Machine learning accurately predicted clinical response to TMS based on pretreatment EEG coherence, and a Support Vector

Machine classifier accurately assigned EEG recordings to either pre- or post-TMS treatment. The accuracy for alpha, beta, theta, and delta bands was 75.4 ± 1.5%, 77.4 ± 1.4%, 73.8 ± 1.5%, and 78.6 ± 1.4%, respectively, all significantly better than chance. Recently, there was a Spectral Event EEG re-analysis of these data to identify novel EEG biomarkers of effective TMS. This approach uses unaveraged EEG to identify individual-level transient increases in power; prior data indicates these may correlate with cognitive functions, particularly in the beta band (Morris et al. 2023). Spectral Events in delta/theta (1–6 Hz), alpha (7–14 Hz), and beta (15–29 Hz) bands were observed in all patients. Pre- to post-treatment changes in fronto-central electrode beta event features, including frontal beta event frequency spans and durations, and central beta event maxima power, were associated with symptom improvement. Additionally, frontal pre-treatment beta event duration negatively correlated with MDD improvement, raising the possibility that beta spectral events might serve as a future biomarker of TMS response.

Recently, two studies investigated the efficacy of left-sided 10 Hz TMS at the DLPFC in individuals with comorbid PTSD and MDD. Because of TMS's mixed results in military populations for MDD (Yesavage et al. 2018), Madore et al. investigated the effects of 10 Hz left DLPFC TMS for PTSD and MDD, using a large TMS registry study of veterans (Madore et al. 2022). Over two-thirds of veterans met the threshold for PTSD (operationally defined as score ≥ 33 on the PCL-5) (Weathers et al. 2013) and 33.6% self-reported having a comorbid anxiety disorder. The PCL-5 and PHQ-9 measured PTSD and MDD symptoms, respectively. TMS was associated with clinically meaningful and statistically significant reductions in PTSD and MDD symptoms. PTSD outcomes were favorable, as 65.3% demonstrated clinically meaningful reduction (>10 points), and 46.1% no longer met PTSD threshold criteria after TMS (PCL-5 < 33; akin to MDD remission). The average reduction in PCL-5 scores was 19 points. This large-scale data supported the effectiveness of standard 10 Hz TMS for PTSD and MDD. Philip et al. conducted a follow-up analysis investigating the efficacy of TMS in veterans with mild traumatic brain injury (mTBI) and found that mTBI status generally did not impact clinical outcomes, even when examining the effect of cumulative head injuries (Philip et al. 2023).

Collectively, these studies indicate that left-sided and right-sided TMS to the DLPFC can improve PTSD and anxiety symptoms. When considering PTSD/MDD comorbidity, left-sided high-frequency protocols (traditionally used for pharmacoresistant MDD) might be the preferred option. When reviewing the literature, we must question how important laterality is in TMS targeting PTSD and GAD. If laterality is important, are there lateralized biological correlates to improvement in specific domains (e.g., P3a amplitude with hyperarousal)? Might early response serve as a predictor for other measures, including (but not limited to) patient retention or duration of effect? That stated, when considering laterality, there is evidence that stimulation on one side may propagate to the other and in other nonlinear ways due to callosal fibers connecting bilateral brain regions, which complicates laterality findings more than might be initially appreciated (Siebner et al. 2022).

4.3 H-Coil TMS

Another TMS technique called "deep" TMS (dTMS) utilizes an H-coil to broadly stimulate cortical regions (Deng et al. 2014). Moraga-Amaro et al. used unblinded dTMS of the insula (H2 coil) and PFC (insular-prefrontal cortex; H4 coil) in participants experiencing occupational stress and/or GAD (Moraga-Amaro et al. 2023). Stimulation decreased anxiety across various scales, including the HAM-A, CAS, and STAI. There was a HAM-A reduction of 88.7% and 70.7% in participants with occupational stress and GAD, respectively, indicating promise of dTMS for these conditions.

Similarly, Pell et al. conducted a retrospective analysis of patients' anxiety scores in studies that administered dTMS (H1 coil) versus sham for MDD (Pell et al. 2022). The primary measure was the adjusted HDRS-A/S factor score, collected prior to stimulation and after 4 weeks. Active stimulation yielded larger improvements in anxiety scores ($d = 0.34$, $p = 0.03$) which were sustained until 16 weeks ($d = 0.35$, $p = 0.04$) versus sham; these findings led to the US Food and Drug Administration's expansion of dTMS clearance to anxious depression.

Overall, these studies demonstrate that dTMS is a promising option for individuals with anxiety symptoms, GAD, and comorbid MDD/GAD. In these studies, higher baseline anxiety scores were associated with greater successful outcomes (Pell et al. 2022). That stated, whether these findings hold for isolated GAD remains unknown.

4.4 TBS and Other Newer Modalities

4.4.1 TBS

Theta burst stimulation (TBS) is a newer TMS approach that delivers triplet bursts of 5 Hz pulses with shorter session times (<10 min) (Huang et al. 2005; Larson and Munkácsy 2015; McDonald 2024). TBS is statistically non-inferior to standard TMS for MDD and has been investigated for the treatment of various psychiatric disorders, including PTSD (Blumberger et al. 2018; Yesavage et al. 2018). In the first RCT of TBS for PTSD, veterans with PTSD were randomized to active TBS or sham for 10 blinded sessions (Philip et al. 2019b). Afterward, participants received 10 additional unblinded sessions of active TBS. When interpreting these outcomes for once-daily TBS, it should be noted that using more recent definitions, this study would be defined as an "accelerated" TBS study, as it delivered the equivalent of three TBS sessions per day. As such, it could be interpreted that the end of 10 days represented 30 sessions of treatment, and 20 days with an equivalence of 60 daily treatment sessions. At the end of the blinded phase, the active group was superior in social/occupational function, with effect sizes indicating superior outcomes in PTSD (PCL-5) and depression (IDS-SR) symptoms. At 1 month, outcomes in the active group were superior on all measures (all *p*s < 0.001). A secondary analysis demonstrated active TBS also reduced anger (van't Wout-Frank et al. 2021). A convenience subset received pretreatment resting-state fMRI and demonstrated two predictors of PTSD response. One was increased positive connectivity within the DMN; the other was anticorrelation (increased negative connectivity) between the DMN and the ECN. Over the following year, a naturalistic clinical outcomes study was conducted and showed a dose-dependent effect on the duration of response. Sham participants, who received 10 fewer active TBS sessions, were 3.5 times more likely to experience PTSD relapse within 1 year (Petrosino et al. 2020). Re-examination of the resting-state fMRI data again found that higher within-DMN connectivity predicted better 1-year outcomes. In conclusion, TBS and other forms of neuromodulation may decrease the DMN hyperconnectivity in patients with PTSD, improving their symptoms. Stimulation may also decrease anticorrelated connectivity between the ECN and the DMN; by decreasing the hypoactivity of the ECN, a biomarker commonly seen in patients with MDD and other psychiatric disorders, TMS may improve PTSD symptoms (Liston et al. 2014).

To explore TBS' mechanism through electrophysiologic functional connectivity, a subset of this study sample had EEG performed before and after blinded stimulation (Zandvakili et al. 2021). Using Support Vector Machine learning, which tracks changes in EEG functional connectivity, the active group was successfully identified against the sham post-treatment group with 75% accuracy, in delta (1–4 Hz) band coherence, with findings related to reduction in coherence around the stimulation site and increased at midline. Not only did this study contribute to the literature on TMS's biological mechanism, but it also raises the possibility of an inexpensive technique for therapeutic monitoring.

In 2020, another TBS pilot study was conducted on veterans with PTSD; participants received unblinded bilateral TBS over 2 weeks and demonstrated significant reductions in PTSD and depressive symptoms (HAM-D) that sustained 3 months later (Nursey et al. 2020), with a large effect size on the CAPS.

In 2022, a retrospective study compared clinical outcomes of TBS versus 5 Hz TMS for PTSD and MDD (Philip et al. 2022). PTSD outcomes were inferior in the TBS group, whereas depression outcomes did not differ. Interestingly effect sizes were smaller in the TBS group regardless of outcome measure, whereas effect sizes in the 5 Hz group were comparable to the 2020 TBS study above, perhaps indicating a dose-response curve where higher doses yield superior outcomes.

In 2023, an RCT compared TBS, TMS, and sham at the right DLPFC (Yuan et al. 2023). Both active groups outperformed sham, with no significant differences between the two active groups (in symptoms or side effects). These results underscore the likely equivalency of these TMS approaches and emphasize that TBS is the preferable option when considering participant and staff burden.

Shenasa et al. also compared TBS and 10 Hz TMS through a retrospective case series that analyzed treatment outcomes in veterans who received 10 Hz TMS or TBS. (Shenasa et al. 2023). The two treatment options were compared through participants' reported PTSD symptom scores on the PCL-5. Unlike the smaller case report comparing TBS and 5 Hz TMS described previously (Philip et al.

2022), this report identified no statistically significant differences in PTSD symptoms between the two protocols, again underscoring the efficaciousness of TBS.

4.4.2 Other TMS Modalities

Synchronized TMS (sTMS) utilizes three rotating magnets to deliver relatively low energy stimulation that synchronizes with the participant's individual alpha frequency (IAF; alpha derived from the patient's EEG), but unlike standard TMS is insufficient to elicit depolarization (Leuchter et al. 2015). In 2019, a study recruited veterans with comorbid PTSD/MDD and randomized them to sTMS or sham (Philip et al. 2019a). While the main outcomes were not significant, active sTMS showed significantly greater reductions in threshold-level PTSD symptoms (threshold operationally defined as score ≥33 on the PCL-5) (Weathers et al. 2013). Compared to prior work, high cumulative dose appeared to lead to superior clinical outcomes.

"Personalized TMS" (PrTMS®) is a form of TMS that aims to modulate alpha EEG activity in PTSD (Makale et al. 2023). This approach adjusts the treatment loci and frequency during treatment based on a proprietary algorithm but delivers suprathreshold TMS. In an unblinded study, veterans treated with PrTMS® demonstrated improved PTSD symptoms, alongside increased alpha frequency and synchronization. However, whether this approach yields superior outcomes compared to sham or standard TMS remains unknown.

As newer TMS modalities for PTSD, both TBS and sTMS are promising options. Although it does not appear that TBS yields superior outcomes compared to standard TMS, TBS appears to have the additional benefit of decreasing patient burden. fMRI and EEG collection within TBS studies allow for the exploration of the mechanism of action and, in return, may reveal predictors of clinical response; increased within-network connectivity of the DMN could help inform treatment and improve outcomes. How and whether this additional information can be used cost-effectively remains unknown; other forms of TBS (i.e., continuous) are largely untested. Clinically relevant questions about treatment effects become apparent by conducting follow-up studies to investigate long-term outcomes, like the previously mentioned 1-year TBS study. Of those who benefit from TMS, how long do these benefits last? Is total dose the most important factor in the duration of response, or are there other parameters that could better impact this? How can we predict who will benefit from TMS and experience sustained effects?

5 Combination of TMS and Psychotherapy

Prolonged exposure (PE) and cognitive processing therapy (CPT) remain first-line psychotherapies for PTSD. PE teaches individuals to gradually approach their trauma-related memories, feelings, and situations, allowing them to learn that these memories and cues are not dangerous or to be avoided (Hembree et al.

2003). Comparatively, traumatic memories are deliberately elicited during CPT; patients are then supported as they directly confront conflicts and maladaptive beliefs associated with these memories. Several studies have explored their efficacy in combination with TMS (Table 3); details regarding each study, including stimulation parameters, are in Table 3. While exposure therapy and CBT are efficacious for GAD symptom reduction (Garakani et al. 2020), there is a larger gap in research examining the combination of neuromodulation and psychotherapy for GAD.

One study randomized participants with PTSD into active or sham 1 Hz right-sided TMS during an imaginal exposure exercise (Osuch et al. 2009). This protocol was based on the theory that right frontal cortex hyperactivation is associated with trauma re-experiencing in PTSD; by dampening this response using low-frequency stimulation, it was predicted that this would enhance fear extinction. Serum cortisol, thyroid hormones, and prolactin, as well as urinary cortisol, dopamine, epinephrine, and norepinephrine were assessed to evaluate arousal. The study team hypothesized that decreased emotional arousal in participants would be associated with decreased cortisol, dopamine, epinephrine, norepinephrine, and thyroid hormone (T4) levels, and increased prolactin levels. While there were no significant between-group differences, larger effect sizes were observed for active TMS on hyperarousal. Counter to the study's hypothesis, active low-frequency TMS yielded increased 24-h norepinephrine and T4 levels, alongside decreased prolactin. No correlations between serum/urinary measures and clinical outcomes were reported; these mixed findings complicated the study's interpretation.

Another study investigated dTMS to the medial PFC (MPFC) combined with an ultra-brief exposure in patients with PTSD (targeting informed by Francati et al.) (Isserles et al. 2013; Francati et al. 2007). The MPFC is part of the ECN and is associated with the fear response regulation. The MPFC is anticorrelated with the amygdala, a part of the limbic system that is hyperactive in individuals with PTSD and anxiety disorders. Hypothetically, dTMS upregulates the MPFC, which downregulates the amygdala, therefore dampening patients' fear response (Petrosino et al. 2021). Participants were randomized into three groups: active 20 Hz dTMS to the bilateral MPFC after trauma exposure, active 20 Hz dTMS to bilateral MPFC after non-trauma exposure, and sham dTMS after trauma exposure; exposure was performed prior to dTMS. The active dTMS + trauma exposure group showed a significant improvement in intrusive symptoms compared to both control groups; the active group also demonstrated a reduction in heart rate response to the traumatic scripts. Significant within-group reductions in avoidance and hyperarousal measures (i.e., traditional anxiety symptoms) and in total CAPS were shown in the active group only, and effects were sustained at 2 months. Of note, this study lacked the fourth control group (sham dTMS after non-trauma exposure). This work was followed by a multisite RCT, albeit using a different coil designed to have greater stimulation of the dorsomedial PFC (Isserles et al. 2021). Participants received brief exposure with active dTMS or sham stimulation over 4 weeks. Counter to their hypothesis, sham dTMS significantly outperformed active stimulation. This finding indicated that active stimulation yielded countertherapeutic effects when paired with ultra-brief exposure. This study did not include neuroimaging or other biomarkers to

Table 3 Randomized controlled trials of combination TMS and psychotherapy for PTSD and GAD

Study	Participants	N	TMS type	Psychotherapy type	Groups (frequency, target)	MT (%)	Total dose	Outcome measures	Main outcomes
Osuch et al. (2009)	PTSD	9	rTMS	*Imaginal exposure:* Personalized list of 10 events or cues related to their trauma. During TMS, subjects spoke events or cues aloud.	1 Hz right DLPFC + exposure Sham + exposure	100%	20 active +20 sham sessions (crossover design) 36,000 active pulses	CAPS HAM-D IES Serum and urine measures	The active group showed reductions in hyperarousal vs. sham. No significant improvement in avoidance, intrusion, and depression scores in either group. The active group had increased 24 h urinary norepinephrine and T4 and decreased prolactin
Isserles et al. (2013)	PTSD	30	dTMS (H1-coil)	*Ultra-brief script-driven exposure:* Subjects described 3 life-events (positive, neutral, and traumatic) and wrote 30 s audio scripts. In trauma exposure groups, traumatic script was followed by a 30-s silent imaginal period, then the neutral script and imaginal period. The non-trauma exposure group played the positive script	20 Hz bilateral MPFC + trauma exposure 20 Hz bilateral MPFC + non-trauma exposure Sham coil + trauma exposure	120%	12 sessions 20,160 pulses	CAPS PSS-SR HAM-D BDI Heart rate Skin conductance	The active exposure group showed significant improvement in intrusive symptoms and reduced heart rate response to traumatic scripts. Significant within-group reductions in total CAPS, avoidance, and hyperarousal only in the active exposure group

Kozel et al. (2018)	PTSD	103	rTMS	*Cognitive processing therapy (CPT):* Traumatic memories are elicited and patients are supported through direct confrontation of conflicts and maladaptive beliefs associated with their memories. CPT was immediately after TMS.	1 Hz right DLPFC + CPT Sham coil + CPT	110%	12–15 sessions 216,000–27,000 pulses	CAPS PCL M-PTSD QIDS IPF	The active group demonstrated greater reductions in PTSD across all timepoints, including 6-month follow-up
Fryml et al. (2019)	PTSD	8	rTMS	*Prolonged exposure (PE):* Subjects received standardized PE. An audio recording of their PE session was played during TMS.	10 Hz right or left DLPFC + PE Sham + PE	120%	8 sessions 48,000 pulses (one session per week)	CAPS HAM-D HAM-A PCL	Feasibility of simultaneous TMS and PE
Isserles et al. (2021)	PTSD	125	dTMS (H7-coil)	*Ultra-brief script-driven exposure:* See Isserles et al. (2013) except with no neutral script	18 Hz bilateral MPFC/ ACC + trauma exposure Sham + trauma exposure	100%	12 sessions 34,560 pulses	CAPS MPSS HAM-D	Active dTMS was significantly inferior to sham
Gao et al. (2021)	GAD	92	rTMS	*Mindfulness therapy:* Subjects trained to complete mindfulness and introspection, breathing, body scanning, yoga, and walking zen. Therapy completed daily, 2–3x per day, for 4 weeks	1 Hz at anterolateral right frontal lobe 1 Hz at anterolateral right frontal lobe + mindfulness therapy	100%	20 sessions 24,000 total pulses	HAM-A CGI	Post-treatment, the TMS + mindfulness group had a higher total effective rate than controls

(continued)

Table 3 (continued)

Study	Participants	N	TMS type	Psychotherapy type	Groups (frequency, target)	MT (%)	Total dose	Outcome measures	Main outcomes
Thierrée et al. (2022)	PTSD	38	rTMS	*Trauma script exposure:* Participants wrote a traumatic event script and read it during TMS	10 Hz rTMS right DLPFC + trauma exposure (HF group) 1 Hz right DLPFC + trauma exposure (LF group)	110% (10 Hz) 70% (1 Hz)	8 sessions 24,000 total pulses (HF) 8 sessions 2,400 total pulses (LF)	CAPS PCL-5 HAM-A HAM-D	No significant group differences

ACC anterior cingulate cortex, dTMS deep transcranial magnetic stimulation, MPFC medial prefrontal cortex, MT motor threshold, PTSD posttraumatic stress disorder, rTMS repetitive transcranial magnetic stimulation disentangle the observed findings; these outcomes could be attributed to off-target effects from stimulation, a detrimental effect of exposure immediately before stimulation, or some unknown factor.

A small pilot study was published in 2021 on simultaneous PE with left or right 10 Hz TMS versus sham in PTSD (Fryml et al. 2019). Participants completed an adapted 8-week PE course with TMS once per week; during TMS, participants listened to a recording of their PE session. Though the study was not powered for clinical outcomes, adherence was high and there was a nonsignificant trend in PTSD symptom improvement favoring the active group.

In 2022, another study sought to facilitate fear extinction by combining trauma script exposure with TMS to reduce PTSD symptoms (Thierrée et al. 2022). Participants were randomized to receive 1 Hz or 10 Hz TMS to the right DLPFC with trauma script exposure. In both groups, CAPS scores were reduced, with larger (nonsignificant) effects in the 10 Hz group with a medium effect size. Thus, while not significant, this data also supports the finding that higher frequency TMS combined with trauma script exposure can be efficacious for PTSD.

In the largest RCT to date that combined TMS with CPT for PTSD, veterans were randomized into active or sham 1 Hz TMS at the right DLPFC (Kozel et al. 2018). TMS, followed by a standardized CPT session, occurred weekly for 12 to 15 weeks. Both groups had significant reductions in PTSD symptoms as measured by the CAPS and PCL, but the active group was statistically superior with a clinically meaningful difference. Interestingly, these effects were specific to PTSD, as both groups demonstrated reduced depressive symptoms. 57% of veterans completed the 6-month follow-up and effects were sustained. In total, the authors concluded that TMS with CPT produced meaningfully greater treatment results for PTSD symptoms.

Comparatively, there is a gap in the literature regarding TMS with psychotherapy for GAD. However, one study explored the clinical effects of mindfulness therapy with TMS for GAD. Participants were randomized into TMS with mindfulness therapy or TMS without mindfulness (Gao et al. 2021). 1 Hz TMS was administered at the right prefrontal cortex. At endpoint, the TMS plus mindfulness therapy group had a statistically significant reduction in HAM-A scores compared to the control group; this indicates the promise of potential combined use of stimulation plus mindfulness for GAD.

Collectively, these studies demonstrate the tolerability and feasibility of combining TMS with psychotherapy, although present a somewhat mixed story. Therapy outcomes appear to be improved as an adjunct to standard TMS; however, with broader dTMS coils, the ideal approach is less clear. Individual variability in therapy is another complicating factor, as is the use of abbreviated elements (e.g., brief exposure) of gold-standard therapy; how this might impact clinical trial results remains unclear. Whether there is an ideal timing of stimulation and therapy also remains unknown. However, recent data may provide guidance to this effect – a recent RCT using combined tDCS (a non-depolarizing form of electrical stimulation) at the VMPFC with virtual reality exposure demonstrated that simultaneous administration yielded improvements in PTSD symptoms and autonomic arousal.

Procedures were informed by the mechanistic theory that stimulating the VMPFC may enhance safety learning in patients when combined with exposure to fear or anxiety-inducing cues (van't Wout-Frank et al. 2024).

Furthermore, while other mechanisms for enhancing the effects of TMS with psychotherapy have been proposed (supplementing fear extinction by decreasing right frontal cortex hyperactivity or increasing hypoactivity with downstream effects), these studies have not fully assessed the corresponding biomarkers. Future research should be done to investigate the efficacy of exposure and CBT in combination with TMS for GAD, as current research is limited. Furthermore, use of biological markers in future studies is important to interpret observed clinical outcomes.

6 Discussion: Designing Future Studies for TMS in PTSD and GAD

This review of TMS for PTSD and GAD focused on evidence for its use while highlighting pertinent questions meant to guide future research. In all cases, RCTs have demonstrated that TMS is safe and tolerable. TMS appears to be efficacious for PTSD and GAD, and higher doses of stimulation may be required for improvement. However, there is still no clear indicator that any one protocol or target is superior to another for PTSD or GAD. Furthermore, there are few studies targeting brain regions outside the DLPFC. Additional guidance from clinical registry studies provides a clear indication that left-sided high-frequency protocols can be efficacious for PTSD and comorbid MDD. Results from GAD studies indicate promise but with reliance on secondary data analyses. Efficacy of newer TMS modalities remains mixed, but at least are safe. Overall, TBS does not appear to meaningfully differ from standard TMS; as the quicker option, TBS may decrease patient burden and be more cost-effective. Whether and how this information can be leveraged in newer accelerated TMS studies with higher doses remains to be seen, although case reports appear promising (Zuschlag et al. 2023).

When combining psychotherapy with TMS, results are more mixed. While there is modest data on the combination of TMS plus therapy for GAD, it appears that brief exposure to a traumatic script, at least when stimulating the DMPFC, can be countertherapeutic for PTSD. This contrasts with the more successful pairing of DLPFC stimulation with manualized cognitive processing (which is consistent with the recent tDCS plus virtual reality for PTSD study) (van't Wout-Frank et al. 2024). While combined approaches are at least tolerable and feasible, it is also important to consider that combining TMS and therapy may increase patient burden and increase staffing costs, and if pursued will need to yield outcomes that are superior to TMS monotherapy.

Caveats aside, the studies reviewed here indicate that TMS is safe and can improve symptoms in patients with PTSD and GAD. The research to date has not yet indicated that specific individual parameters are able to yield superior outcomes, at least on the group level. Precision targeting (e.g., Siddiqi et al.) (Siddiqi et al. 2024), tailored to an individual, may provide superior outcomes, although that remains a hypothesis with the current state of the literature. More clearly, higher cumulative administrations of TMS appear to yield increased efficacy. When aiming to reduce patient burden and provide cost-effective care, continuing to investigate TMS-related biomarkers is needed to fully understand clinical outcomes (and may serve as a barrier to interpret unanticipated results). Due to its mixed results and lack of literature, investigating the combination of TMS and various forms of psychotherapy provides an exciting, new opportunity for future studies. In closing, TMS for PTSD and GAD appears to be safe and likely effective. However, further work must be done to confirm efficacy and biomarkers of response in TMS, particularly when combined with psychotherapy.

Author Contributions MG performed the initial literature search, SLC and NSP wrote the first draft of the article; ARA, AZV, YAB, and CC provided critical review and editing. We express our appreciation to Nicholas Petrosino, MD for his assistance with the tables.

Conflict of Interest Statement In the past 3 years, NSP reports clinical trial support (through US Dept of Veterans Affairs contracts) with Neurolief and Wave Neuro. He is a consultant for Motif Neurotech and is on the Scientific Advisory board of Pulvinar Neuro.

Funding and Support Effort on this paper was supported in part by VA grants I50 RX002864, I01 CX002088, IK2 CX002115, and IK2 CX002603, and NIH grants R01 MH120126 and P20 GM130452. The views expressed in this article are those of the authors and do not necessarily reflect the position or policy of the US Department of Veterans Affairs or the National Institutes of Health.

All figures in this piece were generated by the author team using an academic BioRender license (https://www.biorender.com/).

References

American Psychiatric Association (2013) Desk reference to the diagnostic criteria from DSM-5, 5th edn. American Psychiatric Association, Arlington, VA

Shalev A, Liberzon I, Marmar C (2017) Posttraumatic stress disorder. N Engl J Med 376:2459–2469

Goldstein RB, Smith SM, Chou SP et al (2016) The epidemiology of DSM-5 posttraumatic stress disorder in the United States: results from the National Epidemiologic Survey on alcohol and related conditions–III. Soc Psychiatry Psychiatr Epidemiol 51:1137–1148

Hoge CW, Castro CA, Messer SC et al (2004) Combat duty in Iraq and Afghanistan, mental health problems, and barriers to care. N Engl J Med 351:13–22

Flory JD, Yehuda R (2015) Comorbidity between post-traumatic stress disorder and major depressive disorder: alternative explanations and treatment considerations. Dialogues Clin Neurosci 17:141–150

Pietrzak RH, Goldstein RB, Southwick SM et al (2011) Prevalence and Axis I comorbidity of full and partial posttraumatic stress disorder in the United States: results from wave 2 of the National Epidemiologic Survey on alcohol and related conditions. J Anxiety Disord 25:456–465

Solomon SD, Davidson JR (1997) Trauma: prevalence, impairment, service use, and cost. J Clin Psychiatry 58:5–11

Lee DJ, Schnitzlein CW, Wolf JP et al (2016) Psychotherapy versus pharmacotherapy for posttraumatic stress disorder: systemic review and meta-analyses to determine first-line treatments. Depress Anxiety 33:792–806

Steenkamp MM, Litz BT, Hoge CW et al (2015) Psychotherapy for military-related PTSD: a review of randomized clinical trials. JAMA 314:489–500

Imel ZE, Laska K, Jakupcak M et al (2013) Meta-analysis of dropout in treatments for posttraumatic stress disorder. J Consult Clin Psychol 81:394–404

Watts BV, Schnurr PP, Mayo L et al (2013) Metaanalysis of the efficacy of treatments for posttraumatic stress disorder. J Clin Psychiatry 74:e541–e550

Koek RJ, Roach J, Athanasiou N et al (2019) Neuromodulatory treatments for posttraumatic stress disorder (PTSD). Prog Neuro-Psychopharmacol Biol Psychiatry 92:148–160

George MS, Lisanby SH, Avery D et al (2010) Daily left prefrontal transcranial magnetic stimulation therapy for major depressive disorder: a sham-controlled randomized trial. Arch Gen Psychiatry 67:507–516

O'Reardon JP, Solvason HB, Janicak PG et al (2007) Efficacy and safety of transcranial magnetic stimulation in the acute treatment of major depression: a multisite randomized controlled trial. Biol Psychiatry 62:1208–1216

Cosmo C, Zandvakili A, Petrosino NJ, Berlow YA, Philip NS (2021) Repetitive transcranial magnetic stimulation for treatment-resistant depression: recent critical advances in patient care. Curr Treat Options Psychiatry 8(2):47–63. https://doi.org/10.1007/s40501-021-00238-y

Barker AT, Freeston IL, Jabinous R et al (1986) Clinical evaluation of conduction time measurements in central motor pathways using magnetic stimulation of human brain. Lancet 1:1325–1326

Perera T, George MS, Grammer G et al (2016) The clinical TMS society consensus review and treatment recommendations for TMS therapy for major depressive disorder. Brain Stimul 9: 336–346

Pascual-Leone A, Valls-Solé J, Wassermann EM et al (1994) Responses to rapid-rate transcranial magnetic stimulation of the human motor cortex. Brain 117:847–858

Chen R, Classen J, Gerloff C et al (1997) Depression of motor cortex excitability by low-frequency transcranial magnetic stimulation. Neurology 48:1398–1403

Huang YZ, Edwards MJ, Rounis E et al (2005) Theta burst stimulation of the human motor cortex. Neuron 45:201–206

Larson J, Munkácsy E (2015) Theta-burst LTP. Brain Res 1621:38–50

Blumberger DM, Vila-Rodriguez F, Thorpe KE et al (2018) Effectiveness of theta burst versus high-frequency repetitive transcranial magnetic stimulation in patients with depression (THREE-D): a randomised non-inferiority trial. Lancet 391:1683–1692

Van Der Kolk BA, Burbridge JA, Suzuki J (1997) The psychobiology of traumatic memory. Clinical implications of neuroimaging studies. Ann N Y Acad Sci 821:99–113

George MS, Wassermann EM, Williams WA et al (1995) Daily repetitive transcranial magnetic stimulation (rTMS) improves mood in depression. Neuroreport 6:1853–1856

McCann UD, Kimbrell TA, Morgan CM et al (1998) Repetitive transcranial magnetic stimulation for posttraumatic stress disorder. Arch Gen Psychiatry 55:276–279

Grisaru N, Amir M, Cohen H et al (1998) Effect of transcranial magnetic stimulation in posttraumatic stress disorder: a preliminary study. Biol Psychiatry 44:52–55

Rosenberg PB, Mehndiratta RB, Mehndiratta YP et al (2002) Repetitive transcranial magnetic stimulation treatment of comorbid posttraumatic stress disorder and major depression. J Neuropsychiatry Clin Neurosci 14:270–276

Cohen H, Kaplan Z, Kotler M et al (2004) Repetitive transcranial magnetic stimulation of the right dorsolateral prefrontal cortex in posttraumatic stress disorder: a double-blind, placebo-controlled study. Am J Psychiatry 161:515–524

Blake DD, Weathers FW, Nagy LM et al (1995) The development of a clinician-administered PTSD scale. J Trauma Stress 8:75–90

Boggio PS, Rocha M, Oliveira MO et al (2010) Noninvasive brain stimulation with high frequency and low-intensity repetitive transcranial magnetic stimulation treatment for posttraumatic stress disorder. J Clin Psychiatry 71:992–999
Watts BV, Landon B, Groft A et al (2012) A sham controlled study of repetitive transcranial magnetic stimulation for posttraumatic stress disorder. Brain Stimul 5:38–43
Nam DH, Pae CU, Chae JH (2013) Low-frequency, repetitive transcranial magnetic stimulation for the treatment of patients with posttraumatic stress disorder: a double-blind, sham-controlled study. Clin Psychopharmacol Neurosci 11:96–102
Ahmadizadeh MJ, Rezaei M (2018) Unilateral right and bilateral dorsolateral prefrontal cortex transcranial magnetic stimulation in treatment post-traumatic stress disorder: a randomized controlled study. Brain Res Bull 140:334–340
Kozel FA, Van Trees K, Larson V et al (2019) One hertz versus ten hertz repetitive TMS treatment of PTSD: a randomized clinical trial. Psychiatry Res 273:153–162
Philip NS, Aiken EE, Kelley ME et al (2019a) Synchronized transcranial magnetic stimulation for posttraumatic stress disorder and comorbid major depression. Brain Stimul 12:1335–1337
Philip NS, Barredo J, Aiken E et al (2019b) Thetaburst transcranial magnetic stimulation for posttraumatic stress disorder. Am J Psychiatry 176:939–948
Leong K, Chan P, Ong L et al (2020) A randomized sham-controlled trial of 1-Hz and 10-Hz repetitive transcranial magnetic stimulation (rTMS) of the right dorsolateral prefrontal cortex in civilian post-traumatic stress disorder: Un essai randomisé contrôlé simulé de stimulation magnétique transcrânienne repetitive (SMTr) de 1 Hz et 10 Hz du cortex préfrontal dorsolatéral droit dans le trouble de stress post-traumatique chez des civils. Can J Psychiatr 65:770–778
Carpenter LL, Conelea C, Tyrka AR et al (2018) 5 Hz repetitive transcranial magnetic stimulation for posttraumatic stress disorder comorbid with major depressive disorder. J Affect Disord 235: 414–420
Tillman GD, Kimbrell TA, Calley CS et al (2011) Repetitive transcranial magnetic stimulation and threat memory: selective reduction of combat threat memory p300 response after right frontal lobe stimulation. J Neuropsychiatry Clin Neurosci 23:40–47
Nakama H, Garcia A, O'Brien K et al (2014) Case report of a 24-year-old man with resolution of treatment resistant major depressive disorder and comorbid PTSD using rTMS. J ECT 30:e9–e10
Oznur T, Akarsu S, Celik C et al (2014) Is transcranial magnetic stimulation effective in treatment resistant combat related posttraumatic stress disorder. Neurosciences (Riyadh) 19:29–32
Philip NS, Ridout SJ, Albright SE et al (2016) 5-Hz transcranial magnetic stimulation for comorbid posttraumatic stress disorder and major depression. J Trauma Stress 29:93–96
Nursey J, Sbisa A, Knight H et al (2020) Exploring theta burst stimulation for post-traumatic stress disorder in Australian veterans – a pilot study. Mil Med 185:e1770–e1778
Weiss DS, Marmar C (1997) The impact of events scale–revised. In: Wilson JP, Keane T (eds) Assessing psychological trauma and PTSD. The Guilford Press, New York, pp 399–411
Wortmann JH, Jordan AH, Weathers FW et al (2016) Psychometric analysis of the PTSD checklist–5 (PCL-5) among treatment-seeking military service members. Psychol Assess 28:1392–1403
Montgomery SA, Asberg M (1979) A new depression scale designed to be sensitive to change. Br J Psychiatry 134:382–389
Rush AJ, Trivedi MH, Ibrahim HM et al (2003) The 16-item quick inventory of depressive symptomatology (QIDS), clinician rating (QIDS-C), and self-report (QIDS-SR): a psychometric evaluation in patients with chronic major depression. Biol Psychiatry 54:573–583
Beck AT, Ward CH, Mendelson M et al (1961) An inventory for measuring depression. Arch Gen Psychiatry 4:561–571
Philip NS, Barredo J, van't Wout-Frank M et al (2018) Network mechanisms of clinical response to transcranial magnetic stimulation in posttraumatic stress disorder and major depressive disorder. Biol Psychiatry 83:263–272
Zandvakili A, Philip NS, Jones SR et al (2019) Use of machine learning in predicting clinical response to transcranial magnetic stimulation in comorbid posttraumatic stress disorder and major depression: a resting state electroencephalography study. J Affect Disord 252:47–54

Leuchter AF, Cook IA, Feifel D et al (2015) Efficacy and safety of low-field synchronized transcranial magnetic stimulation (sTMS) for treatment of major depression. Brain Stimul 8: 787–794

van't Wout-Frank M, Shea MT, Sorensen DO et al (2021) A secondary analysis on effects of theta burst transcranial magnetic stimulation to reduce anger in veterans with posttraumatic stress disorder. Neuromodulation 24:870–878

Petrosino NJ, Wout-Frank MV, Aiken E et al (2020) One-year clinical outcomes following theta burst stimulation for post-traumatic stress disorder. Neuropsychopharmacology 45:940–946

Zandvakili A, Swearingen HR, Philip NS (2021) Changes in functional connectivity after thetaburst transcranial magnetic stimulation for posttraumatic stress disorder: a machine-learning study. Eur Arch Psychiatry Clin Neurosci 271:29–37

Osuch EA, Benson BE, Luckenbaugh DA et al (2009) Repetitive TMS combined with exposure therapy for PTSD: a preliminary study. J Anxiety Disord 23:54–59

Isserles M, Shalev AY, Roth Y et al (2013) Effectiveness of deep transcranial magnetic stimulation combined with a brief exposure procedure in post-traumatic stress disorder–a pilot study. Brain Stimul 6:377–383

Kozel FA, Motes MA, Didehbani N et al (2018) Repetitive TMS to augment cognitive processing therapy in combat veterans of recent conflicts with PTSD: a randomized clinical trial. J Affect Disord 229:506–514

Fryml LD, Pelic CG, Acierno R et al (2019) Exposure therapy and simultaneous repetitive transcranial magnetic stimulation: a controlled pilot trial for the treatment of posttraumatic stress disorder. J ECT 35:53–60

Isserles M, Tendler A, Roth Y et al (2021) Deep transcranial magnetic stimulation combined with brief exposure for posttraumatic stress disorder: a prospective multisite randomized trial. Biol Psychiatry 90(10):721–728. https://doi.org/10.1016/j.biopsych.2021.04.019

Francati V, Vermetten E, Bremner JD (2007) Functional neuroimaging studies in posttraumatic stress disorder: review of current methods and findings. Depress Anxiety 24:202–218

Guy W (1976) ECDEU assessment manual for psychopharmacology, revised. US Department of Health, Education, and Welfare Publication (ADM), National Institute of Mental Health, Rockville

Bystritsky A, Kaplan JT, Feusner JD et al (2008) A preliminary study of fMRI-guided rTMS in the treatment of generalized anxiety disorder. J Clin Psychiatry 69(7):1092–1098. https://doi.org/10.4088/jcp.v69n0708

Bystritsky A, Kerwin LE, Feusner JD (2009) A preliminary study of fMRI-guided rTMS in the treatment of generalized anxiety disorder: 6-month follow-up. J Clin Psychiatry 70(3):431–432. https://doi.org/10.4088/JCP.08l04641

Diefenbach GJ, Assaf M, Goethe JW, Gueorguieva R, Tolin DF (2016) Improvements in emotion regulation following repetitive transcranial magnetic stimulation for generalized anxiety disorder. J Anxiety Disord 43:1–7. https://doi.org/10.1016/j.janxdis.2016.07.002

Dilkov D, Hawken ER, Kaludiev E, Milev R (2017) Repetitive transcranial magnetic stimulation of the right dorsal lateral prefrontal cortex in the treatment of generalized anxiety disorder: a randomized, double-blind sham controlled clinical trial. Prog Neuro-Psychopharmacol Biol Psychiatry 78:61–65. https://doi.org/10.1016/j.pnpbp.2017.05.018

Maier W, Buller R, Philipp M, Heuser I (1988) The Hamilton anxiety scale: reliability, validity and sensitivity to change in anxiety and depressive disorders. J Affect Disord 14(1):61–68. https://doi.org/10.1016/0165-0327(88)90072-9

Philip NS, Doherty RA, Faucher C, Aiken E, van 't Wout-Frank M. (2022) Transcranial magnetic stimulation for posttraumatic stress disorder and major depression: comparing commonly used clinical protocols. J Trauma Stress 35(1):101–108. https://doi.org/10.1002/jts.22686

Moraga-Amaro R, Muñoz P, Villalobos T et al (2023) Real-world data of non-invasive stimulation of the human insula-prefrontal cortices using deep TMS to treat anxiety for occupational stress and generalized anxiety disorder. Psychiatry Res 320:115036. https://doi.org/10.1016/j.psychres.2022.115036

Pell GS, Harmelech T, Zibman S, Roth Y, Tendler A, Zangen A (2022) Efficacy of deep TMS with the H1 coil for anxious depression. J Clin Med 11(4):1015. Published 2022 Feb 15. https://doi.org/10.3390/jcm11041015

Yuan H, Liu B, Li F et al (2023) Effects of intermittent theta-burst transcranial magnetic stimulation on post-traumatic stress disorder symptoms: a randomized controlled trial. Psychiatry Res 329: 115533. https://doi.org/10.1016/j.psychres.2023.115533

Shenasa MA, Ellerman-Tayag E, Canet P, Martis B, Mishra J, Ramanathan DS (2023) Theta burst stimulation is not inferior to high-frequency repetitive transcranial magnetic stimulation in reducing symptoms of posttraumatic stress disorder in veterans with depression: a retrospective case series. Neuromodulation 26(4):885–891. https://doi.org/10.1016/j.neurom.2023.02.082

Thierrée S, Raulin-Briot M, Legrand M et al (2022) Combining trauma script exposure with rTMS to reduce symptoms of post-traumatic stress disorder: randomized controlled trial. Neuromodulation 25(4):549–557. https://doi.org/10.1111/ner.13505

Makale MT, Abbasi S, Nybo C et al (2023) Personalized repetitive transcranial magnetic stimulation (prtms®) for post-traumatic stress disorder (ptsd) in military combat veterans. Heliyon 9(8): e18943. Published 2023 Aug 8. https://doi.org/10.1016/j.heliyon.2023.e18943

Morris AT, Temereanca S, Zandvakili A et al (2023) Fronto-central resting-state 15-29 Hz transient beta events change with therapeutic transcranial magnetic stimulation for posttraumatic stress disorder and major depressive disorder. Sci Rep 13(1):6366. Published 2023 Apr 19. https://doi.org/10.1038/s41598-023-32801-3

Madore MR, Kozel FA, Williams LM et al (2022) Prefrontal transcranial magnetic stimulation for depression in US military veterans – a naturalistic cohort study in the veterans health administration. J Affect Disord 297:671–678. https://doi.org/10.1016/j.jad.2021.10.025

Philip NS, Ramanathan D, Gamboa B et al (2023) Repetitive transcranial magnetic stimulation for depression and posttraumatic stress disorder in veterans with mild traumatic brain injury. Neuromodulation 26(4):878–884. https://doi.org/10.1016/j.neurom.2022.11.015

Garakani A, Murrough JW, Freire RC et al (2020) Pharmacotherapy of anxiety disorders: current and emerging treatment options. Front Psychiatry 11:595584. Published 2020 Dec 23. https://doi.org/10.3389/fpsyt.2020.595584

Curtiss J, Andrews L, Davis M, Smits J, Hofmann SG (2017) A meta-analysis of pharmacotherapy for social anxiety disorder: an examination of efficacy, moderators, and mediators. Expert Opin Pharmacother 18(3):243–251. https://doi.org/10.1080/14656566.2017.1285907

Newman MG, Zainal NH, Hoyer J (2020) Cognitive-behavioral therapy (CBT) for generalized anxiety disorder (GAD). Generalized anxiety disorder and worrying, vol 19, pp 203–230. https://doi.org/10.1002/9781119189909.ch10

Erickson TM, Newman MG (2005) Cognitive behavioral psychotherapy for generalized anxiety disorder: a primer. Expert Rev Neurother 5(2):247–257

Reinhold JA, Rickels K (2015) Pharmacological treatment for generalized anxiety disorder in adults: an update. Expert Opin Pharmacother 16(11):1669–1681. https://doi.org/10.1517/14656566.2015.1059424

Saccenti D, Lodi L, Moro AS, Scaini S, Forresi B, Lamanna J, Ferro M (2024) Novel approaches for the treatment of post-traumatic stress disorder: a systematic review of non-invasive brain stimulation interventions and insights from clinical trials. Brain Sci 14(3):210. https://doi.org/10.3390/brainsci14030210

Kan RLD, Zhang BBB, Zhang JJQ, Kranz GS (2020) Non-invasive brain stimulation for posttraumatic stress disorder: a systematic review and meta-analysis. Transl Psychiatry 10(1): 168. https://doi.org/10.1038/s41398-020-0851-5

Magnuson J, Ozdemir MA, Mathieson E, Kirkman S, Passera B, Rampersad S, Dufour AB, Brooks D, Pascual-Leone A, Fried PJ, Shafi MM, Ozdemir RA (2023) Neuromodulatory effects and reproducibility of the most widely used repetitive transcranial magnetic stimulation protocols. PLoS One 18(6). https://doi.org/10.1371/journal.pone.0286465

McDonald WM (2024) Theta burst TMS technology: great promise and a lot to learn. Am J Psychiatry 181(1):14–15. https://doi.org/10.1176/appi.ajp.20230879

Duecker F, Sack AT (2015) Rethinking the role of sham TMS. Front Psychol 6:210. Published 2015 Feb 26. https://doi.org/10.3389/fpsyg.2015.00210

Rytwinski NK, Scur MD, Feeny NC, Youngstrom EA (2013) The co-occurrence of major depressive disorder among individuals with posttraumatic stress disorder: a meta-analysis. J Trauma Stress 26(3):299–309. https://doi.org/10.1002/jts.21814

Yesavage JA, Fairchild JK, Mi Z et al (2018) Effect of repetitive transcranial magnetic stimulation on treatment-resistant major depression in US veterans: a randomized clinical trial. JAMA Psychiatry 75(9):884–893. https://doi.org/10.1001/jamapsychiatry.2018.1483

Deng ZD, Lisanby SH, Peterchev AV (2014) Coil design considerations for deep transcranial magnetic stimulation. Clin Neurophysiol 125(6):1202–1212. https://doi.org/10.1016/j.clinph.2013.11.038

Gao L, Xie J, Huang T, Shang Y, Gao Z (2021) Effects of mindfulness decompression therapy combined with transcranial magnetic stimulation in generalized anxiety disorder. Am J Transl Res 13(6):6827–6836. Published 2021 Jun 15

van't Wout-Frank M, Arulpragasam AR, Faucher C et al (2024) Virtual reality and transcranial direct current stimulation for posttraumatic stress disorder: a randomized clinical trial. JAMA Psychiatry. https://doi.org/10.1001/jamapsychiatry.2023.5661

Carpenter L, Aaronson S, Hutton TM et al (2021) Comparison of clinical outcomes with two transcranial magnetic stimulation treatment protocols for major depressive disorder. Brain Stimul 14(1):173–180. https://doi.org/10.1016/j.brs.2020.12.003

Zuschlag ZD, Bajor L, Van Trees K et al (2023) An accelerated course of TMS using intermittent theta burst for veterans with major depressive disorder: a case series. Ann Clin Psychiatry 35(2): 110–117. https://doi.org/10.12788/acp.0110

Siddiqi S, Philip NS, Palm S et al (2024) A potential neuromodulation target for PTSD in veterans derived from focal brain lesions. Res Sq:rs.3.rs-3132332. Published 2024 Mar 19. https://doi.org/10.21203/rs.3.rs-3132332/v1

Weathers FW, Litz BT, Keane TM et al (2013) The PTSD checklist for DSM-5 (PCL-5) – standard [measurement instrument]. Nat Cent PTSD. https://www.ptsd.va.gov/professional/assessment/adult-sr/ptsd-checklist.asp

Hembree EA, Rauch SAM, Foa EB (2003) Beyond the manual: the insider's guide to prolonged exposure therapy for PTSD. Cogn Behav Pract 10(1):22–30. https://doi.org/10.1016/S1077-7229(03)80005-6

Petrosino NJ, Cosmo C, Berlow YA, Zand Vakili A, van 't Wout-Frank M, Philip NS. (2021) Transcranial magnetic stimulation for post-traumatic stress disorder. Ther Adv Psychopharmacol 11:20451253211049921. Published 2021 Oct 28. https://doi.org/10.1177/20451253211049921

Rauch SL, van der Kolk BA, Fisler RE, Alpert NM, Orr SP, Savage CR et al (1996) A symptom provocation study of posttraumatic stress disorder using positron emission tomography and script-driven imagery. Arch Gen Psychiatry 53:380–387

Liston C, Chen AC, Zebley BD et al (2014) Default mode network mechanisms of transcranial magnetic stimulation in depression. Biol Psychiatry 76(7):517–526. https://doi.org/10.1016/j.biopsych.2014.01.023

White LK, Makhoul W, Teferi M, Sheline YI, Balderston NL (2023) The role of dlPFC laterality in the expression and regulation of anxiety. Neuropharmacology 224:109355. https://doi.org/10.1016/j.neuropharm.2022.109355

Siebner HR, Funke K, Aberra AS et al (2022) Transcranial magnetic stimulation of the brain: what is stimulated? – a consensus and critical position paper. Clin Neurophysiol 140:59–97. https://doi.org/10.1016/j.clinph.2022.04.022

Part V
Co-occurring Disorders

Sensory Processing and Anxiety: Within and Beyond the Autism Spectrum

Alisa R. Zoltowski, Caitlin A. Convery, Ekomobong Eyoh, Emily Plump, Molly Sullivan, Eshani R. Arumalla, Jennifer M. Quinde-Zlibut, Bahar Keceli-Kaysili, Brianna Lewis, and Carissa J. Cascio

Contents

A. R. Zoltowski
Psychiatry and Behavioral Sciences, Vanderbilt University Medical Center, Nashville, TN, USA

Life Span Institute, University of Kansas, Lawrence, KS, USA

C. A. Convery, J. M. Quinde-Zlibut, and B. Lewis
Psychiatry and Behavioral Sciences, Vanderbilt University Medical Center, Nashville, TN, USA

E. Eyoh
Institute of Child Development, University of Minnesota, Minneapolis, MN, USA

E. Plump and E. R. Arumalla
Vanderbilt University, Nashville, TN, USA

M. Sullivan
Vanderbilt University School of Nursing, Nashville, TN, USA

B. Keceli-Kaysili
Hearing and Speech Sciences, Vanderbilt University Medical Center, Nashville, TN, USA

C. J. Cascio (✉)
Psychiatry and Behavioral Sciences, Vanderbilt University Medical Center, Nashville, TN, USA

Life Span Institute, University of Kansas, Lawrence, KS, USA

Department of Psychology, University of Kansas, Lawrence, KS, USA
e-mail: carissa.cascio@ku.edu

© The Author(s), under exclusive license to Springer Nature Switzerland AG 2024

Curr Topics Behav Neurosci (2025) 73: 557–580
https://doi.org/10.1007/7854_2024_557
Published Online: 14 December 2024

Abstract In this article, we briefly overview how the expression, measurement, and treatment of anxiety in autism may be different from the general population. We review the literature on links between sensory processing differences and anxiety, which show transdiagnostic patterns but are an especially prominent feature of anxiety in autism. Specifically, we focus on how the sense of interoception, i.e., how we perceive sensory information from within our bodies, contributes to anxiety in autism. We present new findings integrating multimodal interoceptive measures and total anxiety symptoms in a sample of $n = 38$ non-autistic and $n = 43$ autistic individuals, ages 8–55 years. Using principal components analysis, we found two components relating to interoceptive confusion (i.e., self-reported ability to localize and interpret interoceptive cues): one component that closely relates to anxiety symptoms and one component that is distinct from anxiety. Interoceptive perception (i.e., performance on a lab-based task) was uniformly related to interoceptive confusion when distinguished from anxiety but showed complex relations with total anxiety symptoms. Combined, these findings suggest meaningful subtypes of interoceptive difficulties and their interrelationship with anxiety. We present conclusions and future directions for consideration of individual differences, toward creating a personalized understanding of anxiety-interoception links.

Keywords Anxiety · Autism · Interoception · Sensory processing

1 Characterizing Anxiety in Autism

1.1 Presentation of Anxiety in Autism

Autism is characteristically defined by differences in social communication and sensorimotor behaviors (American Psychiatric Association 2013). Along with these core autism features, autistic individuals often experience clinically significant levels of anxiety (White et al. 2009). While exact estimates vary, most studies suggest elevated rates of anxiety symptoms and diagnosed anxiety disorders in autistic individuals, with co-occurring anxiety disorder diagnoses estimated to impact at least 40–50% (De Bruin et al. 2007; Vasa and Mazurek 2015), and perhaps up to 79% of autistic individuals (Kent and Simonoff 2017); whereas, the lifetime prevalence of an anxiety disorder by age 18 is estimated to 32% in the general population (Merikangas et al. 2010). These high levels of anxiety are often reported to be equally or more impairing to autistic individuals as challenges relating to core autism features (Ozsivadjian et al. 2012). Thus, understanding how anxiety impacts autistic individuals, and how this impact may be shared versus distinct from that of non-autistic individuals, is an important part of supporting autistic individuals' well-being.

Evidence on the presentation of anxiety in autistic individuals suggests that there are both common and unique features compared to how anxiety presents in non-autistic individuals (White et al. 2009; Kerns and Kendall 2012; Kerns et al. 2014). Though the major anxiety disorder types (e.g., social, separation, generalized, specific phobia) are observed in both autistic and non-autistic individuals, the focus of specific phobias in autism may include different items or situations than what commonly occurs for non-autistic individuals (Kerns and Kendall 2012). There are also several facets of autism that may overlap with common anxiety features or influence how anxiety features present (White et al. 2009; Kerns et al. 2014). For example, behaviors such as social avoidance, a strong preference for routines, insistence on sameness, compulsive and repetitive behavior, and motor overflow overlap between core autism features and features of anxiety (White et al. 2009; Kerns et al. 2014). Clinicians may need more information to disentangle these factors. One important factor that can help differentiate features of anxiety from autism is the onset of symptoms during the clinical encounter, e.g., a change in social features compared to baseline presentation can indicate the presence of anxiety. In the social domain, assessing social motivation may also be useful for understanding links between observed social behaviors and social anxiety (Pallathra et al. 2018). Differences in language and communication in autism also complicate how anxiety is understood in this population. For example, measures of anxiety that heavily rely on self-report or caregiver report of verbalized worry limit the extent to which individuals with cognitive and/or language disabilities may be considered anxious (Hallett et al. 2013).

1.2 Measuring and Treating Anxiety in Autism

Researchers have considered whether common anxiety measures appropriately capture the same construct in autistic individuals (Lecavalier et al. 2014). A review completed by Lecavalier et al. (2014) of the available anxiety measures at the time suggested that of 10 measures considered, 4 were considered conditionally appropriate for use in autism and 3 were considered potentially appropriate, with caution. Since then, new autism-focused anxiety scales have been developed to address some of these limitations. One is the Parent-Rated Anxiety Scale for Youth with Autism Spectrum Disorder (PRAS-ASD, Scahill et al. 2019).[1] This measure includes a total anxiety score using items designed to capture how anxiety more commonly presents in autistic individuals, including an enhanced focus on behavioral expressions of anxiety. An additional pair of child and adult anxiety measures which include anxiety subtype scores is the Anxiety Scale for Children with Autism Spectrum Disorder (ASC-ASD, Rodgers et al. 2016) and the Anxiety Scale for Adults with Autism Spectrum Disorder (ASA-ASD, Rodgers et al. 2020). These domain scores are comparable to anxiety subtypes common in the general population (e.g., these scales include domain scores for separation and social anxiety, respectively, dependent on participant age), but domain scores for uncertainty and anxious arousal include items that reflect potentially distinct types of anxiety more commonly present in autistic individuals.

Additionally, treatments of anxiety as developed for non-autistic populations may also be effective in autism, including some (but not all) anxiety medications and cognitive behavioral therapy (Vasa et al. 2014). For cognitive behavioral therapy (CBT), autism-specific modifications including increased structure, expanded timing, and specific skills to target are generally recommended (White et al. 2009; Walters et al. 2016). In practice, autistic adults are more likely to be prescribed multiple concurrent medications for anxiety and less likely to receive talk therapy than non-autistic adults, though work on optimizing the balance between medication and psychosocial approaches in this population is still needed (Maddox et al. 2018). Thus, an evolving understanding of what is unique to anxiety in autism, versus what features of anxiety are more broadly shared, will continue to inform measurement of anxiety presentation and ideal treatments in this population.

[1] These measure titles are listed as published, though in general our terminology choices are guided by Bottema-Beutel et al. (2021) to preferentially use "autism" over "autism spectrum disorder."

2 Sensory Processing Differences: A Distinct Basis of Anxiety?

2.1 Sensory Differences and Their General Relationship to Anxiety

The overlap between autistic features and anxiety has led researchers to search for factors that may predispose autistic individuals to higher anxiety. There are likely multiple intrinsic and extrinsic factors that contribute, including differences in physiological and emotional regulation as well as increased negative feedback from others (White et al. 2014). One factor that may be particularly distinctive of autism is sensory processing. While sensory processing differences are not unique to autism (Isaacs et al. 2022; Van Den Boogert et al. 2022), they are now included in the core diagnostic criteria for the condition (American Psychiatric Association 2013). Notably, though early sensory issues occur in both autistic and non-autistic individuals, autistic individuals are more likely to show sharper increases in reported sensory issues between infancy and school age (Chen et al. 2022). The evidence to date suggests broad differences in how autistic individuals experience and respond to sensory information. Autistic sensory differences have been reported across different modes of inquiry (e.g., by sensory testing or self-report) and different sensory modalities, especially auditory and tactile (Altemus et al. 2014; Baranek et al. 2006, 2022; Minshew and Hobson 2008). Substantial heterogeneity exists in how these differences present across individuals. Efforts to organize sensory processing patterns have commonly focused on the nature of the observable behavioral response. Most prominently, "*hyper*-responsiveness" or "sensory over-responsivity" (SOR) refer to strong, affectively negative behavioral reactions to typical sensations. "*Hypo*-responsiveness" refers to reduced reactions to salient sensations. "Sensory interests, repetition and seeking" (SIRS) refers to behaviors including increased intentional engagement with sensory stimuli (Baranek et al. 2006; Boyd et al. 2010; Kirby et al. 2015; Lee et al. 2022). On the surface, these might seem mutually exclusive such that an individual would be either *hyper*- or *hypo*-responsive to sensory stimuli. However, many individuals show both of these types of patterns (Ausderau et al. 2016; Williams et al. 2018; Baranek et al. 2022; Lee et al. 2022). The potential for sensory processing to be a source of anxiety for autistic individuals has been shown in features of autism-specific anxiety, such that certain sensations are more likely to be classified as specific phobias for autistic individuals than for the general population (Kerns et al. 2014). This is further reflected in items in newer autism-focused anxiety scales (Rodgers et al. 2016), including items such as "I worry about being in places that are too loud, or too bright or too busy." Thus, sensory differences are becoming increasingly embedded into the conceptualization and measurement of anxiety in autism.

2.2 *Explaining the Link: Sensory Hyper-Responsiveness and Anxiety*

Of the three primary sensory responsiveness patterns, sensory *hyper*-responsiveness has been the most thoroughly investigated and has shown strong correspondence with higher anxiety levels (Green and Ben-Sasson 2010; Green et al. 2017, 2018). Green and Ben-Sasson (2010) discuss multiple theories to explain this link. One theory suggests that anxiety may cause sensory *hyper*-responsiveness, such that sustained hyper-vigilance over time leads to increasing sensitivity to incoming sensations. Alternatively, another possibility is that sensory *hyper*-responsiveness may cause anxiety, perhaps driving an increasing intolerance of uncertainty as one continues to experience the sensory environment as difficult to predict and often aversive (Wigham et al. 2015; South and Rodgers 2017). This matches how autistic individuals report their perceptions of this relationship, such that they perceive their sensitivity to certain sensations as a cause rather than a result of their anxiety (Verhulst et al. 2022). Lane et al. (2012) found that the magnitude of electrodermal responses to a sensory challenge fully mediated the relationship between baseline arousal and reported anxiety symptoms, in their sample of individuals ages 8–10 spanning autistic, ADHD, and typically developing groups. Green et al. (2012) provide support in longitudinal studies that sensory *hyper*-responsiveness precedes and predicts anxiety levels in autistic toddlers; similar findings have emerged in other longitudinal samples (Dwyer et al. 2020; Dellapiazza et al. 2022). Other evidence suggests that a causal relationship between early sensory *hyper*-responsiveness and later anxiety is not exclusive to autism. Carpenter et al. (2019) also provide compelling longitudinal evidence that sensory *hyper*-responsivity precedes and predicts anxiety in a more general cohort as well, including individuals without clinical diagnoses as well as diagnoses other than autism, suggesting a transdiagnostic relationship.

Findings that sensory *hyper*-responsiveness predicts anxiety transdiagnostically could suggest that this relationship occurs on a single continuum, and any average differences in autistic samples might reflect the degree to which both traits are experienced. On the other hand, early sensory *hyper*-responsivity may interact with other core aspects of autism, including that such sensory differences tend to more sharply increase over early development for autistic than non-autistic individuals (Chen et al. 2022), which could result in a stronger coupling of early sensory responsiveness to the manifestation of later anxiety in autistic compared to non-autistic individuals. Researchers have used magnetic resonance imaging (MRI) to examine the volume and growth of the amygdala in autism, given its key role in fear and anxiety. These studies have suggested that early increases in amygdala volume are associated with autism status alone, but whether this area grows more quickly or more slowly than typical development depends on whether individuals have co-occurring anxiety that presents as more "typical" or more "autism-specific", respectively (Herrington et al. 2017; Andrews et al. 2022). Andrews et al. (2022) found that the faster amygdala growth profiles were most associated with global sensory differences, though autistic individuals with the slow growth profile also showed greater sensory processing differences than did the intermediate growth group (who did not exhibit clinically significant anxiety). This suggests a potential role of early amygdala overgrowth in reinforcing the sensory-anxiety link in autistic individuals, though the existence of different subgroups suggests multiple potential mechanisms.

Functional studies link sensory over-responsivity in autism with reduced habituation to sensory stimulation, in both sensory cortices and amygdala, combined with reduced amygdala-prefrontal coupling (Green et al. 2015). Further, such amygdala responses during sensory stimulation do suggest some divergence between anxious individuals who are autistic versus non-autistic (Cummings et al. 2024). Cummings et al. (2024) found that amygdala responses to mildly aversive sensations relate most closely to sensory responsivity (rather than anxiety) in autistic individuals and most closely to reported anxiety levels (rather than sensory responsivity) in non-autistic individuals. These findings thus suggest that sensory processing differences may play more of a primary role in autistic versus non-autistic experiences of anxiety, consistent with autistic individuals' reported experiences (Verhulst et al. 2022).

2.3 Breaking Down Sensory Heterogeneity: Other Responsiveness Patterns and Specific Modalities

Most prior work on links between sensory processing and anxiety has either focused on global sensory differences or the *hyper*-responsiveness pattern. There is some justification for these two areas of focus, as studies that examine the three different responsiveness patterns have found that sensory *hyper*-responsiveness is indeed most closely related to increased anxiety (MacLennan et al. 2020), but there is added utility in considering total severity of sensory symptoms across all three responsiveness patterns (Uljarević et al. 2016). However, MacLennan et al. (2020) found interesting relationships as to how each responsiveness pattern relates to anxiety subtypes. They found that sensory *hyper*-responsiveness was most likely to relate to physical injury fears and specific phobias. On the other hand, sensory *hypo*-responsiveness showed somewhat of a protective effect, such that increased *hypo*-responsiveness corresponded with reduced social anxiety. In addition to the three responsiveness pattern types, sensory processing may also differ by distinct sensory modalities, such as visual versus auditory systems. Though there is still much to explore in these relationships, auditory difficulties, compared to other sensory modalities, have been tentatively found to especially correlate with anxiety in both adult (Bang and Igelström 2023) and youth (ages 6–11; Bang et al. 2024) samples.

Additionally, global sensory processing differences in autism may pervade other areas of functioning and thus can impact anxiety in ways not thought to be traditionally sensory. Sensory difficulties may exacerbate the challenges of social environments for autistic individuals, since difficulties such as with sound tolerance (Williams et al. 2021) and with integrating information between multiple senses (Feldman et al. 2018) may add to the difficulties that autistic individuals experience in navigating social environments. Corbett et al. (2016) found that the diagnostic group (autism versus not) significantly moderated the association between sensory processing differences and cortisol levels during social play, suggesting specificity in that sensory differences may show additive effects to other stressors for autistic but not non-autistic individuals. Though stress and anxiety are not synonymous (Simon and Corbett 2013), experiencing this type of heightened arousal in social situations may be one factor leading to higher social anxiety (Bellini 2006).

3 Specific Sensory Systems and Anxiety: Interoception

3.1 Frameworks of Interoception and Anxiety

Though most prior attention to sensory processing in autism has been paid to sensory modalities originating outside of the body (broadly, exteroception, such as vision and hearing), individuals in the general population also vary in their sensory processing of internal, bodily signals, i.e., interoception (Craig 2002). Compared to exteroceptive systems, interoception has been posited to uniquely link to anxiety. This idea first originated outside of the autism literature. Paulus and Stein (2006) proposed their theory of anxiety as disordered interoceptive predictions, i.e., predicting a "threat" to bodily state even when interoceptive sensations were within homeostatic ranges. This idea was tested by early figures in interoceptive measurement, including Schandry (1981), whose developed heartbeat counting task has been one of the most popular interoceptive measures to date. Mean effect sizes do suggest heightened ability to detect and discriminate heartbeats in individuals with high anxiety sensitivity, high trait anxiety, and especially panic disorder (Domschke et al. 2010). However, the interoception-anxiety link has remained in question as researchers have attempted to resolve measurement issues regarding these signals (Brener and Ring 2016), which are less able to be directly controlled by researchers and often are processed below conscious awareness. Notably, non-interoceptive factors such as intelligence quotient (IQ) explain substantial variation in the types of cardiac perception paradigms that were used in many of these prior studies (Desmedt et al. 2018). A few studies have leveraged approaches such as isoproterenol administration to experimentally manipulate interoceptive signaling; these studies have shown that such alterations increase anxiety-related behaviors in rodent research (Hsueh et al. 2023) and in humans with lesions (Khalsa et al. 2016). Thus, researchers have continued to study links between interoception and anxiety with increasingly nuanced conditions for manipulating and measuring interoceptive signals.

Advanced computational frameworks have helped to propose and test neural models for filtering and responding to interoceptive signals. One such model is the

Embodied Predictive Interoception Coding (EPIC) model (Barrett and Simmons 2015). In terms of the underlying neural computations, this type of predictive coding can be characterized using Bayesian statistics (Huang and Rao 2011; Barrett and Simmons 2015; Ainley et al. 2016). Bayesian modeling provides a way to estimate sensory expectations based on prior experience (priors) and to update responses to incoming signals (posteriors), particularly when these diverge from expectations. Estimating prior and posterior distributions necessitates two experimental conditions: baseline and altered. In contrast to other sensory modalities, the "altered" condition in interoceptive studies involves manipulating signals within the participant's body, rather than the external environment. Approaches include medical interventions (e.g., isoprotenerol) or targeted manipulations, such as instructing individuals to hold their breath. Khalsa et al. (2009, 2016) and Smith et al. (2020) have used these approaches to demonstrate that anxious individuals show a blunted ability to adapt their cardiac perception between the baseline versus altered condition. Pang et al. (2019) found similar condition effects for the heartbeat-evoked potential (HEP), an electroencephalogram (EEG) measure of cardiac interoceptive processing, such that non-anxious individuals showed differences in HEP between eyes-open and eyes-closed conditions, whereas this condition effect was not present for highly anxious individuals. Newer generations of tasks have begun to expand the study of interoception beyond cardiac sensation, including new measures of respiration (Walsh et al. 2019; Harrison et al. 2021) and gastrointestinal perception (Khalsa et al. 2022). Initial findings from these respiration tasks have shown further links with anxiety in non-autistic individuals (Walsh et al. 2019; Harrison et al. 2021).

3.2 *Interoceptive Differences in Autism and Association with Anxiety*

Qualitative self-report data suggests that many autistic individuals are concerned with their interoception, providing statements such as "The best way I can describe this to health professionals is that I receive a signal from somewhere I'm not exactly sure, and I have difficulties interpreting what they might mean" or "I'm super sensitive to little changes in how my body feels" (Trevisan et al. 2021). Perceptual studies of interoception in autism have been mixed but provide some evidence for how interoceptive processing may differ in autism. Interoception studies in autism so far have typically used either heartbeat counting or heartbeat discrimination paradigms, reflecting a strong cardiac focus in the general interoception literature. Williams et al. (2022) meta-analysis of these studies found evidence for significantly reduced heartbeat counting, but not heartbeat discrimination accuracy, in autistic compared to non-autistic samples, though they note that the lack of correspondence between these two tasks warrants further exploration of these effects.

Building from the predictive coding framework of interoception, Garfinkel et al. (2016) first investigated difference scores between cardiac task-based interoceptive accuracy and self-reported body awareness in autistic adults and found that the extent of this mismatch (generally, higher reported body awareness relative to lower interoceptive accuracy) correlated with higher trait anxiety levels. Palser et al. (2018) replicated Garfinkel et al. (2016) findings in a youth sample, ages 6–18, using the same measures and approach. Preliminary intervention work building upon these findings suggests that interoceptive-focused interventions, including interoceptive awareness training, show promising effects on reducing anxiety in autistic individuals (Quadt et al. 2021). In Quadt et al. (2021) intervention, participants complete the two common cardiac interoception tasks (heartbeat counting and discrimination) with structured feedback, tied in with brief bouts of physical activity that serve to i) boost these signals and ii) provide an alternative condition in which participants may also practice tuning and adapting their interoceptive processing (e.g., as in Smith et al. 2020). Mindfulness approaches, which typically include interoceptive attention, have also been of increasing interest and promise toward reducing anxiety in autistic individuals (Hatfield et al. 2023). These studies suggest that blending sensory neuroscience approaches with such existing anxiety interventions may particularly serve to benefit autistic individuals. However, there is remaining work to integrate these findings into precision medicine-type approaches. In Quadt et al. (2021) work, they found that about a third of their participants responded optimally to their intervention, leaving open the question of whether there are specific subgroups of autistic participants that respond best to interoceptive interventions and relevant characteristics of responders compared to non-responders.

3.3 A Unified Theory: Linking Predictive Coding Across Exteroception and Interoception

The predictive coding framework of sensory processing may provide further cohesion of the mechanism for how internal and external sensory systems each contribute to anxiety in autism. Predictive coding theories of autism suggest that, across the various sensory systems, autistic individuals maintain less precise predictions of incoming sensory information than non-autistic individuals, altering the balance toward veridical, bottom-up sensory input which results in an increasingly detail-focused processing style (Pellicano and Burr 2012; Pellicano 2013). Under this model, autistic individuals may be especially accurate in detecting low-level sensory information, such as simple auditory or visual stimuli, but have difficulty with integrating and filtering their wealth of sensory information. This provides an alternative explanation that the "sensory overload" commonly experienced by autistic individuals may have as much to do with the amount and complexity of stimulation as does individual reactivity to a single stimulus (Pellicano 2013). In line with this theory, there is also a relatively robust literature that autistic individuals have difficulty integrating information between their different senses (Feldman et al. 2018).

Thus, altered predictive coding in both internal and external sensory systems may interact in a loop to generate and maintain anxious feelings. Experiencing the internal sensory environment within this framework, such that internal sensations are experienced as more volatile even when within homeostatic ranges, may prime diffuse feelings of uncertainty and stress in the external world. This may then cue hypervigilance, amplify external sensory over-responsiveness, and/or make it more difficult to regulate when external sensory aversion and uncertainty are experienced. These may then be maintained by intolerance of uncertainty and insistence on sameness behaviors (Wigham et al. 2015; South and Rodgers 2017; Normansell-Mossa et al. 2021). Related to predictive coding, active inference accounts suggest that we interact with our environments in a way to minimize prediction error signaling. Thus, sensory factors may particularly drive autistic individuals' strong preference for routines and sameness (South and Rodgers 2017). This then limits the type of naturalistic exposure that ameliorates anxiety in general samples. Thus, attention toward the many parts of this loop, including how interoceptive and exteroceptive information are integrated within an individual, may be fruitful for understanding autistic versus non-autistic experiences of anxiety.

Reciprocal relationships between interoceptive sensitivity, exteroceptive sensitivity, and anxiety may extend beyond diagnostic boundaries as well. Aron and Aron (1997) proposed the phenotype of "highly sensitive people," who experience high levels of sensory and socio-affective sensitivities. The clustering of these sensitivities within an individual is consistent with the close links between interoceptive and emotional brain systems, as also found in autistic samples (Mul et al. 2018). Acevedo et al. (2018) examined the highly sensitive phenotype across diagnostic bounds and found that this phenotype occurred within the different clinical conditions studied and was not exclusive to autism. Further, highly sensitive people tend to experience higher rates of anxiety, lending support for transdiagnostic sensory-anxiety effects that span interoceptive and exteroceptive systems (Liss et al. 2005).

3.4 Bridging Findings: Studying Interoception and Anxiety with Broader Tasks and Samples

As previously overviewed, there are several remaining areas in which to extend our understanding of interoception and anxiety in autistic and non-autistic individuals. One area is going beyond the most common cardiac tasks and including tasks that account for the integration between internal and external stimuli. The perception and processing of respiration has shown links with anxiety in non-autistic samples (Walsh et al. 2019; Harrison et al. 2021), and such findings may be readily translated into existing interventions. Specifically, a respiration integration task developed by Walsh et al. (2019) was designed to measure the interoceptive perceptual changes associated with a mindfulness intervention conducted by the study team and did indeed capture increased perceptual integration after completing the intervention, compared to baseline. Another area may be to continue to build our understanding of how interoceptive perception relates to different aspects of self-reported interoceptive difficulties, since reported interoceptive difficulties are likely multidimensional (Desmedt et al. 2022). Lastly, there has been very limited work including autistic and anxious non-autistic samples to understand interoception-anxiety links transdiagnostically. Thus, new work from our lab presented in the following sections investigates how interoceptive perception as measured by Walsh et al. (2019) respiration integration task relates to newer measures of reported interoceptive concerns and anxiety across a broad autistic and non-autistic sample.

4 New Findings: Respiration Integration in an Autistic and Non-autistic Sample

4.1 *Materials and Methods*

4.1.1 Participants

Our sample included $n = 43$ autistic and $n = 38$ non-autistic participants, ages 8–55. Study procedures were approved by the Vanderbilt Institutional Review Board, under protocol #191677. All individuals gave informed consent and/or assent for study participation. Autism diagnoses were confirmed using research-reliable administration of the Autism Diagnostic Observation Scale-2 (ADOS-2; Gotham et al. 2009) and judgment of a licensed clinical psychologist. Exclusion criteria for both groups included factors that could limit the completion and interpretation of perceptual tasks, e.g., atypical cardiac rhythms, an intelligence quotient (IQ) < 35, genetic, or neurologic conditions. See Table 1 for participant characteristics.

Table 1 Participant characteristics by group

Group	*n*	Gender % male	Age groups: C, ADO, ADU	IQ	ISQ	SCA (A)RED	ASA/ C-ASD
Non-autistic	38	39%	8, 3, 27	117 (94–137)	38.8 (20–116)	0.44 (0–1.39)	0.76 (0–2.41)
Autistic	43	65%	9, 9, 25	106 (60–143)	56.2 (20–93)	0.73 (0.07–1.54)	1.11 (0.08–2.5)

Note: Sample sizes (*n*) are summarized per group. Characteristics of the total sample per group are summarized as percent for categorical variables and mean (range) for continuous variables. Sample sizes per age group are divided into child (C, 8–12), adolescent (ADO, 13–17), and adult (ADU, 18+). Clinical cut-offs using rescaled scores are: SCA(A)RED: 0.52 Adults, 0.61 Youth; ASA/C-ASD: 1.4 Adults; 0.83 Youth. Abbreviations: IQ = full-scale intelligence quotient from the Weschler Abbreviated Scale of Intelligence (WASI-2; Wechsler 2011), ISQ = Interoception Sensory Questionnaire, ASA/C-ASD: Anxiety Scale for Adults(A)/Children(C) with ASD, SCA (A)RED: Screen for Child/Adult(A) Anxiety Related Disorders

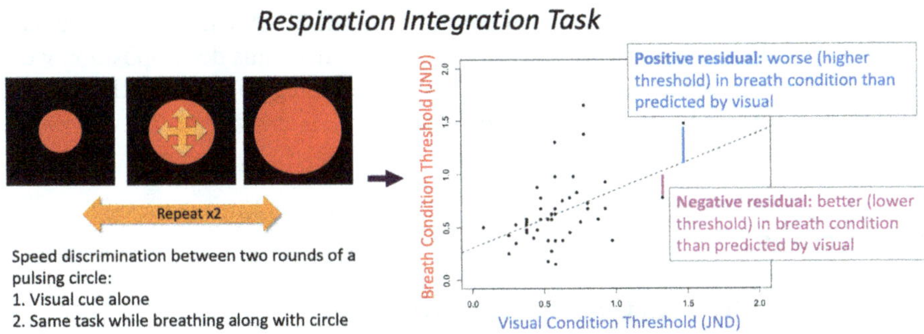

Fig. 1 Overview of respiration integration task. The experimental design (left) and analytic parameters (right) of Walsh et al. (2019) respiration integration task are shown. In this task, participants complete two rounds of a visual speed discrimination task, with and without matched breathing. This task results in a respiratory integration score reflecting the degree to which participants improved their visual speed judgments (lower scores) when using matched breathing

4.1.2 Questionnaire Measures

Participants completed self- and/or parent-report surveys to characterize their reported confusion about interoceptive sensations and anxiety levels, specifically the Interoception Sensory Questionnaire (ISQ, Fiene et al. 2018), the Anxiety Scale for Adults(A)/Children(C) with ASD (ASA/C-ASD; Rodgers et al. 2016; 2020), and the Screen for Child/Adult(A) Anxiety Related Disorders (SCA(A)RED; Birmaher et al. 1997; Angulo et al. 2017). ASC/A-ASD and SCA(A)RED scores have been rescaled by the total number of items.

4.1.3 Respiration Integration Task (RIT)

This task was developed by Walsh et al. (2019). Participants are asked to indicate whether the target circle is expanding/contracting faster or slower than a reference circle. A staircase procedure is used to converge on a speed discrimination threshold, first using vision alone. Once this threshold is established, participants practice matching their breathing to the movement of the circle. Participants then repeat the task while continuing this matched breathing. A respiratory integration score is calculated as the residual value from a linear model to predict breath condition thresholds from visual alone thresholds. See Fig. 1 for task overview.

4.1.4 Analytic Approach

We studied the shared versus distinct variability captured by our questionnaire measures by completing a principal components decomposition on our three parameters of interest: (i) ASA/C-ASD total score, (ii) SC(A)RED total score, and (iii) ISQ total. Individuals' scores on the components from this decomposition were then analyzed in a linear model including respiration integration task (RIT) scores, diagnostic group, and a group by RIT interaction.

4.2 Results

4.2.1 Patterns Between Questionnaire Measures

The principal components derived from these measures suggest a primary component of shared anxiety symptoms and interoceptive confusion, since about 84% of individual variation was explained by a first component loading strongly with all three measures (ASA/C-ASD: 0.60; SCA(A)RED: 0.60; and ISQ: 0.53). Reflecting a much smaller degree of total variation, a second component reflected interoceptive confusion that did not correspond with higher anxiety levels (ISQ loading: 0.84), explaining 13% of individual variation. A last, very small component (3% variance) reflected diverging anxiety scores on autistic (ASA/C-ASD: 0.69 loading) versus non-autistic (SCA(A)RED loading: −0.71) measures.

4.2.2 Relationships Between Anxiety, Interoceptive Confusion, and Respiration Integration

Group was a significant predictor of the first combined anxiety symptoms/interoceptive confusion component, with lower of these combined scores in the non-autistic group ($t = -3.59, p < 0.001, \mathrm{BF}_{\mathrm{inclusion}} = 39.34$). However, the pattern between respiration integration and the anxiety symptoms/interoceptive confusion score was neither clearly the same nor clearly different between groups (Group x RIT interaction term: $t = 1.53, p = 0.13, \mathrm{BF}_{\mathrm{inclusion}} = 0.75$). On the other hand, the second principal component as related to specific interoceptive confusion, distinct from anxiety, showed a strong and consistent relationship to RIT performance, showing a positive relationship such that individuals with lower respiratory integration (and thus higher task scores) reported more interoceptive confusion ($t = 2.55$, $p = 0.01, \mathrm{BF}_{\mathrm{inclusion}} = 29.12$). This relationship was consistent between groups (See Fig. 2). In supplemental analyses, neither gender ($t = 0.57, p = 0.57$) nor IQ ($t = -0.19, p = 0.85$) was significantly related to RIT task scores but increased age related to lower RIT scores ($t = -2.56, p = 0.01$).

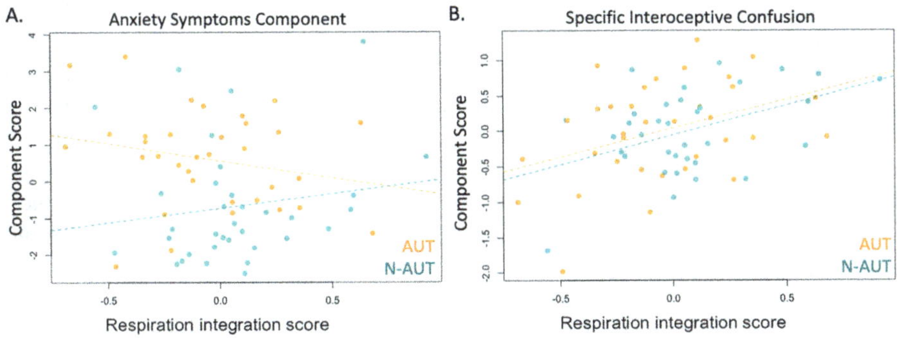

Fig. 2 Plot of anxiety symptoms component (**a**) and specific interoceptive confusion component (**b**) with respiration integration scores by group. For both panels, the *x*-axis reflects individual scores on the respiration integration task. Positive scores = less integration and negative scores = more integration. (**a**) The *y*-axis shows the first principal component scores; higher scores reflect higher levels of anxiety symptoms combined with higher interoceptive confusion (Anxiety Symptoms Component). (**b**) The *y*-axis reflects the second principal component scores; higher scores reflect higher interoceptive confusion but not necessarily higher anxiety (Specific Interoceptive Confusion). Colors: orange = autistic and teal = non-autistic group

5 Discussion

5.1 Interoceptive Confusion and Anxiety: Composite Findings

When composite patterns across our three anxiety and interoceptive questionnaires were considered, all three measures (autism-focused anxiety, traditional anxiety, and interoceptive confusion) corresponded strongly to a general anxiety symptoms factor. This provides some initial context for how to interpret the more recently developed ISQ scale. Interestingly, this scale was developed with autistic populations in mind, to capture the types of interoceptive confusion in autism as found in recent literature (Fiene et al. 2018; Trevisan et al. 2021). However, we found that ISQ scores strongly related to total anxiety symptoms in both our autistic and non-autistic groups. At least at the self-report level, this provides evidence consistent with theories linking anxiety and interoception. This also further highlights the pervasiveness of anxiety in autistic experiences and difficulties in disentangling the effects of autism versus anxiety, given that a scale designed to detect interoceptive sensory patterns in autism is so highly linked to anxiety levels.

5.2 Relationships Between Perceptual Tasks, Self-Reported Anxiety, and Interoceptive Confusion

Examining the second principal component that reflects specific interoceptive confusion, which may be less anxiety-linked, we found an expected relationship such that individuals who showed lower respiration integration effects also reported more confusion about their bodily sensations. This finding was very consistent across groups. On the other hand, the relationship between respiration integration and total anxiety symptoms showed a negative trend for autistic individuals and positive trend for non-autistic individuals. In non-autistic individuals, this relationship went in a direction that is more intuitive for understanding the close links between anxiety and interoceptive confusion: that individuals who showed less ability to integrate their breathing with a visual stimulus also reported higher levels of both anxiety and interoceptive confusion.

In the autistic group, the negative relation between integration scores and anxiety-related interoceptive confusion is reminiscent of findings from Simon and Corbett (2013) who found that autistic children's self-reported anxiety levels corresponded substantially less with their measured cortisol than for their non-autistic peers. This may represent a distinction between low-level sensory integration and high-level contextual interpretation of how interoceptive stimuli relate to bodily needs, which prior studies suggest diverges in autism (Garfinkel et al. 2016; Palser et al. 2018). Aligning such discrepancies within a predictive coding framework, interoception may be another example in which autistic individuals may be perceiving the "trees" at the expense of the "forest" (Pellicano and Burr 2012), e.g., heightened attention toward moment-to-moment sensations may come at the expense of an ability to integrate and adapt to these sensations. Paulus and Stein (2006) provide an initial framework for how enhanced cardioception in anxious individuals may impede rather than aid the effective interpretation of bodily needs. Our results suggest that this effect may occur for respiration as well, for at least some individuals. However, the results do not suggest a statistically significant interaction effect between the two groups. Thus, it is difficult to conclude based on our single experimental task (respiration integration) whether the relationship between interoceptive perception and anxiety symptoms is the same or different between autistic and non-autistic groups. Our current results, of neither a clear across-group correlation or clear interaction effects, suggest this may not be a single or direct relationship.

Still, the identified relationships between interoceptive perception, reported interoceptive confusion, and anxiety are consistent with prior accounts that interoceptive interventions may be especially helpful for certain autistic individuals (Quadt et al. 2021). However, they also lend further context in defining the goal of these interventions. These findings suggest two potential ways in which interoceptive perception may link to interoceptive confusion and considering anxiety levels may help distinguish between the two. For interoceptive confusion that is reported to be much higher than expected from anxiety levels, interoceptive interventions may have a straightforward goal of increasing attention and accurate perception of these cues. For individuals whose interoceptive confusion is in proportion to anxiety levels, it may be especially helpful for autistic individuals to receive support in interpreting the bigger picture from their wealth of internal signals.

6 Areas of Future Research

6.1 *Age, Sex, and Gender Effects*

One aspect of understanding relationships between interoception and anxiety that may be expanded upon in future work is how to better measure and interpret these patterns across age and development. Palser et al. (2018) and Garfinkel et al. (2016) present complementary evidence that suggests consistency in how similar types of predictive-coding-related interoceptive difficulties correlate with anxiety in both children and adults. However, there may still be interesting ways in which these patterns develop over time; further, there may be different types of measures that best capture anxiety-related interoceptive difficulties at different ages. Though age was not a focus of our presented results, we did find increased respiration integration effects in older versus younger individuals. One idea for future studies to consider is whether the absolute measure of interoceptive processing ability versus its deviation from an expected developmental trajectory is a better predictor of anxiety. Additionally, there are substantial differences in how anxiety scores from paired youth and adult measures may be interpreted at different ages. This is notable in the differences in clinical concern cut-offs, as cut-offs for the rescaled scores we used are higher in adults than children (Birmaher et al. 1997; Rodgers et al. 2016, 2020; Angulo et al. 2017). Further, the interpretation of self-reported ratings in children may be different than in adults as children continue to develop internal self-awareness and may show differences with age in types and expression of anxiety (e.g., Nauta et al. 2004). There is room to further explore these issues both at the measurement level and how these measures relate to the underlying construct, i.e., how anxiety and its presentation may develop over time. Thus, a detailed investigation of developmental trajectories of interoceptive processing relative to anxiety presentation would provide meaningful context for the interpretation of interoception-anxiety correlations at different ages. Since the work considering age thus far has generally done so in cross-sectional samples, longitudinal designs are also needed to test these hypotheses.

In addition, there may be meaningful sex and gender differences to explore in further understanding the relations between interoception and anxiety. In the broader literature, there is evidence that sex and gender do each contribute to experiences of anxiety (Altemus et al. 2014) and autism (McQuaid et al. 2022), but the literature has been less conclusive on the extent to which sex or gender differences in interoception exist. The differences that do exist seem to be specific to interoceptive sub-modalities and plausibly, though not conclusively, linked to sex differences in relevant physiology, e.g., cardiac interoception relating to body mass around the heart (Prentice and Murphy 2022). Thus, it will be interesting for future studies to consider whether specific aspects of the relationships between autism, interoception, and anxiety further interact with sex or gender effects.

6.2 A Multifaceted System

Another direction for future studies to take are designs that include additional interoceptive measures, since this system is multifaceted and findings may be distinct across interoceptive sub-modalities, e.g., cardioception, respiration, gastro-intestinal perception, etc. (Crucianelli et al. 2022). Other ways to further pinpoint the specificity of relationships between domains of interoception and anxiety are to further explore anxiety sub-types, potentially using the subscales of the measures described here (SC(A)RED and ASA/C-ASD), given that prior studies have found distinctive patterns for how sensory responsiveness may relate to panic disorder versus social anxiety (MacLennan et al. 2020). Such patterns are likely to exist for interoception as this sense may differentially influence processing in situations that are more internal (such as panic) versus external (such as social). Lastly, there is much further room to explore the heterogeneity within autism. We found that the type of paired interoceptive versus exteroceptive condition design used in the respiration integration task did not seem to show a meaningful association with IQ, compared to what has been found in common cardioception tasks (Williams et al. 2022). However, individual variation in the social and repetitive behavior domains as well as how individual sensory differences may relate between interoception and exteroception are all important areas to explore, particularly in even larger samples that may have enhanced room to explore this variability.

6.3 Advanced Approaches: The Neural Link

The increasingly feasible use of ultra-high-resolution MRI also lends promise toward the future of understanding the mechanistic basis of sensory-anxiety relationships. Advanced computational approaches applied to behavioral paradigms have lent support to predictive coding theories of interoception and anxiety. However, the corresponding brain bases have been difficult to test using the resolution of typical neuroimaging approaches used in human research, in which feedforward and feedback information flow are difficult to distinguish. Harrison et al. (2021) provide initial evidence of how combined Bayesian statistical approaches with ultra-high-resolution MRI imaging may be used to precisely localize the basis of anxiety-related processing differences, combining these tools during a manipulated respiration task to pinpoint the exact location within the insula in which Bayesian processing parameters were best reflected. The ability to resolve layers may be expanded in other ways, given the known patterns of information flow within the laminar gradient of the insula (Barrett and Simmons 2015). Thus, we may be able to build a more detailed understanding of how interoceptive system architecture contributes to anxiety and, even in cases where anxious autistic and non-autistic individuals perform similarly on behavioral measures, whether the mechanism beneath these phenotypes varies by diagnostic group.

7 Conclusions

We have reviewed growing literature about how sensory perception, including a particular focus on interoceptive perception and its integration with external information, relates to anxiety in autistic and non-autistic individuals. We provided new findings that suggest that there may be two components related to interoceptive confusion: one that is highly correlated with anxiety symptoms and a smaller component that is distinct from anxiety symptoms, the latter of which shows a clear, transdiagnostic relationship with interoceptive perceptual integration. These results suggest that interoceptive confusion is indeed related to anxiety, but may have a complex perceptual basis, warranting further investigation at different points of the interoceptive processing stream and of differing interoceptive sub-modalities. Other future directions of research include the exploration of individual heterogeneity, including individual differences that may emerge based on sex, gender, language levels, or differential autistic features. Ultra-high-resolution MRI may also provide new opportunities to explore the relevant information flow with sensory circuits that may contribute to anxiety levels. Overall, this work highlights how sensory processing is an important facet to consider in both autistic and non-autistic experiences of anxiety.

Declaration of interest None.

Acknowledgments This work was supported by the National Institute of Mental Health (R01MH102272 to C.J.C.).

References

Acevedo B, Aron E, Pospos S, Jessen D (2018) The functional highly sensitive brain: a review of the brain circuits underlying sensory processing sensitivity and seemingly related disorders. Philos Trans R Soc B 373:20170161. https://doi.org/10.1098/rstb.2017.0161

Ainley V, Apps MAJ, Fotopoulou A, Tsakiris M (2016) 'Bodily precision': a predictive coding account of individual differences in interoceptive accuracy. Philos Trans R Soc B Biol Sci 371: 20160003. https://doi.org/10.1098/rstb.2016.0003

Altemus M, Sarvaiya N, Neill Epperson C (2014) Sex differences in anxiety and depression clinical perspectives. Front Neuroendocrinol 35:320–330. https://doi.org/10.1016/j.yfrne.2014.05.004

American Psychiatric Association (2013) Diagnostic and statistical manual of mental disorders, 5th edn. American Psychiatric Association, Arlington

Andrews DS, Aksman L, Kerns CM et al (2022) Association of Amygdala development with different forms of anxiety in autism spectrum disorder. Biol Psychiatry 91:977–987. https://doi.org/10.1016/j.biopsych.2022.01.016

Angulo M, Rooks BT, Gill M et al (2017) Psychometrics of the screen for adult anxiety related disorders (SCAARED) – a new scale for the assessment of DSM-5 anxiety disorders. Psychiatry Res 253:84–90. https://doi.org/10.1016/j.psychres.2017.02.034

Aron EN, Aron A (1997) Sensory-processing sensitivity and its relation to introversion and emotionality. J Pers Soc Psychol 73:345–368. https://doi.org/10.1037/0022-3514.73.2.345

Ausderau KK, Sideris J, Little LM et al (2016) Sensory subtypes and associated outcomes in children with autism spectrum disorders: sensory subtypes and outcomes in children with ASD. Autism Res 9:1316–1327. https://doi.org/10.1002/aur.1626

Bang P, Igelström K (2023) Modality-specific associations between sensory differences and autistic traits. Autism 27:2158–2172. https://doi.org/10.1177/13623613231154349

Bang P, Andemichael DK, Pieslinger JF, Igelström K (2024) Sensory symptoms associated with autistic traits and anxiety levels in children aged 6–11 years. J Neurodevelop Disord 16:45. https://doi.org/10.1186/s11689-024-09562-9

Baranek GT, David FJ, Poe MD et al (2006) Sensory experiences questionnaire: discriminating sensory features in young children with autism, developmental delays, and typical development: SEQ. J Child Psychol Psychiatry 47:591–601. https://doi.org/10.1111/j.1469-7610.2005.01546.x

Baranek GT, Sideris J, Chen Y et al (2022) Early measurement of autism risk constructs in the general population: a new factor structure of the first years inventory (FYIv3.1) for ages 6–16 months. Autism Res 15:915–928. https://doi.org/10.1002/aur.2691

Barrett LF, Simmons WK (2015) Interoceptive predictions in the brain. Nat Rev Neurosci 16:419–429. https://doi.org/10.1038/nrn3950

Bellini S (2006) The development of social anxiety in adolescents with autism spectrum disorders. Focus Autism Other Dev Disabl 21:138–145. https://doi.org/10.1177/10883576060210030201

Birmaher B, Khetarpal S, Brent D et al (1997) The screen for child anxiety related emotional disorders (SCARED): scale construction and psychometric characteristics. J Am Acad Child Adolesc Psychiatry 36:545–553. https://doi.org/10.1097/00004583-199704000-00018

Bottema-Beutel K, Kapp SK, Lester JN et al (2021) Avoiding Ableist Language: Suggestions for Autism Researchers. Autism in Adulthood 3:18–29. https://doi.org/10.1089/aut.2020.0014

Boyd BA, Baranek GT, Sideris J et al (2010) Sensory features and repetitive behaviors in children with autism and developmental delays. Autism Res. https://doi.org/10.1002/aur.124

Brener J, Ring C (2016) Towards a psychophysics of interoceptive processes: the measurement of heartbeat detection. Philosophical Transactions of the Royal Society B: Biological Sciences 371:20160015. https://doi.org/10.1098/rstb.2016.0015

Carpenter KLH, Baranek GT, Copeland WE et al (2019) Sensory over-responsivity: an early risk factor for anxiety and behavioral challenges in young children. J Abnorm Child Psychol 47: 1075–1088. https://doi.org/10.1007/s10802-018-0502-y

Chen Y, Sideris J, Watson LR et al (2022) Developmental trajectories of sensory patterns from infancy to school age in a community sample and associations with autistic traits. Child Dev 93. https://doi.org/10.1111/cdev.13745

Corbett BA, Muscatello RA, Blain SD (2016) Impact of sensory sensitivity on physiological stress response and novel peer interaction in children with and without autism Spectrum disorder. Front Neurosci 10. https://doi.org/10.3389/fnins.2016.00278

Craig AD (2002) How do you feel? Interoception: the sense of the physiological condition of the body. Nat Rev Neurosci 3:655–666. https://doi.org/10.1038/nrn894

Crucianelli L, Enmalm A, Ehrsson HH (2022) Interoception as independent cardiac, thermosensory, nociceptive, and affective touch perceptual submodalities. Biol Psychol 172: 108355. https://doi.org/10.1016/j.biopsycho.2022.108355

Cummings KK, Jung J, Zbozinek TD et al (2024) Shared and distinct biological mechanisms for anxiety and sensory over-responsivity in youth with autism versus anxiety disorders. J Neurosci Res 102:e25250. https://doi.org/10.1002/jnr.25250

De Bruin EI, Ferdinand RF, Meester S et al (2007) High rates of psychiatric co-morbidity in PDD-NOS. J Autism Dev Disord 37:877–886. https://doi.org/10.1007/s10803-006-0215-x

Dellapiazza F, Michelon C, Picot M-C, Baghdadli A (2022) Early risk factors for anxiety disorders in children with autism spectrum disorders: results from the ELENA cohort. Sci Rep 12:10914. https://doi.org/10.1038/s41598-022-15165-y

Desmedt O, Luminet O, Corneille O (2018) The heartbeat counting task largely involves non-interoceptive processes: evidence from both the original and an adapted counting task. Biol Psychol 138:185–188. https://doi.org/10.1016/j.biopsycho.2018.09.004

Desmedt O, Heeren A, Corneille O, Luminet O (2022) What do measures of self-report interoception measure? Insights from a systematic review, latent factor analysis, and network approach. Biol Psychol 169:108289. https://doi.org/10.1016/j.biopsycho.2022.108289

Domschke K, Stevens S, Pfleiderer B, Gerlach AL (2010) Interoceptive sensitivity in anxiety and anxiety disorders: an overview and integration of neurobiological findings. Clin Psychol Rev 30:1–11. https://doi.org/10.1016/j.cpr.2009.08.008

Dwyer P, Saron CD, Rivera SM (2020) Identification of longitudinal sensory subtypes in typical development and autism spectrum development using growth mixture modelling. Res Autism Spectr Disord 78:101645. https://doi.org/10.1016/j.rasd.2020.101645

Feldman JI, Dunham K, Cassidy M et al (2018) Audiovisual multisensory integration in individuals with autism spectrum disorder: a systematic review and meta-analysis. Neurosci Biobehav Rev 95:220–234. https://doi.org/10.1016/j.neubiorev.2018.09.020

Fiene L, Ireland MJ, Brownlow C (2018) The interoception sensory questionnaire (ISQ): a scale to measure interoceptive challenges in adults. J Autism Dev Disord 48:3354–3366. https://doi.org/10.1007/s10803-018-3600-3

Garfinkel SN, Tiley C, O'Keeffe S et al (2016) Discrepancies between dimensions of interoception in autism: implications for emotion and anxiety. Biol Psychol 114:117–126. https://doi.org/10.1016/j.biopsycho.2015.12.003

Gotham K, Pickles A, Lord C (2009) Standardizing ADOS scores for a measure of severity in autism spectrum disorders. J Autism Dev Disord 39:693–705. https://doi.org/10.1007/s10803-008-0674-3

Green SA, Ben-Sasson A (2010) Anxiety disorders and sensory over-responsivity in children with autism spectrum disorders: is there a causal relationship? J Autism Dev Disord 40:1495–1504. https://doi.org/10.1007/s10803-010-1007-x

Green SA, Ben-Sasson A, Soto TW, Carter AS (2012) Anxiety and sensory over-responsivity in toddlers with autism spectrum disorders: bidirectional effects across time. J Autism Dev Disord 42:1112–1119. https://doi.org/10.1007/s10803-011-1361-3

Green SA, Hernandez L, Tottenham N et al (2015) Neurobiology of sensory overresponsivity in youth with autism spectrum disorders. JAMA Psychiatry 72:778. https://doi.org/10.1001/jamapsychiatry.2015.0737

Green SA, Hernandez LM, Bowman HC et al (2017) Sensory over-responsivity and social cognition in ASD: effects of aversive sensory stimuli and attentional modulation on neural responses to social cues. Dev Cogn Neurosci. https://doi.org/10.1016/j.dcn.2017.02.005

Green SA, Hernandez LM, Bowman HC et al (2018) Sensory over-responsivity and social cognition in ASD: effects of aversive sensory stimuli and attentional modulation on neural responses to social cues. Dev Cogn Neurosci 29:127–139. https://doi.org/10.1016/j.dcn.2017.02.005

Hallett V, Lecavalier L, Sukhodolsky DG et al (2013) Exploring the manifestations of anxiety in children with autism spectrum disorders. J Autism Dev Disord 43:2341–2352. https://doi.org/10.1007/s10803-013-1775-1

Harrison OK, Köchli L, Marino S et al (2021) Interoception of breathing and its relationship with anxiety. Neuron 109:4080–4093.e8. https://doi.org/10.1016/j.neuron.2021.09.045

Hatfield MK, Ashcroft E, Maguire S et al (2023) "Stop and just breathe for a minute": perspectives of children on the autism spectrum and their caregivers on a mindfulness group. J Autism Dev Disord 53:3394–3405. https://doi.org/10.1007/s10803-022-05542-x

Herrington JD, Maddox BB, Kerns CM et al (2017) Amygdala volume differences in autism spectrum disorder are related to anxiety. J Autism Dev Disord 47:3682–3691. https://doi.org/10.1007/s10803-017-3206-1

Hsueh B, Chen R, Jo Y et al (2023) Cardiogenic control of affective behavioural state. Nature. https://doi.org/10.1038/s41586-023-05748-8

Huang Y, Rao RPN (2011) Predictive coding. WIREs Cogn Sci 2:580–593. https://doi.org/10.1002/wcs.142

Isaacs D, Key AP, Cascio CJ et al (2022) Cross-disorder comparison of sensory over-responsivity in chronic tic disorders and obsessive-compulsive disorder. Compr Psychiatry 113:152291. https://doi.org/10.1016/j.comppsych.2021.152291

Kent R, Simonoff E (2017) Prevalence of anxiety in autism spectrum disorders. In: Anxiety in children and adolescents with autism spectrum disorder. Elsevier, pp 5–32

Kerns CM, Kendall PC (2012) The presentation and classification of anxiety in autism spectrum disorder. Clin Psychol Sci Pract 19:323–347. https://doi.org/10.1111/cpsp.12009

Kerns CM, Kendall PC, Berry L et al (2014) Traditional and atypical presentations of anxiety in youth with autism spectrum disorder. J Autism Dev Disord 44:2851–2861. https://doi.org/10.1007/s10803-014-2141-7

Khalsa SS, Rudrauf D, Sandesara C et al (2009) Bolus isoproterenol infusions provide a reliable method for assessing interoceptive awareness. Int J Psychophysiol 72:34–45. https://doi.org/10.1016/j.ijpsycho.2008.08.010

Khalsa SS, Feinstein JS, Li W et al (2016) Panic anxiety in humans with bilateral amygdala lesions: pharmacological induction via cardiorespiratory interoceptive pathways. J Neurosci 36:3559–3566. https://doi.org/10.1523/JNEUROSCI.4109-15.2016

Khalsa SS, Berner LA, Anderson LM (2022) Gastrointestinal interoception in eating disorders: charting a new path. Curr Psychiatry Rep 24:47–60. https://doi.org/10.1007/s11920-022-01318-3

Kirby AV, Little LM, Schultz B, Baranek GT (2015) Observational characterization of sensory interests, repetitions, and seeking behaviors. Am J Occup Ther 69: 6903220010p1–6903220010p9. https://doi.org/10.5014/ajot.2015.015081

Lane SJ, Reynolds S, Dumenci L (2012) Sensory overresponsivity and anxiety in typically developing children and children with autism and attention deficit hyperactivity disorder: cause or coexistence? Am J Occup Ther 66:595–603. https://doi.org/10.5014/ajot.2012.004523

Lecavalier L, Wood JJ, Halladay AK et al (2014) Measuring anxiety as a treatment endpoint in youth with autism spectrum disorder. J Autism Dev Disord 44:1128–1143. https://doi.org/10.1007/s10803-013-1974-9

Lee H, Chen Y-J, Sideris J et al (2022) Sensory features of young children from a large community sample: latent factor structures of the sensory experiences questionnaire (version 2.1, short form). Am J Occup Ther 76:7603205140. https://doi.org/10.5014/ajot.2022.046995

Liss M, Timmel L, Baxley K, Killingsworth P (2005) Sensory processing sensitivity and its relation to parental bonding, anxiety, and depression. Personal Individ Differ 39:1429–1439. https://doi.org/10.1016/j.paid.2005.05.007

MacLennan K, Roach L, Tavassoli T (2020) The relationship between sensory reactivity differences and anxiety subtypes in autistic children. Autism Res 13:785–795. https://doi.org/10.1002/aur.2259

Maddox BB, Kang-Yi CD, Brodkin ES, Mandell DS (2018) Treatment utilization by adults with autism and co-occurring anxiety or depression. Res Autism Spectr Disord 51:32–37. https://doi.org/10.1016/j.rasd.2018.03.009

McQuaid GA, Lee NR, Wallace GL (2022) Camouflaging in autism spectrum disorder: examining the roles of sex, gender identity, and diagnostic timing. Autism 26:552–559. https://doi.org/10.1177/13623613211042131

Merikangas KR, He J, Burstein M et al (2010) Lifetime prevalence of mental disorders in U.-S. adolescents: results from the national comorbidity survey replication–adolescent supplement (NCS-A). J Am Acad Child Adolesc Psychiatry 49:980–989. https://doi.org/10.1016/j.jaac.2010.05.017

Minshew NJ, Hobson JA (2008) Sensory sensitivities and performance on sensory perceptual tasks in high-functioning individuals with Autism. Journal of Autism and Developmental Disorders 38:1485–1498. https://doi.org/10.1007/s10803-007-0528-4

Mul C, Stagg SD, Herbelin B, Aspell JE (2018) The feeling of me feeling for you: interoception, alexithymia and empathy in autism. J Autism Dev Disord 48:2953–2967. https://doi.org/10.1007/s10803-018-3564-3

Nauta MH, Scholing A, Rapee RM et al (2004) A parent-report measure of children's anxiety: psychometric properties and comparison with child-report in a clinic and normal sample. Behav Res Ther 42:813–839. https://doi.org/10.1016/S0005-7967(03)00200-6

Normansell-Mossa KM, Top DN, Russell N et al (2021) Sensory sensitivity and intolerance of uncertainty influence anxiety in autistic adults. Front Psychol 12:731753. https://doi.org/10.3389/fpsyg.2021.731753

Ozsivadjian A, Knott F, Magiati I (2012) Parent and child perspectives on the nature of anxiety in children and young people with autism spectrum disorders: a focus group study. Autism 16:107–121. https://doi.org/10.1177/1362361311431703

Pallathra AA, Calkins ME, Parish-Morris J et al (2018) Defining behavioral components of social functioning in adults with autism spectrum disorder as targets for treatment. Autism Res 11:488–502. https://doi.org/10.1002/aur.1910

Palser ER, Fotopoulou A, Pellicano E, Kilner JM (2018) The link between interoceptive processing and anxiety in children diagnosed with autism spectrum disorder: extending adult findings into a developmental sample. Biol Psychol 136:13–21. https://doi.org/10.1016/j.biopsycho.2018.05.003

Pang J, Tang X, Li H et al (2019) Altered interoceptive processing in generalized anxiety disorder – a heartbeat-evoked potential research. Front Psych 10:616. https://doi.org/10.3389/fpsyt.2019.00616

Paulus MP, Stein MB (2006) An insular view of anxiety. Biol Psychiatry 60:383–387. https://doi.org/10.1016/j.biopsych.2006.03.042

Pellicano E (2013) Sensory symptoms in autism: a blooming, buzzing confusion? Child Dev Perspect 7:143–148. https://doi.org/10.1111/cdep.12031

Pellicano E, Burr D (2012) When the world becomes 'too real': a Bayesian explanation of autistic perception. Trends Cogn Sci 16:504–510. https://doi.org/10.1016/j.tics.2012.08.009

Prentice F, Murphy J (2022) Sex differences in interoceptive accuracy: a meta-analysis. Neurosci Biobehav Rev 132:497–518. https://doi.org/10.1016/j.neubiorev.2021.11.030

Quadt L, Garfinkel SN, Mulcahy JS et al (2021) Interoceptive training to target anxiety in autistic adults (ADIE): a single-center, superiority randomized controlled trial. EClinicalMedicine 39:101042. https://doi.org/10.1016/j.eclinm.2021.101042

Rodgers J, Wigham S, McConachie H et al (2016) Development of the anxiety scale for children with autism spectrum disorder (ASC-ASD): measuring anxiety in ASD. Autism Res 9:1205–1215. https://doi.org/10.1002/aur.1603

Rodgers J, Farquhar K, Mason D et al (2020) Development and initial evaluation of the anxiety scale for autism-adults. Autism Adulthood 2:24–33. https://doi.org/10.1089/aut.2019.0044

Scahill L, Lecavalier L, Schultz RT et al (2019) Development of the parent-rated anxiety scale for youth with autism spectrum disorder. J Am Acad Child Adolesc Psychiatry 58:887–896.e2. https://doi.org/10.1016/j.jaac.2018.10.016

Schandry R (1981) Heart beat perception and emotional experience. Psychophysiology 18:483–488. https://doi.org/10.1111/j.1469-8986.1981.tb02486.x

Simon DM, Corbett BA (2013) Examining associations between anxiety and cortisol in high functioning male children with autism. Journal of Neurodevelopmental Disorders 5:32

Smith R, Kuplicki R, Feinstein J et al (2020) A Bayesian computational model reveals a failure to adapt interoceptive precision estimates across depression, anxiety, eating, and substance use disorders. PLoS Comput Biol 16:e1008484. https://doi.org/10.1371/journal.pcbi.1008484

South M, Rodgers J (2017) Sensory, emotional and cognitive contributions to anxiety in autism spectrum disorders. Front Hum Neurosci 11. https://doi.org/10.3389/fnhum.2017.00020

Trevisan DA, Parker T, McPartland JC (2021) First-hand accounts of interoceptive difficulties in autistic adults. J Autism Dev Disord 51:3483–3491. https://doi.org/10.1007/s10803-020-04811-x

Uljarević M, Lane A, Kelly A, Leekam S (2016) Sensory subtypes and anxiety in older children and adolescents with autism spectrum disorder. Autism Res 9:1073–1078. https://doi.org/10.1002/aur.1602

Van Den Boogert F, Klein K, Spaan P et al (2022) Sensory processing difficulties in psychiatric disorders: a meta-analysis. J Psychiatr Res 151:173–180. https://doi.org/10.1016/j.jpsychires.2022.04.020

Vasa RA, Mazurek MO (2015) An update on anxiety in youth with autism spectrum disorders. Curr Opin Psychiatry 28:83–90. https://doi.org/10.1097/YCO.0000000000000133

Vasa RA, Carroll LM, Nozzolillo AA et al (2014) A systematic review of treatments for anxiety in youth with autism spectrum disorders. J Autism Dev Disord 44:3215–3229. https://doi.org/10.1007/s10803-014-2184-9

Verhulst I, MacLennan K, Haffey A, Tavassoli T (2022) The perceived causal relations between sensory reactivity differences and anxiety symptoms in autistic adults. Autism Adulthood 4:183–192. https://doi.org/10.1089/aut.2022.0018

Walsh KM, Saab BJ, Farb NA (2019) Effects of a mindfulness meditation app on subjective well-being: active randomized controlled trial and experience sampling study. JMIR Mental Health 6:e10844. https://doi.org/10.2196/10844

Walters S, Loades M, Russell A (2016) A systematic review of effective modifications to cognitive behavioural therapy for young people with autism spectrum disorders. Rev J Autism Dev Disord 3:137–153. https://doi.org/10.1007/s40489-016-0072-2

Wechsler D (2011) Wechsler abbreviated scales of intelligence (WASI-II), 2nd edn. Pearson, Toronto

White SW, Oswald D, Ollendick T, Scahill L (2009) Anxiety in children and adolescents with autism spectrum disorders. Clin Psychol Rev 29:216–229. https://doi.org/10.1016/j.cpr.2009.01.003

White SW, Mazefsky CA, Dichter GS et al (2014) Social-cognitive, physiological, and neural mechanisms underlying emotion regulation impairments: understanding anxiety in autism spectrum disorder. Int J Dev Neurosci 39:22–36. https://doi.org/10.1016/j.ijdevneu.2014.05.012

Wigham S, Rodgers J, South M et al (2015) The interplay between sensory processing abnormalities, intolerance of uncertainty, anxiety and restricted and repetitive Behaviours in autism spectrum disorder. J Autism Dev Disord 45:943–952. https://doi.org/10.1007/s10803-014-2248-x

Williams KL, Kirby AV, Watson LR et al (2018) Sensory features as predictors of adaptive behaviors: a comparative longitudinal study of children with autism spectrum disorder and other developmental disabilities. Res Dev Disabil 81:103–112. https://doi.org/10.1016/j.ridd.2018.07.002

Williams ZJ, He JL, Cascio CJ, Woynaroski TG (2021) A review of decreased sound tolerance in autism: definitions, phenomenology, and potential mechanisms. Neurosci Biobehav Rev 121:1–17. https://doi.org/10.1016/j.neubiorev.2020.11.030

Williams ZJ, Suzman E, Bordman SL et al (2022) Characterizing interoceptive differences in autism: a systematic review and meta-analysis of case–control studies. J Autism Dev Disord. https://doi.org/10.1007/s10803-022-05656-2

Fear and Anxiety in Schizophrenia: A Focus on Development, Assessment, and Mechanisms

Brandee Feola, Marren Jenkins, Julia M. Sheffield, and Jennifer Urbano Blackford

Contents

Abstract In people with schizophrenia, anxiety is highly prevalent and related to numerous negative outcomes; unfortunately, anxiety is both underreported and understudied in schizophrenia. The current review highlights the importance and utility of assessing anxiety in schizophrenia by addressing four main questions: (1) What does anxiety look like throughout the development of schizophrenia?; (2) How do we measure anxiety in schizophrenia?; (3) What are the mechanisms underlying anxiety in schizophrenia; (4) How do we treat anxiety in schizophrenia?

B. Feola (✉), M. Jenkins, and J. M. Sheffield
Department of Psychiatry and Behavioral Sciences, Vanderbilt University Medical Center, Nashville, TN, USA
e-mail: brandee.feola@vumc.org

J. U. Blackford
Munroe-Meyer Institute for Genetics and Rehabilitation, University of Nebraska Medical Center, Omaha, NE, USA

Department of Psychiatry and Behavioral Sciences, Vanderbilt University Medical Center, Nashville, TN, USA

© The Author(s), under exclusive license to Springer Nature Switzerland AG 2024

Curr Topics Behav Neurosci (2025) 73: 581–607
https://doi.org/10.1007/7854_2024_558
Published Online: 17 December 2024

We also provide take-home points and propose future directions for the field. We hope this emphasis on the critical role of anxiety in schizophrenia will help researchers appropriately identify the presence of anxiety, better address these symptoms, and improve the lives of people at risk for or experiencing psychosis.

Keywords Anxiety disorders · Negative affect · Negative emotion · Psychosis · Schizophrenia spectrum disorders

1 Introduction

Anxiety is highly relevant to the development and maintenance of schizophrenia. Anxiety symptoms are often present throughout all stages of schizophrenia including the prodrome (Shioiri et al. 2007), early stages of schizophrenia (Strålin and Hetta 2021), and chronic schizophrenia (Mulugeta et al. 2023; Achim et al. 2011). A pooled meta-analysis of 52 studies estimated that almost 40% of individuals with schizophrenia experience a co-occurring anxiety disorder (Achim et al. 2011), and closer to 50% or more people with schizophrenia experience heightened anxiety symptoms (Achim et al. 2011; Michail and Birchwood 2009; Goodwin et al. 2002; Cosoff and Julian 1998). Moreover, anxiety profoundly impacts the lives of people with schizophrenia; having anxiety is associated with an increased risk of attempting suicide, greater lethality of suicide attempts, higher risk of substance and alcohol abuse, higher number of hospitalizations, lower quality of life, poorer social functioning, higher levels of shame, and lower self-esteem (Cosoff and Julian 1998; Taiminen et al. 2001; Pallanti et al. 2004; Blanchard et al. 1998; Birchwood et al. 2007; Lysaker and Salyers 2007; Gumley et al. 2004). Importantly, people with schizophrenia identify anxiety and its related symptoms, such as fear and uncertainty, as integral components of their experience (Fusar-Poli et al. 2022).

Many theories of schizophrenia highlight the impact of negative affect, including stress and anxiety, on the development and maintenance of the disorder. For example, the neural diathesis-stress model proposes that individuals with stress vulnerability or heightened stress sensitivity are more likely to develop schizophrenia when a stressor occurs later in development (Walker and Diforio 1997; Pruessner et al. 2017). Another model, the "affective pathway to psychosis", proposes that people with heightened stress sensitivity who develop schizophrenia are more likely to have prominent symptoms of anxiety and distress (Myin-Germeys and van Os 2007), compared to people with more prominent negative or cognitive symptoms. Rodent models of schizophrenia also support a role for stress and anxiety. For example, the prenatal methylazoxymethanol acetate (MAM) rodent model of schizophrenia shows increased behavioral stress sensitivity and anxiety-like behaviors in adolescence (Gomes et al. 2019).

Anxiety is also proposed to be a major component of delusional thinking, particularly persecutory delusions and paranoia (Freeman 2016). The prominent threat anticipation model includes worry as a contributing and precipitous factor of the development and maintenance of persecutory delusions (Freeman 2016). This model proposes that worrying brings threat beliefs to mind and expands on them, fueling the fear that one is under threat (Freeman 2016). At a neurobiological level, the Systems Neuroscience of Psychosis (SyNoPsis) model highlights the importance of a limbic loop giving rise to threat-related symptoms that underly paranoia and delusions (Strik et al. 2018). The limbic loop includes brain regions involved in fear and anxiety such as the amygdala, bed nucleus of the stria terminalis (BNST), hippocampus, and ventral striatum.

Anxiety remains understudied and underreported in schizophrenia – despite anxiety being extremely common in psychotic disorders and associated with negative outcomes. Multiple factors may contribute to the lack of focus on anxiety in schizophrenia, including an emphasis on other prominent symptoms of schizophrenia (delusions, hallucinations, cognitive deficits), lack of recognition and assessment of anxiety, and challenges associated with assessing anxiety in schizophrenia when other diagnoses or symptoms are present (e.g., differentiating between social anxiety and paranoia; Roy et al. 2018a; Temmingh and Stein 2015). Another barrier is that anxiety is not a part of the Diagnostic and Statistical Manual of Mental Disorders (DSM-5-T)R (American Psychiatric Association 2013) diagnostic criteria or the Clinician-Rated Dimensions of Psychosis Symptom Severity (CRDPSS), which was proposed for use in research and clinical evaluation. The dimensions include hallucinations, delusions, disorganized speech, abnormal psychomotor behavior, negative symptoms, impaired cognition, depression, and mania. However, anxiety was not included.

The goal of this review is to highlight the high prevalence of anxiety disorders in schizophrenia and the importance of recognizing and assessing anxiety in schizophrenia. We will address four key questions about anxiety in schizophrenia, including: (1) What does anxiety look like throughout the development of schizophrenia?; (2) How do we measure anxiety in schizophrenia?; (3) What are the mechanisms underlying anxiety in schizophrenia?; (4) How do we treat anxiety in schizophrenia? Finally, we propose future directions for the field to address anxiety in schizophrenia.

2 What Does Anxiety Look Like Throughout the Development of Schizophrenia?

Early childhood: Risk factors for anxiety, such as temperament, arise early in childhood. For example, children with inhibited temperament are typically very shy and cautious and display heightened threat sensitivity and physiological responses to novelty (Garcia Coll et al. 1984). Inhibited children are at a much higher risk for developing anxiety disorders (Clauss and Blackford 2012) and individuals with schizophrenia have higher levels of inhibited temperament relative to controls (Feola et al. 2019; Jetha et al. 2011). A study from our group in 166 people with a schizophrenia spectrum disorder found that higher inhibited temperament scores were associated with higher likelihood of co-occurring anxiety disorder, higher levels of anxiety symptoms, higher levels of distress, and lower quality of life (Feola et al. 2019). Therefore, early emerging traits that confer risk for anxiety may also contribute to risk for developing schizophrenia.

Adolescence: Anxiety disorders commonly arise early in adolescence (Solmi et al. 2022; De Lijster et al. 2017). Of the different types of anxiety disorders, separation anxiety, social anxiety, and specific phobias all have an average onset of 15 years old (Solmi et al. 2022; De Lijster et al. 2017). Generalized anxiety disorder (GAD) tends to emerge latest, with a median onset of 32 years of age, although there is substantial variability (Solmi et al. 2022; De Lijster et al. 2017). Thus, most anxiety disorders emerge before schizophrenia, which has an average onset in early adulthood (Solmi et al. 2022).

Prior to the onset of schizophrenia, many individuals experience some level of psychotic symptoms and are classified as being at clinical high risk (CHR) that includes either attenuated or subsyndromal symptoms of schizophrenia (Kline and Schiffman 2014). While anxiety is not part of the CHR criteria, anxiety is a common symptom experienced by people at CHR and in the prodromal phase of schizophrenia (Granö et al. 2014). For example, rates of anxiety disorders in CHR populations range from 15% to 51% (Fusar-Poli et al. 2014) and people at CHR with anxiety have worse clinical outcomes and functioning (Hasmi et al. 2021; Fulford et al. 2013; McAusland et al. 2017). Of the anxiety disorders, social anxiety is particularly prevalent in CHR samples. In two CHR studies, 40–51% met the criteria for social anxiety disorder and social anxiety symptoms were associated with increased negative symptoms and poorer social functioning (Rietdijk et al. 2013; Kuhney et al. 2021). Data from a prospective longitudinal study of 5,362 people found that those who developed schizophrenia rated themselves as less socially confident at age 13 and teachers reported them as being more socially anxious at age 15, suggesting that social anxiety may also confer risk for developing schizophrenia (Jones et al. 1994).

Persistently high anxiety during adolescence appears to be an important risk factor for developing psychosis. For example, a study using the Avon Longitudinal Study of Parents and Children (ALSPAC) showed that children with persistently high levels of anxiety at 8, 10, and 13 years of age were more likely to develop psychotic-like experiences and psychotic disorders (Morales-Muñoz et al. 2022). Another study demonstrated that people at CHR with consistently high levels of anxiety had poorer clinical outcomes, increased polygenic risk score, and higher levels of trauma exposure than the people at CHR without persistently high levels of anxiety (Deng et al. 2023).

Adulthood: Anxiety is prevalent in adults with schizophrenia and is associated with negative clinical and functional outcomes (Fusar-Poli et al. 2022). Anxiety disorders are very common in people with schizophrenia; a pooled meta-analysis demonstrated that almost 40% of people with schizophrenia are diagnosed with a co-occurring anxiety disorder (Achim et al. 2011). A large study of over 300,000 people using the UK biobank demonstrated that schizophrenia is linked with higher odds of having an anxiety disorder (odds ratio 1.41; Mulugeta et al. 2023). However, it is important to note that the rates of anxiety disorders vary considerably across studies of schizophrenia, including people in the early (Wilson et al. 2020) versus chronic states of illness (Roy et al. 2018a, b). The high prevalence of anxiety disorders is clinically important because numerous negative outcomes are associated with anxiety in people with schizophrenia, including increased risk of substance use and dependence, increased risk of suicide, higher number of hospitalizations, lower quality of life, poorer social function, higher levels of shame, and lower self-esteem (Taiminen et al. 2001; Pallanti et al. 2004; Blanchard et al. 1998; Birchwood et al. 2007; Gumley et al. 2004; Ritsner et al. 2014; Romm et al. 2012; Priebe et al. 2000). A year-long study in people with schizophrenia showed that anxiety was the most significant predictor of quality of life, where higher anxiety predicted lower quality of life (Huppert and Smith 2001).

Social anxiety is the most common anxiety disorder in adults with schizophrenia. In a meta-analysis of people with schizophrenia based on DSM-IV criteria, social anxiety disorder was the most prominent anxiety disorder (14.4%) (Achim et al. 2011). In another study, Roy, Achim, and colleagues (Roy et al. 2015) assessed anxiety disorders in 80 people with psychosis and found that 60% of people with psychosis met the criteria for a lifetime anxiety disorder. Of those with an anxiety disorder, 80% had social anxiety disorder. Other studies demonstrate that co-occurring social anxiety is related to higher levels of positive symptoms, poorer social functioning, and lower quality of life in people with schizophrenia (Romm et al. 2012; Huppert and Smith 2005; Penn et al. 1994). In our current study of anxiety in schizophrenia ($n = 40$; R01MH127018), we also replicate the previous findings of social anxiety as the most prominent anxiety disorder (Fig. 1). Although we are currently using the DSM-5 anxiety disorders categories, they do not currently include PTSD and OCD as primary anxiety disorders.

3 How Do we Measure Anxiety in Schizophrenia?

There is significant heterogeneity in the rates of co-occurring anxiety disorders in schizophrenia across studies (Achim et al. 2011; Roy et al. 2018a, b; Wilson et al. 2020). This may be due to multiple factors including under-recognition of anxiety, lack of assessment of anxiety, and an "applied diagnostic hierarchy" or the notion that anxiety symptoms are due to schizophrenia and not a separate co-occurring disorder (Achim et al. 2011; Roy et al. 2015, 2018a, b; Temmingh and Stein 2015). Measuring anxiety in schizophrenia can be challenging as the lines are often blurry, making the symptoms hard to disentangle. The first step to accurately capturing anxiety in schizophrenia is to assess both anxiety and schizophrenia symptoms.

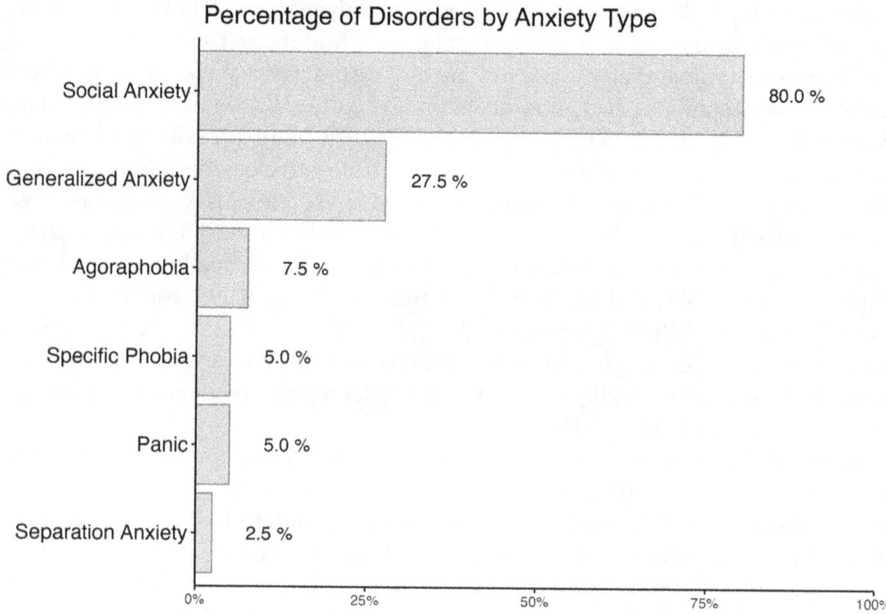

Fig. 1 Distribution of types of anxiety disorders in ongoing research study of $n = 40$ adults with schizophrenia and co-occurring anxiety disorders

3.1 Dissecting Anxiety and Paranoia

Anxiety and paranoia are tightly linked constructs that exist through similar cognitive biases (e.g., threat biases) and emotions (e.g., worry, stress, fear). Anxiety and paranoia share many commonalities – including heightened anticipation of threat or fear of harm – and have overlapping predictive and maintenance factors, such as worry and safety behaviors (Freeman 2007, 2016; Freeman et al. 2002; Green and Phillips 2004). For example, evidence suggests that anxiety and worry predict paranoia and persistence of paranoid thinking (Freeman et al. 2012). One difference is that paranoia, but not anxiety, involves the perception of intent to harm and perceptual anomalies or a feeling that "something is off" or "does not feel right" (Freeman 2007; Freeman et al. 2008). When a threat belief is activated, paranoid thinking differs from anxiety in its attribution and intent of the threat. For example, an individual with social anxiety may have a fear of interacting with a group of people at a party and may have thoughts of the individuals laughing at them. On the other hand, an individual with paranoia may similarly fear others will laugh at them at the party but experience an additional feeling that something is "off", causing them to worry that people at the party are out to get them. Therefore, an individual with paranoia is processing the situation under a lens of threat and the intent of the others fueled by misattributions, jumping to conclusions, and strong confirmatory bias (Freeman 2007). Careful assessment of both anxiety and paranoia can help capture these subtle differences and aid in deciphering the distinction between anxiety and schizophrenia.

3.2 Measurement of Anxiety in Schizophrenia

There are two main approaches to measuring anxiety in schizophrenia: interview-based and self-report measures. Interview-based measures provide an assessment of anxiety, and potentially a diagnosis, by a trained individual. Self-report measures of anxiety provide the first-person perspective of experiencing anxiety symptoms. The goal of this section is to highlight relevant interview-based and self-report measures of anxiety that are applicable to studying anxiety in schizophrenia. The measures are summarized in Table 1.

Interview-based measures: Two common goals for interview-based measures include making a diagnostic assessment and capturing the level of symptom severity. The gold-standard psychiatric research diagnostic tool, the Structured Clinical Interview for DSM-5 Disorders (SCID-5) (First et al. 2015), provides a comprehensive method for assessing all psychiatric diagnoses in the DSM-5. The comprehensive assessment considers a person's entire psychiatric history, which is helpful when identifying disorders that co-occur with psychosis, like anxiety disorders. For example, an interviewer may need the full picture of the individual's psychotic symptoms and history of psychotic illness to distinguish primary anxiety symptoms from those that are secondary to the psychosis. However, one risk in using the SCID-5 is that the anxiety module may only be used if the screener question is endorsed, which could potentially lead to underdiagnosis of anxiety. In addition, the SCID-5 is time-intensive and may not be feasible in all settings or studies. A shorter diagnostic interview is the Mini International Neuropsychiatric Interview (MINI), a brief structured interview designed particularly for its efficiency (Mordal and Gundersen 2010). Multiple studies have used the MINI to assess co-occurring anxiety disorders in psychosis (Seedat et al. 2007; Aikawa et al. 2018; Tibbo et al. 2003). However, it is unclear how accurately the MINI assesses symptoms in people with severe psychopathology.

Another option is to use anxiety-specific interview-based measures, such as the Anxiety Disorders Interview Schedule for DSM-5 (ADIS-5) (Brown and Barlow 2016). The ADIS is a semi-structured interview that assesses both categorical diagnostic ratings and dimensional ratings of symptom severity for each type of anxiety disorder. Importantly, the interviewers must rate the current anxiety symptoms even if the person denies the initial screening questions for that disorder, which reduces the risk of missing an anxiety diagnosis. Using the ADIS for in-depth anxiety assessment also provides the information necessary for the critical distinction between primary and secondary anxiety symptoms. The Hamilton Anxiety Scale (HAM-A) (Hamilton 1959; Maier et al. 1988) is another commonly used measure for assessing the presence and severity of mental and somatic aspects of anxiety. The HAM-A has been widely used, but it is criticized for the lack of ability

Table 1 Interview-based and self-report measures of anxiety relevant for schizophrenia

Measure	Anxiety-related constructs assessed	Outcome	Time period aassessed	Length/ # of items	Cronbach's alpha	Test-retest reliability	Sample items
Interview-based Measures							
SCID-5-RV (First et al. 2015)	DSM-5 anxiety disorders, OCD, PTSD	Diagnostic, categorical	Current, lifetime	30–120 min			
ADIS (Brown and Barlow, 2016)	DSM-5 anxiety disorders, OCD, PTSD	Diagnostic, dimensional	Current, lifetime	30–90 min			
MINI (Mordal and Gundersen 2010)	DSM-5 anxiety disorders, OCD, PTSD	Diagnostic	Current, lifetime	15 min			
HAM-A (Hamilton 1959)	Anxiety	Dimensional	Current	10–15 min			
PANSS (Kay et al. 1987)	Anxiety item, tension item, distress item	Dimensional	Past 2 weeks	30–40 min			
BPRS (Brown and Barlow 2016; Overall and Gorham 1988)	Anxiety item	Dimensional	Past 2–3 days	20–30 min			
S-SARS (Van Staden et al. 2022)	Anxiety in schizophrenia	Dimensional	Past week	10 items			
SAES (Llorca et al. 2014)	Anxiety in schizophrenia	Dimensional	Past week	29 items			
Self-report Measures							
DASS (Lovibond and Lovibond 1995; Antony et al. 1998; Brown et al. 1997)	Depression, anxiety, stress	Dimensional	Past week	42 items	0.92–0.97	0.71–0.81	"I felt I was close to panic"

MASQ (Watson et al. 1995a, b; Boschen and Oei 2007)	Depression, anxiety, general distress	Dimensional	Past week	90 items	0.86–0.91		"Felt nervous" "Felt restless"
STAI (Spielberger et al. 1983; Barnes et al. 2002; Stojanović et al. 2020)	State anxiety, trait anxiety	Dimensional	At the moment (state), generally (trait)	40 items	0.71–0.86	0.93	"I feel nervous and restless" "I am worried"
BAI (Beck et al. 1988)	Anxiety	Dimensional	Generally	21 items	0.92	0.75	"Unable to Relax" "Nervous"
SAS (Zung 1971)	Anxiety	Dimensional	Past several days	20 items	0.82		"I feel more nervous and anxious than normal" "My face gets hot and blushes"
LSAS (Liebowitz 1987)	Social anxiety	Dimensional	Past week	24 items	0.95	0.83	"Going to a party" "Giving a report to a group"
SIAS (Mattick and Clarke 1998; Le Blanc et al. 2014; Heimberg et al. 1992)	Social anxiety	Dimensional	Generally	20 items	0.94	0.92	"I am at ease meeting people at parties, etc." "I find myself worrying that I won't know what to say in social situations"
GAD-7 (Spitzer et al. 2006; Johnson et al. 2019)	Generalized anxiety	Dimensional	Past 2 weeks	7 items	0.91	0.83	"Feeling nervous, anxious, or on edge" "Worrying too much about different things"
PSWQ (Meyer et al. 1990; Brown et al. 1992)	Worry, generalized anxiety	Dimensional	Generally	16 items	0.93	0.92	"I do not tend to worry about things" "I have been a worrier all my life"
DWQ (Freeman et al. 2020)	Worry, paranoia	Dimensional	Past month	15 items	0.95–0.99	0.96–0.97	"I've been worried a lot" "I've been worrying about someone trying to harm me"

to discriminate between anxiety and medication side effects (Hamilton 1959; Maier et al. 1988).

Another group of interview-based measures aims to assess severity of psychopathology symptoms rather than a categorical diagnosis, including psychosis and anxiety symptoms. Commonly used measures of psychosis symptoms, such as the Positive and Negative Syndrome Scale (PANSS; Kay et al. 1987) and Brief Psychiatric Rating Scale (BPRS; Overall and Gorham 1988), contain very few items assessing anxiety. For example, the PANSS (Kay et al. 1987) includes a single anxiety item that assesses physical anxiety symptoms. Despite the narrow description of anxiety in psychosis assessments, researchers sometimes use single items to capture anxiety symptoms, leading to a very limited reporting of anxiety symptoms. In addition, the use of a single item lacks validation, as most psychometric analyses of psychosis instruments are conducted with factors, rather than a single item (Jerrell and Hrisko 2013; Van Den Oord et al. 2006). If researchers are limited in measures of anxiety and only have access to the PANSS, we suggest using a five-factor model that includes a distress factor that contains items for anxiety, depression, guilt, and tension. The distress factor is a more stable and reliable assessment of anxiety than the single anxiety item (Dickinson et al. 2018; Wallwork et al. 2012; van der Gaag et al. 2006).

More recently, measures were created to specifically assess anxiety in schizophrenia. For example, the Scale of Anxiety Evaluation in Schizophrenia (SAES) (Llorca et al. 2014) was created to capture a broader scope of anxiety in schizophrenia by compiling items from four other anxiety scales, including the HAM-A. Unlike other anxiety measures, the scale was validated in individuals with schizophrenia. The Staden Anxiety Scale (Van Staden et al. 2022) was also recently developed to assess general and specific anxiety in people with schizophrenia. The 10-item scale has strong psychometric properties within people with schizophrenia and correlates with other measures of anxiety, including the HAM-A (Van Staden et al. 2022). Measures like the Staden Anxiety Scale may capture the unique experience of anxiety in schizophrenia. More research should be conducted on the utility of schizophrenia-specific measures of anxiety.

3.3 Self-Report Measures of Anxiety

Self-report measures provide an important perspective of an individual's experience of anxiety. However, it can be challenging to decide which self-report measures of anxiety to include in a study. There are pros and cons of including a broader assessment of anxiety versus specific anxiety symptoms, such as social anxiety symptoms or worry. In a research setting, the selection of an anxiety measure may depend on several factors, including specific research questions and time constraints. The goal of this section is to provide a list of self-report measures that are the most relevant for psychosis populations (see Table 1).

Several questionnaires measure negative affect more broadly and include an anxiety subscale such as the Mood and Anxiety Symptom Questionnaire (MASQ; Watson et al. 1995a, b) and the Depression, Anxiety, and Stress Scale (DASS; Lovibond and Lovibond 1995). The MASQ is a longer self-report measure that captures anxiety, depression, and general distress (shared between anxiety and depression). The DASS captures a broader aspect of negative emotions and includes three subscales that separately assess anxiety, depression, and stress. However, the seven items included in the anxiety scale assess mostly somatic symptoms of anxiety.

Other self-report measures assess anxiety more broadly, including the State-Trait Anxiety Inventory (STAI; Spielberger et al. 1983), Self-report of Anxiety Scale (SAS; Zung 1971), and Beck Anxiety Inventory (BAI; Beck et al. 1988). The STAI contains two subscales to separately assess state anxiety (the level of anxiety "at this moment") and trait anxiety (the level of anxiety "generally"). The SAS assesses anxiety broadly within the past few days. The BAI also assesses anxiety broadly, but most questions focus on somatic symptoms.

Another group of questionnaires focuses on anxiety symptoms specific to a type of anxiety, such as social anxiety, generalized anxiety disorder, or worry. Social anxiety is one of the most widely studied and prevalent types of anxiety in schizophrenia. The Leibowitz Social Anxiety Scale (LSAS; Liebowitz 1987) is one of the most common social anxiety measures. One of the strengths of the LSAS is that it is highly specific to the presence and severity of social anxiety, separately assesses social interaction and performance situations, and demonstrates relatively good psychometric properties in people with schizophrenia (Smith et al. 2021). The Social Interaction Anxiety Scale (SIAS; Mattick and Clarke 1998) is another measure that assesses social anxiety specifically. The Generalized Anxiety Disorder-7 (GAD-7; Spitzer et al. 2006) is a short measure used to specifically capture GAD symptoms. The Penn State Worry Questionnaire (PSWQ; Meyer et al. 1990) assesses worry symptoms, which are a core feature of GAD. The Dunn Worry Questionnaire (DWQ; Freeman et al. 2020) was developed to assess both general worry and paranoid worry, which is especially useful among people with schizophrenia. The general worry subscale is shown to be sensitive to worry across general and clinical populations, while the paranoid worry subscale is most sensitive in clinical populations (Freeman et al. 2020).

Despite the important perspective provided by self-report measures, there is a concern that people with schizophrenia may not be reliable self-reporters. We believe that self-report measures provide critical information about an individual's experience that is essential for accurately capturing psychiatric disorders. In addition, emerging evidence suggests that self-report measures are reliable and valid in people with schizophrenia, as self-report measures of anxiety converge with diagnoses measured by the ADIS (Huppert et al. 2002). Self-report measures, including the DASS and PSWQ, were highly reliable with high internal consistency (0.85–0.95) and high test-retest reliability (0.58–0.88) 3 months later. Of note, most individuals completing self-report measures were considered "stable", and it is less clear whether self-reports are less reliable and/or valid when someone is experiencing an acute psychotic episode.

4 What Are the Mechanisms Underlying Anxiety in Schizophrenia?

4.1 Neurobiological Basis of Fear and Anxiety in Schizophrenia

Fear and anxiety involve two key brain regions: the amygdala and bed nucleus of the stria terminalis (BNST). Although fear and anxiety are overlapping constructs, there are data to suggest that fear and anxiety are distinct and mediated by different brain regions (for reviews see Avery et al. 2016; Fox et al. 2015; Goode and Maren 2017; Lebow and Chen 2016). Fear is a response to an imminent, close, certain threat and mediated by the amygdala, while anxiety is a sustained response to a potential, distant, and uncertain threat that is mediated by the BNST. Although it was initially challenging to image the BNST in humans, technological advances have led to numerous human neuroimaging studies of the BNST over the past decade (Theiss et al. 2017). For example, our group characterized the structural and functional connectivity of the BNST and developed a BNST mask for use by the field (Theiss et al. 2017; Avery et al. 2014). Our group and others identified a reliable and robust BNST network in healthy individuals (Avery et al. 2014; McMenamin et al. 2014; Tillman et al. 2018; Torrisi et al. 2015) and demonstrated that BNST function is correlated with anxiety symptoms, trait-like symptoms, and anxiety disorders (Somerville et al. 2010; Clauss et al. 2019).

Most research examining the neurobiological basis of fear and anxiety in schizophrenia has focused on the amygdala (Feola et al. 2024a). The amygdala was first thought to be hypoactive to emotional stimuli in individuals with schizophrenia. However, those data were based on neuroimaging studies that compared emotional versus neutral stimuli. More recent studies clarified that the reported differences were driven by heightened amygdala responses to neutral stimuli and not group differences in responses to the emotional stimuli (Anticevic et al. 2012; Dugré et al. 2019). Studies also show that individuals with psychosis have altered amygdala connectivity, including decreased amygdala-prefrontal connectivity relative to controls at rest (Anticevic et al. 2014; Liu et al. 2014) and during emotion processing (Bjorkquist et al. 2016; Mukherjee et al. 2012, 2014; Feola et al. 2021). These findings suggest that amygdala activation and connectivity with regulation regions may be altered at baseline in individuals with schizophrenia.

In contrast to the amygdala, the BNST has received considerably less attention in psychosis research. Our group has examined the brain basis of anxiety in psychosis utilizing two approaches: a hypothesis-driven region of interest approach and a data-driven approach. Our first project examined BNST and amygdala responses to unpredictable threat in people with schizophrenia with co-occurring anxiety, individuals with schizophrenia without co-occurring anxiety, and healthy controls (Feola et al. 2021). We used a threat anticipation task to investigate brain activation and connectivity to unpredictable and predictable threat anticipation. Our study found that individuals with schizophrenia exhibited alterations in BNST and amygdala connectivity, but no differences in activation. The most prominent differences in connectivity were between the two subgroups of people with schizophrenia, with and without co-occurring anxiety disorders. The individuals with schizophrenia and co-occurring anxiety disorders had increased BNST connectivity with regions of the salience network (i.e., dorsal anterior cingulate and insula) during unpredictable threat anticipation and increased BNST connectivity with regions of a fear network (i.e., sublenticular extended amygdala and insula) during predictable threat anticipation, compared to the individuals with schizophrenia without co-occurring anxiety disorders. Our study also replicated previous amygdala findings of decreased amygdala-prefrontal connectivity in schizophrenia; however, the amygdala connectivity did not distinguish between the two subgroups of schizophrenia. These findings suggest that individuals with schizophrenia and co-occurring anxiety disorders exhibit distinct patterns of BNST connectivity compared to those without co-occurring anxiety disorders, implying potential differences in the pathophysiology of these subgroups.

Our second project used the Human Connectome Project Early Psychosis (HCP-EP) dataset to determine how self-reported fear and anxiety symptoms relate to resting state connectivity in people in the early stages of psychosis and controls (Feola et al. 2024b). Most previous studies have examined the amygdala or BNST as a priori brain regions of interest. We utilized an alternative, data-driven approach to identify brain regions that correlated with fear and anxiety symptoms. We found that higher levels of fear and anxiety were correlated with stronger activation in a cluster comprising the amygdala and hippocampus across the entire sample. In addition, connectivity between the amygdala/hippocampus cluster and the insula was significantly correlated with fear/anxiety scores across the whole sample. One caveat is that the self-report measure did not separately assess fear and anxiety, so the results represent the brain basis of fear and anxiety combined. Our findings validated decades of hypothesis-driven studies demonstrating that the amygdala, hippocampus, and insula are pivotal for fear and anxiety. Importantly, our study also discovered a psychosis-specific connection between the amygdala/hippocampus cluster and the ventromedial prefrontal cortex (vmPFC). For people with early psychosis, higher fear/anxiety scores were correlated with stronger connectivity between the vmPFC and amygdala/hippocampus cluster. Our findings suggest that people with psychosis with anxiety have sustained and heightened connectivity of fear and anxiety networks at rest, potentially identifying future targets for treatment interventions.

4.2 Neurobiological Basis of Fear and Anxiety in Paranoia

Researchers have also studied the neurobiological basis of fear and anxiety in paranoia, with a focus on the amygdala (Phillips et al. 1999; Williams et al. 2004; Park and Ichinose 2015; Fan et al. 2021; Walther et al. 2022). For example, people with paranoia have increased cerebral blood flow (CBF) in the amygdala during resting state and inducing paranoia increases amygdala CBF in healthy individuals (Pinkham et al. 2022). Studies also show hypoactivation of the amygdala during tasks with emotional stimuli in people with paranoia (Williams et al. 2004; Pinkham et al. 2015). People with paranoia show stronger amygdala connectivity with the prefrontal cortex, insula, and hippocampus at rest relative to controls (Fan et al. 2021; Walther et al. 2022). Together, the two lines of research in people with schizophrenia and paranoia highlight disruption in fear and anxiety networks at baseline. These findings may further align with the aberrant salience theory, which proposes that elevated presynaptic dopamine in psychotic disorders marks mundane or non-threatening information as salient, therefore increasing stress or anxiety (Nudelman and Waltz 2022; Howes and Nour 2016; Kapur 2003). Critically, this may result in a feedback loop, as heightened stress sensitizes the dopamine system, further promoting aberrant salience.

Although these studies highlight important alterations in fear and anxiety networks in schizophrenia, future studies will be crucial to further understand the brain basis of anxiety in schizophrenia. A few important questions that need to be clarified include how the anxiety network is altered in schizophrenia, whether anxiety networks help us understand heterogeneity in schizophrenia, when do alterations in the anxiety network arise in the development of schizophrenia, and whether we can modulate fear and anxiety networks in schizophrenia.

4.3 Trajectories of Anxiety in Psychosis

A critical question is through what mechanisms do anxiety and schizophrenia co-occur? The affective pathway model proposes that there is a unique developmental trajectory, characterized by early trauma or stress sensitivity, for people with schizophrenia and high levels of negative affect (Myin-Germeys and van Os 2007). We aim to build on the affective pathway and neural diathesis-stress models to propose three specific pathways to psychosis characterized by high levels of anxiety. Components of these pathways have been proposed by other researchers that suggest anxiety can arise independent of the psychosis or arise in tandem with the onset of psychotic symptoms (Temmingh and Stein 2015). We aim to integrate multiple theories of psychosis by proposing three different pathways to anxiety in psychosis: (1) Risk Factor; (2) Stressor; (3) Concurrent (Fig. 2). These pathways are the beginning of our conceptualization of the development of psychosis with high levels of anxiety and future research will be fundamental for understanding the mechanisms driving anxiety in psychosis.

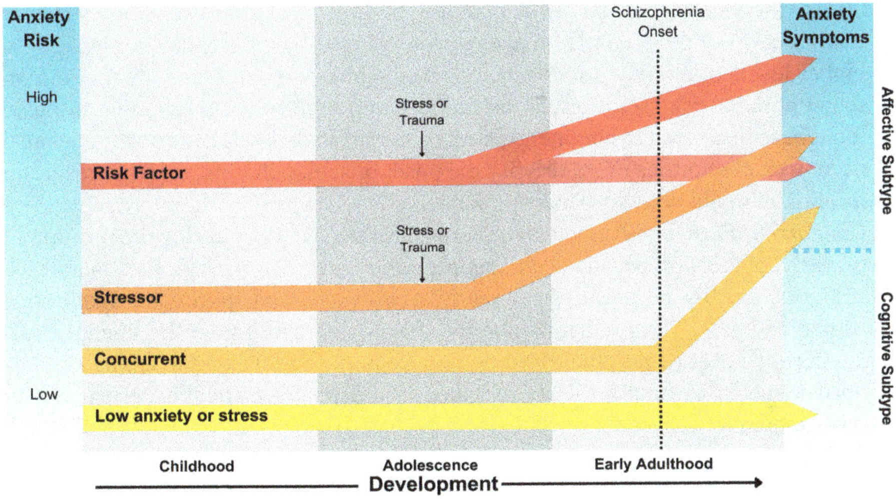

Fig. 2 Pathways to schizophrenia with anxiety across development. The figure shows three pathways to an affective subtype of schizophrenia with high anxiety, and one pathway with low anxiety that results in a cognitive subtype of schizophrenia with low anxiety. The top pathway (in red) shows the Risk Factor pathway that is characterized by risk factors for anxiety that are present early in development, leading to high anxiety and stress sensitivity before the onset of schizophrenia. The second pathway from the top (in orange) represents the Stressor pathway where the occurrence of trauma or life stressors results in high levels of anxiety and stress sensitivity prior to the onset of schizophrenia. The third pathway (in lighter orange) displays the Concurrent pathway where anxiety arises with the onset of schizophrenia. The fourth pathway (in yellow) shows a low anxiety pathway, highlighting the subgroup of people with schizophrenia that do not have anxiety and may fall into the cognitive subtype of schizophrenia

1. *Risk Factor Pathway*: We propose that risk factors for anxiety also increase the risk for schizophrenia with co-occurring anxiety. For example, one proposed mechanism for increased risk for schizophrenia is that children with inhibited temperament also experience heightened physiological arousal and stress sensitivity that may leave them more vulnerable to the impact of life stressors (Aas et al. 2014). Therefore, this pathway represents the impact of early anxiety risk factors, such as inhibited temperament, on heightened stress sensitivity and anxiety that lead to increased risk for schizophrenia. In addition, the experience of trauma or life stressor can exacerbate the risk for schizophrenia with high anxiety. In the Risk Factor pathway, anxiety emerges early and is persistent across development and into schizophrenia.
2. *Stressor Pathway*: We propose that the experience of significant stress during development leads to heightened stress sensitivity and/or anxiety arising prior to the onset of psychosis. This pathway aligns with prominent theories of schizophrenia that highlight the role of stress, such as the neural diathesis-stress model (Walker and Diforio 1997; Pruessner et al. 2017). Trauma is one of the most prominent risk factors for developing schizophrenia and is associated with higher symptom severity (Aas et al. 2014, 2019; Isvoranu et al. 2017; Lardinois et al.

2011; van Nierop et al. 2018). A potential biological mechanism is that stress or trauma early in development makes stress systems hyper-responsive, leading to higher levels of anxiety prior to the onset of schizophrenia. Many other common schizophrenia risk factors, such as immigration and urban upbringing, are also thought to be related to heightened stress (Anglin et al. 2021). A stressor may also occur as a precipitous event driving the onset of schizophrenia, together resulting in schizophrenia with co-occurring anxiety.

3. *Concurrent Pathway*: Anxiety symptoms may also develop during the prodromal or early phases of psychosis in people who have no anxiety risk factors or stressors. In this pathway, the anxiety may result from psychotic symptoms, stigma associated with a schizophrenia diagnosis, or negative life changes following the onset of psychotic symptoms. Regardless of when the anxiety developed in an individual's life, it remains a core piece of their experience with schizophrenia.

While we proposed three different pathways to psychosis with anxiety, it is important to acknowledge that not all people with schizophrenia experience anxiety. Theories such as the affective pathway to psychosis propose that anxiety and stress will only be experienced by an affective subgroup (Myin-Germeys and van Os 2007). The other subgroup includes people with schizophrenia that have prominent cognitive deficits or persistent negative symptoms, consistent with cognitive and deficit subgroups (Dickinson et al. 2018; Voineskos et al. 2013; Kirkpatrick et al. 2001). While anxiety is a significant aspect of psychosis for many individuals, it is essential to recognize the heterogeneity. To highlight the importance of this caveat, the cognitive pathway is shown on the bottom of Fig. 2.

Another caveat is that the three pathways to schizophrenia with high anxiety are shown as distinct, but these pathways share commonalities and may overlap. For example, all pathways may include multiple risk factors for psychosis that impact stress responses leading to heightened stress sensitivity and anxiety such as immigration status, trauma, and discrimination. In addition, studies have highlighted that anxiety disorders are more prevalent in women (Bangasser and Cuarenta 2021) and the affective pathway to psychosis is proposed to be more common in women (Myin-Germeys and van Os 2007). Therefore, it may be more likely that women would display anxiety in the proposed pathways to psychosis with anxiety. In addition, a meta-analysis of associations with psychosis and common risk factors highlighted specific stressors, including childhood trauma and urbanicity, are associated with different symptom presentations in women versus men (Center for Mental Health Services SA and MHSA 2021). These studies highlight the complexity in sex and gender effects and the need for more research examining sex and gender differences in psychosis.

5 How Do We Treat Anxiety in Schizophrenia?

The next important question is how do we treat anxiety in schizophrenia? As we previously highlighted, anxiety has largely been ignored in schizophrenia research and this is also true for interventions, despite evidence of the benefits of treating anxiety in schizophrenia (Kingsep et al. 2003; Montreuil et al. 2016; Lepage et al. 2023; Halperin et al. 2000). From a pharmacological standpoint, the first-line treatment is antipsychotics. Approximately 14–22% of patients are treated with an antipsychotic plus an anti-anxiety medication (e.g., anxiolytics), and 31–38% are treated with an antipsychotic plus an antidepressant medication (e.g., selective serotonin reuptake inhibitors), which often helps treat co-morbid anxiety (Chakos et al. 2006; Pickar et al. 2008) .

Psychotherapy can provide an effective treatment for anxiety in psychosis. Cognitive Behavioral Therapy for psychosis (CBTp) is the gold standard for schizophrenia and other psychotic disorders (Center for Mental Health Services SA and MHSA 2021) and meta-analyses report that CBTp yields small to moderate effects (0.35–0.44) on symptoms (Wykes et al. 2008). Interestingly, CBTp was developed from CBT for anxiety, using the same methods of formulation and treatment, including cognitive restructuring and behavioral experiments (i.e., exposure therapy). Psychotherapy may be particularly effective for schizophrenia through its impact on anxiety and stress. In youth at CHR, psychotherapy focused on stress management and coping reduced conversion to a primary psychotic disorder in a randomized clinical trial (Bechdolf et al. 2012). In addition, group-based CBT improves anxiety symptoms in individuals with early psychosis and those at high risk with social anxiety disorders (Kingsep et al. 2003; Montreuil et al. 2016; Lepage et al. 2023; Halperin et al. 2000).

A newer approach to CBTp involves the direct targeting of specific symptoms, instead of psychosis more broadly. This symptom-specific approach has been most successfully applied to persecutory delusions and paranoia (Freeman et al. 2021). For example, an eight-session CBTp-based psychotherapy was developed to specifically target worry in individuals with severe persecutory delusions and paranoia, using techniques commonly used for treating GAD. The treatment had a large impact on reducing the severity of persecutory delusions and paranoia and worry accounted for 66% of the change in delusion severity (Freeman et al. 2015), directly linking the mechanism of anxiety symptom underlying paranoia. Future studies should continue to examine the potential benefit of targeting worry and anxiety in psychosis.

Other approaches used to manage anxiety in psychosis include mindfulness and yoga. Studies of mindfulness for treating anxiety in psychosis show some potential benefits including decreasing distress and negative emotions and increasing wellbeing (Bosanac and Castle 2015; Chadwick et al. 2005; Perlini et al. 2020; Brown et al. 2010). One study of a yoga intervention in 20 people with early psychosis showed improvements in anxiety including individuals more relaxed and less worried (McGuire et al. 2022). Although these findings are promising, more work is needed to clarify the potential utility and benefits of yoga and mindfulness approaches to target anxiety in psychosis.

6 Future Directions

This review highlighted the current state of assessment and research of anxiety in psychosis. Below we provide suggestions for future directions to advance our knowledge and interventions for anxiety in psychosis.

1. **Increased screening and intervention for anxiety early in development.** Anxiety disorders arise early in development and may be associated with increased risk of developing schizophrenia. Therefore, early identification and intervention on anxiety may reduce the risk of developing schizophrenia.
2. **Improvements in assessing and reporting of anxiety in psychosis.** A broader, consistent assessment of anxiety in schizophrenia is crucial to understand true prevalence of anxiety in schizophrenia. While interview-based and self-report measures are available, we suggest that the field create a more formal assessment of anxiety in schizophrenia. One potential is adding anxiety to the Clinician-Rated Dimensions of Psychosis Symptom Severity (CRDPSS) to improve the broader recognition and assessment of anxiety.
3. **Increased training and discussion of anxiety in schizophrenia.** Anxiety in schizophrenia is underrecognized and underreported. To make a significant impact, the field must adopt more training and education about anxiety in schizophrenia. This can occur at multiple levels including in medical or graduate school education, as well as broader education for senior investigators. Please contact our group with any questions about educational resources.
4. **More research examining anxiety in schizophrenia.** Theories such as the affective pathway to psychosis and our proposed pathways are a start; however, experimental studies of anxiety in schizophrenia are crucial to fully understand how anxiety impacts the mechanisms driving schizophrenia and to improve our current treatments.
5. **Integrating lived experiences of anxiety in schizophrenia.** First-hand accounts of individuals' experiences of schizophrenia provide a crucial perspective. Most of the research on anxiety in schizophrenia is based on measures and reports of anxiety outside of the field of schizophrenia. However, anxiety experienced by individuals with schizophrenia may be different than individuals that experience anxiety without psychosis. Efforts are underway to create measures to improve assessment of anxiety in people with schizophrenia, such as the Staden Anxiety Scale (Van Staden et al. 2022). More research will be helpful in validating these schizophrenia-specific measures and future studies will be crucial for understanding the lived experience of anxiety in people with schizophrenia.
6. **Examining additional factors influencing anxiety.** There are additional factors that are shown to influence anxiety across psychopathologies such as gender/sex,

trauma, cognition, insight, co-occurring disorders, and substance and alcohol use. Increased reporting of these variables is a first step in understanding the role of these factors in anxiety in people with schizophrenia. Future research should aim to examine the interaction of these factors in relation to anxiety in schizophrenia, which will help advance our understanding of the mechanisms driving schizophrenia.

7 Conclusions

This review served to highlight the importance, prevalence, and clinical significance of anxiety in schizophrenia. Advancing the recognition and knowledge of anxiety in schizophrenia will help providers recognize anxiety, leading to increased treatment of anxiety and improvement of our current treatments of schizophrenia. The goal of understanding anxiety in schizophrenia is to help alleviate some of the burden for people with schizophrenia and their families.

Acknowledgments The authors would also like to acknowledge the funding sources that supported this work: the National Institute of Mental Health (R01MH127018-01 to JUB/BF; K23MH126313 to JMS).

References

Aas M, Dazzan P, Mondelli V, Melle I, Murray RM, Pariante CM (2014) A systematic review of cognitive function in first-episode psychosis, including a discussion on childhood trauma, stress, and inflammation. Front Psychiatry 4:1–13. https://doi.org/10.3389/fpsyt.2013.00182

Aas M, Pizzagalli DA, Laskemoen JF et al (2019) Elevated hair cortisol is associated with childhood maltreatment and cognitive impairment in schizophrenia and in bipolar disorders. Schizophr Res 213:65–71. https://doi.org/10.1016/j.schres.2019.01.011

Achim AM, Maziade M, Raymond É, Olivier D, Mérette C, Roy MA (2011) How prevalent are anxiety disorders in schizophrenia? A meta-analysis and critical review on a significant association. Schizophr Bull 37(4):811–821. https://doi.org/10.1093/schbul/sbp148

Aikawa S, Kobayashi H, Nemoto T et al (2018) Social anxiety and risk factors in patients with schizophrenia: relationship with duration of untreated psychosis. Psychiatry Res 263:94–100. https://doi.org/10.1016/j.psychres.2018.02.038

American Psychiatric Association (2013) Diagnostic and statistical manual of mental disorders, 5th edn. American Psychiatric Association. https://doi.org/10.1176/appi.books.9780890425596

Anglin DM, Ereshefsky S, Klaunig MJ et al (2021) From womb to neighborhood: a racial analysis of social determinants of psychosis in the United States. Am J Psychiatry 178(7):599–610. https://doi.org/10.1176/appi.ajp.2020.20071091

Anticevic A, Van Snellenberg JX, Cohen RE, Repovs G, Dowd EC, Barch DM (2012) Amygdala recruitment in schizophrenia in response to aversive emotional material: a meta-analysis of neuroimaging studies. Schizophr Bull 38(3):608–621. https://doi.org/10.1093/schbul/sbq131

Anticevic A, Tang Y, Cho YT et al (2014) Amygdala connectivity differs among chronic, early course, and individuals at risk for developing schizophrenia. Schizophr Bull 40(5):1105–1116. https://doi.org/10.1093/schbul/sbt165

Antony M, Bieling P, Cox B, Enns M, Swinson R (1998) Psychometric properties of the 42-item and 21-item versions of the depression anxiety stress scales in clinical groups and a community sample. Psychol Assess 10:176–181. https://doi.org/10.1037/1040-3590.10.2.176

Avery SN, Clauss JA, Winder DG, Woodward N, Heckers S, Blackford JU (2014) BNST neurocircuitry in humans. Neuroimage 91:311–323. https://doi.org/10.1016/j.neuroimage.2014.01.017

Avery SNN, Clauss JAA, Blackford JUU (2016) The human BNST: functional role in anxiety and addiction. Neuropsychopharmacology 41(1):126–141. https://doi.org/10.1038/npp.2015.185

Bangasser DA, Cuarenta A (2021) Sex differences in anxiety and depression: circuits and mechanisms. Nat Rev Neurosci 22(11):674–684. https://doi.org/10.1038/s41583-021-00513-0

Barnes LLB, Harp D, Jung WS (2002) Reliability generalization of scores on the Spielberger state-trait anxiety inventory. Educ Psychol Meas 62(4):603–618

Bechdolf A, Wagner M, Ruhrmann S et al (2012) Preventing progression to first-episode psychosis in early initial prodromal states. Br J Psychiatry 200(1):22–29. https://doi.org/10.1192/bjp.bp.109.066357

Beck AT, Epstein N, Brown G, Steer RA (1988) An inventory for measuring clinical anxiety: psychometric properties. J Consult Clin Psychol 56(6):893–897

Birchwood M, Trower P, Brunet K, Gilbert P, Iqbal Z, Jackson C (2007) Social anxiety and the shame of psychosis: a study in first episode psychosis. Behav Res Ther 45(5):1025–1037. https://doi.org/10.1016/j.brat.2006.07.011

Bjorkquist OA, Olsen EK, Nelson BD, Herbener ES (2016) Altered amygdala-prefrontal connectivity during emotion perception in schizophrenia. Schizophr Res 175(1–3):35–41. https://doi.org/10.1016/j.schres.2016.04.003

Blanchard JJ, Mueser KT, Bellack AS (1998) Anhedonia, positive and negative affect, and social functioning in schizophrenia. Schizophr Bull 24(3):413–424. https://doi.org/10.1093/oxfordjournals.schbul.a033336

Bosanac P, Castle D (2015) How should we manage anxiety in patients with schizophrenia? Australas Psychiatry 23(4):374–377. https://doi.org/10.1177/1039856215588207

Boschen MJ, Oei TPS (2007) Discriminant validity of the MASQ in a clinical sample. Psychiatry Res 150(2):163–171. https://doi.org/10.1016/j.psychres.2006.03.008

Brown TA, Barlow DH (2016) Anxiety and related disorders interview schedule for DSM-5 (ADIS-5) – adult version. Published online

Brown TA, Antony MM, Barlow DH (1992) Psychometric properties of the Penn State Worry Questionnaire in a clinical anxiety disorders sample. Behav Res Ther 30(1):33–37

Brown TA, Chorpita BF, Korotitscw W, Barlow DH (1997) Psychometric properties of the depression anxiety stress scales (DASS) in clinical samples. Behav Res Ther 35(I):79–89

Brown LF, Davis LW, Larocco VA, Strasburger A (2010) Participant perspectives on mindfulness meditation training for anxiety in schizophrenia. Am J Psychiatr Rehabil 13(3):224–242. https://doi.org/10.1080/15487768.2010.501302

Center for Mental Health Services SA and MHSA (2021) Substance Abuse and Mental Health Services Administration: routine administration of cognitive behavioral therapy for psychosis as the standard of care for individuals seeking treatment for psychosis: state of the science and implementation considerations for key stakeholders

Chadwick P, Taylor KN, Abba N (2005) Mindfulness groups for people with psychosis. Behav Cogn Psychother 33(3):351–359. https://doi.org/10.1017/S1352465805002158

Chakos MH, Glick ID, Miller AL et al (2006) Baseline use of concomitant psychotropic medications to treat schizophrenia in the CATIE trial. Psychiatr Serv 57(8):1094–1101

Clauss JA, Blackford JU (2012) Behavioral inhibition and risk for developing social anxiety disorder: a meta-analytic study. J Am Acad Child Adolesc Psychiatry 51(10):1066–1075.e1. https://doi.org/10.1016/j.jaac.2012.08.002

Clauss JA, Avery SN, Benningfield MM, Blackford JU (2019) Social anxiety is associated with BNST response to unpredictability. Depress Anxiety 36:666–675. https://doi.org/10.1002/da.22891

Cosoff SJ, Julian HR (1998) The prevalence of comorbid anxiety in schizophrenia, schizoaffective disorder and bipolar disorder. Aust N Z J Psychiatry 32(1):67–72. https://doi.org/10.1046/j.1440-1614.1998.00374.x

De Lijster JM, Dierckx B, Utens EMWJ et al (2017) The age of onset of anxiety disorders: a meta-analysis. Can J Psychiatry 62(4):237–246. https://doi.org/10.1177/0706743716640757

Deng W, Addington J, Bearden CE et al (2023) Characterizing sustained social anxiety in individuals at clinical high risk for psychosis: trajectory, risk factors, and functional outcomes. Psychol Med 53(8):3644–3651. https://doi.org/10.1017/S0033291722000277

Dickinson D, Pratt DN, Giangrande EJ et al (2018) Attacking heterogeneity in schizophrenia by deriving clinical subgroups from widely available symptom data. Schizophr Bull 44(1): 101–113. https://doi.org/10.1093/schbul/sbx039

Dugré JR, Bitar N, Dumais A, Potvin S (2019) Limbic hyperactivity in response to emotionally neutral stimuli in schizophrenia: a neuroimaging meta-analysis of the hypervigilant mind. Am J Psychiatry 176:1021–1029. https://doi.org/10.1176/appi.ajp.2019.19030247

Fan L, Klein H, Bass E, Springfield C, Pinkham A (2021) Amygdala hyperconnectivity in the paranoid state: a transdiagnostic study. J Psychiatr Res 138:117–124. https://doi.org/10.1016/j.jpsychires.2021.03.049

Feola B, Armstrong K, Woodward ND, Heckers S, Blackford JU (2019) Childhood temperament is associated with distress, anxiety and reduced quality of life in schizophrenia spectrum disorders. Psychiatry Res 275:196–203. https://doi.org/10.1016/j.psychres.2019.03.016

Feola B, McHugo M, Armstrong K et al (2021) BNST and amygdala connectivity are altered during threat anticipation in schizophrenia. Behav Brain Res 412:113428. https://doi.org/10.1016/j.bbr.2021.113428

Feola B, Moussa-Tooks AB, Sheffield JM, Heckers S, Woodward ND, Blackford JU (2024a) Threat responses in schizophrenia: a negative valence systems framework. Curr Psychiatry Rep 26(1):9–25. https://doi.org/10.1007/s11920-023-01479-9

Feola B, Beermann A, Manzanarez Felix K et al (2024b) Data-driven, connectome-wide analysis identifies psychosis-specific brain correlates of fear and anxiety. Mol Psychiatry 29:2601–2610. https://doi.org/10.1038/s41380-024-02512-w

First MB, Williams JBW, Karg RS, Spitzer RL (2015) Structured clinical interview for DSM-5, research version. American Psychiatric Publishing, Inc

Fox AS, Oler JA, Tromp DPM, Fudge JL, Kalin NH (2015) Extending the amygdala in theories of threat processing. Trends Neurosci 38(5):319–329. https://doi.org/10.1016/j.tins.2015.03.002

Freeman D (2007) Suspicious minds: the psychology of persecutory delusions. Clin Psychol Rev 27(4):425–457. https://doi.org/10.1016/j.cpr.2006.10.004

Freeman D (2016) Persecutory delusions: a cognitive perspective on understanding and treatment. Lancet 3:685–692

Freeman D, Garety PA, Kuipers E, Fowler D, Bebbington PE (2002) A cognitive model of persecutory delusions. Br J Clin Psychol 41(4):331–347. https://doi.org/10.1348/014466502760387461

Freeman D, Gittins M, Pugh K, Antley A, Slater M, Dunn G (2008) What makes one person paranoid and another person anxious? The differential prediction of social anxiety and persecutory ideation in an experimental situation. Psychol Med 38(8):1121–1132. https://doi.org/10.1017/S0033291708003589

Freeman D, Stahl D, McManus S et al (2012) Insomnia, worry, anxiety and depression as predictors of the occurrence and persistence of paranoid thinking. Soc Psychiatry Psychiatr Epidemiol 47(8):1195–1203. https://doi.org/10.1007/s00127-011-0433-1

Freeman D, Dunn G, Startup H et al (2015) Effects of cognitive behaviour therapy for worry on persecutory delusions in patients with psychosis (WIT): a parallel, single-blind, randomised controlled trial with a mediation analysis. Lancet Psychiatry 2(4):305–313. https://doi.org/10.1016/S2215-0366(15)00039-5

Freeman D, Bird JC, Loe BS et al (2020) The Dunn Worry Questionnaire and the Paranoia Worries Questionnaire: New assessments of worry. Psychol Med 50(5):771–780. https://doi.org/10.1017/S0033291719000588

Freeman D, Emsley R, Diamond R et al (2021) Comparison of a theoretically driven cognitive therapy (the Feeling Safe Programme) with befriending for the treatment of persistent persecutory delusions: a parallel, single-blind, randomised controlled trial. Lancet Psychiatry 8(8): 696–707. https://doi.org/10.1016/S2215-0366(21)00158-9

Fulford D, Niendam TA, Floyd EG et al (2013) Symptom dimensions and functional impairment in early psychosis: more to the story than just negative symptoms. Schizophr Res 147(1):125–131. https://doi.org/10.1016/j.schres.2013.03.024

Fusar-Poli P, Nelson B, Valmaggia L, Yung AR, McGuire PK (2014) Comorbid depressive and anxiety disorders in 509 individuals with an at-risk mental state: impact on psychopathology and transition to psychosis. Schizophr Bull 40(1):120–131. https://doi.org/10.1093/schbul/sbs136

Fusar-Poli P, Estradé A, Stanghellini G et al (2022) The lived experience of psychosis: a bottom-up review co-written by experience and academics. World Psychiatry 21:168–188. https://www.gamian.eu

Garcia Coll C, Kagan J, Reznick JS (1984) Behavioral inhibition in young children. Child Dev 55(3):1005–1019

Gomes FV, Zhu X, Grace AA (2019) Stress during critical periods of development and risk for schizophrenia. Schizophr Res 213:107–113. https://doi.org/10.1016/j.schres.2019.01.030

Goode TD, Maren S (2017) Role of the bed nucleus of the stria terminalis in aversive learning and memory. Learn Mem 24(9):480–491. https://doi.org/10.1101/lm.044206.116

Goodwin R, Lyons JS, Mcnally RJ (2002) Panic attacks in schizophrenia. Schizophr Res 58:213–220. www.elsevier.com/locate/schres

Granö N, Karjalainen M, Edlund V et al (2014) Anxiety symptoms in adolescents at risk for psychosis: a comparison among help seekers. Child Adolesc Ment Health 19(2):97–101. https://doi.org/10.1111/camh.12012

Green MJ, Phillips ML (2004) Social threat perception and the evolution of paranoia. Neurosci Biobehav Rev 28(3):333–342. https://doi.org/10.1016/j.neubiorev.2004.03.006

Gumley A, O'grady M, Power K, Schwannauer M (2004) Negative beliefs about self and illness: a comparison of individuals with psychosis with or without comorbid social anxiety disorder. Aust N Z J Psychiatry 38(11–12):960–964

Halperin S, Nathan P, Drummond P, Castle D (2000) A cognitive-behavioural, group-based intervention for social anxiety in schizophrenia. Aust N Z J Psychiatry 34(5):809–813. https://doi.org/10.1046/j.1440-1614.2000.00820.x

Hamilton M (1959) The assessment of anxiety state by ratings. Br J Med Psychol 32(1):50–55. https://doi.org/10.1111/j.2044-8341.1959.tb00467.x

Hasmi L, Pries LK, ten Have M et al (2021) What makes the psychosis 'clinical high risk' state risky: psychosis itself or the co-presence of a non-psychotic disorder? Epidemiol Psychiatr Sci:30. https://doi.org/10.1017/S204579602100041X

Heimberg RG, Mueller GP, Holt CS, Hope DA, Liebowitz MR (1992) Assessment of anxiety in social interaction and being observed by others: the social interaction anxiety scale and the social phobia scale. Behav Therapy 23:53–73.

Howes OD, Nour MM (2016) Dopamine and the aberrant salience hypothesis of schizophrenia. World Psychiatry 15(1):3–4. https://doi.org/10.1002/wps.20276

Huppert JD, Smith TE (2001) Longitudinal analysis of subjective quality of life in schizophrenia: anxiety as the best symptom predictor. J Nerv Ment Dis 189(10) http://journals.lww.com/jonmd

Huppert JD, Smith TE (2005) Anxiety and schizophrenia: the interaction of subtypes of anxiety and psychotic symptoms. CNS Spectr 10(9):721–731. https://doi.org/10.1017/S1092852900019714

Huppert JD, Smith TE, Apfeldorf WJ (2002) Use of self-report measures of anxiety and depression in outpatients with schizophrenia: reliability and validity. J Psychopathol Behav Assess 24(4)

Isvoranu AM, Van Borkulo CD, Lou BL et al (2017) A network approach to psychosis: pathways between childhood trauma and psychotic symptoms. Schizophr Bull 43(1):187–196. https://doi.org/10.1093/schbul/sbw055

Jerrell JM, Hrisko S (2013) Utility of two PANSS 5-factor models for assessing psychosocial outcomes in clinical programs for persons with schizophrenia. Schizophr Res Treat 2013:1–9. https://doi.org/10.1155/2013/705631

Jetha MK, Goldberg JO, Schmidt LA (2011) Temperament and its relation to social functioning in schizophrenia. Int J Soc Psychiatry 59(3):254–263. https://doi.org/10.1177/0020764011433639

Johnson SU, Ulvenes PG, Øktedalen T, Hoffart A (2019) Psychometric properties of the GAD-7 in a heterogeneous psychiatric sample. Front Psychol 10. https://doi.org/10.3389/fpsyg.2019.01713

Jones P, Murray R, Jones P, Rodgers B, Marmot M (1994) Child developmental risk factors for adult schizophrenia in the British 1946 birth cohort. Lancet 344(8934):1398–1402. https://doi.org/10.1016/S0140-6736(94)90569-X

Kapur S (2003) Psychosis as a state of aberrant salience: a framework linking biology, phenomenology, and pharmacology in schizophrenia. Am J Psychiatry 160:1. http://ajp.psychiatryonline.org

Kay SR, Fiszbein A, Opler LA (1987) The positive and negative syndrome scale (PANSS) for schizophrenia. Schizophr Bull 13(2):261–276. https://doi.org/10.1093/schbul/13.2.261

Kingsep P, Nathan P, Castle D (2003) Cognitive behavioural group treatment for social anxiety in schizophrenia. Schizophr Res 63(1–2):121–129. https://doi.org/10.1016/S0920-9964(02)00376-6

Kirkpatrick B, Buchanan RW, Ross DE, Carpenter WT Jr (2001) A separate disease within the syndrome of schizophrenia. Arch Gen Psychiatry 58(2):165–171. https://doi.org/10.1001/archpsyc.58.2.165

Kline E, Schiffman J (2014) Psychosis risk screening: a systematic review. Schizophr Res 158(1–3):11–18. https://doi.org/10.1016/J.SCHRES.2014.06.036

Kuhney FS, Damme KSF, Pelletier-Baldelli A et al (2021) Prevalence and functional consequences of social anxiety in individuals at clinical high-risk for psychosis: perspective from a community sample comparison. Schizophr Bull Open 2(1). https://doi.org/10.1093/schizbullopen/sgab025

Lardinois M, Lataster T, Mengelers R, van Os J, Myin-Germeys I (2011) Childhood trauma and increased stress sensitivity in psychosis. Acta Psychiatr Scand 123(1):28–35. https://doi.org/10.1111/j.1600-0447.2010.01594.x

Le Blanc AL, Bruce LC, Heimberg RG et al (2014) Evaluation of the psychometric properties of two short forms of the social interaction anxiety scale and the social phobia scale. Assessment 21(3):312–323. https://doi.org/10.1177/1073191114521279

Lebow MA, Chen A (2016) Overshadowed by the amygdala: the bed nucleus of the stria terminalis emerges as key to psychiatric disorders. Mol Psychiatry 21:450–463. https://doi.org/10.1038/mp.2016.1

Lepage M, Bowie CR, Montreuil T et al (2023) Manualized group cognitive behavioral therapy for social anxiety in first-episode psychosis: a randomized controlled trial. Psychol Med 53(8):3335–3344. https://doi.org/10.1017/S0033291721005328

Liebowitz MR (1987) Liebowitz social anxiety scale (LSAS). APA PsycTest. https://doi.org/10.1037/t07671-000

Liu H, Tang Y, Womer F et al (2014) Differentiating patterns of amygdala-frontal functional connectivity in schizophrenia and bipolar disorder. Schizophr Bull 40(2):469–477. https://doi.org/10.1093/schbul/sbt044

Llorca PM, Lancon C, Blanc O et al (2014) A composite scale applied to evaluate anxiety in schizophrenic patients (SAES). Eur Arch Psychiatry Clin Neurosci 264(2):171–178. https://doi.org/10.1007/s00406-013-0416-1

Lovibond PF, Lovibond SH (1995) Depression anxiety and stress scales (DASS-42). APA PsycTests. Published online

Lysaker PH, Salyers MP (2007) Anxiety symptoms in schizophrenia spectrum disorders: associations with social function, positive and negative symptoms, hope and trauma history. Acta Psychiatr Scand 116(4):290–298. https://doi.org/10.1111/j.1600-0447.2007.01067.x
Maier W, Buller R, Philipp M, Heuser I (1988) The Hamilton anxiety scale: reliability, validity and sensitivity to change in anxiety and depressive disorders. J Affect Disord 14(1):61–68. https://doi.org/10.1016/0165-0327(88)90072-9
Mattick RP, Clarke JC (1998) Development and validation of measures of social phobia scrutiny fear and social interaction anxiety. Behav Res Ther 36(4):455–470. https://doi.org/10.1016/S0005-7967(97)10031-6
McAusland L, Buchy L, Cadenhead KS et al (2017) Anxiety in youth at clinical high risk for psychosis. Early Interv Psychiatry 11(6):480–487. https://doi.org/10.1111/eip.12274
McGuire D, Shannon A, Somaiya J, Brown E, O'Donoghue B (2022) A pilot study of a yoga intervention for the treatment of anxiety in young people with early psychosis. Early Interv Psychiatry 16(2):200–204. https://doi.org/10.1111/eip.13151
McMenamin BW, Langeslag SJE, Sirbu M, Padmala S, Pessoa XL (2014) Network organization unfolds over time during periods of anxious anticipation. J Neurosci 34(34):11261–11273. https://doi.org/10.1523/JNEUROSCI.1579-14.2014
Meyer TJ, Miller ML, Metzger RL, Borkovec TD (1990) Development and validation of the Penn State Worry Questionnaire. Behav Res Ther 28(6):487–495. https://doi.org/10.1016/0005-7967(90)90135-6
Michail M, Birchwood M (2009) Social anxiety disorder in first-episode psychosis: incidence, phenomenology and relationship with paranoia. Br J Psychiatry 195(3):234–241. https://doi.org/10.1192/bjp.bp.108.053124
Montreuil TC, Malla AK, Joober R, Bélanger C, Myhr G, Lepage M (2016) Manualized group cognitive-behavioral therapy for social anxiety in at-risk mental state and first episode psychosis: a pilot study of feasibility and outcomes. Int J Group Psychother 66(2):225–245. https://doi.org/10.1080/00207284.2015.1106190
Morales-Muñoz I, Palmer ER, Marwaha S, Mallikarjun PK, Upthegrove R (2022) Persistent childhood and adolescent anxiety and risk for psychosis: a longitudinal birth cohort study. Biol Psychiatry 92(4):275–282. https://doi.org/10.1016/j.biopsych.2021.12.003
Mordal J, Gundersen BJG (2010) Norwegian version of the mini-international neuropsychiatric interview: feasibility, acceptability and test-retest reliability in an acute psychiatric ward. Eur Psychiatry 25(3):172–177. https://doi.org/10.1016/j.eurpsy.2009.02.004
Mukherjee P, Whalley HC, McKirdy JW et al (2012) Lower effective connectivity between amygdala and parietal regions in response to fearful faces in schizophrenia. Schizophr Res 134(2–3):118–124. https://doi.org/10.1016/j.schres.2011.09.033
Mukherjee P, Whalley HC, McKirdy JW et al (2014) Altered amygdala connectivity within the social brain in schizophrenia. Schizophr Bull 40(1):152–160. https://doi.org/10.1093/schbul/sbt086
Mulugeta A, Suppiah V, Hyppönen E (2023) Schizophrenia and co-morbidity risk: evidence from a data driven phenomewide association study. J Psychiatr Res 162:1–10. https://doi.org/10.1016/j.jpsychires.2023.04.010
Myin-Germeys I, van Os J (2007) Stress-reactivity in psychosis: evidence for an affective pathway to psychosis. Clin Psychol Rev 27(4):409–424. https://doi.org/10.1016/j.cpr.2006.09.005
Nudelman JL, Waltz JA (2022) Acute and lifetime stress and psychotic illness: the roles of reward and salience networks. J Psychiatr Brain Sci 7(6):e220012. https://doi.org/10.20900/jpbs.20220012
Overall JE, Gorham DR (1988) The brief psychiatric rating scale (BPRS): recent developments in ascertainment and scaling. Psychopharmacol Bull 24(1):97–99
Pallanti S, Quercioli L, Hollander E (2004) Social anxiety in outpatients with schizophrenia: a relevant cause of disability. Am J Psychiatry 161(1):53–58. https://doi.org/10.1176/appi.ajp.161.1.53

Park S, Ichinose M (2015) Amygdala on the lookout. Am J Psychiatry 172(8):704–705. https://doi.org/10.1176/appi.ajp.2015.15050646

Penn DL, Hope DA, Spaulding W, Kucera J (1994) Social anxiety in schizophrenia. Schizophr Res 11(3):277–284. https://doi.org/10.1016/0920-9964(94)90022-1

Perlini C, Bellani M, Rossetti MG et al (2020) Mindfulness-based interventions in the early phase of affective and non-affective psychoses: special section on "Translational and Neuroscience Studies in Affective Disorders" section editor, Maria Nobile MD, PhD. J Affect Disord 263: 747–753. https://doi.org/10.1016/j.jad.2019.10.011

Phillips ML, Williams L, Senior C et al (1999) A differential neural response to threatening and non-threatening negative facial expressions in paranoid and non-paranoid schizophrenics. Psychiatry Res 92:1131

Pickar D, Vinik J, Bartko JJ (2008) Pharmacotherapy of schizophrenic patients: preponderance of off-label drug use. PLoS One 3(9). https://doi.org/10.1371/journal.pone.0003150

Pinkham AE, Liu P, Lu H, Kriegsman M, Simpson C, Tamminga C (2015) Amygdala hyperactivity at rest in paranoid individuals with schizophrenia. Am J Psychiatry 172(8):784–792. https://doi.org/10.1176/appi.ajp.2014.14081000

Pinkham A, Bass E, Klein H, Springfield C, Kent J, Aslan S (2022) Inducing paranoia is linked to increased resting amygdala cerebral blood flow in healthy individuals. J Exp Psychopathol 13(3). https://doi.org/10.1177/20438087221130693

Priebe S, Roeder-Wanner UU, Kaiser W (2000) Quality of life in first-admitted schizophrenia patients: a follow-up study. Psychol Med 30(1):225–230. https://doi.org/10.1017/S0033291798008253

Pruessner M, Cullen AE, Aas M, Walker EF (2017) The neural diathesis-stress model of schizophrenia revisited: an update on recent findings considering illness stage and neurobiological and methodological complexities. Neurosci Biobehav Rev 73:191–218. https://doi.org/10.1016/J.NEUBIOREV.2016.12.013

Rietdijk J, Ising HK, Dragt S et al (2013) Depression and social anxiety in help-seeking patients with an ultra-high risk for developing psychosis. Psychiatry Res 209(3):309–313. https://doi.org/10.1016/j.psychres.2013.01.012

Ritsner MS, Lisker A, Grinshpoon A (2014) Predicting 10-year quality-of-life outcomes of patients with schizophrenia and schizoaffective disorders. Psychiatry Clin Neurosci 68(4):308–317. https://doi.org/10.1111/pcn.12135

Romm KL, Melle I, Thoresen C, Andreassen OA, Rossberg JI (2012) Severe social anxiety in early psychosis is associated with poor premorbid functioning, depression, and reduced quality of life. Compr Psychiatry 53(5):434–440. https://doi.org/10.1016/j.comppsych.2011.06.002

Roy MA, Achim AM, Vallières C et al (2015) Comorbidity between anxiety disorders and recent-onset psychotic disorders. Schizophr Res 166(1–3):353–354. https://doi.org/10.1016/j.schres.2015.04.041

Roy MA, Demers MF, Achim AM (2018a) Social anxiety disorder in schizophrenia: a neglected, yet potentially important comorbidity. J Psychiatry Neurosci 43(4):287–288. https://doi.org/10.1503/jpn.180081

Roy MA, Vallières C, Lehoux C, Leclerc LD, Demers MF, Achim AM (2018b) More intensive probing increases the detection of social anxiety disorders in schizophrenia. Psychiatry Res 268: 358–360. https://doi.org/10.1016/j.psychres.2018.07.032

Seedat S, Fritelli V, Oosthuizen P, Emsley RA, Stein DJ (2007) Measuring anxiety in patients with schizophrenia. J Nerv Ment Dis 195(4):320–324. https://doi.org/10.1097/01.nmd.0000253782.47140.ac

Shioiri T, Shinada K, Kuwabara H, Someya T (2007) Early prodromal symptoms and diagnoses before first psychotic episode in 219 inpatients with schizophrenia. Psychiatry Clin Neurosci 61(4):348–354. https://doi.org/10.1111/j.1440-1819.2007.01685.x

Smith EL, Garety PA, Harding H, Hardy A (2021) Are there reliable and valid measures of anxiety for people with psychosis? A systematic review of psychometric properties. Psychol Psychother Theory Res Pract 94(1):173–198. https://doi.org/10.1111/papt.12265

Solmi M, Radua J, Olivola M et al (2022) Age at onset of mental disorders worldwide: large-scale meta-analysis of 192 epidemiological studies. Mol Psychiatry 27(1):281–295. https://doi.org/10.1038/s41380-021-01161-7

Somerville LH, Whalen PJ, Kelley WM (2010) Human bed nucleus of the stria terminalis indexes hypervigilant threat monitoring. Biol Psychiatry 68(5):416–424. https://doi.org/10.1016/j.biopsych.2010.04.002

Spielberger CD, Gorsuch RL, Lushene R, Vagg PR, Jacobs GA (1983) Manual for the state-trait anxiety inventory. Consulting Psychologists Press

Spitzer RL, Kroenke K, Williams JBW, Löwe B (2006) A brief measure for assessing generalized anxiety disorder: the GAD-7. Arch Intern Med 166(10):1092–1097. https://doi.org/10.1001/archinte.166.10.1092

Stojanović NM, Randjelović PJ, Nikolić G et al (2020) Reliability and validity of the spielberger's state-trait anxiety inventory (STAI) in serbian university student and psychiatric non-psychotic outpatient populations. Acta Facultatis Medicae Naissensis 327(2):149–159. https://doi.org/10.5937/afmnai37-25011

Strålin P, Hetta J (2021) First episode psychosis: register-based study of comorbid psychiatric disorders and medications before and after. Eur Arch Psychiatry Clin Neurosci 271(2):303–313. https://doi.org/10.1007/s00406-020-01139-6

Strik W, Stegmayer K, Walther S, Dierks T (2018) Systems neuroscience of psychosis: mapping schizophrenia symptoms onto brain systems. Neuropsychobiology 75(3):100–116. https://doi.org/10.1159/000485221

Taiminen T, Huttunen J, Heilä H et al (2001) The schizophrenia suicide risk scale (SSRS): development and initial validation. Schizophr Res 47(2–3):199–213. https://doi.org/10.1016/S0920-9964(00)00126-2

Temmingh H, Stein DJ (2015) Anxiety in patients with schizophrenia: epidemiology and management. CNS Drugs 29(10):819–832. https://doi.org/10.1007/s40263-015-0282-7

Theiss JD, Ridgewell C, McHugo M, Heckers S, Blackford JU (2017) Manual segmentation of the human bed nucleus of the stria terminalis using 3 T MRI. Neuroimage 146:288-292. https://doi.org/10.1016/j.neuroimage.2016.11.047

Tibbo P, Swainson J, Chue P, LeMelledo JM (2003) Prevalence and relationship to delusions and hallucinations of anxiety disorders in schizophrenia. Depress Anxiety 17(2):65–72. https://doi.org/10.1002/da.10083

Tillman RM, Stockbridge MD, Nacewicz BM et al (2018) Intrinsic functional connectivity of the central extended amygdala. Hum Brain Mapp 39(3):1291–1312. https://doi.org/10.1002/hbm.23917

Torrisi S, O'Connell K, Andrew D et al (2015) Resting state connectivity of the bed nucleus of the stria terminalis at ultra-high field. Hum Brain Mapp 36(10):4076–4088. https://doi.org/10.1016/j.bbi.2017.04.008

Van Den Oord EJCG, Rujescu D, Robles JR et al (2006) Factor structure and external validity of the PANSS revisited. Schizophr Res 82(2–3):213–223. https://doi.org/10.1016/j.schres.2005.09.002

van der Gaag M, Hoffman T, Remijsen M et al (2006) The five-factor model of the positive and negative syndrome scale II: a ten-fold cross-validation of a revised model. Schizophr Res 85(1–3):280–287. https://doi.org/10.1016/j.schres.2006.03.021

van Nierop M, Lecei A, Myin-Germeys I et al (2018) Stress reactivity links childhood trauma exposure to an admixture of depressive, anxiety, and psychosis symptoms. Psychiatry Res 260:451–457. https://doi.org/10.1016/j.psychres.2017.12.012

Van Staden W, Dlagnekova A, Naidu K (2022) Validity and reliability of the Staden schizophrenia anxiety rating scale. Diagnostics 12(4). https://doi.org/10.3390/diagnostics12040831

Voineskos AN, Foussias G, Lerch J et al (2013) Neuroimaging evidence for the deficit subtype of schizophrenia. JAMA Psychiatry 70(5):472–480. https://doi.org/10.1001/jamapsychiatry.2013.786

Walker EF, Diforio D (1997) Schizophrenia: a neural diathesis-stress model. Psychol Rev 104(4): 667–685. http://www.ncbi.nlm.nih.gov/pubmed/9337628

Wallwork RS, Fortgang R, Hashimoto R, Weinberger DR, Dickinson D (2012) Searching for a consensus five-factor model of the positive and negative syndrome scale for schizophrenia. Schizophr Res 137(1–3):246–250. https://doi.org/10.1016/j.schres.2012.01.031

Walther S, Lefebvre S, Conring F et al (2022) Limbic links to paranoia: increased resting-state functional connectivity between amygdala, hippocampus and orbitofrontal cortex in schizophrenia patients with paranoia. Eur Arch Psychiatry Clin Neurosci 272(6):1021–1032. https://doi.org/10.1007/s00406-021-01337-w

Watson D, Weber K, Assenheimer JS, Clark LA, Strauss ME, McCormick RA (1995a) Testing a tripartite model: I. Evaluating the convergent and discriminant validity of anxiety and depression symptom scales. J Abnorm Psychol 104(1):3–14

Watson D, Clark LA, Weber K, Assenheimer JS, Strauss ME, McCormick RA (1995b) Testing a tripartite model: II. Exploring the symptom structure of anxiety and depression in student, adult, and patient samples. J Abnorm Psychol 104(1):15–25. https://doi.org/10.1037/0021-843X.104.1.3

Williams LM, Das P, Harris AW et al (2004) Dysregulation of arousal and amygdala-prefrontal systems in paranoid schizophrenia. Am J Psychiatry 161:3. http://ajp.psychiatryonline.org

Wilson RS, Yung AR, Morrison AP (2020) Comorbidity rates of depression and anxiety in first episode psychosis: a systematic review and meta-analysis. Schizophr Res 216:322–329. https://doi.org/10.1016/j.schres.2019.11.035

Wykes T, Steel C, Everitt B, Tarrier N (2008) Cognitive behavior therapy for schizophrenia: effect sizes, clinical models, and methodological rigor. Schizophr Bull 34(3):523–537. https://doi.org/10.1093/schbul/sbm114

Zung WW (1971) A rating scale for anxiety disorders. Psychosomatics 12(6):371-9.

Wyman [illegible]

Insights into Overlapping Brain Networks for Anxiety and Alcohol Use Disorders

Nicole L. Zabik and Jennifer Urbano Blackford

Contents

Abstract Alcohol use disorder (AUD) is a debilitating condition affecting over 30 million Americans. AUD commonly co-occurs with other disorders, like other substance use disorders, trauma-related disorders, and anxiety disorders. Of the

N. L. Zabik
Munroe-Meyer Institute, University of Nebraska Medical Center, Omaha, NE, USA

J. U. Blackford (✉)
Munroe-Meyer Institute, University of Nebraska Medical Center, Omaha, NE, USA

Department of Psychiatry and Behavioral Sciences, Vanderbilt University Medical Center, Nashville, TN, USA
e-mail: jblackford@unmc.edu

© The Author(s), under exclusive license to Springer Nature Switzerland AG 2025

Curr Topics Behav Neurosci (2025) 73: 609–635
https://doi.org/10.1007/7854_2025_592
Published Online: 15 May 2025

numerous co-occurring disorders, anxiety disorders are the most pervasive: anxiety disorders serve as a risk factor for developing AUD, emerge as co-occurring disorders that maintain alcohol drinking, and impede the effectiveness of treatments for AUD. Anxiety, therefore, shapes the development, course, and treatment of AUDs. AUDs can also increase anxiety, suggesting a complex, bidirectional relation between alcohol use and anxiety. The intersection of AUDs and anxiety is also supported by their overlapping neural circuits, specifically neural circuits involved in stress responding, reward processing, and cognitive control. The current review highlights findings from several decades of research on how anxiety impacts the brain and treatment outcomes in AUDs. We also provide important considerations for future research, with the goal of reducing the shame and burden of alcohol use for individuals with AUD and their families.

Keywords Alcohol use disorders · Anxiety disorders · Negative affect · Neuroimaging

1 Introduction

Alcohol use disorder (AUD) is a debilitating, chronic condition that is characterized by excessive or compulsive use of alcohol despite social, financial, or professional consequences. Prevalence rates of AUD have doubled in the past five years, with nearly 30 million adults qualifying for an AUD in 2022 (SAMHSA 2022). Alcohol use disorder remains an elusive and costly disorder in the US, with researchers exerting tremendous effort to keep pace with understanding the development of AUD and potential treatments. The complexity of alcohol use extends further, as chronic alcohol use is closely connected to disorders like other substance use disorders, depressive disorders, and anxiety disorders (Grant et al. 2015). Of the co-occurring disorders, anxiety disorders represent the most pervasive: anxiety disorders serve as a risk factor for developing AUD, emerge as co-occurring disorders that maintain alcohol drinking, and impede effectiveness of treatments for AUD. Thus, identifying the role of anxiety in AUD serves as an important step towards identifying novel preventions and treatments.

A leading model of addiction recognizes negative affect (i.e., anxiety) as a cornerstone of AUD (Koob and Volkow 2010, 2016). To date, most research in AUDs has focused on reward, craving, and cognition in chronic alcohol use, leading to important discoveries about the neurobiology and psychology of AUD. Thus, there remains a substantial gap in our knowledge about how acute and chronic anxiety may impact AUDs. Greater understanding of the role of anxiety in the development, maintenance, and relapse of alcohol use has great potential to inform the development of preventions and interventions.

The goal of this review is to characterize the role of anxiety in AUD, with a focus on how anxiety-related factors contribute to the development of AUD, maintenance of alcohol use, and relapse following treatment. This review will focus on the impact of anxiety and AUD on brain function in humans. The first section introduces how anxiety and stress impact the development, prevalence, and maintenance of alcohol use. The second section summarizes the distinct, but overlapping, neural networks involved in anxiety and AUD. The third section reviews current neuroimaging literature that investigates how anxiety and AUD shape brain function. Finally, the last section discusses how anxiety may impede the effectiveness of alcohol treatment. This review is guided by three primary principles: (1) anxiety shapes the development and maintenance of chronic alcohol use, (2) anxiety and AUD have significant overlap in their underlying neural networks, and (3) anxiety should be a clinical focal point in AUD recovery.

2 How Anxiety Shapes Alcohol Use Disorder

Anxiety contributes to the development and course of AUD. The impact of anxiety on AUD has been characterized by numerous longitudinal studies. The longitudinal studies we highlight below have three primary foci: (1) developmental impact of anxiety on AUD, (2) anxiety that co-occurs with AUD, and (3) anxiety's role in maintaining alcohol use and impeding alcohol abstinence.

2.1 *Developmental Impact of Anxiety on Alcohol Use*

Adolescence Adolescence is a critical period of development when anxiety disorders are emerging, but is a period before AUDs typically emerge, making it ideal for studying longitudinal relations between anxiety and alcohol use. Klein et al. (2022) investigated whether lifetime or current adolescent anxiety predicted lifetime substance experimentation. Based on an initial wave of data collected from the Adolescent Brain and Cognitive Development dataset (n = 11,785), the authors found that children with an anxiety disorder were 31% more likely to have experimented with alcohol. Similarly, McCabe et al. (2023) investigated the relation between negative affect symptoms (anxiety) and binge drinking over five years using data from the National Consortium on Alcohol and Neurodevelopment in Adolescence-Adulthood (n = 831). Girls reported higher anxiety across the study period and had significant increases in binge drinking at follow-up visits compared with boys, but anxiety was not a significant predictor of binge drinking. Findings from these two studies suggest that adolescent anxiety predicted greater alcohol experimentation, but not binge drinking. One potential explanation for these findings is that anxious adolescents might be more cautious about alcohol use, such that they will use alcohol but not engage in binge drinking, whereas adolescents without anxiety may be more likely to be more extraverted and risk-takers—traits associated with binge drinking.

Emerging Young Adults Emerging adulthood is an important stage of development for alcohol use – the average age of first drink is 18 years, consistent with the transition to college and other major life events (Maggs and Schulenberg 2004; SAMHSA 2023). Zimmerman et al. (2003) prospectively followed emerging adults in the Early Developmental Stages of Psychopathology study ($n = 3,021$). Participants with an anxiety disorder at baseline had higher levels of regular or hazardous alcohol use four years later, which was largely driven by social anxiety disorder, panic disorder, and agoraphobia. Fröjd et al. (2011) prospectively followed adolescents into adulthood in the Adolescent Mental Health Cohort Study ($n = 2,070$). Levels of general anxiety at baseline, but not social anxiety, were associated with a threefold higher incidence of frequent alcohol use and drunkenness. At the two-year follow-up, baseline levels of general anxiety predicted a twofold higher incidence of frequent alcohol use and drunkenness. Wolitzky-Taylor and colleagues (2012) utilized data from the Northwestern-UCLA Youth Emotion Project to determine whether adolescent emotional disorders predicted alcohol use, as well as whether alcohol use predicted adolescent emotional disorders ($n = 627$). Young adults with a baseline anxiety disorder were nearly three times more likely to develop an AUD, while baseline AUD did not predict future anxiety disorders. These data in emerging young adults give us greater insight into the direction of interaction between anxiety and alcohol – anxiety *predicts* alcohol use and AUD, while alcohol use and AUD do not predict anxiety disorders.

Adulthood Studies of adulthood offer a unique view on both lifetime history and current anxiety and alcohol use. Marquenie et al. (2006) aimed to determine the order of onset of anxiety disorders and AUD through the Netherlands Mental Health Survey and Incidence Study, a retrospective and prospective study conducted over three years. Participants were grouped according to baseline diagnoses ($n = 7,076$): anxiety disorder with an AUD (1.4%), anxiety disorder without an AUD (14.7%), AUD without an anxiety disorder (3.5%), and no anxiety disorder or AUD (80.4%). Anxiety disorders preceded the development of AUD in 68% of people with both anxiety and AUD. People with a current anxiety disorder at baseline were over two times more likely to develop AUD at a follow-up assessment. AUD preceded anxiety disorders in 24% of people, while anxiety disorders and AUD co-emerged in 8%. Torvik et al. (2019) investigated whether anxiety disorders predicted development of an AUD in a twin cohort study ($n = 2,284$). Social anxiety disorder at baseline predicted the development of an AUD ten years later, and social anxiety disorder was a stronger predictor of AUD than any other type of anxiety disorder. AUD at baseline did not predict the development of a social anxiety disorder but moderately predicted other anxiety disorders at follow-up. Like studies in emerging adulthood, studies in adults suggest that anxiety disorders largely precede the development of alcohol use.

Together, these findings suggest that anxiety disorders largely precede the development of AUD. Interestingly, the data are most consistent in studies of emerging young adults and adults, with mixed findings in adolescence. While these differences may be the result of methodological variations, such as questionnaires vs. clinical interviews, they might also be the result of relatively lower rates of anxiety disorders and drinking during adolescence. Of interest, both generalized anxiety disorders and social anxiety disorders were predictors of later AUD, consistent with population-level studies investigating co-occurrence of anxiety disorder types and AUD (e.g., Grant et al. 2015).

2.2 Anxiety Co-occurring with Alcohol Use

Prevalence of Co-occurring Anxiety Disorders and AUD Recognition of anxiety as a common co-occurring disorder existed colloquially, but the relation between anxiety and AUD was only demonstrated with standard diagnostic methods 25 years ago using the DSM-III. Kessler et al. (1997) and Kushner et al. (1999) used epidemiological datasets to assess co-occurring disorders across the DSM-III, revealing a two-to-fivefold increase in the likelihood of having an anxiety disorder and an AUD. Additional epidemiological studies confirmed the co-occurrence of anxiety and AUD using later versions of the DSM: DSM-IV (Hasin et al. 2007; Schneier et al. 2010) and DSM-5 (Grant et al. 2015). These studies found that individuals with an anxiety disorder are four times more likely to have a co-occurring AUD (Hasin et al. 2007; Kessler et al. 1997). Notably, criteria for both anxiety disorders and AUDs changed across DSM-III, IV, and 5, which may influence the increase in co-occurrence rates – or suggests increased accuracy of capturing complex disorders. Despite changes across diagnostic criteria, co-occurrence of anxiety disorders and AUD is now a well-established association.

The most recent National Epidemiologic Survey on Alcohol and Related Conditions collected data from over 36,000 US adults on past-year and lifetime prevalence of AUD and psychiatric disorders using DSM-5 criteria (Grant et al. 2015). Individuals with any lifetime AUD diagnosis were 30% more likely to have an anxiety disorder, while those with severe lifetime AUD were 40% more likely to have an anxiety disorder. Controlling for sociodemographic variables (e.g., race, sex, income level) weakened the association between alcohol use and anxiety, ultimately suggesting important underlying factors that increase the likelihood of developing both AUD and anxiety disorders. Interestingly, controlling for other co-occurring disorders did not impact the relation between AUD and anxiety disorders – suggesting a specific link between anxiety and AUD.

Anxiety Maintaining Alcohol Use In addition to anxiety co-occurring with AUD, anxiety can impact how and why an individual drinks. Motivations for alcohol use generally reflect positive and negative reinforcement: initially, individuals may drink for the rewarding properties of alcohol (positive reinforcement); however, as chronic drinking develops, individuals may drink to avoid negative emotions and/or withdrawal symptoms (negative reinforcement) (Koob 2021). Considering that anxiety disorders often precede development of AUD, it is posited that those with anxiety disorders use alcohol for its negative reinforcing effects and transition faster to using alcohol for its negatively reinforcing properties. Several studies in humans support the negative reinforcing properties of alcohol in anxiety disorders. Kushner et al. (2011) found that individuals with an anxiety disorder transitioned more rapidly from first-time regular drinking and getting drunk to severe AUD compared with those without an anxiety disorder. Menary et al. (2011) found that among adults with an anxiety disorder, approximately 20% report self-medicating with alcohol. Adults who self-medicate with alcohol also drink significantly more than those who do not self-medicate, and those who self-medicate with alcohol are more likely to develop an AUD (Menary et al. 2011). Importantly, self-medication with alcohol in people with anxiety disorders also leads to greater utilization of mental health services and lower quality of life (Robinson et al. 2009). While not all people with anxiety engage in the negative reinforcing aspects of alcohol, motivations for alcohol drinking may be an important factor when looking toward improving treatments for AUD.

Anxiety Development During Alcohol Withdrawal and Abstinence Anxiety is common during alcohol withdrawal, as cessation of alcohol use results in increased stress reactivity and the emergence of anxiety symptoms (Koob 2013; Koob and Volkow 2010, 2016). During acute withdrawal, higher drinking frequency in adults with AUD predicts higher anxiety, craving, and cortisol levels (McCaul et al. 2017). In addition, adults with AUD experiencing alcohol withdrawal exhibit higher anxiety levels over a three-week study period compared with healthy controls (Petit et al. 2020). Individuals with an AUD who have a co-occurring anxiety disorder also experience more intense withdrawal symptoms than those without a co-occurring anxiety disorder (Johnston et al. 1991). Beyond acute withdrawal symptoms, anxiety generally decreases during early abstinence (2–4 weeks; Zywiak et al. 1996). However, there is significant heterogeneity, with many people continuing to experience anxiety months to years into abstinence (Heilig et al. 2010). For example, individuals with an AUD and co-occurring anxiety disorder have higher anxiety and depression scores after 18 months of abstinence (Wolitzky-Taylor et al. 2015). Finally, and most importantly, several studies have shown that anxiety contributes to high relapse rates; individuals with high trait anxiety or an anxiety disorder are 2–5 times more likely to relapse within one year of abstinence (e.g., Schellekens et al. 2015; Willinger et al. 2002; Zywiak et al. 1996). The relation between anxiety and alcohol use continues during alcohol abstinence – anxiety increases during acute withdrawal and early alcohol abstinence.

2.3 Conclusion

Together, these data suggest anxiety shapes the development, prevalence, and maintenance of AUD. Findings from longitudinal studies reveal that anxiety disorders increase the risk for developing an AUD, which is consistent across several diagnostic measures and study designs. Epidemiological studies also show a high co-occurrence between anxiety disorders and AUDs. Finally, anxiety plays an important role in the maintenance of alcohol use and symptoms experienced during alcohol withdrawal abstinence. Understanding alcohol use through the lens of anxiety reveals a complex and cyclical relation – anxiety disorders are a significant predictor of later alcohol use, and alcohol use can increase anxiety symptoms – suggesting an important interaction between alcohol and anxiety. To better understand the relation between alcohol use and anxiety and to improve treatment for AUDs, we need to investigate their intersection in the brain.

3 Neural Networks Involved in Anxiety and Alcohol Use Disorder

Studies investigating the neural networks underlying anxiety disorders or AUD suggest distinct, yet overlapping, neural networks. Comparing the anxiety and AUD networks highlights overlap in both stress and reward brain systems. We previously proposed an intersection of anxiety and addiction, based on evidence that there is substantial overlap between anxiety and addiction neurocircuitry (Avery et al. 2016). Here, we expand that review to discuss major overlaps between anxiety and AUD neural networks, including those related to reward and cognition.

3.1 Brain Networks in Anxiety

Anxiety is characterized by hypervigilance related to distant, vague, or unpredictable threats (Brown et al. 1998; Zinbarg and Barlow 1996). Processing and coordinating responses to unpredictable threats involves a broad network of brain regions, including the bed nucleus of the stria terminalis (BNST), amygdala, hippocampus, hypothalamus, insula, ventromedial prefrontal cortex (vmPFC), dorsomedial and dorsolateral prefrontal cortex (PFC), and mid-cingulate and anterior cingulate cortex (ACC) (Abend 2023; Chavanne and Robinson 2021). Within this network, different brain regions coordinate anticipation, responses, and processing of unpredictable threats (Abend 2023; Somerville and Abend 2024). Many studies have investigated brain differences between anxious and healthy people; we review a select number of meta- and mega-analyses to provide a broad understanding of anxiety's impact on the brain.

Structural Brain Measures Measures of brain structure, like gray matter volume, can provide clues about structural abnormalities that may be associated with anxiety. The Enhancing NeuroImaging Genetics through Meta-Analysis (ENIGMA)-Anxiety Working Group reported a mega-analysis of structural neuroimaging data (Harrewijn et al. 2021). The sample ($n = 4{,}394$) included people with a lifetime or current generalized anxiety disorder (25%) and healthy controls. The mega-analysis revealed a group x sex interaction, such that men with a generalized anxiety disorder had a larger gray matter volume in the ventral diencephalon than control men; no differences were detected between women. The ventral diencephalon is comprised of multiple small brain regions, including the hypothalamus, lateral and medial geniculate, and subthalamic nuclei, that are involved in reward and coordinating responses to anxiety. An additional mega-analysis from the ENIGMA-Anxiety Working Group (n = 3,890) compared people with a lifetime or current social anxiety disorder (29%) and healthy controls (Groenewold et al. 2023). The mega-analysis focused on volumetric differences between subcortical structures, revealing that individuals with social anxiety disorder have smaller putamen. The putamen is part of the dorsal striatum, which helps coordinate reward-related learning and movements associated with the reward learning. A post-hoc analysis by age (adolescents vs. adults) revealed that adults with social anxiety drove the effects in the putamen; adolescents showed no significant differences in subcortical volumes. These structural brain findings were intriguing because the differences did not emerge in the putative anxiety brain network, but instead were in brain regions involved in processing stimuli and coordinating responses to the environment.

Resting-State Functional Connectivity Measures Resting state functional connectivity provides a measure of correlated activity between brain regions in the absence of a task and is thought to provide a measure of the intrinsic structure of brain networks. Zugman et al. (2023) conducted a meta-analysis on 14 resting-state connectivity studies that used the amygdala as a seed region, based on the amygdala's role in fear and anxiety disorders (n = 572 subjects, 51% anxiety disorder). The primary finding revealed that people with an anxiety disorder had lower connectivity between the amygdala and the medial PFC, cingulate gyrus, and ACC. The authors also did an additional meta-analysis of 29 resting-state connectivity studies using any brain region as a seed, revealing no significant connectivity differences between groups (n = 1,506, 49% anxiety disorder). Xu et al. (2019) conducted a meta-analysis with 20 studies comparing healthy controls to individuals with an anxiety disorder (n = 974, 48% anxiety disorder). There were multiple group differences in resting-state functional connectivity across the brain. Relevant to anxiety-related brain regions, individuals with an anxiety disorder exhibited weaker connectivity between the following regions: cerebellum-dorsolateral PFC, posterior cingulate cortex-dorsolateral PFC, anterior insula-ACC, and anterior insula-dorsal ACC. Overall, resting-state functional connectivity studies reveal weaker connectivity between corticolimbic regions, which may reflect impaired ability to regulate fear and anxiety.

Task-Based Measures Task-based neuroimaging studies are an important avenue for understanding how people process and respond to stimuli. While responses to tasks likely result from differences in brain structure and resting state connectivity, tasks provide a direct probe of psychological processing and thus may reveal additional insights. A seminal meta-analysis from Etkin and Wager (2007) revealed that anxiety disorders were associated with higher activation of the amygdala, insula, hippocampus, and mid-cingulate cortex during negative emotional conditions. A meta-analysis by McTeague et al. (2020) similarly showed that people with anxiety disorders exhibited greater activation in the amygdala and hippocampus during emotional tasks.

In addition to emotion-processing tasks, tasks that provoke anxiety in humans (e.g., threat of shock, N-P-U tasks, CO_2 challenges) consistently recruit the BNST, amygdala, thalamus, hippocampus, insula, dorsal ACC, and dorsomedial PFC – all displaying greater activation in those with higher anxiety (e.g., Geng et al. 2018; Radoman et al. 2021; Somerville et al. 2010). These studies also report stronger functional connectivity from the amygdala to the dorsal ACC and dorsomedial PFC, as well as stronger connectivity from the dorsomedial PFC to the insula and hippocampus (e.g., Geng et al. 2018). Taken together, these findings suggest that altered anxiety network interactions underlie anxious processing.

Anxiety disorders are also characterized by avoidance behaviors related to potential threats or cues that trigger anxiety. Therefore, additional processes are at play when the brain interprets and responds to its environment, like evaluation of risk versus reward related to approaching or avoiding anxiogenic stimuli. Risk is overestimated in anxiety, causing an "inflation" of potential threat (Grupe and Nitschke 2013). Tasks that tap into risk versus reward evaluation include monetary incentive delay tasks and approach-avoidance conflict tasks. Risk versus reward evaluations require prefrontal cortex integration (dorsomedial PFC, dorsal ACC) with reward-processing regions, like the nucleus accumbens, caudate, and putamen; anxious individuals display less activation of reward-related regions (nucleus accumbens, caudate, putamen) when receiving a reward (e.g., Auerbach et al. 2021). In approach-avoidance conflict tasks, anxious individuals exhibit lower medial PFC activation (Aupperle and Paulus 2010). Overall, reward-related processing is altered in anxiety disorders, which may lead people with anxiety to be more susceptible to seeking reward through alcohol.

In summary, weaker corticolimbic connectivity at rest, coupled with stronger connectivity during tasks, suggests that there may be compensatory mechanisms at large. Specifically, weaker connectivity at rest may be due to poor baseline emotion regulation, while stronger connectivity during tasks may result from overcompensation to regulate emotion.

3.2 *Brain Networks in Alcohol Use Disorder*

Chronic alcohol use results in substantial, widespread impact across the brain. Alcohol is a neurotoxin that degrades gray and white matter structure and impairs coordination between brain regions associated with cognition (lateral and medial PFC, hippocampus) and emotion processing and regulation (BNST, amygdala, striatum, hypothalamus) (Koob and Volkow 2010, 2016). Many studies have investigated brain differences between people with AUDs and controls; we review a select number of seminal studies and meta-analyses to provide a broad understanding of alcohol's impact on the brain.

Structural Brain Measures Chronic alcohol use results in widespread structural changes in the brain, which was demonstrated in some of the earliest studies on alcohol use in humans (e.g., Courville 1955; Jernigan et al. 1982; Pfefferbaum et al. 1988). Xiao et al. (2015) conducted a meta-analysis of nine studies assessing gray matter differences between people with AUD and controls ($n = 655$, 45% AUD). People with AUD had significantly lower gray matter in the dorsal and rostral ACC, posterior ACC, striatum, and insula, compared with controls. Yang et al. (2016) conducted a meta-analysis of 12 studies that revealed more widespread gray matter changes ($n = 931$, 47% AUD). AUD was associated with lower gray matter volume in the insula, superior temporal gyrus, striatum, precentral gyrus, dorsolateral PFC, dorsal ACC, thalamus, and hippocampus, compared with controls. Recently, Li et al. (2021) conducted a meta-analysis of 23 studies comparing people with AUD with controls ($n = 1{,}724$, 49% AUD) revealing lower gray matter volume of the dorsolateral PFC, insula, and dorsal ACC in people with AUD. These findings suggest that chronic alcohol use is associated with consistent gray matter loss in regions associated with emotion processing and regulation.

Resting-State Functional Connectivity Measures Resting-state functional connectivity is also altered in AUD. Rodríguez et al. (2025) systematically reviewed studies that analyzed resting-state data using seed-based functional connectivity ($n = 7$). Overall, these studies found that people with AUD, compared with controls, exhibited stronger connectivity of the anterior insula with the hippocampus and medial orbitofrontal cortex, and between the posterior and anterior cingulate cortices to the inferior frontal gyrus and supramarginal gyrus. In addition, AUD was associated with weaker connectivity of the anterior insula with the superior parietal lobule, the thalamus with the anterior insula, and the ACC with the thalamus, medial frontal gyrus, and orbitofrontal cortex. Recently, Bottino et al. (2025) conducted a systematic review of 39 studies of AUD using a resting-state functional connectivity relevance matrix that quantified the frequency of brain regions associated with significant results in the literature. Overall, AUD was positively associated with connectivity between the insula, ACC, and superior temporal gyrus and between the inferior frontal gyrus and middle temporal pole. AUD was negatively associated with a connection between the superior temporal cortex and lingual gyrus, and between the middle temporal cortex and precuneus. The brain regions most associated with significant results in AUD were corticolimbic brain regions: the medial PFC, dorsolateral PFC, putamen, ACC (subgenual, pregenual, supracallosal), and insula. Overall, these findings suggest consistent differences in people with AUD across studies of brain structure and resting state functional connectivity, especially in corticolimbic brain regions relevant for anxiety processing and responding.

Task-Based Measures The most common tasks for determining the impact of alcohol on brain function relate to the three-stage models of AUD, which highlights the plasticity of brain networks continually adapting and shifting with continued alcohol use. The three-stage model outlines the stages of addiction: binge/intoxication, withdrawal/negative affect, and preoccupation/anticipation. More recently, the Addictions Neuroclinical Assessment, building on the National Institute of Mental

Health's Research Domain Criteria framework, has suggested measurable domains for each of the three stages: incentive salience, negative emotionality, and executive function (Cuthbert and Insel 2013; Kwako et al. 2016, 2017). Binge/intoxication (incentive salience) is characterized by increased reactivity to alcohol-related cues and impaired reward learning – higher striatum and thalamus activation are reported and are associated with a decreased threshold for reward. Withdrawal/negative affect (negative emotionality) is characterized by negative affect and stress resulting from the cessation of alcohol use – higher activation of the BNST, amygdala, nucleus accumbens, and hypothalamus are reported during alcohol withdrawal and abstinence and are associated with a decreased threshold for stress responses. Preoccupation/anticipation (executive function) is characterized by alcohol craving in response to stress- and alcohol-related cues and deficits in executive functioning – higher insula activation is combined with lower medial PFC activation, and these differences are associated with increased craving and decreased executive functioning.

The earliest task-based fMRI studies of AUD focused on assessing brain alterations to alcohol cues. These studies found higher activation of the striatum, dorsal ACC, and medial and lateral PFC to alcohol cues in individuals with AUD compared with controls (Braus et al. 2001; George et al. 2001; Grüsser et al. 2004; Modell and Mountz 1995). Additionally, these early studies found a relation between alcohol cue-induced activity and relapse in people receiving treatment, where higher activation of the striatum and PFC in response to alcohol cues predicted relapse three months later (Braus et al. 2001; Grüsser et al. 2004). A recent meta-analysis from Zeng et al. (2021) investigated 21 studies of alcohol cue reactivity in people with an AUD ($n = 795$), showing that people with AUD had higher cue-related activation of the insula and posterior cingulate cortex. The authors also investigated 17 studies comparing people with AUD with controls ($n = 817$, 56% AUD), revealing higher activation of the dorsal ACC, mid-cingulate cortex, and the ventromedial PFC in people with an AUD. In addition, in 14 studies of people receiving treatment for an AUD ($n = 338$), following treatment (relative to placebo or pre-treatment) there was higher activation of the precentral gyrus and lower activation of the rostral medial PFC, dorsolateral PFC, caudate, and insula.

Studies of negative emotionality have largely used emotional face tasks to probe emotion processing. Both acute and chronic alcohol use are associated with lesser activation of the amygdala, dorsal ACC, and hippocampus to both positive and negative emotional faces (Gilman et al. 2012; Marinkovic et al. 2009; Salloum et al. 2007). Further, chronic alcohol use impacts task-based functional connectivity when responding to emotional stimuli, such that individuals with AUD display less connectivity between the amygdala and dorsal ACC and BNST, and between the insula and ventromedial PFC, ventrolateral PFC, and medial orbitofrontal cortex (Kienast et al. 2013; O'Daly et al. 2012). To our knowledge, there are no meta-analyses that investigate studies of emotion processing in AUD. However, across the select studies reviewed, there are consistencies: lesser activation and connectivity of brain regions involved in emotion processing and regulation in AUDs.

Studies of executive function in AUD use a variety of cognitive tasks related to set/goal shifting, working memory, and processing speed. Quaglieri et al. (2020) conducted a meta-analysis of 13 neuroimaging studies investigating decision-making (monetary incentive delay tasks, delay discounting tasks, and Go/NoGo tasks) in AUD ($n = 654$, 47% AUD). During decision-making, people with AUD exhibited higher activation of the putamen, dorsomedial PFC, and precuneus and lower activation of dorsolateral PFC, dorsal ACC, mid-cingulate cortex, and putamen, compared with controls. Cao et al. (2023) performed a meta-analysis of 13 studies investigating response inhibition comparing people with AUD with controls ($n = 584$, 43% AUD). People with AUD exhibited higher activation of the medial orbitofrontal cortex, dorsal ACC, paracingulate gyrus, striatum, and postcentral gyrus. People with AUD also exhibited lower activation of the dorsolateral PFC, inferior parietal gyrus, middle temporal gyrus, and superior temporal gyrus.

Chronic alcohol use results in widespread and pervasive changes to the brain, spanning fundamental regions involved in memory and executive function, to regions necessary for emotional processing and regulation. Notably, impacts of alcohol are consistently reported across imaging modalities (structural, resting-state, task-based).

3.3 *Overlapping Neural Network in Anxiety and Alcohol Use*

Anxiety and alcohol are associated with distinct, yet overlapping, brain networks (Fig. 1). Overlap in the neural networks associated with anxiety and alcohol use is of particular interest and concern. First, the overlap is of interest considering the prominent longitudinal data that suggests that having an anxiety disorder confers risk for developing an AUD. Second, the overlap is of concern considering that anxiety disorders, stress processing, and the ability to regulate stress may impact an individual's ability to remain abstinent. Anxiety and alcohol use shape brain function in similar ways, suggesting an important lens through which to understand the development, maintenance, and treatment of AUDs.

4 Role of Anxiety and Alcohol Use Disorder on Brain Function

Few studies have directly investigated the specific impact of anxiety on AUD. However, a related area – the relation between stress and AUD – has received more attention. Stress and anxiety are similar physiological responses that differ in

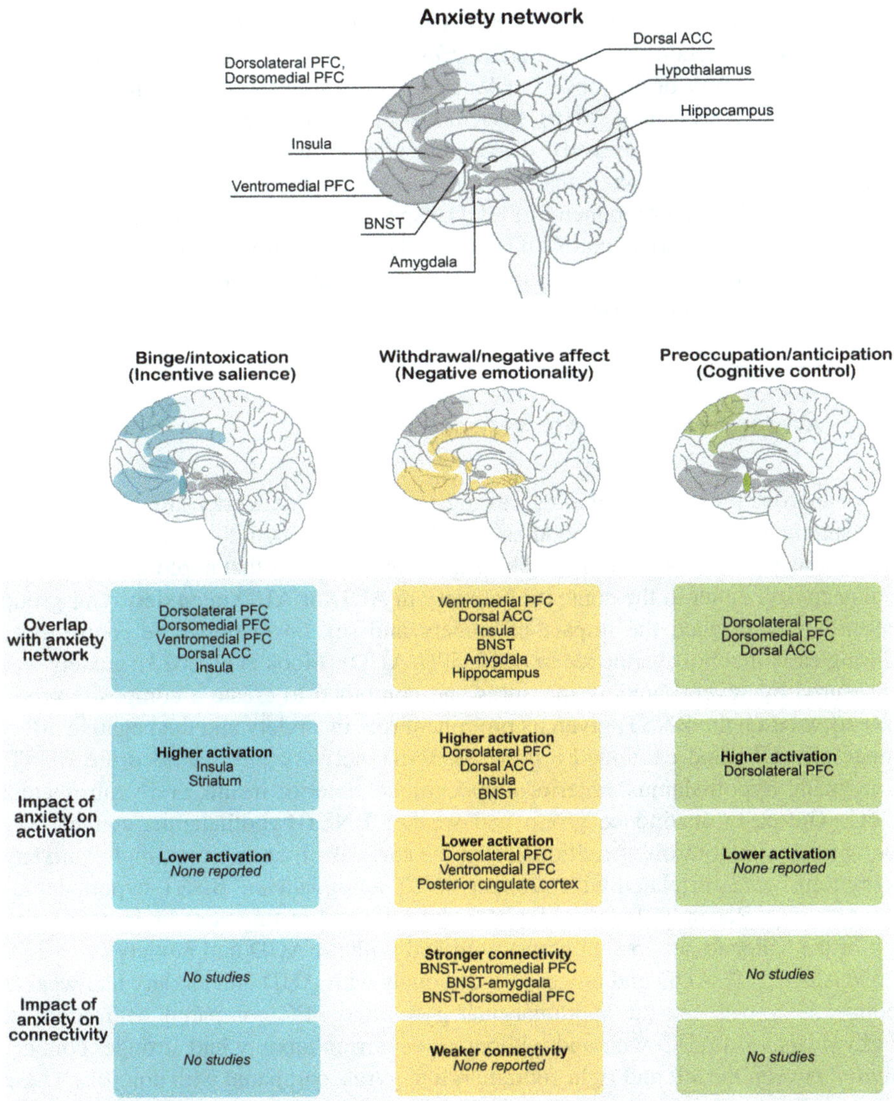

Fig. 1 Overlapping brain networks in anxiety and alcohol use disorders. Top panel ("Anxiety network") displays the canonical regions associated with anxiety and anxiety disorders. Middle panel displays the three stages of AUDs with regions associated with each stage highlighted; regions unshared with the anxiety network are gray or solid color. Bottom table shows the overlapping regions and highlights research reviewed in this paper determining how anxiety impacts brain activation and connectivity in each stage of AUD. *Regions are shown at approximate anatomical locations for illustrative purposes. The only non-shared region is the striatum (binge/intoxication and preoccupation/anticipation); however, note that the striatum is frequently associated with anxiety disorders but is not in the canonical anxiety network. PFC* prefrontal cortex, *ACC* anterior cingulate cortex

their scope and definition: anxiety refers to the physiological response related to unknown or uncertain threats, whereas stress is a physiological experience that can result from anxiety or any short- or long-term exposure to real or perceived threat (i.e., stressor) (Daviu et al. 2019). Both anxiety and stress result in altered responses to emotionally and cognitively demanding stimuli and tasks, leading to similar impacts on brain function. Thus, we will also cover literature outlining the impact of stress on different components of AUD, especially in cases where the literature on the intersection of anxiety and AUD is scant. The intersection of anxiety and alcohol is at the forefront of alcohol research, resulting in a limited amount of research that focuses on their intersection.

4.1 Resting-State Neural Networks

Two approaches to investigating resting-state neural networks are (1) including measurements of anxiety in studies of AUD and (2) including measurements of alcohol use in studies of anxiety. There are limited reports that explore resting-state connectivity, either in the context of anxiety in AUD or AUD in anxiety. Our group recently investigated the impact of anxiety and sex on resting-state connectivity during early alcohol abstinence ($n = 40$, 50% AUD) (Flook et al. 2023); anxiety was measured by several anxiety measures and combined to create a composite score. We focused on the BNST, given its prominent role in anxiety and the negative affect stage of AUD, and examined a putative BNST network that included the BNST, amygdala, hypothalamus, anterior hippocampus, anterior insula, and ventromedial PFC. The early abstinence group had weaker BNST-hypothalamus connectivity compared with controls. Additionally, in the early abstinence group, higher anxiety symptoms were correlated with stronger BNST-amygdala and BNST-hypothalamus connectivity. Recently, Patel et al. (2025) investigated resting-state brain connectivity of the salience network in adults with and without AUD and anxiety ($n = 264$, 25% AUD, 25% AUD and anxiety). Individuals with AUD and anxiety had weaker connectivity between the supramarginal gyrus and PFC compared with controls. Individuals with AUD with and without co-occurring anxiety had stronger connectivity between the left and right supramarginal gyrus, compared with controls. These limited data suggest that people with AUD in early abstinence have alterations in brain regions at the intersection of alcohol and anxiety networks.

4.2 Tasks Probing General Emotion Processing

Cognition and Reward Bach et al. (2024) investigated stress induction on brain responses to alcohol cues in people with AUD ($n = 91$, 100% AUD). People with AUD who underwent psychosocial stress (vs. control condition) exhibited higher insula activation to alcohol cues. Tenekedjieva et al. (2024) investigated the impact of different transdiagnostic factors on brain function in three cognition and reward tasks in people with AUD ($n = 86$, 73% men). Participants completed tasks probing reward (monetary incentive delay task), craving (cue reactivity task), and cognition (Go/NoGo task). A factor analysis of the Depression Anxiety and Stress Scale-42 and Posttraumatic Stress Disorder Scale-5 revealed four factors: trauma distress, negative affect, hyperarousal, and somatic anxiety – thus, three out of four symptom factors described anxiety (negative affect, hyperarousal, and somatic anxiety). During the reward task, all anxiety factors were associated with lower posterior parietal cortex activation. There were also some factor-specific associations during the reward task: (1) higher negative affect was associated with decreased caudate and putamen activation, (2) higher hyperarousal was associated with decreased caudate activation, and (3) higher somatic anxiety was associated with decreased putamen activation. During the craving task, negative affect and hyperarousal were associated with lower dorsolateral PFC activation, negative affect was associated with lower ACC activation, and hyperarousal was associated with lower posterior cingulate cortex activation. Finally, during the cognition task, negative affect was associated with higher dorsolateral PFC activation. Thus, anxiety was associated with alterations in multiple brain regions during reward, craving, and cognition, suggesting that anxiety impacts many underlying components of AUD.

Emotion Processing Kienast et al. (2013) investigated the impact of anxiety on brain activation to aversive (vs. neutral) images in men recently abstinent from an AUD and controls ($n = 24$, 46% AUD); anxiety was measured using the State-Trait Anxiety Inventory (Spielberger et al. 1983). When viewing aversive images, men with AUD had lower activation of the posterior cingulate cortex, cuneus/precuneus, and caudate, compared with controls. Anxiety was not associated with differences in brain activation. Men with AUD had higher trait anxiety and stronger amygdala-dorsal ACC connectivity, compared with controls, but there was no correlation between anxiety and brain connectivity. MacIlvane and colleagues (2020) investigated brain responses during an emotional face-matching task in three groups ($n = 232$): AUD and co-occurring anxiety (39%), AUD without co-occurring anxiety (17%), and controls (44%). The co-occurring AUD and anxiety group had higher face-matching accuracy and stronger temporoparietal junction and supplementary motor area activation (neutral faces), compared with controls. The AUD without anxiety group exhibited higher supramarginal gyrus activation (fearful faces > control). When comparing the AUD groups with and without co-occurring anxiety, the anxiety group exhibited lower cerebellum and inferior frontal gyrus activation (fearful > neutral).

These data suggest that there are complex associations between AUD with and without a co-occurring anxiety disorder, and these associations impact fundamental brain regions related to stimuli processing.

4.3 Tasks Probing Anxiety-Related Brain Networks

Stress Lee et al. (2023) investigated the impact of social stress on brain activation in adults with AUD and high trait anxiety ($n = 39$) and whether pexacerfont, a stress system antagonist, would impact brain activation. Anxiety was measured with the State-Trait Anxiety Inventory-Trait version (cutoff score = 40). Participants completed the Trier Social Stress Task, then watched video recordings of themselves and other participants completing the social stress task while in the scanner. Adults with AUD and high trait anxiety had increased activation of the insula, ACC, ventrolateral PFC, and dorsolateral PFC when viewing themselves complete the social stress task; there was no effect of pexacerfont.

In addition to social stress, several studies have used two well-validated tasks that induce stress by either developing individualized stress scripts (e.g., retelling a recently stressful event the participant experienced) or showing stressful images. Seo et al. (2013) investigated brain activation to individualized stress vs. neutral script ($n = 60$, 50% AUD). Compared with controls, individuals with AUD exhibited higher activation during stress imagery (vs. neutral) in the dorsomedial PFC, ventromedial PFC, dorsal ACC, posterior cingulate cortex, hippocampus, amygdala, caudate, temporal lobe, and cerebellum. Blaine et al. (2020) investigated brain activation to stress images in two samples. The first sample included individuals with an AUD during acute abstinence (2–4 weeks) and controls ($n = 87$, 51% AUD). Compared with controls, individuals with AUD in acute abstinence showed lower activation during stress (vs. neutral) images in the dorsomedial PFC, ventromedial PFC, rostral ACC, and striatum. The second sample included individuals in AUD treatment ($n = 69$). Shorter abstinence length predicted lower activation during stress (vs. neutral) images in the dorsomedial PFC, ventromedial PFC, rostral ACC, insula, hippocampus, amygdala, and striatum. Goldfarb et al. (2022) investigated whole-brain connectivity networks during a stressful images task between light and risky drinkers ($n = 104$, 51% risky drinkers). While both groups exhibited similar networks, risky drinkers exhibited a network with stronger connections with the dorsolateral PFC, dorsal ACC, insula, primary visual and motor cortices, superior temporal gyrus, cerebellum, and ventral tegmental area. Radoman et al. (2024) investigated brain activation during a task with three conditions: alcohol images, stress images, and neutral images in treatment-seeking individuals with AUD ($n = 77$). Individuals with AUD showed higher activation of the orbitofrontal cortex, amygdala, hippocampus, dorsal caudate, and putamen during stress (vs. neutral) images. Individuals with AUD also showed lesser activation of the dorsomedial PFC, dorsolateral PFC, ventromedial PFC, ventrolateral PFC, and ventral caudate during stress (vs. neutral) images. Finally, Seo et al. (2024) investigated brain activation during the stressful images task in light and risky drinkers ($n = 48$, 54% risky drinkers). Risky drinkers exhibited lower activation during stress images in the ventromedial PFC, dorsal ACC, motor cortex, insula, striatum, and temporal gyrus, compared with the light drinkers.

Across these data, there is consistent recruitment of key stress and anxiety brain regions: the dorsomedial PFC, ventromedial PFC, dorsal ACC, insula, amygdala, and striatum. However, direction of effects differs across task type (script vs. imagery) and sample (current drinkers, risky drinkers, or abstinent), suggesting a complex relation between responding to stress and stage of AUD.

Uncertainty and Anticipation Wilcox et al. (2020) investigated brain responses to threat and the association with anxiety in adults with an AUD ($n = 52$). Participants completed an anticipatory anxiety task that varied the threat of a heat stimulus, where brain activation was measured during early, middle, and late phases of the task. Overall, people with AUD reported higher anxiety during high (vs. low) threat stimuli. During the early phase of threat, people with AUD exhibited lower activation of the subgenual ACC/ventromedial PFC (vs. low threat). During the middle phase of threat, people with AUD also exhibited lower activation of the posterior cingulate cortex, dorsolateral PFC, middle and superior frontal gyri, and precuneus (vs. low threat).

Gorka et al. (2020) used the "N-P-U" task to study people with lifetime AUD and controls ($n = 95$, 38% AUD). The AUD group displayed greater activation of the insula and dorsal ACC to unpredictable threat, compared with controls. Later, Gorka et al. (2023) investigated whether brain activation during the "N-P-U" task predicted later drinking habits in youth at high risk for AUD ($n = 91$). Greater activation in the anterior insula and dorsal ACC to unpredictable threat at baseline predicted higher binge drinking one year later.

Our group recently investigated brain activation and BNST functional connectivity during an unpredictable threat task in early abstinence from an AUD and controls ($n = 40$, 50% AUD) (Zabik et al. 2024). In the early abstinence group, higher anxiety was correlated with higher BNST activation during unpredictable threat, compared with controls. There was an interaction with sex; men in early abstinence with higher anxiety exhibited higher activation in the insula, dorsal ACC, and rostral ACC during unpredictable threat, compared with control men. For BNST functional connectivity during unpredictable threat, adults in early abstinence with higher anxiety displayed stronger BNST connectivity with the amygdala, ventromedial PFC, and dorsomedial PFC. Alcohol-abstinent men also showed stronger BNST-ventromedial PFC connectivity than control men. Women in early abstinence had weaker BNST-ventromedial PFC and BNST-thalamus connectivity than control women. Thus, both sex and anxiety impacted brain responses to unpredictable threat in early abstinence.

Several conclusions can be drawn from these data. First, studies of anxiety using unpredictable threat tasks are a recent development in AUD research and offer a new insight into brain processes associated with chronic alcohol use. Second, emerging evidence suggests there are consistent alterations in anxiety brain networks, including heightened insula and dorsal ACC activation. Third, sex differences are seen in AUDs and affect anxiety-related brain networks.

4.4 Conclusion

In summary, findings from task-based fMRI studies converge on alterations in the BNST, insula, dorsal ACC, and medial and lateral PFC subregions (dorsomedial, ventromedial, and dorsolateral). The BNST is the hub of anxiety in the brain and is impacted by chronic alcohol use, causing hyperactivation during alcohol abstinence (Avery et al. 2016). The insula is largely responsible for processing internal information – like anxiety – and is altered by chronic alcohol use (Uddin et al. 2017). The dorsal ACC also has numerous unique roles relevant to anxiety and alcohol use; the dorsal ACC coordinates responses to threatening and anticipatory information, monitors for errors and rewards in the environment, and regulates motor control related to threat and reward (Bush et al. 2002; Knutson et al. 2000; Mesulam 1990; Murtha et al. 1996). Finally, the medial and lateral PFC are the primary integrators of emotional and cognitive information, where they balance reward processing and decision-making to guide behavior (Lindquist et al. 2012). Together, these regions represent critical hubs for emotional states and behaviors resulting from them – ultimately making them susceptible to anxiety and alcohol use. Of note, while the studies reviewed included people with current AUD, with lifetime AUD, and recently abstinent, no simple differences emerge in brain activation across these groups. The direction of effects was largely task-dependent: lower activation was observed during cognitive and emotion processing tasks, and higher activation was observed for alcohol cue and unpredictable threat tasks.

5 Anxiety in the Treatment of Alcohol Use Disorder

Thus far, we have discussed the impact anxiety has on brain measures in AUD. These findings suggest that anxiety plays a significant role in acute and chronic alcohol use, as well as abstinence. Importantly, anxiety shapes the course of AUD beyond development – the following sections will review how anxiety impacts alcohol treatment and the need for anxiety-focused treatments in AUD.

5.1 Anxiety as a Barrier to Recovery

In the early 1990s, Johnston et al. (1991) characterized AUD treatment differences between people with and without a co-occurring anxiety disorder (n = 33, 52% anxiety). During the first 21 days of abstinence, men with AUD and co-occurring anxiety had more severe withdrawal symptoms. A follow-up study in the same sample revealed that the AUD with anxiety group also had significantly elevated levels of self-reported anxiety across the first 21 days of abstinence (Thevos et al. 1991).

Severe withdrawal symptoms and anxiety may also impact recovery from an AUD. Kushner et al. (2005) studied adults with AUD, with or without a co-occurring anxiety disorder, at treatment entry and 120 days after treatment onset ($n = 53$, 55% anxiety). Adults with an AUD and co-occurring anxiety disorder had more severe withdrawal symptoms, a higher number of drinks consumed 90 days after abstinence, fewer days to first drink, and fewer days to first binge-drinking occasion.

5.2 *Anxiety's Impact on Alcohol Use Disorder Treatments*

Milivojevic et al. (2020) investigated whether prazosin treatment for AUD, as well as lifetime anxiety diagnosis, would impact acute measurements of anxiety and stress. People with AUD received prazosin or placebo and completed a stress-induced imagery task to capture acute anxiety, stress-related alcohol cravings, and cortisol levels ($n = 40$). People with AUD who had an anxiety disorder exhibited higher anxiety ratings across the task, with no effect of medication on those ratings. A treatment x anxiety diagnosis interaction was detected on cortisol measures: adults with AUD without anxiety who received placebo had higher cortisol responses to stress- and alcohol-related cues, with no cortisol changes in the prazosin group. These findings suggest that a lifetime diagnosis of anxiety may not impact prazosin treatment.

A meta-analysis from Agabio et al. (2021) explored the impact of anxiety on randomized control trials of baclofen treatment (GABA agonist) for AUD. For studies that included participants with clinically significant levels of anxiety, baclofen resulted in more days abstinent than in the placebo group. For studies that included participants with subclinical levels of anxiety, no difference was detected between baclofen and placebo on days abstinent. Thus, baclofen was more effective for people with significant anxiety.

5.3 *Anxiety-Based Interventions for Alcohol Use Disorder*

Schadé et al. (2005) investigated whether additional anxiety treatment (cognitive behavioral therapy and medication) would improve outcomes for adults with AUD and co-occurring anxiety disorders ($n = 96$). Participants who received dual AUD and anxiety treatment had a significant reduction in anxiety levels but no differences in relapse rates, compared with the AUD treatment-only group. Kushner et al. (2009) used a similar clinical trial design where participants either received AUD treatment as usual or AUD treatment and cognitive behavioral therapy for panic disorder ($n = 54$). The combination of AUD treatment and cognitive behavioral therapy was associated with a decrease in anxiety/panic symptoms and AUD symptoms, compared with the AUD treatment as usual group. Kushner et al. (2013) investigated whether mindfulness-based treatment (vs. cognitive behavioral therapy) would decrease anxiety symptoms and motivations to drink in adults with AUD and co-occurring anxiety disorder ($n = 247$). Cognitive behavioral therapy was associated with lower relapse rates, compared with the mindfulness-based therapy group.

5.4 Conclusions

Anxiety continues to shape AUD, even when individuals are abstinent, by impacting the effectiveness of treatments. These data highlight the importance of measuring anxiety in people seeking treatment to determine who may benefit more from treatments that target anxiety. While many novel treatments are being tested in clinical trials, it is worth noting that many individuals with co-occurring anxiety disorders are largely excluded from early-phase trials (Blanco et al. 2008; Hoertel et al. 2014). Thus, the studies reviewed represent a small sample of focused effort determining how to manage the intersection of anxiety and AUDs.

6 Conclusions and Future Directions

In this review, we described how anxiety shapes the development and maintenance of alcohol use. By highlighting the intersection of anxiety and AUD through longitudinal and brain imaging studies, we hope to underscore the critical importance and cyclical relation of anxiety and AUDs. By highlighting how anxiety shapes alcohol use, we hope to provide researchers and clinicians with important insights to inspire future research and clinical care – ultimately easing the shame and burden of alcohol use for individuals with AUDs and their families.

Below, we provide suggestions for future research and potential avenues for intervention.

1. *What can we do to halt the progress from anxiety to alcohol use?* Anxiety disorders develop during critical years of adolescence. Early identification and treatment could diminish the association between anxiety and alcohol use, as well as introduce positive coping mechanisms for young adults, therefore preventing the development of alcohol use.
2. *How can clinicians and researchers identify anxiety?* Measuring anxiety during alcohol abstinence will be a key avenue for gathering epidemiological data on the course of alcohol use beyond developmental milestones. Formal assessments of anxiety and anxiety screening in primary care will be necessary to understand the course and shaping of alcohol use and treatment.
3. *How can we examine the negative affect component of AUD in the clinic and lab?* An important knowledge gap in AUD is understanding how anxiety impacts each part of the addiction cycle. Researchers are beginning to collect and analyze data in the three domains proposed by the Addictions Neuroclinical Assessment

(Kwako et al. 2016, 2017). State anxiety measures before and after tasks will be a simple – yet effective – way to gauge how anxiety may shape different behaviors and brain processes in AUD. Additionally, studying interventions that target anxiety symptoms will be critical for people with co-occurring anxiety disorders. Alcohol craving is a focal point of treatment and has been successful; we believe the next step is to identify ways to treat anxiety in AUD. Negative affect is a chronic, lingering symptom that can drive craving and impulsive alcohol use.

Acknowledgments The authors would also like to acknowledge the funding sources that supported this work: the National Institute of Alcohol Abuse and Alcoholism (R01AA029127 to JUB, P60AA031124 to JUB, F32AA032170 to NLZ).

References

Abend R (2023) Understanding anxiety symptoms as aberrant defensive responding along the threat imminence continuum. Neurosci Biobehav Rev 152:105305. https://doi.org/10.1016/j.neubiorev.2023.105305

Agabio R, Baldwin DS, Amaro H, Leggio L, Sinclair JMA (2021) The influence of anxiety symptoms on clinical outcomes during baclofen treatment of alcohol use disorder: a systematic review and meta-analysis. Neurosci Biobehav Rev 125:296–313. https://doi.org/10.1016/j.neubiorev.2020.12.030

Auerbach RP, Pagliaccio D, Hubbard NA, Frosch I, Kremens R, Cosby E, Jones R, Siless V, Lo N, Henin A, Hofmann SG, Gabrieli JDE, Yendiki A, Whitfield-Gabrieli S, Pizzagalli DA (2021) Reward-related neural circuitry in depressed and anxious adolescents: a human connectome project. J Am Acad Child Adolesc Psychiatry 61(2):308. https://doi.org/10.1016/j.jaac.2021.04.014

Aupperle RL, Paulus MP (2010) Neural systems underlying approach and avoidance in anxiety disorders. Dialogues Clin Neurosci 12(4):517. https://doi.org/10.31887/DCNS.2010.12.4/raupperle

Avery SN, Clauss JA, Blackford JU (2016) The human BNST: functional role in anxiety and addiction. Neuropsychopharmacology 41(1):Article 1. https://doi.org/10.1038/npp.2015.185

Bach P, Zaiser J, Zimmermann S, Gessner T, Hoffmann S, Gerhardt S, Berhe O, Bekier NK, Abel M, Radler P, Langejürgen J, Tost H, Lenz B, Vollstädt-Klein S, Stallkamp J, Kirschbaum C, Kiefer F (2024) Stress-induced sensitization of insula activation predicts alcohol craving and alcohol use in alcohol use disorder. Biol Psychiatry 95(3):245–255. https://doi.org/10.1016/j.biopsych.2023.08.024

Blaine SK, Wemm S, Fogelman N, Lacadie C, Seo D, Scheinost D, Sinha R (2020) Prefrontal-striatal functional pathology is associated with alcohol abstinence days at treatment initiation and heavy drinking after treatment initiation. Am J Psychiatry 177(11):1048–1059. https://doi.org/10.1176/appi.ajp.2020.19070703

Blanco C, Olfson M, Okuda M, Nunes EV, Liu S-M, Hasin DS (2008) Generalizability of clinical trials for alcohol dependence to community samples. Drug Alcohol Depend 98(0):123–128. https://doi.org/10.1016/j.drugalcdep.2008.05.002

Bottino M, Bocková N, Poller NW, Smolka MN, Böhmer J, Walter H, Marxen M (2025) Relating functional connectivity and alcohol use disorder: a systematic review and derivation of relevance maps for regions and connections. Hum Brain Mapp 46(2):e70156. https://doi.org/10.1002/hbm.70156

Braus DF, Wrase J, Grüsser S, Hermann D, Ruf M, Flor H, Mann K, Heinz A (2001) Alcohol-associated stimuli activate the ventral striatum in abstinent alcoholics. J Neural Transm 108(7): 887–894. https://doi.org/10.1007/s007020170038

Brown TA, Chorpita BF, Barlow DH (1998) Structural relationships among dimensions of the DSM-IV anxiety and mood disorders and dimensions of negative affect, positive affect, and autonomic arousal. J Abnorm Psychol 107(2):179–192. https://doi.org/10.1037/0021-843x.107.2.179

Bush G, Vogt BA, Holmes J, Dale AM, Greve D, Jenike MA, Rosen BR (2002) Dorsal anterior cingulate cortex: a role in reward-based decision making. Proc Natl Acad Sci USA 99(1): 523–528. https://doi.org/10.1073/pnas.012470999

Cao Y, Tian F, Zeng J, Gong Q, Yang X, Jia Z (2023) The brain activity pattern in alcohol-use disorders under inhibition response task. J Psychiatr Res 163:127–134. https://doi.org/10.1016/j.jpsychires.2023.05.009

Chavanne AV, Robinson OJ (2021) The overlapping neurobiology of induced and pathological anxiety: a meta-analysis of functional neural activation. Am J Psychiatry 178(2):156–164. https://doi.org/10.1176/appi.ajp.2020.19111153

Courville CB (1955) Effects of alcohol on the nervous system of man. San Lucas Press, Romeo, p 102

Cuthbert BN, Insel TR (2013) Toward the future of psychiatric diagnosis: the seven pillars of RDoC. BMC Med 11(1):126. https://doi.org/10.1186/1741-7015-11-126

Daviu N, Bruchas MR, Moghaddam B, Sandi C, Beyeler A (2019) Neurobiological links between stress and anxiety. Neurobiol Stress 11:100191. https://doi.org/10.1016/j.ynstr.2019.100191

Etkin A, Wager TD (2007) Functional neuroimaging of anxiety: a meta-analysis of emotional processing in PTSD, social anxiety disorder, and specific phobia. Am J Psychiatry 164(10): 1476–1488. https://doi.org/10.1176/appi.ajp.2007.07030504

Flook EA, Feola B, Benningfield MM, Silveri MM, Winder DG, Blackford JU (2023) Alterations in BNST intrinsic functional connectivity in early abstinence from alcohol use disorder. Alcohol Alcoholism (Oxford, Oxfordshire) 58(3):298–307. https://doi.org/10.1093/alcalc/agad006

Fröjd S, Ranta K, Kaltiala-Heino R, Marttunen M (2011) Associations of social phobia and general anxiety with alcohol and drug use in a community sample of adolescents. Alcohol Alcohol 46(2):192–199. https://doi.org/10.1093/alcalc/agq096

Geng H, Wang Y, Gu R, Luo Y-J, Xu P, Huang Y, Li X (2018) Altered brain activation and connectivity during anticipation of uncertain threat in trait anxiety. Hum Brain Mapp 39(10): 3898. https://doi.org/10.1002/hbm.24219

George MS, Anton RF, Bloomer C, Teneback C, Drobes DJ, Lorberbaum JP, Nahas Z, Vincent DJ (2001) Activation of prefrontal cortex and anterior thalamus in alcoholic subjects on exposure to alcohol-specific cues. Arch Gen Psychiatry 58(4):345–352. https://doi.org/10.1001/archpsyc.58.4.345

Gilman JM, Ramchandani VA, Crouss T, Hommer DW (2012) Subjective and neural responses to intravenous alcohol in young adults with light and heavy drinking patterns. Neuropsychopharmacology 37(2):467–477. https://doi.org/10.1038/npp.2011.206

Goldfarb EV, Scheinost D, Fogelman N, Seo D, Sinha R (2022) High-risk drinkers engage distinct stress-predictive brain networks. Biol Psychiatry Cognit Neurosci Neuroimaging 7(8):805–813. https://doi.org/10.1016/j.bpsc.2022.02.010

Gorka SM, Kreutzer KA, Petrey KM, Radoman M, Phan KL (2020) Behavioral and neural sensitivity to uncertain threat in individuals with alcohol use disorder: associations with drinking behaviors and motives. Addict Biol 25(3):e12774. https://doi.org/10.1111/adb.12774

Gorka SM, Radoman M, Jimmy J, Kreutzer KA, Manzler C, Culp S (2023) Behavioral and brain reactivity to uncertain stress prospectively predicts binge drinking in youth. Neuropsychopharmacology 48(8):1194–1200. https://doi.org/10.1038/s41386-023-01571-x

Grant BF, Goldstein RB, Saha TD, Chou SP, Jung J, Zhang H, Pickering RP, Ruan WJ, Smith SM, Huang B, Hasin DS (2015) Epidemiology of DSM-5 alcohol use disorder. JAMA Psychiatry 72(8):757–766. https://doi.org/10.1001/jamapsychiatry.2015.0584

Groenewold NA, Bas-Hoogendam JM, Amod AR, Laansma MA, Van Velzen LS, Aghajani M, Hilbert K, Oh H, Salas R, Jackowski AP, Pan PM, Salum GA, Blair JR, Blair KS, Hirsch J, Pantazatos SP, Schneier FR, Talati A, Roelofs K et al (2023) Volume of subcortical brain regions in social anxiety disorder: mega-analytic results from 37 samples in the ENIGMA-Anxiety Working Group. Mol Psychiatry 28(3):1079–1089. https://doi.org/10.1038/s41380-022-01933-9

Grupe DW, Nitschke JB (2013) Uncertainty and anticipation in anxiety: an integrated neurobiological and psychological perspective. Nat Rev Neurosci 14(7):488. https://doi.org/10.1038/nrn3524

Grüsser SM, Wrase J, Klein S, Hermann D, Smolka MN, Ruf M, Weber-Fahr W, Flor H, Mann K, Braus DF, Heinz A (2004) Cue-induced activation of the striatum and medial prefrontal cortex is associated with subsequent relapse in abstinent alcoholics. Psychopharmacology 175(3): 296–302. https://doi.org/10.1007/s00213-004-1828-4

Harrewijn A, Cardinale EM, Groenewold NA, Bas-Hoogendam JM, Aghajani M, Hilbert K, Cardoner N, Porta-Casteràs D, Gosnell S, Salas R, Jackowski AP, Pan PM, Salum GA, Blair KS, Blair JR, Hammoud MZ, Milad MR, Burkhouse KL, Phan KL et al (2021) Cortical and subcortical brain structure in generalized anxiety disorder: findings from 28 research sites in the ENIGMA-Anxiety Working Group. Transl Psychiatry 11:502. https://doi.org/10.1038/s41398-021-01622-1

Hasin DS, Stinson FS, Ogburn E, Grant BF (2007) Prevalence, correlates, disability, and comorbidity of DSM-IV alcohol abuse and dependence in the United States: results from the national epidemiologic survey on alcohol and related conditions. Arch Gen Psychiatry 64(7):830–842. https://doi.org/10.1001/archpsyc.64.7.830

Heilig M, Egli M, Crabbe JC, Becker HC (2010) Acute withdrawal, protracted abstinence and negative affect in alcoholism: are they linked? Addict Biol 15(2):169–184. https://doi.org/10.1111/j.1369-1600.2009.00194.x

Hoertel N, Falissard B, Humphreys K, Gorwood P, Seigneurie A-S, Limosin F (2014) Do clinical trials of treatment of alcohol dependence adequately enroll participants with co-occurring independent mood and anxiety disorders? An analysis of data from the national epidemiologic survey on alcohol and related conditions (NESARC). J Clin Psychiatry 75(3):2963. https://doi.org/10.4088/JCP.13m08424

Jernigan TL, Zatz LM, Ahumada AJ, Pfefferbaum A, Tinklenberg JR, Moses JA (1982) CT measures of cerebrospinal fluid volume in alcoholics and normal volunteers. Psychiatry Res 7(1):9–17. https://doi.org/10.1016/0165-1781(82)90048-8

Johnston AL, Thevos AK, Randall CL, Anton RF (1991) Increased severity of alcohol withdrawal in in-patient alcoholics with a co-existing anxiety diagnosis. Br J Addict 86(6):719–725. https://doi.org/10.1111/j.1360-0443.1991.tb03098.x

Kessler RC, Crum RM, Warner LA, Nelson CB, Schulenberg J, Anthony JC (1997) Lifetime co-occurrence of DSM-III-R alcohol abuse and dependence with other psychiatric disorders in the National Comorbidity Survey. Arch Gen Psychiatry 54(4):313–321. https://doi.org/10.1001/archpsyc.1997.01830160031005

Kienast T, Schlagenhauf F, Rapp MA, Wrase J, Daig I, Buchholz H-G, Smolka MN, Gründer G, Kumakura Y, Cumming P, Charlet K, Bartenstein P, Hariri AR, Heinz A (2013) Dopamine-modulated aversive emotion processing fails in alcohol-dependent patients. Pharmacopsychiatry 46(4):130–136. https://doi.org/10.1055/s-0032-1331747

Klein RJ, Gyorda JA, Jacobson NC (2022) Anxiety, depression, and substance experimentation in childhood. PLoS One 17(5):e0265239. https://doi.org/10.1371/journal.pone.0265239

Knutson B, Westdorp A, Kaiser E, Hommer D (2000) FMRI visualization of brain activity during a monetary incentive delay task. NeuroImage 12(1):20–27. https://doi.org/10.1006/nimg.2000.0593

Koob GF (2013) Addiction is a reward deficit and stress surfeit disorder. Front Psychiatry 4:72. https://doi.org/10.3389/fpsyt.2013.00072

Koob GF (2021) Drug addiction: hyperkatifeia/negative reinforcement as a framework for medications development. Pharmacol Rev 73(1):163–201. https://doi.org/10.1124/pharmrev.120.000083

Koob GF, Volkow ND (2010) Neurocircuitry of addiction. Neuropsychopharmacology 35(1): Article 1. https://doi.org/10.1038/npp.2009.110

Koob GF, Volkow ND (2016) Neurobiology of addiction: a neurocircuitry analysis. Lancet Psychiatry 3(8):760–773. https://doi.org/10.1016/S2215-0366(16)00104-8

Kushner MG, Sher KJ, Erickson DJ (1999) Prospective analysis of the relation between DSM-III anxiety disorders and alcohol use disorders. Am J Psychiatry 156(5):723–732. https://doi.org/10.1176/ajp.156.5.723

Kushner MG, Abrams K, Thuras P, Hanson KL, Brekke M, Sletten S (2005) Follow-up study of anxiety disorder and alcohol dependence in comorbid alcoholism treatment patients. Alcohol Clin Exp Res 29(8):1432–1443. https://doi.org/10.1097/01.alc.0000175072.17623.f8

Kushner MG, Sletten S, Donahue C, Thuras P, Maurer E, Schneider A, Frye B, Van Demark J (2009) Cognitive-behavioral therapy for panic disorder in patients being treated for alcohol dependence: moderating effects of alcohol outcome expectancies. Addict Behav 34(6–7): 554–560. https://doi.org/10.1016/j.addbeh.2009.03.011

Kushner MG, Maurer E, Menary K, Thuras P (2011) Vulnerability to the rapid ("telescoped") development of alcohol dependence in individuals with anxiety disorder. J Stud Alcohol Drugs 72(6):1019–1027

Kushner MG, Maurer EW, Thuras P, Donahue C, Frye B, Menary KR, Hobbs J, Haeny AM, Van Demark J (2013) Hybrid cognitive behavioral therapy versus relaxation training for co-occurring anxiety and alcohol disorder: a randomized clinical trial. J Consult Clin Psychol 81(3):429–442. https://doi.org/10.1037/a0031301

Kwako LE, Momenan R, Litten RZ, Koob GF, Goldman D (2016) Addictions neuroclinical assessment: a neuroscience-based framework for addictive disorders. Biol Psychiatry 80(3): 179–189. https://doi.org/10.1016/j.biopsych.2015.10.024

Kwako LE, Momenan R, Grodin EN, Litten RZ, Koob GF, Goldman D (2017) Addictions neuroclinical assessment: a reverse translational approach. Neuropharmacology 122:254–264. https://doi.org/10.1016/j.neuropharm.2017.03.006

Lee MR, Rio D, Kwako L, George DT, Heilig M, Momenan R (2023) Corticotropin-Releasing Factor receptor 1 (CRF1) antagonism in patients with alcohol use disorder and high anxiety levels: effect on neural response during trier social stress test video feedback. Neuropsychopharmacology 48(5):816–820. https://doi.org/10.1038/s41386-022-01521-z

Li L, Yu H, Liu Y, Meng Y, Li X, Zhang C, Liang S, Li M, Guo W, Wang Q, Deng W, Ma X, Coid J, Li T (2021) Lower regional grey matter in alcohol use disorders: evidence from a voxel-based meta-analysis. BMC Psychiatry 21:247. https://doi.org/10.1186/s12888-021-03244-9

Lindquist KA, Wager TD, Kober H, Bliss-Moreau E, Barrett LF (2012) The brain basis of emotion: a meta-analytic review. Behav Brain Sci 35(3):121–143. https://doi.org/10.1017/S0140525X11000446

MacIlvane N, Fede SJ, Pearson EE, Diazgranados N, Momenan R (2020) A distinct neurophenotype of fearful face processing in alcohol use disorder with and without comorbid anxiety. Alcohol Clin Exp Res 44(11):2212–2224. https://doi.org/10.1111/acer.14465

Maggs JL, Schulenberg JE (2004) Trajectories of alcohol use during the transition to adulthood. Alcohol Res Health 28(4):195–201

Marinkovic K, Oscar-Berman M, Urban T, O'Reilly CE, Howard JA, Sawyer K, Harris GJ (2009) Alcoholism and dampened temporal limbic activation to emotional faces. Alcohol Clin Exp Res *33*(11):1880. https://doi.org/10.1111/j.1530-0277.2009.01026.x

Marquenie LA, Schadé A, van Balkom AJLM, Comijs HC, de Graaf R, Vollebergh W, van Dyck R, van den Brink W (2006) Origin of the comorbidity of anxiety disorders and alcohol dependence: findings of a general population study. Eur Addict Res 13(1):39–49. https://doi.org/10.1159/000095814

McCabe CJ, Brumback T, Brown SA, Meruelo AD (2023) Assessing cross-lagged associations between depression, anxiety, and binge drinking in the national consortium on alcohol and neurodevelopment in adolescence (NCANDA) study. Drug Alcohol Depend 243:109761. https://doi.org/10.1016/j.drugalcdep.2022.109761

McCaul ME, Hutton HE, Stephens MAC, Xu X, Wand GS (2017) Anxiety, anxiety sensitivity, and perceived stress as predictors of recent drinking, alcohol craving, and social stress response in heavy drinkers. Alcohol Clin Exp Res 41(4):836–845. https://doi.org/10.1111/acer.13350

McTeague LM, Rosenberg BM, Lopez JW, Carreon DM, Huemer J, Jiang Y, Chick CF, Eickhoff SB, Etkin A (2020) Identification of common neural circuit disruptions in emotional processing across psychiatric disorders. Am J Psychiatry 177(5):411–421. https://doi.org/10.1176/appi.ajp.2019.18111271

Menary KR, Kushner MG, Maurer E, Thuras P (2011) The prevalence and clinical implications of self-medication among individuals with anxiety disorders. J Anxiety Disord 25(3):335–339. https://doi.org/10.1016/j.janxdis.2010.10.006

Mesulam M-M (1990) Large-scale neurocognitive networks and distributed processing for attention, language, and memory. Ann Neurol 28(5):597–613. https://doi.org/10.1002/ana.410280502

Milivojevic V, Angarita GA, Hermes G, Sinha R, Fox HC (2020) Effects of Prazosin on provoked alcohol craving and autonomic and neuroendocrine response to stress in alcohol use disorder. Alcohol Clin Exp Res 44(7):1488–1496. https://doi.org/10.1111/acer.14378

Modell JG, Mountz JM (1995) Focal cerebral blood flow change during craving for alcohol measured by SPECT. J Neuropsychiatry Clin Neurosci 7(1):15–22. https://doi.org/10.1176/jnp.7.1.15

Murtha S, Chertkow H, Beauregard M, Dixon R, Evans A (1996) Anticipation causes increased blood flow to the anterior cingulate cortex. Hum Brain Mapp 4(2):103–112. https://doi.org/10.1002/(SICI)1097-0193(1996)4:2<103::AID-HBM2>3.0.CO;2-7

O'Daly OG, Trick L, Scaife J, Marshall J, Ball D, Phillips ML, Williams SS, Stephens DN, Duka T (2012) Withdrawal-associated increases and decreases in functional neural connectivity associated with altered emotional regulation in alcoholism. Neuropsychopharmacology 37(10):2267–2276. https://doi.org/10.1038/npp.2012.77

Patel DM, Poblete GF, Castellanos A, Salas R (2025) Functional brain connectivity of the salience network in alcohol use and anxiety disorders. J Affect Disord 377:124–133. https://doi.org/10.1016/j.jad.2025.02.045

Petit G, Deschietere G, Loas G, Luminet O, de Timary P (2020) Link between anhedonia and depression during early alcohol abstinence: gender matters. Alcohol Alcoholism (Oxford, Oxfordshire) *55*(1):71–77. https://doi.org/10.1093/alcalc/agz090

Pfefferbaum A, Rosenbloom M, Crusan K, Jernigan TL (1988) Brain CT changes in alcoholics: effects of age and alcohol consumption. Alcohol Clin Exp Res 12(1):81–87. https://doi.org/10.1111/j.1530-0277.1988.tb00137.x

Quaglieri A, Mari E, Boccia M, Piccardi L, Guariglia C, Giannini AM (2020) Brain network underlying executive functions in gambling and alcohol use disorders: an activation likelihood estimation meta-analysis of fMRI studies. Brain Sci 10(6):353. https://doi.org/10.3390/brainsci10060353

Radoman M, Lieberman L, Jimmy J, Gorka SM (2021) Shared and unique neural circuitry underlying temporally unpredictable threat and reward processing. Soc Cogn Affect Neurosci 16(4):370–382. https://doi.org/10.1093/scan/nsab006

Radoman M, Fogelman N, Lacadie C, Seo D, Sinha R (2024) Neural correlates of stress and alcohol cue-induced alcohol craving and of future heavy drinking: evidence of sex differences. Am J Psychiatry 181(5):412–422. https://doi.org/10.1176/appi.ajp.20230849

Robinson JA, Sareen J, Cox BJ, Bolton JM (2009) Correlates of self-medication for anxiety disorders: results from the national epidemiolgic survey on alcohol and related conditions. J Nerv Ment Dis 197(12):873–878. https://doi.org/10.1097/NMD.0b013e3181c299c2

Rodríguez GC, Russell MA, Claus ED (2025) Systematic review on resting-state fMRI in people with AUD and people who binge drink. Mol Psychiatry 30(2):752–762. https://doi.org/10.1038/s41380-024-02796-y

Salloum JB, Ramchandani VA, Bodurka J, Rawlings R, Momenan R, George D, Hommer DW (2007) Blunted rostral anterior cingulate response during a simplified decoding task of negative emotional facial expressions in alcoholic patients. Alcohol Clin Exp Res 31(9):1490–1504. https://doi.org/10.1111/j.1530-0277.2007.00447.x

SAMHSA (2022) Center for behavioral health statistics and quality. 2022 National survey on drug use and health. Table 5.9A – alcohol use disorder in past year among persons aged 12 or older, by age group and demographic characteristics: numbers in thousands. https://www.samhsa.gov/data/sites/default/files/reports/rpt42728/NSDUHDetailedTabs2022/NSDUHDetailedTabs2022/NSDUHDetTabsSect5pe2022.htm?s=5.9&#tab5.9a

SAMHSA (2023) Center for behavioral health statistics and quality. 2023 National survey on drug use and health. Table 4.11B – past year initiation of substance use: among people aged 12 to 49; mean age at first substance use: among past year initiates aged 12 to 49; by sex, numbers in thousands and averages, 2022 and 2023. https://www.samhsa.gov/data/report/2023-nsduh-detailed-tables

Schadé A, Marquenie LA, van Balkom AJLM, Koeter MWJ, de Beurs E, van den Brink W, van Dyck R (2005) The effectiveness of anxiety treatment on alcohol-dependent patients with a comorbid phobic disorder: a randomized controlled trial. Alcohol Clin Exp Res 29(5):794–800. https://doi.org/10.1097/01.alc.0000163511.24583.33

Schellekens AFA, de Jong CAJ, Buitelaar JK, Verkes RJ (2015) Co-morbid anxiety disorders predict early relapse after inpatient alcohol treatment. Eur Psychiatry 30(1):128–136. https://doi.org/10.1016/j.eurpsy.2013.08.006

Schneier FR, Foose TE, Hasin DS, Heimberg RG, Liu S-M, Grant BF, Blanco C (2010) Social anxiety disorder and alcohol use disorder co-morbidity in the national epidemiologic survey on alcohol and related conditions. Psychol Med 40(6):977–988. https://doi.org/10.1017/S0033291709991231

Seo D, Lacadie CM, Tuit K, Hong K-I, Constable RT, Sinha R (2013) Disrupted ventromedial prefrontal function, alcohol craving, and subsequent relapse risk. JAMA Psychiatry (Chicago, Ill) 70(7):727–739. https://doi.org/10.1001/jamapsychiatry.2013.762

Seo D, Martins JS, Sinha R (2024) Brain correlates and functional connectivity linking stress, autonomic dysregulation, and alcohol motivation. Neurobiol Stress 31:100645. https://doi.org/10.1016/j.ynstr.2024.100645

Somerville Y, Abend R (2024) The organization of anxiety symptoms along the threat imminence continuum. Springer, pp 1–29. https://doi.org/10.1007/7854_2024_548

Somerville LH, Whalen PJ, Kelley WM (2010) Human bed nucleus of the stria terminalis indexes hypervigilant threat monitoring. Biol Psychiatry 68(5):416–424. https://doi.org/10.1016/j.biopsych.2010.04.002

Spielberger CD, Gorsuch RL, Luschene R, Vagg PR, Jacobs GA (1983) Manual for the state-trait anxiety inventory (Form Y). Mind Garden. https://cir.nii.ac.jp/crid/1370285712575158016

Tenekedjieva L-T, McCalley DM, Goldstein-Piekarski AN, Williams LM, Padula CB (2024) Transdiagnostic mood, anxiety, and trauma symptom factors in alcohol use disorder: neural correlates across 3 Brain Networks. Biol Psychiatry Cognit Neurosci Neuroimaging. https://doi.org/10.1016/j.bpsc.2024.01.013

Thevos AK, Johnston AL, Latham PK, Randall CL, Adinoff B, Malcolm R (1991) Symptoms of anxiety in inpatient alcoholics with and without DSM-III-R anxiety diagnoses. Alcohol Clin Exp Res 15(1):102–105. https://doi.org/10.1111/j.1530-0277.1991.tb00525.x

Torvik FA, Rosenström TH, Gustavson K, Ystrom E, Kendler KS, Bramness JG, Czajkowski N, Reichborn-Kjennerud T (2019) Explaining the association between anxiety disorders and alcohol use disorder: a twin study. Depress Anxiety 36(6):522–532. https://doi.org/10.1002/da.22886

Uddin LQ, Nomi JS, Hebert-Seropian B, Ghaziri J, Boucher O (2017) Structure and function of the human insula. J Clin Neurophysiol 34(4):300–306. https://doi.org/10.1097/WNP.0000000000000377

Wilcox CE, Adinoff B, Clifford J, Ling J, Witkiewitz K, Mayer AR, Boggs KM, Eck M, Bogenschutz M (2020) Brain activation and subjective anxiety during an anticipatory anxiety task is related to clinical outcome during prazosin treatment for alcohol use disorder. NeuroImage Clin 26:102162. https://doi.org/10.1016/j.nicl.2020.102162

Willinger U, Lenzinger E, Hornik K, Fischer G, Schönbeck G, Aschauer HN, Meszaros K, European Fluvoxamine in Alcoholism Study Group (2002) Anxiety as a predictor of relapse in detoxified alcohol-dependent patients. Alcohol Alcoholism (Oxford, Oxfordshire) 37(6): 609–612. https://doi.org/10.1093/alcalc/37.6.609

Wolitzky-Taylor K, Bobova L, Zinbarg RE, Mineka S, Craske MG (2012) Longitudinal investigation of the impact of anxiety and mood disorders in adolescence on subsequent substance use disorder onset and vice versa. Addict Behav 37(8):982–985. https://doi.org/10.1016/j.addbeh.2012.03.026

Wolitzky-Taylor K, Brown LA, Roy-Byrne P, Sherbourne C, Stein MB, Sullivan G, Bystritsky A, Craske MG (2015) The impact of alcohol use severity on anxiety treatment outcomes in a large effectiveness trial in primary care. J Anxiety Disord 30:88–93. https://doi.org/10.1016/j.janxdis.2014.12.011

Xiao P, Dai Z, Zhong J, Zhu Y, Shi H, Pan P (2015) Regional gray matter deficits in alcohol dependence: a meta-analysis of voxel-based morphometry studies. Drug Alcohol Depend 153: 22–28. https://doi.org/10.1016/j.drugalcdep.2015.05.030

Xu J, Van Dam NT, Feng C, Luo Y, Ai H, Gu R, Xu P (2019) Anxious brain networks: a coordinate-based activation likelihood estimation meta-analysis of resting-state functional connectivity studies in anxiety. Neurosci Biobehav Rev 96:21–30. https://doi.org/10.1016/j.neubiorev.2018.11.005

Yang X, Tian F, Zhang H, Zeng J, Chen T, Wang S, Jia Z, Gong Q (2016) Cortical and subcortical gray matter shrinkage in alcohol-use disorders: a voxel-based meta-analysis. Neurosci Biobehav Rev 66:92–103. https://doi.org/10.1016/j.neubiorev.2016.03.034

Zabik NL, Flook EA, Feola B, Benningfield MM, Silveri MM, Winder DG, Blackford JU (2024) Bed nucleus of the stria terminalis network responses to unpredictable threat in early alcohol abstinence. Alcohol Clin Exp Res 48:1716

Zeng J, Yu S, Cao H, Su Y, Dong Z, Yang X (2021) Neurobiological correlates of cue-reactivity in alcohol-use disorders: a voxel-wise meta-analysis of fMRI studies. Neurosci Biobehav Rev 128: 294–310. https://doi.org/10.1016/j.neubiorev.2021.06.031

Zimmermann P, Wittchen H-U, Höfler M, Pfister H, Kessler RC, Lieb R (2003) Primary anxiety disorders and the development of subsequent alcohol use disorders: a 4-year community study of adolescents and young adults. Psychol Med 33(7):1211–1222. https://doi.org/10.1017/S0033291703008158

Zinbarg RE, Barlow DH (1996) Structure of anxiety and the anxiety disorders: a hierarchical model. J Abnorm Psychol 105(2):181–193. https://doi.org/10.1037/0021-843x.105.2.181

Zugman A, Jett L, Antonacci C, Winkler AM, Pine DS (2023) A systematic review and meta-analysis of resting-state fMRI in anxiety disorders: need for data sharing to move the field forward. J Anxiety Disord 99:102773. https://doi.org/10.1016/j.janxdis.2023.102773

Zywiak WH, Connors GJ, Maisto SA, Westerberg VS (1996) Relapse research and the reasons for drinking questionnaire: a factor analysis of Marlatt's relapse taxonomy. Addiction 91(Suppl): S121–S130

Part VI
Conclusion

Synopsis of Current Trends in Behavioral Neuroscience Volume __

Murray B. Stein

Contents

M. B. Stein (✉)
University of California San Diego, La Jolla, CA, USA
e-mail: mstein@health.ucsd.edu

© The Author(s), under exclusive license to Springer Nature Switzerland AG 2025
Curr Topics Behav Neurosci (2025) 73: 639–653
https://doi.org/10.1007/7854_2025_594
Published Online: 11 June 2025

Abstract Following is a synopsis of the reviews contained within this volume. Each of these reviews is written by experts in their field, and each provides an authoritative treatise on its subject matter. The topics cover the gamut of anxiety-relevant research, from basic mechanistic and animal models through to clinical studies. The authors do a splendid job of introducing key concepts and terms relevant to their fields of study, thereby opening vistas of information for interested readers. The authors—and clearly some credit must also go to the editors—have made great efforts not only to review what has been done, but to imagine where the field(s) will go and how those may, eventually if not immediately, impact the prevention and treatment of anxiety disorders. I hope you enjoy reading and learning from this cadre of experts as much as I did.

1 Basic Mechanisms of Fear and Anxiety

1.1 Disentangling the Neural Circuits of Arousal and Anxiety-Like Behavior

Nicolas Pégard and Jose Rodriguez-Romaguera

Summary This review explores the role of arousal in anxiety disorders, highlighting its impact on alertness and stimulus salience. Dysregulation of arousal can lead to hypo- or hyperarousal, affecting response selection and threat perception. This review highlights challenges in studying arousal in preclinical models and discusses novel methodologies that integrate physiological measurements with neuron activity tracking in awake animals.

Commentary Technological advancements have enabled the detection of rapid processes that determine an organism's readiness to respond to stimuli. Arousal involves multiple processes orchestrated by brainstem nuclei and various neuropeptide systems, interacting with deep brain nuclei and cortical regions. This review emphasizes the importance of integrating manipulation and neural recording techniques with physiological measurements to advance anxiety disorder research.

1.2 The Emerging Role of Brain Mitochondria in Fear and Anxiety

Carmen Sandi

Summary This review explores the critical role of mitochondria in brain function and behavior, particularly in relation to fear and anxiety. Mitochondria, beyond their traditional role in energy metabolism, are essential for linking cellular energy processes with neural circuits governing these behaviors. This review evaluates current knowledge about fear and anxiety neural circuits and summarizes the fundamental roles of mitochondria, including energy production, calcium buffering, ROS generation, and signaling.

Commentary This review highlights how mitochondrial function and dysfunction impact the neural substrates of fear and anxiety, with implications for PTSD and anxiety disorders. It proposes that mitochondrial pathways could be new therapeutic targets, integrating insights from genetic, biochemical, neurobiological, behavioral, and clinical studies. Microglia are now believed to play important roles in synaptic plasticity, neurodevelopment, and stress response mechanisms (Chen et al. 2023). By merging mitochondrial biology with behavioral and circuit neuroscience, this review uncovers promising avenues for therapeutic intervention, suggesting that targeting mitochondrial dysfunction could improve treatment outcomes for anxiety and related disorders.

1.3 Behavioral and Brain Mechanisms of Active Avoidance and Their Relevance to Anxiety Disorders

Mohammed R. Milad

Summary This review examines avoidance behaviors through behavioral and neural lenses, including current treatments and their impact on brain restructuring. Avoidance is an adaptive behavior crucial for survival, helping individuals avoid true threats. However, it can become maladaptive, interfering with daily functioning and engagement in value-driven behaviors. Anxiety disorders involve unrestrained fear or avoidance of perceived threats, leading to behavioral modifications and physical symptoms. Avoidance behaviors, negatively reinforced, significantly impact social, interpersonal, and occupational functioning. Research focuses on neural circuits involved in fear and avoidance, aiming to develop medications and therapies to interrupt maladaptive neural activity.

Commentary This is a key review for understanding the role of avoidance in anxiety disorders. Maladaptive avoidance behaviors are critical in anxiety disorders' development and maintenance. While avoidance offers immediate relief, it reinforces maladaptive patterns, exacerbating anxiety. Key brain regions involved include the amygdala, hippocampus, striatum, and ACC. Therapeutic approaches like cognitive behavioral therapy and pharmacological interventions are promising. A multifaceted approach combining neural understanding, psychotherapy, and pharmacotherapy may improve outcomes for anxiety disorder patients.

1.4 Current State of the Neuroscience of Fear Extinction and Its Relevance to Anxiety Disorders

Raffael Kalisch

Summary This review examines the functional neuroanatomy of human fear extinction, initially explored through fMRI studies translating rodent research findings. A critical meta-analysis (Fullana et al, 2018) revealed inconsistent activation in key brain areas, challenging previous assumptions. This review provides new insights into extinction-related brain activation, emphasizing the involvement of the reward learning system, cerebellum, vmPFC, hippocampus, and a broader brain network, along with dopamine and noradrenaline.

Commentary Extinction learning and memory retrieval are thought to be impaired in individuals with anxiety disorders, including PTSD. Methodological challenges, such as low test-retest reliability of conditioning measures and high MRI costs, complicate the identification of abnormalities in fear extinction circuitry. Functional connectivity (FC) metrics offer more reliable insights, revealing general features of anxiety disorders. This review discusses potential therapeutic approaches, including pharmacological treatments, transcranial magnetic stimulation (TMS), and imaging-based biofeedback, to enhance fear extinction and improve treatment outcomes for anxiety disorders.

1.5 The Organization of Anxiety Symptoms Along the Threat Imminence Continuum

Rany Abend

Summary Despite extensive research, understanding and translating the mechanisms of anxiety disorders into clinical practice remains limited. This review highlights two key complications: reliance on phenomenology-driven definitions and limited understanding of symptom chronicity. It proposes organizing anxiety symptoms along a temporal dimension based on threat imminence, viewing them as aberrant expressions of normative defensive responses.

Commentary Grounding anxiety symptoms in threat imminence-dependent defensive responses can theoretically address diagnostic challenges and improve clinical application. This approach may reduce comorbidity and highlight core anxiety symptomatology, enabling better anticipation and targeting of symptoms. It complements existing models across disciplines, integrating biological, evolutionary, cognitive, and psychodynamic perspectives. Experimentally accounting for distinct defensive response patterns can enhance mechanistic research and facilitate the development of targeted therapeutic interventions.

1.6 Structural Brain Correlates of Anxiety During Development

Antonia N. Kaczkurkin

Summary This review explores the associations between neurodevelopmental patterns and anxiety in youth, focusing on brain structure (volume, cortical thickness, and surface area). It synthesizes existing literature, revealing more research on brain volume than other modalities, and highlights variability in implicated brain regions and association directions. This review discusses gaps and limitations in the current body of work, suggesting future research directions.

Commentary This review covers neurostructural correlates of various anxiety phenotypes, including generalized anxiety disorder, social anxiety disorder, OCD, and PTSD. It notes a lack of studies on panic, specific phobia, agoraphobia, and selective mutism. The variability in findings across studies is attributed to differences in sample characteristics, symptom measurement, and neuroimaging approaches. Addressing these gaps is crucial for advancing our understanding of anxiety's neurobiological underpinnings and improving clinical interventions.

1.7 Computational Mechanisms of Information-Seeking in Anxiety

Ryan Smith

Summary This review evaluates the relationship of anxiety to information-seeking behavior, focusing on studies using cognitive computational models and behavioral tasks. It highlights the adaptive and maladaptive aspects of exploratory behaviors in reducing uncertainty.

Commentary This review examines altered information-seeking in anxiety, noting mixed results depending on the type of exploration and measurement methods. It discusses the neural basis of exploration, highlighting brain regions linked to anxiety and exploration. This review suggests future research directions, including the need for consistent experimental approaches and the potential for clinical interventions targeting exploratory behavior. Understanding the links between anxiety and information-seeking can inform novel therapeutic strategies and improve treatment outcomes.

1.8 Emotion Regulation Under Stress: A Social Processing and Memory Perspective

Mauricio R. Delgado

Summary This review explores alternative strategies for emotion regulation under stress, focusing on social processing and memory recall. Ethical considerations regarding emotional memory updating are also discussed, highlighting the importance of both positive and negative memories in shaping our interactions and personal identity.

Commentary Autobiographical memory, particularly positive social memories, can serve as a regulatory resource under stress. These memories buffer the physiological stress response and free cognitive resources to reframe perspectives. However, challenges exist for individuals with anxiety and depression, who may struggle with recalling specific positive memories. Interventions aimed at improving positive memory recall and social connections show promise.

2 Risk Factors and Moderating Factors

2.1 Developmental Shifts in Amygdala Function

J. Amiel Rosenkranz

Summary This review discusses the socio-emotional developmental trajectory in mammals, focusing on the transition from reliance on others to independent navigation of the social environment. The amygdala plays a key role in social and emotional processing, displaying developmental changes during early juvenile, adolescent, and adult stages. This review proposes a framework where the developmental stage of intra-amygdala circuits and their regulation by cortical inputs biases animals toward developmentally appropriate social and emotional behavior.

Commentary The amygdala is highly sensitive to the environment, particularly during early post-natal and adolescent periods. This sensitivity allows for optimization of social and emotional function by tuning the amygdala's developmental trajectory. Future research should focus on integrating how environmental challenges during development exert coordinated actions across the amygdala and align with other neural regions to contribute to the developmental trajectory of social and emotional behaviors. Though difficult to conduct, this line of research offers promising insights into the role of the amygdala in anxiety and related disorders.

2.2 *Intergenerational Effects of Stress: A Focus on Learning and Memory*

Brian G. Dias

Summary This review explores how stress, in various forms and durations, affects learning and memory in both the directly exposed generation and across generations. It emphasizes associative fear conditioning in humans and rodents, discussing the relevance of extinction training and recall as indicators of normative and disrupted learning. This review also examines neuroanatomical substrates and the influences of post-natal, in utero, and pre-conceptional stress on learning and memory across generations.

Commentary This review highlights how early life stress increases the likelihood of developing trauma and anxiety-related disorders like PTSD. It notes that gestational and pre-conceptional stress also reverberate across generations. The discussion includes mechanisms through which stress at different developmental stages disrupts learning and memory, emphasizing the importance of understanding these processes for addressing intergenerational effects of stress. This line of investigation provides tremendous insight into the long-lived impact of stress and trauma.

2.3 *Insights into the Neurobiology of Behavioral Inhibition from Nonhuman Primate Models*

Andrew S. Fox

Summary Children exhibiting extreme behavioral inhibition (BI) are at risk for anxiety and depressive disorders (Schwartz et al. 2003). Understanding the neurobiology of BI can guide early-life interventions to prevent these disorders. Nonhuman primates (NHPs) are valuable models for studying BI, as they exhibit similar behaviors and neural circuits.

Commentary Rhesus monkey models of BI have implicated a distributed brain network, including the extended amygdala complex (EAc), in early-life BI. These studies complement human research by identifying neural substrates and molecular pathways involved in BI. Manipulating specific brain regions and molecules in NHPs has shown that altering one molecule can affect the entire BI-related brain network. Despite progress, challenges remain in bridging the gap between rodent and human studies. NHP models continue to provide insights into the neurobiology of BI, promising to inform new strategies for treating and preventing anxiety and depressive disorders.

2.4 *The Impact of Early Life Experiences on Stress Neurobiology and the Development of Anxiety*

Yanbin Niu

Summary This review examines how early life stress, from the prenatal period to age two, influences the development of anxiety disorders. Stress during critical periods of brain development can shape neural circuits, increasing the risk of anxiety. This review covers the association between early life stress and anxiety risk, the neurobiology of anxiety, and the vulnerability of anxiety-related neural regions to early stress. It also discusses the neural mechanisms linking early stress to anxiety and highlights the importance of considering other contributing factors like genetics and environmental influences.

Commentary Early life stress significantly increases the risk of developing anxiety disorders (Bandoli et al. 2017). Changes in brain regions associated with salience/arousal and top-down regulation systems, particularly the amygdala and prefrontal cortex, may underlie this association. Despite substantial evidence, more research is needed to establish causal pathways from early stress exposure to anxiety, considering limitations like small sample sizes and cross-sectional designs.

2.5 *Genetic Vulnerability to Social Anxiety Disorder*

Janna Marie Bas-Hoogendam

Summary This review explores the genetic factors contributing to social anxiety disorder (SAD), one of the most prevalent anxiety disorders. It reviews family, twin, and population-based studies to understand genetic vulnerability. This review discusses the endophenotype approach, highlighting findings from the Leiden Family Lab study on SAD, and examines the relationship between behavioral inhibition (BI) and SAD development. The relevance of this work for improving preventative and therapeutic interventions for socially anxious youth is emphasized.

Commentary Genetics significantly impact the development of SAD. Recent research on endophenotypes and inhibited temperament has advanced our understanding of genetic vulnerability to SAD. Whereas the genetics of anxiety—writ large—has seen recent advances (Strom et al. 2025), the particular genetics of SAD remains largely unexplored. Neuroimaging and neuromodulation techniques offer potential for novel interventions. More research in large samples is needed to identify reliable targets for these interventions.

2.6 *Neurodevelopmental Pathways from Temperamental Fear to Anxiety*

Eunkyung Shin

Summary his review discusses the link between fearful temperament and anxiety disorders, highlighting how biological and environmental factors shape vulnerability from early life. It describes fearful temperament as a marker for anxiety risk, neural mechanisms underlying this temperament, and factors that moderate the association between fearful temperament and anxiety.

Commentary Fearful temperament and anxiety disorders share neural processes, reflecting early deviations from typical brain development. This review examines how children with fearful temperament process environmental information and regulate arousal, and how these processes interact with environmental factors. Understanding these mechanisms can inform early intervention and prevention strategies. The heterogeneity in developmental trajectories suggests that internal and external factors can buffer or exacerbate the risk associated with fearful temperament.

3 Sex Differences

3.1 Sex Differences in Neural Circuits Underlying Fear Processing

Raül Andero Galí

Summary This review explores how sex differences influence neural circuitry involved in anxiety and fear-related disorders. Studies integrating female and male data reveal distinct interconnectivity between brain areas, highlighting the significant impact of sex as a biological variable on fear processing.

Commentary Impairments in fear memory processing are central to the development of fear-related disorders like PTSD, panic disorder, and specific phobias. Understanding the interconnectivity between brain regions is crucial for developing targeted treatments. Females and males exhibit distinct neural responses and interconnectivity patterns, emphasizing the importance of considering sex differences in fear processing research.

3.2 Sex Differences in the Neurobiology of Fear and Anxiety

Jennifer S. Stevens

Summary This review evaluates the neural, cognitive, and physiological mechanisms contributing to sex differences in fear and anxiety. Women are diagnosed with anxiety and stress-related disorders at twice the rate of men (Kessler et al. 2010). This review focuses on human studies among healthy populations and patients with various anxiety disorders, revealing nuanced differences in stimuli that evoke anxiety and stress responses in women and men.

Commentary Women generally show heightened neural responses in threat-related regions, while men show greater responses in emotion regulatory regions. Sex hormones like progesterone, estradiol, and testosterone play significant roles in these differences. Women with anxiety disorders exhibit hyperactivity in emotional response regions and reduced connectivity between the amygdala and the prefrontal cortex. More research is needed to explore these differences, considering factors like cycling sex hormones and including diverse populations in studies.

4 Treatment

4.1 Neuromodulation of Fear and Anxiety Circuits

Wen Li

Summary This review assesses the complex neural underpinnings of anxiety, highlighting the expansion of anxiety and fear circuitry beyond the amygdala to a distributed network. It discusses noninvasive brain stimulation (NIBS) as a promising clinical intervention to modulate anxiety-related neural circuits. NIBS methodologies, including transcranial magnetic, direct-current, and alternating-current stimulation (TMS, tDCS, tACS), are explored for their potential to test, target, and normalize the pathophysiology of fear and anxiety circuits.

Commentary Significant progress has been made in understanding the neural basis of fear and anxiety disorders. The fear circuit is now thought to include the sensory cortex, insula, and dorsal ACC, along with large-scale networks like the salience network (SN) and default mode network (DMN). Neuromodulation has shown promise in treating these disorders, a subject further addressed in the review by Noah Philip and colleagues (chapter "Transcranial Magnetic Stimulation for Posttraumatic Stress Disorder and Generalized Anxiety Disorder").

4.2 Using Inhibitory Learning Theories to Optimize Exposure Treatment for Children with Anxiety Disorders

Jennifer L. Hudson

Summary Cognitive behavioral therapy (CBT) is the recommended first-line treatment for anxiety disorders in children. However, relapse rates are high, indicating the need for improved efficacy. Exposure therapy, a core component of CBT, involves facing feared situations. This review evaluates inhibitory learning theory, which may enhance exposure therapy by focusing on expectancy violation and developing nonthreat associations.

Commentary Adapting inhibitory learning strategies for children has the potential to improve treatment outcomes and reduce long-term mental health burdens. Future studies should focus on optimizing these strategies in clinical practice and understanding factors that influence their effectiveness.

4.3 Neural Markers of Treatment Response in Pediatric Anxiety and PTSD

Kate D. Fitzgerald

Summary The authors review the literature on neural predictors of treatment response as well as changes that occur in response to treatment. Their review points to changes in activity in the amygdala in response to threatening stimuli and pretreatment activity within regions of the cingulate cortex are predictive of response to cognitive behavioral therapy (CBT). They also indicate that activation within and connectivity between other regions of the brain can contribute to prediction of therapeutic response.

Commentary Pediatric anxiety disorders and posttraumatic stress disorder can be disabling and difficult to treat, and many children do not achieve remission with existing therapies. Understanding how brain systems work together to enhance positive outcomes from psychological (and possibly other) therapies may lead to new advances in therapy as well as inform ways to improve upon currently available treatments.

4.4 Fears Worth Testing Out: A Systematic Review of the Neural Mechanisms of Treatment Outcome for Anxiety-Related Disorders

Robin Aupperle

Summary This review conducts a systematic analysis of task-based fMRI studies examining brain activation changes with pharmacologic or psychotherapy interventions for various anxiety-related disorders. This review focuses on prefrontal-insula-amygdala networks, identifying treatment-related decreases in amygdala and anterior insula activation as consistent findings. This review emphasizes the need for larger sample sizes, nuanced computational approaches, and increased focus on replication to enhance understanding and development of effective mental health treatments.

Commentary Human neuroimaging holds promise for integrating psychological and neurobiological research to inspire new treatment approaches. Despite inconsistent findings, abandoning treatment-related fMRI research would be premature. Larger, multi-site studies and data sharing are crucial for identifying consistent results. Future research should explore additional cognitive and emotional processes, leveraging neuroimaging to develop treatments targeting specific mechanisms.

While fMRI research has yet to profoundly impact clinical care, it has subtly influenced treatment development, suggesting continued exploration is warranted.

4.5 Transcranial Magnetic Stimulation for Posttraumatic Stress Disorder and Generalized Anxiety Disorder

Noah S. Philip

Summary This review comprehensively reviews the use of transcranial magnetic stimulation (TMS) for treating PTSD and GAD.

Commentary TMS is safe and appears efficacious for PTSD and GAD, though no single protocol has proven superior. Higher doses of stimulation may improve outcomes. Combining TMS with psychotherapy shows mixed results, highlighting the need for further investigation. Precision targeting and higher cumulative administrations of TMS may yield better efficacy. Continued research on TMS-related biomarkers and combination therapies is crucial for advancing treatment.

5 Co-occurring Disorders

5.1 Sensory Processing and Anxiety: Within and Beyond the Autism Spectrum

Carissa J. Cascio

Summary This review explores how anxiety in autism differs from the general population, focusing on sensory processing differences. It highlights the role of interoception, the perception of internal sensory information, in contributing to anxiety in autism. New findings from a study of autistic and non-autistic individuals suggest two components of interoceptive confusion: one closely related to anxiety symptoms and another distinct from anxiety. These findings indicate meaningful subtypes of interoceptive difficulties and their relationship with anxiety.

Commentary Sensory perception, particularly interoceptive perception, is crucial in understanding anxiety in autistic and non-autistic individuals. This review suggests further investigation into interoceptive processing and individual differences, including sex, gender, and autistic features. Ultra-high-resolution MRI may offer insights into sensory circuits contributing to anxiety. Overall, sensory processing is an important aspect of anxiety experiences across different populations.

5.2 *Fear and Anxiety in Schizophrenia: A Focus on Development, Assessment, and Mechanisms*

Brandee Feola

Summary Anxiety is highly prevalent in people with schizophrenia and is related to numerous negative outcomes. However, it is often underreported and understudied. This review addresses four main questions: the development of anxiety in schizophrenia, measurement methods, underlying mechanisms, and treatment approaches. This review emphasizes the importance of assessing anxiety in schizophrenia to improve identification, treatment, and overall quality of life for those affected.

Commentary Recognizing and understanding anxiety in schizophrenia is crucial for improving treatment outcomes. Increased awareness and knowledge can help providers better address anxiety symptoms, ultimately alleviating the burden for individuals with schizophrenia and their families. Future research should focus on enhancing the recognition and treatment of anxiety in schizophrenia.

5.3 *Insights into Overlapping Brain Networks for Anxiety and Alcohol Use Disorders*

Jennifer Urbano Blackford

Summary Alcohol use disorder (AUD) affects over 30 million Americans and often co-occurs with anxiety disorders. Anxiety can increase the risk of developing AUD, maintain alcohol use, and hinder treatment effectiveness (Craske and Stein 2016). The relationship between AUD and anxiety is bidirectional and involves overlapping neural circuits related to stress, reward, and cognitive control. This review examines how anxiety impacts the brain and treatment outcomes in AUD, aiming to reduce the burden of alcohol use.

Commentary Anxiety significantly influences the development and maintenance of AUD. Understanding this intersection through research can improve clinical care and reduce the stigma of alcohol use. Future research should focus on early identification and treatment of anxiety to prevent AUD, and determining how anxiety affects different stages of addiction.

References

Bandoli G, Campbell-Sills L, Kessler RC, Heeringa SG, Nock MK, Rosellini AJ, Sampson NA, Schoenbaum M, Ursano RJ, Stein MB (2017) Childhood adversity, adult stress, and the risk of major depression or generalized anxiety disorder in US soldiers: a test of the stress sensitization hypothesis. Psychol Med 47(13):2379–2392. https://doi.org/10.1017/S0033291717001064

Chen W, Zhao H, Li Y (2023) Mitochondrial dynamics in health and disease: mechanisms and potentialtargets. Signal Transduct Target Ther 8:333. https://doi.org/10.1038/s41392-023-01547-9

Cill SL, Goldberg MA, Cosmo C, Arulpragasam AR, Zandvakili A, Berlow YA, Philip NS (in press) Transcranial magnetic stimulation for posttraumatic stress disorder and generalized anxiety disorder. Update to information from this volume. Craske MG, Stein MB (2016) Anxiety. Lancet 388(10063):3048–3059. https://doi.org/10.1016/S0140-6736(16)30381-6

Fullana MA, Albajes-Eizagirre A, Soriano-Mas C, Vervliet B, Cardoner N, Benet O, Radua J, Harrison BJ (2018) Fear extinction in the human brain: a meta-analysis of fMRI studies in healthy participants. Neurosci Biobehav Rev 88:16–25

Kessler RC, Ruscio AM, Shear K, Wittchen HU (2010) Epidemiology of anxiety disorders. Curr Top Behav Neurosci 2:21–35

Schwartz, C. E., Wright, C. I., Shin, L. M., Kagan, J., & Rauch, S. L. (2003). Inhibited and uninhibited infants "grown up": adult amygdalar response to novelty. Science (New York.), 300(5627), 1952–1953. https://doi.org/10.1126/science.1083703

Strom NI, Verhulst B, Bacanu SA, Cheesman R, Purves KL et al (2025) Genome-wide association study of major anxiety disorders in 122,341 European-ancestry cases identifies 58 loci and highlights GABAergic signaling. medRxiv [Preprint]. 2024 Jul 5:2024.07.03.24309466. https://doi.org/10.1101/2024.07.03.24309466.

GPSR Compliance
The European Union's (EU) General Product Safety Regulation (GPSR) is a set of rules that requires consumer products to be safe and our obligations to ensure this.

If you have any concerns about our products, you can contact us on

ProductSafety@springernature.com

In case Publisher is established outside the EU, the EU authorized representative is:

Springer Nature Customer Service Center GmbH
Europaplatz 3
69115 Heidelberg, Germany

www.ingramcontent.com/pod-product-compliance
Lightning Source LLC
Chambersburg PA
CBHW062351140426
43165CB00003B/32
9783031984013